Try Me in Lectures!

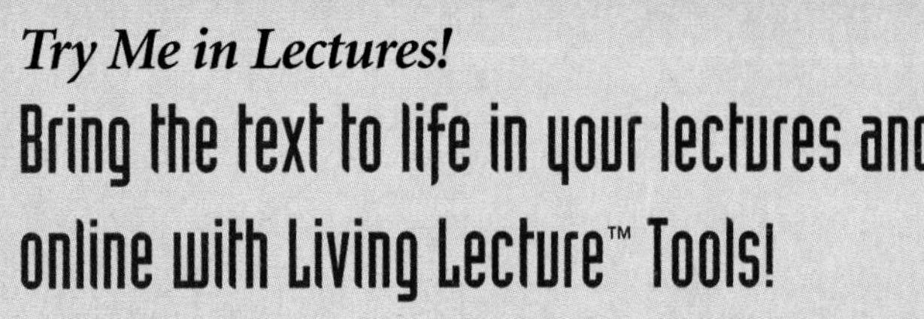

FREE to qualified adopters, **Multimedia Manager with Living Lecture™ Tools** animates the Active Figures from the text, allowing you to integrate three-dimensional, dynamic models of meteorological processes into Microsoft® PowerPoint® presentations. You can also incorporate traditional art from all of Brooks/Cole's meteorology texts, lecture notes, and your own audio and other media clips. In short, you can create fluid lectures to stimulate the imagination and improve the understanding of even the least-engaged students.

Living Lecture Tools are also at the heart of **MeteorologyNow™**, the sophisticated, assessment-driven learning system for students that accompanies this text. One click animates the **Living Lecture Tools'** superb three-dimensional models of the text's Active Figures, making the lessons intuitive and exciting as students work to master the concepts, complete their assignments, and demonstrate their understanding.

THOMSON™
BROOKS/COLE

Learn more about MeteorologyNow ▶
on pages P-2 and P-3.

What do your students need to learn NOW?

Meteorology Now

Help your students make the best possible use of their study time with **MeteorologyNow,** the first assessment-centered student learning tool for meteorology!

This powerful and interactive resource helps students gauge their unique study needs, then gives them a *Personalized Learning Plan* that focuses their studies on the concepts they most need to master. Web-based and free with every new copy of the text, **MeteorologyNow** is seamlessly tied to the text and to your lectures. That's because it includes animations of the text's Active Figures—the same ones available on **Multimedia Manager with Living Lecture Tools.**

What does an integrated learning system do?

This unmatched resource and the new edition of this text were developed in concert, to enhance one another and provide students with a seamless, integrated learning system. As they work through the text, students will see notes (like the one at the bottom of this page) that direct them to the dynamic, media-enhanced activities and helpful tutorials in **MeteorologyNow.** This precise, page-by-page integration means they'll spend less time flipping through pages or navigating Web sites and more time learning.

MeteorologyNow is available through this text's Book Companion Web Site at **http://earthscience.brookscole.com/ahrens/ess4e.** In addition to **MeteorologyNow,** students using the Web site will have access to maps, Web links, Internet and **InfoTrac® College Edition** exercises, learning objectives, discussion questions, chapter outlines, and much more.

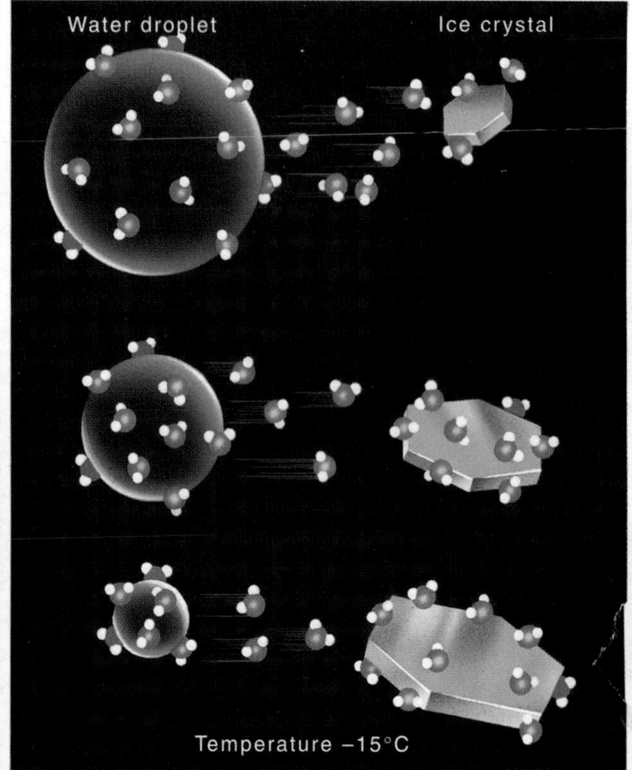

MeteorologyNow™ ACTIVE FIGURE 5.20
The ice-crystal process. The greater number of water vapor molecules around the liquid droplets causes water molecules to diffuse from the liquid drops toward the ice crystals. The ice crystals absorb the water vapor and grow larger, while the water droplets grow smaller.
Watch this Active Figure at http://earthscience.brookscole.com/ahrens/ess4e.

MeteorologyNow™

How does it work?

Students log on to **MeteorologyNow** at **http://earthscience.brookscole.com/ahrens/ess4e** by using the free access code packaged with their text. They'll immediately notice the system's simple, browser-based format—as easy to use as surfing the Web. Just a click of the mouse gives students the freedom to enter and explore the system at any point. Students can build a complete, *Personalized Learning Plan* for themselves by taking advantage of all three of **MeteorologyNow's** powerful components:

What Do I Know?

A *Pre-Test* is your students' first step. Each chapter's *Pre-Test* includes approximately 20 multiple-choice questions that help students identify any conceptual weaknesses. The questions concentrate on the most common mistakes students make. Students' results may be e-mailed to the professor so that both can work together on the best learning strategy.

Once they've completed the *Pre-Test,* students are presented with a detailed *Personalized Learning Plan* that outlines the concepts they need to review, identifying where the concepts can be found in the text as well as in the program.

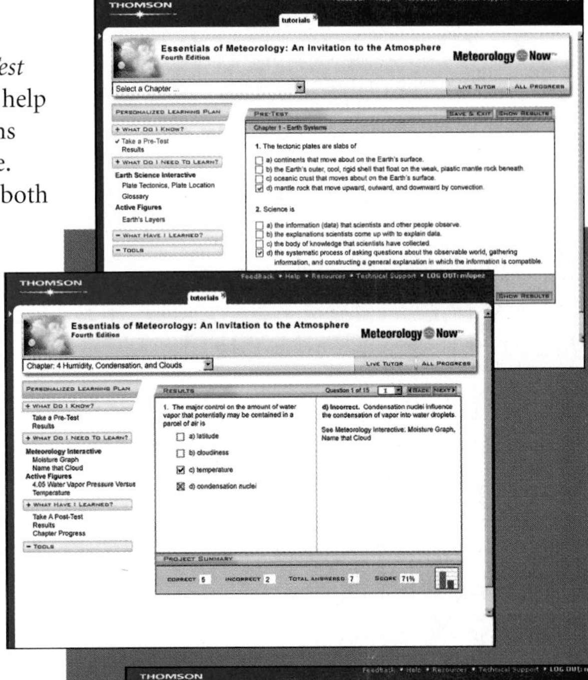

What Do I Need to Learn?

Working from their *Personalized Learning Plan,* students encounter activities, movies, and interactive tutorials—including numerous animations—that function together to help them master key concepts and visualize meteorological processes. Among these are animated versions of Active Figures from the text, each of which demonstrates a key concept and is paired with conceptual questions to help students focus on what is happening in the animated model of the figure.

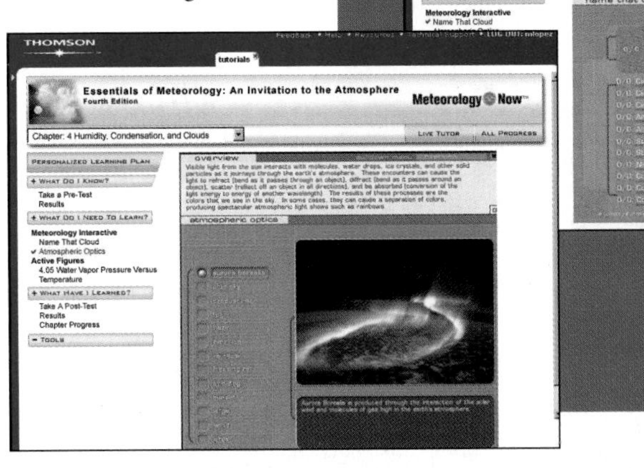

What Have I Learned?

An optional 10-question *Post-Test* ensures that the student has mastered the concepts in each chapter. As with the *Pre-Test,* students' results may be e-mailed to the instructor to help both assess the student's progress. If students need to improve their score, **MeteorologyNow** works with them as they continue to build their knowledge.

This "essentials" version of Ahrens' best seller is back—with updated content and new interactive resources that make learning about meteorology more fascinating than ever.

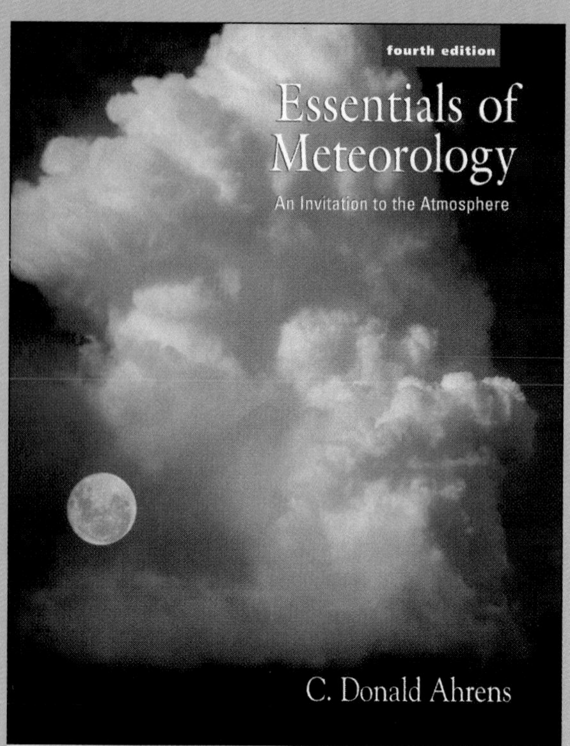

Essentials of Meteorology, Fourth Edition

C. Donald Ahrens
Modesto Junior College

There are many good reasons why Donald Ahrens is the most widely read and trusted author in introductory meteorology. With his accessible style, practical pedagogy, and ideal blend of theory and application, Ahrens successfully engages students in exploring topics in this dynamic field. His ability to explain ideas in a student-friendly, manageable fashion allows even non-science students to immediately apply the text material to the world around them—and understand the underlying meteorological principles.

This Fourth Edition of ***Essentials of Meteorology*** includes updated material throughout as well as newly revised art, maps, and figures. Add the supplemental resource package that accompanies the text—including a powerful presentation tool featuring all of the text's art and photos as well as numerous animations and video clips—and you and your students will have all the teaching and learning support that you need.

Highlights of the Fourth Edition

- **Numerous updates** on meteorological advances and phenomena (see page P-5).
- **Significantly revised art program** with new photos and figures (see page P-6).
- **Updated *Focus on* and *Did You Know?* boxes** that expand on material in the text (see page P-7).
- **A superior ancillary package, anchored by the FREE *MeteorologyNow* online tutorial** (see pages P-1 through P-3 for information on **MeteorologyNow** and page P-8 for details on other supplements).
- **A removable Cloud Chart** in the text that allows students to identify clouds wherever they go.

From current Air Quality Index information to new material on thunderstorms, the book is completely up-to-date.

Certainly, the likelihood of a thunderstorm producing a tornado increases when the storm becomes a supercell, but not all supercells produce tornadoes. And not all tornadoes come from rotating thunderstorms (supercells).

Nonsupercell Tornadoes Tornadoes that do not occur in association with a pre-existing wall cloud of a supercell are called **nonsupercell tornadoes.** These tornadoes may occur with intense multicell storms as well as with ordinary thunderstorms, even relatively weak ones. Some nonsupercell tornadoes extend from the base of a thunderstorm as a visible funnel cloud, as shown in Fig. 10.36, whereas others may begin on the ground and build upwards in the absence of a condensation funnel.

Nonsupercell tornadoes may form along a gust front where the cool downdraft of the thunderstorm forces warm, humid air upwards. Tornadoes that form along a gust front are commonly called **gustnadoes.** These relatively weak tornadoes normally are short-lived and rarely inflict significant damage. Gustnadoes are often seen as a rotating cloud of dust or debris rising above the surface.

Occasionally, rather weak, short-lived tornadoes will occur with rapidly building cumulus congestus clouds. Tornadoes such as these commonly form over east-central Colorado. Because they look similar to waterspouts that form over water, they are sometimes called **landspouts.**

Figure 10.37 illustrates how a landspout can form. Suppose, for example, that the winds at the surface converge along a boundary, as illustrated in Fig. 10.37a. (The wind may converge due to topographic irregularities or any number of other factors, including temperature and moisture variations.) Notice that along the boundary, the air is rising, condensing, and forming into a cumulus congestus cloud. Notice also that along the surface at the boundary there is horizontal rotation (spin) created by the wind blowing in opposite directions along the boundary. If the developing cloud should move over the region of rotating air, the spinning air may be drawn up into the cloud by the storm's updraft. As the spinning, rising air shrinks in diameter, it produces a tornado-like structure called a *landspout.* As with a tornado, the landspout increases in rotational speed much like a spinning skater increases in speed when the arms are brought in close to the body. Landspouts usually dissipate when rain falls through the cloud and destroys the updraft. Tornadoes may form in this manner along many types of converging wind boundaries, including sea breezes and gust fronts. Nonsupercell tornadoes and funnel clouds may also form with thunderstorms when cold air aloft (associated with an upper-level trough) moves over a region. Common along the west coast of North America, these short-lived tornadoes are sometimes called *cold-air funnels.*

FIGURE 10.36 A funnel cloud extends downward from the base of a nonsupercell thunderstorm over central California.

© James Tyler

Several models all SRES envelope

Model ensemble all SRES envelope

A1F1
A1B
A1T
A2
B1
B2
IS92e high
IS92a
IS92c low
Third Assessment Report method

Temperature Change, °C

Year

...t the emissions of CO_2 from ...rall, present trends indicate ...ospheric CO_2 increase, the ...sorb a *decreasing* percentage ...dition, it is not known how ...r than CO_2 will increase.

...s and more water vapor is ...udiness might increase as ...s—which come in a variety ...m at different altitudes— ... Clouds reflect incoming ...cess that tends to cool the ...bsorb infrared radiation ... to warm it. Just how the ...anges in cloudiness will ...e of clouds that form and ...ch as liquid water (or ice) ...istribution. For example, ...(composed mostly of ice) tend to promote a net warming effect: They allow a good deal of sunlight to pass through (which warms the earth's surface), yet because they are cold, they warm the atmosphere by absorbing more infrared radiation from the earth than they emit upward. Low stratified clouds, on the other hand, tend to promote a net cooling effect. Composed mostly of water droplets, they reflect much of the sun's incoming energy, and, because their tops are relatively warm, they radiate away much of the infrared energy they receive from the earth. Satellite data confirm that, overall, clouds presently have a *net cooling effect* on our planet, which means that, without clouds, our atmosphere would be warmer.

Additional clouds in a warmer world would not necessarily have a net cooling effect, however. Their influence on the average surface air temperature would depend on their extent and on whether low or high clouds dominate the climate scene. Consequently, the feedback from clouds could potentially enhance or reduce the warming produced by increasing greenhouse gases. Most models show that as the surface air warms, there will be more convection, more convective-type clouds, and an increase in cirrus clouds. This situation would tend to provide a positive feedback on the climate system, and the effect of clouds on cooling the earth would be diminished (see Fig. 14.16).

Updated Content That Reflects Recent Advances, Including:

- The latest information on greenhouse warming (Chapter 2).
- Additional information on air temperature measuring devices and freezes—along with the latest wind chill charts used by the National Weather Service (Chapter 3).
- Current material on dew-point temperature (Chapter 4) and additional information on cloud seeding (Chapter 5).
- A new section on the Pacific Decadal Oscillation (Chapter 7).
- A new section on the formation of landspouts and new material on supercell tornadoes and non-supercell tornadoes in a significantly revised chapter on thunderstorms and tornadoes (Chapter 10).
- The latest material on National Ambient Air Quality Standards and the Air Quality Index (Chapter 12).
- The most up-to-date information on temperature and climate change, including the Intergovernmental Panel on Climate Change (IPCC) Third Assessment Report (TAR), released in 2001 (Chapter 14).

The updated art program showcases the excitement and drama of weather phenomena.

New art and photos ▶▶▶
The text's photos, many of which were taken by the author, provide unmatched images of spectacular weather phenomena. New art has also been added to some of the **Focus on** boxes, and all satellite imagery is now in color.

16 CHAPTER 1 The Earth's Atmosphere

Middle Latitude Storm

Thunderstorms

isible reflected light) shows a variety of cloud patterns and storms in the earth's atmosphere.

y to clearing skies
orthwest after the

e that not only do
ge. Steered by the
e-latitude storm in
rm, which moves

eastward, carrying its clouds and weather with it. In advance of this system, a sunny day in Ohio will gradually cloud over and yield heavy showers and thunderstorms by nightfall. Behind the storm, cool dry northerly winds rushing into eastern Colorado cause an overcast sky to give way to clearing conditions. Farther south, the thunderstorms presently over the Gulf of Mexico (Fig. 1.10)

© Ed Darack

A mass of moist, stable air gliding up and over these mountains condenses into lenticular clouds.

84 CHAPTER 4 Humidity, Condensation, and Clouds

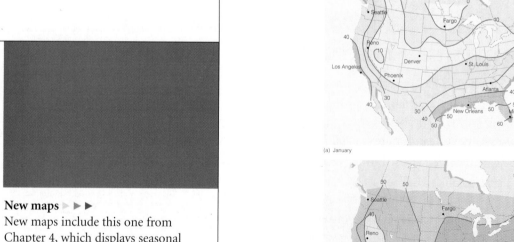

FIGURE 4.7 Average surface dew-point temperatures (°F) for (a) January and for (b) July.

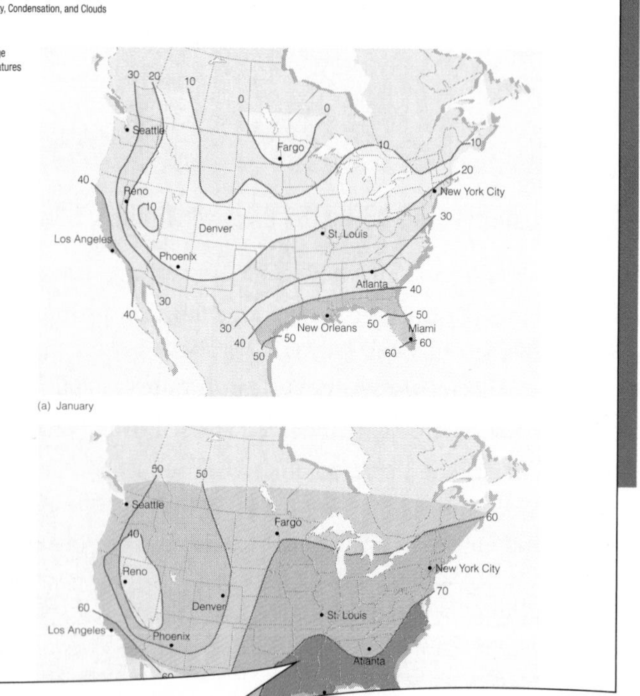

New maps ▶▶▶
New maps include this one from Chapter 4, which displays seasonal differences in dew-point temperatures in North America.

New Visuals

In-text learning aids engage and support students as they read through each chapter.

FOCUS ON AN OBSERVATION

The Aurora—A Dazzling Light Show

At high latitudes after darkness has fallen, a faint, white glow may appear in the sky. Lasting from a few minutes to a few hours, the light may move across the sky as a yellow-green arc much wider than a rainbow; or, it may faintly decorate the sky with flickering draperies of blue, green, and purple light that constantly change in form and location, as if blown by a gentle breeze. This eerie yet beautiful light show is called the **aurora** (see Fig. 2).

The aurora is caused by charged particles from the sun interacting with our atmosphere. From the sun and its tenuous atmosphere comes a continuous discharge of particles. This discharge happens because, at extremely high temperatures, gases become stripped of electrons by violent collisions and acquire enough speed to escape the gravitational pull of the sun. As these charged particles (ions and electrons) travel through space, they are known as the *solar wind*. When the solar wind moves close enough to the earth, it interacts with the earth's magnetic field, disturbing it. This disturbance causes energetic solar wind particles to enter the upper atmosphere, where they collide with atmospheric gases. These gases then become excited and emit visible radiation (light), which causes the sky to glow like a neon light, thus producing the aurora.

In the Northern Hemisphere, the aurora is called the *aurora borealis*, or northern lights; its counterpart in the Southern Hemisphere is the *aurora australis*, or southern lights. The aurora is most frequently seen in the polar regions, where the earth's magnetic field lines emerge from the earth (see Fig. 3). But during active sun periods when there are numerous sunspots (huge cooler regions on the sun's surface) and giant flares (solar eruptions), large quantities of particles travel outward away from the sun at high speeds (hundreds of kilometers a second). These energetic particles are able to penetrate unusually deep into the earth's magnetic field, where they provide sufficient energy to produce auroral displays. During these conditions in North America, we see the aurora much farther south than usual.

FIGURE 2 The aurora borealis is a phenomenon that forms as energetic particles from the sun interact with the earth's atmosphere.

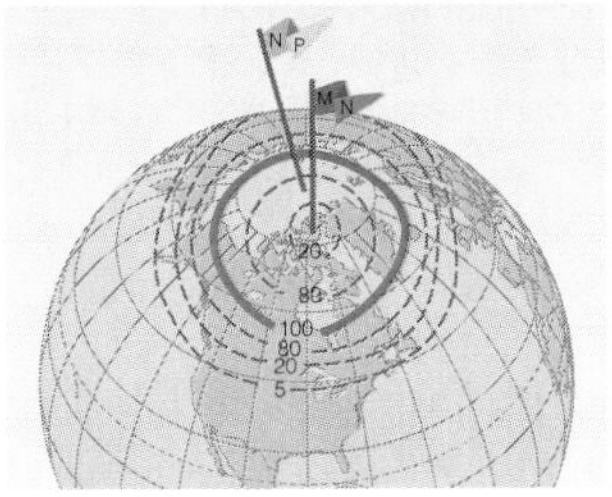

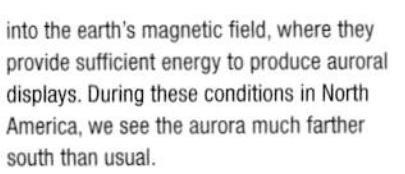

FIGURE 3 The aurora belt (solid red line) represents the region where you would most likely observe the aurora on a clear night. (The numbers represent the average number of nights per year on which you might see an aurora if the sky were clear.) The flag MN denotes the magnetic North Pole, where the earth's magnetic field lines emerge from the earth. The flag NP denotes the geographic North Pole, about which the earth rotates.

Focus on boxes

Focus on boxes expand on the primary text discussions, emphasizing observations, special topics, and advanced topics. For instance, a new box in Chapter 11 gives students a look at how a tropical storm can produce devastating floods over Texas and across the southeastern United States. Some boxes are now entitled **Focus on an Environmental Issue** to better reflect their environmental emphasis.

Did You Know? boxes

At least one new **Did You Know?** box has been added in almost every chapter to provide interesting facts and additional information on recent meteorological events.

DID YOU KNOW?

If all the water vapor in the atmosphere were to suddenly condense and fall as rain, it would cover the earth's surface with about 2.5 cm (1 in.) of water.

DID YOU KNOW?

The large ears of a jackrabbit are efficient emitters of infrared energy. Its ears help the rabbit survive the heat of a summer's day by radiating a great deal of infrared energy to the cooler sky above. Similarly, the large ears of the African elephant greatly increase its radiating surface area and promote cooling of its large mass.

BRIEF REVIEW

Before we go on to the section on clouds, here is a brief review of some of the important concepts and facts we have covered so far:

- Dew, frost, and frozen dew generally form on clear nights when the temperature of objects on the surface cools below the air's dew-point temperature.
- Visible white frost forms in saturated air when the air temperature is at or below freezing. Under these conditions, water vapor can change directly to ice, in a process called *deposition*.
- Condensation nuclei act as surfaces on which water vapor condenses. Those nuclei that have an affinity for water vapor are called *hygroscopic*.
- Fog is a cloud resting on the ground.
- Radiation fog, advection fog, and upslope fog all form as the air cools. Evaporation (mixing) fog, such as steam fog and frontal fog, form as water evaporates and mixes with drier air.

CLOUDS

Clouds are aesthetically appealing and add excitement to the atmosphere. Without them, there would be no rain or snow, thunder or lightning, rainbows or halos. How monotonous if one had only a clear blue sky to look at. A *cloud* is a visible aggregate of tiny water droplets or ice crystals suspended in the air. Some are found ...

naturalist Lamarck (1744–1829) proposed the first system for classifying clouds in 1802; however, his work did not receive wide acclaim. One year later, Luke Howard, an English naturalist, developed a cloud classification system that found general acceptance. In essence, Howard's innovative system employed Latin words to describe clouds as they appear to a ground observer. He named a sheetlike cloud *stratus* (Latin for "layer"); a puffy cloud *cumulus* ("heap"); a wispy cloud *cirrus* ("curl of hair"); and a rain cloud *nimbus* ("violent rain"). In Howard's system, these were the four basic cloud forms. Other clouds could be described by combining the basic types. For example, nimbostratus is a rain cloud that shows layering, whereas cumulonimbus is a rain cloud having pronounced vertical development.

In 1887, Abercromby and Hildebrandsson expanded Howard's original system and published a classification system that, with only slight modification, is still used today. Ten principal cloud forms are divided into four primary cloud groups. Each group is identified by the height of the cloud's base above the surface: high clouds, middle clouds, and low clouds. The fourth group contains clouds showing more vertical than horizontal development. Within each group, cloud types are identified by their appearance. Table 4.2 lists these four groups and their cloud types.

The approximate base height of each cloud group is ... in Table 4.3. Note that the altitude separating the high ...

In-text learning aids

Introductory stories at the beginning of each chapter draw students into the discussion. **Brief Reviews** after major sections reinforce important concepts as they read through the chapter; and a **Summary, Key Terms, Questions for Review,** and **Questions for Thought and Exploration** at the end of each chapter provide additional opportunities to review and assess understanding.

In-Text Learning Aids

For Instructors—FREE to Qualified Adopters

Instructor's Resource Manual and Test Bank
0-534-42267-5

This edition features a completely updated Test Bank and expanded student projects, which can be used for student homework, classwork, etc., and do not require a lab setting. It also includes a section in each chapter with teaching suggestions and demonstrations that you can present in class.

Transparency Acetates
0-534-42271-3

This set of 100 transparency acetates contains photos and images from the text.

Multimedia Manager with Living Lecture™ Tools
0-534-42269-1

Multimedia Manager with Living Lecture Tools animates selected figures from the text (the same ones available to students in **MeteorologyNow**), letting you integrate three-dimensional, dynamic models of meteorological processes into Microsoft PowerPoint presentations. It also includes all art and figures from the text, CNN® video clips, and the Instructor's Resource Manual and Test Bank in Microsoft® Word format. See page P-1 for more information on **Living Lecture Tools.**

CNN® Today Video: Meteorology
Vol. I: 0-534-53770-7 • Vol. II: 0-534-37206-6
Vol. III: 0-534-38906-6 • Vol. IV: 0-534-39087-0
Vol. V: 0-534-39778-6 • Vol. VI: 0-534-42280-2

These videos contain short, two- to four-minute clips of high-interest news stories about topics that will spark class discussion and generate a deeper understanding of the importance of meteorology.

New!
WebTutor™ ToolBox on WebCT and Blackboard
On WebCT packaged with the text: 0-534-65243-3
On Blackboard packaged with the text: 0-534-65198-4

Available FREE with this text (if requested), **WebTutor™ ToolBox** is preloaded with content and available via access code. **WebTutor ToolBox** pairs the content of the text's rich Book Companion Web Site with the course management functionality of a WebCT or Blackboard product. Assign materials (including online quizzes) and have the results flow automatically to your grade book. **ToolBox** is ready to use as soon as you log on—or, you can customize its content by uploading images and other resources, adding Web links, or creating your own practice materials. Students have access only to student resources, while instructors have the option to access password-protected instructor resources.

ExamView®
0-534-42268-3

Create, deliver, and customize tests and study guides (both print and online) in minutes with this easy-to-use assessment and tutorial system. **ExamView®** offers both a *Quick Test Wizard* and an *Online Test Wizard* that guide you step-by-step through the process of creating tests, while the unique "WYSIWYG" capability allows you to see the test you are creating on the screen exactly as it will print or display online.

For Students

Workbook/Study Guide
0-534-42266-7 • Packaged with the text: 0-534-60024-7

Written by C. Donald Ahrens, this workbook/study guide is organized by chapter and includes chapter summaries, important concepts, and self-tests with true/false, multiple choice, and essay-type questions and answers. A list of additional suggested reading material is also included.

Exploring Tropical Cyclones: GIS Investigations for the Earth Sciences
0-534-39147-8 • Packaged with the text: 0-534-59997-4

Exploring Water Resources: GIS Investigations for the Earth Sciences
0-534-39156-7 • Packaged with the text: 0-534-63317-X

By Michelle K. Hall-Wallace, C. Scott Walker, Larry P. Kendall, and Christian J. Schaller—all of The University of Arizona®, Tucson

These groundbreaking guides let even novice users tap the power of GIS to explore, manipulate, and analyze large data sets. They come with all the software and data sets needed to complete the exercises, and are available alone or bundled with **Essentials of Meteorology, Fourth Edition. Visit** **http://earthscience.brookscole.com/GISguides** to learn about other guides in the series of four ***GIS Investigations for the Earth Sciences!***

Essential Study Skills for Science Students
By Daniel D. Chiras

0-534-37595-2 •Packaged with the text: 0-534-60060-3

Written specifically for science students, this book discusses how to develop good study habits, prepare for tests, produce excellent term papers, and more.

NEW and FREE with every new copy of the text!
MeteorologyNow™
See pages P-1 through P-3 for details about this assessment-driven student learning tool.

Packaged FREE with every new copy of this text!
InfoTrac® College Edition
http://www.infotrac-college.com

Every new copy of the text includes four months of FREE access to **InfoTrac® College Edition.** This online library presents full-length articles from nearly 5,000 publications, including *Weatherwise, American Scientist,* and *Atmospheric Monitoring & Abatement News.* New! Students now have instant access to critical-thinking and paper-writing tools through **InfoWrite.**

Free!
The Brooks/Cole Earth Sciences Resource Center
http://earthscience.brookscole.com

This site offers maps, Web links, and a range of other resources, including easy access to MeteorologyNow and a direct link to the Book Companion Web Site at **http://earthscience.brookscole.com/ahrens/ess4e** for chapter-by-chapter study tools such as learning objectives and InfoTrac College Edition exercises.

Chapter 1 – The Earth's Atmosphere

Class Preparation / Lecture Tools	Testing Tools / Course Management	Student Mastery / Homework / Tutorials	Beyond the Book
Instructor's Resource Manual and Test Bank Chapter Summary, Key Terms, Teaching Suggestions, Student Projects, and Test Bank **Multimedia Manager with Living Lecture™ Tools** Create a multimedia lecture using the Microsoft® PowerPoint® link tool **Transparency Acetates** Chapter 1	**Instructor's Resource Manual and Test Bank** Multiple-Choice, True/False, Word Choice, Essay, and Short Answer Questions ExamView **ExamView®** Computerized test bank with online capabilities WebTUTOR™ ToolBox **WebTutor™ ToolBox** Online course management tool for WebCT or Blackboard preloaded with text-specific content and media resources Meteorology Now™ **MeteorologyNow™** Use movies, Active Figures, Pre- and Post-Tests to quiz students on what they currently know and what they need to learn	**Study Guide** Chapter Summary, Key Concepts, Terms, and Review Questions Meteorology Now™ **MeteorologyNow™** Create Web-based personalized learning plans using movies, Active Figures, Pre- and Post-Tests that focus on concepts students need to master	**Book Companion Web Site** **http://earthscience.brookscole.com/ahrens/ess4e** Online resources for you and your students **InfoTrac® College Edition** **http://www.infotrac-college.com** *Keywords:* atmosphere, carbon dioxide, ozone, troposphere, stratosphere, weather, climate **Relevant Web Sites** Exploration of Earth's Atmosphere **http://liftoff.msfc.nasa.gov/academy/space/atmosphere.html** NASA Earth Observatory **http://earthobservatory.nasa.gov**

Chapter 2 – Warming the Earth and the Atmosphere

Class Preparation / Lecture Tools	Testing Tools / Course Management	Student Mastery / Homework / Tutorials	Beyond the Book
Instructor's Resource Manual and Test Bank Chapter Summary, Key Terms, Teaching Suggestions, Student Projects, and Test Bank **Multimedia Manager with Living Lecture™ Tools** Create a multimedia lecture using the Microsoft® PowerPoint® link tool **CNN® Today Video**  Vol. III, Solar Storms (1:55) Vol. III, Tornadoes on the Sun (1:30) **Transparency Acetates** Chapter 2	**Instructor's Resource Manual and Test Bank** Multiple-Choice, True/False, Word Choice, Essay, and Short Answer Questions ExamView **ExamView®** Computerized test bank with online capabilities WebTUTOR™ ToolBox **WebTutor™ ToolBox** Online course management tool for WebCT or Blackboard preloaded with text-specific content and media resources Meteorology Now™ **MeteorologyNow™** Use movies, Active Figures, Pre- and Post-Tests to quiz students on what they currently know and what they need to learn	**Study Guide** Chapter Summary, Key Concepts, Terms, and Review Questions Meteorology Now™ **MeteorologyNow™** Create Web-based personalized learning plans using movies, Active Figures, Pre- and Post-Tests that focus on concepts students need to master	**Book Companion Web Site** **http://earthscience.brookscole.com/ahrens/ess4e** Online resources for you and your students **InfoTrac® College Edition** **http://www.infotrac-college.com** *Keywords:* energy, temperature, heat, heat capacity, conduction, convection, radiation **Relevant Web Sites**  Stanford Solar Center **http://solar-center.stanford.edu** SOHO: Exploring the Sun **http://sohowww.nascom.nasa.gov** 

Chapter 3 – Air Temperature

Class Preparation / Lecture Tools	Testing Tools / Course Management	Student Mastery / Homework / Tutorials	Beyond the Book
Instructor's Resource Manual and Test Bank Chapter Summary, Key Terms, Teaching Suggestions, Student Projects, and Test Bank **Multimedia Manager with Living Lecture™ Tools** Create a multimedia lecture using the Microsoft® PowerPoint® link tool **Transparency Acetates** Chapter 3	**Instructor's Resource Manual and Test Bank** Multiple-Choice, True/False, Word Choice, Essay, and Short Answer Questions ExamView **ExamView®** Computerized test bank with online capabilities WebTUTOR™ ToolBox **WebTutor™ ToolBox** Online course management tool for WebCT or Blackboard preloaded with text-specific content and media resources Meteorology Now™ **MeteorologyNow™** Use movies, Active Figures, Pre- and Post-Tests to quiz students on what they currently know and what they need to learn	**Study Guide** Chapter Summary, Key Concepts, Terms, and Review Questions Meteorology Now™ **MeteorologyNow™** Create Web-based personalized learning plans using movies, Active Figures, Pre- and Post-Tests that focus on concepts students need to master	**Book Companion Web Site** **http://earthscience.brookscole.com/ahrens/ess4e** Online resources for you and your students **InfoTrac® College Edition** **http://www.infotrac-college.com** *Keywords:* Sublimination, evaporation, condensation, saturation, hydrologic cycle **Relevant Web Sites** NOAA's Extreme Weather **http://lwf.ncdc.noaa.gov/oa/climate/severeweather/extremes.html** U. Illinois: Hydrologic Cycle **http://ww2010.atmos.uiuc.edu/(Gh)/guides/mtr/hyd/home.rxml**

Chapter 4 – Humidity, Condensation, and Clouds

Class Preparation / Lecture Tools	Testing Tools / Course Management	Student Mastery / Homework / Tutorials	Beyond the Book
Instructor's Resource Manual and Test Bank Chapter Summary, Key Terms, Teaching Suggestions, Student Projects, and Test Bank **Multimedia Manager with Living Lecture™ Tools** Create a multimedia lecture using the Microsoft® PowerPoint® link tool **CNN® Today Video** Vol. I, New Fog Detection System in Tennessee (2:38) Vol. I, Temperature, Humidity, and Tennis (2:04) **Transparency Acetates** Chapter 4	**Instructor's Resource Manual and Test Bank** Multiple-Choice, True/False, Word Choice, Essay, and Short Answer Questions ExamView **ExamView®** Computerized test bank with online capabilities WebTUTOR™ ToolBox **WebTutor™ ToolBox** Online course management tool for WebCT or Blackboard preloaded with text-specific content and media resources Meteorology Now™ **MeteorologyNow™** Use movies, Active Figures, Pre- and Post-Tests to quiz students on what they currently know and what they need to learn	**Study Guide** Chapter Summary, Key Concepts, Terms, and Review Questions **Exploring Water Resources: GIS Investigations** Activities 1.2, 1.4 Meteorology Now™ **MeteorologyNow™** Create Web-based personalized learning plans using movies, Active Figures, Pre- and Post-Tests that focus on concepts students need to master	**Book Companion Web Site** **http://earthscience.brookscole.com/ahrens/ess4e** Online resources for you and your students **InfoTrac® College Edition** **http://www.infotrac-college.com** *Keywords:* dew, frost, condensation nuclei, haze, fog, radiation fog, advection fog, cirrus clouds, stratus clouds, cumulus clouds, lenticular clouds, satellites **Relevant Web Sites**  U. Illinois: Clouds and Precipitation **http://ww2010.atmos.uiuc.edu/(Gh)/guides/mtr/cld/dvlp/wtr.rxml** 

Chapter 5 – Cloud Development and Precipitation

Class Preparation / Lecture Tools	Testing Tools / Course Management	Student Mastery / Homework / Tutorials	Beyond the Book
Instructor's Resource Manual and Test Bank Chapter Summary, Key Terms, Teaching Suggestions, Student Projects, and Test Bank **Multimedia Manager with Living Lecture™ Tools** Create a multimedia lecture using the Microsoft® PowerPoint® link tool **CNN® Today Video** 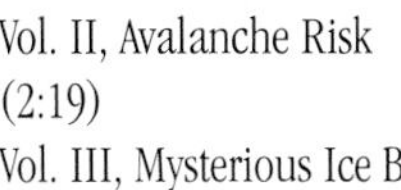 Vol. II, Avalanche Risk (2:19) Vol. III, Mysterious Ice Balls (1:15) **Transparency Acetates** Chapter 5	ExamView **ExamView®** Computerized test bank with online capabilities WebTUTOR™ ToolBox **WebTutor™ ToolBox** Online course management tool for WebCT or Blackboard preloaded with text-specific content and media resources Meteorology Now™ **MeteorologyNow™** Use movies, Active Figures, Pre- and Post-Tests to quiz students on what they currently know and what they need to learn	**Study Guide** Chapter Summary, Key Concepts, Terms, and Review Questions **Exploring Water Resources: GIS Investigations** Activities 2.2, 2.3, 2.4, 3.4, 4.1, 4.2 Meteorology Now™ **MeteorologyNow™** Create Web-based personalized learning plans using movies, Active Figures, Pre- and Post-Tests that focus on concepts students need to master	**Book Companion Web Site** **http://earthscience.brookscole.com/ahrens/ess4e** Online resources for you and your students **InfoTrac® College Edition** **http://www.infotrac-college.com** 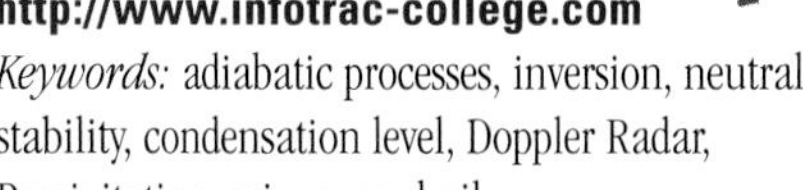 *Keywords:* adiabatic processes, inversion, neutral stability, condensation level, Doppler Radar, Precipitation, rain, snow, hail **Relevant Web Sites** Images of Clouds from Space **http://www.solarviews.com/eng/cloud1.htm** National Drought Mitigation Center **http://drought.unl.edu/ndmc**

Chapter 6 – Air Pressure and Winds

Class Preparation / Lecture Tools	Testing Tools / Course Management	Student Mastery / Homework / Tutorials	Beyond the Book
Instructor's Resource Manual and Test Bank Chapter Summary, Key Terms, Teaching Suggestions, Student Projects, and Test Bank **Multimedia Manager with Living Lecture™ Tools** Create a multimedia lecture using the Microsoft® PowerPoint® link tool **Transparency Acetates** Chapter 6	ExamView **ExamView®** Computerized test bank with online capabilities WebTUTOR™ ToolBox **WebTutor™ ToolBox** Online course management tool for WebCT or Blackboard preloaded with text-specific content and media resources Meteorology Now™ **MeteorologyNow™** Use movies, Active Figures, Pre- and Post-Tests to quiz students on what they currently know and what they need to learn	**Study Guide** Chapter Summary, Key Concepts, Terms, and Review Questions **Exploring Tropical Cyclones: GIS Investigations** Activity 1.3 **Exploring Water Resources: GIS Investigations** Activity 2.3 Meteorology Now™ **MeteorologyNow™** Create Web-based personalized learning plans using movies, Active Figures, Pre- and Post-Tests that focus on concepts students need to master	**Book Companion Web Site** **http://earthscience.brookscole.com/ahrens/ess4e** Online resources for you and your students **InfoTrac® College Edition**  **http://www.infotrac-college.com** *Keywords:* air pressure, barometer, isobar, anticyclone, mid-latitude cyclone, Coriolis force, geostrophic wind, cyclonic flow, anticyclonic flow, zonal flow, meridional flow, Buys-Ballot's law, hydrostatic equilibrium **Relevant Web Sites** U. Illinois: Forces and Winds **http://ww2010.atmos.uiuc.edu/(Gh)/guides/mtr/fw/home.rxml** Coriolis Force **http://www.windpower.org/en/tour/wres/coriolis.htm**

Chapter 7 – Atmospheric Circulations

Class Preparation / Lecture Tools	Testing Tools / Course Management	Student Mastery / Homework / Tutorials	Beyond the Book
Instructor's Resource Manual and Test Bank Chapter Summary, Key Terms, Teaching Suggestions, Student Projects, and Test Bank **Multimedia Manager with Living Lecture™ Tools** Create a multimedia lecture using the Microsoft® PowerPoint® link tool **CNN® Today Video** Vol. I, Air Turbulence (2:34) Vol. I, El Nino Returns (2:20) Vol. III, Pacific Decadal Oscillation (2:05) **Transparency Acetates** Chapter 7	**Instructor's Resource Manual and Test Bank** Multiple-Choice, True/False, Word Choice, Essay, and Short Answer Questions ExamView® **ExamView®** Computerized test bank with online capabilities WebTUTOR™ ToolBox **WebTutor™ ToolBox** Online course management tool for WebCT or Blackboard preloaded with text-specific content and media resources Meteorology Now™ **MeteorologyNow™** Use movies, Active Figures, Pre- and Post-Tests to quiz students on what they currently know and what they need to learn	**Study Guide** Chapter Summary, Key Concepts, Terms, and Review Questions Meteorology Now™ **MeteorologyNow™** Create Web-based personalized learning plans using movies, Active Figures, Pre- and Post-Tests that focus on concepts students need to master	**Book Companion Web Site** **http://earthscience.brookscole.com/ahrens/ess4e** Online resources for you and your students **InfoTrac® College Edition** **http://www.infotrac-college.com** *Keywords:* eddies, wind shear, clear air turbulence, wind waves, swells, wind direction, prevailing wind, wind rose, anemometer, aerovane, land breeze, sea breeze, monsoon, El Niño, La Niña, Southern Oscillation, Hadley cell, trade winds, westerlies **Relevant Web Sites** National Severe Storms Laboratory **http://www.nssl.noaa.gov** NOAA El Niño Theme Page **http://www.elnino.noaa.gov** NOVA Online: Tracking El Niño **http://www.pbs.org/wgbh/nova/elnino**

Chapter 8 – Air Masses, Fronts, and Middle-Latitude Cyclones

Class Preparation / Lecture Tools	Testing Tools / Course Management	Student Mastery / Homework / Tutorials	Beyond the Book
Instructor's Resource Manual and Test Bank Chapter Summary, Key Terms, Teaching Suggestions, Student Projects, and Test Bank **Multimedia Manager with Living Lecture™ Tools** Create a multimedia lecture using the Microsoft® PowerPoint® link tool **Transparency Acetates** Chapter 8	**Instructor's Resource Manual and Test Bank** Multiple-Choice, True/False, Word Choice, Essay, and Short Answer Questions ExamView® **ExamView®** Computerized test bank with online capabilities WebTUTOR™ ToolBox **WebTutor™ ToolBox** Online course management tool for WebCT or Blackboard preloaded with text-specific content and media resources Meteorology Now™ **MeteorologyNow™** Use movies, Active Figures, Pre- and Post-Tests to quiz students on what they currently know and what they need to learn	**Study Guide** Chapter Summary, Key Concepts, Terms, and Review Questions **Exploring Tropical Cyclones: GIS Investigations** Activities 1.2–1.5, 2.1–2.4, 2.6 **Exploring Water Resources: GIS Investigations** Activity 2.3 Meteorology Now™ **MeteorologyNow™** Create Web-based personalized learning plans using movies, Active Figures, Pre- and Post-Tests that focus on concepts students need to master	**Book Companion Web Site** **http://earthscience.brookscole.com/ahrens/ess4e** Online resources for you and your students **InfoTrac® College Edition** **http://www.infotrac-college.com** *Keywords:* air masses, source regions, lake-effect snows, front, stationary front, cold front, warm front, occluded front, upper-air front **Relevant Web Sites** U. Illinois: Air Masses and Fronts **http://ww2010.atmos.uiuc.edu/(Gh)/guides/mtr/af/home.rxml** Weather Underground **http://www.wunderground.com/US/Region/US/Fronts.html** Lake-Effect Snow **http://www.comet.ucar.edu/class/smfaculty/byrd/sld001.htm**

Chapter 9 – Weather Forecasting

Class Preparation / Lecture Tools	Testing Tools / Course Management	Student Mastery / Homework / Tutorials	Beyond the Book
Instructor's Resource Manual and Test Bank Chapter Summary, Key Terms, Teaching Suggestions, Student Projects, and Test Bank **Multimedia Manager with Living Lecture™ Tools** Create a multimedia lecture using the Microsoft® PowerPoint® link tool **CNN® Today Video** 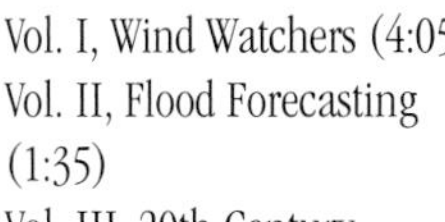 Vol. I, Wind Watchers (4:05) Vol. II, Flood Forecasting (1:35) Vol. III, 20th Century Weather (2:00) **Transparency Acetates** Chapter 9	**Instructor's Resource Manual and Test Bank** Multiple-Choice, True/False, Word Choice, Essay, and Short Answer Questions ExamView **ExamView®** Computerized test bank with online capabilities WebTUTOR™ ToolBox **WebTutor™ ToolBox** Online course management tool for WebCT or Blackboard preloaded with text-specific content and media resources Meteorology Now™ **MeteorologyNow™** Use movies, Active Figures, Pre- and Post-Tests to quiz students on what they currently know and what they need to learn	**Study Guide** Chapter Summary, Key Concepts, Terms, and Review Questions Meteorology Now™ **MeteorologyNow™** Create Web-based personalized learning plans using movies, Active Figures, Pre- and Post-Tests that focus on concepts students need to master	**Book Companion Web Site** **http://earthscience.brookscole.com/ahrens/ess4e** Online resources for you and your students **InfoTrac® College Edition** **http://www.infotrac-college.com** *Keywords:* weather watch, weather warning, numerical weather prediction, atmospheric models, ensemble forecasting, persistence forecast, steady-state method, analogue method, statistical forecast, probability forecast **Relevant Web Sites** NOAA-CIRES Climate Diagnostics Center **http://www.cdc.noaa.gov/index.html** National Weather Service **http://www.nws.noaa.gov**

Chapter 10 – Thunderstorms and Tornadoes

Class Preparation / Lecture Tools	Testing Tools / Course Management	Student Mastery / Homework / Tutorials	Beyond the Book
Instructor's Resource Manual and Test Bank Chapter Summary, Key Terms, Teaching Suggestions, Student Projects, and Test Bank **Multimedia Manager with Living Lecture™ Tools** Create a multimedia lecture using the Microsoft® PowerPoint® link tool **CNN® Today Video**  Vol. I, Flood Tracking with the USGS (2:38) Vol. I, Wind Shear Alert for Aircraft (1:40) Vol. II, Twister Scale (2:16) **Transparency Acetates** Chapter 10	**Instructor's Resource Manual and Test Bank** Multiple-Choice, True/False, Word Choice, Essay, and Short Answer Questions ExamView **ExamView®** Computerized test bank with online capabilities WebTUTOR™ ToolBox **WebTutor™ ToolBox** Online course management tool for WebCT or Blackboard preloaded with text-specific content and media resources Meteorology Now™ **MeteorologyNow™** Use movies, Active Figures, Pre- and Post-Tests to quiz students on what they currently know and what they need to learn	**Study Guide** Chapter Summary, Key Concepts, Terms, and Review Questions Meteorology Now™ **MeteorologyNow™** Create Web-based personalized learning plans using movies, Active Figures, Pre- and Post-Tests that focus on concepts students need to master	**Book Companion Web Site** **http://earthscience.brookscole.com/ahrens/ess4e** Online resources for you and your students **InfoTrac® College Edition** **http://www.infotrac-college.com** *Keywords:* thunderstorm, multicell storms, gust front, shelf cloud, roll clouds, downdraft, microburst, squall line, dryline, flash flood, lightning thunder, sonic boom, St. Elmo's Fire, sferics, tornadoes, funnel cloud, Fujita scale  **Relevant Web Sites**  NOAA's Storm Prediction Center **http://www.spc.noaa.gov** U. Illinois: Severe Storms **http://ww2010.atmos.uiuc.edu/(Gh)/guides/mtr/svr/home.rxml**

Chapter 11 – Hurricanes

Class Preparation / Lecture Tools	Testing Tools / Course Management	Student Mastery / Homework / Tutorials	Beyond the Book
Instructor's Resource Manual and Test Bank Chapter Summary, Key Terms, Teaching Suggestions, Student Projects, and Test Bank **Multimedia Manager with Living Lecture™ Tools** Create a multimedia lecture using the Microsoft® PowerPoint® link tool **CNN® Today Video** Vol. I, Hurricane Hunters (2:37) Vol. II, Hurricanes (1:38) Vol. II, Lightning (1:25) Vol. II, Tsunami Risk (1:35) **Transparency Acetates** Chapter 11	**Instructor's Resource Manual and Test Bank** Multiple-Choice, True/False, Word Choice, Essay, and Short Answer Questions ExamView **ExamView®** Computerized test bank with online capabilities WebTUTOR ToolBox **WebTutor™ ToolBox** Online course management tool for WebCT or Blackboard preloaded with text-specific content and media resources Meteorology Now™ **MeteorologyNow™** Use movies, Active Figures, Pre- and Post-Tests to quiz students on what they currently know and what they need to learn	**Study Guide** Chapter Summary, Key Concepts, Terms, and Review Questions **Exploring Tropical Cyclones: GIS Investigations** Activities 1.1–1.5, 2.1–2.4, 3.1–3.4, 4.1–4.4 Meteorology Now™ **MeteorologyNow™** Create Web-based personalized learning plans using movies, Active Figures, Pre- and Post-Tests that focus on concepts students need to master	**Book Companion Web Site** **http://earthscience.brookscole.com/ahrens/ess4e** Online resources for you and your students **InfoTrac® College Edition** **http://www.infotrac-college.com** *Keywords:* streamlines, hurricane, typhoon, eye, eye wall, trade wind, inversion, tropical disturbance, tropical depression, tropical storm, hurricane watch, hurricane warning, Saffir-Simpson scale, super-typhoon **Relevant Web Sites** NOAA's National Hurricane Center **http://hurricanes.noaa.gov/**  U. Illinois: Hurricanes **http://ww2010.atmos.uiuc.edu/(Gh)/guides/mtr/hurr/home.rxml**

Chapter 12 – Air Pollution

Class Preparation / Lecture Tools	Testing Tools / Course Management	Student Mastery / Homework / Tutorials	Beyond the Book
Instructor's Resource Manual and Test Bank Chapter Summary, Key Terms, Teaching Suggestions, Student Projects, and Test Bank **Multimedia Manager with Living Lecture™ Tools** Create a multimedia lecture using the Microsoft® PowerPoint® link tool **CNN® Today Video** Vol. I, Ozone Hole Healing (2:10) Vol. II, Urban Heat (2:25) Vol. III, Changing Weather (1:35) **Transparency Acetates** Chapter 12	**Instructor's Resource Manual and Test Bank** Multiple-Choice, True/False, Word Choice, Essay, and Short Answer Questions ExamView **ExamView®** Computerized test bank with online capabilities WebTUTOR ToolBox **WebTutor™ ToolBox** Online course management tool for WebCT or Blackboard preloaded with text-specific content and media resources Meteorology Now™ **MeteorologyNow™** Use movies, Active Figures, Pre- and Post-Tests to quiz students on what they currently know and what they need to learn	**Study Guide** Chapter Summary, Key Concepts, Terms, and Review Questions Meteorology Now™ **MeteorologyNow™** Create Web-based personalized learning plans using movies, Active Figures, Pre- and Post-Tests that focus on concepts students need to master	**Book Companion Web Site** **http://earthscience.brookscole.com/ahrens/ess4e** Online resources for you and your students **InfoTrac® College Edition** **http://www.infotrac-college.com** *Keywords:* air pollutants, particulate matter, carbon monoxide (CO), sulfur dioxide (SO_2), hydrocarbons, nitrogen dioxide (NO_2), nitric oxide (NO), smog, photochemical smog, ozone (O_3), ozone hole 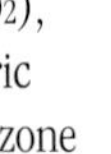 **Relevant Web Sites** EPA: Ozone Depletion **http://www.epa.gov/ozone/index.html** National Academy of Sciences: Ozone Depletion Phenomenon **http://www.beyonddiscovery.org/content/view.article.asp?a=73**

Chapter 13 - Global Climate

Class Preparation / Lecture Tools	Testing Tools / Course Management	Student Mastery / Homework / Tutorials	Beyond the Book
Instructor's Resource Manual and Test Bank Chapter Summary, Key Terms, Teaching Suggestions, Student Projects, and Test Bank **Multimedia Manager with Living Lecture™ Tools** Create a multimedia lecture using the Microsoft® PowerPoint® link tool **CNN® Today Video** Vol. II, Thin Ice (2:32)  **Transparency Acetates** Chapter 13	**Instructor's Resource Manual and Test Bank** Multiple-Choice, True/False, Word Choice, Essay, and Short Answer Questions ExamView **ExamView®** Computerized test bank with online capabilities WebTUTOR™ ToolBox **WebTutor™ ToolBox** Online course management tool for WebCT or Blackboard preloaded with text-specific content and media resources Meteorology Now™ **MeteorologyNow™** Use movies, Active Figures, Pre- and Post-Tests to quiz students on what they currently know and what they need to learn	**Study Guide** Chapter Summary, Key Concepts, Terms, and Review Questions Meteorology Now™ **MeteorologyNow™** Create Web-based personalized learning plans using movies, Active Figures, Pre- and Post-Tests that focus on concepts students need to master	**Book Companion Web Site** **http://earthscience.brookscole.com/ahrens/ess4e** Online resources for you and your students **InfoTrac® College Edition** **http://www.infotrac-college.com** *Keywords:* climatic controls, Koppen classification system, P/E index, tropical rain forest, laterite, tropical wet-and-dry, savanna grass, arid climates, desert, xerophytes, semi arid, steppe, humid subtropical, marine, dry- humid continental, summer subtropical (Mediterranean) **Relevant Web Sites** IRI/LDEO Climate Data Library **http://ingrid.ldgo.columbia.edu** EPA: Ecosystems **http://www.epa.gov/ebtpages/ecosystems.html**

Chapter 14 - Climate Change

Class Preparation / Lecture Tools	Testing Tools / Course Management	Student Mastery / Homework / Tutorials	Beyond the Book
Instructor's Resource Manual and Test Bank Chapter Summary, Key Terms, Teaching Suggestions, Student Projects, and Test Bank **Multimedia Manager with Living Lecture™ Tools** Create a multimedia lecture using the Microsoft® PowerPoint® link tool **CNN® Today Video** Vol. II, Global Warming (2:10) Vol. II, Weird Weather (2:20) Vol. III, Alternative Fuels (2:05) **Transparency Acetates** Chapter 14	**Instructor's Resource Manual and Test Bank** Multiple-Choice, True/False, Word Choice, Essay, and Short Answer Questions ExamView **ExamView®** Computerized test bank with online capabilities WebTUTOR™ ToolBox **WebTutor™ ToolBox** Online course management tool for WebCT or Blackboard preloaded with text-specific content and media resources Meteorology Now™ **MeteorologyNow™** Use movies, Active Figures, Pre- and Post-Tests to quiz students on what they currently know and what they need to learn	**Study Guide** Chapter Summary, Key Concepts, Terms, and Review Questions **Exploring Water Resources: GIS Investigations** Activity 1.4 Meteorology Now™ **MeteorologyNow™** Create Web-based personalized learning plans using movies, Active Figures, Pre- and Post-Tests that focus on concepts students need to master	**Book Companion Web Site** **http://earthscience.brookscole.com/ahrens/ess4e** Online resources for you and your students **InfoTrac® College Edition** **http://www.infotrac-college.com** *Keywords:* dendrochronology, Ice Age, interglacial periods, mid-Holocene maximum, Medieval Climatic Optimum, Little Ice Age, positive feedback mechanism **Relevant Web Sites**  Intergovernmental Panel on Climate Change **http://www.ipcc.ch** Union of Concerned Scientists: Global Warming **http://www.ucsusa.org/global_environment/global_warming/index.cfm** Climate Research Unit **http://www.cru.uea.ac.uk**

Chapter 15 – Light, Color, and Atmospheric Optics

Class Preparation / Lecture Tools	Testing Tools / Course Management	Student Mastery / Homework / Tutorials	Beyond the Book
Instructor's Resource Manual and Test Bank Chapter Summary, Key Terms, Teaching Suggestions, Student Projects, and Test Bank **Multimedia Manager with Living Lecture™ Tools** Create a multimedia lecture using the Microsoft® PowerPoint® link tool **Transparency Acetates** Chapter 15	**Instructor's Resource Manual and Test Bank** Multiple-Choice, True/False, Word Choice, Essay, and Short Answer Questions ExamView **ExamView®** Computerized test bank with online capabilities WebTUTOR™ ToolBox **WebTutor™ ToolBox** Online course management tool for WebCT or Blackboard preloaded with text-specific content and media resources Meteorology Now™ **MeteorologyNow™** Use movies, Active Figures, Pre- and Post-Tests to quiz students on what they currently know and what they need to learn	**Study Guide** Chapter Summary, Key Concepts, Terms, and Review Questions Meteorology Now™ **MeteorologyNow™** Create Web-based personalized learning plans using movies, Active Figures, Pre- and Post-Tests that focus on concepts students need to master	**Book Companion Web Site** **http://earthscience.brookscole.com/ahrens/ess4e** Online resources for you and your students **InfoTrac® College Edition** **http://www.infotrac-college.com** *Keywords:* reflected light, scattered light, crepuscular rays, refraction, mirage, dispersion, sundog, rainbow, corona, diffraction **Relevant Web Sites** U. Illinois: Light and Optics **http://ww2010.atmos.uiuc.edu/(Gh)/guides/mtr/opt/home.rxml** About Rainbows **http://www.unidata.ucar.edu/staff/blynds/rnbw.html** Crepuscular Rays **http://www.ems.psu.edu/~demark/471/CrepuscularRays.html**

Meteorology Now™

http://earthscience.brookscole.com/ahrens/ess4e

Here's your *Media Integration Guide* for **MeteorologyNow**™, the interactive Web-based learning tool that's FREE with this text. After you take a **MeteorologyNow** *Pre-Test* or *Post-Test* for each text chapter, your automatically generated *Personalized Learning Plan* will direct you to one or more of the animations and interactive activities listed here. For example, your plan for Chapter 2 may suggest that you review an animated version of Active Figure 2.16 on the angle of the sun, or an interesting interactive exercise on energy balance. Whatever the activity, you'll learn much more effectively than you would by reading alone!

Chapter	MeteorologyNow™ Active Figures	MeteorologyNow™ Interactive (BlueSkies)	What You'll Learn
1: **The Earth's Atmosphere**		▶ Layers of the atmosphere	Examine the boundaries between different atmospheric layers using actual, current atmospheric measurements anywhere in the world.
2: **Warming the Earth and the Atmosphere**	▶ 2.9: Selective absorption ▶ 2.11: Energy budget for earth/atmosphere ▶ 2.16: Angle of the sun ▶ 2.17: Reasons for seasons	▶ Energy balance	Many different factors interact to determine the temperature of the earth's atmosphere and surface. These activities allow you to interactively observe and experiment with several individual causes of temperature.
3: **Air Temperature**			
4: **Humidity, Condensation, and Clouds**	▶ 4.5: Water vapor pressure vs. temperature	▶ Moisture graph ▶ Name that cloud	These interactive activities provide hands-on experience in exploring the relationships between different moisture variables, and valuable practice in cloud identification.
5: **Cloud Development and Precipitation**	▶ 5.7: Development of a cumulus cloud ▶ 5.12: Adiabatic process ▶ 5.20: Ice crystal (Bergeron) process ▶ 5.26: Formation of sleet, freezing rain ▶ 5.31: Formation of hail	▶ Adiabatic	These animations and interactive figures allow you to examine many of the factors influencing the formation of clouds and precipitation, including air motions, adiabatic warming and cooling, vapor pressures, atmospheric temperature structure, and vertical winds.
6: **Air Pressure and Winds**	▶ 6.2: Formation of direct thermal circulation ▶ 6.14: Coriolis force	▶ Coriolis force ▶ Winds in the two hemispheres	Winds are caused by the interaction of several different forces. These activities and animations allow you to examine these forces, as well as the winds they cause in different parts of the world.
7: **Atmospheric Circulations**	▶ 7.13: Formation of dust devil ▶ 7.16: General circulation ▶ 7.21: Formation of polar front jet stream	▶ Global atmosphere ▶ Global ocean ▶ Southern oscillation	From small dust devils to city-size wind cells to the global atmospheric circulation and ocean currents, wind patterns and vertical motions are related to the horizontal distribution of temperature. These activities provide opportunities to visualize several examples of atmospheric circulations and their causes.

Chapter	MeteorologyNow™ Active Figures	MeteorologyNow™ Interactive (BlueSkies)	What You'll Learn
8: Air Masses, Fronts, and Middle-Latitude Cyclones	▶ 8.13: Cold front ▶ 8.15: Warm front ▶ 8.16: Occluded front ▶ 8.19: Cyclogenesis ▶ 8.23: Wave cyclone formation ▶ Figure 6: Shortwave propagation	▶ Isopleths ▶ Find the front	These animations and activities will help you understand the relationship between temperature, pressure patterns, and wind circulations. You can test your understanding by examining current weather data around the globe, drawing your own isobars, and locating fronts.
9: Weather Forecasting	▶ 9.9a: Visible/IR/Water vapor satellite imagery	▶ Forecasting	These activities allow you to examine a multitude of charts depicting current weather conditions around the globe, make your own weather forecasts, and compare your forecasts with those made by professional meteorologists.
10: Thunderstorms and Tornadoes	▶ 10.6: Role of gust front in severe thunderstorm formation ▶ 10.11: Air mass thunderstorm formation ▶ 10.20: Lightning processes ▶ 10.40: Formation of waterspouts	▶ Microbursts ▶ Lightning	Nature's fury is often unleashed in the form of thunderstorms. These animations and activities help you to explore the details of the wind, moisture, temperature, and electrical charge patterns that cause thunder, lightning, severe winds, waterspouts, and microbursts.
11: Hurricanes	▶ 11.3: Hurricane morphology	▶ Virtual hurricane ▶ Hurricane forecasting	These activities allow you to explore hurricanes in three dimensions to examine the distribution of winds, rain, and vertical motions. You can also try your hand at forecasting hurricane movement.
12: Air Pollution	▶ 12.3: Atmospheric visibility in Grand Canyon ▶ 12.7: Ozone hole formation	▶ Smog	These activities will help you understand the interaction between the chemical precursors of photochemical smog, the effects of air pollution on visibility, and features of the stratospheric ozone hole.
13: Global Climate			
14: Climate Change	▶ 14.9: Orbital fluctuations	▶ Temperature trends	Examine the components of the Milankovitch theory of climate fluctuations as well as future temperatures, including the geographical distribution of warming and cooling, as predicted by a global climate model.
15: Light, Color, and Atmospheric Optics		▶ Atmospheric optics	Test your skill in identifying various atmospheric optical phenomena.

Essentials of Meteorology

Photo by author

ESSENTIALS OF METEOROLOGY

AN INVITATION TO THE ATMOSPHERE

FOURTH EDITION

C. DONALD AHRENS
Modesto Junior College

Australia ■ Canada ■ Mexico ■ Singapore ■ Spain ■ United Kingdom ■ United States

Earth Science Editor: Keith Dodson
Development Editor: Marie Carigma-Sambilay
Assistant Editor: Carol Ann Benedict
Editorial Assistant: Melissa Newt
Technology Project Manager: Ericka Yeoman-Saler
Marketing Manager: Melanie Banfield
Marketing Assistant: Leyla Jowza
Advertising Project Manager: Kelley McAllister
Project Manager, Editorial Production: Hal Humphrey
Art Director: Vernon Boes
Print/Media Buyer: Barbara Britton
Permissions Editor: Sarah Harkrader
Production Service: Janet Bollow Associates

Printed in the United States of America
1 2 3 4 5 6 7 08 07 06 05 04

For more information about our products,
contact us at:
Thomson Learning Academic Resource Center
1-800-423-0563
For permission to use material from this text or product,
submit a request online at
http://www.thomsonrights.com.
Any additional questions about permissions can
be submitted by email to
thomsonrights@thomson.com.

Library of Congress Control Number: 20041722
Student Edition: ISBN 0-534-42264-0
Instructor's Edition: ISBN 0-534-42265-9
International Student Edition: ISBN 0-534-40679-3

Text Designer: Janet Bollow
Art Editor: Janet Bollow
Photo Researchers: Lita Ahrens and Janet Bollow
Copy Editor: Stuart Kenter
Illustrators: Alexander Productions, Myrna Vladic, Richard Sheppard, George Kelvin, Carl Brown, Sue Sellars, Folium, House of Graphics, Joe Medeiros
Cover Designer: Cheryl Carrington
Cover Image: Mary Clay/Getty Images
Compositor: Graphic World, Inc.
Text Printer: Courier Corporation/Kendallville
Cover Printer: Phoenix Color Corp

Thomson Brooks/Cole
10 Davis Drive
Belmont, CA 94002
USA

Asia
Thomson Learning
5 Shenton Way #01-01
UIC Building
Singapore 068808

Australia/New Zealand
Thomson Learning
102 Dodds Street
Southbank, Victoria 3006
Australia

Canada
Nelson
1120 Birchmount Road
Toronto, Ontario M1K 5G4
Canada

Europe/Middle East/Africa
Thomson Learning
High Holborn House
50/51 Bedford Row
London WC1R 4LR
United Kingdom

Latin America
Thomson Learning
Seneca, 53
Colonia Polanco
11560 Mexico D.F.
Mexico

Spain/Portugal
Paraninfo
Calle Magallanes, 25
28015 Madrid Spain

Contents

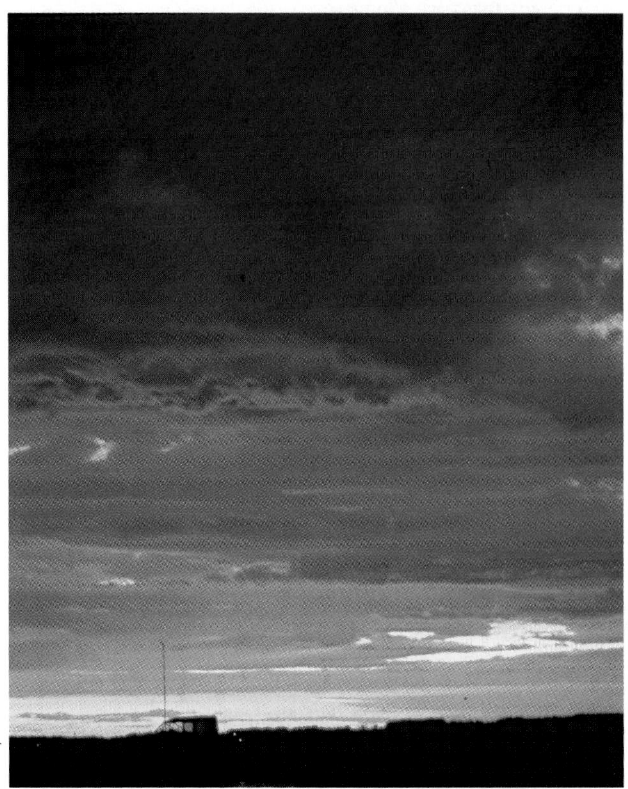
Photo by author

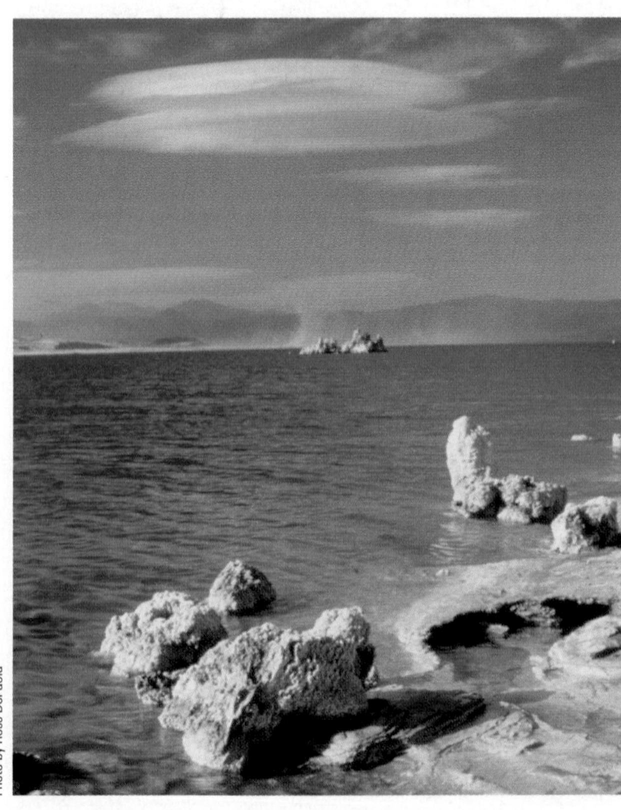
Photo by Ross DePaola

5 CLOUD DEVELOPMENT AND PRECIPITATION 110

6 AIR PRESSURE AND WINDS 140

Photo by author

7 ATMOSPHERIC CIRCULATIONS 166

8 AIR MASSES, FRONTS, AND MIDDLE-LATITUDE CYCLONES 200

9 WEATHER FORECASTING 230

10 THUNDERSTORMS AND TORNADOES 256

11 HURRICANES 292

Photo by author

Photo by author

12 AIR POLLUTION 316

13 GLOBAL CLIMATE 342

14 CLIMATE CHANGE 372

15 LIGHT, COLOR, AND ATMOSPHERIC OPTICS 402

Photo by author

APPENDICES

Preface

The world is an ever-changing picture of naturally occurring events. From drought and famine to devastating floods, some of the greatest challenges we face come in the form of natural disasters created by weather. Yet, dealing with weather and climate is an inevitable part of our lives. Sometimes it is as small as deciding what to wear for the day or how to plan a vacation. But it can also have life-shattering consequences, especially for those who are victims of a hurricane or tornado.

In recent years, weather and climate have become front page news from such environmental issues as greenhouse warming and ozone depletion in the stratosphere to the global weather influences of El Niño. The dynamic nature of the atmosphere seems to demand our attention and understanding more these days than ever before. Almost daily, there are newspaper articles describing some weather event or impending climate change. For this reason, and the fact that weather influences our daily lives in so many ways, interest in meteorology (the study of the atmosphere) has been growing. This rapidly developing and popular science is giving us more information about the workings of the atmosphere than ever before. The atmosphere will always provide challenges for us, but as research and technology advance, our ability to understand our atmosphere improves as well. The information available to you in this book, therefore, is intended to aid in your own personal understanding and appreciation of our earth's dynamic atmosphere.

ABOUT THIS BOOK

Essentials of Meteorology is written for students taking an introductory course on the atmospheric environment. The main purpose of the text is to convey meteorological concepts in a visual, practical, and nonmathematical manner. In addition, the intent of the book is to stimulate curiosity in the reader and to answer questions about weather and climate that arise in our day-to-day lives. Although introductory in nature, this fourth edition maintains scientific integrity and includes up-to-date information on important topics, such as El Niño, ozone depletion, and global warming. Discussion of weather events, such as the tornado outbreak of May, 1999, that brought destruction to parts of Oklahoma City, and the massive flooding wrought by tropical storm Allison over Texas and the southeastern United States during June, 2001, also are included. As was the case with the previous editions, no special prerequisites are necessary.

Written expressly for the student, this book emphasizes the understanding and application of meteorological principles. The text encourages watching the weather so that it becomes "alive," allowing readers to immediately apply textbook material to the world around them. To assist with this endeavor, a color Cloud Chart appears at the back of the text. The Cloud Chart can be separated from the book and used as a learning tool at any place one chooses to observe the sky. To strengthen points and clarify concepts, illustrations are rendered in full color throughout. Color photographs were carefully selected to illustrate features, stimulate interest, and show how exciting the study of weather can be.

This edition, organized into fifteen chapters, is designed to provide maximum flexibility to instructors of weather and climate courses. Thus, chapters can be covered in any desired order. For example, Chapter 12, "Air Pollution," and Chapter 15, "Light, Color, and Atmospheric Optics," are both self-contained and can be covered earlier if so desired. Instructors, then, are able to tailor this text to their particular needs. This book basically follows a traditional approach. After an introductory chapter on the origin, composition, and structure of the atmosphere, it covers solar energy, and air temperature, humidity, clouds, precipitation, and winds. Then comes a chapter on air masses, fronts, and middle-latitude cyclonic storms. Weather prediction and severe storms are next. A chapter on air pollution is followed by a chapter on global climate. A chapter on climate change is next. The final chapter deals with atmospheric optics.

Each chapter contains at least two Focus sections, which either expand on material in the main text or explore a subject closely related to what is being discussed. Focus sections fall into one of three distinct categories: Observations, Special Topics, and Environmental Issues. Some include material that is not always found in introductory meteorology textbooks—subjects such as sun burning and UV rays, fog dispersal, and aircraft icing. Others help bridge theory and practice. A Focus section new to this edition is "A Tropical Storm Named Allison."

Set apart as "Did You Know?" features in each chapter is weather information that may not be commonly known, yet pertains to the topic under discussion. Designed to bring the reader into the text, most of these weather highlights relate to some interesting weather fact or astonishing event. Many new "Did You Know?" items have been added to this edition.

Each chapter incorporates other effective learning aids:

- A major topic outline begins each chapter.
- Interesting introductory pieces draw the reader naturally into the main text.
- Important terms are boldfaced, with their definitions appearing in the glossary or in the text.
- Key phrases are italicized.
- English equivalents of metric units are immediately provided in parentheses.
- A brief review of the main points is placed toward the middle of most chapters.
- Intext callouts direct the student to the Meteorology Now program on the Book Companion web site.
- Summaries at the end of each chapter review the chapter's main ideas.
- A list of key terms following each chapter allows students to review and reinforce their knowledge of the main concepts they have encountered.
- Questions for Review act to check how well students assimilate the material.
- Questions for Thought and Exploration encourage students to synthesize learned concepts for a deeper understanding of the material.

Eight appendices conclude the book. Some are more technical than the main text, such as Appendix B, "Equations and Constants." Others can be used in observing the weather, such as Appendix F, "The Beaufort Wind Scale." In addition, at the end of the book, a compilation of supplementary material, as well as an extensive glossary, is presented.

FOURTH EDITION CHANGES

To help visualize how exciting weather and climate can be, many new color photographs and more than 50 new or revised color illustrations have been added to this edition. To complement the photographs and new art, the fourth edition of *Essentials of Meteorology* has been extensively updated and revised.

Chapter 1, "The Earth's Atmosphere,"still serves as a broad overview of the atmosphere. To help with this endeavor, many new photographs have been added. Chapter 2, "Warming the Earth and the Atmosphere," contains the latest information on greenhouse warming while Chapter 3, "Air Temperature," contains the latest wind-chill charts used by the National Weather Service and additional information on freezes.

Chapter 4, "Humdity, Condensation, and Clouds," now contains average dew-point temperature maps for the United States and southern Canada for January and July. Additional information on cloud seeding has been added to Chapter 5, "Cloud Development and Precipitation." The chapter on atmospheric circulations (Chapter 7) now contains information on the Pacific Decadal Oscillation (PDO). The chapter on air masses, fronts, and middle-latitude cyclones (Chapter 8) has been revised and expanded for clarity. Much of Chapter 9, "Weather Forecasting," has been restructured and revised with a new section added that describes the various types of forecasts.

Chapter 10, "Thunderstorms and Tornadoes," has been significantly reorganized and revised, and includes new art and new photos. Much of the section on thunderstorms has been rewritten. Moreover, new to this chapter are sections on supercell and nonsupercell tornadoes, including material on the formation of landspouts. Several sections of Chapter 11, "Hurricanes," have been revised. A new Focus section on tropical storm Allison now appears in the chapter.

Chapter 12, "Air Pollution," contains information on the new National Ambient Air Quality Standards and the Air Quality Index (AQI). The chapter on climate change (Chapter 14) has been significantly revised and rewritten to include the latest assessment on climate change from the 2001 Third Assessment Report (TAR) of the Intergovernmental Panel on Climate Change (IPCC).

ACKNOWLEDGMENTS

A special thank-you to the many individuals who have contributed to this fourth edition. A tremendous thank-you goes to my wife, Lita, for her invaluable assistance. My special thanks to Mabel Labiak for carefully reading the entire manuscript and to Jan Null for his researching some of the photos. Thanks to Janet Bollow for designing the book and for the time and careful attention she put into ensuring a beautiful product. Thanks to Stuart Kenter for his careful editing.

A special thanks to the many professional people at Brooks/Cole and Thomson who were instrumental in seeing that this fourth edition became a reality. Thanks to my friends and colleagues who provided comments, suggestions, and thoughtful input. I am indebted to those individuals who were kind enough to review all or part of this book, including:

William V. Ackerman
Ohio State University

Stephen K. Cox
Colorado State University

Nancy Dignon
Tallahassee Community College

Tim Doggett
Texas Tech University

Bart Geerts
University of Wyoming

Dennis Hartmann
University of Washington

Dan Johnson
Portland State University

Jon Kahl
University of Wisconsin, Milwaukee

Jeffrey K. Lew
UCLA

Larry McAdam
Seminole Community College

John Monteverdi
San Francisco State University

John Mullins
Florida Community College

Jan Null
Golden Gate Weather Service

Kyaw Tha Paw U
University of California, Davis

Mohan Ramamurthy
University of Illinois

Alan Robock
Rutgers University

Bob Weisman
St. Cloud State University

Andy White
University of Oklahoma

TO THE STUDENT

Learning about the atmosphere can be an enjoyable experience, especially if you become involved. This book is intended to give you some insight into the workings of the atmosphere, but for a real appreciation of your atmospheric environment, you must go outside and observe. Mountains take millions of years to form, while a cumulus cloud can develop into a raging thunderstorm in less than an hour. To help with your observations, a color Cloud Chart is bound at the back of the book for easy reference. Remove it and keep it with you. And, remember, all of the information in this book is out there—please, take the time to look.

Donald Ahrens

Photo by author

We live at the bottom of a swirling ocean of air. Here, air billowing up from the earth's surface forms into clouds and thunderstorms over the warm landmass of North America.

Meteorology Now™ This icon, appearing throughout the book, indicates an opportunity to explore interactive tutorials, animations, or practice problems available on the MeteorologyNow Web site at http://earthscience.brookscole.com/ahrens/ess4e.

The Earth's Atmosphere

I well remember a brilliant red balloon which kept me completely happy for a whole afternoon, until, while I was playing, a clumsy movement allowed it to escape. Spellbound, I gazed after it as it drifted silently away, gently swaying, growing smaller and smaller until it was only a red point in a blue sky. At that moment I realized, for the first time, the vastness above us: a huge space without visible limits. It was an apparent void, full of secrets, exerting an inexplicable power over all the earth's inhabitants. I believe that many people, consciously or unconsciously, have been filled with awe by the immensity of the atmosphere. All our knowledge about the air, gathered over hundreds of years, has not diminished this feeling.

Theo Loebsack, *Our Atmosphere*

CONTENTS

Our *atmosphere* is a delicate life-giving blanket of air that surrounds the fragile earth. In one way or another, it influences everything we see and hear—it is intimately connected to our lives. Air is with us from birth, and we cannot detach ourselves from its presence. In the open air, we can travel for many thousands of kilometers in any horizontal direction, but should we move a mere eight kilometers above the surface, we would suffocate. We may be able to survive without food for a few weeks, or without water for a few days, but, without our atmosphere, we would not survive more than a few minutes. Just as fish are confined to an environment of water, so we are confined to an ocean of air. Anywhere we go, it must go with us.

The earth without an atmosphere would have no lakes or oceans. There would be no sounds, no clouds, no red sunsets. The beautiful pageantry of the sky would be absent. It would be unimaginably cold at night and unbearably hot during the day. All things on the earth would be at the mercy of an intense sun beating down upon a planet utterly parched.

Living on the surface of the earth, we have adapted so completely to our environment of air that we sometimes forget how truly remarkable this substance is. Even though air is tasteless, odorless, and (most of the time) invisible, it protects us from the scorching rays of the sun and provides us with a mixture of gases that allows life to flourish. Because we cannot see, smell, or taste air, it may seem surprising that between your eyes and the pages of this book are trillions of air molecules. Some of these may have been in a cloud only yesterday, or over another continent last week, or perhaps part of the life-giving breath of a person who lived hundreds of years ago.

Warmth for our planet is provided primarily by the sun's energy. At an average distance from the sun of nearly 150 million kilometers (km), or 93 million miles (mi), the earth intercepts only a very small fraction of the sun's total energy output. However, it is this radiant energy* that drives the atmosphere into the patterns of everyday wind and weather, and allows life to flourish.

At its surface, the earth maintains an average temperature of about 15°C (59°F).** Although this temperature is mild, the earth experiences a wide range of temperatures, as readings can drop below −85°C (−121°F) during a frigid Antarctic night and climb during the day, to above 50°C (122°F) on the oppressively hot, subtropical desert.

*Radiant energy, or radiation, is energy transferred in the form of waves that have electrical and magnetic properties. The light that we see is radiation, as is ultraviolet light. More on this important topic is given in Chapter 2.

**The abbreviation °C is used when measuring temperature in degrees Celsius, and °F is the abbreviation for degrees Fahrenheit. More information about temperature scales is given in Appendix A and in Chapter 2.

DID YOU KNOW?

If the earth were to shrink to the size of a large beach ball, its inhabitable atmosphere would be thinner than a piece of paper.

In this chapter, we will examine a number of important concepts and ideas about the earth's atmosphere, many of which will be expanded in subsequent chapters.

OVERVIEW OF THE EARTH'S ATMOSPHERE

The earth's **atmosphere** is a thin, gaseous envelope comprised mostly of nitrogen (N_2) and oxygen (O_2), with small amounts of other gases, such as water vapor (H_2O) and carbon dioxide (CO_2). Nested in the atmosphere are clouds of liquid water and ice crystals.

Although our atmosphere extends upward for many hundreds of kilometers, almost 99 percent of the atmosphere lies within a mere 30 km (about 19 mi) of the earth's surface (see Fig. 1.1). This thin blanket of air constantly shields the surface and its inhabitants from the sun's dangerous ultraviolet radiant energy, as well as from the onslaught of material from interplanetary space. There is no definite upper limit to the atmosphere; rather, it becomes thinner and thinner, eventually merging with empty space, which surrounds all the planets.

Composition of the Atmosphere Table 1.1 shows the various gases present in a volume of air near the earth's surface. Notice that **nitrogen** (N_2) occupies about 78 percent and **oxygen** (O_2) about 21 percent of the total volume of dry air. If all the other gases are removed, these percentages for nitrogen and oxygen hold fairly constant up to an elevation of about 80 km (or 50 mi).

At the surface, there is a balance between destruction (output) and production (input) of these gases. For example, nitrogen is removed from the atmosphere primarily by biological processes that involve soil bacteria. In addition, nitrogen is taken from the air by tiny ocean-dwelling plankton that convert it into nutrients that help fortify the ocean's food chain. It is returned to the atmosphere mainly through the decaying of plant and animal matter. Oxygen, on the other hand, is removed from the atmosphere when organic matter decays and when oxygen combines with other substances, producing oxides. It is also taken from the atmosphere during breathing, as the lungs take in oxy-

NASA photo

FIGURE 1.1 The earth's atmosphere as viewed from space during sunrise. About 90 percent of the earth's atmosphere is within the bright area and about 70 percent lies below the top of the highest cloud.

gen and release carbon dioxide. The addition of oxygen to the atmosphere occurs during photosynthesis, as plants, in the presence of sunlight, combine carbon dioxide and water to produce sugar and oxygen.

The concentration of the invisible gas **water vapor,** however, varies greatly from place to place, and from time to time. Close to the surface in warm, steamy, tropical locations, water vapor may account for up to 4 percent of the atmospheric gases, whereas in colder arctic areas, its concentration may dwindle to a mere fraction of a percent. Water vapor molecules are, of course, invisible. They become visible only when they transform into larger liquid or solid particles, such as cloud droplets and ice crystals. The changing of water vapor into liquid water is called *condensation,* whereas the process of liquid water becoming water vapor is called *evaporation.* In the lower atmosphere, water is everywhere. It is the only substance that exists as a gas, a liquid, and a solid at those temperatures and pressures normally found near the earth's surface (see Fig. 1.2).

Water vapor is an *extremely important* gas in our atmosphere. Not only does it form into both liquid and solid cloud particles that grow in size and fall to earth as precipitation, but it also releases large amounts of heat—called *latent heat*—when it changes from vapor into liquid water or ice. Latent heat is an important

TABLE 1.1 Composition of the Atmosphere Near the Earth's Surface

PERMANENT GASES			VARIABLE GASES			
Gas	*Symbol*	*Percent (by Volume) Dry Air*	*Gas (and Particles)*	*Symbol*	*Percent (by Volume)*	*Parts per Million (ppm)**
Nitrogen	N_2	78.08	Water vapor	H_2O	0 to 4	
Oxygen	O_2	20.95	Carbon dioxide	CO_2	0.037	375*
Argon	Ar	0.93	Methane	CH_4	0.00017	1.7
Neon	Ne	0.0018	Nitrous oxide	N_2O	0.00003	0.3
Helium	He	0.0005	Ozone	O_3	0.000004	0.04†
Hydrogen	H_2	0.00006	Particles (dust, soot, etc.)		0.000001	0.01–0.15
Xenon	Xe	0.000009	Chlorofluorocarbons (CFCs)		0.00000002	0.0002

*For CO_2, 375 parts per million means that out of every million air molecules, 375 are CO_2 molecules.

†Stratospheric values at altitudes between 11 km and 50 km are about 5 to 12 ppm.

Photo by author

FIGURE 1.2 The earth's atmosphere is a rich mixture of many gases, with clouds of condensed water vapor and ice crystals. Here, water evaporates from the ocean's surface. Rising air currents then transform the invisible water vapor into many billions of tiny liquid droplets that appear as puffy cumulus clouds. If the rising air in the cloud should extend to greater heights, where air temperatures are quite low, some of the liquid droplets would freeze into minute ice crystals.

source of atmospheric energy, especially for storms, such as thunderstorms and hurricanes. Moreover, water vapor is a potent *greenhouse gas* because it strongly absorbs a portion of the earth's outgoing radiant energy (somewhat like the glass of a greenhouse prevents the heat inside from escaping and mixing with the outside air). Thus, water vapor plays a significant role in the earth's heat-energy balance.

Carbon dioxide (CO_2), a natural component of the atmosphere, occupies a small (but important) percent of a volume of air, about 0.037 percent. Carbon dioxide enters the atmosphere mainly from the decay of vegetation, but it also comes from volcanic eruptions, the exhalations of animal life, from the burning of fossil fuels (such as coal, oil, and natural gas), and from deforestation. The removal of CO_2 from the atmosphere takes place during *photosynthesis,* as plants consume CO_2 to produce green matter. The CO_2 is then stored in roots, branches, and leaves. The oceans act as a huge reservoir for CO_2, as phytoplankton (tiny drifting plants) in surface water fix CO_2 into organic tissues. Carbon dioxide that dissolves directly into surface water mixes downward and circulates through greater depths. Estimates are that the oceans hold more than 50 times the total atmospheric CO_2 content.

Figure 1.3 reveals that the atmospheric concentration of CO_2 has risen more than 18 percent since 1958, when it was first measured at Mauna Loa Observatory in Hawaii. This increase means that CO_2 is entering the atmosphere at a greater rate than it is being removed. The increase appears to be due mainly to the burning of fossil fuels; however, deforestation also plays a role as cut timber, burned or left to rot, releases CO_2 directly into the air, perhaps accounting for about 20 percent of the observed increase. Measurements of CO_2 also come from ice cores. In Greenland and Antarctica, for example, tiny bubbles of air trapped within the ice sheets reveal that before the industrial revolution, CO_2 levels were stable at about 280 parts per million (ppm). Since the early 1800s, however, CO_2 levels have increased by more than 25 percent. With CO_2 levels presently increasing by about 0.4 percent annually (1.5 ppm/year), scientists now estimate that the concentration of CO_2 will likely rise from its current value of about 375 ppm to a value near 500 ppm toward the end of this century.

Carbon dioxide is another important greenhouse gas because, like water vapor, it traps a portion of the earth's outgoing energy. Consequently, with everything else being equal, as the atmospheric concentration of CO_2 increases, so should the average global surface air temperature. Most of the mathematical model experiments that predict future atmospheric conditions estimate that increasing levels of CO_2 (and other greenhouse gases) will result in a *global warming* of surface air between 1.4°C and 5.8°C (about 2.5°F and 10.5°F) by the year 2100. Such warming (as we will learn in more detail in Chapter 14) could result in a variety of consequences, such as increasing precipitation in certain areas and reducing it in others as the global air currents that guide the major storm systems across the earth begin to shift from their "normal" paths.

Carbon dioxide and water vapor are not the only greenhouse gases. Recently, others have been gaining notoriety, primarily because they, too, are becoming more concentrated. Such gases include *methane* (CH_4), *nitrous oxide* (N_2O), and *chlorofluorocarbons* (CFCs).*

Levels of methane, for example, have been rising over the past century, increasing recently by about one-half of one percent per year. Most methane appears to derive from the breakdown of plant material by certain bacteria in rice paddies, wet oxygen-poor soil, the biological activity of termites, and biochemical reactions in the stomachs of cows. Just why methane should be increasing so rapidly is currently under study. Levels of nitrous oxide—commonly known as laughing gas—have been rising annually at the rate of about one-quarter of a percent. Nitrous oxide forms in the soil through a

*Because these gases (including CO_2) occupy only a small fraction of a percent in a volume of air near the surface, they are referred to collectively as *trace gases.*

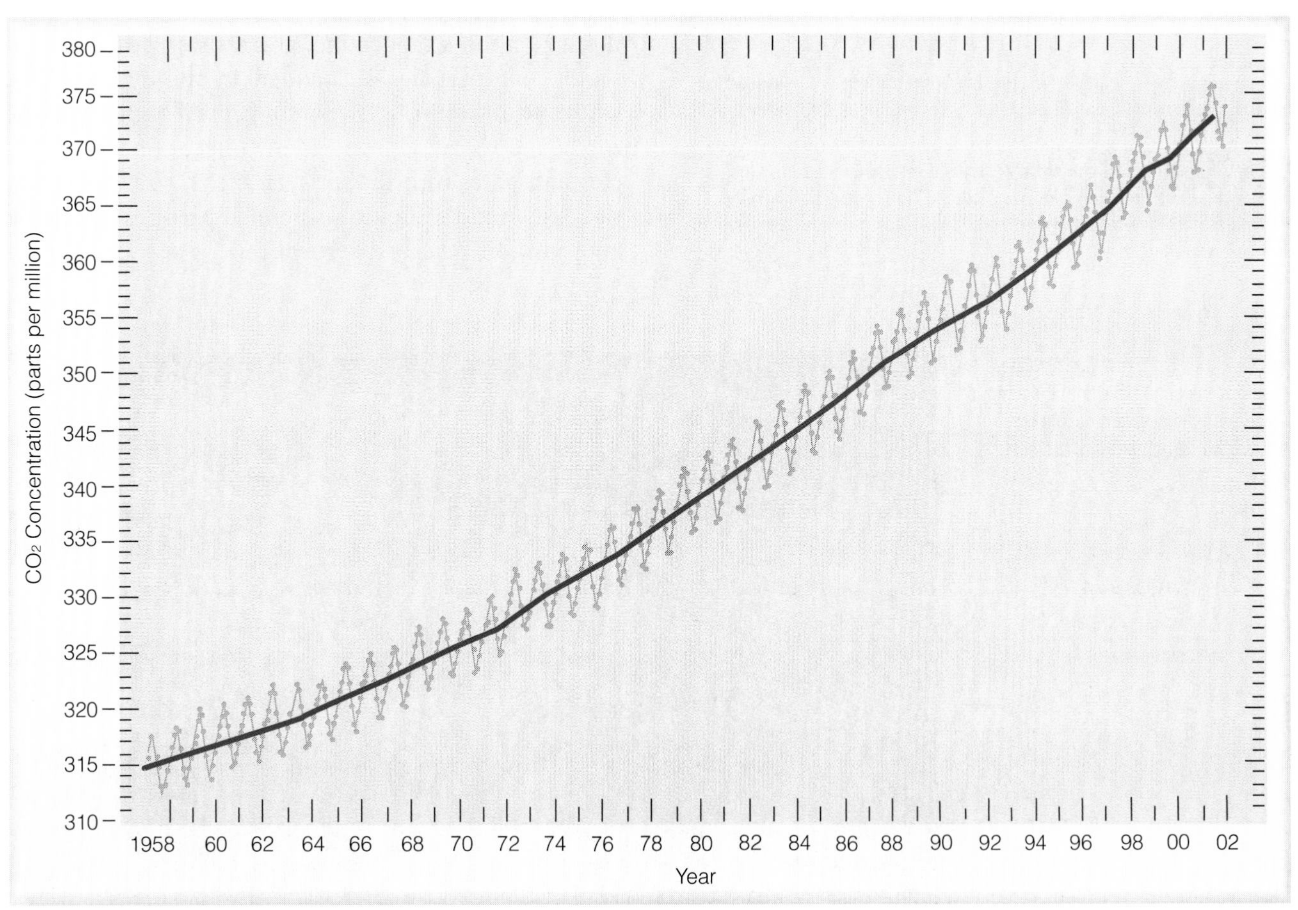

FIGURE 1.3 Measurements of CO_2 in parts per million (ppm) at Mauna Loa Observatory, Hawaii. Higher readings occur in winter when plants die and release CO_2 to the atmosphere. Lower readings occur in summer when more abundant vegetation absorbs CO_2 from the atmosphere. The solid line is the average yearly value.

chemical process involving bacteria and certain microbes. Ultraviolet light from the sun destroys it.

Chlorofluorocarbons represent a group of greenhouse gases that, up until recently, had been increasing in concentration. At one time, they were the most widely used propellants in spray cans. Today, however, they are mainly used as refrigerants, as propellants for the blowing of plastic-foam insulation, and as solvents for cleaning electronic microcircuits. Although their average concentration in a volume of air is quite small (see Table 1.1), they have an important effect on our atmosphere as they not only have the potential for raising global temperatures, they also play a part in destroying the gas ozone in the stratosphere.*

At the surface, **ozone** (O_3) is the primary ingredient of *photochemical smog,*** which irritates the eyes and throat and damages vegetation. But the majority of atmospheric ozone (about 97 percent) is found in the upper atmosphere—in the stratosphere—where it is formed naturally, as oxygen atoms combine with oxygen molecules. Here, the concentration of ozone averages less than 0.002 percent by volume. This small quantity is important, however, because it shields plants, animals, and humans from the sun's harmful ultraviolet rays. It is ironic that ozone, which damages plant life in a polluted environment, provides a natural protective shield in the upper atmosphere so that plants on the surface may survive. We will see in Chapter 12 that when CFCs enter the stratosphere, ultraviolet rays break them apart, and the CFCs release ozone-destroying *chlorine.* Because of this effect, ozone concentration in the stratosphere has been decreasing over parts of the Northern and Southern Hemispheres. The reduction in stratospheric ozone levels over springtime Antarctica has plummeted at such an alarming rate that during September and October, there is an *ozone hole* over the region. (We will examine the ozone hole situation, as well as photochemical ozone, in Chapter 12.)

Impurities from both natural and human sources are also present in the atmosphere: Wind picks up dust and soil from the earth's surface and carries it aloft; small saltwater drops from ocean waves are swept into the air (upon evaporating, these drops leave microscopic salt particles suspended in the atmosphere); smoke from forest fires is often carried high above the earth; and volcanoes spew many tons of fine ash particles and gases into the air (see Fig. 1.4). Collectively, these tiny solid or liquid suspended particles of various composition are called **aerosols.**

*The stratosphere is located at an altitude between about 11 km and 50 km above the earth's surface.

**Originally the word *smog* meant the combining of smoke and fog. Today, however, the word usually refers to the type of smog that forms in large cities, such as Los Angeles, California. Because this type of smog forms when chemical reactions take place in the presence of sunlight, it is termed *photochemical smog.*

FIGURE 1.4 Erupting volcanoes can send tons of particles into the atmosphere, along with vast amounts of water vapor, carbon dioxide, and sulfur dioxide.

Some natural impurities found in the atmosphere are quite beneficial. Small, floating particles, for instance, act as surfaces on which water vapor condenses to form clouds. However, most human-made impurities (and some natural ones) are a nuisance, as well as a health hazard. These we call **pollutants.** For example, automobile engines emit copious amounts of *nitrogen dioxide* (NO_2), *carbon monoxide* (CO), and *hydrocarbons.* In sunlight, nitrogen dioxide reacts with hydrocarbons and other gases to produce ozone. Carbon monoxide is a major pollutant of city air. Colorless and odorless, this poisonous gas forms during the incomplete combustion of carbon-containing fuel. Hence, over 75 percent of carbon monoxide in urban areas comes from road vehicles.

The burning of sulfur-containing fuels (such as coal and oil) releases the colorless gas *sulfur dioxide* (SO_2) into the air. When the atmosphere is sufficiently moist, the SO_2 may transform into tiny dilute drops of sulfuric acid. Rain containing sulfuric acid corrodes metals and painted surfaces, and turns freshwater lakes acidic. *Acid rain* (thoroughly discussed in Chapter 12) is a major environmental problem, especially downwind from major industrial areas. In addition, high concentrations of SO_2 produce serious respiratory problems in humans, such as bronchitis and emphysema, and have an adverse effect on plant life. (More information on these and other pollutants is given in Chapter 12.)

The Early Atmosphere The atmosphere that originally surrounded the earth was probably much different from the air we breathe today. The earth's first atmosphere (some 4.6 billion years ago) was most likely *hydrogen* and *helium*—the two most abundant gases found in the universe—as well as hydrogen compounds, such as methane and ammonia. Most scientists feel that this early atmosphere escaped into space from the earth's hot surface.

A second, more dense atmosphere, however, gradually enveloped the earth as gases from molten rock within its hot interior escaped through volcanoes and steam vents. We assume that volcanoes spewed out the same gases then as they do today: mostly water vapor (about 80 percent), carbon dioxide (about 10 percent), and up to a few percent nitrogen. These gases (mostly water vapor and carbon dioxide) probably created the earth's second atmosphere.

As millions of years passed, the constant outpouring of gases from the hot interior—known as **outgassing**—provided a rich supply of water vapor, which formed into clouds.* Rain fell upon the earth for many thousands of years, forming the rivers, lakes, and oceans of the world. During this time, large amounts of CO_2 were dissolved in the oceans. Through chemical and biological processes, much of the CO_2 became locked up in carbonate sedimentary rocks, such as limestone. With much of the water vapor already condensed and the concentration of CO_2 dwindling, the atmosphere gradually became rich in nitrogen (N_2), which is usually not chemically active.

It appears that oxygen (O_2), the second most abundant gas in today's atmosphere, probably began an extremely slow increase in concentration as energetic rays from the sun split water vapor (H_2O) into hydrogen and oxygen. The hydrogen, being lighter, probably rose and escaped into space, while the oxygen remained in the atmosphere.

This slow increase in oxygen may have provided enough of this gas for primitive plants to evolve, perhaps 2 to 3 billion years ago. Or the plants may have evolved in an almost oxygen-free (anaerobic) environment. At any rate, plant growth greatly enriched our atmosphere with oxygen. The reason for this enrichment is that, during the process of photosynthesis, plants, in the presence of sunlight, combine carbon dioxide and water to produce oxygen. Hence, after plants evolved, the atmospheric oxygen content increased more rapidly, probably reaching its present composition about several hundred million years ago.

BRIEF REVIEW

Before going on to the next several sections, here is a review of some of the important concepts presented so far:

- The earth's atmosphere is a mixture of many gases. In a volume of dry air near the surface, nitrogen (N_2) occupies about 78 percent and oxygen (O_2) about 21 percent.
- Water vapor, which normally occupies less than 4 percent in a volume of air near the surface, can condense into liquid cloud droplets or transform into delicate ice crystals. Water is the only substance in our atmosphere that is found naturally as a gas (water vapor), as a liquid (water), and as a solid (ice).
- Both water vapor and carbon dioxide (CO_2) are important greenhouse gases.
- The majority of water on our planet is believed to have come from its hot interior through outgassing.

*It is now believed that some of the earth's water may have originated from numerous collisions with small meteors and disintegrating comets when the earth was very young.

DID YOU KNOW?

The air density in the mile-high city of Denver, Colorado, is normally about 15 percent less than the air density at sea level. As the air density decreases, the drag force on a hit baseball also decreases. Because of this fact, a baseball will travel farther when hit at Denver's Coors Field compared to one hit at sea level. Hence, a baseball hit for a 340-foot home run down the left field line on a warm, calm day at Coors Field in Denver, Colorado, would simply be a 300-foot out if hit at Camden Yards Stadium in Baltimore, Maryland.

VERTICAL STRUCTURE OF THE ATMOSPHERE

A vertical profile of the atmosphere reveals that it can be divided into a series of layers. Each layer may be defined in a number of ways: by the manner in which the air temperature varies through it, by the gases that comprise it, or even by its electrical properties. At any rate, before we examine these various atmospheric layers, we need to look at the vertical profile of two important variables: air pressure and air density.

A Brief Look at Air Pressure and Air Density Air molecules (as well as everything else) are held near the earth by *gravity.* This strong, invisible force pulling down on the air above squeezes (compresses) air molecules closer together, which causes their number in a given volume to increase. The more air above a level, the greater the squeezing effect or compression. Since **air density** is the number of air molecules in a given space (volume), it follows that air density is greatest at the surface and decreases as we move up into the atmosphere. Notice in Fig. 1.5 that, owing to the fact that the air near the surface is compressed, air density normally decreases rapidly at first, then more slowly as we move farther away from the surface.

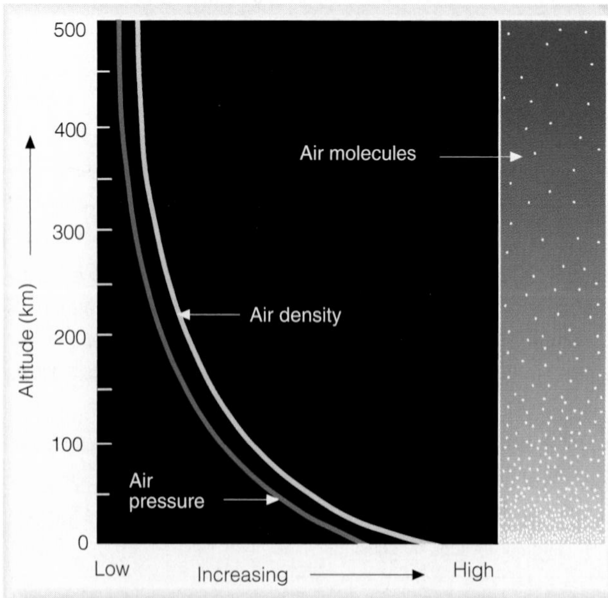

FIGURE 1.5 Both air pressure and air density decrease with increasing altitude.

Air molecules have weight.* In fact, air is surprisingly heavy. The weight of all the air around the earth is a staggering 5600 trillion tons. The weight of the air molecules acts as a force upon the earth. The amount of force exerted over an area of surface is called *atmospheric pressure* or, simply, **air pressure.**** The pressure at any level in the atmosphere may be measured in terms of the total mass of the air above any point. As we climb in elevation, fewer air molecules are above us; hence, *atmospheric pressure always decreases with increasing height.* Like air density, air pressure decreases rapidly at first, then more slowly at higher levels (see Fig. 1.5).

If we weigh a column of air 1 square inch in cross section, extending from the average height of the ocean surface (sea level) to the "top" of the atmosphere, it would weigh very nearly 14.7 pounds. Thus, normal atmospheric pressure near sea level is close to 14.7 pounds per square inch. If more molecules are packed into the column, it becomes more dense, the air weighs more, and the surface pressure goes up. On the other hand, when fewer molecules are in the column, the air weighs less, and the surface pressure goes down. So, a change in air density can bring about a change in air pressure.

Pounds per square inch is, of course, just one way to express air pressure. Presently, the most common unit for air pressure found on surface weather maps is the *millibar* (mb), although the *hectopascal*† (hPa) is gradually replacing the millibar as the preferred unit of pressure on surface maps. Another unit of pressure is *inches*

*The *weight* of an object, including air, is the force acting on the object due to gravity. In fact, weight is defined as the mass of an object times the acceleration of gravity. An object's *mass* is the quantity of matter in the object. Consequently, the mass of air in a rigid container is the same everywhere in the universe. However, if you were to instantly travel to the moon, where the acceleration of gravity is one-sixth that of earth, the mass of air in the container would be the same, but its weight would decrease to one-sixth of its earth value.

**Because air pressure is measured with an instrument called a *barometer,* atmospheric pressure is often referred to as *barometric pressure.*

†One hectopascal equals 1 millibar.

of mercury (Hg), which is commonly used both in the field of aviation and in television and radio weather broadcasts. At sea level, the *standard value* for atmospheric pressure is

$$1013.25 \text{ mb} = 1013.25 \text{ hPa} = 29.92 \text{ in. Hg.}$$

Figure 1.6 (and Fig. 1.5) illustrates how rapidly air pressure decreases with height. Near sea level, atmospheric pressure decreases rapidly, whereas at high levels it decreases more slowly. With a sea-level pressure near 1000 mb, we can see in Fig. 1.6 that, at an altitude of only 5.5 km (or 3.5 mi), the air pressure is about 500 mb, or half of the sea-level pressure. This situation means that, if you were at a mere 5.5 km (which is about 18,000 feet) above the surface, you would be above one-half of all the molecules in the atmosphere.

At an elevation approaching the summit of Mount Everest (about 9 km or 29,000 ft), the air pressure would be about 300 mb. The summit is above nearly 70 percent of all the molecules in the atmosphere. At an altitude of about 50 km, the air pressure is about 1 mb, which means that 99.9 percent of all the air molecules are below this level. Yet the atmosphere extends upwards for many hundreds of kilometers, gradually becoming thinner and thinner until it ultimately merges with outer space.

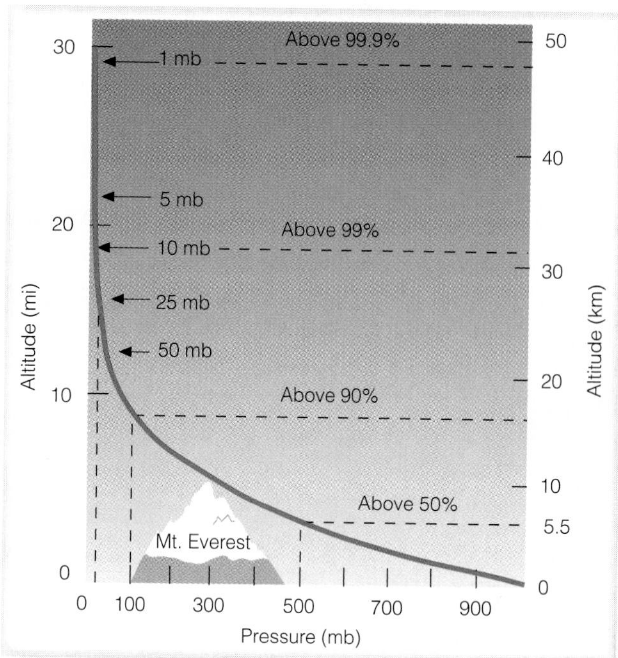

FIGURE 1.6 Atmospheric pressure decreases rapidly with height. Climbing to an altitude of only 5.5 km, where the pressure is 500 mb, would put you above one-half of the atmosphere's molecules.

Layers of the Atmosphere We have seen that both air pressure and density decrease with height above the earth—rapidly at first, then more slowly. *Air temperature,* however, has a more complicated vertical profile.*

Look closely at Fig. 1.7 and notice that air temperature normally decreases from the earth's surface up to an altitude of about 11 km, which is nearly 36,000 ft, or 7 mi. This decrease in air temperature with increasing height is due primarily to the fact (investigated further in Chapter 2) that sunlight warms the earth's surface, and the surface, in turn, warms the air above it. The rate at which the air temperature decreases with height is called the temperature **lapse rate.** The *average* (or *standard) lapse rate* in this region of the lower atmosphere is about 6.5 degrees Celsius (°C) for every 1000 meters (m) or about 3.6 degrees Fahrenheit (°F) for every 1000 ft rise in elevation. Keep in mind that these values are only averages. On some days, the air becomes colder more quickly as we move upward, which would increase or steepen the lapse rate. On other days, the air temperature would decrease more slowly with height, and the lapse rate would be less. Occasionally, the air temperature may actually *increase* with height, producing a condition known as a **temperature inversion.** So the lapse rate fluctuates, varying from day to day and season to season. (The instrument that measures the vertical profile of air temperature in the atmosphere up to an elevation sometimes exceeding 30 km (100,000 ft) is the **radiosonde.** More information on this instrument is given in the Focus section on p. 11.)

The region of the atmosphere from the surface up to about 11 km contains all of the weather we are familiar with on earth. Also, this region is kept well stirred by rising and descending air currents. Here, it is common for air molecules to circulate through a depth of more than 10 km in just a few days. This region of circulating air extending upward from the earth's surface to where the air stops becoming colder with height is called the

Air temperature is the degree of hotness or coldness of the air and, as we will see in Chapter 2, it is also a measure of the average speed of the air molecules.

DID YOU KNOW?

Air temperature normally decreases with increasing height above the surface; thus, if you are flying in a jet aircraft at about 9 km (30,000 ft), the air temperature just outside your window would typically be about −50°C (−58°F)—more than 60°C (108°F) colder than the air at the earth's surface, directly below you.

troposphere—from the Greek *tropein,* meaning to turn, or to change.

Notice in Fig. 1.7 that just above 11 km the air temperature normally stops decreasing with height. Here, the lapse rate is zero. This region, where, on average, the air temperature remains constant with height, is referred to as an *isothermal* (equal temperature) zone. *The bottom of this zone marks the top of the troposphere and the beginning of another layer, the **stratosphere.** The boundary separating the troposphere from the stratosphere is called the **tropopause.** The height of the tropopause varies. It is normally found at higher elevations over equatorial regions, and it decreases in elevation as we travel poleward. Generally, the tropopause is higher in summer and lower in winter at all latitudes. In some regions, the tropopause "breaks" and is difficult to locate and, here, scientists have observed tropospheric air mixing with stratospheric air and vice versa. These breaks also mark the position of *jet streams*—high winds that meander in a narrow channel like an old river, often at speeds exceeding 100 knots.*

From Fig. 1.7 we can see that, in the stratosphere at an altitude near 20 km (12 mi), the air temperature begins to increase with height, producing a *temperature inversion.* The inversion region, along with the lower isothermal layer, tends to keep the vertical currents of the troposphere from spreading into the stratosphere. The inversion also tends to reduce the amount of verti-

*In many instances, the isothermal layer is not present and the air temperature begins to increase with increasing height.

*A knot is a nautical mile per hour. One knot is equal to 1.15 miles per hour (mi/hr), or 1.9 kilometers per hour (km/hr).

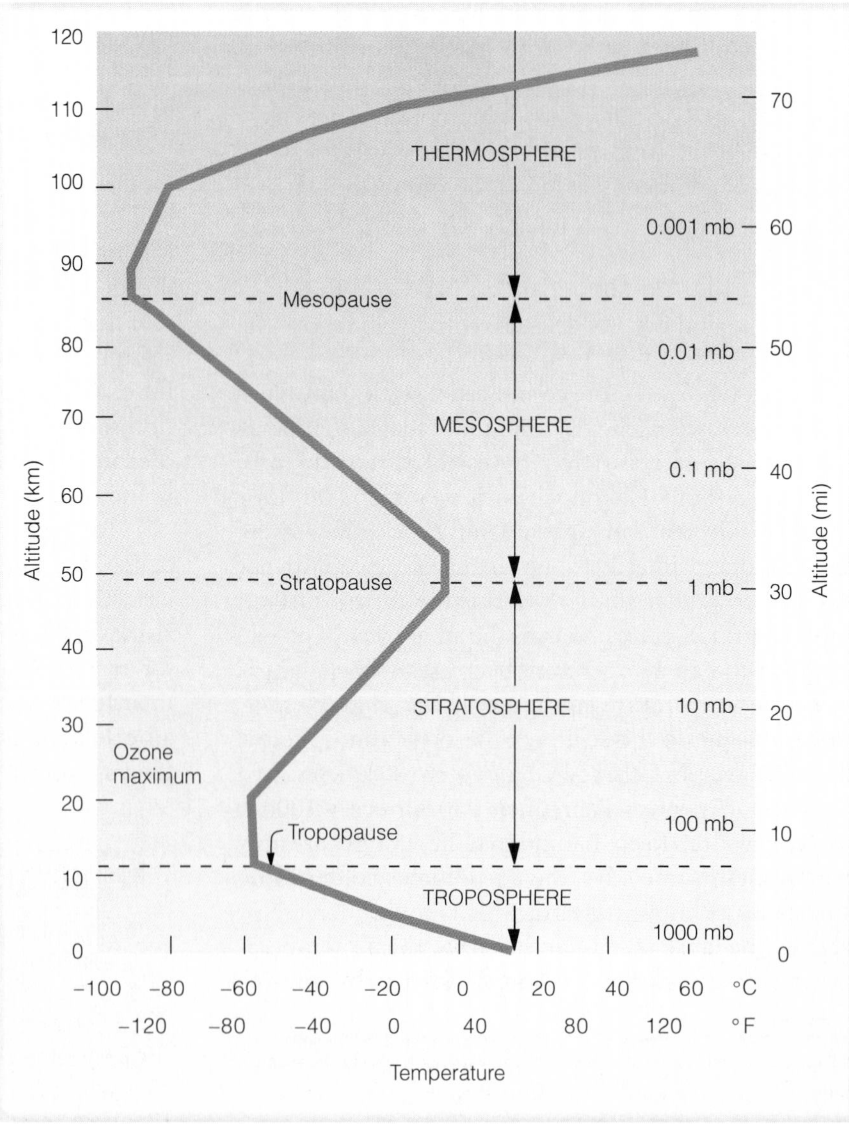

FIGURE 1.7 Layers of the atmosphere as related to the average profile of air temperature above the earth's surface. The heavy line illustrates how the average temperature varies in each layer.

FOCUS ON AN OBSERVATION

The Radiosonde

The vertical distribution of temperature, pressure, and humidity up to an altitude of about 30 km can be obtained with an instrument called a *radiosonde.** The radiosonde is a small, lightweight box equipped with weather instruments and a radio transmitter. It is attached to a cord that has a parachute and a gas-filled balloon tied tightly at the end (see Fig. 1). As the balloon rises, the attached radiosonde measures air temperature with a small electrical thermometer—a thermistor—located just outside the box. The radiosonde measures humidity electrically by sending an electric current across a carbon-coated plate. Air pressure is obtained by a small barometer located inside the box. All of this information is transmitted to the surface by radio. Here, a computer rapidly reconverts the various frequencies into values of temperature, pressure, and moisture. Special tracking equipment at the surface may also be used to provide a vertical profile of winds.* (When winds are added, the observation is called a *rawinsonde.*) When plotted on a graph, the vertical distribution of temperature, humidity, and wind is called a *sounding.* Eventually, the balloon bursts and the radiosonde returns to earth, its descent being slowed by its parachute.

At most sites, radiosondes are released twice a day, usually at the time that corresponds to midnight and noon in Greenwich, England. Releasing radiosondes is an expensive operation because many of the instruments are never retrieved, and many of those that are retrieved are often in poor working condition. To complement the radiosonde, modern geostationary satellites (using instruments that measure radiant energy) are providing scientists with vertical temperature profiles in inaccessible regions.

*A radiosonde that is dropped by parachute from an aircraft is called a *dropsonde.*

*A modern development in the radiosonde is the use of satellite Global Positioning System (GPS) equipment. Radiosondes can be equipped with a GPS device that provides more accurate position data back to the computer for wind computations.

Photo by author

FIGURE 1 The radiosonde with parachute and balloon.

cal motion in the stratosphere itself; hence, it is a stratified layer. Even though the air temperature is increasing with height, the air at an altitude of 30 km is extremely cold, averaging less than −46°C.

The reason for the inversion in the stratosphere is that the gas ozone plays a major part in heating the air at this altitude. Recall that ozone is important because it absorbs energetic ultraviolet (UV) solar energy. Some of this absorbed energy warms the stratosphere, which explains why there is an inversion. If ozone were not present, the air probably would become colder with height, as it does in the troposphere.*

Above the stratosphere is the **mesosphere** (middle sphere). The air here is extremely thin and the atmospheric pressure is quite low (again, refer back to Fig. 1.7). Even though the percentage of nitrogen and oxygen in the mesosphere is about the same as it was at the earth's surface, a breath of mesospheric air contains far fewer oxygen molecules than a breath of tropospheric air. At this level, without proper oxygen-breathing equipment, the brain would soon become oxygen-starved—a condition known as *hypoxia*—and suffocation would result. With an average temperature of −90°C, the top of the mesosphere represents the coldest part of our atmosphere.

The "hot layer" above the mesosphere is the **thermosphere.** Here, oxygen molecules (O_2) absorb energetic solar rays, warming the air. In the thermosphere, there are relatively few atoms and molecules. Consequently, the absorption of a small amount of energetic solar energy can cause a large increase in air temperature that may exceed 500°C, or 900°F (see Fig. 1.8).

*Recall from an earlier discussion that the concentration of stratospheric ozone is decreasing over portions of the globe as chlorofluorocarbons break apart and release ozone-destroying chlorine in the process. Again, additional material on this topic is given in Chapter 12.

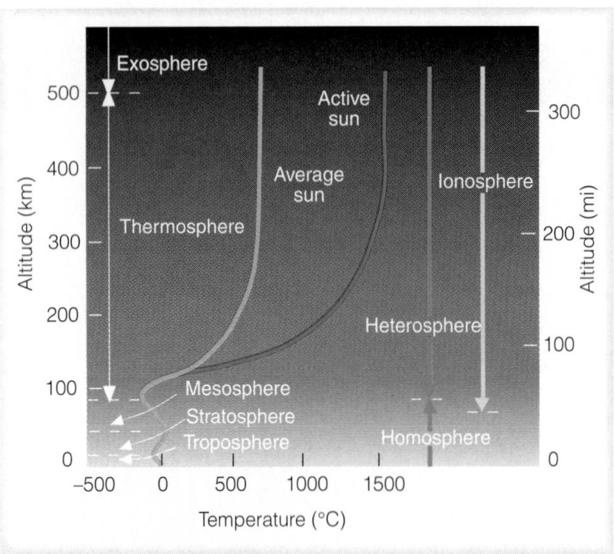

FIGURE 1.8 Layers of the atmosphere based on temperature (red line), composition (green line), and electrical properties (blue line).

Even though the temperature in the thermosphere is exceedingly high, a person shielded from the sun would not necessarily feel hot. The reason for this fact is that there are too few molecules in this region of the atmosphere to bump against something (exposed skin, for example) and transfer enough heat to it to make it feel warm. The low density of the thermosphere also means that an air molecule will move an average distance of over one kilometer before colliding with another molecule. A similar air molecule at the earth's surface will move an average distance of less than one millionth of a centimeter before it collides with another molecule.

At the top of the thermosphere, about 500 km (300 mi) above the earth's surface, molecules can move great distances before they collide with other molecules. Here, many of the lighter, faster-moving molecules traveling in the right direction actually escape the earth's gravitational pull. The region where atoms and molecules shoot off into space is sometimes referred to as the *exosphere,* which represents the upper limit of our atmosphere.

Up to this point, we have examined the atmospheric layers based on the vertical profile of temperature. The atmosphere, however, may also be divided into layers based on its composition. For example, the composition of the atmosphere begins to slowly change in the lower part of the thermosphere. Below the thermosphere, the composition of air remains fairly uniform (78% nitrogen, 21% oxygen) by turbulent mixing. This lower, well-mixed region is known as the *homosphere* (see Fig. 1.8).

In the thermosphere, collisions between atoms and molecules are infrequent, and the air is unable to keep itself stirred. As a result, diffusion takes over as heavier atoms and molecules (such as oxygen and nitrogen) tend to settle to the bottom of the layer, while lighter gases (such as hydrogen and helium) float to the top. The region from about the base of the thermosphere to the top of the atmosphere is often called the *heterosphere.*

The Ionosphere The **ionosphere** is not really a layer, but rather an electrified region within the upper atmosphere where fairly large concentrations of ions and free electrons exist. *Ions* are atoms and molecules that have lost (or gained) one or more electrons. Atoms lose electrons and become positively charged when they cannot absorb all of the energy transferred to them by a colliding energetic particle or the sun's energy.

The lower region of the ionosphere is usually about 60 km above the earth's surface. From here (60 km), the ionosphere extends upward to the top of the atmosphere. Hence, the bulk of the ionosphere is in the thermosphere (see Fig. 1.8).

The ionosphere plays a major role in radio communications. The lower part (called the *D* region) reflects standard AM radio waves back to earth, but at the same time it seriously weakens them through absorption. At night, though, the *D* region gradually disappears and AM radio waves are able to penetrate higher into the ionosphere (into the *E* and *F* regions—see Fig. 1.9), where the waves are reflected back to earth. Because there is, at night, little absorption of radio waves in the higher reaches of the ionosphere, such waves bounce repeatedly from the ionosphere to the earth's surface and back to the ionosphere again. In this way, standard AM radio waves are able to travel for many hundreds of kilometers at night.

Around sunrise and sunset, AM radio stations usually make "necessary technical adjustments" to compensate for the changing electrical characteristics of the *D* region. Because they can broadcast over a greater distance at night, most AM stations reduce their output near sunset. This reduction prevents two stations—both transmitting at the same frequency but hundreds of kilometers apart—from interfering with each other's radio programs. At sunrise, as the *D* region intensifies, the power supplied to AM radio transmitters is normally increased. FM stations do not need to make these adjustments because FM radio waves are shorter than AM waves, and are able to penetrate through the ionosphere without being reflected.

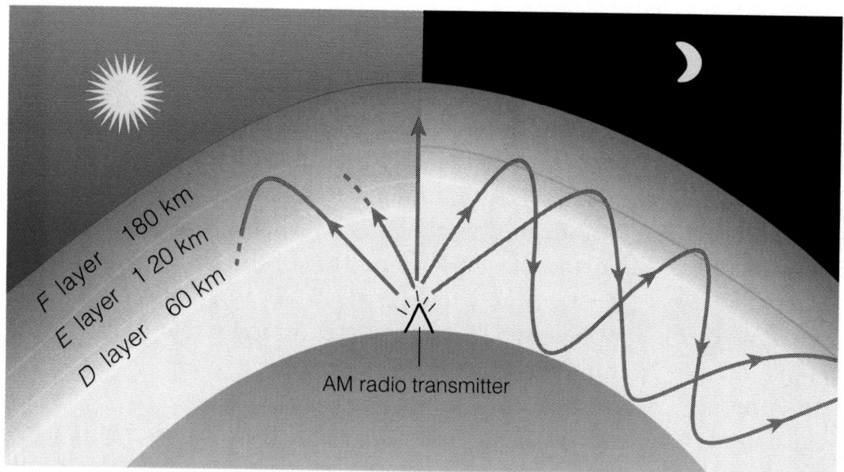

FIGURE 1.9 At night, the higher region of the ionosphere (*F* region) strongly reflects AM radio waves, allowing them to be sent over great distances. During the day, the lower *D* region strongly absorbs and weakens AM radio waves, preventing them from being picked up by distant receivers.

BRIEF REVIEW

We have, in the last several sections, been examining our atmosphere from a vertical perspective. A few of the main points are:

- Atmospheric pressure at any level represents the total mass of air above that level, and atmospheric pressure always decreases with increasing height above the surface.
- The atmosphere may be divided into layers (or regions) according to its vertical profile of temperature, its gaseous composition, or its electrical properties.
- Ozone at the earth's surface is the main ingredient of photochemical smog, whereas ozone in the stratosphere protects life on earth from the sun's harmful ultraviolet rays.

We will now turn our attention to weather events that take place in the lower atmosphere. As you read the remainder of this chapter, keep in mind that the content serves as a broad overview of material to come in later chapters, and that many of the concepts and ideas you encounter are designed to familiarize you with items you might read about in a newspaper or magazine, or see on television.

Meteorology Now™ Click "Layers in the Atmosphere" to examine *current* vertical profiles of temperature, pressure, and other meteorological variables.

WEATHER AND CLIMATE

When we talk about the **weather,** we are talking about the condition of the atmosphere at any particular time and place. Weather—which is always changing—is comprised of the elements of:

1. *air temperature*—the degree of hotness or coldness of the air
2. *air pressure*—the force of the air above an area
3. *humidity*—a measure of the amount of water vapor in the air
4. *clouds*—a visible mass of tiny water droplets and/or ice crystals that are above the earth's surface
5. *precipitation*—any form of water, either liquid or solid (rain or snow), that falls from clouds and reaches the ground
6. *visibility*—the greatest distance one can see
7. *wind*—the horizontal movement of air

If we measure and observe these **weather elements** over a specified interval of time, say, for many years, we would obtain the "average weather" or the **climate** of a particular region. Climate, therefore, represents the accumulation of daily and seasonal weather events (the average range of weather) over a long period of time. The concept of climate is much more than this, for it also includes the extremes of weather—the heat waves of summer and the cold spells of winter—that occur in a particular region. The *frequency* of these extremes is what helps us distinguish among climates that have similar averages.

DID YOU KNOW?

When it rains, it rains pennies from heaven—sometimes. On July 17, 1940, a tornado reportedly picked up a treasure of over 1000 sixteenth-century silver coins, carried them into a thunderstorm, then dropped them on the village of Merchery in the Gorki region of Russia.

If we were able to watch the earth for many thousands of years, even the climate would change. We would see rivers of ice moving down stream-cut valleys and huge glaciers—sheets of moving snow and ice—spreading their icy fingers over large portions of North America. Advancing slowly from Canada, a single glacier might extend as far south as Kansas and Illinois, with ice several thousands of meters thick covering the region now occupied by Chicago. Over an interval of 2 million years or so, we would see the ice advance and retreat several times. Of course, for this phenomenon to happen, the average temperature of North America would have to decrease and then rise in a cyclic manner.

Suppose we could photograph the earth once every thousand years for many hundreds of millions of years. In time-lapse film sequence, these photos would show that not only is the climate altering, but the whole earth itself is changing as well: mountains would rise up only to be torn down by erosion; isolated puffs of smoke and steam would appear as volcanoes spew hot gases and fine dust into the atmosphere; and the entire surface of the earth would undergo a gradual transformation as some ocean basins widen and others shrink.*

In summary, the earth and its atmosphere are dynamic systems that are constantly changing. While major transformations of the earth's surface are completed only after long spans of time, the state of the atmosphere can change in a matter of minutes. Hence, a watchful eye turned skyward will be able to observe many of these changes.

A Satellite's View of the Weather A good view of the weather can be seen from a weather satellite. Figure 1.10 is a satellite image showing a portion of the Pacific Ocean and the North American continent. The photograph was obtained from a *geostationary satellite* situated about 36,000 km (22,300 mi) above the earth. At this elevation, the satellite travels at the same rate as the earth spins, which allows it to remain positioned above the same spot so it can continuously monitor what is taking place beneath it.

The dotted lines running from pole to pole on the satellite picture are called *meridians.* Since the zero meridian (or prime meridian) runs through Greenwich, England, the *longitude* of any place on earth is simply how far east or west, in degrees, it is from the prime meridian. North America is west of Great Britain and most of the United States lies between 75°W and 125°W longitude.

The dotted lines that parallel the equator are called *parallels of latitude.* The latitude of any place is how far north or south, in degrees, it is from the equator. The latitude of the equator is 0°, whereas the latitude of the North Pole is 90°N and that of the South Pole is 90°S. Most of the United States is located between latitude 30°N and 50°N, a region commonly referred to as the **middle latitudes.**

Storms of All Sizes Probably the most dramatic spectacle in Fig. 1.10 is the whirling cloud masses of all shapes and sizes. The clouds appear white because sunlight is reflected back to space from their tops. The dark areas show where skies are clear. The largest of the organized cloud masses are the sprawling storms. One such storm shows as an extensive band of clouds, over 2000 km long, west of the Great Lakes. Superimposed on the satellite image is the storm's center (indicated by the large red L) and its adjoining weather fronts in red, blue, and purple. This **middle-latitude cyclonic storm** system (or *extratropical cyclone*) forms outside the tropics and, in the Northern Hemisphere, has winds spinning counterclockwise about its center, which is presently over Minnesota.

A slightly smaller but more vigorous storm is located over the Pacific Ocean near latitude 12°N and longitude 116°W. This tropical storm system, with its swirling band of rotating clouds and surface winds in excess of 64 knots* (74 mi/hr), is known as a **hurricane.** The diameter of the hurricane is about 800 km (500 mi). The tiny dot at its center is called the *eye.* Near the surface, in the eye, winds are light, skies are generally clear, and the atmospheric pressure is lowest. Around the eye, however, is an extensive region where heavy rain and high surface winds are reaching peak gusts of 100 knots.

Smaller storms are seen as bright spots over the Gulf of Mexico. These spots represent clusters of towering *cumulus* clouds that have grown into **thunderstorms,** that is, tall churning clouds accompanied by lightning, thunder, strong gusty winds, and heavy rain. If you look closely at Fig. 1.10, you will see similar cloud forms in many regions. There were probably thousands of thunderstorms occurring throughout the world at that very moment. Although they cannot be seen indi-

*The movement of the ocean floor and continents is explained in the widely acclaimed theory of *plate tectonics,* formerly called the theory of continental drift.

*Recall from p. 10 that 1 knot equals 1.15 miles per hour.

vidually, there are even some thunderstorms embedded in the cloud mass west of the Great Lakes. Later in the day on which this photograph was taken, a few of these storms spawned the most violent disturbance in the atmosphere—the **tornado.**

A tornado is an intense rotating column of air that extends downward from the base of a thunderstorm. Sometimes called *twisters,* or *cyclones,* they may appear as ropes or as a large circular cylinder. The majority are less than a kilometer wide and many are smaller than a football field. Tornado winds may exceed 200 knots but most probably peak at less than 125 knots. The rotation of some tornadoes never reaches the ground, and the rapidly rotating funnel appears to hang from the base of its parent cloud. Often, they dip down, then rise up before disappearing.

A Look at a Weather Map We can obtain a better picture of the middle-latitude storm system by examining a simplified surface weather map for the same day that the satellite picture was taken. The weight of the air above different regions varies and, hence, so does the atmospheric pressure. In Fig. 1.11, the letter L on the map indicates a region of low atmospheric pressure, often called a *low,* which marks the center of the middle-latitude storm. (Compare the center of the storm in Fig. 1.11 with that in Fig. 1.10.) The two letters H on the map represent regions of high atmospheric pressure, called *highs,* or *anticyclones.* The circles on the map represent individual weather stations. The **wind** is the horizontal movement of air. The **wind direction**—the direction *from which* the wind is blowing*—is given by lines that parallel the wind and extend outward from the center of the station. The *wind speed*—the rate at which the air is moving past a stationary observer—is indicated by barbs.

Notice how the wind blows around the highs and the lows. The horizontal pressure differences create a force that starts the air moving from higher pressure toward lower pressure. Because of the earth's rotation, the winds are deflected toward the right in the Northern Hemisphere.** This deflection causes the winds to blow *clockwise* and *outward* from the center of the highs, and *counterclockwise* and *inward* toward the center of the low.

*If you are facing north and the wind is blowing in your face, the wind would be called a "north wind."

**This deflecting force, known as the *Coriolis force,* is discussed more completely in Chapter 6, as are the winds.

As the surface air spins into the low, it flows together and rises, much like toothpaste does when its open tube is squeezed. The rising air cools, and the water vapor in the air condenses into clouds. Notice in Fig. 1.11 that the area of precipitation (the shaded green area) in the vicinity of the low corresponds to an extensive cloudy region in the satellite image (Fig. 1.10).

Also notice by comparing Figs. 1.10 and 1.11 that, in the regions of high pressure, skies are generally clear. As the surface air flows outward away from the center of a high, air sinking from above must replace the laterally spreading air. Since sinking air does not usually produce clouds, we find generally clear skies and fair weather associated with the regions of high pressure.

The swirling air around the areas of high and low pressure are the major weather producers for the middle latitudes. Look at the middle-latitude storm and the surface temperatures in Fig. 1.11 and notice that, to the southeast of the storm, southerly winds from the Gulf of Mexico are bringing warm, humid air northward over much of the southeastern portion of the nation. On the storm's western side, cool dry northerly breezes combine with sinking air to create generally clear weather over the Rocky Mountains. The boundary that separates the warm and cool air appears as a heavy, dark line on the map—a **front,** across which there is a sharp change in temperature, humidity, and wind direction.

Where the cool air from Canada replaces the warmer air from the Gulf of Mexico, a *cold front* is drawn in blue, with arrowheads showing its general direction of movement. Where the warm Gulf air is replacing cooler air to the north, a *warm front* is drawn in red, with half circles showing its general direction of movement. Where the cold front has caught up to the warm front and cold air is now replacing cool air, an *occluded front* is drawn in purple, with alternating arrowheads and half circles to show how it is moving. Along each of the fronts, warm air is rising, producing clouds and precipitation. In the satellite image (Fig. 1.10), the occluded front and the cold front appear as an elongated, curling cloud band that stretches from the low pressure area over Minnesota into the northern part of Texas.

Notice in Fig. 1.11 that the weather front is to the west of Chicago. As the westerly winds aloft push the front eastward, a person on the outskirts of Chicago might observe the approaching front as a line of towering thunderstorms similar to those in Fig. 1.12. In a few hours, Chicago should experience heavy showers with thunder, lightning, and gusty winds as the front passes.

FIGURE 1.10 This satellite image (taken in visible reflected light) shows a variety of cloud patterns and storms in the earth's atmosphere.

All of this, however, should give way to clearing skies and surface winds from the west or northwest after the front has moved on by.

Observing storm systems, we see that not only do they move but they constantly change. Steered by the upper-level westerly winds, the middle-latitude storm in Fig. 1.11 intensifies into a larger storm, which moves eastward, carrying its clouds and weather with it. In advance of this system, a sunny day in Ohio will gradually cloud over and yield heavy showers and thunderstorms by nightfall. Behind the storm, cool dry northerly winds rushing into eastern Colorado cause an overcast sky to give way to clearing conditions. Farther south, the thunderstorms presently over the Gulf of Mexico (Fig. 1.10)

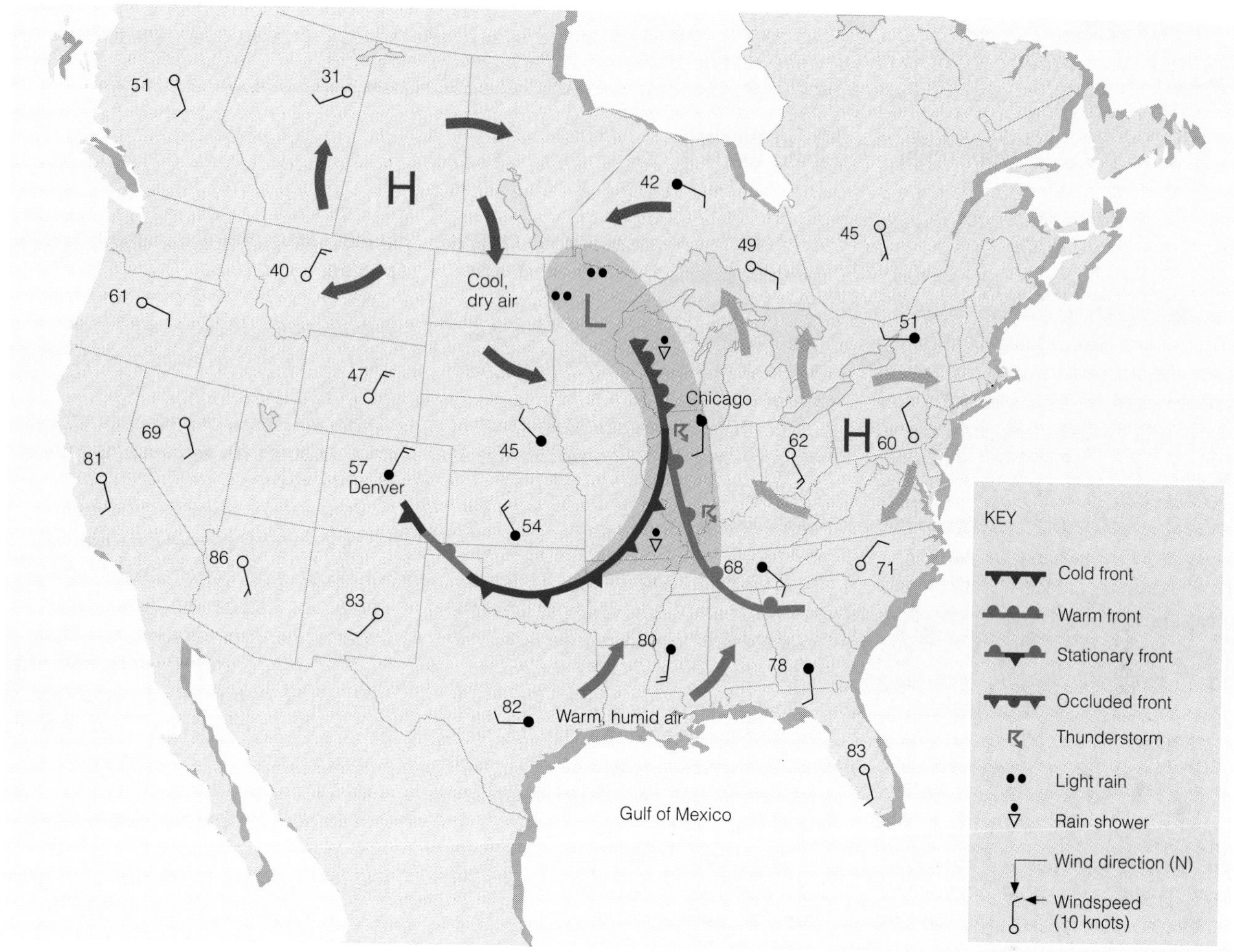

FIGURE 1.11 Simplified surface weather map that correlates with the satellite image shown in Fig. 1.10. The shaded green area represents precipitation. The numbers on the map represent air temperatures in °F.

expand a little, then dissipate as new storms appear over water and land areas. To the west, the hurricane over the Pacific Ocean drifts northwestward and encounters cooler water. Here, away from its warm energy source, it loses its punch; winds taper off, and the storm soon turns into an unorganized mass of clouds and tropical moisture.

Up to this point, we have looked at the concepts of weather and climate without discussing the word **meteorology.** What does this word actually mean, and where did it originate? If you are interested in this information, read the Focus section entitled "Meteorology—A Brief History" on p. 18.

Meteorology Now™ Click "Forecasting" to examine current weather conditions in different parts of the country.

Weather and Climate in Our Lives Weather and climate play a major role in our lives. Weather, for example, often dictates the type of clothing we wear, while climate influences the type of clothing we buy. Climate determines when to plant crops as well as what type of crops can be planted. Weather determines if these same crops will grow to maturity. Although weather and climate affect our lives in many ways, perhaps their most immediate effect is on our comfort. In order to survive the cold of winter and heat of summer, we build homes, heat them, air condition them, insulate them—only to find that when we leave our shelter, we are at the mercy of the weather elements.

Even when we are dressed for the weather properly, wind, humidity, and precipitation can change our perception of how cold or warm it feels. On a cold, windy day the effects of *wind chill* tell us that it feels much colder

FOCUS ON A SPECIAL TOPIC

Meteorology—A Brief History

Meteorology is the study of the atmosphere and its phenomena. The term itself goes back to the Greek philosopher Aristotle who, about 340 B.C., wrote a book on natural philosophy entitled *Meteorologica*. This work represented the sum of knowledge on weather and climate at that time, as well as material on astronomy, geography, and chemistry. Some of the topics covered included clouds, rain, snow, wind, hail, thunder, and hurricanes. In those days, all substances that fell from the sky, and anything seen in the air, were called meteors, hence the term *meteorology*, which actually comes from the Greek word *meteoros*, meaning "high in the air." Today, we differentiate between those meteors that come from extraterrestrial sources outside our atmosphere (meteoroids) and particles of water and ice observed in the atmosphere (hydrometeors).

In *Meteorologica*, Aristotle attempted to explain atmospheric phenomena in a philosophical and speculative manner. Several years later, Theophrastus, a student of Aristotle, compiled a book on weather forecasting called the *Book of Signs*, which attempted to foretell the weather by observing certain weather-related indicators. Even though many of their ideas were found to be erroneous, the work of Aristotle and Theophrastus remained a dominant influence in the field of meteorology for almost 2000 years.

The birth of meteorology as a genuine natural science did not take place until the invention of weather instruments. During the late 1500s, the Italian physicist and astronomer Galileo invented a crude water thermometer. In 1643, Evangelista Torricelli, a student of Galileo, invented the mercury barometer for measuring air pressure. A few years later, French mathematician-philosophers Blaise Pascal and René Descartes, using a barometer, demonstrated that atmospheric pressure decreases with increasing altitude. In 1667, Robert Hooke, a British scientist, invented a swing-type (plate) anemometer for measuring wind speed.

In 1719, German physicist Gabriel Daniel Fahrenheit, working on the boiling and freezing of water, developed a temperature scale. British meteorologist George Hadley, in 1735, explained how the earth's rotation influences the winds in the tropics. In 1742, Swedish astronomer Anders Celsius developed the centigrade (Celsius) temperature scale. By flying a kite in a thunderstorm in 1752, American statesman and scientist Benjamin Franklin demonstrated the electrical nature of lightning. In 1780, Horace deSaussure, a Swiss geologist and meteorologist, invented the hair hygrometer for measuring humidity.

With observations from instruments available, attempts were then made to explain certain weather phenomena employing scientific experimentation and the physical laws that were being developed at the time. French chemist Jacques Charles, in 1787, discovered the relationship between temperature and a volume of air. Enough weather information was available in 1821 that a crude weather map was drawn. In 1835, French physicist Gaspard Coriolis mathematically demonstrated the effect that the earth's rotation has on atmospheric motions.

As more and better instruments were developed, the science of meteorology progressed. By the 1840s, ideas about winds and storms were partially understood. Meteorology got a giant boost in 1843 with the invention of the telegraph. Weather observations and information could now be rapidly disseminated and, in 1869, *isobars* (lines of equal pressure) were placed on a weather map. Around 1920, the concepts of air masses and weather fronts were formulated in Norway. By the 1940s, upper-air balloon observations of temperature, humidity, and pressure gave a three-dimensional view of the atmosphere, and high-flying military aircraft discovered the existence of jet streams.

Meteorology took another step forward in the 1950s, when high-speed computers were developed to solve the mathematical equations that describe the behavior of the atmosphere. At the same time, a group of scientists at Princeton, New Jersey, developed numerical means for predicting the weather. Today, computers plot the observations, draw the lines on the map, and forecast the state of the atmosphere at some desired time in the future.

After World War II, surplus military radars became available, and many were transformed into precipitation-measuring tools. In the mid-1990s, these conventional radars were replaced by the more sophisticated *Doppler radars*, which have the ability to peer into severe thunderstorms and unveil their winds.

In 1960, the first weather satellite, *Tiros 1*, was launched, ushering in space-age meteorology. Subsequent satellites provided a wide range of useful information, ranging from day and night time-lapse images of clouds and storms to images that depict swirling ribbons of water vapor flowing around the globe. Throughout the 1990s and into the twenty-first century, ever more sophisticated satellites were developed to supply computers with a far greater network of data so that more accurate forecasts—perhaps up to two weeks or more—will be available in the future.

---

than it really is, and, if not properly dressed, we run the risk of *frostbite* or even *hypothermia* (the rapid, progressive mental and physical collapse that accompanies the lowering of human body temperature). On a hot, humid day we normally feel uncomfortably warm and blame it on the humidity. If we become too warm, our bodies overheat and *heat exhaustion* or *heat stroke* may result. Those most likely to suffer these maladies are the elderly with impaired circulatory systems and infants, whose heat regulatory mechanisms are not yet fully developed.

Photo by author

FIGURE 1.12 Thunderstorms developing along an approaching cold front.

Weather affects how we feel in other ways, too. Arthritic pain is most likely to occur when rising humidity is accompanied by falling pressures. In ways not well understood, weather does seem to affect our health. The incidence of heart attacks shows a statistical peak after the passage of warm fronts, when rain and wind are common, and after the passage of cold fronts, when an abrupt change takes place as showery precipitation is accompanied by cold gusty winds. Headaches are common on days when we are forced to squint, often due to hazy skies or a thin, bright overcast layer of high clouds.

For some people, a warm, dry wind blowing downslope (a *chinook wind*) adversely affects their behavior (they often become irritable and depressed). Just how and why these winds impact humans physiologically is not well understood. We will take up the question of why these winds are warm and dry in Chapter 7.

When the weather turns colder or warmer than normal, it impacts directly on the lives and pocketbooks of many people. For example, the exceptionally cool summer of 1992 over the eastern two-thirds of North America saved people billions of dollars in air-conditioning costs. On the other side of the coin, the colder-than-normal winter of 2000–2001 over much of North America sent heating costs soaring as demand for heating fuel escalated.

Major cold spells accompanied by heavy snow and ice can play havoc by snarling commuter traffic, curtailing airport services, closing schools, and downing power lines, thereby cutting off electricity to thousands of customers (see Fig. 1.13). For example, a huge ice storm during January, 1998, in northern New England and Canada left millions of people without power and caused over a billion dollars in damages, and a devastating snow storm during March, 1993, buried parts of the East Coast with 14-foot snow drifts and left Syracuse, New York, paralyzed with a snow depth of 36 inches. When the frigid air settles into the Deep South, many millions of dollars worth of temperature-sensitive fruits and vegetables may be ruined, the eventual consequence being higher produce prices in the supermarket.

Prolonged dry spells, especially when accompanied by high temperatures, can lead to a shortage of food and, in some places, widespread starvation. Parts of Africa, for example, have periodically suffered through major droughts and famine. In 1986, the southeastern section of the United States experienced a terrible drought as searing summer temperatures wilted crops, causing losses in excess of a billion dollars. When the climate turns hot and dry, animals suffer too. Over 500,000 chickens perished in Georgia alone during a two-day period at the peak of the

Photo by Judy Champlin

FIGURE 1.13 The ice storm of January, 1998.

summer heat. Severe drought also has an effect on water reserves, often forcing communities to ration water and restrict its use. During periods of extended drought, vegetation often becomes tinder-dry and, sparked by lightning or a careless human, such a dried-up region can quickly become a raging inferno. During the summer of 1998, hundreds of thousands of acres in drought-stricken northern and central Florida were ravaged by wildfires.

Each summer, scorching *heat waves* take many lives. During the past 20 years, an annual average of more than 300 deaths in the United States were attributed to excessive heat exposure. In one particularly devastating heat wave that hit Chicago, Illinois, during July, 1995, high temperatures coupled with high humidity claimed the lives of more than 500 people.

Each year, the violent side of weather influences the lives of millions. It is amazing how many people whose family roots are in the Midwest know the story of someone who was severely injured or killed by a tornado. Tornadoes have not only taken many lives, but annually they cause damage to buildings and property totaling in the hundreds of millions of dollars, as a single large tornado can level an entire section of a town (see Fig. 1.14).

DID YOU KNOW?

On the average, 146 people die each year in the United States from floods and flash floods. Of those who died in flash floods during the past ten years, over half of them were in motor vehicles.

Although the gentle rains of a typical summer thunderstorm are welcome over much of North America, the heavy downpours, high winds, and hail of the *severe thunderstorms* are not. Cloudbursts from slowly moving, intense thunderstorms can provide too much rain too quickly, creating *flash floods* as small streams become raging rivers composed of mud and sand entangled with uprooted plants and trees (see Fig. 1.15). On the average, more people die in the United States from floods and flash floods than from any other natural disaster. Strong downdrafts originating inside an intense thunderstorm (a *downburst*) create turbulent winds that are capable of destroying crops and inflicting damage upon surface structures. Several airline crashes have been attributed to the turbulent *wind shear* zone within the downburst. Annually, hail damages crops worth millions of dollars, and lightning takes the lives of about eighty people in the United States and starts fires that destroy many thousands of acres of valuable timber (see Fig. 1.16).

Even the quiet side of weather has its influence. When winds die down and humid air becomes more tranquil, fog may form. Heavy fog can restrict visibility at airports, causing flight delays and cancellations. Every winter, deadly fog-related auto accidents occur along our busy highways and turnpikes. But fog has a positive side, too, especially during a dry spell, as fog moisture collects on tree branches and drips to the ground, where it provides water for the tree's root system.

Weather and climate have become so much a part of our lives that the first thing many of us do in the

FIGURE 1.14 Tornadoes annually inflict widespread damage and cause the loss of many lives.

morning is to listen to the local weather forecast. For this reason, many radio and television newscasts have their own "weather person" to present weather information and give daily forecasts. More and more of these people are professionally trained in meteorology, and many stations require that the weathercaster obtain a seal of approval from the American Meteorological Society (AMS), or a certificate from the National Weather Association (NWA). To make their weather presentation as up-to-the-minute as possible, an increasing number of stations are taking advantage of the information provided by the National Weather Service (NWS), such as computerized weather forecasts, time-lapse satellite pictures, and color Doppler radar displays.

For more than twenty years now, a staff of trained professionals at "The Weather Channel" have provided weather information twenty-four hours a day on cable television. And finally, the National Oceanic and Atmospheric Administration (NOAA), in cooperation with the National Weather Service, sponsors weather radio broadcasts at selected locations across the United States. Known as *NOAA weather radio* (and transmitted at

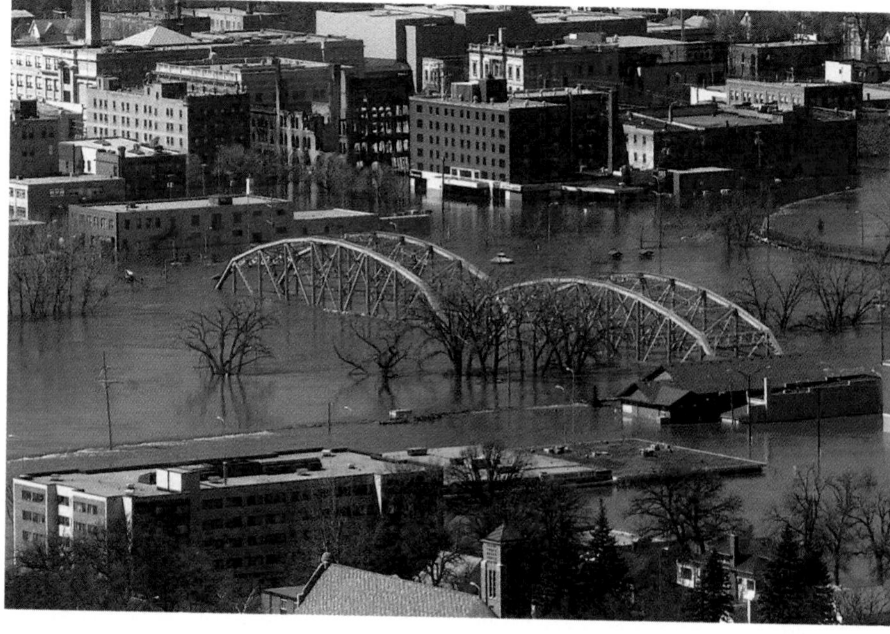

FIGURE 1.15 Flooding during April, 1997, inundates Grand Forks, North Dakota, as flood waters of the Red River extend over much of the city.

FIGURE 1.16 Estimates are that lightning strikes the earth about 100 times every second. About 25 million lightning strikes hit the United States each year. Consequently, lightning is a very common, and sometimes deadly, weather phenomenon.

VHF–FM frequencies), this service provides continuous weather information and regional forecasts (as well as special weather advisories, including watches and warnings) for over 90 percent of the United States.

SUMMARY

This chapter provides an overview of the earth's atmosphere. Our atmosphere is one rich in nitrogen and oxygen as well as smaller amounts of other gases, such as water vapor, carbon dioxide, and other greenhouse gases whose increasing levels may result in global warming. We examined the earth's early atmosphere and found it to be much different from the air we breathe today.

We investigated the various layers of the atmosphere: the troposphere (the lowest layer), where almost all weather events occur, and the stratosphere, where ozone protects us from a portion of the sun's harmful rays. Above the stratosphere lies the mesosphere, where the air temperature drops dramatically with height. Above the mesosphere lies the warmest part of the atmosphere, the thermosphere. At the top of the thermosphere is the exosphere, where collisions between gas molecules and atoms are so infrequent that fast-moving lighter molecules can actually escape the earth's gravitational pull, and shoot off into space. The ionosphere represents that portion of the upper atmosphere where large numbers of ions and free electrons exist.

We looked briefly at the weather map and a satellite image and observed that dispersed throughout the atmosphere are storms and clouds of all sizes and shapes. The movement, intensification, and weakening of these systems, as well as the dynamic nature of air itself, produce a variety of weather events that we described in terms of weather elements. The sum total of weather and its extremes over a long period of time is what we call climate. Although sudden changes in weather may occur in a moment, climatic change takes place gradually over many years. The study of the atmosphere and all of its related phenomena is called *meteorology,* a term whose origin dates back to the days of Aristotle. Finally, we discussed some of many ways weather and climate influence our lives.

MeteorologyNow™ Assess your understanding of this chapter's topics with additional quizzing and tutorials at http://earthscience.brookscole.com/ahrens/ess4e.

KEY TERMS

The following terms are listed in the order they appear in the text. Define each. Doing so will aid you in reviewing the material covered in this chapter.

atmosphere
nitrogen
oxygen
water vapor
carbon dioxide
ozone
aerosols
pollutants
outgassing
air density
air pressure
lapse rate
temperature inversion
radiosonde
troposphere
stratosphere
tropopause
mesosphere
thermosphere
ionosphere
weather
weather elements
climate
middle latitudes
middle-latitude cyclonic storm
hurricane
thunderstorm
tornado
wind
wind direction
front
meteorology

QUESTIONS FOR REVIEW

1. What is the primary source of energy for the earth's atmosphere?
2. List the four most abundant gases in today's atmosphere.
3. Of the four most abundant gases in our atmosphere, which one shows the greatest variation from place to place at the earth's surface?
4. Explain how the atmosphere "protects" inhabitants at the earth's surface.
5. What are some of the important roles that water plays in our atmosphere?
6. Briefly explain the production and natural destruction of carbon dioxide near the earth's surface. Give a reason for the increase of carbon dioxide over the past 100 years.
7. What are some of the aerosols in the atmosphere?
8. What are the two most abundant greenhouse gases in the earth's atmosphere?
9. How has the earth's atmosphere changed over time?
10. (a) Explain the concept of air pressure in terms of weight of air above some level.
    (b) Why does air pressure always decrease with increasing height above the surface?
11. What is standard atmospheric pressure at sea level in
    (a) inches of mercury,
    (b) millibars, and
    (c) hectopascals?
12. On the basis of temperature, list the layers of the atmosphere from the lowest layer to the highest.
13. Briefly describe how the air temperature changes from the earth's surface to the lower thermosphere.
14. (a) What atmospheric layer contains all of our weather?
    (b) In what atmospheric layer do we find the highest concentration of ozone? The highest average air temperature?
15. Even though the actual concentration of oxygen is close to 21 percent (by volume) in the upper stratosphere, explain why, without proper breathing apparatus, you would not be able to survive there.
16. What is the ionosphere and where is it located?
17. List the common weather elements.
18. How does weather differ from climate?
19. Rank the following storms in size from largest to smallest: hurricane, tornado, middle-latitude cyclonic storm, thunderstorm.
20. When someone says that "the wind direction today is south," does this mean that the wind is blowing *toward the south* or *from the south*?
21. Weather in the middle latitudes tends to move in what general direction?
22. Describe some of the features observed on a surface weather map.
23. Define *meteorology* and discuss the origin of this word.
24. Explain how the wind generally blows around areas of low and high pressure in the Northern Hemisphere.
25. Describe some of the ways weather and climate can influence people's lives.

QUESTIONS FOR THOUGHT AND EXPLORATION

1. Why does a radiosonde observation rarely extend above 30 km (100,000 ft) in altitude?
2. Explain how you considered both weather and climate in your choice of the clothing you chose to wear today.
3. Compare a newspaper weather map with a professional weather map obtained from the internet. Discuss any differences in the two maps. Look at both maps and see if you can identify a warm front, a cold front, and a middle-latitude cyclonic storm.
4. Which of the following statements relate more to weather and which relate more to climate?
   (a) The summers here are warm and humid.
   (b) Cumulus clouds presently cover the entire sky.
   (c) Our lowest temperature last winter was −29°C (−18°F).
   (d) The air temperature outside is 22°C (72°F).
   (e) December is our foggiest month.
   (f) The highest temperature ever recorded in Phoenixville, Pennsylvania, was 44°C (111°F) on July 10, 1936.
   (g) Snow is falling at the rate of 5 cm (2 in.) per hour.
   (h) The average temperature for the month of January in Chicago, Illinois, is −3°C (26°F).
5. Keep track of the weather. On an outline map of North America, mark the daily position of fronts and pressure systems for a period of several weeks or more. (This information can be obtained from newspapers, the TV news, or from the Internet.) Plot the general upper-level flow pattern on the map. Observe how the surface systems move. Relate this information to the material on wind, fronts, and cyclones covered in later chapters.
6. Compose a one-week journal, including daily newspaper weather maps and weather forecasts from the newspaper or from the Internet. Provide a commentary for each day regarding the coincidence of actual and predicted weather.

Go to the Brooks/Cole Earth Sciences Resource Center (http://earthscience.brookscole.com) for critical thinking exercises, articles, and additional readings from InfoTrac College Edition, Brooks/Cole's online student library.

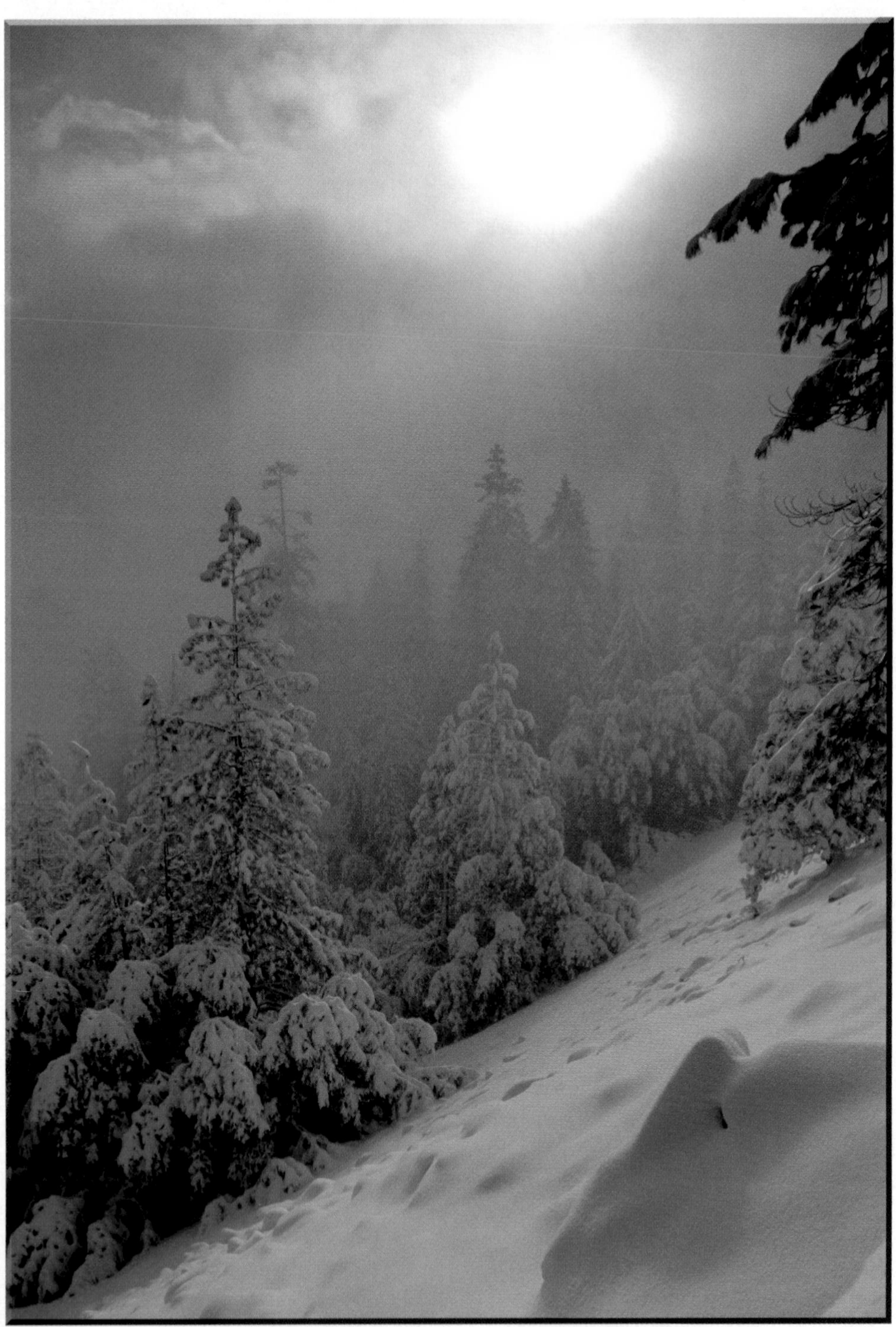

© Brad Perks

The sun peering through a break in the clouds casts a blanket of warmth over a cold, snowy landscape.

Meteorology Now™ This icon, appearing throughout the book, indicates an opportunity to explore interactive tutorials, animations, or practice problems available on the MeteorologyNow Web site at http://earthscience.brookscole.com/ahrens/ess4e.

CHAPTER 2

Warming the Earth and the Atmosphere

The sun doesn't rise or fall: it doesn't move, it just sits there, and we rotate in front of it. Dawn means that we are rotating around into sight of it, while dusk means we have turned another 180 degrees and are being carried into the shadow zone. The sun never "goes away from the sky." It's still there sharing the same sky with us; it's simply that there is a chunk of opaque earth between us and the sun which prevents our seeing it. Everyone knows that, but I really see it now. No longer do I drive down a highway and wish the blinding sun would set; instead I wish we could speed up our rotation a bit and swing around into the shadows more quickly.

Michael Collins, *Carrying the Fire*

CONTENTS

As you sit quietly reading this book, you are part of a moving experience. The earth is speeding around the sun at thousands of miles per hour while, at the same time, it is spinning on its axis. When we look down upon the North Pole, we see that the direction of spin is counterclockwise, meaning that we are moving toward the east at hundreds of miles per hour. We normally don't think of it in that way, but, of course, this is what causes the sun, moon, and stars to rise in the east and set in the west. In fact, it is these motions coupled with energy from the sun, striking a tilted planet, that cause our seasons. But, as we will see later, the sun's energy is not distributed evenly over the earth, as tropical regions receive more energy than polar regions. It is this energy imbalance that drives our atmosphere into the dynamic patterns we experience as wind and weather.

Therefore, we will begin this chapter by examining the concept of energy and heat transfer. Then we will see how our atmosphere warms and cools. Finally, we will examine how the earth's motions and the sun's energy work together to produce the seasons.

TEMPERATURE AND HEAT TRANSFER

Temperature is the quantity that tells us how hot or cold something is relative to some set standard value. But we can look at temperature in another way.

We know that air is a mixture of countless billions of atoms and molecules. If they could be seen, they would appear to be moving about in all directions, freely darting, twisting, spinning, and colliding with one another like an angry swarm of bees. Close to the earth's surface, each individual molecule would travel about a thousand times its diameter before colliding with another molecule. Moreover, we would see that all the atoms and molecules are not moving at the same speed, as some are moving faster than others. The energy associated with this motion is called **kinetic energy,** the energy of motion. The temperature of the air (or any substance) is a measure of its average kinetic energy. Simply stated, **temperature** *is a measure of the average speed of the atoms and molecules,* where higher temperatures correspond to faster average speeds.

Suppose we examine a volume of surface air about the size of a large flexible balloon. If we warm the air inside, the molecules would move faster, but they also would move slightly farther apart—the air becomes less dense. Conversely, if we cool the air, the molecules would slow down, crowd closer together, and the air would become more dense. This molecular behavior is why, in many places throughout the book, we refer to surface air as either *warm, less-dense air* or as *cold, more-dense air.*

Suppose we continue to slowly cool the air. Its atoms and molecules would move slower and slower until the air reaches a temperature of –273°C (–459°F), which is the lowest temperature possible. At this temperature, called **absolute zero,** the atoms and molecules would possess a minimum amount of energy and theoretically no thermal motion.

The atmosphere contains internal energy, which is the total energy stored in its molecules. **Heat,** on the other hand, *is energy in the process of being transferred from one object to another because of the temperature difference between them.* After heat is transferred, it is stored as internal energy. In the atmosphere, heat is transferred by *conduction, convection,* and *radiation.* We will examine these mechanisms of energy transfer after we look at temperature scales and the important concept of latent heat.

Temperature Scales Recall that, theoretically, at a temperature of absolute zero there is no thermal motion. Consequently, at absolute zero, we can begin a temperature scale called the *absolute scale,* or **Kelvin scale,** after Lord Kelvin (1824–1907), a famous British scientist who first introduced it. Since the Kelvin scale contains no negative numbers, it is quite convenient for scientific calculations. Two other temperature scales commonly used today are the Fahrenheit and Celsius (formerly centigrade). The **Fahrenheit scale** was developed in the early 1700s by the physicist G. Daniel Fahrenheit, who assigned the number 32 to the temperature at which water freezes, and the number 212 to the temperature at which water boils. The zero point was simply the lowest temperature that he obtained with a mixture of ice, water, and salt. Between the freezing and boiling points are 180 equal divisions, each of which is called a degree. A thermometer calibrated with this scale is referred to as a Fahrenheit thermometer, for it measures an object's temperature in degrees Fahrenheit (°F).

The **Celsius scale** was introduced later in the eighteenth century. The number 0 (zero) on this scale is assigned to the temperature at which pure water freezes, and the number 100 to the temperature at which pure water boils at sea level. The space between freezing and boiling is divided into 100 equal degrees. Therefore, each Celsius degree (°C) is 180/100 or 1.8 times bigger than a Fahrenheit degree. Put another way, an increase in temperature of 1°C equals an increase of 1.8°F.

A formula for converting °C to °F is

$$°C = \tfrac{5}{9}\,(°F - 32).$$

On the Kelvin scale, degrees Kelvin are called *Kelvins* (abbreviated K). Each degree on the Kelvin scale is exactly the same size as a degree Celsius, and a temperature of 0 K is equal to –273°C. Converting from °C to K can be done by simply adding 273 to the Celsius temperature, as

$$K = °C + 273.$$

Figure 2.1 compares the Kelvin, Celsius, and Fahrenheit scales. Converting a temperature from one scale to another can be done by simply reading the corresponding temperature from the adjacent scale. Thus, 303 on the Kelvin scale is the equivalent of 30°C and 86°F.*

In most of the world, temperature readings are taken in °C. In the United States, however, temperatures above the surface are taken in °C, while temperatures at the surface are typically read in °F. Currently, then, temperatures on upper-level maps are plotted in °C, while, on surface weather maps, they are in °F. Since both scales are in use, temperature readings in this book will, in most cases, be given in °C followed by their equivalent in °F.

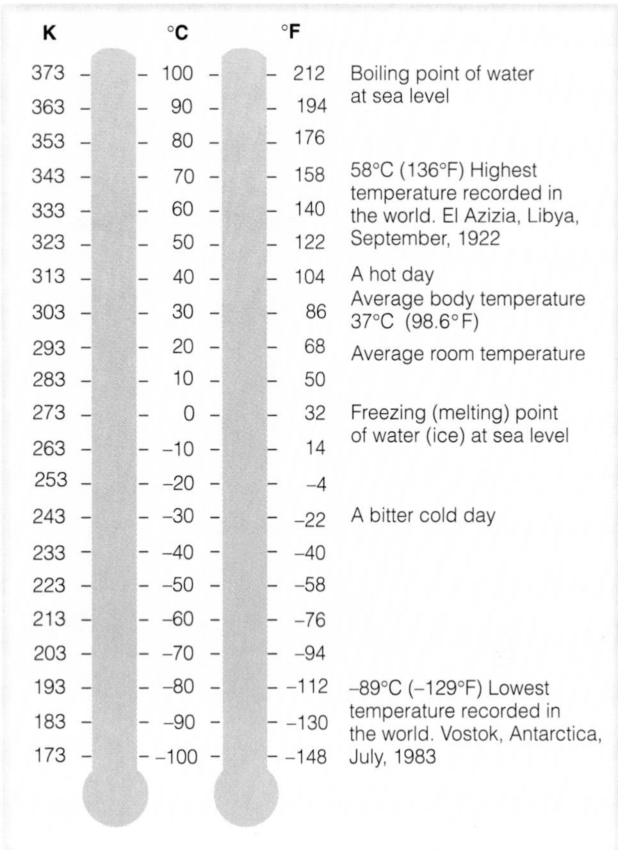

FIGURE 2.1 Comparison of Kelvin, Celsius, and Fahrenheit scales.

Latent Heat—The Hidden Warmth We know from Chapter 1 that water vapor is an invisible gas that becomes visible when it changes into larger liquid or solid (ice) particles. This process of transformation is known as a *change of state* or, simply, a *phase change.* The heat energy required to change a substance, such as water, from one state to another is called **latent heat.** But why is this heat referred to as "latent"? To answer this question, we will begin with something familiar to most of us—the cooling produced by evaporating water.

Suppose we microscopically examine a small drop of pure water. At the drop's surface, molecules are constantly escaping (evaporating). Because the more energetic, faster-moving molecules escape most easily, the average motion of all the molecules left behind decreases as each additional molecule evaporates. Since temperature is a measure of average molecular motion, the slower motion suggests a lower water temperature. *Evaporation is, therefore, a cooling process.* Stated another way, evaporation is a cooling process because the energy needed to evaporate the water—that is, to change its phase from a liquid to a gas—may come from the water or other sources, including the air.

The energy lost by liquid water during evaporation can be thought of as carried away by, and "locked up" within, the water vapor molecule. The energy is thus in a "stored" or "hidden" condition and is, therefore, called *latent heat.* It is latent (hidden) in that the temperature of the substance changing from liquid to vapor is still the same. However, the heat energy will reappear as **sensible heat** (the heat we can feel and measure with a thermometer) when the vapor condenses back into liquid water. Therefore, *condensation* (the opposite of evaporation) *is a warming process.*

The heat energy released when water vapor condenses to form liquid droplets is called *latent heat of condensation.* Conversely, the heat energy used to change liquid into vapor at the same temperature is called *latent heat of evaporation* (vaporization). Nearly 600 calories* are required to evaporate a single gram of water at room temperature. With many hundreds of grams of water evaporating from the body, it is no wonder that after a shower we feel cold before drying off. Figure 2.2 summarizes the concepts examined so far. When the change

*A more complete table of conversions is given in Appendix A.

*By definition, a calorie is the amount of heat required to raise the temperature of 1 gram of water from 14.5°C to 15.5°C. In the International System (SI), the unit of energy is the joule (J), where 1 calorie = 4.186 J. (For pronunciation: joule rhymes with pool.)

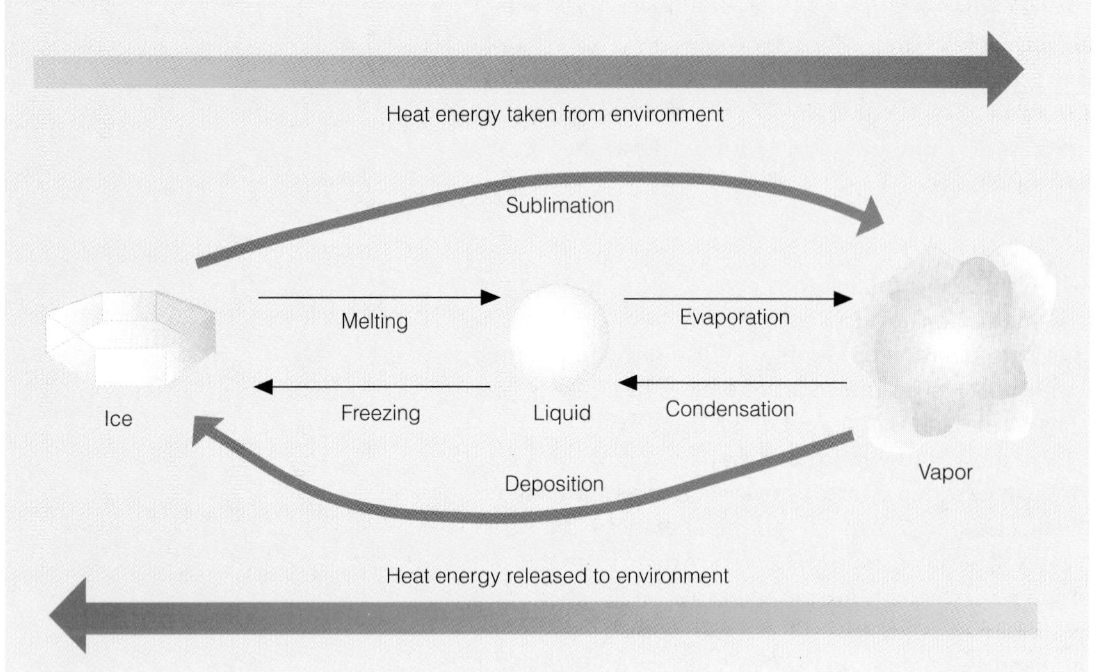

FIGURE 2.2 Heat energy absorbed and released.

of state is from left to right, heat is absorbed by the substance and taken away from the environment. The processes of melting, evaporation, and sublimation (ice to vapor) all cool the environment. When the change of state is from right to left, heat energy is given up by the substance and added to the environment. The processes of freezing, condensation, and deposition (vapor to ice) all warm their surroundings.

Latent heat is an important source of atmospheric energy. Once vapor molecules become separated from the earth's surface, they are swept away by the wind, like dust before a broom. Rising to high altitudes where the air is cold, the vapor changes into liquid and ice cloud particles. During these processes, a tremendous amount of heat energy is released into the environment (see Fig. 2.3).

FIGURE 2.3 Every time a cloud forms, it warms the atmosphere. Inside this developing thunderstorm, a vast amount of stored heat energy (latent heat) is given up to the air, as invisible water vapor becomes countless billions of water droplets and ice crystals. In fact, for the duration of this storm alone, more heat energy is released inside this cloud than is unleashed by a small nuclear bomb.

Photo by author

Water vapor evaporated from warm, tropical water can be carried into polar regions, where it condenses and gives up its heat energy. Thus, as we will see, evaporation-transportation-condensation is an extremely important mechanism for the relocation of heat energy (as well as water) in the atmosphere.

Conduction The transfer of heat from molecule to molecule within a substance is called **conduction.** Hold one end of a metal straight pin between your fingers and place a flaming candle under the other end (see Fig. 2.4). Because of the energy they absorb from the flame, the molecules in the pin vibrate faster. The faster-vibrating molecules cause adjoining molecules to vibrate faster. These, in turn, pass vibrational energy on to their neighboring molecules, and so on, until the molecules at the finger-held end of the pin begin to vibrate rapidly. These fast-moving molecules eventually cause the molecules of your finger to vibrate more quickly. Heat is now being transferred from the pin to your finger, and both the pin and your finger feel hot. If enough heat is transferred, you will drop the pin. The transmission of heat from one end of the pin to the other, and from the pin to your finger, occurs by conduction. Heat transferred in this fashion always flows from *warmer to colder* regions. Generally, the greater the temperature difference, the more rapid the heat transfer.

When materials can easily pass energy from one molecule to another, they are considered to be good conductors of heat. How well they conduct heat depends upon how their molecules are structurally bonded together. Table 2.1 shows that solids, such as metals, are good heat conductors. It is often difficult, therefore, to judge the temperature of metal objects. For example, if you grab a metal pipe at room temperature, it will seem to be much colder than it actually is because the metal conducts heat away from the hand quite rapidly. Conversely, *air is an extremely poor conductor of heat,* which is why most insulating materials have a large number of air spaces trapped within them. Air is such a poor heat conductor that, in calm weather, the hot ground only warms a shallow layer of air a few centimeters thick by conduction. Yet, air can carry this energy rapidly from one region to another. How, then, does this phenomenon happen?

Convection The transfer of heat by the mass movement of a fluid (such as water and air) is called **convection.** This type of heat transfer takes place in liquids and gases because they can move freely and it is possible to set up currents within them.

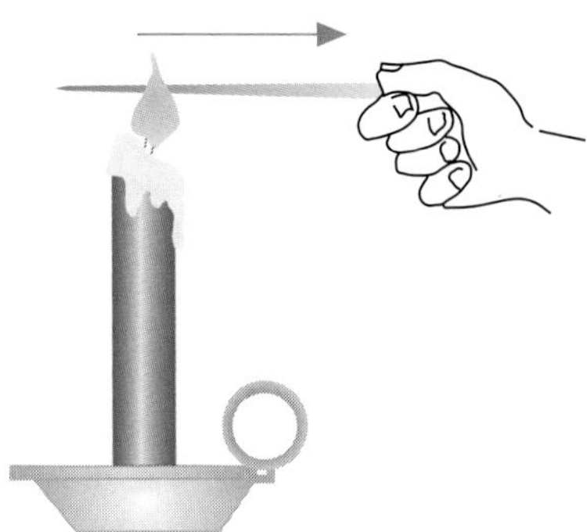

FIGURE 2.4 The transfer of heat from the hot end of the metal pin to the cool end by molecular contact is called *conduction.*

Convection happens naturally in the atmosphere. On a warm, sunny day certain areas of the earth's surface absorb more heat from the sun than others; as a result, the air near the earth's surface is heated somewhat unevenly. Air molecules adjacent to these hot surfaces bounce against them, thereby gaining some extra energy by conduction. The heated air expands and becomes less dense than the surrounding cooler air. The expanded warm air is buoyed upward and rises. In this manner, large bubbles of warm air rise and transfer heat energy upward. Cooler, heavier air flows toward the surface to replace the rising air. This cooler air becomes heated in turn, rises, and the cycle is repeated. In meteorology, this

TABLE 2.1
Heat Conductivity* of Various Substances

SUBSTANCE	HEAT CONDUCTIVITY (WATTS† PER METER PER °C)
Still air	0.023 (at 20°C)
Wood	0.08
Dry soil	0.25
Water	0.60 (at 20°C)
Snow	0.63
Wet soil	2.1
Ice	2.1
Sandstone	2.6
Granite	2.7
Iron	80
Silver	427

*Heat (thermal) conductivity describes a substance's ability to conduct heat as a consequence of molecular motion.

†A watt (W) is a unit of power where one watt equals one joule (J) per second (J/s). One joule equals 0.24 calories.

DID YOU KNOW?

Although we can't see air, there are signs that tell us where the air is rising. On a calm day, you can watch a hawk circle and climb high above level ground while its wings remain motionless. A rising thermal carries the hawk upward as it scans the terrain for prey. If the water vapor inside the rising thermal condenses into liquid cloud droplets, the thermal becomes visible to us as a puffy cumulus cloud. Flying in a light aircraft beneath these clouds usually produces a bumpy ride, as passengers are jostled around by the rising and sinking air associated with convection.

vertical exchange of heat is called *convection,* and the rising air bubbles are known as **thermals** (see Fig. 2.5).

The rising air expands and gradually spreads outward. It then slowly begins to sink. Near the surface, it moves back into the heated region, replacing the rising air. In this way, a *convective circulation,* or thermal "cell," is produced in the atmosphere. In a convective circulation the warm, rising air cools. In our atmosphere, *any air that rises will expand and cool, and any air that sinks is compressed and warms.* This important concept is detailed in the Focus section on p. 31.

Although the entire process of heated air rising, spreading out, sinking, and finally flowing back toward its original location is known as a convective circulation, meteorologists usually restrict the term *convection* to the process of the rising and sinking part of the circulation.

The horizontally moving part of the circulation (called *wind*) carries properties of the air in that particular area with it. The transfer of these properties by horizontally moving air is called **advection.** For example, wind blowing across a body of water will "pick up" water vapor from the evaporating surface and transport it elsewhere in the atmosphere. If the air cools, the water vapor may condense into cloud droplets and release latent heat. In a sense, then, heat is advected (carried) by the water vapor as it is swept along with the wind. Earlier, we saw that this is an important way to redistribute heat energy in the atmosphere.

BRIEF REVIEW

Before moving on to the next section, here is a summary of some of the important concepts and facts we have covered:

- The temperature of a substance is a measure of the average kinetic energy (average speed) of its atoms and molecules.
- Evaporation (the transformation of liquid into vapor) is a cooling process that can cool the air, whereas condensation (the transformation of vapor into liquid) is a warming process that can warm the air.
- Heat is energy in the process of being transferred from one object to another because of the temperature difference between them.
- In conduction, which is the transfer of heat by molecule-to-molecule contact, heat always flows from warmer to colder regions.
- Air is a poor conductor of heat.
- Convection is an important mechanism of heat transfer, as it represents the vertical movement of warmer air upward and cooler air downward.

There is yet another mechanism for the transfer of energy—radiation, or *radiant energy,* which is what we receive from the sun. In this method, energy may be transferred from one object to another without the space between them necessarily being heated.

Radiation On a summer day, you may have noticed how warm and flushed your face feels as you stand

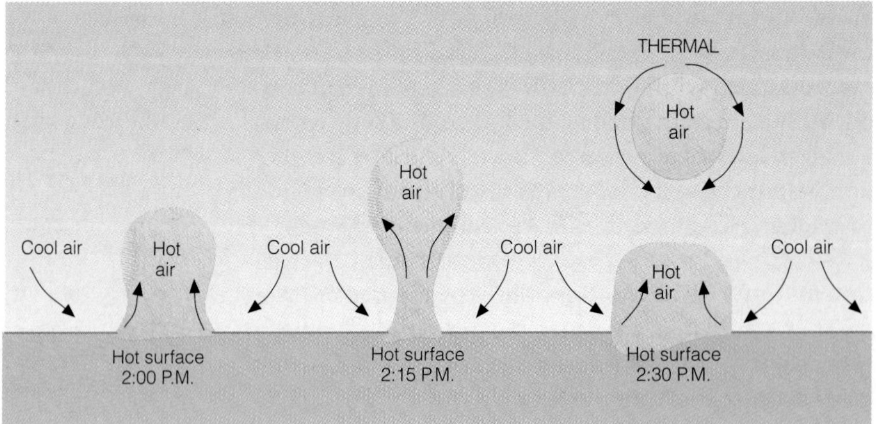

FIGURE 2.5 The development of a thermal. A thermal is a rising bubble of air that carries heat energy upward by *convection.*

FOCUS ON A SPECIAL TOPIC

Rising Air Cools and Sinking Air Warms

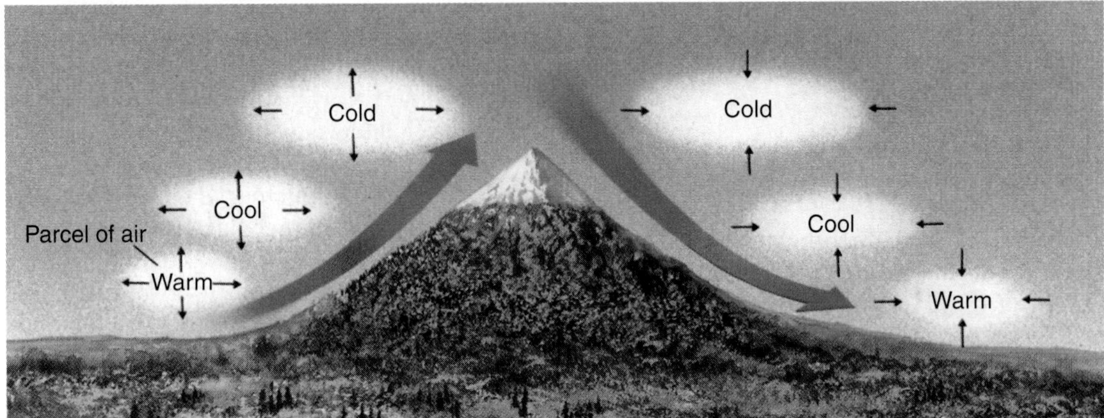

FIGURE 1 Rising air expands and cools; sinking air is compressed and warms.

To understand why rising air cools and sinking air warms we need to examine some air. Suppose we place air in an imaginary thin, elastic wrap about the size of a large balloon (see Fig. 1). This invisible balloonlike "blob" is called a ***parcel.*** The air parcel can expand and contract freely, but neither external air nor heat is able to mix with the air inside. By the same token, as the parcel moves, it does not break apart, but remains as a single unit.

At the earth's surface, the parcel has the same temperature and pressure as the air surrounding it. Suppose we lift the parcel. Recall from Chapter 1 that air pressure always decreases as we move up into the atmosphere. Consequently, as the parcel rises, it enters a region where the surrounding air pressure is lower. To equalize the pressure, the parcel molecules inside push the parcel walls outward, expanding it. Because there is no other energy source, the air molecules inside use some of their own energy to expand the parcel. This energy loss shows up as slower molecular speeds, which represent a lower parcel temperature. Hence, *any air that rises always expands and cools.*

If the parcel is lowered to the earth, it returns to a region where the air pressure is higher. The higher outside pressure squeezes (compresses) the parcel back to its original (smaller) size. Because air molecules have a faster rebound velocity after striking the sides of a collapsing parcel, the average speed of the molecules inside goes up. (A Ping-Pong ball moves faster after striking a paddle that is moving toward it.) This increase in molecular speed represents a warmer parcel temperature. Therefore, *any air that sinks (subsides), warms by compression.*

facing the sun. Sunlight travels through the surrounding air with little effect upon the air itself. Your face, however, absorbs this energy and converts it to thermal energy. Thus, sunlight warms your face without actually warming the air. The energy transferred from the sun to your face is called **radiant energy,** or **radiation.** It travels in the form of waves that release energy when they are absorbed by an object. Because these waves have magnetic and electrical properties, we call them **electromagnetic waves.** Electromagnetic waves do not need molecules to propagate them. In a vacuum, they travel at a constant speed of nearly 300,000 km (186,000 mi) per second—the speed of light.

Figure 2.6 shows some of the different wavelengths of radiation. Notice that the *wavelength* (which is often expressed by the Greek letter lambda, λ) is the distance measured along a wave from one crest to another. Also notice that some of the waves have exceedingly short lengths. For example, radiation that we can see (visible light) has an average wavelength of less than one-millionth of a meter—a distance nearly one-hundredth the diameter of a human hair. To measure these short lengths, we introduce a new unit of measurement called a **micrometer** (abbreviated μm), which is equal to one-millionth of a meter (m); thus,

$$1 \text{ micrometer } (\mu\text{m}) = 0.000001 \text{ m} = 10^{-6} \text{ m}.$$

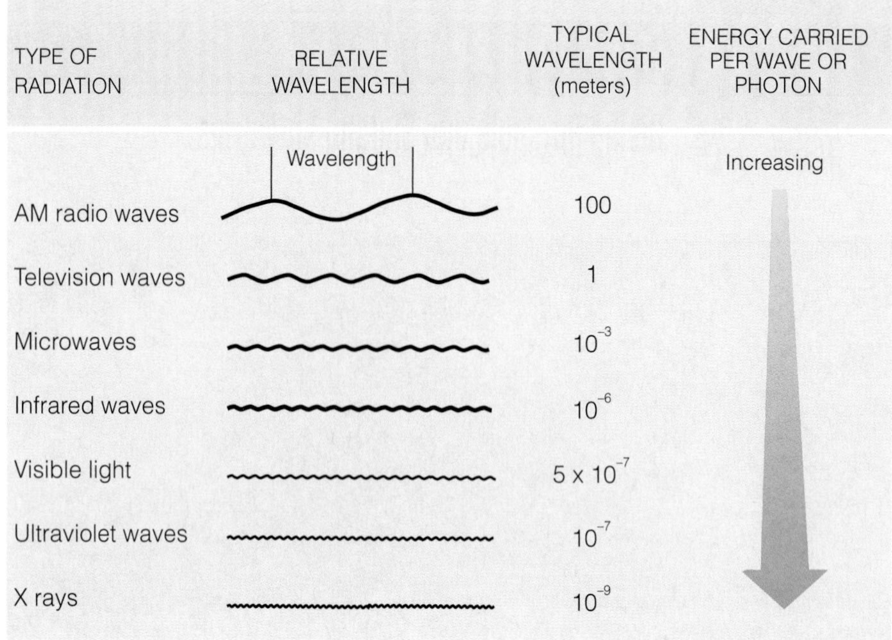

TYPE OF RADIATION
RELATIVE WAVELENGTH
TYPICAL WAVELENGTH (meters)
ENERGY CARRIED PER WAVE OR PHOTON
Wavelength
Increasing
AM radio waves
100
Television waves
1
Microwaves
10^{-3}
Infrared waves
10^{-6}
Visible light
5×10^{-7}
Ultraviolet waves
10^{-7}
X rays
10^{-9}

FIGURE 2.6 Radiation characterized according to wavelength. As the wavelength decreases, the energy carried per wave increases.

In Fig. 2.6, we can see that the average wavelength of visible light is about 0.000005 meters, which is the same as 0.5 μm. To give you a common object for comparison, the average height of a letter on this page is about 2000 μm, or 2 millimeters (2 mm), whereas the thickness of this page is about 100 μm.

We can also see in Fig. 2.6 that the longer waves carry less energy than do the shorter waves. When comparing the energy carried by various waves, it is useful to give electromagnetic radiation characteristics of particles in order to explain some of the wave's behavior. We can actually think of radiation as streams of particles, or **photons,** that are discrete packets of energy.*

An ultraviolet (UV) photon carries more energy than a photon of visible light. In fact, certain ultraviolet photons have enough energy to produce sunburns and penetrate skin tissue, sometimes causing skin cancer. (Additional information on radiant energy and its effect on humans is given in the Focus section on p. 33.)

To better understand the concept of radiation, here are a few important concepts and facts to remember:

1. All things (whose temperature is above absolute zero), no matter how big or small, emit radiation. The air, your body, flowers, trees, the earth, the stars are all radiating a wide range of electromagnetic waves. The energy originates from rapidly vibrating electrons, billions of which exist in every object.
2. The wavelengths of radiation that an object emits depend primarily on the object's temperature. *The higher the object's temperature, the shorter are the wavelengths of emitted radiation.* By the same token, as an object's temperature increases, its peak emission of radiation shifts toward shorter wavelengths. This relationship between temperature and wavelength is called *Wien's law** (or *Wien's displacement law*) after the German physicist Wilhelm Wien (pronounced *Ween,* 1864–1928) who discovered it.
3. Objects that have a high temperature emit radiation at a greater rate or intensity than objects with a lower temperature. Thus, as the temperature of an object increases, more total radiation (over a given surface area) is emitted each second. This relationship between temperature and emitted radiation is known as the *Stefan-Boltzmann law*** after Josef Stefan (1835–1893) and Ludwig Boltzmann (1844–1906), who devised it.

*Packets of photons make up waves, and groups of waves make up a beam of radiation.

*Wien's law:

$$\lambda_{max} = \frac{\text{constant}}{T}.$$

Where λ_{max} is the wavelength at which maximum radiation emission occurs, T is the object's temperature in Kelvins (K) and the constant is 2897 μmK. More information on Wien's law is given in Appendix B.

**Stefan-Boltzmann law:

$$E = \sigma T^4.$$

Where E is the maximum rate of radiation emitted by each square meter of surface of an object, σ (the Greek letter sigma) is a constant, and T is the object's surface temperature in Kelvins (K). Additional information on the Stefan-Boltzmann law is given in Appendix B.

FOCUS ON AN ENVIRONMENTAL ISSUE

Sun Burning and UV Rays

Earlier, we learned that shorter waves of radiation carry much more energy than longer waves, and that a photon of ultraviolet light carries more energy than a photon of visible light. In fact, ultraviolet (UV) wavelengths in the range of 0.20 and 0.29 μm (known as *UV–C radiation)* are harmful to living things, as certain waves can cause chromosome mutations, kill single-celled organisms, and damage the cornea of the eye. Fortunately, virtually all the ultraviolet radiation at wavelengths in the UV–C range is absorbed by ozone in the stratosphere.

Ultraviolet wavelengths between about 0.29 and 0.32 μm (known as *UV–B radiation)* reach the earth in small amounts. Photons in this wavelength range have enough energy to produce sunburns and penetrate skin tissues, sometimes causing skin cancer. About 90 percent of all skin cancers are linked to sun exposure and UV–B radiation. Oddly enough, these same wavelengths activate provitamin D in the skin and convert it into vitamin D, which is essential to health.

Longer ultraviolet waves with lengths of about 0.32 to 0.40 μm (called *UV–A radiation)* are less energetic, but can still tan the skin. Although UV–B is mainly responsible for burning the skin, UV–A can cause skin redness. It can also interfere with the skin's immune system and cause long-term skin damage that shows up years later as accelerated aging and skin wrinkling. Moreover, recent studies indicate that the longer UV–A exposures needed to create a tan pose about the same cancer risk as a UV–B tanning dose.

Upon striking the human body, ultraviolet radiation is absorbed beneath the outer layer of skin. To protect the skin from these harmful rays, the body's defense mechanism kicks in. Certain cells (when exposed to UV radiation) produce a dark pigment *(melanin)* that begins to absorb some of the UV radiation. (It is the production of melanin that produces a tan.) Consequently, a body that produces little melanin—one with pale skin—has little natural protection from UV–B.

Additional protection can come from a sunscreen. Unlike the old lotions that simply moisturized the skin before it baked in the sun, sunscreens today block UV rays from ever reaching the skin. Some contain chemicals (such as zinc oxide) that reflect UV radiation. (These are the white pastes once seen on the noses of lifeguards.) Others consist of a mixture of chemicals that actually absorb ultraviolet radiation, usually UV–B, although new products with UV–A-absorbing qualities are now on the market. The *Sun Protection Factor (SPF)* number on every container of sunscreen dictates how effective the product is in protecting from UV–B—the higher the number, the better the protection.

Protecting oneself from excessive exposure to the sun's energetic UV rays is certainly wise. Estimates are that, in a single year, over 30,000 Americans will be diagnosed with malignant melanoma, the most deadly form of skin cancer. And if the protective ozone shield continues to diminish, there is an ever-increasing risk of problems associated with UV–B. Using a good sunscreen and proper clothing can certainly help. The best way to protect yourself from too much sun, however, is to limit your time in direct sunlight, especially between the hours of 11 A.M. and 3 P.M. when the sun is highest in the sky and its rays are most direct.

Presently, the National Weather Service makes a daily prediction of UV radiation levels for selected cities throughout the United States. The forecast, known as the *UV Index,* gives the UV level at its peak, around noon standard time or 1 P.M. daylight savings time. The 15-point index corresponds to five exposure categories set by the Environmental Protection Agency (EPA). An index value of between 0 and 2 is considered "minimal," whereas a value of 10 or greater is deemed "very high."

Objects at a high temperature (above about 500°C) radiate waves with many lengths, but some of them are short enough to stimulate the sensation of color. We actually see these objects glow red. Objects cooler than this radiate at wavelengths that are too long for us to see. The page of this book, for example, is radiating electromagnetic waves. But because its temperature is only around 20°C (68°F), the waves emitted are much too long to stimulate vision. We are able to see the page, however, because light waves from other sources (such as light bulbs or the sun) are being *reflected* (bounced) off the paper. If this book were carried into a completely dark room, it would continue to radiate, but the pages would appear black because there are no visible light waves in the room to reflect off the page.

DID YOU KNOW?

The large ears of a jackrabbit are efficient emitters of infrared energy. Its ears help the rabbit survive the heat of a summer's day by radiating a great deal of infrared energy to the cooler sky above. Similarly, the large ears of the African elephant greatly increase its radiating surface area and promote cooling of its large mass.

The sun emits radiation at almost all wavelengths, but because its surface is hot—6000 K (10,500°F)—it radiates

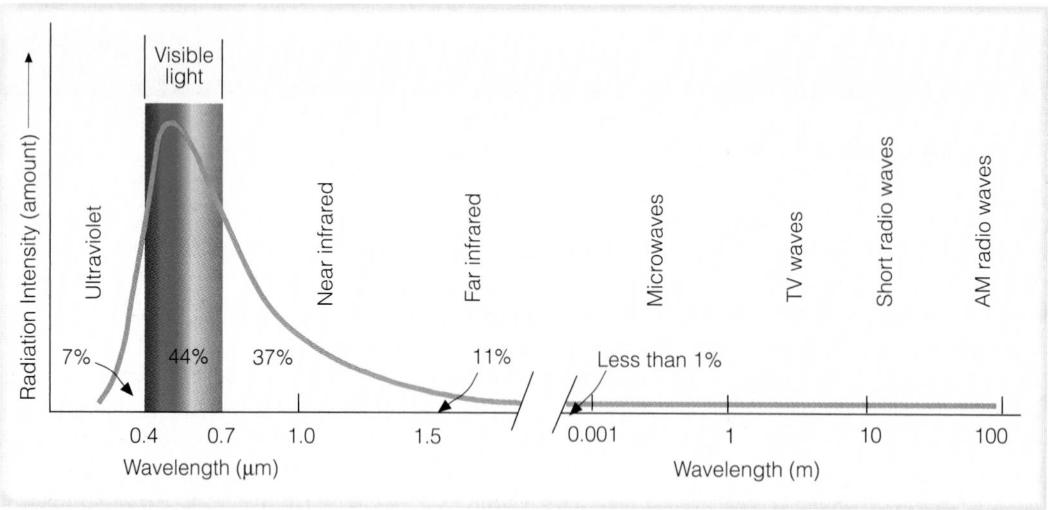

FIGURE 2.7 The sun's electromagnetic spectrum and some of the descriptive names of each region. The numbers underneath the curve approximate the percent of energy the sun radiates in various regions.

FIGURE 2.8 The hotter sun not only radiates more energy than that of the cooler earth (the area under the curve), but it also radiates the majority of its energy at much shorter wavelengths. (The area under the curves is equal to the total energy emitted, and the scales for the two curves differ by a factor of 100,000.)

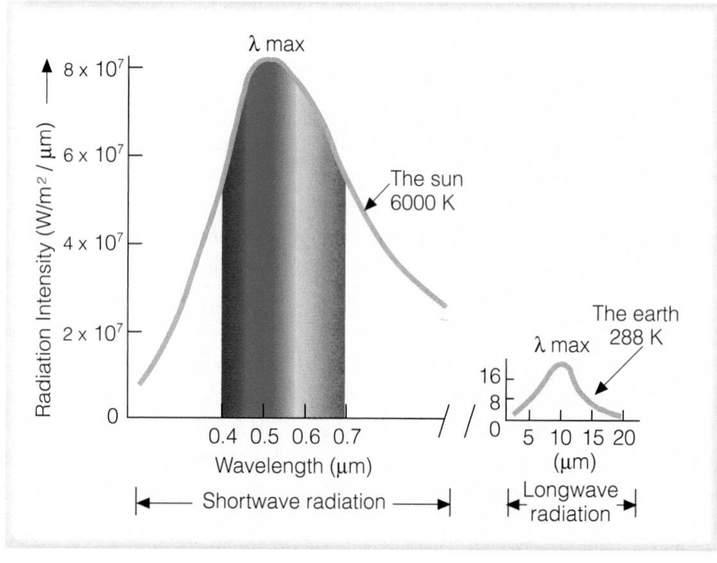

the majority of its energy at relatively short wavelengths. If we look at the amount of radiation given off by the sun at each wavelength, we obtain the sun's *electromagnetic spectrum.* A portion of this spectrum is shown in Fig. 2.7.

Notice that the sun emits a maximum amount of radiation at wavelengths near 0.5 μm. Since our eyes are sensitive to radiation between 0.4 and 0.7 μm, these waves reach the eye and stimulate the sensation of color. This portion of the spectrum is therefore referred to as the **visible region,** and the radiant energy that reaches our eye is called *visible light.* The color violet is the shortest wavelength of visible light. Wavelengths shorter than violet (0.4 μm) are **ultraviolet (UV).** The longest wavelengths of visible light correspond to the color red. Wavelengths longer than red (0.7 μm) are called **infrared (IR).**

Whereas the hot sun emits only a part of its energy in the infrared portion of the spectrum, the relatively cool earth emits almost all of its energy at infrared wavelengths. In fact, the earth, with an average surface temperature near 288 K (15°C, or 59°F), radiates nearly all its energy between 5 and 25 μm, with a peak intensity in the infrared region near 10 μm (see Fig. 2.8). Since the sun radiates the majority of its energy at much shorter wavelengths than does the earth, solar radiation is often called *shortwave radiation,* whereas the earth's radiation is referred to as *longwave* (or *terrestrial*) *radiation.*

BALANCING ACT—ABSORPTION, EMISSION, AND EQUILIBRIUM

If the earth and all things on it are continually radiating energy, why doesn't everything get progressively colder? The answer is that all objects not only radiate energy, they absorb it as well. If an object radiates more energy than it absorbs, it becomes colder; if it absorbs more energy than it emits, it becomes warmer. On a sunny day, the earth's surface warms by absorbing more energy from the sun and the atmosphere than it radiates, whereas at night the earth cools by radiating more energy than it absorbs from its surroundings. When an object emits and absorbs energy at equal rates, its temperature remains constant.

The rate at which something radiates and absorbs energy depends strongly on its surface characteristics, such as color, texture, and moisture, as well as temperature. For example, a black object in direct sunlight is a good absorber of solar radiation. It converts energy from the sun into internal energy, and its temperature ordinarily increases. You need only walk barefoot on a black asphalt road on a summer afternoon to experience this. At night, the blacktop road will cool quickly by emitting infrared energy and, by early morning, it may be cooler than surrounding surfaces.

Any object that is a perfect absorber (that is, absorbs all the radiation that strikes it) and a perfect emitter (emits the maximum radiation possible at its given temperature) is called a **blackbody.** Blackbodies do not have to be colored black, they simply must absorb and emit all possible radiation. Since the earth's surface and the sun absorb and radiate with nearly 100 percent efficiency for their respective temperatures, they both behave as blackbodies.

When we look at the earth from space, we see that half of it is in sunlight, the other half is in darkness. The outpouring of solar energy constantly bathes the earth with radiation, while the earth, in turn, constantly emits infrared radiation. If we assume that there is no other method of transferring heat, then, when the rate of absorption of solar radiation equals the rate of emission of infrared earth radiation, a state of *radiative equilibrium* is achieved. The average temperature at which this occurs is called the **radiative equilibrium temperature.** At this temperature, the earth (behaving as a blackbody) is absorbing solar radiation and emitting infrared radiation at equal rates, and its average temperature does not change. As the earth is about 150 million km (93 million mi) from the sun, the earth's radiative equilibrium temperature is about 255 K (–18°C, 0°F). But this temperature is *much* lower than the earth's observed average surface temperature of 288 K (15°C, 59°F). Why is there such a large difference?

The answer lies in the fact that *the earth's atmosphere absorbs and emits infrared radiation.* Unlike the earth, the atmosphere does *not* behave like a blackbody, as it absorbs some wavelengths of radiation and is transparent to others. Objects that selectively absorb and emit radiation, such as gases in our atmosphere, are known as **selective absorbers.**

Selective Absorbers and the Atmospheric Greenhouse Effect There are many selective absorbers in our environment. Snow, for example, is a good absorber of infrared radiation but a poor absorber of sunlight. Objects that selectively absorb radiation usually selectively emit radiation at the same wavelength. Snow is therefore a good emitter of infrared energy. At night, a snow surface usually emits much more infrared energy than it absorbs from its surroundings. This large loss of infrared radiation (coupled with the insulating qualities of snow) causes the air above a snow surface on a clear, winter night to become extremely cold.

Figure 2.9 shows some of the most important selectively absorbing gases in our atmosphere (the shaded area represents the percent of radiation absorbed by each gas at various wavelengths). Notice that both water vapor (H_2O) and carbon dioxide (CO_2) are strong absorbers of infrared radiation and poor absorbers of visible solar radiation. Other, less important, selective absorbers include nitrous oxide (N_2O), methane (CH_4), and ozone (O_3), which is most abundant in the stratosphere. As these gases absorb infrared radiation emitted from the earth's surface, they gain kinetic energy (energy of motion). The gas molecules share this energy by colliding with neighboring air molecules, such as oxygen and nitrogen (both

DID YOU KNOW?

What an absorber! First detected in the earth's atmosphere in 1999, a greenhouse gas (trifluoromethyl sulfur pentafluoride, SF_5CF_3) pound for pound absorbs about 18,000 times more infrared radiation than CO_2 does. This trace gas, which may form in high-voltage electrical equipment, is increasing in the atmosphere by about 6 percent per year, but it is present in very tiny amounts—about 0.00000012 ppm.

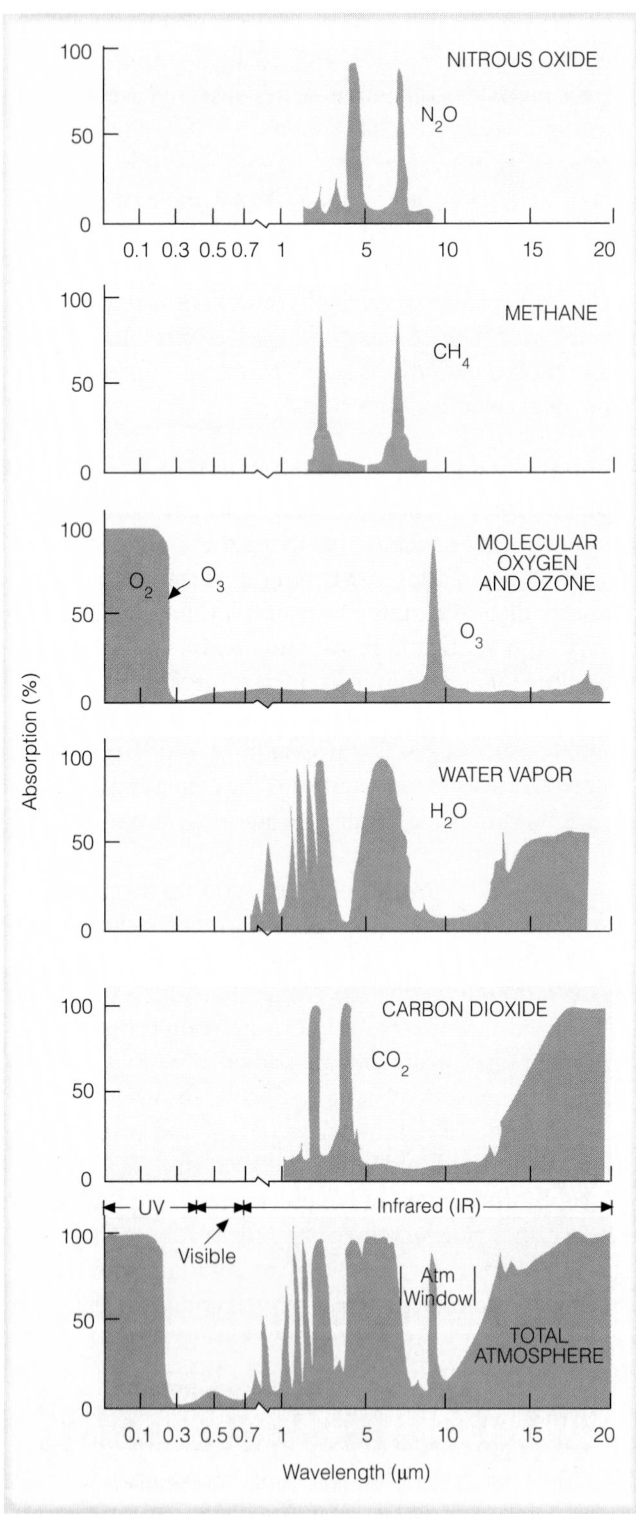

Meteorology Now™ ACTIVE FIGURE 2.9
Absorption of radiation by gases in the atmosphere. The shaded area represents the percent of radiation absorbed. The strongest absorbers of infrared radiation are water vapor and carbon dioxide.
Watch this Active Figure at http://earthscience.brookscole.com/ahrens/ess4e.

of which are poor absorbers of infrared energy). These collisions increase the average kinetic energy of the air, which results in an increase in air temperature. Thus, most of the infrared energy emitted from the earth's surface keeps the lower atmosphere warm.

Besides being selective absorbers, water vapor and CO_2 selectively emit radiation at infrared wavelengths.* This radiation travels away from these gases in all directions. A portion of this energy is radiated toward the earth's surface and absorbed, thus heating the ground. The earth, in turn, radiates infrared energy upward, where it is absorbed and warms the lower atmosphere. In this way, water vapor and CO_2 absorb and radiate infrared energy and act as an insulating layer around the earth, keeping part of the earth's infrared radiation from escaping rapidly into space. Consequently, the earth's surface and the lower atmosphere are much warmer than they would be if these selectively absorbing gases were not present. In fact, as we saw earlier, the earth's mean radiative equilibrium temperature without CO_2 and water vapor would be around –18°C (0°F), or about 33°C (59°F) lower than at present.

The absorption characteristics of water vapor, CO_2, and other gases (such as methane and nitrous oxide depicted in Fig. 2.9) were at one time thought to be similar to the glass of a florist's greenhouse. In a greenhouse, the glass allows visible radiation to come in, but inhibits to some degree the passage of outgoing infrared radiation. For this reason, the absorption of infrared radiation from the earth by water vapor and CO_2 is popularly called the **greenhouse effect.** However, studies have shown that the warm air inside a greenhouse is probably caused more by the air's inability to circulate and mix with the cooler outside air, rather than by the entrapment of infrared energy. Because of these findings, some scientists suggest that the greenhouse effect should be called the *atmosphere effect.* To accommodate everyone, we will usually use the term *atmospheric greenhouse effect* when describing the role that water vapor, CO_2, and other greenhouse gases** play in keeping the earth's mean surface temperature higher than it otherwise would be.

Look again at Fig. 2.9 and observe that, in the bottom diagram, there is a region between about 8 and 11 μm where neither water vapor nor CO_2 readily absorb

*Nitrous oxide, methane, and ozone also emit infrared radiation, but their concentration in the atmosphere is much smaller than water vapor and carbon dioxide (see Table 1.1, p. 3.)

**The term "greenhouse gases" derives from the standard use of "greenhouse effect." Greenhouse gases include, among others, water vapor, carbon dioxide, methane, nitrous oxide, and ozone.

infrared radiation. Because these wavelengths of emitted energy pass upward through the atmosphere and out into space, the wavelength range (between 8 and 11 μm) is known as the **atmospheric window.** Clouds can enhance the atmospheric greenhouse effect. Tiny liquid cloud droplets are selective absorbers in that they are good absorbers of infrared radiation but poor absorbers of visible solar radiation. Clouds even absorb the wavelengths between 8 and 11 μm, which are otherwise "passed up" by water vapor and CO_2. Thus, they have the effect of enhancing the atmospheric greenhouse effect by closing the atmospheric window.

Clouds—especially low, thick ones—are excellent emitters of infrared radiation. Their tops radiate infrared energy upward and their bases radiate energy back to the earth's surface where it is absorbed and, in a sense, radiated back to the clouds. This process keeps calm, cloudy nights warmer than calm, clear ones. If the clouds remain into the next day, they prevent much of the sunlight from reaching the ground by reflecting it back to space. Since the ground does not heat up as much as it would in full sunshine, cloudy, calm days are normally cooler than clear, calm days. Hence, the presence of clouds tends to keep nighttime temperatures higher and daytime temperatures lower.

In summary, the atmospheric greenhouse effect occurs because water vapor, CO_2, and other greenhouse gases are selective absorbers. They allow most of the sun's radiation to reach the surface, but they absorb a good portion of the earth's outgoing infrared radiation, preventing it from escaping into space (see Fig. 2.10). It is the atmospheric greenhouse effect, then, that keeps the temperature of our planet at a level where life can survive. The greenhouse effect is not just a "good thing"; it is essential to life on earth.

Enhancement of the Greenhouse Effect In spite of the inaccuracies that have plagued temperature measurements, studies suggest that for the past 100 years or so, the earth's surface air temperature has been undergoing a slight warming of about 0.6°C (about 1°F). Today, there are scientific computer models, called *general circulation models* (GCMs) that mathematically simulate the physical processes of the atmosphere and oceans. These models (also referred to as *climate models*) predict that if such a warming should continue unabated, we

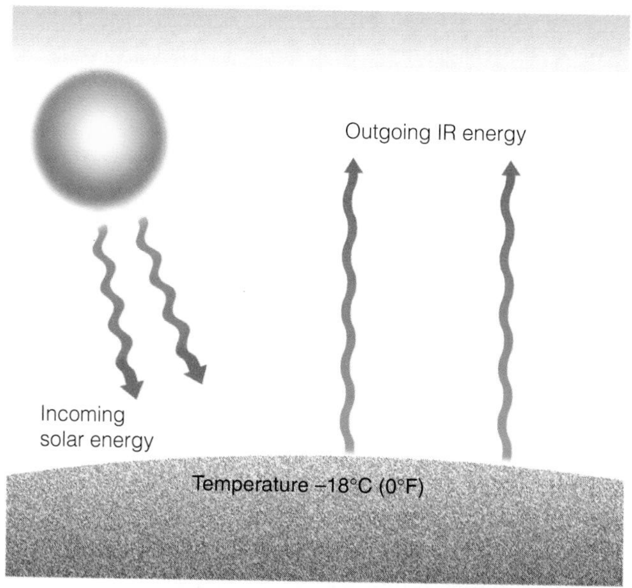

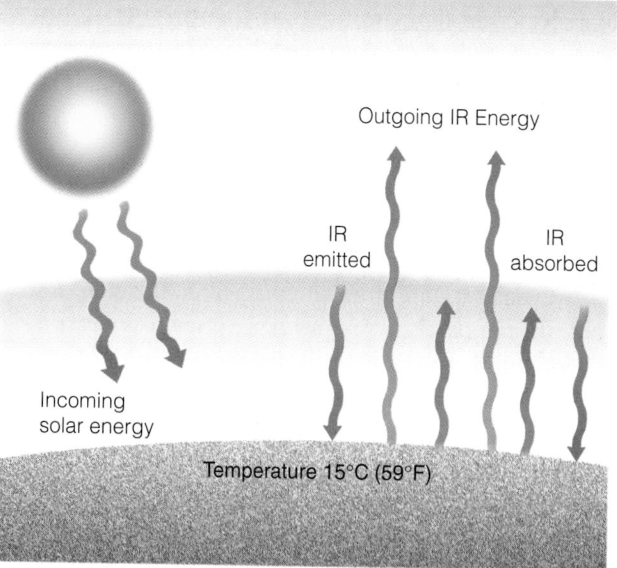

FIGURE 2.10 Sunlight warms the earth's surface only during the day, whereas the surface constantly emits infrared radiation upward during the day and at night. (a) Near the surface *without* water vapor, CO_2, and other greenhouse gases, the earth's surface would constantly emit infrared radiation (IR) energy; incoming energy from the sun would be equal to outgoing IR energy from the earth's surface. Since the earth would receive no IR energy from its lower atmosphere (no atmospheric greenhouse effect), the earth's average surface temperature would be a frigid −18°C (0°F). (b) With greenhouse gases, the earth's surface receives energy from the sun and infrared energy from its atmosphere. Incoming energy still equals outgoing energy, but the added IR energy from the greenhouse gases raises the earth's average surface temperature about 33°C, to a comfortable 15°C (59°F).

would be irrevocably committed to the negative effects of climate change, such as a rise in sea level.

The main cause of this *global warming* appears to be the greenhouse gas CO_2, whose concentration has been increasing primarily due to the burning of fossil fuels and to deforestation. However, in recent years, increasing concentration of other greenhouse gases, such as methane (CH_4), nitrous oxide (N_2O), and chlorofluorocarbons (CFCs), has collectively been shown to have an effect almost equal to that of CO_2. Look back at Fig. 2.9 and notice that both CH_4 and N_2O absorb strongly at infrared wavelengths. Moreover, a particular CFC (CFC-12) absorbs in the region of the atmospheric window between 8 and 11 μm. Thus, in terms of its absorption impact on infrared radiation, the addition of a single CFC-12 molecule to the atmosphere is the equivalent of adding 10,000 molecules of CO_2. Overall, water vapor accounts for about 60 percent of the atmospheric greenhouse effect, CO_2 accounts for about 26 percent, and the remaining greenhouse gases contribute about 14 percent.

Presently, the concentration of CO_2 in a volume of air near the surface is about 0.037 percent. Climate models predict that a continuing increase of CO_2 to an amount more than double its pre-industrial value of 0.028 percent, along with the continued increase of other greenhouse gases, will cause the earth's current average surface air temperature to rise between 1.4°C and 5.8°C (2.5°F and 10.5°F) by the end of this century. How can increasing such a small quantity of CO_2 and adding miniscule amounts of other greenhouse gases bring about such a large temperature increase?

Mathematical climate models predict that rising ocean temperatures will cause an increase in evaporation rates. The added *water vapor*—the primary greenhouse gas—will enhance the atmospheric greenhouse effect and double the temperature rise in what is known as a *positive feedback*. But there are other feedbacks to consider.*

The two potentially largest and least understood feedbacks in the climate system are the clouds and the oceans. Clouds can change area, depth, and radiation properties simultaneously with climatic changes. The net effect of all these changes is not totally clear at this time. Oceans, on the other hand, cover 70 percent of the planet. The response of ocean circulations, ocean temperatures, and sea ice to global warming will determine the global pattern and speed of climate change. Unfortunately, it is not now known how quickly each of these will respond.

Satellite data from the *Earth Radiation Budget Experiment* (ERBE) suggest that clouds overall appear to *cool* the earth's climate, as they reflect and radiate away more energy than they retain. (The earth would be warmer if clouds were not present.) So an increase in global cloudiness (if it were to occur) might offset some of the global warming brought on by an enhanced atmospheric greenhouse effect. Therefore, if clouds were to act on the climate system in this manner, they would provide a *negative feedback* on climate change.*

Uncertainties unquestionably exist about the impact that increasing levels of CO_2 and other greenhouse gases will have on enhancing the atmospheric greenhouse effect. Nonetheless, many scientific studies suggest that increasing the concentration of these gases in our atmosphere will lead to global-scale climatic change by the end of this century. Such change could adversely affect water resources and agricultural productivity. (We will examine this topic further in Chapter 14, where we cover climatic change in more detail.)

BRIEF REVIEW

In the last several sections, we have explored examples of some of the ways radiation is absorbed and emitted by various objects. Before reading the next several sections, let's review a few important facts and principles:

- *All* objects with a temperature above absolute zero emit radiation.
- The higher an object's temperature, the greater the amount of radiation emitted per unit surface area and the shorter the wavelength of maximum emission.
- The earth absorbs solar radiation only during the daylight hours; however, it emits infrared radiation continuously, both during the day and at night.
- The earth's surface behaves as a blackbody, making it a much better absorber and emitter of radiation than the atmosphere.
- Water vapor and carbon dioxide are important atmospheric greenhouse gases that selectively absorb and emit infrared radiation, thereby keeping the earth's average surface temperature warmer than it otherwise would be.
- Cloudy, calm nights are often warmer than clear, calm nights because clouds strongly emit infrared radiation.
- It is *not* the greenhouse effect itself that is of concern, but the *enhancement* of it due to increasing levels of greenhouse gases.

*A feedback is a process whereby an initial change in a process will tend to either reinforce the process (positive feedback) or weaken the process (negative feedback). The *water vapor–temperature rise* feedback is a positive feedback because the initial increase in temperature is reinforced by the addition of more water vapor, which absorbs more of the earth's infrared energy, thus strengthening the greenhouse effect and enhancing the warming.

*Overall, current climate models tend to show that changes in clouds could provide either a net negative or a net positive feedback on climate change.

With these concepts in mind, we will first examine how the air near the ground warms; then we will consider how the earth and its atmosphere maintain a yearly energy balance.

Warming the Air from Below On a clear day, solar energy passes through the lower atmosphere with little effect upon the air. Ultimately it reaches the surface, warming it (see Fig. 2.11). Air molecules in contact with the heated surface bounce against it, gain energy by *conduction,* then shoot upward like freshly popped kernels of corn, carrying their energy with them. Because the air near the ground is very dense, these molecules only travel a short distance before they collide with other molecules. During the collision, these more rapidly moving molecules share their energy with less energetic molecules, raising the average temperature of the air. But air is such a poor heat conductor that this process is only important within a few centimeters of the ground.

As the surface air warms, it actually becomes less dense than the air directly above it. The warmer air rises and the cooler air sinks, setting up thermals, or *free convection cells,* that transfer heat upward and distribute it through a deeper layer of air. The rising air expands and cools, and, if sufficiently moist, the water vapor condenses into cloud droplets, releasing latent heat that warms the air. Meanwhile, the earth constantly emits infrared energy. Some of this energy is absorbed by greenhouse gases (such as water vapor and carbon dioxide) that emit infrared energy upward and downward, back to the surface. Since the concentration of water vapor decreases rapidly above the earth, most of the absorption occurs in a layer near the surface. Hence, the lower atmosphere is mainly heated from below.

DID YOU KNOW?

The atmosphere of Venus, which is mostly carbon dioxide, is considerably more dense than that of Earth. Consequently, the greenhouse effect on Venus is exceptionally strong, producing a surface air temperature of about 500°C, or nearly 950°F.

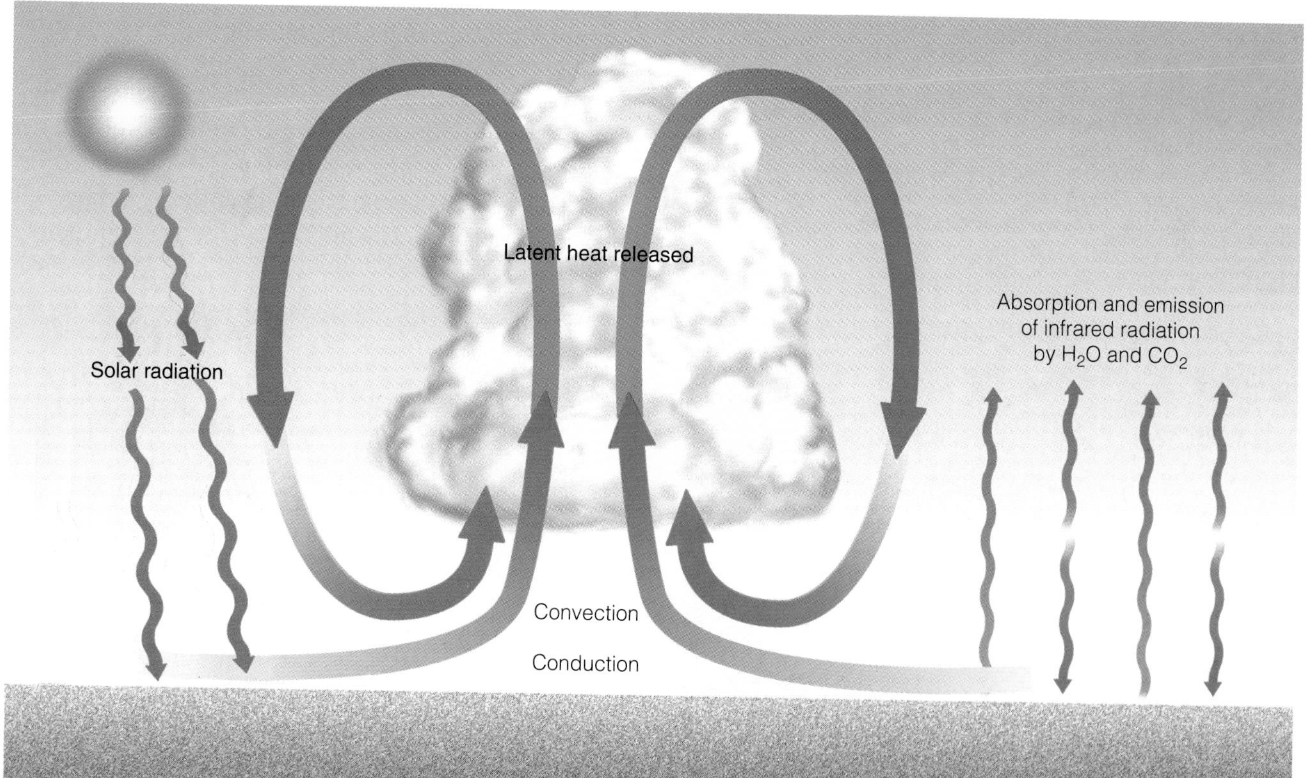

Meteorology Now™ ACTIVE FIGURE 2.11
Air in the lower atmosphere is heated from below. Sunlight warms the ground, and the air above is warmed by conduction, convection, and infrared radiation. Further warming occurs during condensation as latent heat is given up to the air inside the cloud.
Watch this Active Figure at http://earthscience.brookscole.com/ahrens/ess4e.

INCOMING SOLAR ENERGY

As the sun's radiant energy travels through space, essentially nothing interferes with it until it reaches the atmosphere. At the top of the atmosphere, solar energy received on a surface perpendicular to the sun's rays appears to remain fairly constant at nearly two calories on each square centimeter each minute, or 1367 W/m²— a value called the **solar constant.***

Scattered and Reflected Light When solar radiation enters the atmosphere, a number of interactions take place. For example, some of the energy is absorbed by gases, such as ozone, in the upper atmosphere. Moreover, when sunlight strikes very small objects, such as air molecules and dust particles, the light itself is deflected in all directions—forward, sideways, and backwards. The distribution of light in this manner is called **scattering.** (Scattered light is also called *diffuse light.*) Because air molecules are much smaller than the wavelengths of visible light, they are more effective scatterers of the shorter (blue) wavelengths than the longer (red) wavelengths. Hence, when we look away from the direct beam of sunlight, blue light strikes our eyes from all directions, turning the daytime sky blue. At midday, all the wavelengths of visible light from the sun strike our eyes, and the sun is perceived as white. At sunrise and sunset, when the white beam of sunlight must pass through a thick portion of the atmosphere, scattering by air molecules removes the blue light, leaving the longer wavelengths of red, orange, and yellow to pass on through, creating the image of a ruddy or yellowish sun (see Fig. 2.12).

Sunlight can be **reflected** from objects. Generally, reflection differs from scattering in that during the process of reflection more light is sent *backwards.* **Albedo** is the percent of radiation returning from a given surface compared to the amount of radiation initially striking that surface. Albedo, then, represents the *reflectivity* of the surface. In Table 2.2, notice that thick clouds have a higher albedo than thin clouds. On the average, the albedo of clouds is near 60 percent. When solar energy strikes a surface covered with snow, up to 95 percent of the sunlight may be reflected. Most of this energy is in the visible and ultraviolet wavelengths. Consequently, reflected radiation, coupled with direct sunlight, can produce severe sunburns on the exposed skin of unwary snow skiers, and unprotected eyes can suffer the agony of snow blindness.

Water surfaces, on the other hand, reflect only a small amount of solar energy. For an entire day, a smooth water surface will have an average albedo of about 10 percent. Averaged for an entire year, the earth and its atmosphere (including its clouds), will redirect

*By definition, the solar constant (which, in actuality, is *not* "constant") is the rate at which radiant energy from the sun is received on a surface at the outer edge of the atmosphere perpendicular to the sun's rays when the earth is at an average distance from the sun. Satellite measurements from the *Earth Radiation Budget Satellite* suggest the solar constant varies slightly as the sun's radiant output varies. The average is about 1.96 cal/cm²/min, or between 1365 W/m² and 1372 W/m² in the SI system of measurement.

FIGURE 2.12 A brilliant red sunset produced by the process of scattering.

TABLE 2.2
Typical Albedo of Various Surfaces

SURFACE	ALBEDO (PERCENT)
Fresh snow	75 to 95
Clouds (thick)	60 to 90
Clouds (thin)	30 to 50
Venus	78
Ice	30 to 40
Sand	15 to 45
Earth and atmosphere	30
Mars	17
Grassy field	10 to 30
Dry, plowed field	5 to 20
Water	10*
Forest	3 to 10
Moon	7

*Daily average.

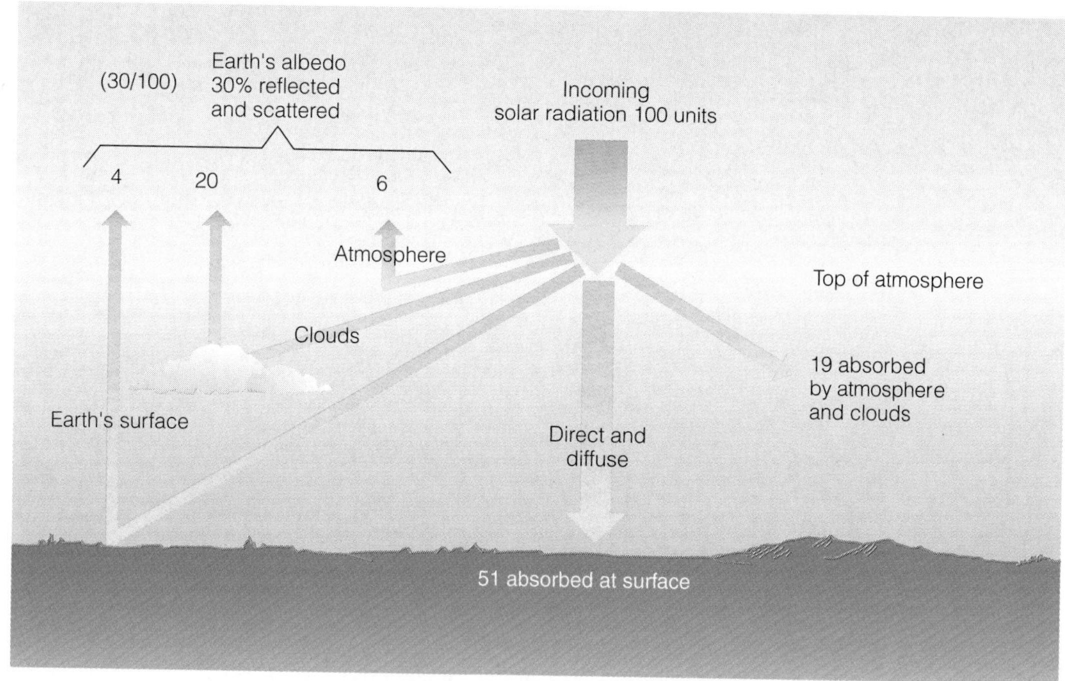

FIGURE 2.13 On the average, of all the solar energy that reaches the earth's atmosphere annually, about 30 percent (30/100) is reflected and scattered back to space, giving the earth and its atmosphere an albedo of 30 percent. Of the remaining solar energy, about 19 percent is absorbed by the atmosphere and clouds, and 51 percent is absorbed at the surface.

about 30 percent of the sun's incoming radiation back to space, which gives the earth and its atmosphere a combined albedo of 30 percent (see Fig. 2.13).

The Earth's Annual Energy Balance Although the average temperature at any one place may vary considerably from year to year, the earth's overall average equilibrium temperature changes only slightly from one year to the next. This fact indicates that, each year, the earth and its atmosphere combined must send off into space just as much energy as they receive from the sun. The same type of energy balance must exist between the earth's surface and the atmosphere. That is, each year, the earth's surface must return to the atmosphere the same amount of energy that it absorbs. If this did not occur, the earth's average surface temperature would change. How do the earth and its atmosphere maintain this yearly energy balance?

Suppose 100 units of solar energy reach the top of the earth's atmosphere. We already know from Fig. 2.13 that, on the average, clouds, the earth, and the atmosphere reflect and scatter 30 units back to space, and that the atmosphere and clouds together absorb 19 units, which leaves 51 units of direct and indirect (diffuse) solar radiation to be absorbed at the earth's surface.

Figure 2.14 shows approximately what happens to the solar radiation that is absorbed by the surface and the atmosphere. Out of 51 units reaching the surface, a large amount (23 units) is used to evaporate water, and about 7 units are lost through conduction and convection, which leaves 21 units to be radiated away as infrared energy. Look closely at Fig. 2.14 and notice that the earth's surface actually radiates upward a whopping 117 units. It does so because, although it receives solar radiation only during the day, it constantly emits infrared energy both during the day and at night. Additionally, the atmosphere above only allows a small fraction of this energy (6 units) to pass through into space. The majority of it (111 units) is absorbed mainly by the greenhouse gases water vapor and CO_2, and by clouds. Much of this energy (96 units) is then radiated back to earth, producing the atmospheric greenhouse effect. Hence, the earth's surface receives nearly twice as much longwave infrared energy from the atmosphere as it does shortwave radiation from the sun. In all these exchanges, notice that the energy lost at the earth's surface (147 units) is exactly balanced by the energy gained there (147 units).

A similar balance exists between the earth's surface and its atmosphere. Again in Fig. 2.14 observe that the

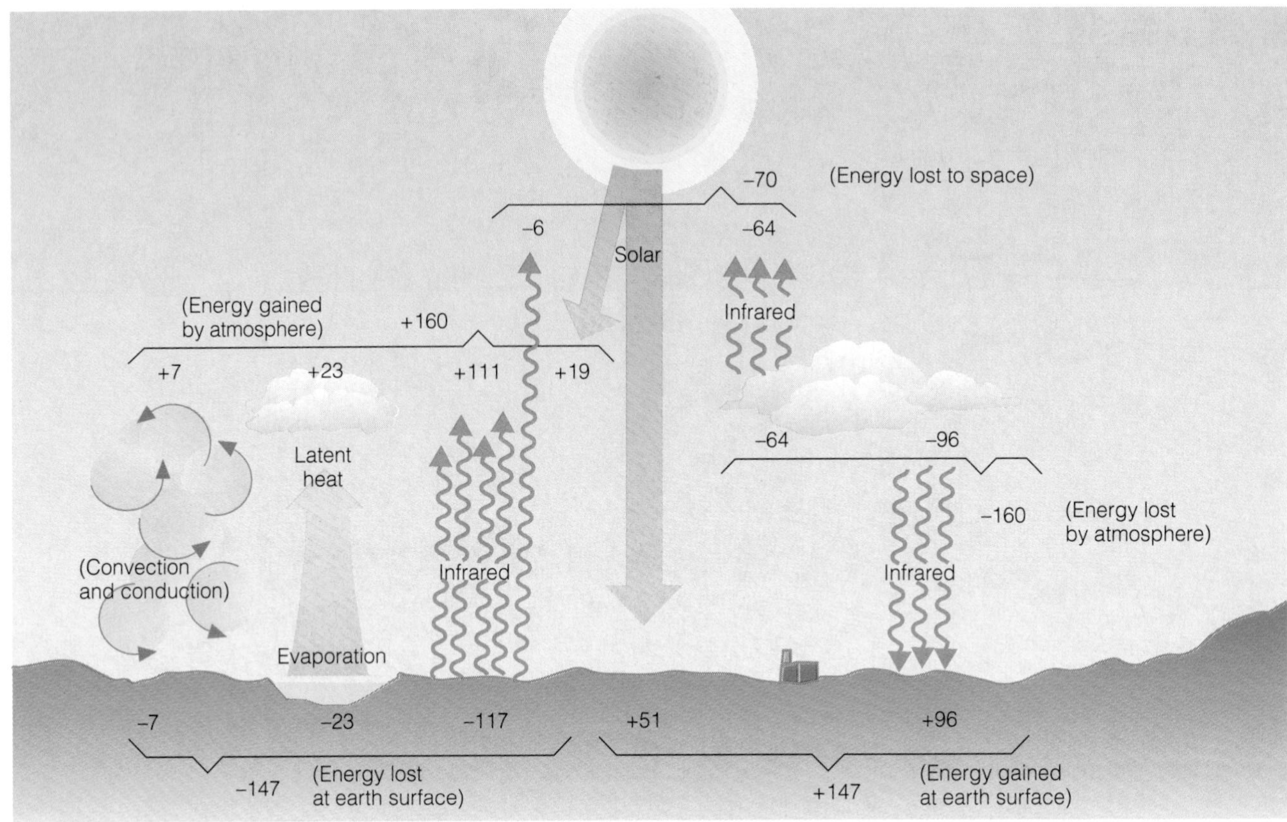

FIGURE 2.14 The earth-atmosphere energy balance. Numbers represent approximations based on surface observations and satellite data. While the actual value of each process may vary by several percent, it is the relative size of the numbers that is important.

energy gained by the atmosphere (160 units) balances the energy lost. Moreover, averaged for an entire year, the solar energy received at the earth's surface (51 units) and that absorbed by the earth's atmosphere (19 units) balances the infrared energy lost to space by the earth's surface (6 units) and its atmosphere (64 units).

We can see the effect that conduction, convection, and latent heat play in the warming of the atmosphere if we look at the energy balance only in radiative terms. The earth's surface receives 147 units of radiant energy from the sun and its own atmosphere, while it radiates away 117 units, producing a *surplus* of 30 units. The atmosphere, on the other hand, receives 130 units (19 units from the sun and 111 from the earth), while it loses 160 units, producing a *deficit* of 30 units. The balance (30 units) is the warming of the atmosphere produced by the heat transfer processes of conduction and convection (7 units) and by the release of latent heat (23 units).

And so, the earth and the atmosphere absorb energy from the sun, as well as from each other. In all of the energy exchanges, a delicate balance is maintained. Essentially, there is no yearly gain or loss of total energy, and the average temperature of the earth and the atmosphere remains fairly constant from one year to the next. This equilibrium does not imply that the earth's average temperature does not change, but that the changes are small from year to year (usually less than one-tenth of a degree Celsius), and become significant only when measured over many years.

We now turn our attention to how incoming solar energy produces the earth's seasons. Before doing so, you may wish to read the Focus section on p. 43, which explains how solar energy, in the form of particles, produces a dazzling light show known as the *aurora.*

Why the Earth Has Seasons The earth revolves completely around the sun in an elliptical path (not quite a circle) in slightly longer than 365 days (one year). As the earth revolves around the sun, it spins on its own axis, completing one spin in 24 hours (one day). The average distance from the earth to the sun is 150 million km (93 million mi). Because the earth's orbit is an ellipse instead of a circle, the actual distance from the earth to the sun varies during the year. The earth comes closer to

FOCUS ON AN OBSERVATION

The Aurora—A Dazzling Light Show

At high latitudes after darkness has fallen, a faint, white glow may appear in the sky. Lasting from a few minutes to a few hours, the light may move across the sky as a yellow-green arc much wider than a rainbow; or, it may faintly decorate the sky with flickering draperies of blue, green, and purple light that constantly change in form and location, as if blown by a gentle breeze. This eerie yet beautiful light show is called the **aurora** (see Fig. 2).

The aurora is caused by charged particles from the sun interacting with our atmosphere. From the sun and its tenuous atmosphere comes a continuous discharge of particles. This discharge happens because, at extremely high temperatures, gases become stripped of electrons by violent collisions and acquire enough speed to escape the gravitational pull of the sun. As these charged particles (ions and electrons) travel through space, they are known as the *solar wind.* When the solar wind moves close enough to the earth, it interacts with the earth's magnetic field, disturbing it. This disturbance causes energetic solar wind particles to enter the upper atmosphere, where they collide with atmospheric gases. These gases then become excited and emit visible radiation (light), which causes the sky to glow like a neon light, thus producing the aurora.

In the Northern Hemisphere, the aurora is called the *aurora borealis,* or northern lights; its counterpart in the Southern Hemisphere is the *aurora australis,* or southern lights. The aurora is most frequently seen in the polar regions, where the earth's magnetic field lines emerge from the earth (see Fig. 3). But during active sun periods when there are numerous sunspots (huge cooler regions on the sun's surface) and giant flares (solar eruptions), large quantities of particles travel outward away from the sun at high speeds (hundreds of kilometers a second). These energetic particles are able to penetrate unusually deep into the earth's magnetic field, where they provide sufficient energy to produce auroral displays. During these conditions in North America, we see the aurora much farther south than usual.

FIGURE 2 The aurora borealis is a phenomenon that forms as energetic particles from the sun interact with the earth's atmosphere.

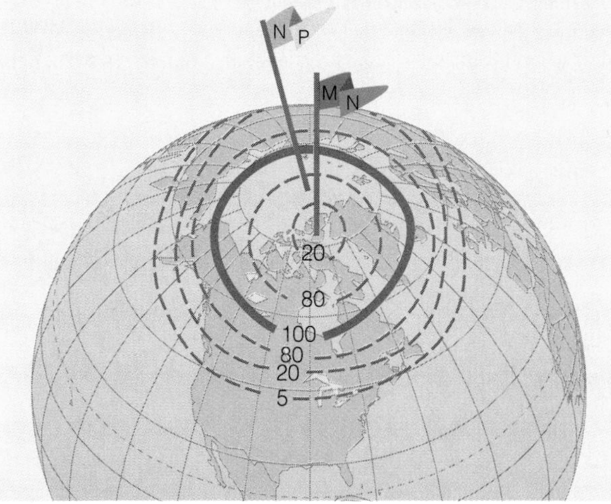

FIGURE 3 The aurora belt (solid red line) represents the region where you would most likely observe the aurora on a clear night. (The numbers represent the average number of nights per year on which you might see an aurora if the sky were clear.) The flag MN denotes the magnetic North Pole, where the earth's magnetic field lines emerge from the earth. The flag NP denotes the geographic North Pole, about which the earth rotates.

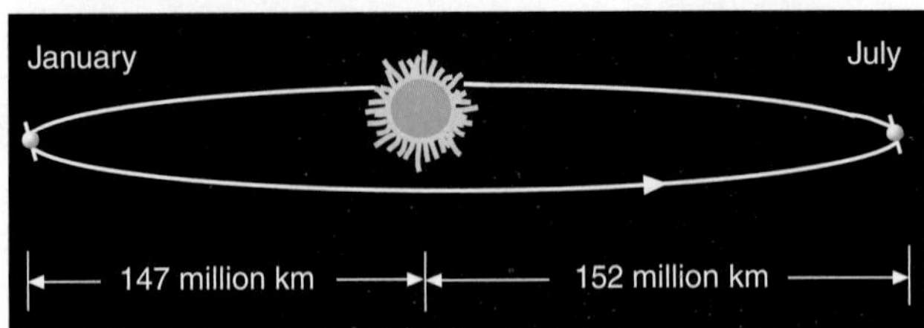

FIGURE 2.15 The elliptical path (highly exaggerated) of the earth about the sun brings the earth slightly closer to the sun in January than in July.

the sun in January (147 million km) than it does in July (152 million km).* (See Fig. 2.15.) From this fact, we might conclude that our warmest weather should occur in January and our coldest weather in July. But, in the Northern Hemisphere, we normally experience cold weather in January when we are closer to the sun and warm weather in July when we are farther away. If nearness to the sun were the primary cause of the seasons then, indeed, January would be warmer than July. However, nearness to the sun is only a small part of the story.

Our seasons are regulated by the amount of solar energy received at the earth's surface. This amount is determined primarily by the angle at which sunlight strikes the surface, and by how long the sun shines on any latitude (daylight hours). Let's look more closely at these factors.

*The time around January 3rd, when the earth is closest to the sun, is called *perihelion* (from the Greek *peri,* meaning "near" and *helios,* meaning "sun"). The time when the earth is farthest from the sun (around July 4th) is called *aphelion* (from the Greek *ap,* meaning "away from").

Solar energy that strikes the earth's surface perpendicularly (directly) is much more intense than solar energy that strikes the same surface at an angle. Think of shining a flashlight straight at a wall—you get a small, circular spot of light (see Fig. 2.16). Now, tip the flashlight and notice how the spot of light spreads over a larger area. The same principle holds for sunlight. Sunlight striking the earth at an angle spreads out and must heat a larger region than sunlight impinging directly on the earth. Everything else being equal, an area experiencing more direct solar rays will receive more heat than the same size area being struck by sunlight at an angle. In addition, the more the sun's rays are slanted from the perpendicular, the more atmosphere they must penetrate. And the more atmosphere they penetrate, the more they can be scattered and absorbed (attenuated). As a consequence, when the sun is high in the sky, it can heat the ground to a much higher temperature than when it is low on the horizon.

The second important factor determining how warm the earth's surface becomes is the length of time the sun shines each day. Longer daylight hours, of course, mean that more energy is available from sunlight. In a given location, more solar energy reaches the earth's surface on a clear, long day than on a day that is clear but much shorter. Hence, more surface heating takes place.

From a casual observation, we know that summer days have more daylight hours than winter days. Also, the noontime summer sun is higher in the sky than is the noontime winter sun. Both of these events occur because our spinning planet is inclined on its axis (tilted) as it revolves around the sun. As Fig. 2.17 illustrates, the angle of

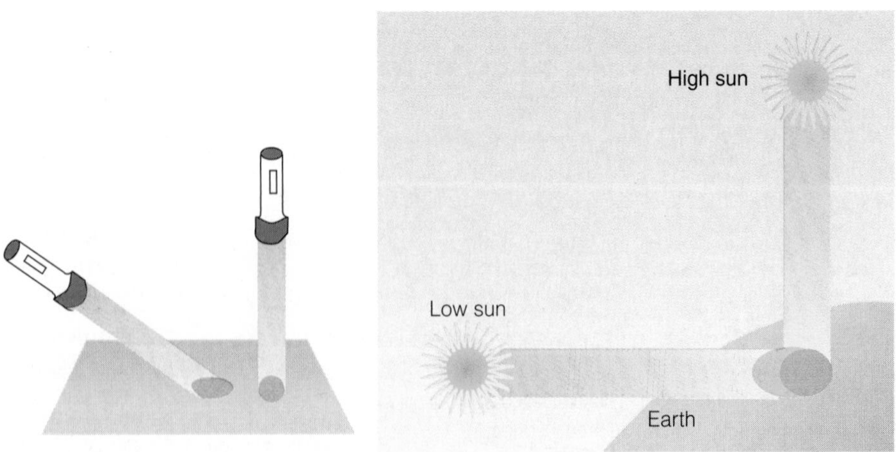

Meteorology Now™ ACTIVE FIGURE 2.16
Sunlight that strikes a surface at an angle is spread over a larger area than sunlight that strikes the surface directly. Oblique sun rays deliver less energy (are less intense) to a surface than direct sun rays.
Watch this Active Figure at http://earthscience.brookscole.com/ahrens/ess4e.

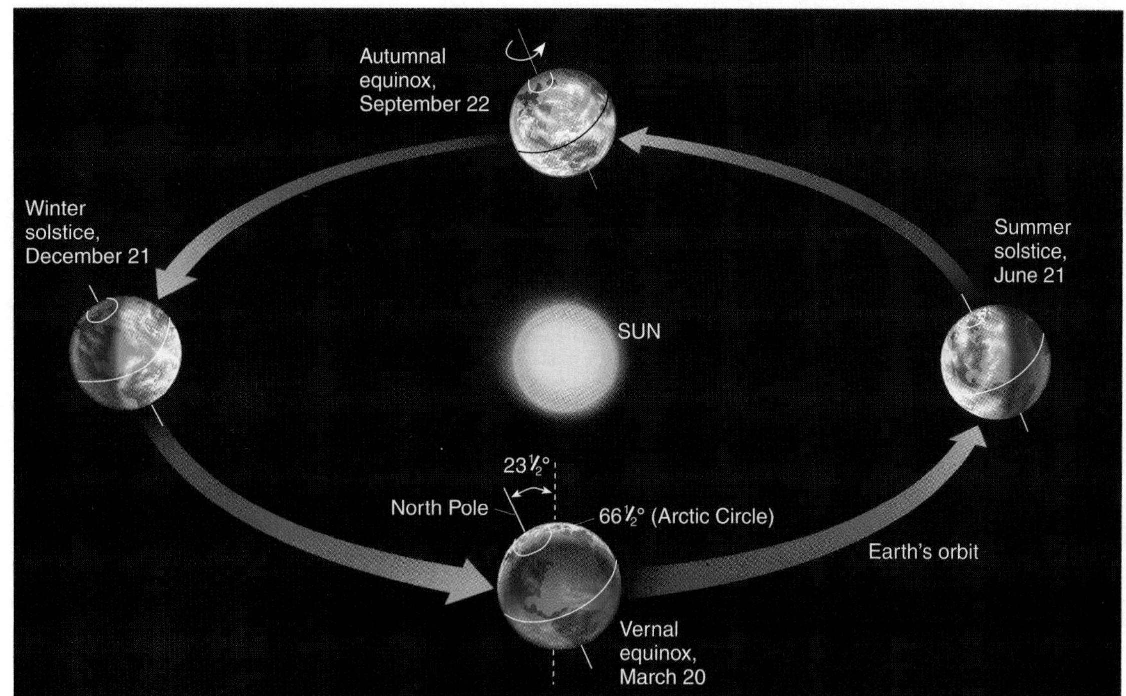

Meteorology Now™ ACTIVE FIGURE 2.17
As the earth revolves about the sun, it is tilted on its axis by an angle of 23½°. The earth's axis always points to the same area in space (as viewed from a distant star). Thus, in June, when the Northern Hemisphere is tipped toward the sun, more direct sunlight and long hours of daylight cause warmer weather than in December, when the Northern Hemisphere is tipped away from the sun. (Diagram, of course, is not to scale.)
Watch this Active Figure at http://earthscience.brookscole.com/ahrens/ess4e.

tilt is 23½° from the perpendicular drawn to the plane of the earth's orbit. The earth's axis points to the same direction in space all year long; thus, the Northern Hemisphere is tilted toward the sun in summer (June), and away from the sun in winter (December).

Seasons in the Northern Hemisphere Notice in Fig. 2.17 that on June 21, the northern half of the world is directed toward the sun. At noon on this day, solar rays beat down upon the Northern Hemisphere more directly than during any other time of year. The sun is at its highest position in the noonday sky, directly above 23½° north (N) latitude (Tropic of Cancer). If you were standing at this latitude on June 21, the sun at noon would be directly overhead. This day, called the **summer solstice,** is the astronomical first day of summer in the Northern Hemisphere.*

Study Fig. 2.17 closely and notice that, as the earth spins on its axis, the side facing the sun is in sunshine and the other side is in darkness. Thus, half of the globe is always illuminated. If the earth's axis were not tilted, the noonday sun would always be directly overhead at the equator, and there would be 12 hours of daylight and 12 hours of darkness at each latitude every day of the year. However, the earth is tilted. Since the Northern Hemisphere faces towards the sun on June 21, each latitude in the Northern Hemisphere will have more than 12 hours of daylight. The farther north we go, the longer are the daylight hours. When we reach the Arctic Circle

DID YOU KNOW?

The Land of Total Darkness. Does darkness (constant night) really occur at the Arctic Circle (66½°N) on the winter solstice? The answer is no. Due to the bending and scattering of sunlight by the atmosphere, the sky is not totally dark at the Arctic Circle on December 21. In fact, on this date, total darkness only happens north of about 82° latitude. Even at the North Pole, total darkness does not occur from September 22 through March 20, but rather from about November 5 through February 5.

*As we will see later in this chapter, the seasons are reversed in the Southern Hemisphere. Hence, in the Southern Hemisphere, this same day is the winter solstice, or the astronomical first day of winter.

FIGURE 2.18 Land of the Midnight Sun. A series of exposures of the sun taken before, during, and after midnight in northern Alaska during July.

(66½°N), daylight lasts for 24 hours, as the sun does not set. Notice in Fig. 2.17 how the region above 66½°N never gets into the "shadow" zone as the earth spins. At the North Pole, the sun actually rises above the horizon on March 20 and has six months until it sets on September 22. No wonder this region is called the "Land of the Midnight Sun"! (See Fig. 2.18.)

Even though in the far north the sun is above the horizon for many hours during the summer (see Table 2.3), the surface air there is not warmer than the air farther south, where days are appreciably shorter. The reason for this fact is shown in Fig. 2.19. When *in*coming *so*lar radi*ation* (called *insolation*) enters the atmosphere, fine dust, air molecules and clouds reflect and scatter it, and some of it is absorbed by atmospheric gases. Generally, the greater the thickness of atmosphere

TABLE 2.3 Length of Time from Sunrise to Sunset for Various Latitudes on Different Dates

NORTHERN HEMISPHERE (READ DOWN)				
Latitude	*March 20*	*June 21*	*Sept. 22*	*Dec. 21*
0°	12 hr	12.0 hr	12 hr	12.0 hr
10°	12 hr	12.6 hr	12 hr	11.4 hr
20°	12 hr	13.2 hr	12 hr	10.8 hr
30°	12 hr	13.9 hr	12 hr	10.1 hr
40°	12 hr	14.9 hr	12 hr	9.1 hr
50°	12 hr	16.3 hr	12 hr	7.7 hr
60°	12 hr	18.4 hr	12 hr	5.6 hr
70°	12 hr	2 months	12 hr	0 hr
80°	12 hr	4 months	12 hr	0 hr
90°	12 hr	6 months	12 hr	0 hr
Latitude	*Sept. 22*	*Dec. 21*	*March 20*	*June 21*
SOUTHERN HEMISPHERE (READ UP)				

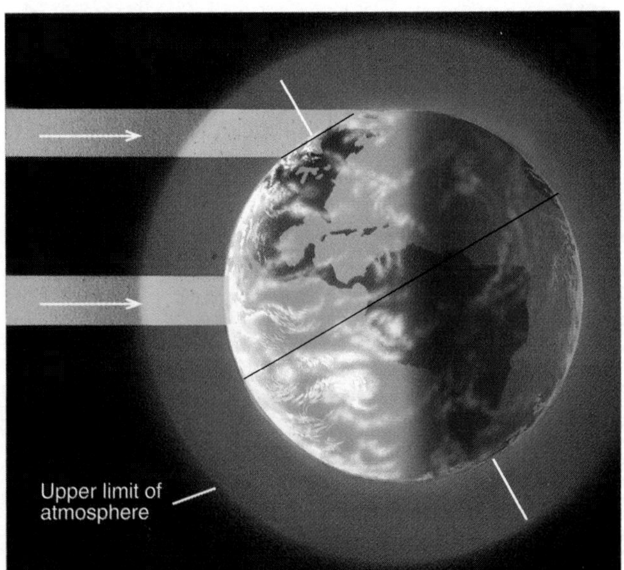

FIGURE 2.19 During the Northern Hemisphere summer, sunlight that reaches the earth's surface in far northern latitudes has passed through a thicker layer of absorbing, scattering, and reflecting atmosphere than sunlight that reaches the earth's surface farther south. Sunlight is lost through both the thickness of the pure atmosphere and by impurities in the atmosphere. As the sun's rays become more oblique, these effects become more pronounced.

that sunlight must penetrate, the greater are the chances that it will be either reflected or absorbed by the atmosphere. During the summer in far northern latitudes, the sun is never very high above the horizon, so its radiant energy must pass through a thick portion of atmosphere before it reaches the earth's surface. Some of the solar energy that does reach the surface melts frozen soil or is reflected by snow or ice. And, that which is absorbed is spread over a large area. So, even though northern cities may experience long hours of sunlight they are not warmer than cities farther south. Overall, they receive less radiation at the surface, and what radiation they do receive does not effectively heat the surface.

Look at Fig. 2.17 again and notice that, by September 22, the earth will have moved so that the sun is directly above the equator. Except at the poles, the days and nights throughout the world are of equal length. This day is called the **autumnal** (fall) **equinox,** and it marks the astronomical beginning of fall in the Northern Hemisphere. At the North Pole, the sun appears on the horizon for 24 hours, due to the bending of light by the atmosphere. The following day (or at least within several days), the sun disappears from view, not to rise again for a long, cold six months. Throughout the northern half of the world on each successive day, there are fewer hours of daylight, and the noon sun is slightly lower in the sky. Less direct sunlight and shorter hours of daylight spell cooler weather for the Northern Hemisphere. Reduced sunlight, lower air temperatures, and cooling breezes stimulate the beautiful pageantry of fall colors (see Fig. 2.20).

DID YOU KNOW?

Contrary to popular belief, it is not the first frost that causes the leaves of deciduous trees to change color. The yellow and orange colors, which are actually in the leaves, begin to show through several weeks before the first frost, as shorter days and cooler nights cause a decrease in the production of the green pigment chlorophyll.

In some years around the middle of autumn, there is an unseasonably warm spell, especially in the eastern two-thirds of the United States. This warm period, referred to as **Indian Summer,** may last from several days up to a week or more. It usually occurs when a large high pressure area stalls near the southeast coast. The clockwise flow of air around this system moves warm air from the Gulf of Mexico into the central or eastern half of the nation. The warm, gentle breezes and smoke from a variety of sources respectively make for mild, hazy days. The warm weather ends abruptly when an outbreak of polar air reminds us that winter is not far away.

On December 21 (three months after the autumnal equinox), the Northern Hemisphere is tilted as far away

FIGURE 2.20 The pageantry of fall colors in New England. The weather most suitable for an impressive display of fall colors is warm, sunny days followed by clear, cool nights with temperatures dropping below 7°C (45°F), but remaining above freezing.

DID YOU KNOW?

On an equinox, is the time between sunrise and sunset exactly 12 hours? The answer depends on how we define "sunrise" and "sunset." For example, sunrise may be defined as the time when the sun's upper edge becomes visible, and sunset when the sun's upper edge disappears. This definition of sunrise and sunset causes the actual date when day and night are of equal length (as, for example, reported in the newspaper) to occur several days before the vernal equinox and several days after the autumnal equinox.

from the sun as it will be all year (see Fig. 2.17, p. 45). Nights are long and days are short. Notice in Table 2.3 that daylight decreases from 12 hours at the equator to 0 (zero) at latitudes above 66½°N. This is the shortest day of the year, called the **winter solstice**—the astronomical beginning of winter in the northern world. On this day, the sun shines directly above latitude 23½°S (Tropic of Capricorn). In the northern half of the world, the sun is at its lowest position in the noon sky. Its rays pass through a thick section of atmosphere and spread over a large area on the surface.

With so little incident sunlight, the earth's surface cools quickly. A blanket of clean snow covering the ground aids in the cooling. In northern Canada and Alaska, arctic air rapidly becomes extremely cold as it lies poised, ready to do battle with the milder air to the south. Periodically, this cold arctic air pushes down into the northern United States, producing a rapid drop in temperature called a *cold wave,* which occasionally reaches far into the south. Sometimes, these cold spells arrive well before the winter solstice—the "official" first day of winter—bringing with them heavy snow and blustery winds. (More information on this "official" first day of winter is given in the Focus section on p. 49.)

Three months past the winter solstice marks the astronomical arrival of spring, which is called the **vernal** (spring) **equinox.** The date is March 20 and, once again, the noonday sun is shining directly on the equator, days and nights throughout the world are of equal length, and, at the North Pole, the sun rises above the horizon after a long six month absence.

At this point it is interesting to note that although sunlight is most intense in the Northern Hemisphere on June 21, the warmest weather in middle latitudes normally occurs weeks later, usually in July or August. This situation (called the *lag in seasonal temperature*) arises because although incoming energy from the sun is greatest in June, it still exceeds outgoing energy from the earth for a period of at least several weeks. When incoming solar energy and outgoing earth energy are in balance, the highest average temperature is attained. When outgoing energy exceeds incoming energy, the average temperature drops. Because outgoing earth energy exceeds incoming solar energy well past the winter solstice (December 21), we normally find our coldest weather occurring in January or February.

Up to now, we have seen that the seasons are controlled by solar energy striking our tilted planet, as it makes its annual voyage around the sun. This tilt of the earth causes a seasonal variation in both the length of daylight and the intensity of sunlight that reaches the surface. Because of these facts, high latitudes tend to lose more energy to space each year than they receive from the sun, while low latitudes tend to gain more energy during the course of a year than they lose. From Fig. 2.21 we can see that only at middle latitudes near 37° does the amount of energy received each year balance the amount lost. From this situation, we might conclude that polar regions are growing colder each year, while tropical regions are becoming warmer. But this does not happen. To compensate for these gains and losses of energy, winds in the atmosphere and currents in the oceans circulate warm air and water toward the poles, and cold air and water toward the equator. Thus, the transfer of heat

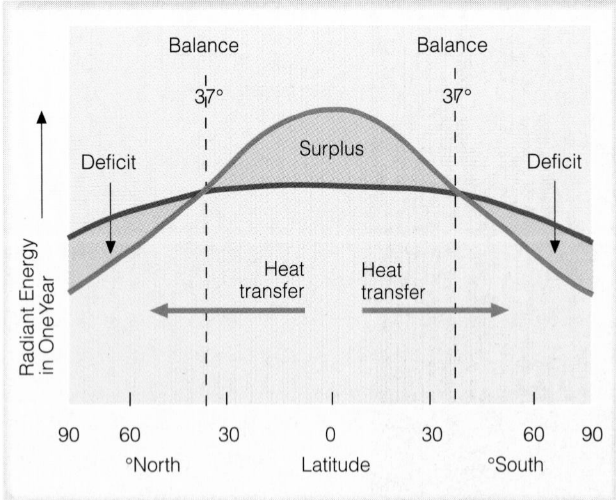

FIGURE 2.21 The average annual incoming solar radiation (red line) absorbed by the earth and the atmosphere along with the average annual infrared radiation (blue line) emitted by the earth and the atmosphere.

FOCUS ON A SPECIAL TOPIC

Is December 21 Really the First Day of Winter?

On December 21 (or 22, depending on the year) after nearly a month of cold weather, and perhaps a snowstorm or two, (see Fig. 4) someone on the radio or television has the audacity to proclaim that "today is the first official day of winter." If during the last several weeks it was not winter, then what season was it?

Actually, December 21 marks the *astronomical* first day of winter in the Northern Hemisphere (NH), just as June 21 marks the *astronomical* first day of summer (NH). The earth is tilted on its axis by 23½° as it revolves around the sun. This fact causes the sun (as we view it from earth) to move in the sky from a point where it is directly above 23½° South latitude on December 21, to a point where it is directly above 23½° North latitude on June 21. The astronomical first day of spring (NH) occurs around March 20 as the sun crosses the equator moving northward and, likewise, the astronomical first day of autumn (NH) occurs around September 22 as the sun crosses the equator moving southward.

In the middle latitudes, summer is defined as the warmest season and winter the coldest season. If the year is divided into four seasons with each season consisting of three months, then the meteorological definition of summer over much of the Northern Hemisphere would be the three warmest months of June, July, and August. Winter would be the three coldest months of December, January, and February. Autumn would be September, October, and November—the transition between summer and winter. And spring would be March, April, and May—the transition between winter and summer.

So, the next time you hear someone remark on December 21 that "winter officially begins today," remember that this is the astronomical definition of the first day of winter. According to the meteorological definition, winter has been around for several weeks.

FIGURE 4 A heavy snowfall covers New York City in early December. Since the snowstorm occurred before the winter solstice, is this a late fall storm or an early winter storm?

energy by atmospheric and oceanic circulations prevents low latitudes from steadily becoming warmer and high latitudes from steadily growing colder. These circulations are extremely important to weather and climate, and will be treated more completely in Chapter 7.

Seasons in the Southern Hemisphere On June 21, the Southern Hemisphere is adjusting to an entirely different season. Because this part of the world is now tilted away from the sun, nights are long, days are short, and solar rays come in at an angle. All of these factors keep air temperatures fairly low. The June solstice marks the astronomical beginning of winter in the Southern Hemisphere. In this part of the world, summer will not "officially" begin until the sun is over the Tropic of Capricorn (23½°S)—remember that this occurs on December 21. So, when it is winter and June in the Southern Hemisphere, it is summer and June in the Northern Hemisphere. Conversely, when it is summer and December in the Southern Hemisphere, it is winter and December in the Northern Hemisphere. So, if you are tired of the cold, December weather in your North-

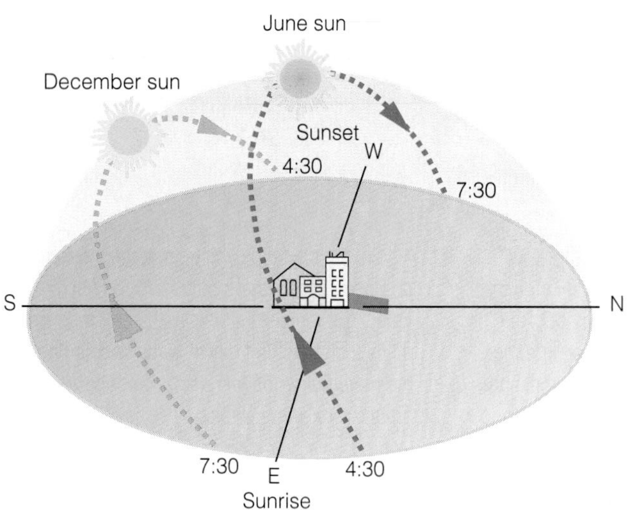

FIGURE 2.22 The changing position of the sun, as observed in middle latitudes in the Northern Hemisphere.

ern Hemisphere city, travel to the summer half of the world and enjoy the warmer weather. The tilt of the earth as it revolves around the sun makes all this possible.

We know the earth comes nearer to the sun in January than in July. Even though this difference in distance amounts to only about 3 percent, the energy that strikes the top of the earth's atmosphere is almost 7 percent greater on January 3 than on July 4. These statistics might lead us to believe that summer should be warmer in the Southern Hemisphere than in the Northern Hemisphere, which, however, is not the case. A close examination of the Southern Hemisphere reveals that nearly 81 percent of the surface is water compared to 61 percent in the Northern Hemisphere. The added solar energy due to the closeness of the sun is absorbed by large bodies of water, becoming well mixed and circulated within them. This process keeps the average summer (January) temperatures in the Southern Hemisphere cooler than the average summer (July) temperatures in the Northern Hemisphere. Because of water's large heat capacity, it also tends to keep winters in the Southern Hemisphere warmer than we might expect.*

DID YOU KNOW?

Seasonal changes can cause depression. For example, some people face each winter with a sense of foreboding, especially at high latitudes where days are short and nights are long and cold. If the depression is lasting and disabling, the problem is called *seasonal affective disorder* (SAD). People with SAD tend to sleep longer, overeat, and feel tired and drowsy during the day. The treatment is usually extra doses of bright light.

Local Seasonal Variations Figure 2.22 shows how the sun's position changes in the middle latitudes of the Northern Hemisphere during the course of one year. Note that, during the winter, the sun rises in the southeast and sets in the southwest. During the summer, it rises in the northeast, reaches a much higher position in the sky at noon, and sets in the northwest. Clearly, objects facing south will receive more sunlight during a year than those facing north. This fact becomes strikingly apparent in hilly or mountainous country.

Hills that face south receive more sunshine and, hence, become warmer than the partially shielded north-facing hills. Higher temperatures usually mean greater rates of evaporation and slightly drier soil conditions. Thus, south-facing hillsides are usually warmer and drier as compared to north-facing slopes at the same elevation. In many areas of the far west, only sparse vegetation grows on south-facing slopes, while, on the same hill, dense vegetation grows on the cool, moist hills that face north (see Fig. 2.23).

In the mountains, snow usually lingers on the ground for a longer time on north slopes than on the warmer south slopes. For this reason, ski runs are built facing north wherever possible. Also, homes and cabins built on the north side of a hill usually have a steep pitched roof, as well as a reinforced deck to withstand the added weight of snow from successive winter storms.

The seasonal change in the sun's position during the year can have an effect on the vegetation around the home. In winter, a large two-story home can shade its own north side, keeping it much cooler than its south side. Trees that require warm, sunny weather should be planted on the south side, where sunlight reflected from the house can even add to the warmth.

The design of a home can be important in reducing heating and cooling costs. Large windows should face south, allowing sunshine to penetrate the home in winter. To block out excess sunlight during the summer, a small eave or overhang should be built. A kitchen with windows facing east will let in enough warm morning sunlight to help heat this area. Because the west side

*For a comparison of January and July temperatures see Figs. 3.8 and 3.9, p. 64.

Photo by author

FIGURE 2.23 In areas where small temperature changes can cause major changes in soil moisture, sparse vegetation on the south-facing slopes will often contrast with lush vegetation on the north-facing slopes.

warms rapidly in the afternoon, rooms having small windows (such as garages) should be placed here to act as a thermal buffer. Deciduous trees planted on the west side of a home provide shade in the summer. In winter, they drop their leaves, allowing the winter sunshine to warm the house. If you like the bedroom slightly cooler than the rest of the home, face it toward the north. Let nature help with the heating and air conditioning. Proper house design, orientation, and landscaping can help cut the demand for electricity, as well as for natural gas and fossil fuels, which are rapidly being depleted.

SUMMARY

In this chapter, we looked at the concepts of heat and temperature and learned that latent heat is an important source of atmospheric heat energy. We also learned that the transfer of heat can take place by conduction, convection, and radiation—the transfer of energy by means of electromagnetic waves.

The hot sun emits most of its radiation as shortwave radiation. A portion of this energy heats the earth, and the earth, in turn, warms the air above. The cool earth emits most of its radiation as longwave infrared energy. Selective absorbers in the atmosphere, such as water vapor and carbon dioxide, absorb some of the earth's infrared radiation and radiate a portion of it back to the surface, where it warms the surface, producing the atmospheric greenhouse effect. The average equilibrium temperature of the earth and the atmosphere remains fairly constant from one year to the next because the amount of energy they absorb each year is equal to the amount of energy they lose.

We examined the seasons and found that the earth has seasons because it is tilted on its axis as it revolves around the sun. The tilt of the earth causes a seasonal variation in both the length of daylight and the intensity of sunlight that reaches the surface. Finally, on a more local setting, we saw that the earth's inclination influences the amount of solar energy received on the north and south side of a hill, as well as around a home.

Meteorology Now™ Assess your understanding of this chapter's topics with additional quizzing and tutorials at http://earthscience.brookscole.com/ahrens/ess4e.

KEY TERMS

The following terms are listed in the order they appear in the text. Define each. Doing so will aid you in reviewing the material covered in this chapter.

kinetic energy	absolute zero
temperature	heat

Kelvin scale
Fahrenheit scale
Celsius scale
latent heat
sensible heat
conduction
convection
thermals
advection
radiant energy (radiation)
electromagnetic waves
micrometer
photons
visible region
ultraviolet radiation (UV)
infrared radiation (IR)
blackbody
radiative equilibrium temperature
selective absorbers
greenhouse effect
atmospheric window
solar constant
scattering
reflected (light)
albedo
aurora
summer solstice
autumnal equinox
Indian summer
winter solstice
vernal equinox

QUESTIONS FOR REVIEW

1. Distinguish between temperature and heat.
2. How does the average speed of air molecules relate to the air temperature?
3. Explain how heat is transferred in our atmosphere by: (a) conduction (b) convection (c) radiation
4. What is latent heat? How is latent heat an important source of atmospheric energy?
5. How does the Kelvin temperature scale differ from the Celsius scale?
6. How does the amount of radiation emitted by the earth differ from that emitted by the sun?
7. How does the temperature of an object influence the radiation it emits?
8. How do the wavelengths of most of the radiation emitted by the sun differ from those emitted by the surface of the earth?
9. When a body reaches a radiative equilibrium temperature, what is taking place?
10. Why are carbon dioxide and water vapor called selective absorbers?
11. Explain how the earth's atmospheric greenhouse effect works.
12. What gases appear to be responsible for the enhancement of the earth's greenhouse effect?
13. Why does the albedo of the earth and its atmosphere average about 30 percent?
14. Explain how the atmosphere near the earth's surface is warmed from below.
15. In the Northern Hemisphere, why are summers warmer than winters even though the earth is actually closer to the sun in January?
16. What are the main factors that determine seasonal temperature variations?
17. If it is winter and January in New York City, what is the season and month in Sydney, Australia?
18. During the Northern Hemisphere's summer, the daylight hours in northern latitudes are longer than in middle latitudes. Explain why northern latitudes are not warmer.
19. Explain why the vegetation on the north-facing side of a hill is frequently different from the vegetation on the south-facing side of the same hill.

QUESTIONS FOR THOUGHT AND EXPLORATION

1. Explain why the bridge in the diagram is the first to become icy.

2. If the surface of a puddle freezes, is heat energy *released to* or *taken from* the air above the puddle? Explain.
3. In houses and apartments with forced-air furnaces, heat registers are usually placed near the floor rather than near the ceiling. Explain why.
4. Which do you feel would have the greatest effect on the earth's greenhouse effect: removing all of the CO_2 from the atmosphere or removing all of the water vapor? Explain your answer.
5. How would the seasons be affected where you live if the tilt of the earth's axis *increased* from $23\frac{1}{2}°$ to 40°?
6. Explain why an increase in cloud cover surrounding the earth would increase the earth's albedo, yet not necessarily lead to a lower earth surface temperature.
7. Why does the surface temperature often increase on a clear, calm night as a low cloud moves overhead?
8. How is heat transferred away from the surface of the moon? (Hint: The moon has no atmosphere.)

9. The Aurora (http://www.exploratorium.edu/learning_studio/auroras/selfguide1.html): Compare the appearance of auroras as viewed from earth and as viewed from space.
10. Ultraviolet Radiation Index (http://www.msc-smc.ec.gc.ca/uv_e.html): On what information do you think the UV Index is based? What are some of the activities that you engage in that might put you at risk for extended exposure to ultraviolet radiation?

- Go to the Brooks/Cole Earth Sciences Resource Center (http://earthscience.brookscole.com) for critical thinking exercises, articles, and additional readings from InfoTrac College Edition, Brooks/Cole's online student library.

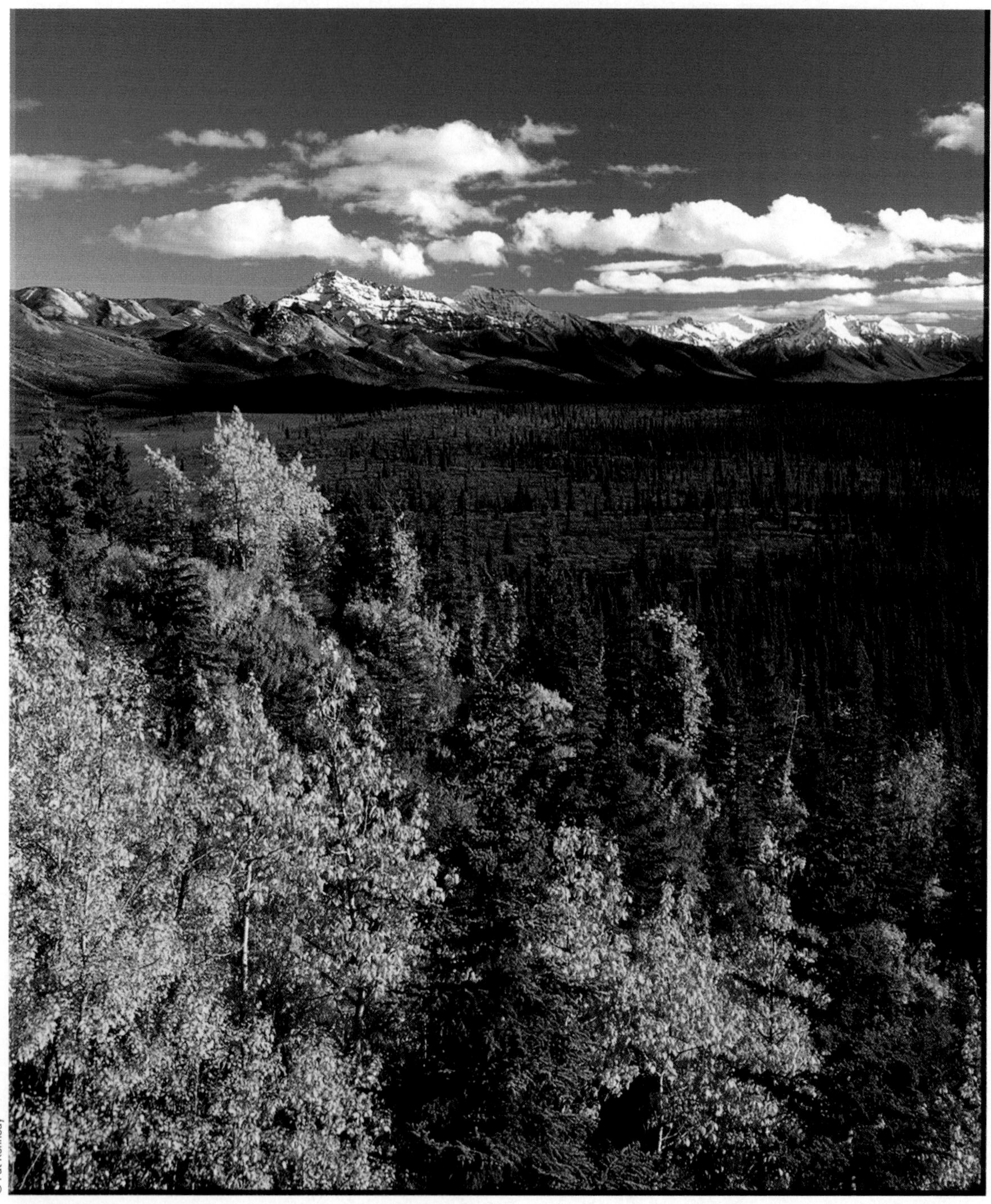

© Pat Kennedy

A warm fall day in Denali National Park, Alaska. Here air temperatures may climb well above freezing during the day and drop to well below freezing at night.

Meteorology Now™ This icon, appearing throughout the book, indicates an opportunity to explore interactive tutorials, animations, or practice problems available on the MeteorologyNow Web site at http://earthscience.brookscole.com/ahrens/ess4e.

Air Temperature

The sun shining full upon the field, the soil of which was sandy, the mouth of a heated oven seemed to me to be a trifle hotter than this ploughed field; it was almost impossible to breathe. . . . The weather was almost too hot to live in, and the British troops in the orchard were forced by the heat to shelter themselves from it under trees. . . . I presume everyone has heard of the heat that day, but none can realize it that did not feel it. Fighting is hot work in cool weather, how much more so in such weather as it was on the twenty-eighth of June 1778.

David M. Ludlum, *The Weather Factor*

CONTENTS

Air temperature is an important weather element. It not only dictates how we should dress for the day, but the careful recording and application of temperature data are tremendously important to us all. For without accurate information of this type, the work of farmers, weather analysts, power company engineers, and many others would be a great deal more difficult. Therefore, we begin this chapter by examining the daily variation in air temperature. Here, we will answer such questions as why the warmest time of the day is normally in the afternoon, and why the coldest is usually in the early morning. And why calm, clear nights are usually colder than windy, clear nights. After we examine the factors that cause temperatures to vary from one place to another, we will look at daily, monthly, and yearly temperature averages and ranges with an eye toward practical applications for everyday living. Near the end of the chapter, we will see how air temperature is measured and how the wind can change our perception of air temperature.

DAILY TEMPERATURE VARIATIONS

In Chapter 2, we learned how the sun's energy coupled with the motions of the earth produce the seasons. In a way, each sunny day is like a tiny season as the air goes through a daily cycle of warming and cooling. The air warms during the morning hours, as the sun gradually rises higher in the sky, spreading a blanket of heat energy over the ground. The sun reaches its highest point around noon, after which it begins its slow journey toward the western horizon. It is around noon when the earth's surface receives the most intense solar rays. However, somewhat surprisingly, noontime is usually not the warmest part of the day. Rather, the air continues to be heated, often reaching a maximum temperature later in the afternoon. To find out why this *lag in temperature* occurs, we need to examine a shallow layer of air in contact with the ground.

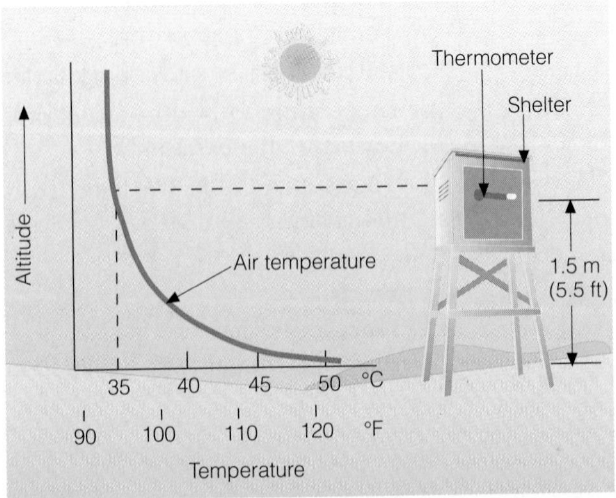

FIGURE 3.1 On a sunny, calm day, the air near the surface can be substantially warmer than the air a meter or so above the surface.

Daytime Warming As the sun rises in the morning, sunlight warms the ground, and the ground warms the air in contact with it by conduction. However, air is such a poor heat conductor that this process only takes place within a few centimeters of the ground. As the sun rises higher in the sky, the air in contact with the ground becomes even warmer, and, on a windless day, a substantial temperature difference usually exists just above the ground. This explains why joggers on a clear, windless, hot summer afternoon may experience air temperatures of over 50°C (122°F) at their feet and only 35°C (95°F) at their waists (see Fig. 3.1).

Near the surface, convection begins, and rising air bubbles (thermals) help to redistribute heat. In calm weather, these thermals are small and do not effectively mix the air near the surface. Thus, large vertical temperature differences are able to exist. On windy days, however, turbulent eddies are able to mix hot, surface air with the cooler air above. This form of mechanical stirring, sometimes called *forced convection,* helps the thermals to transfer heat away from the surface more efficiently. Therefore, on sunny, windy days the temperature difference between the surface air and the air directly above is not as great as it is on sunny, calm days.

We can now see why the warmest part of the day is usually in the afternoon. Around noon, the sun's rays are most intense. However, even though incoming solar radiation decreases in intensity after noon, it still exceeds outgoing heat energy from the surface for a time. This yields an energy surplus for two to four hours after noon and substantially contributes to a lag between the time of maximum solar heating and the time of maximum air temperature several feet above the surface (see Fig. 3.2).

The exact time of the highest temperature reading varies somewhat. Where the summer sky remains cloud-free all afternoon, the maximum temperature may occur sometime between 3:00 and 5:00 P.M. Where there is afternoon cloudiness or haze, the temperature maximum occurs an hour or two earlier. If clouds persist throughout the day, the overall daytime temperatures are usually lower, as clouds reflect a great deal of incoming sunlight.

Adjacent to large bodies of water, cool air moving inland may modify the rhythm of temperature change such that the warmest part of the day occurs at noon or

DID YOU KNOW?

Death Valley, California, had a high temperature of 38°C (100°F) on 134 days during 1974. During July, 1998, the temperature in Death Valley reached a scorching 54°C (129°F)—only 4°C (7°F) below the world record high temperature of 58°C (136°F) measured in El Azizia, Libya, in 1922.

before. In winter, atmospheric storms circulating warm air northward can even cause the highest temperature to occur at night.

Just how warm the air becomes depends on such factors as the type of soil, its moisture content, and vegetation cover. When the soil is a poor heat conductor (as loosely packed sand is), heat energy does not readily transfer into the ground. This allows the surface layer to reach a higher temperature, availing more energy to warm the air above. On the other hand, if the soil is moist or covered with vegetation, much of the available energy evaporates water, leaving less to heat the air. As you might expect, the highest summer temperatures usually occur over desert regions, where clear skies coupled with low humidities and meager vegetation permit the surface and the air above to warm up rapidly.

Where the air is humid, haze and cloudiness lower the maximum temperature by preventing some of the sun's rays from reaching the ground. In humid Atlanta, Georgia, the average maximum temperature for July is 30.5°C (87°F). In contrast, Phoenix, Arizona—in the desert southwest at the same latitude as Atlanta—experiences an average July maximum of 40.5°C (105°F). (Additional information on high daytime temperatures is given in the Focus section on p. 58.)

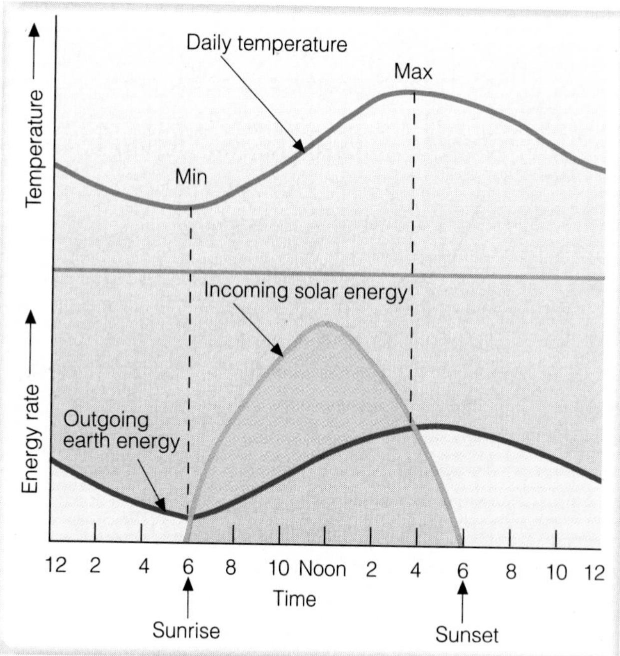

FIGURE 3.2 The daily variation in air temperature is controlled by incoming energy (primarily from the sun) and outgoing energy from the earth's surface. Where incoming energy exceeds outgoing energy (orange shade), the air temperature rises. Where outgoing energy exceeds incoming energy (blue shade), the air temperature falls.

Nighttime Cooling As the sun lowers, its energy is spread over a larger area, which reduces the heat available to warm the ground. Observe in Fig. 3.2 that sometime in late afternoon or early evening, the earth's surface and air above begin to lose more energy than they receive; hence, they start to cool.

Both the ground and air above cool by radiating infrared energy, a process called **radiational cooling.** The ground, being a much better radiator than air, is able to cool more quickly. Consequently, shortly after sunset, the earth's surface is slightly cooler than the air directly above it. The surface air transfers some energy to the ground by conduction, which the ground, in turn, quickly radiates away.

As the night progresses, the ground and the air in contact with it continue to cool more rapidly than the air a few meters higher. The warmer upper air does transfer *some* heat downward, a process that is slow due to the air's poor thermal conductivity. Therefore, by late night or early morning, the coldest air is next to the ground, with slightly warmer air above (see Fig. 3.3).

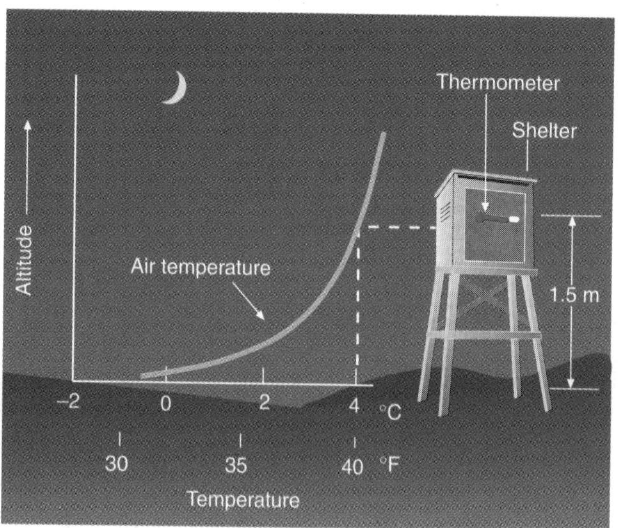

FIGURE 3.3 On a clear, calm night, the air near the surface can be much colder than the air above. The increase in air temperature with increasing height above the surface is called a radiation temperature inversion.

FOCUS ON A SPECIAL TOPIC

Record High Temperatures

Most people are aware of the extreme heat that exists during the summer in the desert southwest of the United States. But how hot does it get there? On July 10, 1913, Greenland Ranch in Death Valley, California, reported the highest temperature ever observed in North America: 57°C (134°F) (Fig. 1). Here, air temperatures are persistently hot throughout the summer, with the average maximum for July being 47°C (116°F). During the summer of 1917, there was an incredible period of 43 consecutive days when the maximum temperature reached 120°F or higher.

Probably the hottest urban area in the United States is Yuma, Arizona. Located along the California–Arizona border, Yuma's high temperature during July averages 42°C (108°F). In 1937, the high reached 100°F or more for 101 consecutive days.

In a more humid climate, the maximum temperature rarely climbs above 41°C (106°F). However, during the record heat wave of 1936, the air temperature reached 121°F near Alton, Kansas. And during the heat wave of 1983, which destroyed about $7 billion in crops and increased the nation's air-conditioning bill by an estimated $1 billion, Fayetteville reported North Carolina's all-time record high temperature when the mercury hit 110°F.

These readings, however, do not hold a candle to the hottest place in the world. That distinction probably belongs to Dallol, Ethiopia. Dallol is located south of the Red Sea, near latitude 12°N, in the hot, dry Danakil Depression. A prospecting company kept weather records at Dallol from 1960 to 1966. During this time, the average daily maximum temperature exceeded 38°C (100°F) every month of the year, except during December and January, when the average maximum lowered to 98°F and 97°F, respectively. On many days, the air temperature exceeded 120°F. The average annual temperature for the six years at Dallol was 34°C (94°F). In comparison, the average annual temperature in Yuma is 23°C (74°F) and at Death Valley, 24°C (76°F). The highest temperature reading on earth (under standard conditions) occurred northeast of Dallol at El Azizia, Libya (32°N), when, on September 13, 1922, the temperature reached a scorching 58°C (136°F). Table 1 gives record high temperatures throughout the world.

TABLE 1 Some Record High Temperatures Throughout the World

LOCATION (LATITUDE)	RECORD HIGH TEMPERATURE (°C)	(°F)	RECORD FOR:	DATE
El Azizia, Libya (32°N)	58	136	The world	September 13, 1922
Death Valley, Calif. (36°N)	57	134	Western Hemisphere	July 10, 1913
Tirat Tsvi, Israel (32°N)	54	129	Middle East	June 21, 1942
Cloncurry, Queensland (21°S)	53	128	Australia	January 16, 1889
Seville, Spain (37°N)	50	122	Europe	August 4, 1881
Rivadavia, Argentina (35°S)	49	120	South America	December 11, 1905
Midale, Saskatchewan (49°N)	45	113	Canada	July 5, 1937
Fort Yukon, Alaska (66°N)	38	100	Alaska	June 27, 1915
Pahala, Hawaii (19°N)	38	100	Hawaii	April 27, 1931
Esparanza, Antarctica (63°S)	14	58	Antarctica	October 20, 1956

FIGURE 1 The hottest place in North America, Death Valley, California, where the air temperature reached 57°C (134°F).

This measured increase in air temperature just above the ground is known as a **radiation inversion** because it forms mainly through radiational cooling of the surface. Because radiation inversions occur on most clear, calm nights, they are also called *nocturnal inversions.**

Meteorology Now™ Click "Energy Balance" to observe energy changes over the course of a day, and to predict when the maximum temperature will occur.

Cold Air Near the Surface A strong radiation inversion occurs when the air near the ground is much colder than the air higher up. Ideal conditions for a strong inversion and, hence, very low nighttime temperatures exist when the air is calm, the night is long, and the air is fairly dry and cloud-free. Let's examine these ingredients one by one.

A windless night is essential for a strong radiation inversion because a stiff breeze tends to mix the colder air at the surface with the warmer air above. This mixing, along with the cooling of the warmer air as it comes in contact with the cold ground, causes a vertical temperature profile that is almost isothermal (constant temperature) in a layer several feet thick. In the absence of wind, the cooler, more-dense surface air does not readily mix with the warmer, less-dense air above, and the inversion is more strongly developed as illustrated in Fig. 3.3.

A long night also contributes to a strong inversion. Generally, the longer the night, the longer the time of radiational cooling and the better are the chances that the air near the ground will be much colder than the air above. Consequently, winter nights provide the best conditions for a strong radiation inversion, other factors being equal.

Finally, radiation inversions are more likely with a clear sky and dry air. Under these conditions, the ground is able to radiate its energy to outer space and thereby cool rapidly. However, with cloudy weather and moist air, much of the outgoing infrared energy is absorbed and radiated to the surface, retarding the rate of cooling. Also, on humid nights, condensation in the form of fog or dew will release latent heat, which warms the air. So, radiation inversions may occur on any night. But, during long winter nights, when the air is still, cloud-free, and relatively dry, these inversions can become strong and deep.

It should now be apparent that how cold the night air becomes depends primarily on the length of the night, the moisture content of the air, cloudiness, and the wind. Even though wind may initially bring cold air into a region, the coldest nights usually occur when the air is clear and relatively calm. (Additional information on very low nighttime temperatures is given in the Focus section on p. 60.)

*Radiation (nocturnal) inversions are also called *surface inversions.*

DID YOU KNOW?

When the surface air temperature dipped to its all-time record low of −88°C (−127°F) on the Antarctic Plateau of Vostok Station, a drop of saliva falling from the lips of a person taking an observation would have frozen solid before reaching the ground.

Look back at Fig. 3.2 (p. 57) and observe that the lowest temperature on any given day is usually observed around sunrise. However, the cooling of the ground and surface air may even continue beyond sunrise for a half hour or so, as outgoing energy can exceed incoming energy. This situation happens because light from the early morning sun passes through a thick section of atmosphere and strikes the ground at a low angle. Consequently, the sun's energy does not effectively warm the surface. Surface heating may be reduced further when the ground is moist and available energy is used for evaporation. Hence, the lowest temperature may occur shortly after the sun has risen.

Cold, heavy surface air slowly drains downhill during the night and eventually settles in low-lying basins and valleys. Valley bottoms are thus colder than the surrounding hillsides (see Fig. 3.4). In middle latitudes, these warmer hillsides, called **thermal belts,** are less likely to experience freezing temperatures than the valley below. This encourages farmers to plant on hillsides those trees unable to survive the valley's low temperature.

On the valley floor, the cold, dense air is unable to rise. Smoke and other pollutants trapped in this heavy air restrict visibility. Therefore, valley bottoms are not only colder, but are also more frequently polluted than nearby hillsides. Even when the land is only gently sloped, cold air settles into lower-lying areas, such as river basins and floodplains. Because the flat floodplains are agriculturally rich areas, cold air drainage often forces farmers to seek protection for their crops.

Protecting Crops from the Cold Night Air On cold nights, many plants may be damaged by low temperatures. To protect small plants or shrubs, cover them with straw, cloth, or plastic sheeting. This prevents ground heat from being radiated away to the colder surroundings. If you are a household gardener concerned about

FOCUS ON A SPECIAL TOPIC

Record Low Temperatures

One city in the United States that experiences very low temperatures is International Falls, Minnesota, where the average temperature for January is −16°C (3°F). Located several hundred miles to the south, Minneapolis–St. Paul, with an average temperature of −9°C (16°F) for the three winter months, is the coldest major urban area in the nation. For duration of extreme cold, Minneapolis reported 186 consecutive hours of temperatures below 0°F during the winter of 1911–1912. Within the forty-eight adjacent states, however, the record for the longest duration of severe cold belongs to Langdon, North Dakota, where the thermometer remained below 0°F for 41 consecutive days during the winter of 1936. The official record for the lowest temperature in the forty-eight adjacent states belongs to Rogers Pass, Montana, where on the morning of January 20, 1954, the mercury dropped to −57°C (−70°F). The lowest official temperature for Alaska, −62°C (−80°F), occurred at Prospect Creek on January 23, 1971.

The coldest areas in North America are found in the Yukon and Northwest Territories of Canada. Resolute, Canada (latitude 75°N), has an average temperature of −32°C (−26°F) for the month of January.

The lowest temperatures and coldest winters in the Northern Hemisphere are found in the interior of Siberia and Greenland. For example, the average January temperature in Yakutsk, Siberia (latitude 62°N), is −43°C (−46°F). There, the mean temperature for the entire year is a bitter cold −11°C (12°F). At Eismitte, Greenland, the average temperature for February (the coldest month) is −47°C (−53°F), with the mean annual temperature being a frigid −30°C (−22°F). Even though these temperatures are extremely low, they do not come close to the coldest area of the world: the Antarctic.

At the geographical South Pole, over nine thousand feet above sea level, where the Amundsen-Scott scientific station has been keeping records for more than forty years, the average temperature for the month of July (winter) is –59°C (−74°F) and the mean annual temperature is −49°C (−57°F). The lowest temperature ever recorded there (−83°C or −117°F) occurred under clear skies with a light wind on the morning of June 23, 1983. Cold as it was, it was not the record low for the world. That belongs to the Russian station at Vostok, Antarctica (latitude 78°S), where the temperature plummeted to −89°C (−129°F) on July 21, 1983. (See Table 2 for record low temperatures throughout the world.)

TABLE 2 Some Record Low Temperatures Throughout the World

LOCATION (LATITUDE)	RECORD LOW TEMPERATURE (°C)	(°F)	RECORD FOR:	DATE
Vostok, Antarctica (78°S)	–89	–129	The world	July 21, 1983
Verkhoyansk, Russia (67°N)	–68	–90	Northern Hemisphere	February 7, 1892
Northice, Greenland (72°N)	–66	–87	Greenland	January 9, 1954
Snag, Yukon (62°N)	–63	–81	North America	February 3, 1947
Prospect Creek, Alaska (66°N)	–62	–80	Alaska	January 23, 1971
Rogers Pass, Montana (47°N)	–57	–70	U.S. (excluding Alaska)	January 20, 1954
Sarmiento, Argentina (34°S)	–33	–27	South America	June 1, 1907
Ifrane, Morocco (33°N)	–24	–11	Africa	February 11, 1935
Charlotte Pass, Australia (36°S)	–22	–8	Australia	July 22, 1949
Mt. Haleakala, Hawaii (20°N)	–10	14	Hawaii	January 2, 1961

outside flowers and plants during cold weather, simply wrap them in plastic or cover each with a paper cup.

Fruit trees are particularly vulnerable to cold weather in the spring when they are blossoming. The protection of such trees presents a serious problem to the farmer. Since the lowest temperatures on a clear, still night occur near the surface, the lower branches of a tree are the most susceptible to damage. Therefore, increasing the air temperature close to the ground may prevent damage. One way this increase can be achieved is to use **orchard heaters,** or "smudge pots," which warm the air around the trees by setting up convection currents close to the ground (see Fig. 3.5). In addition, heat energy radiated from oil or gas-fired orchard heaters is intercepted by the buds of the trees, which raises their temperature.

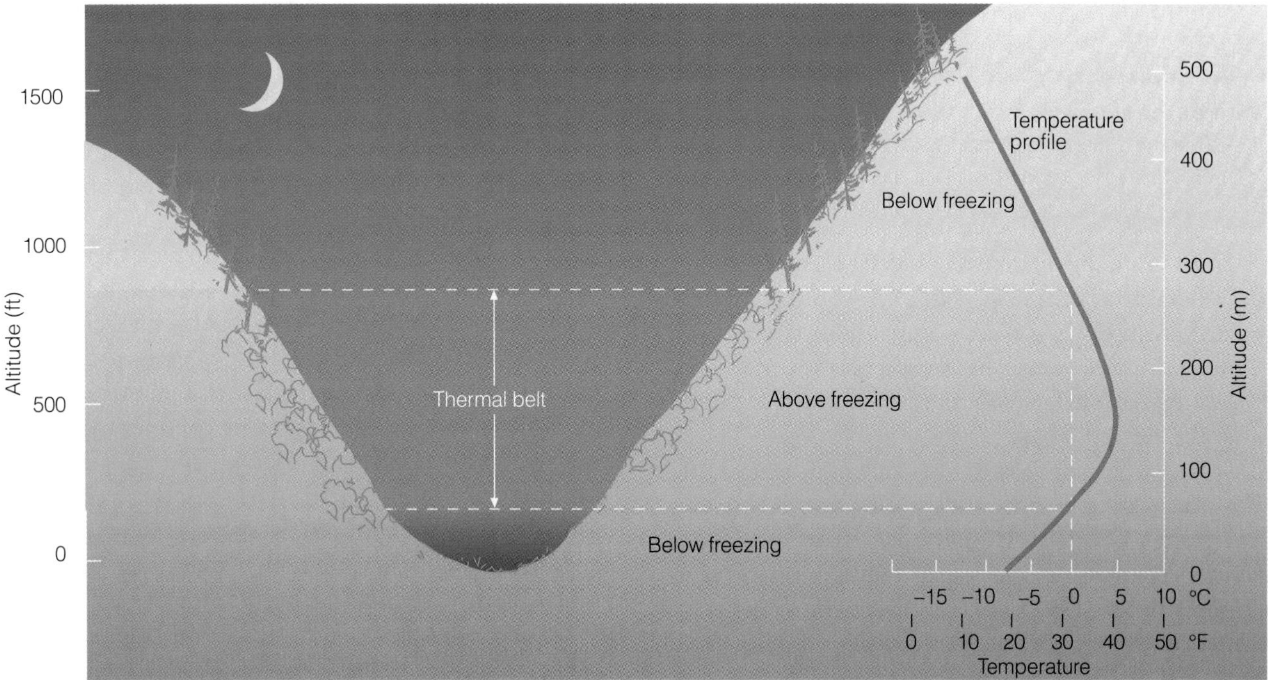

FIGURE 3.4 On cold, clear nights, the settling of cold air into valleys makes them colder than surrounding hillsides. The region along the side of the hill where the air temperature is above freezing is known as a *thermal belt.*

Another way to protect trees is to mix the cold air at the ground with the warmer air above, thus raising the temperature of the air next to the ground. Such mixing can be accomplished by using **wind machines** (see Fig. 3.6), which are power-driven fans that resemble airplane propellers. One significant benefit of wind machines is that they can be thermostatically controlled to turn off and on at prescribed temperatures. Farmers without their own wind machines can rent air mixers in the form of helicopters. Although helicopters are effective in mixing the air, they are expensive to operate.

If sufficient water is available, trees can be protected by irrigation. On potentially cold nights, farmers might flood the orchard. Because water has a high heat capac-

Photo by author

FIGURE 3.5 Orchard heaters circulate the air by setting up convection currents.

Photo by author

FIGURE 3.6 Wind machines mix cooler surface air with warmer air above.

DID YOU KNOW?

Talk about cold turkey! In Fairbanks, Alaska, on Thanksgiving day in 1990, the air temperature plummeted to −42°C (−44°F), only 3°C (5°F) above Fairbanks' all-time record low for November.

ity, it cools more slowly than dry soil. Consequently, the surface does not become as cold as it would if it were dry. Furthermore, wet soil has a higher thermal conductivity than dry soil. Hence, in wet soil heat is conducted upward from subsurface soil more rapidly, which helps to keep the surface warmer.

So far, we have discussed protecting trees against the cold air near the ground during a radiation inversion. Farmers often face another nighttime cooling problem. For instance, when subfreezing air blows into a region, the coldest air is not found at the surface; the air actually becomes colder with height. This condition is known as a **freeze.*** A single freeze in California or Florida can cause several million dollars damage to citrus crops. As a case in point, several freezes during the spring of 2001 caused millions of dollars in damage to California's north coast vineyards, which resulted in higher wine prices.

Protecting an orchard from the damaging cold air blown by the wind can be a problem. Wind machines will not help because they would only mix cold air at the surface with the colder air above. Orchard heaters and irrigation are of little value as they would only protect the branches just above the ground. However, there is one form of protection that does work: An orchard's sprinkling system may be turned on so that it emits a fine spray of water. In the cold air, the water freezes around the branches and buds, coating them with a thin veneer of ice (see Fig. 3.7). As long as the spraying continues, the latent heat—given off as the water changes into ice—keeps the ice temperature at 0°C (32°F). The ice acts as a protective coating against the subfreezing air by keeping the buds (or fruit) at a temperature higher than their damaging point. Care must be taken since too much ice can cause the branches to break. The fruit may be saved from the cold air, while the tree itself may be damaged by too much protection. Sprinklers work well when the air is fairly humid. They do not work well when the air is dry, as a good deal of the water may be lost through evaporation.

*A freeze occurs over a widespread area when the surface air temperature remains below freezing for a long enough time to damage certain agricultural crops. The terms *frost* and *freeze* are often used interchangeably by various segments of society. However, to the grower of perennial crops (such as apples and citrus) who has to protect the crop against damaging low temperatures, it makes no difference if visible "frost" is present or not. The concern is whether or not the plant tissue has been exposed to temperatures equal to or below 32°F. The actual freezing point of the plant, however, can vary because perennial plants can develop hardiness in the fall that usually lasts through the winter, then wears off gradually in the spring.

BRIEF REVIEW

Up to this point we have examined temperature variations on a daily basis. Before going on, here is a review of some of the important concepts and facts we have covered:

- During the day, the earth's surface and air above will continue to warm as long as incoming energy (mainly sunlight) exceeds outgoing energy from the surface.

FIGURE 3.7 A coating of ice protects these almond trees from damaging low temperatures, as an early spring freeze drops air temperatures well below freezing.

- At night, the earth's surface cools, mainly by giving up more infrared radiation than it receives—a process called radiational cooling.
- The coldest nights of winter normally occur when the air is calm, fairly dry (low water-vapor content), and cloud free.
- The highest temperatures during the day and the lowest temperatures at night are normally observed at the earth's surface.
- Radiation inversions exist usually at night when the air near the ground is colder than the air above.

THE CONTROLS OF TEMPERATURE

The main factors that cause variations in temperature from one place to another are called the **controls of temperature.** In the previous chapter, we saw that the greatest factor in determining temperature is the amount of solar radiation that reaches the surface. This, of course, is determined by the length of daylight hours and the intensity of incoming solar radiation. Both of these factors are a function of latitude; hence, latitude is considered an important control of temperature. The main controls are listed below.

1. latitude
2. land and water distribution
3. ocean currents
4. elevation

We can obtain a better picture of these controls by examining Figs. 3.8 and 3.9, which show the average monthly temperatures throughout the world for January and July. The lines on the map are **isotherms**—lines connecting places that have the same temperature. Because air temperature normally decreases with height, cities at very high elevations are much colder than their sea level counterparts. Consequently, the isotherms in Figs. 3.8 and 3.9 are corrected to read at the same horizontal level (sea level) by adding to each station above sea level an amount of temperature that would correspond to an average temperature change with height.*

Figures 3.8 and 3.9 show the importance of latitude on temperature. Note that, on the average, temperatures decrease poleward from the tropics and subtropics in both January and July. However, because there is a greater variation in solar radiation between low and high latitudes in winter than in summer, the isotherms in January are closer together (a tighter gradient)* than they are in July. This means that if you travel from New Orleans to Detroit in January, you are more likely to experience greater temperature variations than if you make the same trip in July. Notice also in Figs. 3.8 and 3.9 that the isotherms do not run horizontally; rather, in many places they bend, especially where they approach an ocean-continent boundary.

On the January map, the temperatures are much lower in the middle of continents than they are at the same latitude near the oceans; on the July map, the reverse is true. The reason for these temperature variations can be attributed to the unequal heating and cooling properties of land and water. For one thing, solar energy reaching land is absorbed in a thin layer of soil; reaching water, it penetrates deeply. Because water is able to circulate, it distributes its heat through a much deeper layer. Also, some of the solar energy striking the water is used to evaporate it rather than heat it.

Another important reason for the temperature contrasts is that water has a higher *specific heat* than land. The **specific heat** of a substance *is the amount of heat needed to raise the temperature of one gram of a substance by one degree Celsius.* It takes a great deal more heat (about five times more) to raise the temperature of a given amount of water by one degree than it does to raise the temperature of the same amount of soil or rock by one degree. Consequently, water has a much higher specific heat than either of these substances. Water not only heats more slowly than land, it cools more slowly as well, and so the oceans act like huge heat reservoirs. Thus, mid-ocean surface temperatures change relatively little from summer to winter compared to the much larger annual temperature changes over the middle of continents.

Along the margin of continents, ocean currents often influence air temperatures. For example, along the eastern margins, warm ocean currents transport warm water poleward, while, along the western margins, they transport cold water equatorward. As we will see in Chapter 7, some coastal areas also experience upwelling, which brings cold water from below to the surface.

Even large lakes can modify the temperature around them. In summer, the Great Lakes remain cooler than the land. As a result, refreshing breezes blow inland, bringing relief from the sometimes sweltering heat. As winter approaches, the water cools more slowly than the land. The first blast of cold air from Canada is modified as it crosses the lakes, and so the first freeze is delayed on the eastern shores of Lake Michigan.

*The amount of change is usually less than the standard temperature lapse rate of 3.6°F per 1000 feet (6.5°C per 1000 meters). The reason is that the standard lapse rate is computed for altitudes above the earth's surface in the "free" atmosphere. In the less-dense air at high elevations, the absorption of solar radiation by the ground causes an overall slightly higher temperature than that of the free atmosphere at the same level.

*Gradient represents the rate of change of some quantity (in this case, temperature) over a given distance.

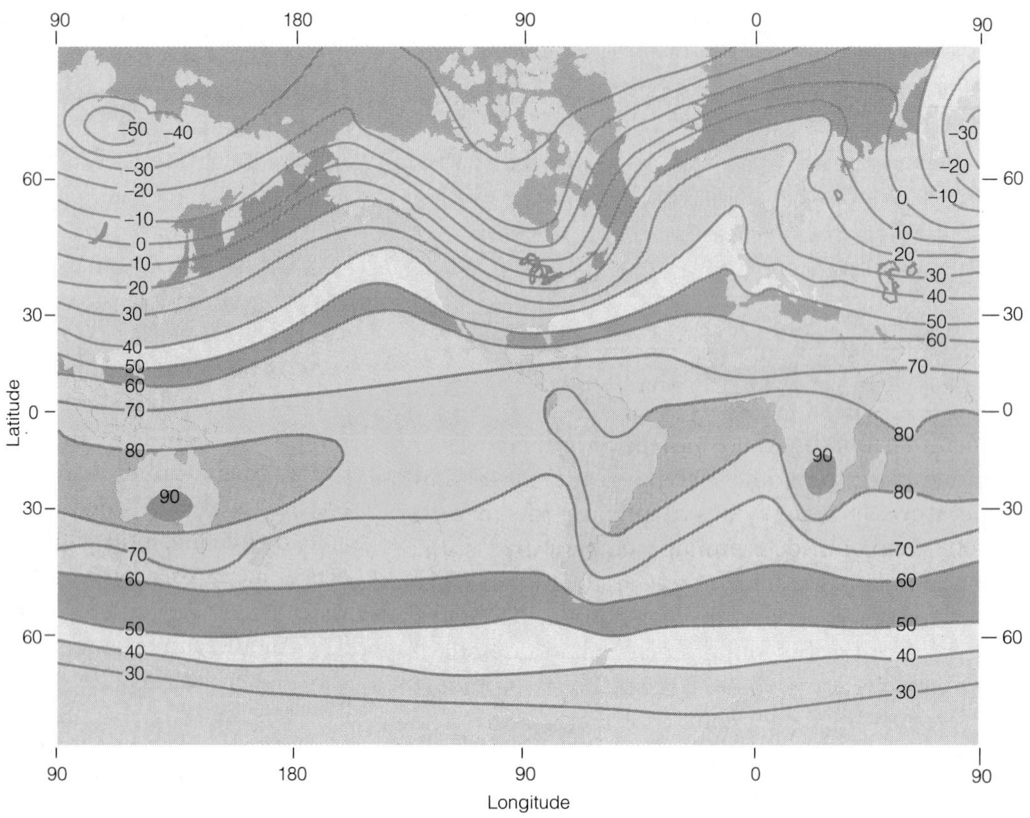

FIGURE 3.8 Average air temperature near sea level in January (°F).

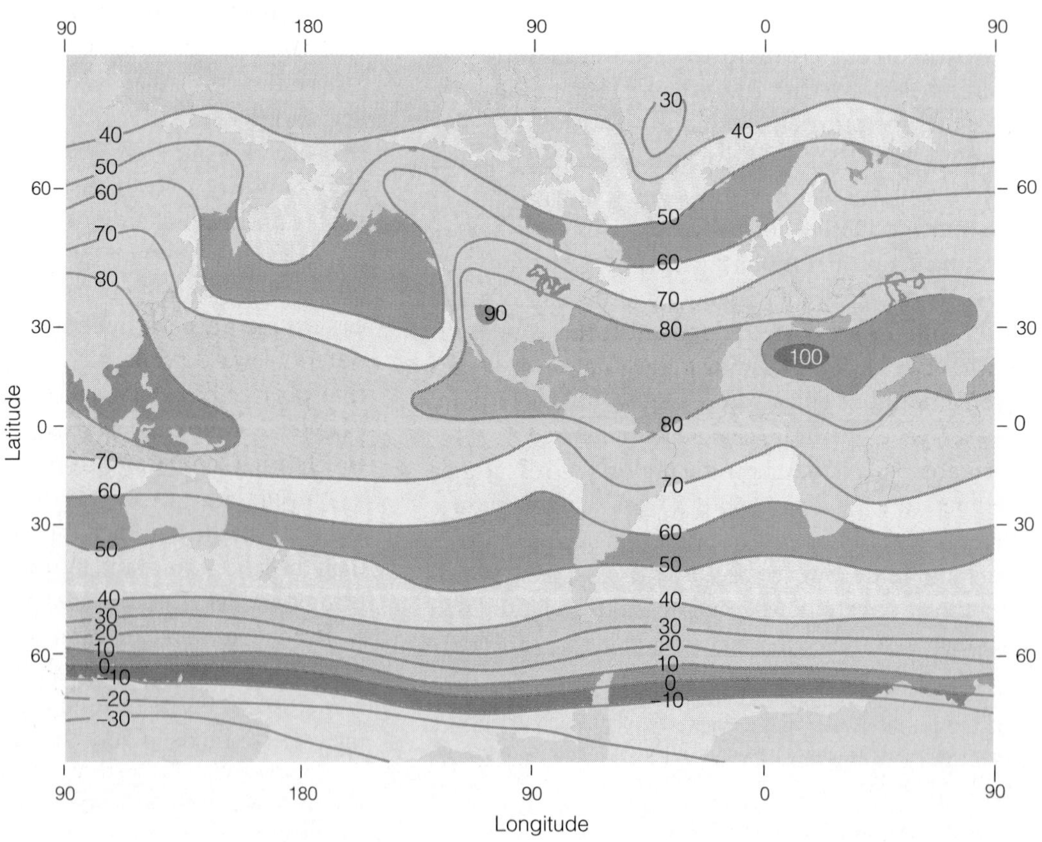

FIGURE 3.9 Average air temperature near sea level in July (°F).

AIR TEMPERATURE DATA

In the previous sections, we considered how air temperature varies on a daily basis and from one place to another. We will now focus on the ways temperature data are organized and used.

Daily, Monthly, and Yearly Temperatures The greatest variation in daily temperature occurs at the earth's surface. In fact, the difference between the daily maximum and minimum temperature—called the **daily (diurnal) range of temperature**—is greatest next to the ground and becomes progressively smaller as we move away from the surface (see Fig. 3.10). This daily variation in temperature is also much larger on clear days than on cloudy ones.

The largest diurnal range of temperature occurs on high deserts, where the air is fairly dry, often cloud-free, and there is little water vapor to radiate much infrared energy back to the surface. By day, clear summer skies allow the sun's energy to quickly warm the ground which, in turn, warms the air above to a temperature sometimes exceeding 35°C (95°F). At night, the ground cools rapidly by radiating infrared energy to space, and the minimum temperature in these regions occasionally dips below 5°C (41°F), thus giving a daily temperature range of more than 30°C (54°F).

In humid regions, the diurnal temperature range is usually small. Here, haze and clouds lower the maximum temperature by preventing some of the sun's energy from reaching the surface. At night, the moist air keeps the minimum temperature high by absorbing the earth's infrared radiation and radiating a portion of it to the ground. An example of a humid city with a small summer diurnal temperature range is Charleston, South Carolina, where the average July maximum temperature is 32°C (90°F), the average minimum is 22°C (72°F), and the diurnal range is only 10°C (18°F).

Cities near large bodies of water typically have smaller diurnal temperature ranges than cities further inland. This phenomenon is caused in part by the additional water vapor in the air and by the fact that water warms and cools much more slowly than land.

Moreover, cities whose temperature readings are obtained at airports often have larger diurnal temperature ranges than those whose readings are obtained in downtown areas. The reason for this fact is that nighttime temperatures in cities tend to be warmer than those in outlying rural areas. This nighttime city warmth—called the *urban heat island*—is due to industrial and urban development, a topic that will be treated more completely in Chapter 12.

DID YOU KNOW?

One of the greatest temperature ranges ever recorded in the Northern Hemisphere (56°C or 100°F) occurred at Browning, Montana, on January 23, 1916, when the air temperature plummeted from 7°C (44°F) to −49°C (−56°F) in less than 24 hours. This huge temperature range, however, would represent a rather typical day on the planet Mars, where the average high temperature reaches about −12°C (10°F) and the average low drops to −79°C (−110°F), producing a daily temperature range of 67°C, or 120°F.

The average of the highest and lowest temperature for a 24-hour period is known as the **mean (average) daily temperature.** Most newspapers list the mean daily temperature along with the highest and lowest temperatures for the preceding day. The average of the mean daily temperatures for a particular date averaged for a 30-year period gives the average (or "*normal*") temperatures for that date. The average temperature for each month is the average of the daily mean temperatures for that month. Additional information on the concept of "normal" temperature is given in the Focus section on p. 66.

At any location, the difference between the average temperature of the warmest and coldest months is called

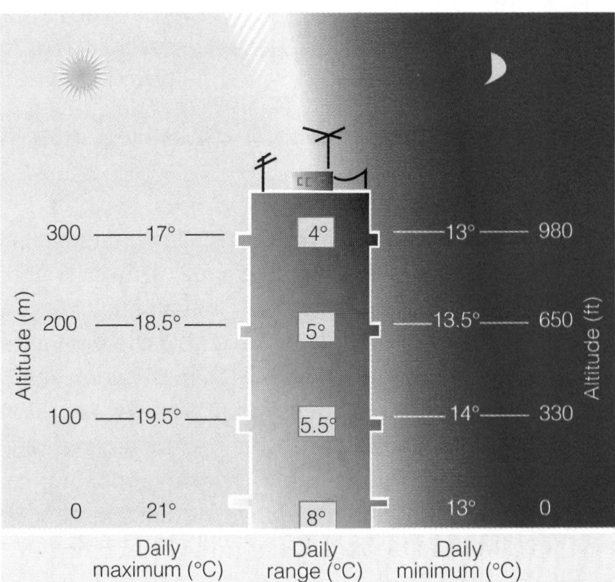

FIGURE 3.10 The daily range of temperature decreases as we climb away from the earth's surface. Hence, there is less day-to-night variation in air temperature near the top of a high-rise apartment complex than at the ground level.

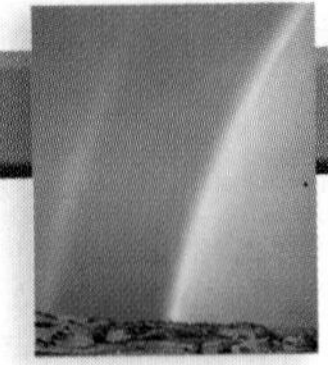

FOCUS ON A SPECIAL TOPIC

When It Comes to Temperature, What's Normal?

When the weathercaster reports that "the normal high temperature for today is 68°F," does this mean that the high temperature on this day is usually 68°F? Or does it mean that we should expect a high temperature near 68°F? Actually, we should expect neither one.

Remember that the word *normal*, or *norm*, refers to weather data averaged over a period of 30 years. For example, Fig. 2 shows the high temperature measured for 30 years in a southwestern city on March 15. The average (mean) high temperature for this period is 68°F; hence, the normal high temperature for this date is 68°F (dashed line). Notice, however, that only on one day during this 30-year period did the high temperature actually measure 68°F (large red dot). In fact, the most common high temperature (called the *mode*) was 60°F, and occurred on 4 days (blue dots).

So what would be considered a typical high temperature for this date? Actually, any high temperature that lies between about 47°F and 89°F (two standard deviations* on either side of 68°F) would be considered typical for this day. While a high temperature of 80°F may be quite warm and a high temperature of 47°F may be quite cool, they are both no more uncommon (unusual) than a high temperature of 68°F, which is the *normal* (average) high temperature for the 30-year period. This same type of reasoning applies to *normal rainfall*, as the actual amount of precipitation will likely be greater or less than the 30-year average.

*A standard deviation is a statistical measure of the spread of the data. Two standard deviations for this set of data mean that 95 percent of the time the high temperature occurs between 47°F and 89°F.

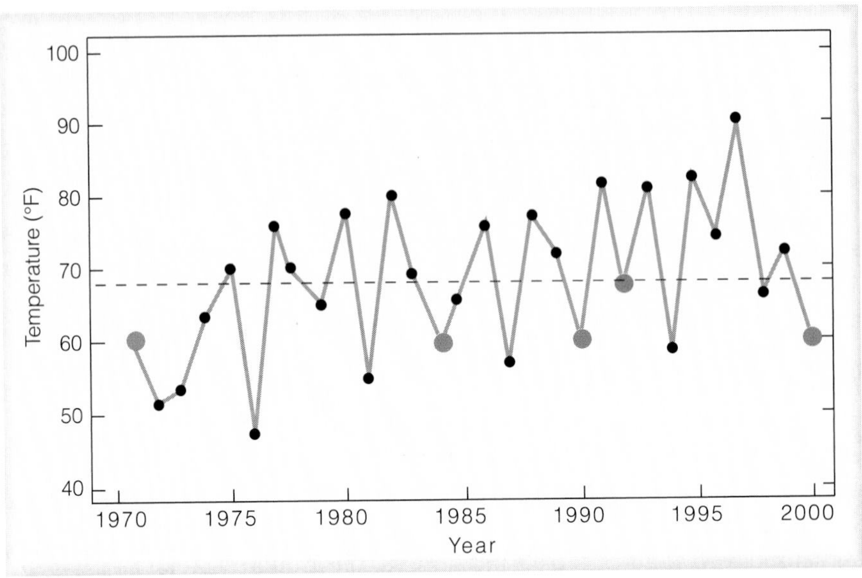

FIGURE 2 The high temperature measured (for 30 years) on March 15 in a city located in the southwestern United States. The dashed line represents the *normal* temperature for the 30-year period.

the **annual range of temperature.** Usually the largest annual ranges occur over land, the smallest over water. Hence, inland cities have larger annual ranges than coastal cities. Near the equator (because daylight length varies little and the sun is always high in the noon sky), annual temperature ranges are small, usually less than 3°C (5°F). Quito, Ecuador—on the equator at an elevation of 2850 m (9350 ft)—experiences an annual range of less than 1°C. In middle and high latitudes, large seasonal variations in the amount of sunlight reaching the surface produce large temperature contrasts between winter and summer. Here, annual ranges are large, especially in the middle of a continent. Yakutsk, in northeastern Siberia near the Arctic Circle, has an extremely large annual temperature range of 62°C (112°F).

The average temperature of any station for the entire year is the **mean (average) annual temperature,** which represents the average of the twelve monthly average temperatures.* When two cities have the same mean annual temperature, it might first seem that their temperatures throughout the year are quite similar. However, often this is not the case. For example, San Francisco, California, and Richmond, Virginia, are at the same latitude (37°N). Both have similar hours of daylight during the year; both have the same mean annual temperature—14°C (57°F). Here, the similarities end. The temperature differences between the two cities are

*The mean annual temperature may be obtained by taking the sum of the 12 monthly means and dividing that total by 12, or by obtaining the sum of the daily means and dividing that total by 365.

apparent to anyone who has traveled to San Francisco during the summer with a suitcase full of clothes suitable for summer weather in Richmond.

Figure 3.11 summarizes the average temperatures for San Francisco and Richmond. Notice that the coldest month for both cities is January. Even though January in Richmond averages only 8°C (14°F) colder than January in San Francisco, people in Richmond awaken to an average January minimum temperature of −6°C (21°F), which is much colder than the lowest temperature ever recorded in San Francisco. Trees that thrive in San Francisco's weather would find it difficult surviving a winter in Richmond. So, even though San Francisco and Richmond have the same mean annual temperature, the behavior and range of their temperatures differ greatly.

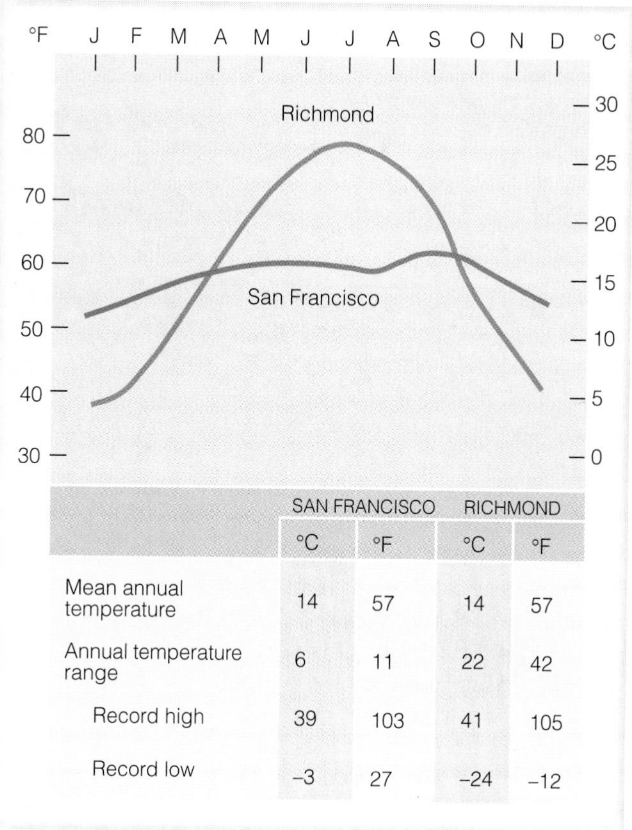

	SAN FRANCISCO		RICHMOND	
	°C	°F	°C	°F
Mean annual temperature	14	57	14	57
Annual temperature range	6	11	22	42
Record high	39	103	41	105
Record low	-3	27	-24	-12

FIGURE 3.11 Temperature data for San Francisco, California (37°N) and Richmond, Virginia (37°N)—two cities with the same mean annual temperature.

The Use of Temperature Data An application of daily temperature developed by heating engineers in estimating energy needs is the **heating degree-day.** The heating degree-day is based on the assumption that people will begin to use their furnaces when the mean daily temperature drops below 65°F. Therefore, heating degree-days are determined by subtracting the mean temperature for the day from 65°F. Thus, if the mean temperature for a day is 64°F, there would be 1 heating degree-day on this day.*

On days when the mean temperature is above 65°F, there are no heating degree-days. Hence, the lower the average daily temperature, the more heating degree-days and the greater the predicted consumption of fuel. When the number of heating degree-days for a whole year is calculated, the heating fuel requirements for any location can be estimated. Figure 3.12 shows the yearly average number of heating degree-days in various locations throughout the United States.

As the mean daily temperature climbs above 65°F, people begin to cool their indoor environment. Consequently, an index, called the **cooling degree-day,** is used during warm weather to estimate the energy needed to cool indoor air to a comfortable level. The forecast of mean daily temperature is converted to cooling degree-days by subtracting 65°F from the mean. The remaining value is the number of cooling degree-days for that day. For example, a day with a mean temperature of 70°F would correspond to (70−65), or 5 cooling degree-days. High values indicate warm weather and high power production for cooling (see Fig. 3.13).

Knowledge of the number of cooling degree-days in an area allows a builder to plan the size and type of equipment that should be installed to provide adequate air conditioning. Also, the forecasting of cooling degree-days during the summer gives power companies a way of predicting the energy demand during peak energy periods. A composite of heating plus cooling degree-days would give a practical indication of the energy requirements over the year.

Farmers use an index called **growing degree-days** as a guide to planting and for determining the approximate dates when a crop will be ready for harvesting. A growing degree-day for a particular crop is defined as a day on which the mean daily temperature is one degree above the *base temperature* (also known as the *zero temperature*)—the minimum temperature required for growth of that crop. For sweet corn, the base temperature is 50°F and, for peas, it is 40°F.

On a summer day in Iowa, the mean temperature might be 80°F. From Table 3.1, we can see that, on this day, sweet corn would accumulate (80−50), or 30 growing degree-days. Theoretically, sweet corn can be harvested when it accumulates a total of 2200 growing degree-days. So, if sweet corn is planted in early April and

*In the United States, the National Weather Service and the Department of Agriculture use degrees Fahrenheit in their computations.

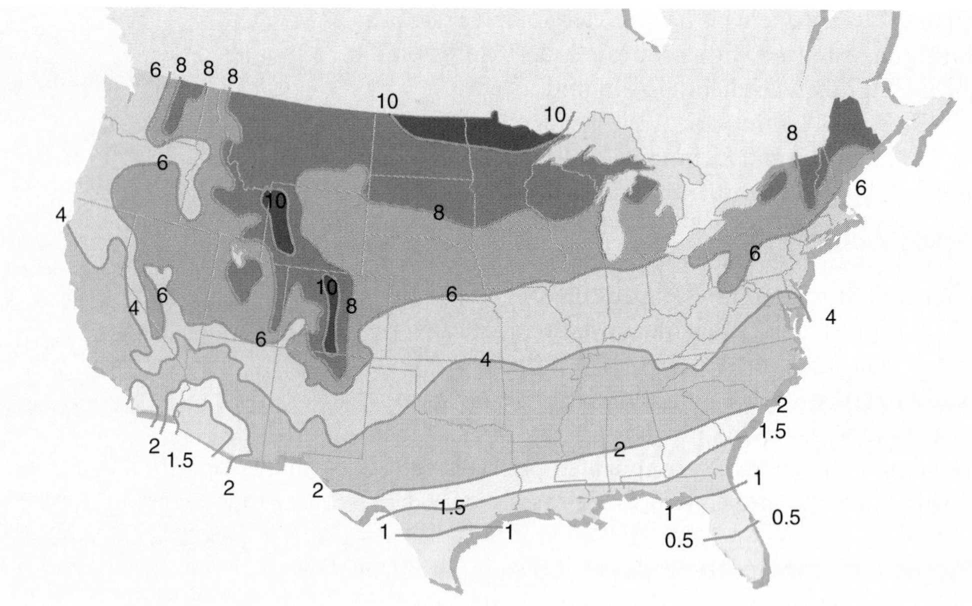

FIGURE 3.12 Mean annual total heating degree-days in thousands of °F, where the number 4 on the map represents 4000 (base 65°F).

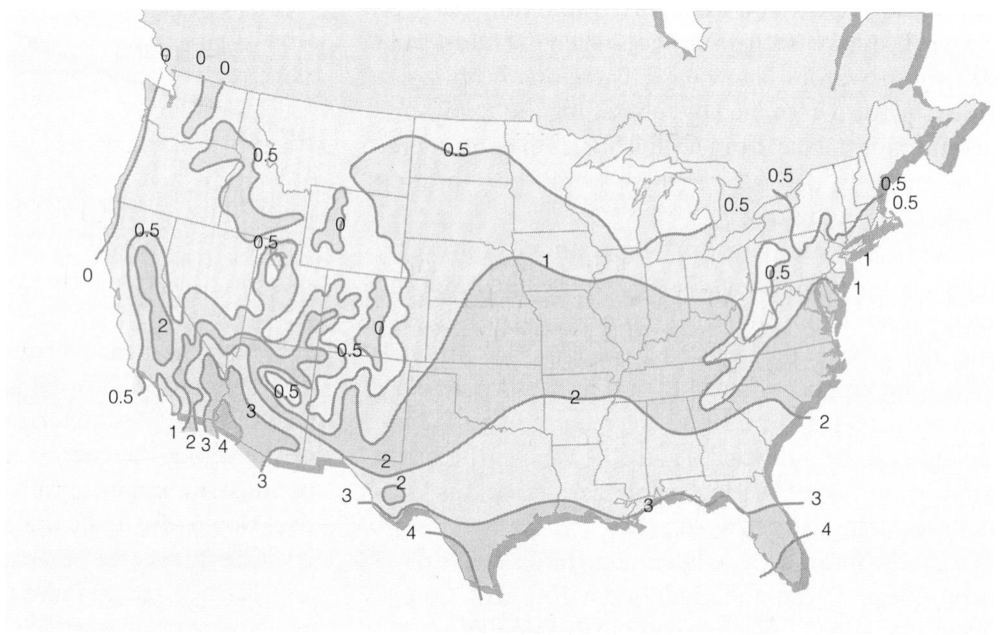

FIGURE 3.13 Mean annual total cooling degree-days in thousands of °F, where the number 1 on the map represents 1000 (base 65°F).

each day thereafter averages about 20 growing degree-days, the corn would be ready for harvest about 110 days later, or around the middle of July.*

At one time, corn varieties were rated in terms of "days to maturity." This rating system was unsuccessful because, in actual practice, corn took considerably longer in some areas than in others. This discrepancy was the reason for defining "growing degree-days." Hence, in humid Iowa, where summer nighttime temperatures are high, growing degree-days accumulate much faster. Consequently, the corn matures in considerably fewer days than in the drier west, where summer nighttime temperatures are lower, and each day accumulates fewer growing degree-days. Although moisture and other conditions are not taken into account, growing degree-days nevertheless serve as a useful guide in forecasting approximate dates of crop maturity.

*As a point of interest, in the corn belt when the air temperature climbs above 86°F, the hot air puts added stress on the growth of the corn. Consequently, the corn grows more slowly. Because of this fact, any maximum temperature over 86°F is reduced to 86°F when computing the mean air temperature.

FOCUS ON A SPECIAL TOPIC

A Thousand Degrees and Freezing to Death

Is there somewhere in our atmosphere where the air temperature can be exceedingly high (say above 500°C or 900°F) yet a person might feel extremely cold? There is a region, but it's not at the earth's surface.

You may recall from Chapter 1 (see Fig. 1.8, p. 12), that in the upper reaches of our atmosphere (in the middle and upper thermosphere), air temperatures may exceed 500°C. However, a thermometer shielded from the sun in this region of the atmosphere would indicate an extremely low temperature. This apparent discrepancy lies in the meaning of air temperature and how we measure it.

In Chapter 2, we learned that the air temperature is directly related to the average speed at which the air molecules are moving—faster speeds correspond to higher temperatures. In the middle and upper thermosphere at altitudes approaching 300 km (200 mi), air molecules are zipping about at speeds corresponding to extremely high temperatures. However, in order to transfer enough energy to heat something up by conduction (exposed skin or a thermometer bulb), an extremely large number of molecules must collide with the object. In the "thin" air of the upper atmosphere, air molecules are moving extraordinarily fast, but there are simply not enough of them bouncing against the thermometer bulb for it to register a high temperature. In fact, when properly shielded from the sun, the thermometer bulb loses far more energy than it receives and indicates a temperature near absolute zero. This explains why an astronaut, when space walking, will not only survive temperatures exceeding 500°C, but will also feel a profound coldness when shielded from the sun's radiant energy. At these high altitudes, the traditional meaning of air temperature (that is, regarding how "hot" or "cold" something feels) is no longer applicable.

AIR TEMPERATURE AND HUMAN COMFORT

Probably everyone realizes that the same air temperature can feel differently on different occasions. For example, a temperature of 20°C (68°F) on a clear, windless March afternoon in New York City can almost feel balmy after a long, hard winter. Yet, this same temperature may feel uncomfortably cool on a summer afternoon in a stiff breeze. The human body's perception of temperature obviously changes with varying atmospheric conditions. The reason for these changes is related to how we exchange heat energy with our environment.

TABLE 3.1 Estimated Growing Degree-Days for Certain Agricultural Crops to Reach Maturity

CROP (VARIETY, LOCATION)	BASE TEMPERATURE (°F)	GROWING DEGREE-DAYS TO MATURITY
Beans (Snap/South Carolina)	50	1200–1300
Corn (Sweet/Indiana)	50	2200–2800
Cotton (Delta Smooth Leaf/Arkansas)	60	1900–2500
Peas (Early/Indiana)	40	1100–1200
Rice (Vegold/Arkansas)	60	1700–2100
Wheat (Indiana)	40	2100–2400

The body stabilizes its temperature primarily by converting food into heat *(metabolism)*. To maintain a constant temperature, the heat produced and absorbed by the body must be equal to the heat it loses to its surroundings. There is, therefore, a constant exchange of heat—especially at the surface of the skin—between the body and the environment.

One way the body loses heat is by emitting infrared energy. But we not only emit radiant energy, we absorb it as well. Another way the body loses and gains heat is by conduction and convection, which transfer heat to and from the body by air motions. On a cold day, a thin layer of warm air molecules forms close to the skin, protecting it from the surrounding cooler air and from the rapid transfer of heat. Thus, in cold weather, when the air is calm, the temperature we perceive—called the **sensible temperature**—is often higher than a thermometer might indicate. (Could the opposite effect occur where the air temperature is very high and a person might feel exceptionally cold? If you are unsure, read the Focus section above.)

Once the wind starts to blow, the insulating layer of warm air is swept away, and heat is rapidly removed from the skin by the constant bombardment of cold air. When all other factors are the same, the faster the wind blows, the greater the heat loss, and the colder we feel. How cold the wind makes us feel is usually expressed as a **wind-chill index** (WCI).

DID YOU KNOW?

During the Korean War, over one-quarter of the United States' troop casualties were caused by frostbite during the winter campaign of 1950–1951.

The modern wind-chill index (see Tables 3.2 and 3.3) was formulated in 2001 by a joint action group of the National Weather Service and other agencies. The new index takes into account the wind speed at about 1.5 m (5 ft) above the ground instead of the 10 m (33 ft) where "official" readings are usually taken. In addition, it translates the ability of the air to take heat away from a person's face (the air's cooling power) into a wind-chill equivalent temperature.* For example, notice in Table 3.2 that an air temperature of 10°F with a wind speed of 10 mi/hr produces a wind-chill equivalent temperature of −4°F. Under these conditions, the skin of a person's exposed face would lose as much heat in one minute in air with a temperature of 10°F and a wind speed of 10 mi/hr as it would in calm air with a temperature of −4°F. Of course, how cold we feel actually depends on a number of factors, including the fit and type of clothing we wear, and the amount of sunshine striking the body, and the actual amount of exposed skin.

*The wind-chill equivalent temperature formulas are as follows: Wind chill (°F) = $35.74 + 0.6215T - 35.75 (V^{0.16}) + 0.4275T (V^{0.16})$, where T is the air temperature in °F and V is the wind speed in mi/hr. Wind chill (°C) = $13.12 + 0.6215T - 11.37 (V^{0.16}) + 0.3965T (V^{0.16})$, where T is the air temperature in °C, and V is the wind speed in km/hr.

TABLE 3.2 Wind-Chill Equivalent Temperature (°F). A 20-mi/hr Wind Combined with an Air Temperature of 20°F Produces a Wind-Chill Equivalent Temperature of 4°F.*

WIND SPEED (MI/HR)	AIR TEMPERATURE (°F)																
Calm	40	35	30	25	20	15	10	5	0	−5	−10	−15	−20	−25	−30	−35	−40
5	36	31	25	19	13	7	1	−5	−11	−16	−22	−28	−34	−40	−46	−52	−57
10	34	27	21	15	9	3	−4	−10	−16	−22	−28	−35	−41	−47	−53	−59	−66
15	32	25	19	13	6	0	−7	−13	−19	−26	−32	−39	−45	−51	−58	−64	−71
20	30	24	17	11	4	−2	−9	−15	−22	−29	−35	−42	−48	−55	−61	−68	−74
25	29	23	16	9	3	−4	−11	−17	−24	−31	−37	−44	−51	−58	−64	−71	−78
30	28	22	15	8	1	−5	−12	−19	−26	−33	−39	−46	−53	−60	−67	−73	−80
35	28	21	14	7	0	−7	−14	−21	−27	−34	−41	−48	−55	−62	−69	−76	−82
40	27	20	13	6	−1	−8	−15	−22	−29	−36	−43	−50	−57	−64	−71	−78	−84
45	26	19	12	5	−2	−9	−16	−23	−30	−37	−44	−51	−58	−65	−72	−79	−86
50	26	19	12	4	−3	−10	−17	−24	−31	−38	−45	−52	−60	−67	−74	−81	−88
55	25	18	11	4	−3	−11	−18	−25	−32	−39	−46	−54	−61	−68	−75	−82	−89
60	25	17	10	3	−4	−11	−19	−26	−33	−40	−48	−55	−62	−69	−76	−84	−91

TABLE 3.3 Wind-Chill Equivalent Temperature (°C)*

WIND SPEED (KM/HR)	AIR TEMPERATURE (°C)												
Calm	10	5	0	−5	−10	−15	−20	−25	−30	−35	−40	−45	−50
10	8.6	2.7	−3.3	−9.3	−15.3	−21.1	−27.2	−33.2	−39.2	−45.1	−51.1	−57.1	−63.0
15	7.9	1.7	−4.4	−10.6	−16.7	−22.9	−29.1	−35.2	−41.4	−47.6	−51.6	−59.9	−66.1
20	7.4	1.1	−5.2	−11.6	−17.9	−24.2	−30.5	−36.8	−43.1	−49.4	−55.7	−62.0	−68.3
25	6.9	0.5	−5.9	−12.3	−18.8	−25.2	−31.6	−38.0	−44.5	−50.9	−57.3	−63.7	−70.2
30	6.6	0.1	−6.5	−13.0	−19.5	−26.0	−32.6	−39.1	−45.6	−52.1	−58.7	−65.2	−71.7
35	6.3	−0.4	−7.0	−13.6	−20.2	−26.8	−33.4	−40.0	−46.6	−53.2	−59.8	−66.4	−73.1
40	6.0	−0.7	−7.4	−14.1	−20.8	−27.4	−34.1	−40.8	−47.5	−54.2	−60.9	−67.6	−74.2
45	5.7	−1.0	−7.8	−14.5	−21.3	−28.0	−34.8	−41.5	−48.3	−55.1	−61.8	−68.6	−75.3
50	5.5	−1.3	−8.1	−15.0	−21.8	−28.6	−35.4	−42.2	−49.0	−55.8	−62.7	−69.5	−76.3
55	5.3	−1.6	−8.5	−15.3	−22.2	−29.1	−36.0	−42.8	−49.7	−56.6	−63.4	−70.3	−77.2
60	5.1	−1.8	−8.8	−15.7	−22.6	−29.5	−36.5	−43.4	−50.3	−57.2	−64.2	−71.1	−78.0

*Dark blue shaded areas represent conditions where frostbite occurs in 30 minutes or less.

High winds, in below-freezing air, can remove heat from exposed skin so quickly that the skin may actually freeze and discolor. The freezing of skin, called **frostbite,** usually occurs on the body extremities first because they are the greatest distance from the source of body heat.

In cold weather, wet skin can be a factor in how cold we feel. A cold, rainy day (drizzly, or even foggy) often feels colder than a "dry" one because water on exposed skin conducts heat away from the body better than air does. In fact, in cold, wet, and windy weather a person may actually lose body heat faster than the body can produce it. This may even occur in relatively mild weather with air temperatures as high as 10°C (50°F). The rapid loss of body heat may lower the body temperature below its normal level and bring on a condition known as **hypothermia**—the rapid, progressive mental and physical collapse that accompanies the lowering of human body temperature.

The first symptom of hypothermia is exhaustion. If exposure continues, judgment and reasoning power begin to disappear. Prolonged exposure, especially at temperatures near or below freezing, produces stupor, collapse, and death when the internal body temperature drops to about 26°C (79°F).

In cold weather, heat is more easily dissipated through the skin. To counteract this rapid heat loss, the peripheral blood vessels of the body constrict, cutting off the flow of blood to the outer layers of the skin. In hot weather, the blood vessels enlarge, allowing a greater loss of heat energy to the surroundings. In addition to this, we perspire. As evaporation occurs, the skin cools. When the air contains a great deal of water vapor and it is close to being saturated, perspiration does not readily evaporate from the skin. Less evaporational cooling causes most people to feel hotter than it really is, and a number of people start to complain about the "heat and humidity." (A closer look at how we feel in hot weather will be given in Chapter 4, after we have examined the concepts of relative humidity and wet-bulb temperature.)

MEASURING AIR TEMPERATURE

Thermometers were developed to measure air temperature. Each thermometer has a definite scale and is calibrated so that a thermometer reading of 0°C in Vermont will indicate the same temperature as a thermometer with the same reading in North Dakota. If a particular reading were to represent different degrees of hot or cold, depending on location, thermometers would be useless.

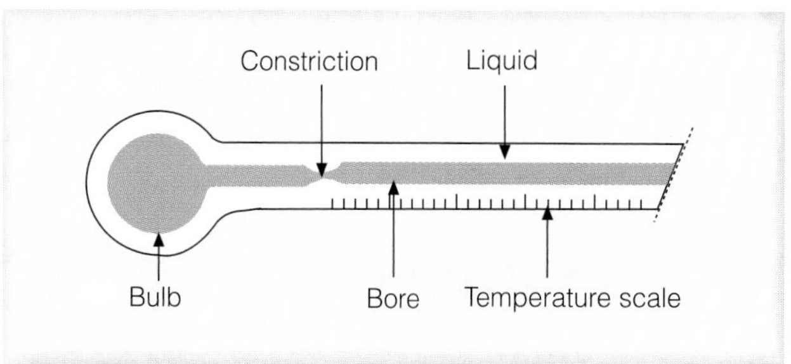

FIGURE 3.14 A section of a maximum thermometer.

Liquid-in-glass thermometers are often used for measuring surface air temperature because they are easy to read and inexpensive to construct. These thermometers have a glass bulb attached to a sealed, graduated tube about 25 cm (10 in.) long. A very small opening, or bore, extends from the bulb to the end of the tube. A liquid in the bulb (usually mercury or red-colored alcohol) is free to move from the bulb up through the bore and into the tube. When the air temperature increases, the liquid in the bulb expands, and rises up the tube. When the air temperature decreases, the liquid contracts, and moves down the tube. Hence, the length of the liquid in the tube represents the air temperature. Because the bore is very narrow, a small temperature change will show up as a relatively large change in the length of the liquid column.

Maximum and minimum thermometers are liquid-in-glass thermometers used for determining daily maximum and minimum temperatures. The **maximum thermometer** looks like any other liquid-in-glass thermometer with one exception: It has a small constriction within the bore just above the bulb (see Fig. 3.14). As the air temperature increases, the mercury expands and freely moves past the constriction up the tube, until the maximum temperature occurs. However, as the air temperature begins to drop, the small constriction prevents the mercury from flowing back into the bulb. Thus, the end of the stationary mercury column indicates the maximum temperature for the day. The mercury will stay at this position until either the air warms to a higher reading or the thermometer is reset by whirling it on a special holder and pivot. Usually, the whirling is sufficient to push the mercury back into the bulb past the constriction until the end of the column indicates the present air temperature.*

A **minimum thermometer** measures the lowest temperature reached during a given period. Most mini-

*Liquid-in-glass thermometers that measure body temperature are maximum thermometers, which is why they are shaken both before and after you take your temperature.

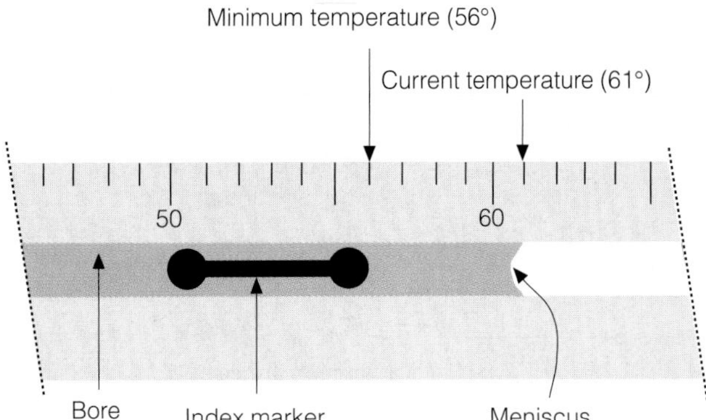

FIGURE 3.15 A section of a minimum thermometer showing both the current air temperature and the minimum temperature.

mum thermometers use alcohol as a liquid, since it freezes at a temperature of –130°C compared to –39°C for mercury. The minimum thermometer is similar to other liquid-in-glass thermometers except that it contains a small barbell-shaped index marker in the bore (see Fig. 3.15). The small index marker is free to slide back and forth within the liquid. It cannot move out of the liquid because the surface tension at the end of the liquid column (the *meniscus*) holds it in.

A minimum thermometer is mounted horizontally. As the air temperature drops, the contracting liquid moves back into the bulb and brings the index marker down the bore with it. When the air temperature stops decreasing, the liquid and the index marker stop moving down the bore. As the air warms, the alcohol expands and moves freely up the tube past the stationary index marker. Because the index marker does not move as the air warms, the minimum temperature is read by observing the upper end of the marker.

To reset a minimum thermometer, simply tip it upside down. This allows the index marker to slide to the upper end of the alcohol column, which is indicating the current air temperature. The thermometer is then remounted horizontally, so that the marker will move toward the bulb as the air temperature decreases.

Highly accurate temperature measurements may be made with **electrical thermometers,** such as the *thermistor* and the *electrical resistance thermometer.* Both of these instruments measure the electrical resistance of a particular material. Since the resistance of the material chosen for these thermometers changes as the temperature changes, the resistance can be calibrated to represent air temperature.

Electrical resistance thermometers are the type of thermometers used in the measurement of air temperature at the over 900 fully automated surface weather stations (known as *ASOS* for *A*utomated *S*urface *O*bserving *S*ystem) that exist at airports and military facilities throughout the United States. Hence, many of the liquid-in-glass thermometers have been replaced with electrical thermometers. At this point it should be noted that the replacement of liquid-in-glass thermometers with electrical thermometers has raised concern among climatologists. For one thing, the response of the electrical thermometers to temperature change is faster. Thus, electrical thermometers may reach a brief extreme reading, which could have been missed by the slower-responding liquid-in-glass thermometer. In addition, many temperature readings, which were taken at airport weather offices, are now taken at ASOS locations that sit near or between runways at the airport. This change in instrumentation and relocation of the measurement site can sometimes introduce a small, but significant, temperature change at the reporting station.

Air temperature may also be obtained with instruments called *infrared sensors,* or **radiometers.** Radiometers do not measure temperature directly; rather, they measure emitted radiation (usually infrared). By measuring both the intensity of radiant energy and the wavelength of maximum emission of a particular gas (either water vapor or carbon dioxide), radiometers in orbiting satellites are now able to estimate the air temperature at selected levels in the atmosphere.

A **bimetallic thermometer** consists of two different pieces of metal (usually brass and iron) welded together to form a single strip. As the temperature changes, the brass expands more than the iron, causing the strip to bend. The small amount of bending is amplified through a system of levers to a pointer on a calibrated scale. The bimetallic thermometer is usually the temperature-sensing part of the **thermograph,** an instrument that measures and records temperature (see Fig. 3.16).

Thermographs are gradually being replaced with *data loggers.* These small instruments have a thermistor connected to a circuit board inside the logger. A computer programs the interval at which readings are taken. The loggers are not only more responsive to air temperature than are thermographs, they are less expensive.

Chances are, you may have heard someone exclaim something like, "Today the thermometer measured 90 degrees in the shade!" Does this mean that the air temperature is sometimes measured in the sun? If you are unsure of the answer, read the Focus section on p. 73 before reading the next section on instrument shelters.

FOCUS ON AN OBSERVATION

Thermometers Should Be Read in the Shade

When we measure air temperature with a common liquid thermometer, an incredible number of air molecules bombard the bulb, transferring energy either to or away from it. When the air is warmer than the thermometer, the liquid gains energy, expands, and rises up the tube; the opposite will happen when the air is colder than the thermometer. The liquid stops rising (or falling) when equilibrium between incoming and outgoing energy is established. At this point, we can read the temperature by observing the height of the liquid in the tube.

It is *impossible* to measure *air temperature* accurately in direct sunlight because the thermometer absorbs radiant energy from the sun in addition to energy from the air molecules. The thermometer gains energy at a much faster rate than it can radiate it away, and the liquid keeps expanding and rising until there is equilibrium between incoming and outgoing energy. Because of the direct absorption of solar energy, the level of the liquid in the thermometer indicates a temperature much higher than the actual air temperature, and so a statement that says, "Today the air temperature measured 100 degrees in the sun," has no meaning. Hence, a thermometer must be kept in a shady place to measure the temperature of the air accurately.

Thermometers and other instruments are usually housed in an **instrument shelter.** The shelter completely encloses the instruments, protecting them from rain, snow, and the sun's direct rays. It is painted white to reflect sunlight, faces north to avoid direct exposure to sunlight, and has louvered sides, so that air is free to flow through it. This construction helps to keep the air inside the shelter at the same temperature as the air outside.

The thermometers inside a standard shelter are mounted about 1.5 to 2 m (5 to 6 ft) above the ground. As we saw in an earlier section, on a clear, calm night the air at ground level may be much colder than the air at the level of the shelter. As a result, on clear winter mornings it is possible to see ice or frost on the ground even though the minimum thermometer in the shelter did not reach the freezing point.

The older instrument shelters are gradually being replaced by the *Max-Min Temperature Shelter* (see Fig. 3.17). The shelter is mounted on a pipe, and wires from the electrical temperature sensor inside are run to a building. A readout inside the building displays the current air temperature and stores the maximum and minimum temperatures for later retrieval. This type of shelter is now used with the automated (ASOS) system.

Because air temperatures vary considerably above different types of surfaces, shelters are usually placed over grass to ensure that the air temperature is measured at the same elevation over the same type of surface. Unfortunately, some shelters are placed on asphalt, others sit on concrete, while others are located on the tops of tall buildings, making it difficult to compare air temperature measurements from different locations. In fact, if either the maximum or minimum air temperature in your area seems suspiciously different from those of nearby towns, find out where the instrument shelter is situated.

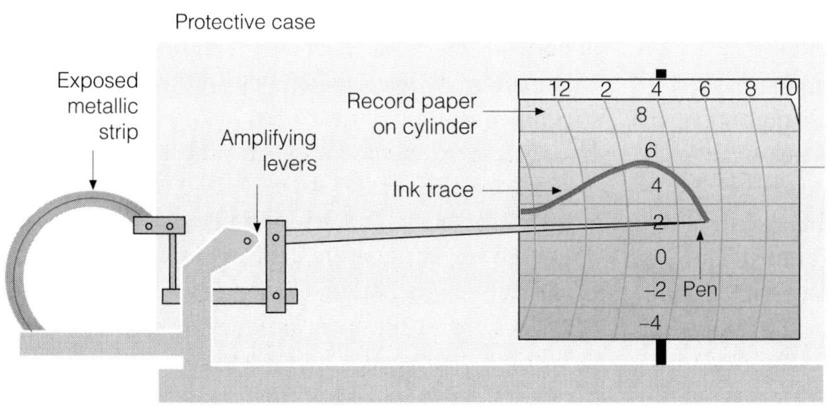

FIGURE 3.16 The thermograph with a bimetallic thermometer.

Photo by Jan Null

FIGURE 3.17 The max-min instrument shelter (middle box) and other weather instruments that comprise the ASOS system.

SUMMARY

The daily variation in air temperature near the earth's surface is controlled mainly by the input of energy from the sun and the output of energy from the surface. On a clear, calm day, the surface air warms, as long as heat input (mainly sunlight) exceeds heat output (mainly convection and radiated infrared energy). The surface air cools at night, as long as heat output exceeds input. Because the ground at night cools more quickly than the air above, the coldest air is normally found at the surface where a radiation inversion usually forms. When the air temperature in agricultural areas drops to dangerously low readings, fruit trees and grape vineyards can be protected from the cold by a variety of means, from mixing the air to spraying the trees and vines with water.

The greatest daily variation in air temperature occurs at the earth's surface. Both the diurnal and annual range of temperature are greater in dry climates than in humid ones. Even though two cities may have similar average annual temperatures, the range and extreme of their temperatures can differ greatly. Temperature information influences our lives in many ways, from deciding what clothes to take on a trip to providing critical information for energy-use predictions and agricultural planning. We reviewed some of the many types of thermometers in use: maximum, minimum, bimetallic, electrical, radiometer. Those designed to measure air temperatures near the surface are housed in instrument shelters to protect them from direct sunlight and precipitation.

Meteorology Now™ Assess your understanding of this chapter's topics with additional quizzing and tutorials at http://earthscience.brookscole.com/ahrens/ess4e.

KEY TERMS

The following terms are listed in the order they appear in the text. Define each. Doing so will aid you in reviewing the material covered in this chapter.

radiational cooling
radiation inversion
thermal belt
orchard heater
wind machine
freeze
controls of temperature
isotherm
specific heat
daily (diurnal) range of temperature
mean (average) daily temperature
annual range of temperature
mean annual temperature
heating degree-day
cooling degree-day
growing degree-day
sensible temperature
wind-chill index
frostbite
hypothermia
liquid-in-glass thermometer
maximum thermometer
minimum thermometer
electrical thermometer
radiometer
bimetallic thermometer
thermograph
instrument shelter

QUESTIONS FOR REVIEW

1. Explain why the warmest time of the day is usually in the afternoon, even though the sun's rays are most direct at noon.
2. On a calm, sunny day, why is the air next to the ground normally much warmer than the air several feet above?
3. Explain how incoming energy and outgoing energy regulate the daily variation in air temperature.
4. Draw a vertical profile of air temperature from the ground to an elevation of 3 m (10 ft) on a clear, windless (a) afternoon and (b) early morning just before sunrise. Explain why the temperature curves are different.
5. Explain how radiational cooling at night produces a radiation temperature inversion.
6. What weather conditions are best suited for the formation of a cold night and a strong radiation inversion?

7. Explain why thermal belts are found along hillsides at night.
8. List some of the measures farmers use to protect their crops against the cold. Explain the physical principle behind each method.
9. Why are the lower branches of trees most susceptible to damage from low temperatures?
10. Describe each of the controls of temperature.
11. Look at Fig. 3.8 (temperature map for January) and explain why the isotherms dip southward (equatorward) over the Northern Hemisphere continents.
12. During the winter, frost can form on the ground when the minimum thermometer indicates a low temperature above freezing. Explain.
13. Why do the first freeze in autumn and the last freeze in spring occur in bottomlands?
14. Explain why the daily range of temperature is normally greater
    (a) in dry regions than in humid regions and
    (b) on clear days than on cloudy days.
15. Why are the largest annual range of temperatures normally observed over continents away from large bodies of water?
16. Two cities have the same mean annual temperature. Explain why this fact does not mean that their temperatures throughout the year are similar.
17. What is a heating degree-day? A cooling degree-day? How are these units calculated?
18. During a cold, calm, sunny day, why do we usually feel warmer than a thermometer indicates?
19. (a) Assume the wind is blowing at 30 mi/hr and the air temperature is 5°F. Determine the wind chill equivalent temperature in Table 3.2, p. 70.
    (b) Under the conditions listed in (a) above, explain why an ordinary thermometer would measure a temperature of 5°F.
20. What atmospheric conditions can bring on hypothermia?
21. Someone says, "Today, the air temperature measured 99°F in the sun." Why does this statement have no meaning?
22. Explain why the minimum thermometer is the one with a small barbell-shaped index marker in the bore.
23. Briefly describe how the following thermometers measure air temperature:
    (a) liquid-in-glass
    (b) bimetallic
    (c) electrical
    (d) radiometer

QUESTIONS FOR THOUGHT AND EXPLORATION

1. How do you feel cloud cover would affect the lag in daily temperature?
2. Which location is most likely to have the greater daily temperature range: a tropical rainforest near the equator or a desert site in Nevada? Explain.
3. Explain why putting on a heavy winter jacket would be effective in keeping you warm, even if the jacket had been outside in sub-freezing temperatures for several hours.
4. Why is the air temperature displayed on a bank or building marquee usually inaccurate?
5. If you were forced to place a meteorological instrument shelter over asphalt rather than over grass, what modification(s) would you have to make so that the temperature measurements inside the shelter were more representative of the actual air temperature?
6. The average temperature in San Francisco, California, for December, January, and February is 11°C (52°F). During the same three-month period the average temperature in Richmond, Virginia, is 4°C (39°F). Yet, San Francisco and Richmond have nearly the same yearly total of heating degree-days. Explain why. (Hint: See Fig. 3.11, p. 67.)
7. How would the lag in daily temperature experienced over land compare to the daily temperature lag over water?
8. Historical Weather Data (http://www.wunderground.com): Enter the name of any city in the United States and find the current weather forecast for that city. Using the "History & Almanac" section of the page, find the record high and low temperatures for three different dates.
9. World Climate Data (http://www.worldclimate.com/climate/index.htm): Enter the name of any city and find the average monthly temperatures for that location. Compare the average monthly temperatures for a coastal city with those of a city located in the interior of a large continent. Discuss the differences you find.

Go to the Brooks/Cole Earth Sciences Resource Center (http://earthscience.brookscole.com) for critical thinking exercises, articles, and additional readings from InfoTrac College Edition, Brooks/Cole's online student library.

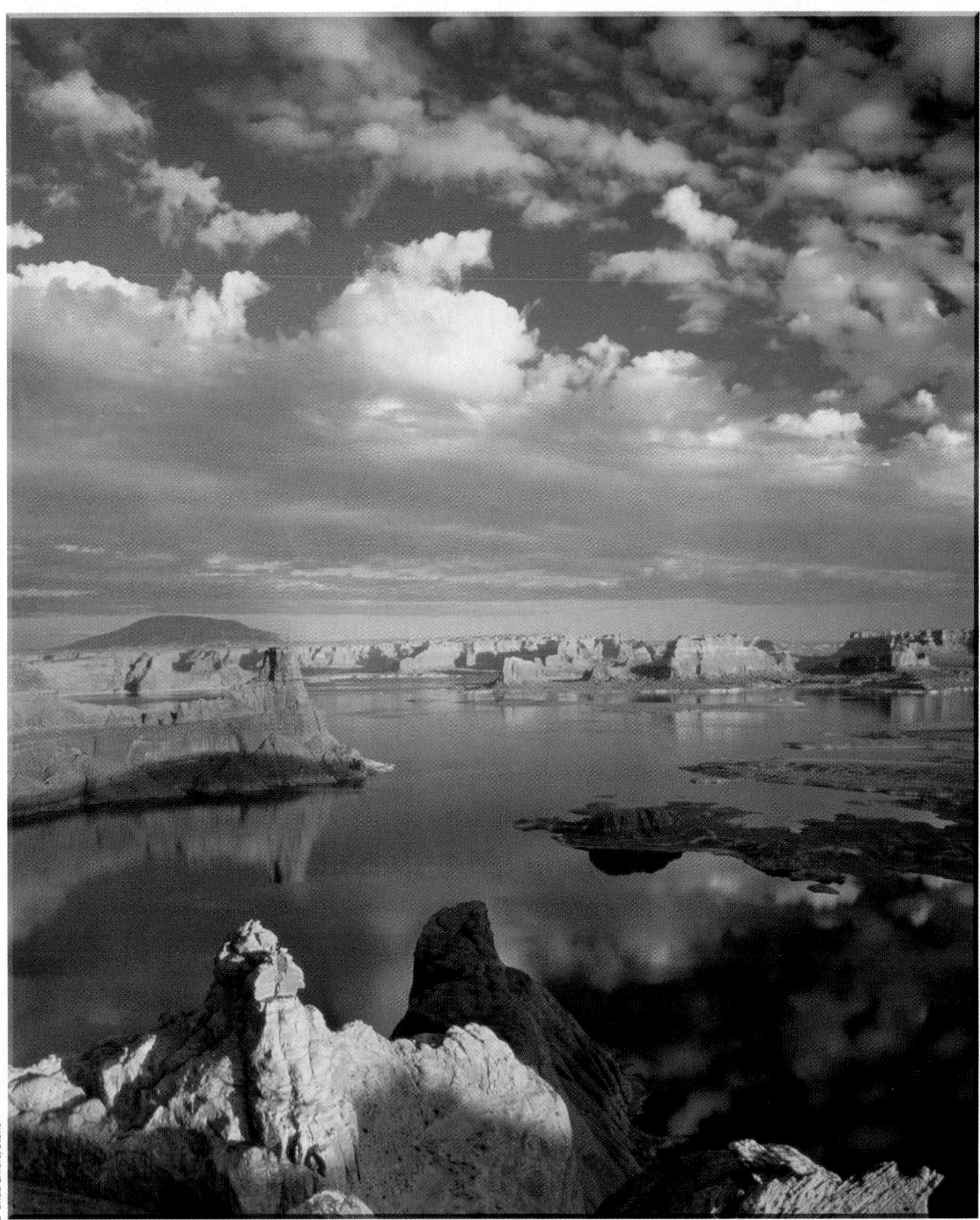

© Chad Ehlers/Stone

Cumuliform clouds develop as air slowly rises over Lake Powell in Utah.

Meteorology Now™ This icon, appearing throughout the book, indicates an opportunity to explore interactive tutorials, animations, or practice problems available on the MeteorologyNow Web site at http://earthscience.brookscole.com/ahrens/ess4e.

CHAPTER 4

Humidity, Condensation, and Clouds

Sometimes it rains and still fails to moisten the desert—the falling water evaporates halfway down between cloud and earth. Then you see curtains of blue rain dangling out of reach in the sky while the living things wither below for want of water. Torture by tantalizing, hope without fulfillment. And the clouds disperse and dissipate into nothingness. . . . The sun climbed noon-high, the heat grew thick and heavy on our brains, the dust clouded our eyes and mixed with our sweat. My canteen is nearly empty and I'm afraid to drink what little water is left—there may never be any more. I'd like to cave in for a while, crawl under yonder cottonwood and die peacefully in the shade, drinking dust.

Edward Abbey, *Desert Solitaire—A Season in the Wilderness*

CONTENTS

We know from Chapter 1 that, in our atmosphere, the concentration of the invisible gas water vapor is normally less than a few percent of all the atmospheric molecules. Yet water vapor is exceedingly important, for it transforms into cloud droplets and ice crystals—particles that grow in size and fall to the earth as precipitation. The term *humidity* is used to describe the amount of water vapor in the air. To most of us, a moist day suggests high humidity. However, there is usually more water vapor in the hot, "dry" air of the Sahara Desert than in the cold, "damp" polar air in New England, which raises an interesting question: Does the desert air have a higher humidity? As we will see later in this chapter, the answer to this question is both yes and no, depending on the type of humidity we mean.

So that we may better understand the concept of humidity, we will begin this chapter by examining the circulation of water in the atmosphere. Then we will look at different ways to express humidity. Near the end of the chapter, we will investigate various forms of condensation, including dew, fog, and clouds.

CIRCULATION OF WATER IN THE ATMOSPHERE

Within the atmosphere, there is an unending circulation of water. Since the oceans occupy over 70 percent of the earth's surface, we can think of this circulation as beginning over the ocean. Here, the sun's energy transforms enormous quantities of liquid water into water vapor in a process called **evaporation.** Winds then transport the moist air to other regions, where the water vapor changes back into liquid, forming clouds, in a process called **condensation.** Under certain conditions, the liquid (or solid) cloud particles may grow in size and fall to the surface as **precipitation**—rain, snow, or hail. If the precipitation falls into an ocean, the water is ready to begin its cycle again. If, on the other hand, the precipitation falls on a continent, a great deal of the water returns to the ocean in a complex journey. This cycle of moving and transforming water molecules from liquid to vapor and back to liquid again is called the **hydrologic** (water) **cycle.** In the most simplistic form of this cycle, water molecules travel from ocean to atmosphere to land and then back to the ocean.

Figure 4.1 illustrates the complexities of the hydrologic cycle. For example, before falling rain ever reaches the ground, a portion of it evaporates back into the air. Some of the precipitation may be intercepted by vegetation, where it evaporates or drips to the ground long after a storm has ended. Once on the surface, a portion of the water soaks into the ground by percolating downward through small openings in the soil and rock, forming groundwater that can be tapped by wells. What does not soak in collects in puddles of standing water or runs off into streams and rivers, which find their way back to

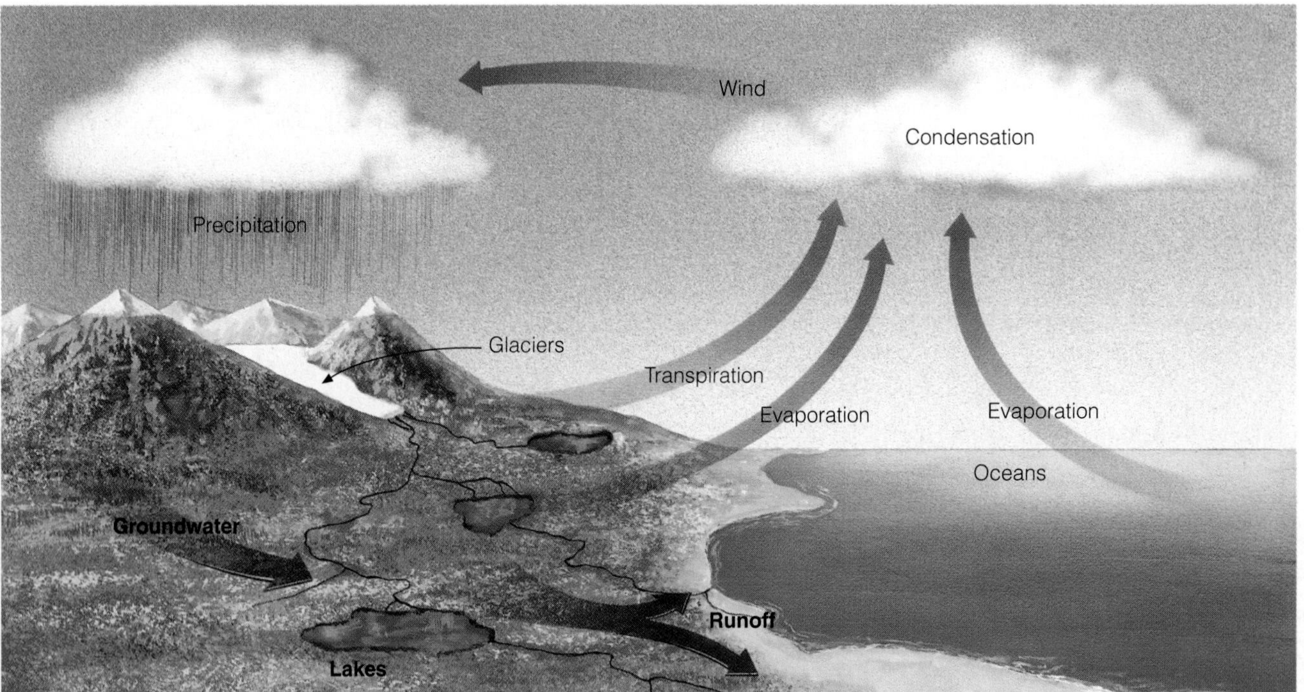

FIGURE 4.1 The hydrologic cycle.

the ocean. Even the underground water moves slowly and eventually surfaces, only to evaporate or be carried seaward by rivers.

Over land, a considerable amount of vapor is added to the atmosphere through evaporation from the soil, lakes, and streams. Even plants give up moisture by a process called *transpiration.* The water absorbed by a plant's root system moves upward through the stem and emerges from the plant through numerous small openings on the underside of the leaf. In all, evaporation and transpiration from continental areas amount to only about 15 percent of the nearly 1.5 billion billion gallons of water vapor that annually evaporate into the atmosphere; the remaining 85 percent evaporates from the oceans. The total mass of water vapor stored in the atmosphere at any moment adds up to only a little over a week's supply of the world's precipitation. Since this amount varies only slightly from day to day, the hydrologic cycle is exceedingly efficient in circulating water in the atmosphere.

EVAPORATION, CONDENSATION, AND SATURATION

To obtain a slightly different picture of water in the atmosphere, suppose we examine water in a beaker similar to the one shown in Fig. 4.2a. If we were able to magnify the surface water about a billion times, we would see water molecules fairly close together, jiggling, bouncing, and moving about. We would also see that the molecules are not all moving at the same speed—some are moving much faster than others. Recall from Chapter 2 that the *temperature* of the water is a measure of the average speed of its molecules. At the surface, molecules with enough speed (and traveling in the right direction) would occasionally break away from the liquid surface and enter into the air above. These molecules, changing from the *liquid state into the vapor state,* are evaporating. While some water molecules are leaving the liquid, others are returning. Those returning are condensing as they are changing from a *vapor state to a liquid state.*

When a cover is placed over the dish (Fig. 4.2b), after a while the total number of molecules escaping from the liquid (evaporating) would be balanced by the number returning (condensing). When this condition exists, the air is said to be **saturated** with water vapor. For every molecule that evaporates, one must condense, and no net loss of liquid or vapor molecules results.

If we remove the cover and blow across the top of the water, some of the vapor molecules already in the air

DID YOU KNOW?

If all the water vapor in the atmosphere were to suddenly condense and fall as rain, it would cover the earth's surface with about 2.5 cm (1 in.) of water.

above would be blown away, creating a difference between the actual number of vapor molecules and the total number required for *saturation.* This would help prevent saturation from occurring and would allow for a greater amount of evaporation. Wind, therefore, enhances evaporation.

The temperature of the water also influences evaporation. All else being equal, warm water will evaporate more readily than cool water. The reason for this phenomenon is that, when heated, the water molecules will speed up. At higher temperatures, a greater fraction of the molecules have sufficient speed to break through the surface tension of the water and zip off into the air above. Consequently, the warmer the water, the greater the rate of evaporation.

If we could examine the air above the water in Fig. 4.2, we would observe the water vapor molecules freely darting about and bumping into each other as well as neighboring molecules of oxygen and nitrogen. We would also observe that mixed in with all of the air molecules are microscopic bits of dust, smoke, and salt from ocean spray. Since many of these serve as surfaces on which water vapor may condense, they are called

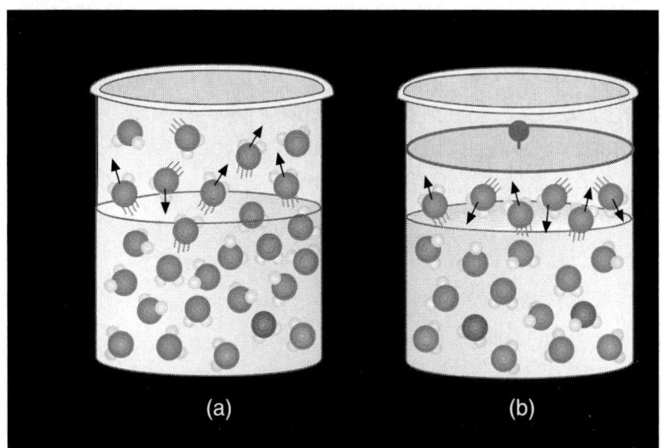

FIGURE 4.2 (a) Water molecules at the surface of the water are evaporating (changing from liquid into vapor) and condensing (changing from vapor into liquid). Since more molecules are evaporating than condensing, net evaporation is occurring. (b) When the number of water molecules escaping from the liquid (evaporating) balances those returning (condensing), the air above the liquid is saturated with water vapor. (For clarity, only water molecules are illustrated.)

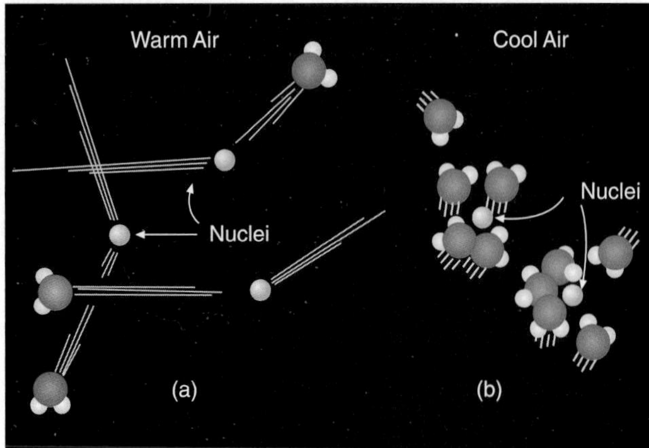

FIGURE 4.3 Condensation is more likely to occur as the air cools. (a) In the warm air, fast-moving H_2O vapor molecules tend to bounce away after colliding with nuclei. (b) In the cool air, slow-moving vapor molecules are more likely to join together on nuclei. The condensing of many billions of water molecules produces tiny liquid water droplets.

condensation nuclei. In the warm air above the water, fast-moving vapor molecules strike the nuclei with such impact that they simply bounce away (see Fig. 4.3). However, if the air is chilled, the molecules move more slowly and are more apt to stick and condense to the nuclei. When many billions of these vapor molecules condense onto the nuclei, tiny liquid cloud droplets form.

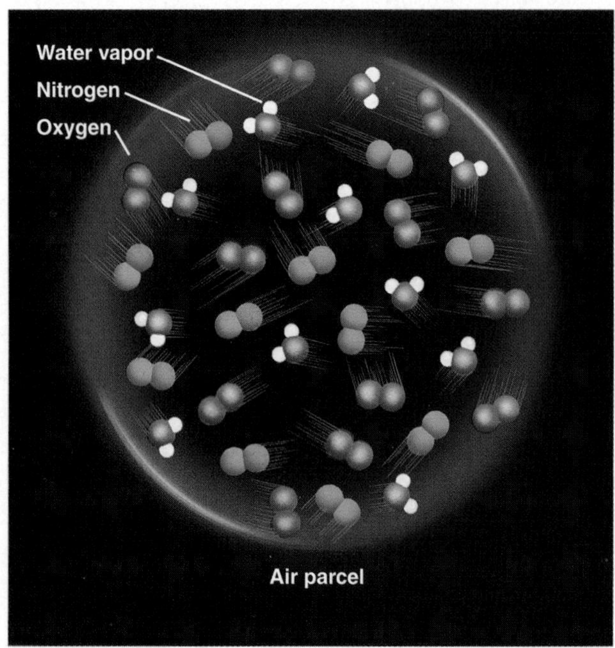

FIGURE 4.4 The water vapor content (humidity) inside this air parcel can be expressed in a number of ways.

We can see then that condensation is more likely to happen as the air cools and the speed of the vapor molecules decreases. As the air temperature increases, condensation is less likely because most of the molecules have sufficient speed (sufficient energy) to remain as a vapor. As we will see in this and other chapters, *condensation occurs primarily when the air is cooled.*

Even though condensation is more likely to occur when the air cools, it is important to note that no matter how cold the air becomes, there will always be a few molecules with sufficient speed (sufficient energy) to remain as a vapor. It should be apparent, then, that with the same number of water vapor molecules in the air, saturation is more likely to occur in cool air than in warm air. This idea often leads to the statement that "warm air can hold more water vapor molecules before becoming saturated than can cold air" or, simply, "warm air has a greater capacity for water vapor than does cold air." At this point, it is important to realize that although these statements are correct, the use of such words as "hold" and "capacity" are misleading when describing water vapor content, as air does not really "hold" water vapor in the sense of making "room" for it.

HUMIDITY

Humidity refers to any one of a number of ways of specifying the amount of water vapor in the air. Since there are several ways to express atmospheric water vapor content, there are several meanings for the concept of humidity.

Imagine, for example, that we enclose a volume of air (about the size of a large balloon) in a thin elastic container—a *parcel*—as illustrated in Fig. 4.4. If we extract the water vapor from the parcel, we would specify the humidity in the following ways:

1. We could compare the weight (mass) of the water vapor with the volume of air in the parcel and obtain the *water vapor density,* or *absolute humidity.*
2. We could compare the weight (mass) of the water vapor in the parcel with the total weight (mass) of all the air in the parcel (including vapor) and obtain the *specific humidity.*
3. Or, we could compare the weight (mass) of the water vapor in the parcel with the weight (mass) of the remaining dry air and obtain the *mixing ratio.*

Absolute humidity is normally expressed as grams of water vapor per cubic meter of air (g/m^3), whereas both specific humidity and mixing ratio are expressed as grams of water vapor per kilogram of air (g/kg).

Look at Fig. 4.4 and notice that we could also express the humidity of the air in terms of *water vapor pressure*—the push (force) that the water vapor molecules are exerting against the inside walls of the parcel.

Vapor Pressure Suppose the air parcel in Fig. 4.4 is near sea level and the air pressure inside the parcel is 1000 millibars (mb).* The total air pressure inside the parcel is due to the collision of all the molecules against the walls of the parcel. In other words, the total pressure inside the parcel is equal to the sum of the pressures of the individual gases. Since the total pressure inside the parcel is 1000 millibars, and the gases inside include nitrogen (78 percent), oxygen (21 percent), and water vapor (1 percent), the partial pressure exerted by nitrogen would then be 780 mb and, by oxygen, 210 mb. The partial pressure of water vapor, called the **actual vapor pressure,** would be only 10 mb (one percent of 1000).** It is evident, then, that because the number of water vapor molecules in any volume of air is small compared to the total number of air molecules in the volume, the actual vapor pressure is normally a small fraction of the total air pressure.

Everything else being equal, the more air molecules in a parcel, the greater the total air pressure. When you blow up a balloon, you increase its pressure by putting in more air. Similarly, an increase in the number of water vapor molecules will increase the total vapor pressure. Hence, the actual vapor pressure is a fairly good measure of the total amount of water vapor in the air: *High actual vapor pressure indicates large numbers of water vapor molecules, whereas low actual vapor pressure indicates comparatively small numbers of vapor molecules.*

Actual vapor pressure indicates the air's total water vapor content, whereas **saturation vapor pressure** describes how much water vapor is necessary to make the air saturated at any given temperature. Put another way, *saturation vapor pressure is the pressure that the water vapor molecules would exert if the air were saturated with vapor at a given temperature.*

We can obtain a better picture of the concept of saturation vapor pressure by imagining molecules evaporating from a water surface. (Look back at Fig. 4.2b.) Recall that when the air is saturated, the number of molecules escaping from the water's surface equals the number returning. Since the number of "fast-moving" molecules increases as the temperature increases, the number of water molecules escaping per second increases also. In order to maintain equilibrium, this situation causes an increase in the number of water vapor molecules in the air above the liquid. Consequently, *at higher air temperatures, it takes more water vapor to saturate the air.* And more vapor molecules exert a greater pressure. Saturation vapor pressure, then, depends primarily on the air temperature. From the graph in Fig. 4.5, we can see that at 10°C, the saturation vapor pressure is about 12 mb, whereas at 30°C it is about 42 mb.

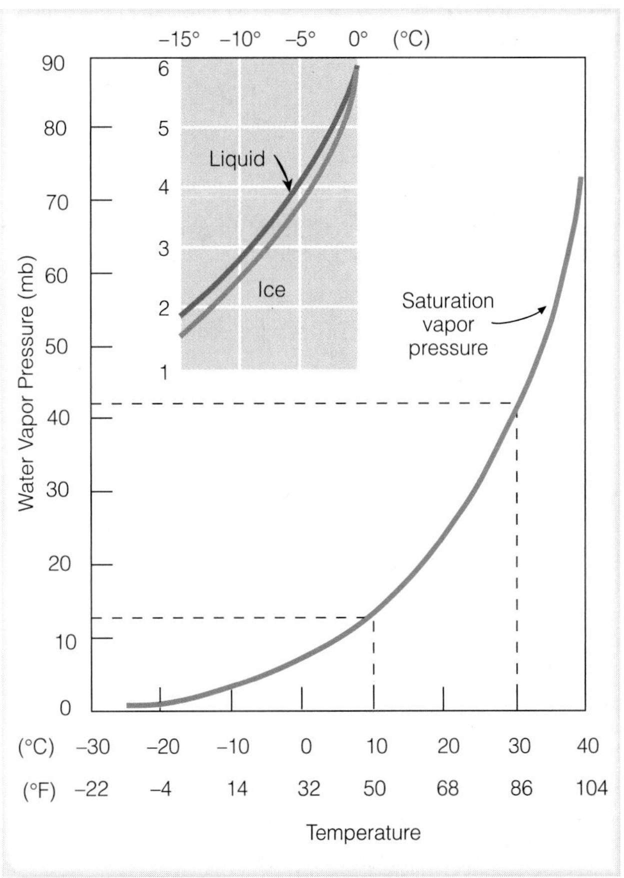

Meteorology Now™ ACTIVE FIGURE 4.5
Saturation vapor pressure increases with increasing temperature. At a temperature of 10°C, the saturation vapor pressure is about 12 mb, whereas at 30°C it is about 42 mb. The insert illustrates that the saturation vapor pressure over water is greater than the saturation vapor pressure over ice. Watch this Active Figure at http://earthscience.brookscole.com/ahrens/ess4e.

*You may recall from Chapter 1 that the millibar is the unit of pressure most commonly found on surface weather maps. The millibar expresses atmospheric pressure as a force over a given area.

**When we use the percentages of various gases in a volume of air, these percentages only give us an approximation of the actual vapor pressure. The point here is that, near the earth's surface, the actual vapor pressure is often close to 10 mb.

Relative Humidity While relative humidity is the most commonly used way of describing atmospheric moisture, it is also, unfortunately, the most misunderstood. The concept of relative humidity may at first seem confusing because it does not indicate the actual amount of water vapor in the air. Instead, it tells us how close the air is to being saturated. The **relative humidity** (RH) is the *ratio of the amount of water vapor actually in the air to the maximum amount of water vapor required for saturation at that particular temperature (and pressure).* It is the *ratio* of the air's water vapor *content* to its *capacity;* thus

$$\text{RH} = \frac{\text{water vapor content}}{\text{water vapor capacity}}.$$

We can think of the actual vapor pressure as a measure of the air's actual water vapor content, and the saturation vapor pressure as a measure of air's total capacity for water vapor. Hence, the relative humidity can be expressed as

$$\text{RH} = \frac{\text{actual vapor pressure}}{\text{saturation vapor pressure}} \times 100 \text{ percent}.$$

Relative humidity is given as a percent.* Air with a 50 percent relative humidity actually contains one-half the amount required for saturation. Air with a 100 percent relative humidity is said to be *saturated* because it is filled to capacity with water vapor. Air with a relative humidity greater than 100 percent is said to be **supersaturated.**

A change in relative humidity can be brought about in two primary ways:

1. by changing the air's water vapor content
2. by changing the air temperature

A change in the air's water vapor content can change the air's actual vapor pressure. If the air temperature remains constant, an increase in the air's water vapor content increases the air's actual vapor pressure and raises the relative humidity. The relative humidity increases as the actual vapor pressure approaches the saturation vapor pressure and the air approaches saturation. Conversely, if the air temperature remains constant, a decrease in the air's water vapor content decreases the air's actual vapor pressure and lowers the relative humidity. In summary, *as water vapor is added to the air (with no change in air temperature), the relative humidity increases, and, as water vapor is removed from the air, the relative humidity lowers.*

A change in the air temperature can bring about a change in the relative humidity. This phenomenon happens because a change in air temperature alters the air's saturation vapor pressure. If the air temperature increases, the saturation vapor pressure also increases, which raises the air's water vapor capacity. If there is no change in the air's actual water vapor content, the relative humidity lowers. If, on the other hand, the air temperature decreases, so does the air's saturation vapor pressure. As the saturation vapor pressure approaches the actual vapor pressure, the relative humidity increases as the air approaches saturation. In summary, *with no change in water vapor content, an increase in air temperature lowers the relative humidity, while a decrease in air temperature raises the relative humidity.*

In many places, the air's total vapor content varies only slightly during an entire day, and so it is the changing air temperature that primarily regulates the daily variation in relative humidity (see Fig. 4.6). As the air cools during the night, the relative humidity increases. Normally, the highest relative humidity occurs in the early morning, during the coolest part of the day. As the air warms during the day, the relative humidity decreases, with the lowest values usually occurring during the warmest part of the afternoon.

These changes in relative humidity are important in determining the amount of evaporation from vegetation and wet surfaces. If you water your lawn on a hot afternoon, when the relative humidity is low, much of the water will evaporate quickly from the lawn, instead

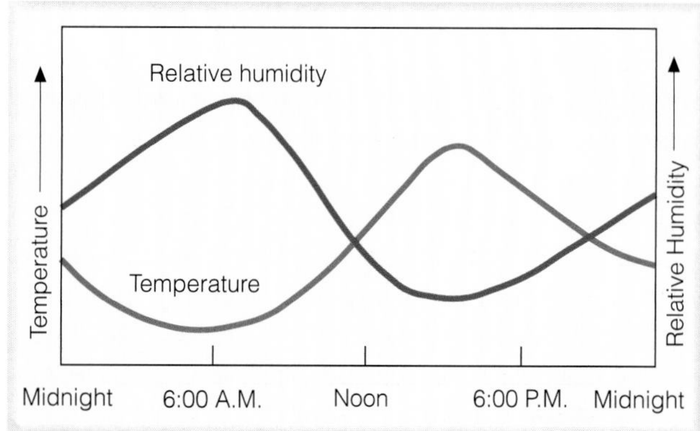

FIGURE 4.6 When the air is cool (morning), the relative humidity is high. When the air is warm (afternoon), the relative humidity is low. These conditions exist in clear weather when the air is calm or of constant wind speed.

*Additional information on relative humidity and vapor pressure is given in Appendix B.

of soaking into the ground. Watering the same lawn in the evening, when the relative humidity is higher, will cut down the evaporation and increase the effectiveness of the watering.

Very low relative humidities in a house can have an adverse effect on things living inside. For example, house plants have a difficult time surviving because the moisture from their leaves and the soil evaporates rapidly. People suffer, too, when the relative humidity is quite low. The rapid evaporation of moisture from exposed flesh causes skin to crack, dry, flake, or itch. These low humidities also irritate the mucous membranes in the nose and throat, producing an "itchy" throat. Similarly, dry nasal passages permit inhaled bacteria to incubate, causing persistent infections.

The relative humidity in a home can be increased just by heating water and allowing it to evaporate into the air. The added water vapor raises the relative humidity to a more comfortable level. In modern homes, a humidifier, installed near the furnace, adds moisture to the air at a rate of about one gallon per room per day. The air, with its increased water vapor, is circulated throughout the home by a forced air heating system. In this way, all rooms get their fair share of moisture—not just the room where the vapor is added.

Relative Humidity and Dew Point Suppose it is early morning and the outside air is saturated. The air temperature is 10°C (50°F) and the relative humidity is 100 percent. We know from the previous section that relative humidity can be expressed as

$$\text{RH} = \frac{\text{actual vapor pressure}}{\text{saturation vapor pressure}} \times 100 \text{ percent.}$$

Looking back at Fig. 4.5 (p. 81), we can see that air with a temperature of 10°C has a saturation vapor pressure of 12 mb. Since the air is saturated and the relative humidity is 100 percent, the actual vapor pressure *must* be the same as the saturation vapor pressure (12 mb), since

$$\text{RH} = \frac{12 \text{ mb}}{12 \text{ mb}} \times 100\% = 100 \text{ percent.}$$

Suppose during the day the air warms to 30°C (86°F), with no change in water vapor content (or air pressure). Because there is no change in water vapor content, the actual vapor pressure must be the same (12 mb) as it was in the early morning when the air was saturated. The saturation vapor pressure, however, has increased because the air temperature has increased. From Fig. 4.5, note that air with a temperature of 30°C has a saturation vapor pressure of 42 mb. The relative humidity of this unsaturated, warmer air is now much lower, as

$$\text{RH} = \frac{12 \text{ mb}}{42 \text{ mb}} \times 100\% = 29 \text{ percent.}$$

To what temperature must the outside air, with a temperature of 30°C, be cooled so that it is once again saturated? The answer, of course, is 10°C. For this amount of water vapor in the air, 10°C is called the **dew-point temperature** or, simply, the **dew point.** It represents *the temperature to which air would have to be cooled (with no change in air pressure or moisture content) for saturation to occur.* Since atmospheric pressure varies only slightly at the earth's surface, *the dew point is a good indicator of the air's actual water vapor content. High dew points indicate high water vapor content; low dew points, low water vapor content.* Addition of water vapor to the air increases the dew point; removing water vapor lowers it.

Figure 4.7a shows the average dew-point temperatures across the United States for January. Notice that the dew points are highest (the greatest amount of water vapor in the air) over the Gulf Coast states and lowest over the interior. Compare New Orleans with Fargo. Cold, dry winds from northern Canada flow relentlessly into the Central Plains during the winter, keeping this area dry. But warm, moist air from the Gulf of Mexico helps maintain a higher dew-point temperature in the southern states.

Figure 4.7b is a similar diagram showing the average dew-point temperatures for July. Again, the highest dew points are observed along the Gulf Coast, with some areas experiencing average dew-point temperatures near 75°F. Note, too, that the dew points over the eastern and central portion of the country are much higher in July, meaning that the July air contains between 3 and 6 times more water vapor than the January air. The reason for the high dew points is that this region is almost constantly receiving humid air from the warm Gulf of

DID YOU KNOW?

On a hot, muggy summer day in the eastern half of the United States, it is common to hear someone complain that "the air temperature today is 90 degrees and the relative humidity is 90 percent." Although this weather situation is remotely possible, it is highly unlikely, as a temperature of 90°F and a relative humidity of 90 percent can occur only if the dew-point temperature is incredibly high—nearly 87°F.

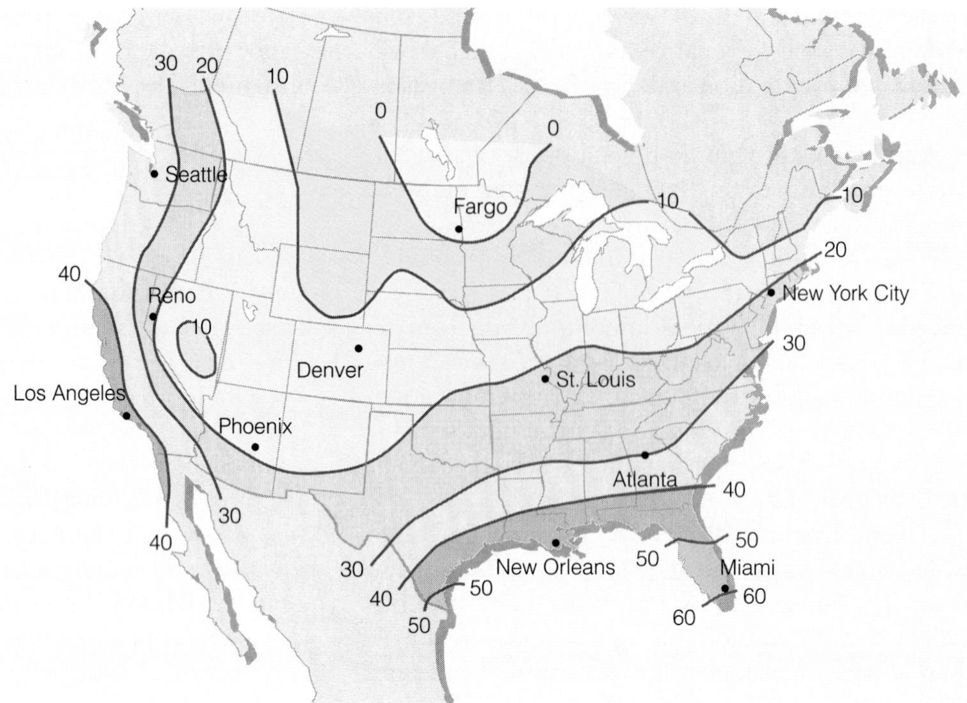

(a) January

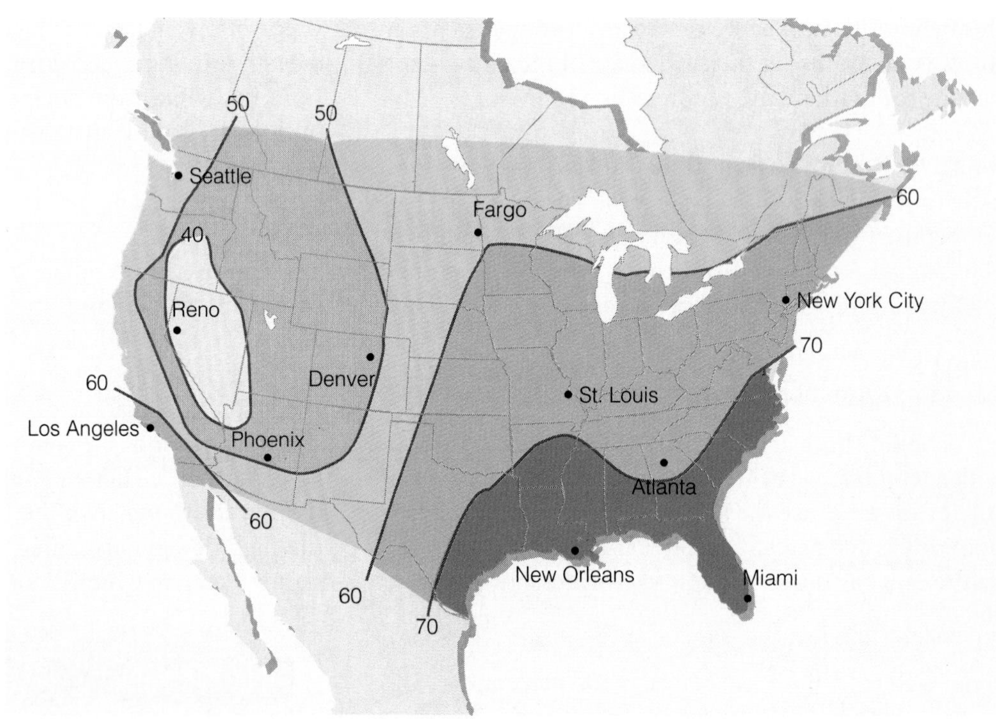

(b) July

FIGURE 4.7 Average surface dew-point temperatures (°F) for (a) January and for (b) July.

Mexico. The lowest dew point, and hence the driest air, is found in the West, with Nevada experiencing the lowest values—a region surrounded by mountains that effectively shield it from significant amounts of moisture moving in from the southwest and northwest.

The difference between air temperature and dew point can indicate whether the relative humidity is low or high. When the air temperature and dew point are far apart, the relative humidity is low; when they are close to the same value, the relative humidity is high. When

(a) POLAR AIR: Air temperature –2°C (28°F)
Dew point –2°C (28°F)
Relative humidity 100 percent

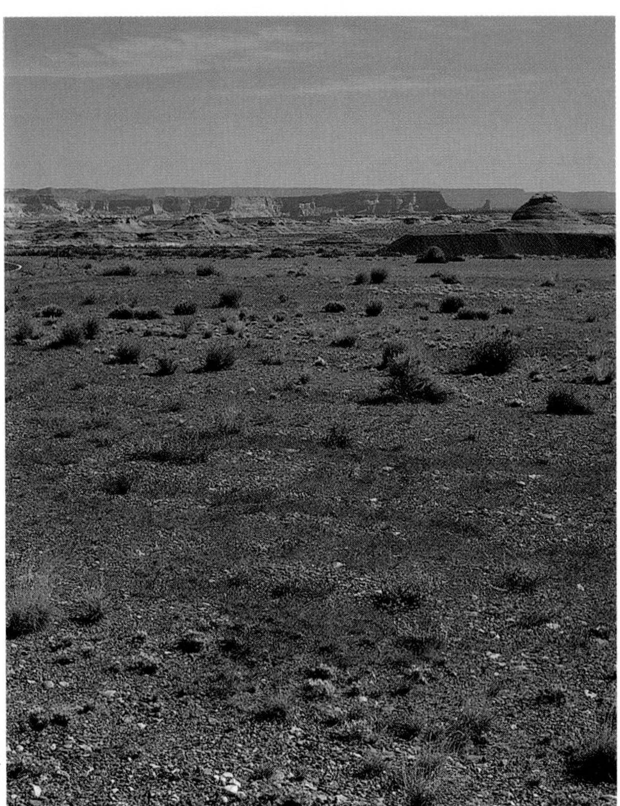

(b) DESERT AIR: Air temperature 35°C (95°F)
Dew point 5°C (41°F)
Relative humidity 16 percent

FIGURE 4.8 The polar air has the higher relative humidity, whereas the desert air, with the higher dew point, contains more water vapor.

the air temperature and dew point are equal, the air is saturated and the relative humidity is 100 percent. Even though the relative humidity may be 100 percent, the air, under certain conditions, may be considered "dry."

Observe, for example, in Fig. 4.8a that because the air temperature and dew point are the same in the polar air, the relative humidity must be 100 percent. On the other hand, the desert air (Fig. 4.8b), with a large separation between air temperature and dew point, has a much lower relative humidity, 16 percent.* However, since dew point is a measure of the amount of water vapor in the air, the desert air (with a higher dew point) must contain *more* water vapor. So even though the polar air has a higher relative humidity, the desert air that contains more water vapor has a higher water vapor density, or *absolute humidity.* (The specific humidity and mixing ratio are also higher in the desert air.)

Now we can see why polar air is often described as being "dry" when the relative humidity is high (often close to 100 percent). In cold, polar air, the dew point and air temperature are normally close together. But the low dew-point temperature means that there is little water vapor in the air. Consequently, the air is "dry" even though the relative humidity is high.

Meteorology Now™ Click "Moisture Graph" to test the relationship between temperature, dew point, vapor pressure, and relative humidity.

BRIEF REVIEW

Up to this point we have looked at the different ways of describing humidity. Before going on, here is a review of some of the important concepts and facts we have covered:

- Relative humidity does not tell us how much water vapor is actually in the air; rather, it tells us how close the air is to being saturated.

*The relative humidity can be computed from Fig. 4.5 (p. 81). The desert air with an air temperature of 35°C has a saturation vapor pressure of about 56 mb. A dew-point temperature of 5°C gives the desert air an actual vapor pressure of about 9 mb. These values produce a relative humidity of 9⁄56 × 100, or 16 percent.

- Relative humidity can change when the air's water-vapor content changes, or when the air temperature changes.
- With a constant amount of water vapor, cooling the air raises the relative humidity and warming the air lowers it.
- The dew-point temperature is a good indicator of the air's water-vapor content.
- Dry air can have a high relative humidity. In polar air, the dew-point temperature is low and the air is considered dry. But the air temperature is close to the dew point, and so the relative humidity is high.

Relative Humidity and Human Discomfort On a hot, muggy day when the relative humidity is high, it is common to hear someone exclaim (often in exasperation), "It's not so much the heat, it's the humidity." Actually, this statement is valid. In warm weather, the main source of body cooling is through evaporation of perspiration. Recall from Chapter 2 that evaporation is a cooling process, so when the air temperature is high and the relative humidity low, perspiration on the skin evaporates quickly, often making us feel that the air temperature is lower than it really is. However, when both the air temperature and relative humidity are high and the air is nearly saturated with water vapor, body moisture does not readily evaporate; instead, it collects on the skin as beads of perspiration. Less evaporation means less cooling, and so we usually feel warmer than we did with a similar air temperature, but a lower relative humidity.

A good measure of how cool the skin can become is the **wet-bulb temperature**—*the lowest temperature that can be reached by evaporating water into the air.* On a hot day when the wet-bulb temperature is low, rapid evaporation (and, hence, cooling) takes place at the skin's surface. As the wet-bulb temperature approaches the air temperature, less cooling occurs, and the skin temperature may begin to rise. When the wet-bulb temperature exceeds the skin's temperature, no net evaporation occurs, and the body temperature can rise quite rapidly. Fortunately, most of the time, the wet-bulb temperature is considerably below the temperature of the skin.

When the weather is hot and muggy, a number of heat-related problems may occur. For example, in hot weather when the human body temperature rises, the *hypothalamus* gland (a gland in the brain that regulates body temperature) activates the body's heat-regulating mechanism, and over ten million sweat glands wet the body with as much as two liters of liquid per hour. As this perspiration evaporates, rapid loss of water and salt can result in a chemical imbalance that may lead to painful *heat cramps.* Excessive water loss through perspiring coupled with an increasing body temperature may result in *heat exhaustion*—fatigue, headache, nausea, and even fainting. If one's body temperature rises above about 41°C (106°F), *heatstroke* can occur, resulting in complete failure of the circulatory functions. If the body temperature continues to rise, death may result. In fact, each year across North America, hundreds of people die from heat-related maladies. Even strong, healthy individuals can succumb to heatstroke, as did the Minnesota Vikings' all-pro offensive lineman, Korey Stringer, who collapsed after practice on July 31, 2001, and died 15 hours later. Before Korey fainted, temperatures on the practice field were in the 90s (°F) with the relative humidity above 55 percent.

In an effort to draw attention to this serious weather-related health hazard, an index called the **heat index (HI)** is used by the National Weather Service. The index combines air temperature with relative humidity to determine an **apparent temperature**—what the air temperature "feels like" to the average person for various combinations of air temperature and relative humidity. For example, in Fig. 4.9, an air temperature of 100°F and a relative humidity of 60 percent produce an apparent temperature of 132°F. Heatstroke or sunstroke is imminent when the index reaches this level. However, as we can see from the preceding paragraph, heatstroke related deaths can occur when the heat index value is considerably lower than 130°F (see Table 4.1).

During hot, humid weather some people remark about how "heavy" or how dense, the air feels. Is hot, humid air really more dense than hot, dry air? If you are interested in the answer, read the Focus section on p. 88.

Measuring Humidity The common instrument used to obtain dew point and relative humidity is a **psychrometer,** which consists of two liquid-in-glass thermometers mounted side by side and attached to a piece

DID YOU KNOW?

Tragically, many hundreds of people died of heat-related maladies during the great Chicago heat wave of July, 1995. On July 13, the afternoon air temperature reached 104°F. With a dew-point temperature of 76°F and a relative humidity near 40 percent, the apparent temperature soared to 119°F. In a van, with the windows rolled up, two small toddlers fell asleep and an hour later were found dead of heat exhaustion. Estimates are that, on a day like this one, temperatures inside a closed vehicle could approach 190°F within half an hour.

Heat Index (or apparent temperature)

Air Temperature (°F) \ Relative Humidity (%)	0	5	10	15	20	25	30	35	40	45	50	55	60	65	70	75	80	85	90	95	100
140	125																				
135	120	128																			
130	117	122	131																		
125	111	116	123	131	141																
120	107	111	116	123	130	139	148														
115	103	107	111	115	120	127	135	143	151												
110	99	102	105	108	112	117	123	130	137	143	150										
105	95	97	100	102	105	109	113	118	123	129	135	142	149								
100	91	93	95	97	99	101	104	107	110	115	120	126	132	138	144						
95	87	88	90	91	93	94	96	98	101	104	107	110	114	119	124	130	136				
90	83	84	85	86	87	88	90	91	93	95	96	98	100	102	106	109	113	117	122		
85	78	79	80	81	82	83	84	85	86	87	88	89	90	91	93	95	97	99	102	105	108
80	73	74	75	76	77	77	78	79	79	80	81	81	82	83	85	86	86	87	88	89	91
75	69	69	70	71	72	72	73	73	74	74	75	75	76	76	77	77	78	78	79	79	80
70	64	64	65	65	66	66	67	67	68	68	69	69	70	70	70	70	71	71	71	71	72

FIGURE 4.9 Air temperature (°F) and relative humidity are combined to determine an apparent temperature or heat index (HI). An air temperature of 95°F with a relative humidity of 55 percent produces an apparent temperature (HI) of 110°F.

of metal that has either a handle or chain at one end (see Fig. 4.10). The thermometers are exactly alike except that one has a piece of cloth (wick) covering the bulb. The wick-covered thermometer—called the *wet bulb*—is dipped in clean water, whereas the other thermometer is kept dry. Both thermometers are ventilated for a few minutes, either by whirling the instrument *(sling psychrometer)*, or by drawing air past it with an electric fan *(aspirated psychrometer)*. Water evaporates from the wick and that thermometer cools. The drier the air, the greater the amount of evaporation and cooling. After a few minutes, the wick-covered thermometer will cool to the lowest value possible. Recall from an earlier section that this is the *wet-bulb temperature*—the lowest temperature that can be attained by evaporating water into the air.

The dry thermometer (commonly called the *dry bulb*) gives the current air temperature, or *dry-bulb tem-*

TABLE 4.1 The Heat Index (HI)

CATEGORY	APPARENT TEMPERATURE (°F)	HEAT SYNDROME
I	130° or higher	Heatstroke or sunstroke *imminent*
II	105°–130°	Sunstroke, heat cramps, or heat exhaustion *likely*, heatstroke *possible* with prolonged exposure and physical activity
III	90°–105°	Sunstroke, heat cramps, and heat exhaustion *possible* with prolonged exposure and physical activity
IV	80°–90°	Fatigue *possible* with prolonged exposure and physical activity

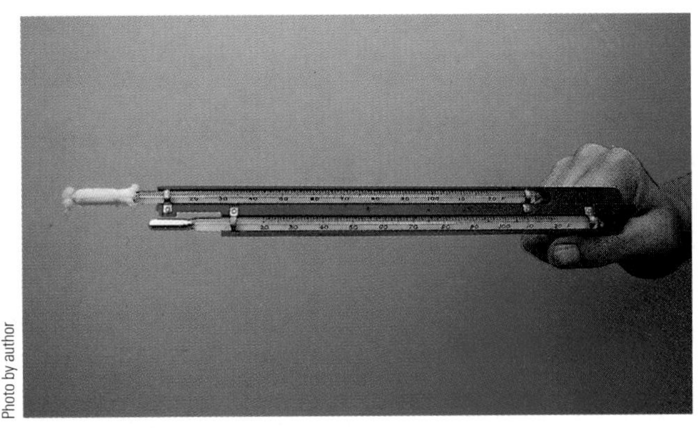

Photo by author

FIGURE 4.10 The sling psychrometer.

FOCUS ON A SPECIAL TOPIC

Humid Air and Dry Air Do Not Weigh the Same

Does a volume of hot, humid air really weigh more than a similar size volume of hot, dry air? The answer is no! At the same temperature and at the same level in the atmosphere, hot, humid air is lighter (less dense) than hot, dry air. The reason for this fact is that a molecule of water vapor (H_2O) weighs appreciably less than a molecule of either nitrogen (N_2) or oxygen (O_2). (Keep in mind that we are referring strictly to water vapor—a gas—and not suspended liquid droplets.)

Consequently, in a given volume of air, as lighter water vapor molecules replace either nitrogen or oxygen molecules one for one, the number of molecules in the volume does not change, but the total weight of the air becomes slightly less. Since air density is the mass of air in a volume, the more humid air must be lighter than the drier air. Hence, ***hot, humid air at the surface is lighter (less dense) than hot, dry air.***

This fact can have an important influence in the weather. The lighter the air becomes, the more likely it is to rise. All other factors being equal, hot, humid (less-dense) air will rise more readily than hot, dry (more-dense) air. It is, of course, the water vapor in the rising air that changes into liquid cloud droplets and ice crystals, which, in turn, grow large enough to fall to the earth as precipitation.

Of lesser importance to weather but of greater importance to sports is the fact that a baseball will "carry" farther in less-dense air. Consequently, without the influence of wind, a ball will travel slightly farther on a hot, humid day than it will on a hot, dry day. So when the sports announcer proclaims "the air today is heavy because of the high humidity" remember that this statement is not true and, in fact, a 404-foot home run on this humid day might simply be a 400-foot out on a very dry day.

perature. The temperature difference between the dry bulb and the wet bulb is known as the *wet-bulb depression.* A large depression indicates that a great deal of water can evaporate into the air and that the relative humidity is low. A small depression indicates that little evaporation of water vapor is possible, so the air is close to saturation and the relative humidity is high. If there is no depression, the dry bulb, the wet bulb, and the dew point are the same; the air is saturated and the relative humidity is 100 percent. (Tables used to compute relative humidity and dew point are given in Appendix D.)

Instruments that measure humidity are commonly called **hygrometers.** One type—called the *hair hygrometer*—uses human (or horse) hair to measure relative humidity. It is constructed on the principle that, as the relative humidity increases, the length of hair increases and, as the relative humidity decreases, so does the hair length. A number of strands of hair (with oils removed) are attached to a system of levers. A small change in hair length is magnified by a linkage system and transmitted to a dial (Fig. 4.11) calibrated to show relative humidity, which can then be read directly or

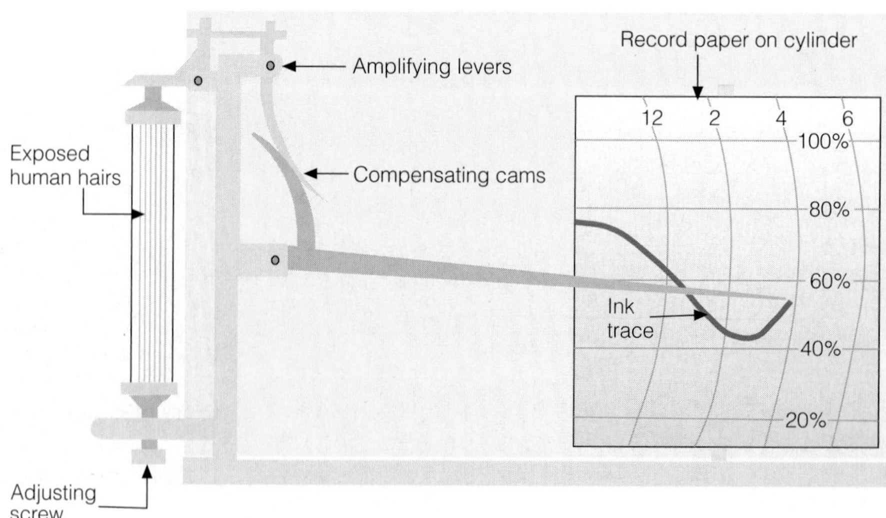

FIGURE 4.11 The hair hygrometer measures relative humidity by amplifying and measuring changes in the length of human (or horse) hair.

recorded on a chart. (Often, the chart is attached to a clock-driven rotating drum that gives a continuous record of relative humidity.) Because the hair hygrometer is not as accurate as the psychrometer (especially at very high and very low relative humidities), it requires frequent calibration, principally in areas that experience large daily variations in relative humidity.

The *electrical hygrometer* is another instrument that measures humidity. It consists of a flat plate coated with a film of carbon. An electric current is sent across the plate. As water vapor is absorbed, the electrical resistance of the carbon coating changes. These changes are translated into relative humidity. This instrument is commonly used in the radiosonde, which gathers atmospheric data at various levels above the earth. Still another instrument—the *infrared hygrometer*—measures atmospheric humidity by measuring the amount of infrared energy absorbed by water vapor in a sample of air. The *dew-point hygrometer* measures the dew-point temperature by cooling the surface of a mirror until condensation (dew) forms. This sensor is the type that measures dew-point temperature in the hundreds of fully automated weather stations—Automated Surface Observing System (ASOS)—that exist throughout the United States. Finally, the *dew cell* determines the amount of water vapor in the air by measuring the air's actual vapor pressure.

Over the last several sections we saw that, as the air cools, the air temperature approaches the dew-point temperature and the relative humidity increases. When the air temperature reaches the dew point, the air is saturated with water vapor and the relative humidity is 100 percent. Continued cooling, however, causes some of the water vapor to condense into liquid water. The cooling may take place in a thick portion of the atmosphere, or it may occur near the earth's surface. In the next section, we will examine condensation that forms near the ground.

DEW AND FROST

On clear, calm nights, objects near the earth's surface cool rapidly by emitting infrared radiation. The ground and objects on it often become much colder than the surrounding air. Air that comes in contact with these cold surfaces cools by conduction. Eventually, the air cools to the dew point. As surfaces (such as twigs, leaves, and blades of grass) cool below this temperature, water vapor begins to condense upon them, forming tiny visible specks of water called **dew** (see Fig. 4.12). If the air temperature should drop to freezing or below, the dew will freeze, becoming tiny beads of ice called *frozen dew*. Because the coolest air is usually at ground level, dew is more likely to form on blades of grass than on objects several feet above the surface. This thin coating of dew not only dampens bare feet, but it also is a valuable source of moisture for many plants during periods of low rainfall.

Dew is more likely to form on nights that are clear and calm than on nights that are cloudy and windy. Clear nights allow objects near the ground to cool rapidly, and calm winds mean that the coldest air will be located at ground level. These atmospheric conditions are usually associated with large fair-weather, high-pressure systems. On the other hand, the cloudy, windy weather that inhibits rapid cooling near the ground and the forming of dew often signifies the approach of a rain-producing storm system. These observations inspired the following folk-rhyme:

When the dew is on the grass,
rain will never come to pass.
When grass is dry at morning light,
look for rain before the night!

DID YOU KNOW?

In hot, muggy weather, people with naturally curly hair often experience the "frizzes" as their hair increases in length. People with long, straight hair often experience a bad hair day as their hair goes "limp" in the hot, humid weather.

Photo by author

FIGURE 4.12 Dew forms on clear nights when objects on the surface cool to a temperature below the dew point. If these beads of water should freeze, they would become frozen dew.

DID YOU KNOW?

Dew can help sensitive vegetation survive a drought because, during the course of a year, dew can yield a blanket of water between one-half inch and two inches thick.

Visible white frost forms on cold, clear, calm mornings when the dew-point temperature is at or below freezing. When the air temperature cools to the dew point (now called the *frost point*) and further cooling occurs, water vapor can change directly to ice without becoming a liquid first—a process called *deposition.** The delicate, white crystals of ice that form in this manner are called *hoarfrost, white frost,* or simply **frost.** Frost has a treelike branching pattern that easily distinguishes it from the nearly spherical beads of frozen dew (see Fig. 4.13).

In very dry weather, the air may become quite cold and drop below freezing without ever reaching the frost point, and no visible frost forms. *Freeze* and *black frost* are words denoting this situation—a situation that can severely damage certain crops (see Chapter 3, pp. 59–62).

As a deep layer of air cools during the night, its relative humidity increases. When the air's relative humidity reaches about 75 percent, some of its water vapor may begin to condense onto tiny floating particles of sea salt and other substances—*condensation nuclei*—that are *hygroscopic* ("water seeking") in that they allow water vapor to condense onto them when the relative humidity is considerably below 100 percent. As water collects onto these nuclei, their size increases and the particles, although still small, are now large enough to scatter visible light in all directions, becoming **haze**—a layer of particles dispersed through a portion of the atmosphere (see Fig. 4.14).

As the relative humidity gradually approaches 100 percent, the haze particles grow larger, and condensation begins on the less-active nuclei. Now a large fraction of the available nuclei have water condensing onto them, causing the droplets to grow even bigger, until eventually they become visible to the naked eye. The increasing size and concentration of droplets further restrict visibility. When the visibility lowers to less than 1 km (or 0.62 mi), and the air is wet with millions of tiny floating water droplets, the haze becomes a cloud resting near the ground, which we call **fog.***

*When the ice changes back into vapor without melting, the process is called *sublimation.*

FIGURE 4.13 These are the delicate ice-crystal patterns that frost exhibits on a window during a cold winter morning.

FOG

Fog, like any cloud, usually forms in one of two ways: (1) by cooling—air is cooled below its saturation point (dew point); and (2) by evaporation and mixing—water vapor is added to the air by evaporation, and the moist air mixes with relatively dry air. Once fog forms it is maintained by new fog droplets, which constantly form on available nuclei. In other words, the air must maintain its degree of saturation either by continual cooling or by evaporation and mixing of vapor into the air.

Fog produced by the earth's radiational cooling is called **radiation fog,** or *ground fog.* It forms best on clear nights when a shallow layer of moist air near the ground is overlain by drier air. Under these conditions, the ground cools rapidly since the shallow, moist layer does not absorb much of the earth's outgoing infrared radiation. As the ground cools, so does the air directly above it, and a surface inversion forms, with colder air at the surface and warmer air above. The moist, lower layer (chilled rapidly by the cold ground) quickly becomes saturated, and fog forms. The longer the night, the longer the time of cooling and the greater the likelihood of fog. Hence, radiation fogs are most common over land in late fall and winter.

Another factor promoting the formation of radiation fog is a light breeze of less than five knots. Although radiation fog may form in calm air, slight air movement

*This is the official international definition of *fog.* The United States Weather Service reports fog as a restriction to visibility when fog restricts the visibility to 6 miles or less and the spread between the air temperature and dew point is 5°F or less. When the visibility is less than one-quarter of a mile, the fog is considered dense.

Photo by author

FIGURE 4.14 The high relative humidity of the cold air above the lake is causing a layer of haze to form on a still winter morning.

brings more of the moist air in direct contact with the cold ground and the transfer of heat occurs more rapidly. A strong breeze tends to prevent a radiation fog from forming by mixing the air near the surface with the drier air above. The ingredients of clear skies and light winds are associated with large high-pressure areas (anticyclones). Consequently, during the winter, when a high becomes stagnant over an area, radiation fog may form on consecutive days.

Because cold, heavy air drains downhill and collects in valley bottoms, we normally see radiation fog forming in low-lying areas. Hence, radiation fog is frequently called *valley fog.* The cold air and high moisture content in river valleys make them susceptible to radiation fog. Since radiation fog normally forms in lowlands, hills may be clear all day long, while adjacent valleys are fogged in (see Fig. 4.15).

Radiation fogs are normally deepest around sunrise. Usually, however, a shallow fog layer will dissipate or *burn off* by afternoon. Of course, the fog does not "burn"; rather, sunlight penetrates the fog and warms the ground, causing the temperature of the air in contact with the ground to increase. The warm air rises and mixes with the foggy air above, which increases the temperature of the foggy air. In the slightly warmer air, some of the fog droplets evaporate, allowing more sunlight to reach the ground, which produces more heating, and soon the fog completely disappears. If the fog layer is quite thick, it may not completely dissipate and a layer of low clouds (called *stratus*) covers the region. This type of fog is sometimes called *high fog.*

When warm, moist air moves over a sufficiently colder surface, the moist air may cool to its saturation point, forming advection fog. A good example of **advection fog** may be observed along the Pacific Coast during summer. The main reason fog forms in this region is that the surface water near the coast is much colder than the surface water farther offshore. Warm, moist air from the Pacific Ocean is carried (advected) by westerly winds over the cold coastal waters. Chilled from below, the air temperature drops to the dew point, and fog is produced. Advection fog, unlike radiation fog, always involves the movement of air, so when there is a stiff summer breeze in San Francisco, it's common to watch advection fog roll in past the Golden Gate Bridge (see Fig. 4.16).

Photo by author

FIGURE 4.15 Radiation fog nestled in a valley.

© Russel D. Curtis/Photo Researchers

FIGURE 4.16 Advection fog rolling in past the Golden Gate Bridge in San Francisco. As fog moves inland, the air warms and the fog lifts above the surface. Eventually, the air becomes warm enough to totally evaporate the fog.

As summer winds carry the fog inland over the warmer land, the fog near the ground dissipates, leaving a sheet of low-lying gray clouds that block out the sun. Further inland, the air is sufficiently warm, so that even these low clouds evaporate and disappear.

Because they provide moisture to the coastal redwood trees, advection fogs are important to the scenic beauty of the Pacific Coast. Much of the fog moisture collected by the needles and branches of the redwoods drips to the ground *(fog drip),* where it is utilized by the tree's shallow root system. Without the summer fog, coastal redwood trees would have trouble surviving the dry California summers. Hence, we find them nestled in the fog belt along the coast.

Advection fogs also prevail where two ocean currents with different temperatures flow next to one another. Such is the case in the Atlantic Ocean off the coast of Newfoundland, where the cold southward-flowing Labrador Current lies almost parallel to the warm northward-flowing Gulf Stream. Warm southerly air moving over the cold water produces fog in that region—so frequently that fog occurs on about two out of three days during summer.

Advection fog also forms over land. In winter, warm, moist air from the Gulf of Mexico moves northward over progressively colder and slightly elevated land. As the air cools to its saturation point, a fog forms in the southern or central United States. Because the cold ground is often the result of radiation cooling, fog that forms in this manner is sometimes called *advection-radiation fog.* During this same time of year, air moving across the warm Gulf Stream encounters the colder land of the British Isles and produces the thick fogs of England. Similarly, fog forms as marine air moves over an ice or snow surface. In extremely cold arctic air, ice crystals form instead of water droplets, producing an *ice fog.*

Fog that forms as moist air flows up along an elevated plain, hill, or mountain is called **upslope fog.** Typically, upslope fog forms during the winter and spring on the eastern side of the Rockies, where the eastward-sloping plains are nearly a kilometer higher than the land further east. Occasionally, cold air moves from the lower eastern plains westward. The air gradually rises, expands, becomes cooler, and—if sufficiently moist—a fog forms. Upslope fogs that form over an extensive area may last for many days.

So far, we have seen how the cooling of air produces fog. But remember that fog may also form by the mixing of two unsaturated masses of air. Fog that forms in this manner is usually called *evaporation fog* because evaporation initially enriches the air with water vapor.

DID YOU KNOW?

Ever hear of caribou fog? No, it's not the fog that forms in Caribou, Maine, but the fog that forms around herds of caribou. In very cold weather, just a little water vapor added to the air will saturate it. Consequently, the perspiration and breath from large herds of caribou add enough water vapor to the air to create a blanket of fog that hovers around the herd.

Probably, a more appropriate name for the fog is **evaporation (mixing) fog.** On a cold day, you may have unknowingly produced evaporation (mixing) fog. When moist air from your mouth or nose meets the cold air and mixes with it, the air becomes saturated, and a tiny cloud forms with each exhaled breath.

A common form of evaporation-mixing fog is the *steam fog,* which forms when cold air moves over warm water. This type of fog forms above a heated outside swimming pool in winter. As long as the water is warmer than the unsaturated air above, water will evaporate from the pool into the air. The increase in water vapor raises the dew point, and, if mixing is sufficient, the air above becomes saturated. The colder air directly above the water is heated from below and becomes warmer than the air directly above it. This warmer air rises and, from a distance, the rising condensing vapor appears as "steam."

It is common to see steam fog forming over lakes on autumn mornings, as cold air settles over water still warm from the long summer. On occasion, over the Great Lakes, columns of condensed vapor rise from the fog layer, forming whirling *steam devils,* which appear similar to the dust devils on land. If you travel to Yellowstone National Park, you will see steam fog forming above thermal ponds all year long (see Fig. 4.17). Over the ocean in polar regions, steam fog is referred to as *arctic sea smoke.*

Steam fog may form above a wet surface on a sunny day. This type of fog is commonly observed after a rain shower as sunlight shines on a wet road, heats the asphalt, and quickly evaporates the water. This added vapor mixes with the air above, producing steam fog. Fog that forms in this manner is short-lived and disappears as the road surface dries.

A warm rain falling through a layer of cold, moist air can produce fog. As a warm raindrop falls into a cold layer of air, some of the water evaporates from the raindrop into the air. This process may saturate the air, and if mixing occurs, fog forms. Fog of this type is often associated with warm air riding up and over a mass of colder

FIGURE 4.17 Even in summer, warm air rising above thermal pools in Yellowstone National Park condenses into a type of steam fog.

surface air. The fog usually develops in the shallow layer of cold air just ahead of an approaching warm front or behind a cold front, which is why this type of evaporation fog is also known as *precipitation fog*, or *frontal fog*.

FOGGY WEATHER

The foggiest regions in the United States are shown in Fig. 4.18. Notice that heavy fog is more prevalent in coastal margins (especially those regions lapped by cold ocean currents) than in the center of the continent. In fact, the foggiest spot near sea level in the United States is Cape Disappointment, Washington. Located at the mouth of the Columbia River, it averages 2556 hours of heavy fog each year. Anyone who travels to this spot hoping to enjoy the sun during August and September would find its name appropriate indeed.

Extremely limited visibility exists while driving at night in heavy fog with the high-beam lights on. The light scattered back to the driver's eyes from the fog droplets makes it difficult to see very far down the road. Along a gently sloping highway, the elevated sections may have excellent visibility, while in lower regions—only a few miles away—fog may cause poor visibility. Driving from the clear area into the fog on a major freeway can be extremely dangerous. In fact, every winter many people are involved in fog-related auto accidents. These usually oc-

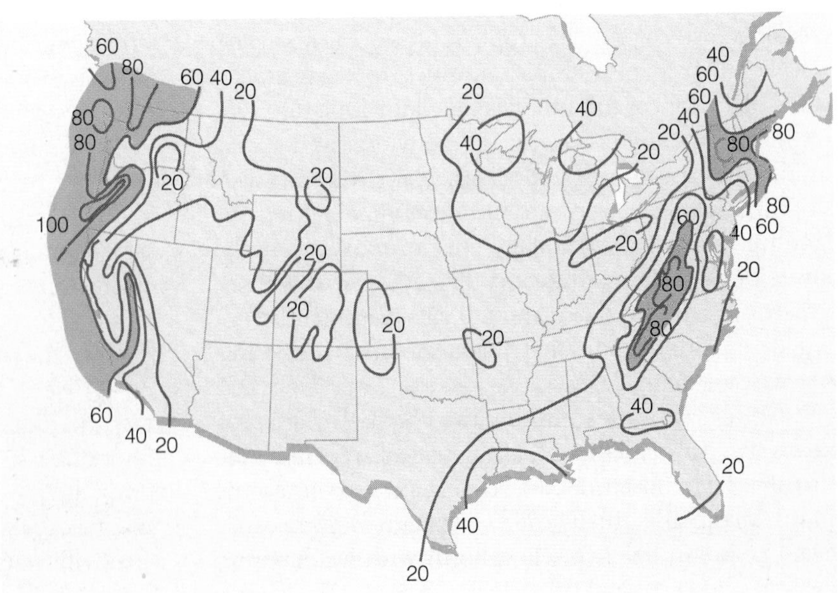

FIGURE 4.18 Average annual number of days with dense fog throughout the United States.

FOCUS ON AN ENVIRONMENTAL ISSUE

Fog Dispersal

In any airport fog-clearing operation the problem is to improve visibility so that aircraft can take off and land. Experts have tried various methods, which can be grouped into four categories: (1) increase the size of the fog droplets, so that they become heavy and settle to the ground as a light drizzle; (2) seed cold fog with dry ice (solid carbon dioxide), so that fog droplets are converted into ice crystals; (3) heat the air, so that the fog evaporates; and (4) mix the cooler saturated air near the surface with the warmer unsaturated air above.

To date, only one of these methods has been reasonably successful—the seeding of cold fog. *Cold fog* forms when the air temperature is below freezing, and most of the fog droplets remain as liquid water. (Liquid fog in below-freezing air is also called *supercooled fog.*) The fog can be cleared by injecting several hundred pounds of dry ice into it. As the tiny pieces of cold (–78°C) dry ice descend, they freeze some of the supercooled fog droplets in their path, producing ice crystals. As we will see in Chapter 5, these crystals then grow larger at the expense of the remaining liquid fog droplets. Hence, the fog droplets evaporate and the larger ice crystals fall to the ground, which leaves a "hole" in the fog for aircraft takeoffs and landings.

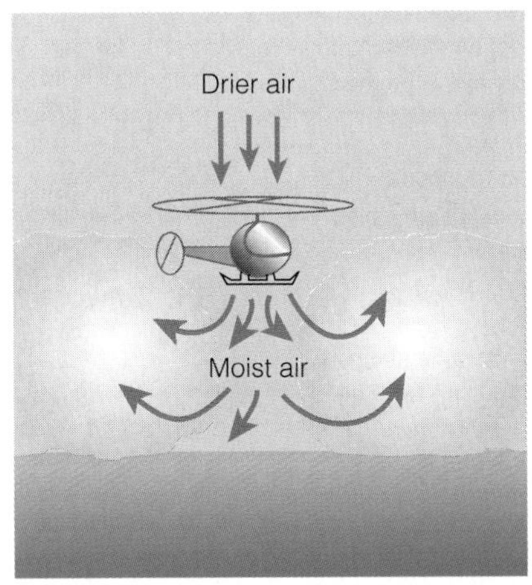

FIGURE 1 Helicopters hovering above an area of shallow fog can produce a clear area by mixing the drier air into the foggy air below.

Unfortunately, most of the fogs that close airports in the United States are *warm fogs* that form when the air temperature is above freezing. Since dry ice seeding does not work in warm fog, other techniques must be tried.

One method involves injecting hygroscopic particles into the fog. Large salt particles and other chemicals absorb the tiny fog droplets and form into larger drops. More large drops and fewer small drops improve the visibility; plus, the larger drops are more likely to fall as a light drizzle. Since the chemicals are expensive and the fog clears for only a short time, this method of fog dispersal is not economically feasible.

Another technique for fog dispersal is to warm the air enough so that the fog droplets evaporate and visibility improves. Tested at Los Angeles International Airport in the early 1950s, this technique was abandoned because it was smoky, expensive, and not very effective. In fact, the burning of hundreds of dollars worth of fuel only cleared the runway for a short time. And the smoke particles, released during the burning of the fuel, provided abundant nuclei for the fog to recondense upon.

A final method of warm fog dispersal uses helicopters to mix the air. The chopper flies across the fog layer, and the turbulent downwash created by the rotor blades brings drier air above the fog into contact with the moist fog layer (see Fig. 1). The aim, of course, is to evaporate the fog. Experiments show that this method works well, as long as the fog is a shallow radiation fog with a relatively low liquid water content. But many fogs are thick, have a high liquid water content, and form by other means. An inexpensive and practical method of dispersing warm fog has yet to be discovered.

cur when a car enters the fog and, because of the reduced visibility, the driver puts on the brakes to slow down. The car behind then slams into the slowed vehicle, causing a chain-reaction accident with many cars involved.

Airports suspend flight operations when fog causes visibility to drop below a prescribed minimum. The resulting delays and cancellations become costly to the airline industry and irritate passengers. With fog-caused problems such as these, it is no wonder that scientists have been seeking ways to disperse, or at least "thin," fog. (For more information on fog-thinning techniques, read the Focus section entitled "Fog Dispersal," above.)

Up to this point, we have looked at the different forms of condensation that occur on or near the earth's surface. In particular, we learned that fog is simply many millions of tiny liquid droplets (or ice crystals) that form near the ground. In the following sections, we will see how these same particles, forming well above the ground, are classified and identified as clouds.

DID YOU KNOW?

On the morning of February 14, 2000, dense fog, limiting visibility to less than 30 m (100 ft), caused a 63-vehicle pileup along Interstate 15 in San Bernardino County, California. This accident left 17 people injured, 4 seriously, and a freeway strewn with twisted cars and big-rigs.

BRIEF REVIEW

Before we go on to the section on clouds, here is a brief review of some of the important concepts and facts we have covered so far:

- Dew, frost, and frozen dew generally form on clear nights when the temperature of objects on the surface cools below the air's dew-point temperature.
- Visible white frost forms in saturated air when the air temperature is at or below freezing. Under these conditions, water vapor can change directly to ice, in a process called *deposition.*
- Condensation nuclei act as surfaces on which water vapor condenses. Those nuclei that have an affinity for water vapor are called *hygroscopic.*
- Fog is a cloud resting on the ground.
- Radiation fog, advection fog, and upslope fog all form as the air cools. Evaporation (mixing) fog, such as steam fog and frontal fog, form as water evaporates and mixes with drier air.

CLOUDS

Clouds are aesthetically appealing and add excitement to the atmosphere. Without them, there would be no rain or snow, thunder or lightning, rainbows or halos. How monotonous if one had only a clear blue sky to look at. A *cloud* is a visible aggregate of tiny water droplets or ice crystals suspended in the air. Some are found only at high elevations, whereas others nearly touch the ground. Clouds can be thick or thin, big or little—they exist in a seemingly endless variety of forms. To impose order on this variety, we divide clouds into ten basic types. With a careful and practiced eye, you can become reasonably proficient in correctly identifying them.

Classification of Clouds Although ancient astronomers named the major stellar constellations about 2000 years ago, clouds were not formally identified and classified until the early nineteenth century. The French naturalist Lamarck (1744–1829) proposed the first system for classifying clouds in 1802; however, his work did not receive wide acclaim. One year later, Luke Howard, an English naturalist, developed a cloud classification system that found general acceptance. In essence, Howard's innovative system employed Latin words to describe clouds as they appear to a ground observer. He named a sheetlike cloud *stratus* (Latin for "layer"); a puffy cloud *cumulus* ("heap"); a wispy cloud *cirrus* ("curl of hair"); and a rain cloud *nimbus* ("violent rain"). In Howard's system, these were the four basic cloud forms. Other clouds could be described by combining the basic types. For example, nimbostratus is a rain cloud that shows layering, whereas cumulonimbus is a rain cloud having pronounced vertical development.

In 1887, Abercromby and Hildebrandsson expanded Howard's original system and published a classification system that, with only slight modification, is still used today. Ten principal cloud forms are divided into four primary cloud groups. Each group is identified by the height of the cloud's base above the surface: high clouds, middle clouds, and low clouds. The fourth group contains clouds showing more vertical than horizontal development. Within each group, cloud types are identified by their appearance. Table 4.2 lists these four groups and their cloud types.

The approximate base height of each cloud group is given in Table 4.3. Note that the altitude separating the high and middle cloud groups overlaps and varies with latitude. Large temperature changes cause most of this latitudinal variation. For example, high cirriform clouds are composed almost entirely of ice crystals. In subtropical regions, air temperatures low enough to freeze all liquid water usually occur only above about 20,000 feet. In polar regions, however, these same temperatures may be found at altitudes as low as 10,000 feet. Hence, while you may observe cirrus clouds at 12,000 feet over northern Alaska, you will not see them at that elevation above southern Florida.

Clouds cannot be accurately identified strictly on the basis of elevation. Other visual clues are necessary. Some of these are explained in the following section.

TABLE 4.2
The Four Major Cloud Groups and Their Types

1. High clouds	3. Low clouds
Cirrus (Ci)	Stratus (St)
Cirrostratus (Cs)	Stratocumulus (Sc)
Cirrocumulus (Cc)	Nimbostratus (Ns)
2. Middle clouds	4. Clouds with vertical development
Altostratus (As)	Cumulus (Cu)
Altocumulus (Ac)	Cumulonimbus (Cb)

Cloud Identification

High Clouds High clouds in middle and low latitudes generally form above 20,000 ft (or 6000 m). Because the air at these elevations is quite cold and "dry," high clouds are composed almost exclusively of ice crystals and are also rather thin.* High clouds usually appear white, ex-

*Small quantities of liquid water in cirrus clouds at temperatures as low as −36°C (−33°F) were discovered during research conducted above Boulder, Colorado.

TABLE 4.3 Approximate Height of Cloud Bases above the Surface for Various Locations

CLOUD GROUP	TROPICAL REGION	MIDDLE-LATITUDE REGION	POLAR REGION
High Ci, Cs, Cc	20,000 to 60,000 ft (6000 to 18,000 m)	16,000 to 43,000 ft (5000 to 13,000 m)	10,000 to 26,000 ft (3000 to 8000 m)
Middle As, Ac	6500 to 26,000 ft (2000 to 8000 m)	6500 to 23,000 ft (2000 to 7000 m)	6500 to 13,000 ft (2000 to 4000 m)
Low St, Sc, Ns	surface to 6500 ft (0 to 2000 m)	surface to 6500 ft (0 to 2000 m)	surface to 6500 ft (0 to 2000 m)

cept near sunrise and sunset, when the unscattered (red, orange, and yellow) components of sunlight are reflected from the underside of the clouds.

The most common high clouds are the **cirrus,** which are thin, wispy clouds blown by high winds into long streamers called *mares' tails.* Notice in Fig. 4.19 that they can look like a white, feathery patch with a faint wisp of a tail at one end. Cirrus clouds usually move across the sky from west to east, indicating the prevailing winds at their elevation.

Cirrocumulus clouds, seen less frequently than cirrus, appear as small, rounded, white puffs that may occur individually, or in long rows (see Fig. 4.20). When in rows, the cirrocumulus cloud has a rippling appearance that distinguishes it from the silky look of the cirrus and the sheetlike cirrostratus. Cirrocumulus seldom cover more than a small portion of the sky. The dappled cloud elements that reflect the red or yellow light of a setting sun make this one of the most beautiful of all clouds. The small ripples in the cirrocumulus strongly resemble the scales of a fish; hence, the expression "*mackerel sky*" commonly describes a sky full of cirrocumulus clouds.

The thin, sheetlike, high clouds that often cover the entire sky are **cirrostratus** (Fig. 4.21), which are so thin that the sun and moon can be clearly seen through them. The ice crystals in these clouds bend the light passing through them and will often produce a halo. In fact, the veil of cirrostratus may be so thin that a halo is the only clue to its presence. Thick cirrostratus clouds give the sky a glary white appearance and frequently form ahead of an advancing storm; hence, they can be used to predict rain or snow within twelve to twenty-four hours, especially if they are followed by middle-type clouds.

Middle Clouds The middle clouds have bases between about 6500 and 23,000 ft (2000 and 7000 m) in the middle latitudes. These clouds are composed of water droplets and—when the temperature becomes low enough— some ice crystals.

Altocumulus clouds are middle clouds that appear as gray, puffy masses, sometimes rolled out in parallel waves or bands (see Fig. 4.22). Usually, one part of the cloud is darker than another, which helps to separate it from the higher cirrocumulus. Also, the individual puffs of the altocumulus appear larger than those of the cirrocumulus. A layer of altocumulus may sometimes be

Photo by author

FIGURE 4.19 Cirrus clouds.

FIGURE 4.20 Cirrocumulus clouds.

Photo by author

Photo by author

FIGURE 4.21 Cirrostratus clouds with a faint halo.

confused with altostratus; in case of doubt, clouds are called altocumulus if there are rounded masses or rolls present. Altocumulus clouds that look like "little castles" *(castellanus)* in the sky indicate the presence of rising air at cloud level. The appearance of these clouds on a warm, humid summer morning often portends thunderstorms by late afternoon.

The **altostratus** is a gray or blue-gray cloud that often covers the entire sky over an area that extends over many hundreds of square kilometers. In the thinner section of the cloud, the sun (or moon) may be *dimly visible* as a round disk, which is sometimes referred to as a "watery sun" (see Fig. 4.23). Thick cirrostratus clouds are occasionally confused with thin altostratus clouds.

The gray color, height, and dimness of the sun are good clues to identifying an altostratus. The fact that halos only occur with cirriform clouds also helps one distinguish them. Another way to separate the two is to look at the ground for shadows. If there are none, it is a good bet that the cloud is altostratus because cirrostratus are usually transparent enough to produce them. Altostratus clouds often form ahead of storms having widespread and relatively continuous precipitation. If precipitation falls from an altostratus, its base usually lowers. If the precipitation reaches the ground, the cloud is then classified as *nimbostratus.*

Low Clouds Low clouds, with their bases lying below 6500 ft (or 2000 m) are almost always composed of water droplets; however, in cold weather, they may contain ice particles and snow.

The **nimbostratus** is a dark gray, "wet"-looking cloud layer associated with more or less continuously falling rain or snow (see Fig. 4.24). The intensity of this precipitation is usually light or moderate—it is never of the heavy, showery variety, unless well-developed cumulus clouds are embedded within the nimbostratus cloud. The base of the nimbostratus cloud is normally impossible to identify clearly and is easily confused with the altostratus.

Photo by author

FIGURE 4.22 Altocumulus clouds.

Photo by author

FIGURE 4.23 Altostratus clouds. The appearance of a dimly visible "watery sun" through a deck of gray clouds is usually a good indication that the clouds are altostratus.

FIGURE 4.24 The nimbostratus is the sheetlike cloud from which light rain is falling. The ragged-appearing cloud beneath the nimbostratus is stratus fractus, or scud.

Photo by author

Thin nimbostratus is usually darker gray than thick altostratus, and you cannot see the sun or moon through a layer of nimbostratus. Visibility below a nimbostratus cloud deck is usually quite poor because rain will evaporate and mix with the air in this region. If this air becomes saturated, a lower layer of clouds or fog may form beneath the original cloud base. Since these lower clouds drift rapidly with the wind, they form irregular shreds with a ragged appearance called *stratus fractus,* or *scud.*

A low, lumpy cloud layer is the **stratocumulus.** It appears in rows, in patches, or as rounded masses with blue sky visible between the individual cloud elements (see Fig. 4.25). Often they appear near sunset as the spreading remains of a much larger cumulus cloud. The color of stratocumulus ranges from light to dark gray. It differs from altocumulus in that it has a lower base and larger individual cloud elements. (Compare Fig. 4.22 with Fig. 4.25.) To distinguish between the two, hold your hand at arm's length and point toward the cloud. Altocumulus cloud elements will generally be about the size of your thumbnail; stratocumulus cloud elements will usually be about the size of your fist. Rain or snow rarely fall from stratocumulus.

Stratus is a uniform grayish cloud that often covers the entire sky. It resembles a fog that does not reach the ground (see Fig. 4.26). Actually, when a thick fog "lifts," the resulting cloud is a deck of low stratus. Normally, no precipitation falls from the stratus, but sometimes it is

FIGURE 4.25 Stratocumulus clouds. Notice that the rounded masses are larger than those of the altocumulus.

Photo by author

Photo by author

FIGURE 4.26 A layer of low-lying stratus clouds.

accompanied by a light mist or drizzle. This cloud commonly occurs over Pacific and Atlantic coastal waters in summer. A thick layer of stratus might be confused with nimbostratus, but the distinction between them can be made by observing the base of the cloud. Often, stratus has a more uniform base than does nimbostratus. Also, a deck of stratus may be confused with a layer of altostratus. However, if you remember that stratus clouds are lower and darker gray, the distinction can be made.

Clouds with Vertical Development Familiar to almost everyone, the puffy **cumulus** cloud takes on a variety of shapes, but most often it looks like a piece of floating cotton with sharp outlines and a flat base (see Fig. 4.27).

Photo by author

FIGURE 4.27 Cumulus clouds. Small cumulus clouds such as these are sometimes called *fair weather cumulus,* or *cumulus humilis.*

Photo by author

FIGURE 4.28 Cumulus congestus. This line of cumulus congestus clouds is building along Maryland's eastern shore.

The base appears white to light gray, and, on a humid day, may be only a few thousand feet above the ground and a half a mile or so wide. The top of the cloud—often in the form of rounded towers—denotes the limit of rising air and is usually not very high. These clouds can be distinguished from stratocumulus by the fact that cumulus clouds are detached (usually a great deal of blue sky between each cloud) whereas stratocumulus usually occur in groups or patches. Also, the cumulus has a dome- or tower-shaped top as opposed to the generally flat tops of the stratocumulus. Cumulus clouds that show only slight vertical growth *(cumulus humilis)* are associated with fair weather; therefore, we call these clouds "fair weather cumulus." If the cumulus clouds are small and appear as broken fragments of a cloud with ragged edges, they are called *cumulus fractus.*

Harmless-looking cumulus often develop on warm summer mornings and, by afternoon, become much larger and more vertically developed. When the growing cumulus resembles a head of cauliflower, it becomes a *cumulus congestus,* or *towering cumulus.* Most often, it is a single large cloud, but, occasionally, several grow into each other, forming a line of towering clouds, as shown in Fig. 4.28. Precipitation that falls from a cumulus congestus is always showery.

If a cumulus congestus continues to grow vertically, it develops into a giant **cumulonimbus**—a thunderstorm cloud (see Fig. 4.29). While its dark base may be no more than 2000 ft above the earth's surface, its top may extend upward to the tropopause, over 35,000 ft higher. A cumulonimbus can occur as an isolated cloud or as part of a line or "wall" of clouds.

Tremendous amounts of energy are released by the condensation of water vapor within a cumulonimbus and result in the development of violent up- and downdrafts, which may exceed fifty knots. The lower (warmer) part of the cloud is usually composed of only water droplets. Higher up in the cloud, water droplets and ice crystals both abound, while, toward the cold top, there are only ice crystals. Swift winds at these higher altitudes can reshape the top of the cloud into a huge flattened *anvil.** These great thunderheads may contain all forms of precipitation—large raindrops,

*An anvil is a heavy block of iron or steel with a smooth, flat top on which metals are shaped by hammering.

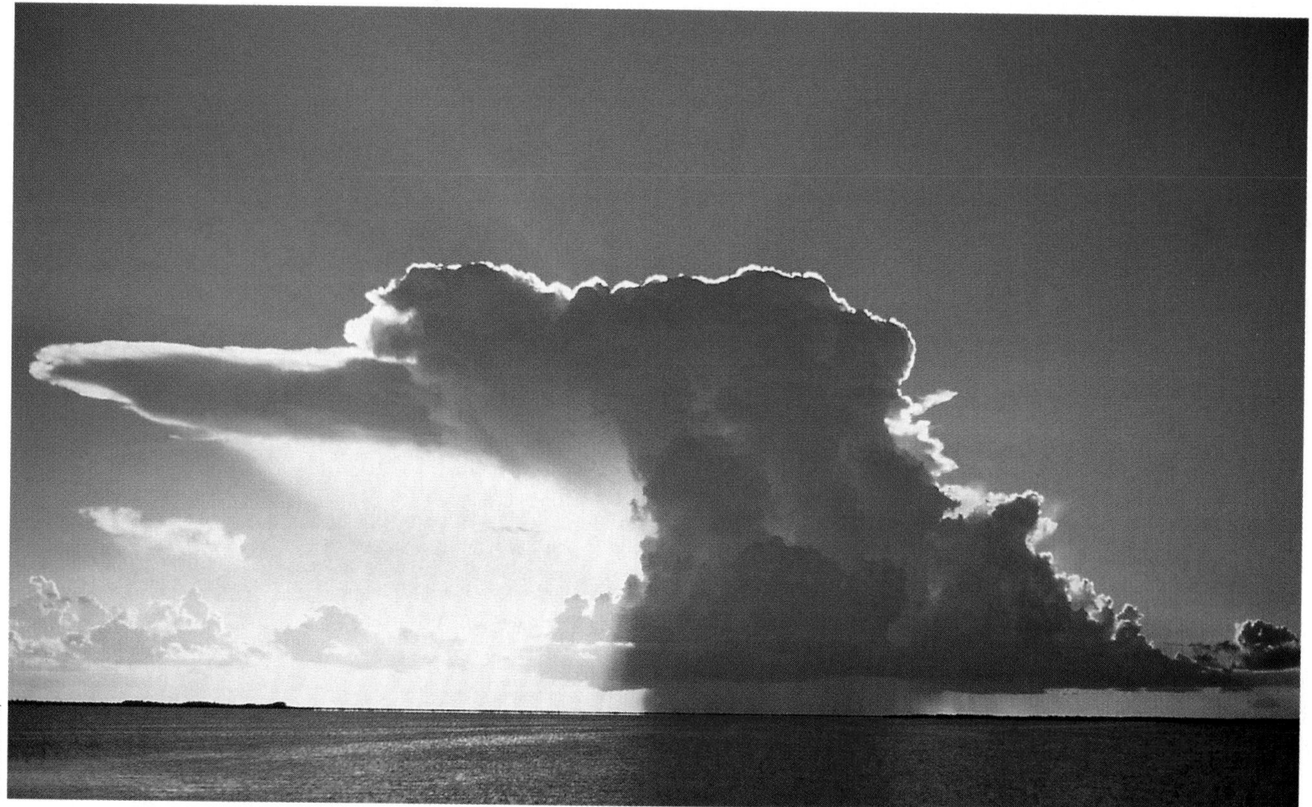

FIGURE 4.29 A cumulonimbus cloud. Strong upper-level winds blowing from right to left produce a well-defined anvil. Sunlight scattered by falling ice crystals produces the white (bright) area beneath the anvil. Notice the heavy rain shower falling from the base of the cloud.

snowflakes, snow pellets, and sometimes hailstones—all of which can fall to earth in the form of heavy showers. Lightning, thunder, and even tornadoes are associated with the cumulonimbus. (More information on the violent nature of thunderstorms and tornadoes is given in Chapter 10.)

Cumulus congestus and cumulonimbus frequently look alike, making it difficult to distinguish between them. However, you can usually distinguish them by looking at the top of the cloud. If the sprouting upper part of the cloud is sharply defined and not fibrous, it is usually a cumulus congestus; conversely, if the top of the cloud loses its sharpness and becomes fibrous in texture, it is usually a cumulonimbus. (Compare Fig. 4.28 with Fig. 4.29.) The weather associated with these clouds also differs: lightning, thunder, and large hail typically occur with cumulonimbus.

So far, we have discussed the ten primary cloud forms, summarized pictorially in Fig. 4.30. This figure, along with the cloud photographs and descriptions, should help you identify the more common cloud forms. Don't worry if you find it hard to estimate cloud heights. This is a difficult procedure, requiring much practice. You can use local objects (hills, mountains, tall buildings) of known height as references on which to base your height estimates.

To better describe a cloud's shape and form, a number of descriptive words may be used in conjunction with its name. We mentioned a few in the previous section; for example, a stratus cloud with a ragged appearance is a stratus fractus, and a cumulus cloud with marked vertical growth is a cumulus congestus. Table 4.4 lists some of the more common terms that are used in cloud identification.

DID YOU KNOW?

On July 26, 1959, Colonel William A. Rankin took a wild ride inside a huge cumulonimbus cloud. Bailing out of his disabled military aircraft inside a thunderstorm at 14.5 km (about 47,500 ft), Rankin free-fell for about 3 km (10,000 ft). When his parachute opened, surging updrafts carried him higher into the cloud, where he was pelted by heavy rain and hail, and nearly struck by lightning.

TABLE 4.4 Common Terms Used in Identifying Clouds

TERM	LATIN ROOT AND MEANING	DESCRIPTION
Lenticularis	(*lens, lenticula,* lentil)	Clouds having the shape of a lens; often elongated and usually with well-defined outlines. This term applies mainly to cirrocumulus, altocumulus, and stratocumulus
Fractus	(*frangere,* to break or fracture)	Clouds that have a ragged or torn appearance; applies only to stratus and cumulus
Humilis	(*humilis,* of small size)	Cumulus clouds with generally flattened bases and slight vertical growth
Congestus	(*congerere,* to bring together; to pile up)	Cumulus clouds of great vertical extent that, from a distance, may resemble a head of cauliflower
Undulatus	(*unda,* wave; having waves)	Clouds in patches, sheets, or layers showing undulations
Translucidus	(*translucere,* to shine through; transparent)	Clouds that cover a large part of the sky and are sufficiently translucent to reveal the position of the sun or moon
Mammatus	(*mamma,* mammary)	Baglike clouds that hang like a cow's udder on the underside of a cloud; may occur with cirrus, altocumulus, altostratus, stratocumulus, and cumulonimbus
Pileus	(*pileus,* cap)	A cloud in the form of a cap or hood above or attached to the upper part of a cumuliform cloud, particularly during its developing stage
Castellanus	(*castellum,* a castle)	Clouds that show vertical development and produce towerlike extensions, often in the shape of small castles

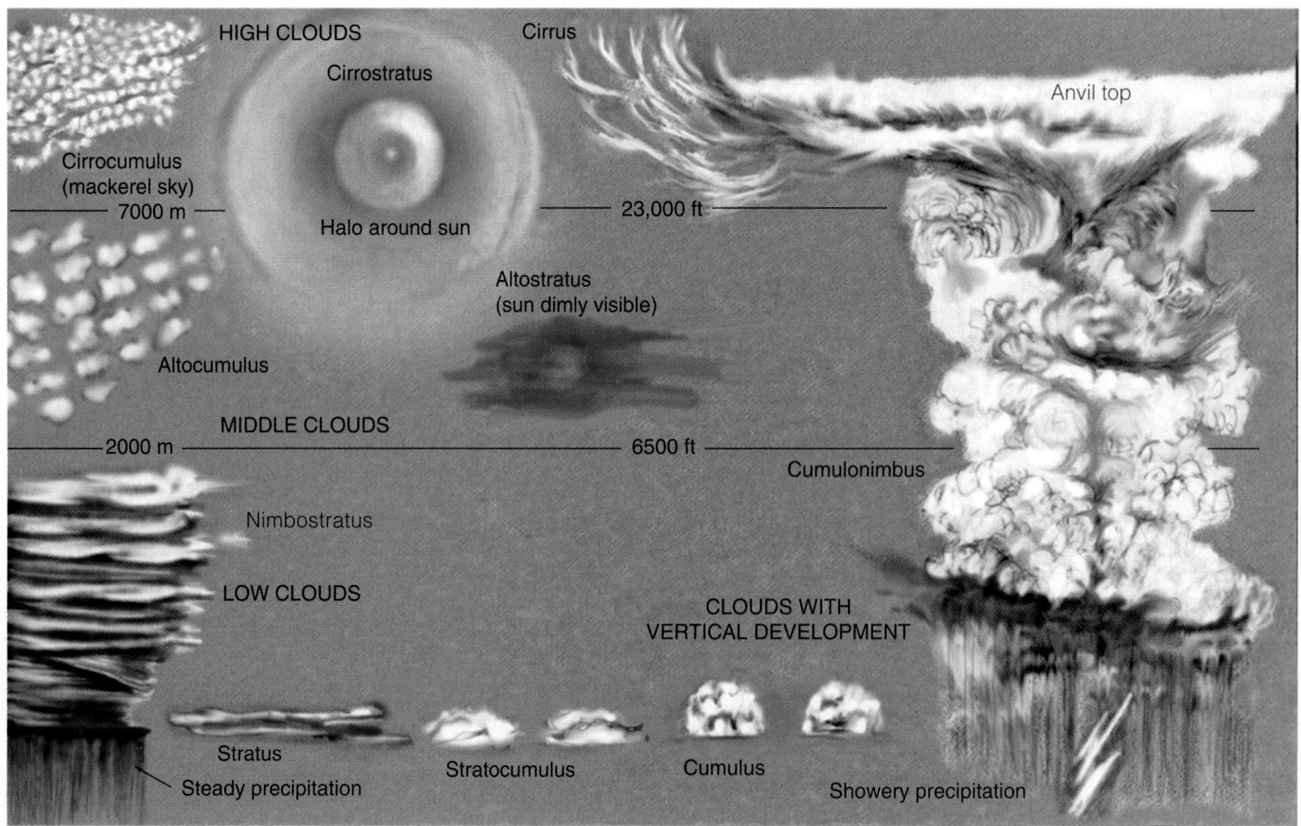

FIGURE 4.30 A generalized illustration of basic cloud types based on height above the surface and vertical development.

© J. L. Medeiros

FIGURE 4.31 Lenticular clouds forming one on top of the other on the eastern side of the Sierra Nevada.

Photo by author

FIGURE 4.32 A pileus cloud forming above a developing cumulus cloud.

Some Unusual Clouds Although the ten basic cloud forms are the most frequently seen, there are some unusual clouds that deserve mentioning. For example, moist air crossing a mountain barrier often forms into waves. The clouds that form in the wave crest usually have a lens shape and are, therefore, called **lenticular clouds** (see Fig. 4.31). Frequently, they form one above the other like a stack of pancakes, and at a distance they may resemble a fleet of hovering spacecraft. Hence, it is no wonder a large number of UFO sightings take place when lenticular clouds are present.

Similar to the lenticular is the *cap cloud,* or **pileus,** that usually resembles a silken scarf capping the top of a sprouting cumulus cloud (see Fig. 4.32). Pileus clouds form when moist winds are deflected up and over the top of a building cumulus congestus or cumulonimbus. If the air flowing over the top of the cloud condenses, a pileus often forms.

Most clouds form in rising air, but the mammatus forms in sinking air. **Mammatus clouds** derive their name from their appearance—baglike sacks that hang beneath the cloud and resemble a cow's udder (see Fig. 4.33). Although mammatus most frequently form on the underside of cumulonimbus, they may develop beneath cirrus, cirrocumulus, altostratus, altocumulus, and stratocumulus.

National Center for Atmospheric Research (NCAR)

FIGURE 4.33 Mammatus clouds forming beneath a thunderstorm.

Jet aircraft flying at high altitudes often produce a cirruslike trail of condensed vapor called a *condensation trail* or **contrail** (see Fig. 4.34). The condensation may come directly from the water vapor added to the air from engine exhaust. In this case, there must be sufficient mixing of the hot exhaust gases with the cold air to produce saturation. Contrails evaporate rapidly when the relative humidity of the surrounding air is low. If the relative humidity is high, however, contrails may persist for many hours. Contrails may also form by a cooling process as the reduced pressure produced by air flowing over the wing causes the air to cool.

Aside from the cumulonimbus cloud that sometimes penetrates into the stratosphere, all of the clouds described so far are observed in the lower atmosphere—in the troposphere. Occasionally, however, clouds may be seen above the troposphere. For example, soft pearly looking clouds called **nacreous clouds,** or *mother-of-pearl*

FIGURE 4.34 A contrail forming behind a jet aircraft.

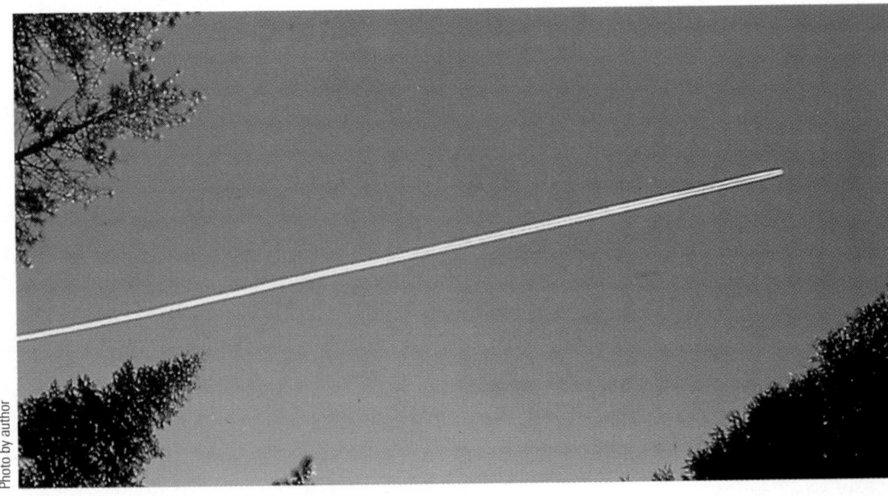

Photo by author

FIGURE 4.35
The clouds in this photograph are nacreous clouds. They form in the stratosphere and are most easily seen at high latitudes.

FIGURE 4.36
The wavy clouds in this photograph are noctilucent clouds. They are usually observed at high latitudes, at altitudes between 75 and 90 km above the earth's surface.

clouds, form in the stratosphere at altitudes above 30 km or 100,000 ft (see Fig. 4.35). They are best viewed in polar latitudes during the winter months when the sun, being just below the horizon, is able to illuminate them because of their high altitude. Their exact composition is not known, although they appear to be composed of water in either solid or liquid (supercooled) form.

Wavy bluish-white clouds, so thin that stars shine brightly through them, may sometimes be seen in the upper mesosphere, at altitudes above 75 km (46 mi). The best place to view these clouds is in polar regions at twilight. At this time, because of their altitude, the clouds are still in sunshine. To a ground observer, they appear bright against a dark background and, for this reason, they are called **noctilucent clouds,** meaning "luminous night clouds" (see Fig. 4.36). Studies reveal that these clouds are composed of tiny ice crystals. The water to make the ice may originate in meteoroids that disintegrate when entering the upper atmosphere or from the chemical breakdown of methane gas at high levels in the atmosphere.

Meteorology Now™ Click "Name that Cloud" to practice identifying clouds.

SUMMARY

In this chapter, we examined the hydrologic cycle and saw how water is circulated within our atmosphere. We then looked at some of the ways of describing humidity and found that relative humidity does not tell us how much water vapor is in the air but, rather, how close the air is to being saturated. A good indicator of the air's actual water vapor content is the dew-point temperature. When the air temperature and dew point are close together, the relative humidity is high; and, when they are far apart, the relative humidity is low.

When the air temperature drops below the dew point in a shallow layer of air near the surface, dew forms. If the dew freezes, it becomes frozen dew. Visible white frost forms when the air cools to a below freezing dew-point temperature. As the air cools in a deeper layer near the surface, the relative humidity increases and water vapor begins to condense upon "water seeking" hygroscopic condensation nuclei, forming haze. As the relative humidity approaches 100 percent, the air can become filled with tiny liquid droplets (or ice crystals) called fog. Upon examining fog, we found that it forms in two primary ways: cooling the air and evaporating and mixing water vapor into the air.

Condensation above the earth's surface produces clouds. When clouds are classified according to their height and physical appearance, they are divided into four main groups: high, middle, low, and clouds with vertical development. Since each cloud has physical characteristics that distinguish it from all the others, careful observation normally leads to correct identification.

Meteorology Now™ Assess your understanding of this chapter's topics with additional quizzing and tutorials at http://earthscience.brookscole.com/ahrens/ess4e.

KEY TERMS

The following terms are listed in the order they appear in the text. Define each. Doing so will aid you in reviewing the material covered in this chapter.

evaporation
condensation
precipitation
hydrologic cycle
saturated air
condensation nuclei
humidity
actual vapor pressure
saturation vapor pressure
relative humidity
supersaturated air
dew-point temperature (dew point)
wet-bulb temperature
heat index (HI)
apparent temperature
psychrometer
hygrometer
dew
frost
haze
fog
radiation fog
advection fog
upslope fog
evaporation (mixing) fog
cirrus clouds
cirrocumulus clouds
cirrostratus clouds
altocumulus clouds
altostratus clouds
nimbostratus clouds
stratocumulus clouds
stratus clouds
cumulus clouds
cumulonimbus clouds
lenticular clouds
pileus clouds
mammatus clouds
contrail
nacreous clouds
noctilucent clouds

QUESTIONS FOR REVIEW

1. Briefly explain the movement of water in the hydrologic cycle.
2. How does condensation differ from precipitation?
3. What are condensation nuclei and why are they important in our atmosphere?
4. In a volume of air, how does the actual vapor pressure differ from the saturation vapor pressure? When are they the same?
5. What does saturation vapor pressure primarily depend upon?
6. (a) What does the relative humidity represent?
   (b) When the relative humidity is given, why is it also important to know the air temperature?
   (c) Explain two ways the relative humidity may be changed.
   (d) During what part of the day is the relative humidity normally lowest? Normally highest?
7. Why do hot and humid summer days usually feel hotter than hot and dry summer days?
8. Why is cold polar air described as "dry" when the relative humidity of that air is very high?
9. Why is the wet-bulb temperature a good measure of how cool human skin can become?
10. (a) What is the dew-point temperature?
    (b) How is the difference between dew point and air temperature related to the relative humidity?
11. How can you obtain both the dew point and the relative humidity using a sling psychrometer?
12. Explain how dew, frozen dew, and visible frost form.
13. List the two primary ways in which fog forms.
14. Describe the conditions that are necessary for the formation of:
    (a) radiation fog
    (b) advection fog

15. How does evaporation (mixing) fog form?
16. Clouds are most generally classified by height above the earth's surface. List the major height categories and the cloud types associated with each.
17. How can you distinguish altostratus clouds from cirrostratus clouds?
18. Which clouds are associated with each of the following characteristics:
    (a) mackerel sky
    (b) lightning
    (c) halos
    (d) hailstones
    (e) mares' tails
    (f) anvil top
    (g) light continuous rain or snow
    (h) heavy rain showers

QUESTIONS FOR THOUGHT AND EXPLORATION

1. Use the concepts of condensation and saturation to explain why eyeglasses often fog up after coming indoors on a cold day.
2. After completing a grueling semester of meteorological course work, you call your travel agent to arrange a much-needed summer vacation. When your agent suggests a trip to the desert, you decline because of a concern that the dry air will make your skin feel uncomfortable. The travel agent assures you that almost daily "desert relative humidities are above 90 percent." Could the agent be correct? Explain.
3. Can the actual vapor pressure ever be greater than the saturation vapor pressure? Explain.
4. Suppose while measuring the relative humidity using a sling psychrometer, you accidently moisten both the dry-bulb and the wet-bulb thermometers. Will the relative humidity you determine be higher or lower than the air's true relative humidity?
5. Why is advection fog more common on the west coast of the United States than on the east coast?
6. With all other factors being equal, would you expect a lower minimum temperature on a night with cirrus clouds or on a night with stratocumulus clouds? Explain your answer.
7. Explain why icebergs are frequently surrounded by fog.
8. While driving from cold air (well below freezing) into much warmer air (well above freezing), frost forms on the windshield of the car. Does the frost form on the inside or outside of the windshield? How can the frost form when the air is so warm?
9. Why do relative humidities seldom reach 100 percent in polluted air?
10. If all fog droplets gradually settle earthward, explain how fog can last (without disappearing) for many days at a time.
11. The air temperature during the night cools to the dew point in a deep layer, producing fog. Before the fog formed, the air temperature cooled each hour about 2°C. After the fog formed, the air temperature cooled by only 0.5°C each hour. Give *two* reasons why the air cooled more slowly after the fog formed.
12. Why can you see your breath on a cold morning? Does the air temperature have to be below freezing for this to occur?
13. The sky is overcast and it is raining. Explain how you could tell if the cloud above you is a nimbostratus or a cumulonimbus.
14. You are sitting inside your house on a sunny afternoon. The shades are drawn and you look at the window and notice the sun disappears for about 10 seconds. The alternate light and dark period lasts for nearly 30 minutes. Are the clouds passing in front of the sun cirrocumulus, altocumulus, stratocumulus, or cumulus? Give a reasonable explanation for your answer.

• Go to the Brooks/Cole Earth Sciences Resource Center (http://earthscience.brookscole.com) for critical thinking exercises, articles, and additional readings from InfoTrac College Edition, Brooks/Cole's online student library.

© Ed Darack

A mass of moist, stable air gliding up and over these mountains condenses into lenticular clouds.

Meteorology Now™ This icon, appearing throughout the book, indicates an opportunity to explore interactive tutorials, animations, or practice problems available on the MeteorologyNow Web site at http://earthscience.brookscole.com/ahrens/ess4e.

CHAPTER 5

Cloud Development and Precipitation

CONTENTS

The weather is an ever-playing drama before which we are a captive audience. With the lower atmosphere as the stage, air and water as the principal characters, and clouds for costumes, the weather's acts are presented continuously somewhere about the globe. The script is written by the sun; the production is directed by the earth's rotation; and, just as no theater scene is staged exactly the same way twice, each weather episode is played a little differently, each is marked with a bit of individuality.

Clyde Orr, Jr., *Between Earth and Space*

Clouds, spectacular features in the sky, add beauty and color to the natural landscape. Yet, clouds are important for nonaesthetic reasons, too. As they form, vast quantities of heat are released into the atmosphere. Clouds help regulate the earth's energy balance by reflecting and scattering solar radiation and by absorbing the earth's infrared energy. And, of course, without clouds there would be no precipitation. But clouds are also significant because they visually indicate the physical processes taking place in the atmosphere; to a trained observer, they are signposts in the sky. In the beginning of this chapter, we will look at the atmospheric processes these signposts point to, the first of which is atmospheric stability. Later, we will examine the different mechanisms responsible for the formation of most clouds. Toward the end of the chapter, we will peer into the tiny world of cloud droplets to see how rain, snow, and other types of precipitation form.

ATMOSPHERIC STABILITY

We know that most clouds form as air rises, expands, and cools. But why does the air rise on some occasions and not on others? And why does the size and shape of clouds vary so much when the air does rise? To answer these questions, let's focus on the concept of atmospheric stability.

When we speak of atmospheric stability, we are referring to a condition of equilibrium. For example, rock A resting in the depression in Fig. 5.1 is in *stable* equilibrium. If the rock is pushed up along either side of the hill and then let go, it will quickly return to its original position. On the other hand, rock B, resting on the top of the hill, is in a state of *unstable* equilibrium, as a slight push will set it moving away from its original position.

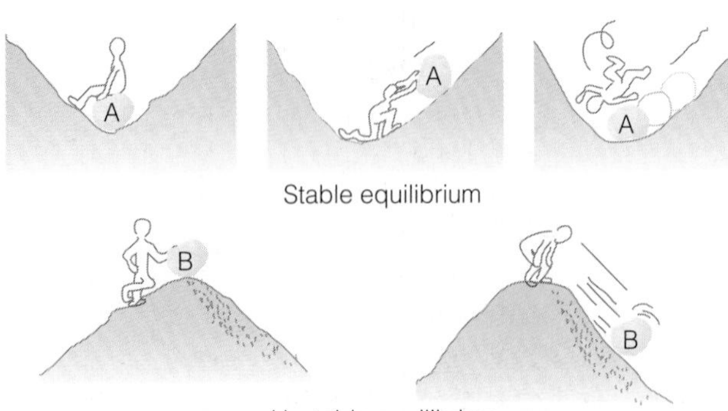

FIGURE 5.1 When rock A is disturbed, it will return to its original position; rock B, however, will accelerate away from its original position.

Applying these concepts to the atmosphere, we can see that air is in stable equilibrium when, after being lifted or lowered, it tends to return to its original position—it resists upward and downward air motions. Air that is in unstable equilibrium will, when given a little push, move farther away from its original position—it favors vertical air currents.

In order to explore the behavior of rising and sinking air, we must first review some concepts we learned in earlier chapters. Recall that a balloonlike blob of air is called an *air parcel.* (The concept of air parcel is illustrated in Fig. 4.4, p. 80.) When an air parcel rises, it moves into a region where the air pressure surrounding it is lower. This situation allows the air molecules inside to push outward on the parcel walls, expanding it. As the air parcel expands, the air inside cools. If the same parcel is brought back to the surface, the increasing pressure around the parcel squeezes (compresses) it back to its original volume, and the air inside warms. Hence, *a rising parcel of air expands and cools, while a sinking parcel is compressed and warms.*

If a parcel of air expands and cools, or compresses and warms, with no interchange of heat with its outside surroundings, this situation is called an **adiabatic process.** As long as the air in the parcel is unsaturated (the relative humidity is less than 100 percent), the rate of adiabatic cooling or warming remains constant and is about 10°C for every 1000 meters of change in elevation, or about 5.5°F for every 1000 feet. Since this rate of cooling or warming only applies to unsaturated air, it is called the **dry adiabatic rate*** (see Fig. 5.2).

As the rising air cools, its relative humidity increases as the air temperature approaches the dew-point temperature. If the air cools to its dew-point temperature, the relative humidity becomes 100 percent. Further lifting results in condensation, a cloud forms, and latent heat is released into the rising air. Because the heat added during condensation offsets some of the cooling due to expansion, the air no longer cools at the dry adiabatic rate but at a lesser rate called the **moist adiabatic rate.** (Because latent heat is added to the rising saturated air, the process is not really adiabatic.**) If a saturated parcel containing water droplets were to sink, it would compress and warm at the moist adiabatic rate because evaporation of the liquid droplets would offset the rate of compressional warming. Hence, the rate at which ris-

*For aviation purposes, the dry adiabatic rate is sometimes expressed as 3°C per 1000 ft.

**If condensed water or ice is removed from the rising saturated parcel, the cooling process is called an *irreversible pseudoadiabatic process.*

ing or sinking saturated air changes temperature—the moist adiabatic rate—is less than the dry adiabatic rate.

Unlike the dry adiabatic rate, the moist adiabatic rate is not constant, but varies greatly with temperature and, hence, with moisture content—as warm saturated air produces more liquid water than cold saturated air. The added condensation in warm, saturated air liberates more latent heat. Consequently, the moist adiabatic rate is much less than the dry adiabatic rate when the rising air is quite warm; however, the two rates are nearly the same when the rising air is very cold. Although the moist adiabatic rate does vary, to make the numbers easy to deal with we will use an average of 6°C per 1000 m (3.3°F per 1000 ft) in most of our examples and calculations.

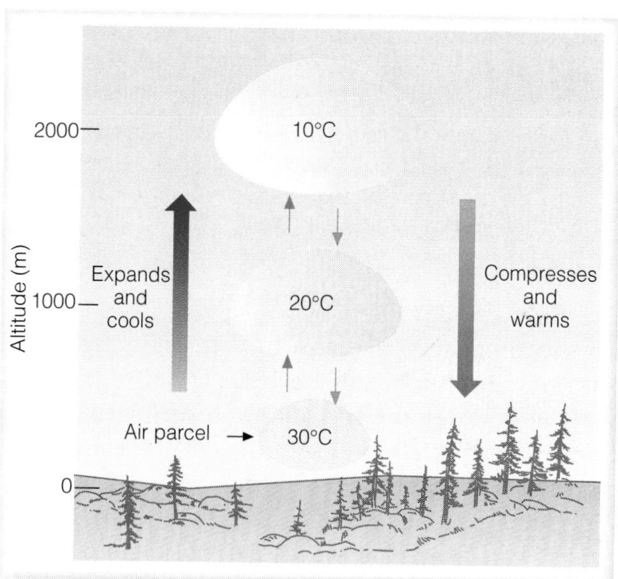

FIGURE 5.2 The dry adiabatic rate. As long as the air parcel remains unsaturated, it expands and cools by 10°C per 1000 m; the sinking parcel compresses and warms by 10°C per 1000 m.

DETERMINING STABILITY

We determine the stability of the air by comparing the temperature of a rising parcel to that of its surroundings. If the rising air is colder than its environment, it will be more dense* (heavier) and tend to sink back to its original level. In this case, the air is *stable* because it resists upward displacement. If the rising air is warmer and, therefore, less dense (lighter) than the surrounding air, it will continue to rise until it reaches the same temperature as its environment. This is an example of *unstable* air. To figure out the air's stability, we need to measure the temperature both of the rising air and of its environment at various levels above the earth.

Stable Air Suppose we release a balloon-borne instrument—a radiosonde (see Fig. 1, p. 11)—and it sends back temperature data as shown in Fig. 5.3. We measure the air temperature in the vertical and find that it decreases by 4°C for every 1000 m. Remember from Chapter 1 that the rate at which the air temperature changes with elevation is called the *lapse rate*. Because this is the rate at which the air temperature surrounding us would be changing if we were to climb upward into the atmosphere, we refer to it as the **environmental lapse rate.**

Notice in Fig. 5.3a that (with an environmental lapse rate of 4°C per 1000 m) a rising parcel of unsaturated, "dry" air is colder and heavier than the air surrounding it at all levels. Even if the parcel is initially saturated (Fig. 5.3b), as it rises it, too, would be colder than

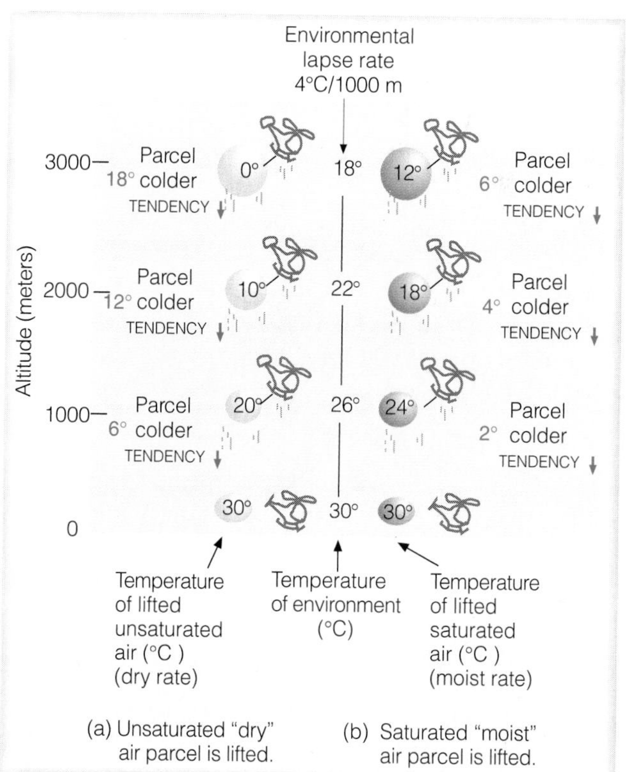

FIGURE 5.3 A stable atmosphere. An *absolutely stable atmosphere* exists when a rising air parcel is colder and heavier (i.e., more dense) than the air surrounding it. If given the chance (i.e., released), the air parcel in both situations would return to its original position, the surface.

*When, at the same level in the atmosphere, we compare parcels of air that are equal in size but vary in temperature, we find that cold air parcels are more dense than warm air parcels; that is, in the cold parcel, there are more molecules that are crowded closer together.

DID YOU KNOW?

If you take a walk on a bitter cold, yet clear, winter morning, when the air is calm and a strong subsidence inversion exists, the air aloft—thousands of meters above you—may be more than 17°C (30°F) warmer than the air at the surface.

its environment at all levels. In both cases, the atmosphere is **absolutely stable** because the lifted parcel of air is colder and heavier than the air surrounding it. If released, the parcel would have a tendency to return to its original position.

Since stable air strongly resists upward vertical motion, it will, *if forced to rise,* tend to spread out horizontally. If clouds form in this rising air, they, too, will spread horizontally in relatively thin layers and usually have flat tops and bases. We might expect to see clouds—such as cirrostratus, altostratus, nimbostratus, or stratus—forming in stable air.

The atmosphere is stable when the environmental lapse rate is small; that is, when there is a relatively small difference in temperature between the surface air and the air aloft. Consequently, the atmosphere tends to become more stable—it *stabilizes*—as the air aloft warms or the surface air cools. The *cooling* of the *surface air* may be due to:

1. nighttime radiational cooling of the surface
2. an influx of cold surface air brought in by the wind
3. air moving over a cold surface

It should be apparent that, on any given day, the air is generally most stable in the early morning around sunrise, when the lowest surface air temperature is recorded.

The air aloft may warm as winds bring in warmer air or as the air slowly sinks over a large area. Recall that sinking (subsiding) air warms as it is compressed. The warming may produce an inversion, where the air aloft is actually warmer than the air at the surface.* An inversion that forms by slow, sinking air is termed a *subsidence inversion.* Because inversions represent a very stable atmosphere, they act as a lid on vertical air motion. When an inversion exists near the ground, stratus, fog, haze, and pollutants are all kept close to the surface (see Fig. 5.4).

Unstable Air The atmosphere is unstable when the air temperature decreases rapidly as we move up into the atmosphere. For example, in Fig. 5.5, notice that the

*Recall from Chapter 3 that an inversion represents an atmospheric condition where the air becomes warmer with height.

FIGURE 5.4 Cold surface air, on this morning, produces a stable atmosphere that inhibits vertical air motions and allows the fog and haze to linger close to the ground.

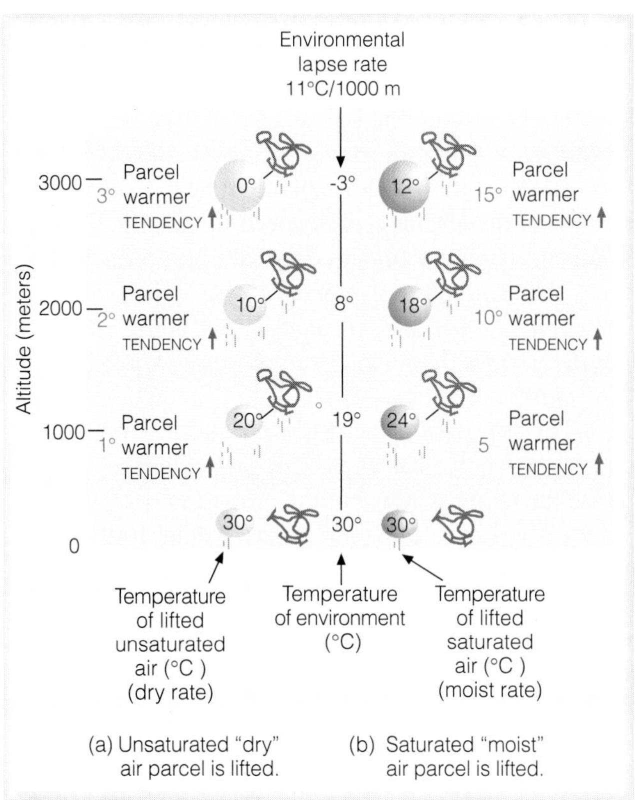

FIGURE 5.5 An unstable atmosphere. An *absolutely unstable atmosphere* exists when a rising air parcel is warmer and lighter (i.e., less dense) than the air surrounding it. If given the chance (i.e., released), the lifted parcel in both (a) and (b) would continue to move away (accelerate) from its original position.

FIGURE 5.6 Unstable air. The warmth from the forest fire heats the air, causing instability near the surface. Warm, less-dense air (and smoke) bubbles upward, expanding and cooling as it rises. Eventually the rising air cools to its dew point, condensation begins, and a cumulus cloud forms.

measured air temperature decreases by 11°C for every 1000-meter rise in elevation, which means that the environmental lapse rate is 11°C per 1000 meters. Also notice that a lifted parcel of unsaturated "dry" air in Fig. 5.5a, as well as a lifted parcel of saturated "moist" air in Fig. 5.5b, will, at each level above the surface, be warmer than the air surrounding them. Since, in both cases, the rising air is warmer and less dense than the air around them, once the parcels start upward, they will continue to rise on their own, away from the surface. Thus, we have an **absolutely unstable atmosphere.**

The atmosphere becomes more unstable as the environmental lapse rate steepens; that is, as the temperature of the air drops rapidly with increasing height. This circumstance may be brought on by either the air aloft becoming colder or the surface air becoming warmer (see Fig. 5.6). The *warming* of the *surface air* may be due to:

1. daytime solar heating of the surface
2. an influx of warm surface air brought in by the wind
3. air moving over a warm surface

Generally, then, as the surface air warms during the day, the atmosphere becomes more unstable—it *destabilizes.* The air aloft may cool as winds bring in colder air or as the air (or clouds) emit infrared radiation to space (radiational cooling). Just as sinking air produces warming and a more stable atmosphere, rising air, especially an entire layer where the top is dry and the bottom is humid, produces cooling and a more unstable atmo-

DID YOU KNOW?

Nature can produce its own fire extinguisher. Forest fires generate atmospheric instability by heating the air near the surface. The hot, rising air above the fire contains tons of tiny smoke particles that act as cloud condensation nuclei. As the air rises and cools, water vapor in the atmosphere as well as water vapor released during the burning of the timber, will often condense onto the nuclei, producing a cumuliform cloud, sometimes called a *pyrocumulus.* If the cloud builds high enough, and remains over the fire area, its heavy showers may actually help to extinguish the fire.

sphere. The lifted layer becomes more unstable as it rises and stretches out vertically in the less dense air aloft. This stretching effect steepens the environmental lapse rate as the top of the layer cools more than the bottom. Instability brought on by the lifting of air is often associated with the development of severe weather, such as thunderstorms and tornadoes, which are investigated more thoroughly in Chapter 10.

It should be noted, however, that deep layers in the atmosphere are seldom, if ever, absolutely unstable. Absolute instability is usually limited to a very shallow layer near the ground on hot, sunny days. Here, the environmental lapse rate can exceed the dry adiabatic rate, and the lapse rate is called *superadiabatic.*

Conditionally Unstable Air Suppose an unsaturated (but humid) air parcel is somehow forced to rise from the surface, as shown in Fig. 5.7. As the parcel rises, it expands, and cools at the *dry adiabatic rate* until its air temperature cools to its dew point. At this level, the air is saturated, the relative humidity is 100 percent, and further lifting results in condensation and the formation of a cloud. The elevation above the surface where the cloud first forms (in this example, 1000 meters) is called the **condensation level.**

In Fig. 5.7, notice that above the condensation level, the rising saturated air cools at the *moist adiabatic rate.* Notice also that from the surface up to a level near 2000 meters, the rising, lifted air is colder than the air surrounding it. The atmosphere up to this level is *stable.* However, due to the release of latent heat, the rising air near 2000 meters has actually become warmer than the air around it. Since the lifted air can rise on its own accord, the atmosphere is now *unstable.* The level in the atmosphere where the air parcel, after being lifted, becomes warmer than the air surrounding it, is called the *level of free convection.*

The atmospheric layer from the surface up to 4000 meters in Fig. 5.7 has gone from stable to unstable because the rising air was humid enough to become saturated, form a cloud, and release latent heat, which warms the air. Had the cloud not formed, the rising air would have remained colder at each level than the air surrounding it. From the surface to 4000 meters, we have what is said to be a **conditionally unstable atmosphere**—the condition for instability being whether or not the rising air becomes saturated. Therefore, *conditional instability* means that, if unsaturated stable air is somehow lifted to a level where it becomes saturated, instability may result.

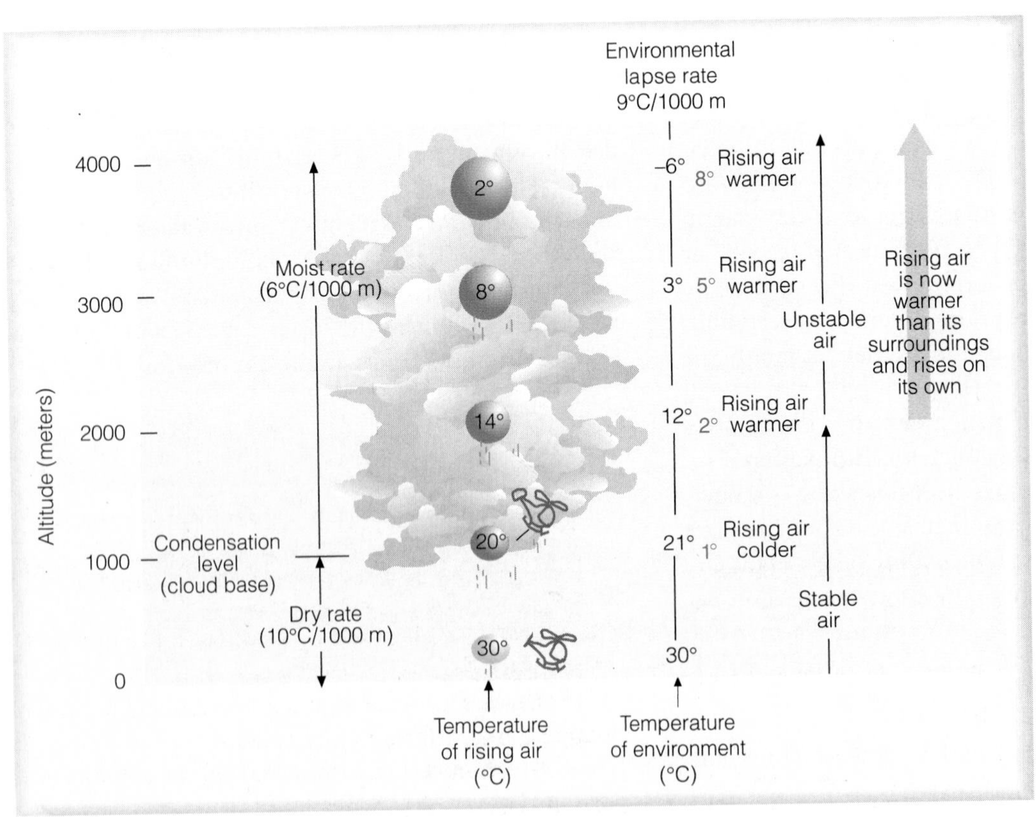

Meteorology Now™

ACTIVE FIGURE 5.7
Conditionally unstable air. The atmosphere is conditionally unstable when unsaturated, stable air is lifted to a level where it becomes saturated and warmer than the air surrounding it. If the atmosphere remains unstable, vertical developing cumulus clouds can build to great heights.
Watch this Active Figure at http://earthscience.brookscole.com/ahrens/ess4e.

In Fig. 5.7, we can see that the environmental lapse rate is 9°C per 1000 meters. This value is between the dry adiabatic rate (10°C/1000 m) and the moist adiabatic rate (6°C/1000 m). Consequently, *conditional instability exists whenever the environmental lapse rate is between the dry and moist adiabatic rates.* Recall from Chapter 1 that the average lapse rate in the troposphere is about 6.5°C per 1000 m (3.6°F per 1000 ft). Since this value lies between the dry adiabatic rate and the average moist rate, *the atmosphere is ordinarily in a state of conditional instability.*

At this point, it should be apparent that the stability of the atmosphere changes during the course of a day. In clear, calm weather around sunrise, surface air is normally colder than the air above it, a radiation inversion exists, and the atmosphere is quite stable, as indicated by smoke or haze lingering close to the ground. As the day progresses, sunlight warms the surface and the surface warms the air above. As the air temperature near the ground increases, the lower atmosphere gradually becomes more unstable, with maximum instability usually occurring during the hottest part of the day. On a humid summer afternoon this phenomenon can be witnessed by the development of cumulus clouds.

BRIEF REVIEW

Up to this point we have looked briefly at stability as it relates to cloud development. The next section describes how atmospheric stability influences the physical mechanisms responsible for the development of individual cloud types. However, before going on, here is a brief review of some of the facts and concepts concerning stability:

- The air temperature in a rising parcel of *unsaturated* air decreases at the dry adiabatic rate, whereas the air temperature in a rising parcel of *saturated* air decreases at the moist adiabatic rate.
- The dry adiabatic rate and moist adiabatic rate of cooling are different due to the fact that latent heat is released in a rising parcel of *saturated* air.
- In a *stable atmosphere,* a lifted parcel of air will be colder (heavier) than the air surrounding it. Because of this fact, the lifted parcel will tend to sink back to its original position.
- In an *unstable atmosphere,* a lifted parcel of air will be warmer (lighter) than the air surrounding it, and thus will continue to rise upward, away from its original position.
- The atmosphere becomes more stable (stabilizes) as the surface air cools, the air aloft warms, or a layer of air sinks (subsides) over a vast area.
- The atmosphere becomes more unstable (destabilizes) as the surface air warms, the air aloft cools, or a layer of air is lifted.
- Layered clouds tend to form in a stable atmosphere, whereas cumuliform clouds tend to form in a conditionally unstable atmosphere.

CLOUD DEVELOPMENT AND STABILITY

Most clouds form as air rises, expands, and cools. Basically, the following mechanisms are responsible for the development of the majority of clouds we observe:

1. surface heating and free convection
2. uplift along topography
3. widespread ascent due to the flowing together (convergence) of surface air
4. uplift along weather fronts (see Fig. 5.8)

Convection and Clouds Some areas of the earth's surface are better absorbers of sunlight than others and, therefore, heat up more quickly. The air in contact with these "hot spots" becomes warmer than its surroundings. A hot "bubble" of air—a *thermal*—breaks away from the warm surface and rises, expanding and cooling as it ascends. As the thermal rises, it mixes with the cooler, drier air around it and gradually loses its identity. Its upward movement now slows. Frequently, before it is completely diluted, subsequent rising thermals penetrate it and help the air rise a little higher. If the rising air cools to its saturation point, the moisture will condense, and the thermal becomes visible to us as a cumulus cloud.

Observe in Fig. 5.9 that the air motions are downward on the outside of the cumulus cloud. The downward motions are caused in part by evaporation around the outer edge of the cloud, which cools the air, making it heavy. Another reason for the downward motion is the completion of the convection current started by the thermal. Cool air slowly descends to replace the rising warm air. Therefore, we have rising air in the cloud and sinking air around it. Since subsiding air greatly inhibits the growth of thermals beneath it, small cumulus clouds usually have a great deal of blue sky between them (see Fig. 5.10).

As the cumulus clouds grow, they shade the ground from the sun. This, of course, cuts off surface heating and upward convection. Without the continual supply of rising air, the cloud begins to erode as its droplets evaporate. Unlike the sharp outline of a growing cumulus, the cloud now has indistinct edges, with cloud fragments extending from its sides. As the cloud dissipates (or moves along with the wind), surface heating begins again and regenerates another thermal, which becomes a new cumulus. This is why you often see cumulus clouds form, gradually disappear, then reform in the same spot.

The stability of the atmosphere plays an important part in determining the vertical growth of cumulus

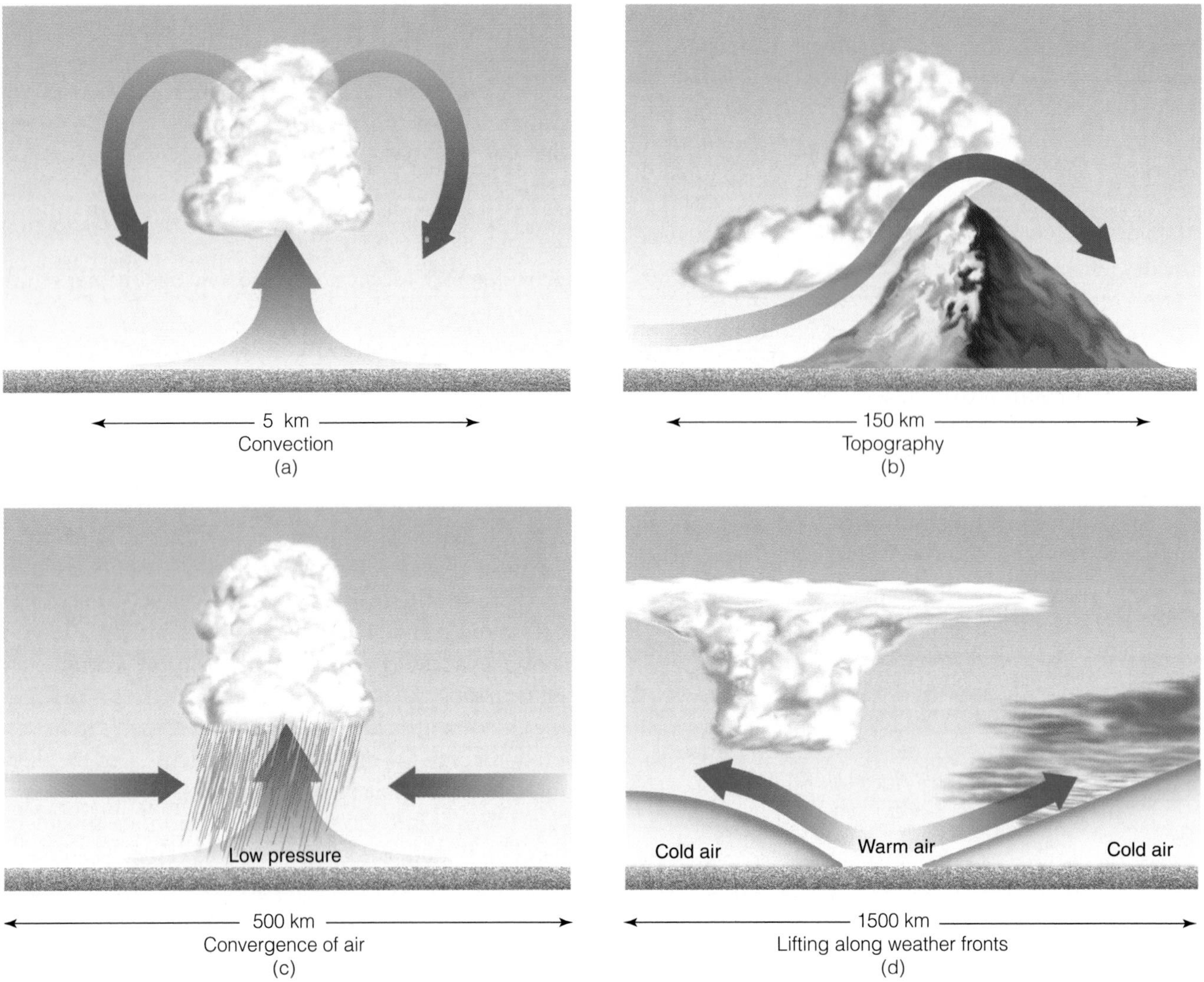

5 km
Convection
(a)

150 km
Topography
(b)

Low pressure

500 km
Convergence of air
(c)

Cold air
Warm air
Cold air

1500 km
Lifting along weather fronts
(d)

FIGURE 5.8 The primary ways clouds form: (a) surface heating and convection; (b) forced lifting along topographic barriers; (c) convergence of surface air; (d) forced lifting along weather fronts.

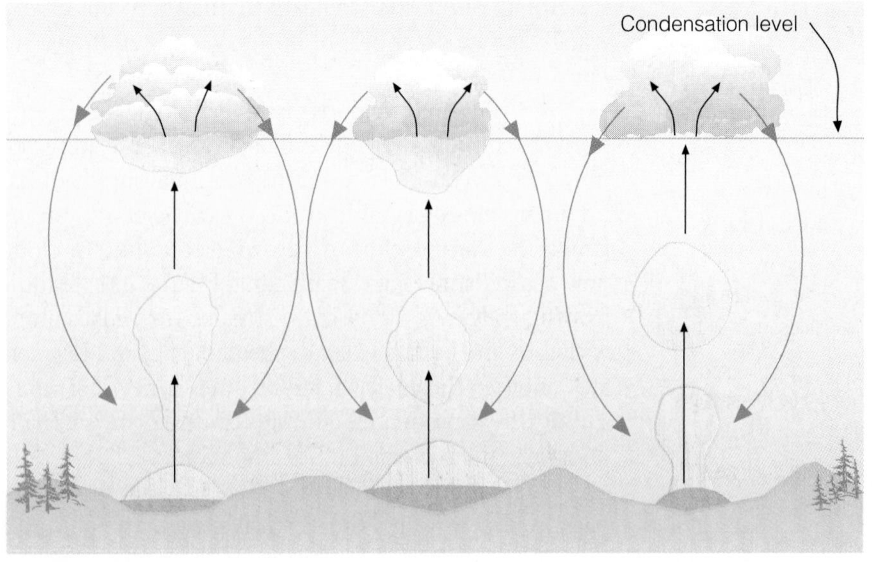

FIGURE 5.9 Cumulus clouds form as hot, invisible air bubbles detach themselves from the surface, then rise and cool to the condensation level. Below and within the cumulus clouds, the air is rising. Around the cloud, the air is sinking.

FIGURE 5.10 Cumulus clouds building on a warm summer afternoon. Each cloud represents a region where thermals are rising from the surface. The clear areas between the clouds are regions where the air is sinking.

FIGURE 5.11 Cumulus clouds developing into thunderstorms in a conditionally unstable atmosphere over the Great Plains. Notice that, in the distance, the cumulonimbus with the anvil top has reached the stable part of the atmosphere.

clouds. For example, if a stable layer (such as an inversion) exists near the top of the cumulus cloud, the cloud would have a difficult time rising much higher, and it would remain as a "fair-weather" cumulus cloud. However, if a deep, conditionally unstable layer exists above the cloud, then the cloud may develop vertically into a towering cumulus congestus with a cauliflowerlike top. When the unstable air is several miles deep, the cumulus congestus may even develop into a cumulonimbus (see Fig. 5.11).

Notice in Fig. 5.11 that the distant thunderstorm has a flat anvil-shaped top. The reason for this shape is due to the fact that the cloud has reached the stable part of the atmosphere, and the rising air is unable to puncture very far into this stable layer. Consequently, the top of the cloud spreads laterally as high winds at this altitude (usually above 10,000 m or 33,000 ft) blow the cloud's ice crystals horizontally.

Topography and Clouds Horizontally moving air obviously cannot go through a large obstacle, such as a mountain, so the air must go over it. Forced lifting along a topographic barrier is called **orographic uplift.** Often, large masses of air rise when they approach a long chain of mountains such as the Sierra Nevada and Rockies. This lifting produces cooling, and if the air is humid, clouds form. Clouds produced in this manner are called *orographic clouds.*

An example of orographic uplift and cloud development is given in Fig. 5.12. Notice that, after having risen over the mountain, the air at the surface on the leeward (downwind) side is considerably warmer than it was at the surface on the windward (upwind) side. The higher air temperature on the leeward side is the result of latent heat being converted into sensible heat during condensation on the windward side. In fact, the rising air at the top of the mountain is considerably warmer than it would have been had condensation not occurred.

Notice also in Fig. 5.12 that the dew-point temperature of the air on the leeward side is lower than it was before the air was lifted over the mountain. The lower dew point and, hence, drier air on the leeward side is the result of water vapor condensing and then remaining as liquid cloud droplets and precipitation on the windward

DID YOU KNOW?

Clouds are heavy—they can easily weigh many tons. Even a relatively small "fair weather" cumulus humilis cloud (about 3000 ft high and 3000 ft in diameter) weighs nearly 400,000 pounds, which is equivalent to the weight of water in two large backyard swimming pools. The cloud, of course, does not fall to the ground because cloud droplets and ice crystals are so tiny that it takes only the slightest updraft to keep these particles suspended.

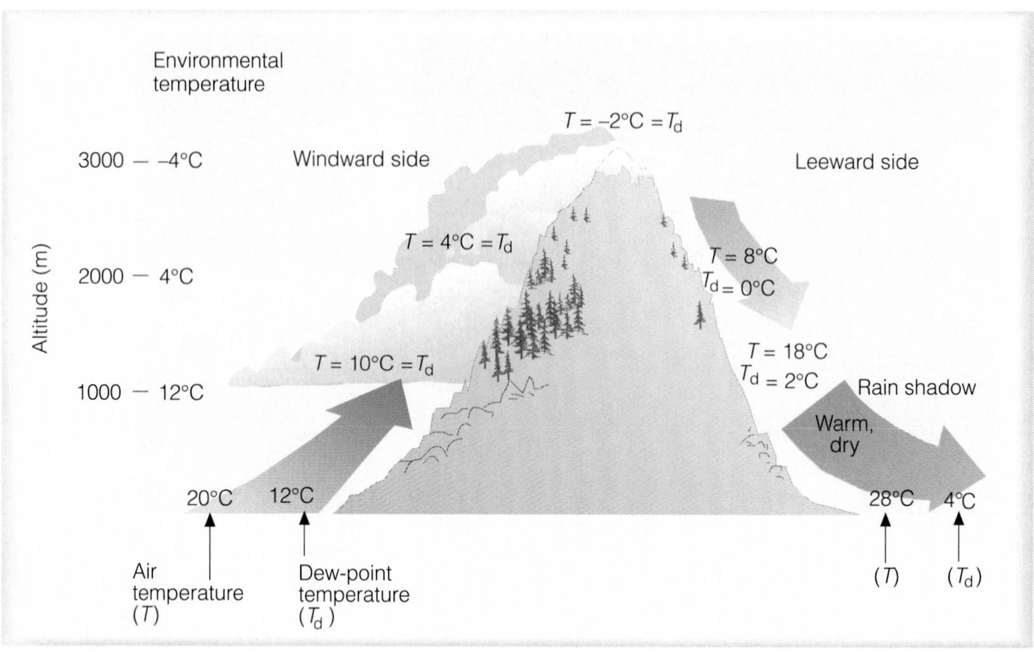

Meteorology Now™ ACTIVE FIGURE 5.12 Orographic uplift, cloud development, and the formation of a rain shadow. Watch this Active Figure at http://earthscience.brookscole.com/ahrens/ess4e.

side. This region on the leeward side of a mountain, where precipitation is noticeably low, and the air is often drier, is called a **rain shadow.**

Although clouds are more prevalent on the windward side of mountains, they may, under certain atmospheric conditions, form on the leeward side as well. For example, stable air flowing over a mountain often moves in a series of waves that may extend for several hundred miles on the leeward side. Such waves often resemble the waves that form in a river downstream from a large boulder. Recall from Chapter 4 that wave clouds often have a characteristic lens shape and are called *lenticular clouds.*

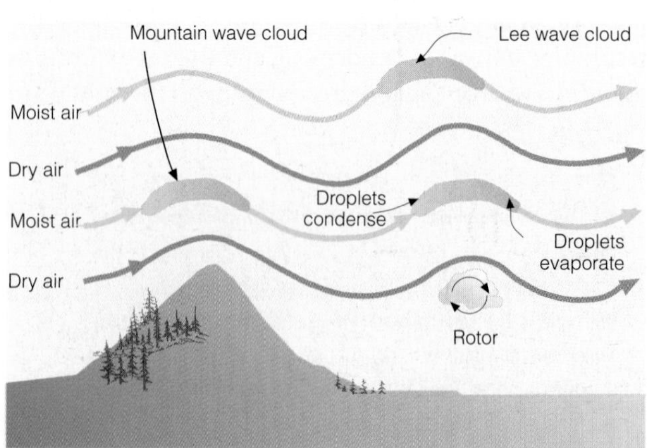

FIGURE 5.13 The formation of lenticular clouds.

The formation of lenticular clouds is shown in Fig. 5.13. As moist air rises on the upwind side of the wave, it cools and condenses, producing a cloud. On the downwind side, the air sinks and warms—the cloud evaporates. Viewed from the ground, the clouds appear motionless as the air rushes through them. When the air between the cloud-forming layers is too dry to produce clouds, lenticular clouds will form one above the other, sometimes extending into the stratosphere and appearing as a fleet of hovering spacecraft. Lenticular clouds that form in the wave directly over the mountain are called *mountain wave clouds* (see Fig. 5.14).

Notice in Fig. 5.13 that beneath the lenticular cloud, a large swirling eddy forms. The rising part of the eddy may cool enough to produce *rotor clouds.* The air in the rotor is extremely turbulent and presents a major hazard to aircraft in the vicinity. Dangerous flying conditions also exist near the lee side of the mountain, where strong downward air motions are present.

Now, having examined the concept of stability and the formation of clouds, we are ready to see how minute cloud particles are transformed into rain and snow. The next section, therefore, takes a look at the processes that produce precipitation.

Meteorology Now™ Click "Adiabatic" to examine changes in temperature, relative humidity, and clouds as air crosses a mountain.

FIGURE 5.14 Lenticular clouds (mountain wave clouds) forming over Mt. Rainier, Washington.

PRECIPITATION PROCESSES

As we all know, cloudy weather does not necessarily mean that it will rain or snow. In fact, clouds may form, linger for many days, and never produce precipitation. In Eureka, California, the August daytime sky is overcast more than 50 percent of the time, yet the average precipitation there for August is merely one-tenth of an inch. How, then, do cloud droplets grow large enough to produce rain? And why do some clouds produce rain, but not others?

In Fig. 5.15, we can see that an ordinary cloud droplet is extremely small, having an average diameter of 0.02 millimeters (mm), which is less than one-thousandth of an inch. Also, notice in Fig. 5.15 that the diameter of a typical cloud droplet is 100 times smaller than a typical raindrop. Clouds, then, are composed of many small droplets—too small to fall as rain. These minute droplets require only slight upward air currents to keep them suspended. Those droplets that do fall, descend slowly and evaporate in the drier air beneath the cloud.

In Chapter 4, we learned that condensation begins on tiny particles called *condensation nuclei.* The growth of cloud droplets by condensation is slow and, even under ideal conditions, it would take several days for this process alone to create a raindrop. It is evident, then, that the condensation process by itself is entirely too slow to produce rain. Yet, observations show that clouds can develop and begin to produce rain in less than an hour. Since it takes about 1 million average size cloud droplets to make an average size raindrop, there must be some other process by which cloud droplets grow large and heavy enough to fall as precipitation.

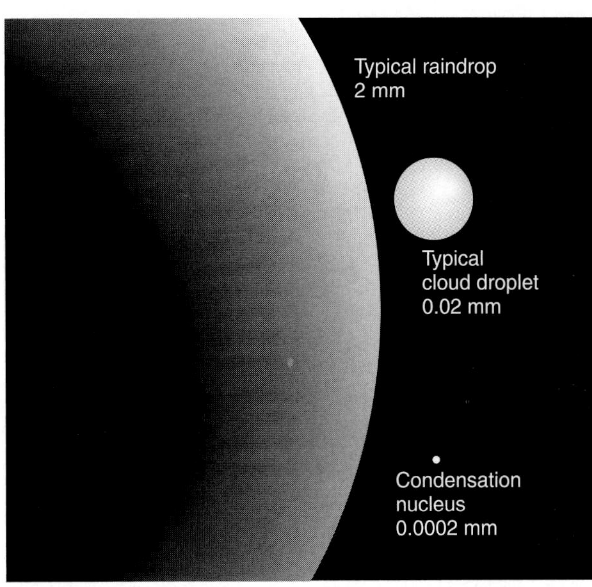

FIGURE 5.15 Relative sizes of raindrops, cloud droplets, and condensation nuclei.

DID YOU KNOW?

Maybe it has never rained cats and dogs, but it has rained maggots. In Acapulco, Mexico, during October, 1968, swarms of maggots (about an inch in length) fell from the sky during a heavy rain shower, covering everything, even people who had gathered there to witness a yachting event. Apparently, the maggots were swept into a thunderstorm by strong vertical air currents.

Even though all the intricacies of how rain is produced are not yet fully understood, two important processes stand out: (1) the collision-coalescence process and (2) the ice-crystal (or Bergeron) process.

Collision and Coalescence Process In clouds with tops warmer than −15°C (5°F), collisions between droplets can play a significant role in producing precipitation. To produce the many collisions necessary to form a raindrop, some cloud droplets must be larger than others. Larger drops may form on large condensation nuclei, such as salt particles, or through random collisions of droplets. Recent studies also suggest that turbulent mixing between the cloud and its drier environment may play a role in producing larger droplets.

As cloud droplets fall, air retards the falling drops. The amount of air resistance depends on the size of the drop and on its rate of fall: The greater its speed, the more air molecules the drop encounters each second. The speed of the falling drop increases until the air resistance equals the pull of gravity. At this point, the drop continues to fall, but at a constant speed, which is called its *terminal velocity.* Because larger drops have a smaller surface-area-to-weight ratio, they must fall faster before reaching their terminal velocity. Thus, *larger drops fall faster than smaller drops.*

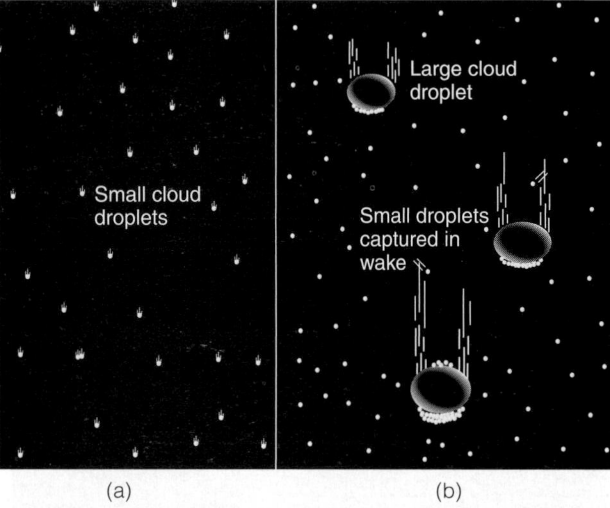

FIGURE 5.16 Collision and coalescence. (a) In a warm cloud composed only of small cloud droplets of uniform size, the droplets are less likely to collide as they all fall very slowly at about the same speed. Those droplets that do collide, frequently do not coalesce because of the strong surface tension that holds together each tiny droplet. (b) In a cloud composed of different size droplets, larger droplets fall faster than smaller droplets. Although some tiny droplets are swept aside, some collect on the larger droplet's forward edge, while others (captured in the wake of the larger droplet) coalesce on the droplet's backside.

Large droplets overtake and collide with smaller drops in their path. This merging of cloud droplets by collision is called **coalescence.** Laboratory studies show that collision does not always guarantee coalescence; sometimes the droplets actually bounce apart during collision. For example, the forces that hold a tiny droplet together *(surface tension)* are so strong that if the droplet were to collide with another tiny droplet, chances are they would not stick together (coalesce) (see Fig. 5.16). Coalescence appears to be enhanced if colliding droplets have opposite (and, hence, attractive) electrical charges.* An important factor influencing cloud droplet growth by the collision process is the amount of time the droplet spends in the cloud. Since rising air currents slow the rate at which droplets fall, a thick cloud with strong updrafts will maximize the time cloud droplets spend in a cloud and, hence, the size to which they grow.

Clouds that have above-freezing temperatures at all levels are called *warm clouds.* In tropical regions, where warm cumulus clouds build to great heights, strong convective updrafts frequently occur. In Fig. 5.17, suppose a cloud droplet is caught in a strong updraft. As the droplet rises, it collides with and captures smaller drops in its path, and grows until it reaches a size of about 1 mm. At this point, the updraft in the cloud is just able to balance the pull of gravity on the drop. Here, the drop remains suspended until it grows just a little bigger. Once the fall velocity of the drop is greater than the updraft velocity in the cloud, the drop slowly descends. As the drop falls, some of the smaller droplets get caught in the airstream around it, and are swept aside. Larger cloud droplets are captured by the falling drop, which then grows larger. By the time this drop reaches the bottom of the cloud, it will be a large raindrop with a diameter of over 5 mm. Because raindrops of this size fall faster and reach the ground first, they typically occur at

*It was once thought that atmospheric electricity played a significant role in the production of rain. Today, many scientists feel that the difference in electrical charge that exists between cloud droplets results from the bouncing collisions between them. It is felt that the weak separation of charge and the weak electrical fields in developing, relatively warm clouds are not significant in initiating precipitation. However, studies show that coalescence is often enhanced in thunderstorms where strongly charged droplets exist in a strong electrical field.

the beginning of a rain shower originating in these warm, convective cumulus clouds.

So far, we have examined the way cloud droplets in warm clouds (that is, those clouds with temperatures above freezing) grow large enough by the collision-coalescence process to fall as raindrops. The most important factor in the production of raindrops is the cloud's liquid water content. In a cloud with sufficient water, other significant factors are:

1. the range of droplet sizes
2. the cloud thickness
3. the updrafts of the cloud
4. the electric charge of the droplets and the electric field in the cloud

Relatively thin stratus clouds with slow, upward air currents are, at best, only able to produce drizzle (the lightest form of rain), whereas the towering cumulus clouds associated with rapidly rising air can cause heavy showers. Now, let's turn our attention to the ice-crystal process of rain formation.

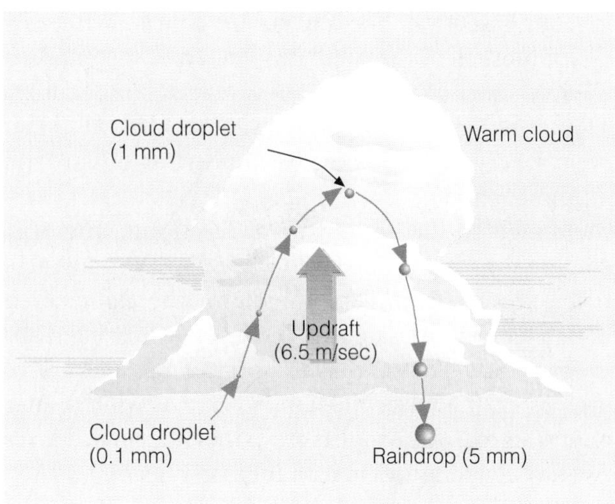

FIGURE 5.17 A cloud droplet rising then falling through a warm cumulus cloud can grow by collision and coalescence and emerge from the cloud as a large raindrop.

Ice-Crystal Process The ice-crystal (or *Bergeron**) process of rain formation proposes that both ice crystals and liquid cloud droplets must co-exist in clouds at temperatures below freezing. Consequently, this process of rain formation is extremely important in middle and high latitudes, where clouds are able to extend upwards into regions where air temperatures are below freezing. Such clouds are called *cold clouds.* Figure 5.18 illustrates a typical cumulonimbus cloud that has formed over the Great Plains of North America.

In the warm region of the cloud (below the freezing level) where only water droplets exist, we might expect to observe cloud droplets growing larger by the collision and coalescence process described in the previous sec-

*The ice-crystal process is also known as the *Bergeron process* after the Swedish meteorologist Tor Bergeron, who proposed that essentially all raindrops begin as ice crystals.

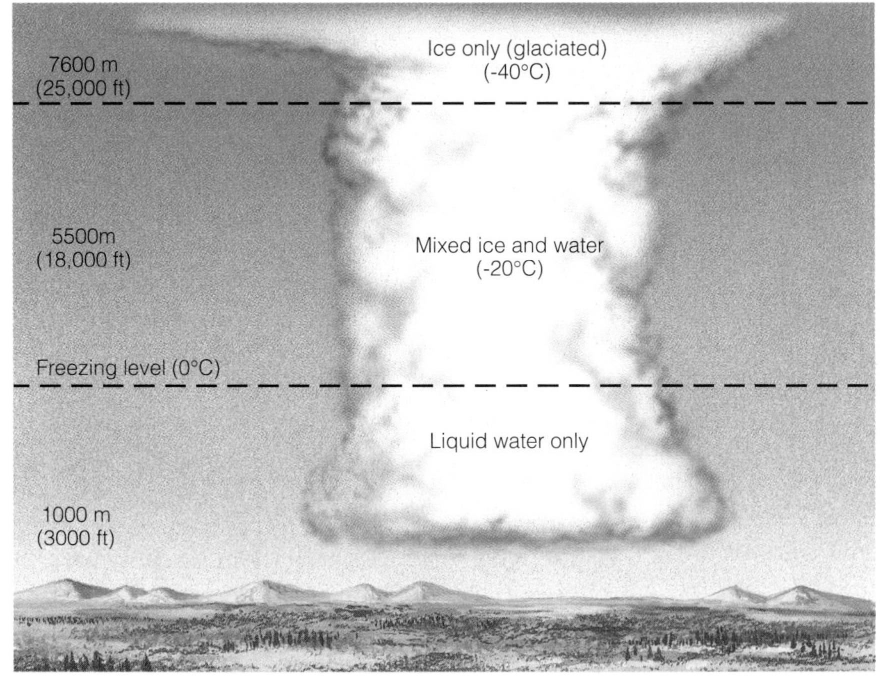

FIGURE 5.18 The distribution of ice and water in a cumulonimbus cloud.

tion. Surprisingly, in the cold air just above the freezing level, almost all of the cloud droplets are still composed of liquid water. Water droplets existing at temperatures below freezing are referred to as **supercooled.** At higher levels, ice crystals become more numerous, but are still outnumbered by water droplets. Ice crystals exist overwhelmingly in the upper part of the cloud, where air temperatures drop to well below freezing. Why are there so few ice crystals in the middle of the cloud, even though temperatures there, too, are below freezing? Laboratory studies reveal that the smaller the amount of pure water, the lower the temperature at which water freezes. Since cloud droplets are extremely small, it takes very low temperatures to turn them into ice.

Just as liquid cloud droplets form on condensation nuclei, ice crystals may form in subfreezing air if there are ice-forming particles present called **ice nuclei.** The number of ice-forming nuclei available in the atmosphere is small, especially at temperatures above −10°C (14°F). Although some uncertainty exists regarding the principal source of ice nuclei, it is known that certain clay minerals, bacteria in decaying plant leaf material, and ice crystals themselves are excellent ice nuclei. Moreover, particles serve as excellent ice-forming nuclei if their geometry resembles that of an ice crystal.

We can now understand why there are so few ice crystals in the subfreezing region of some clouds. Liquid cloud droplets may freeze, but only at very low temperatures. Ice nuclei may initiate the growth of ice crystals, but they do not abound in nature. Therefore, we are left with a cold cloud that contains many more liquid droplets than ice particles, even at low temperatures. Neither the tiny liquid nor solid particles are large enough to fall as precipitation. How, then, does the ice-crystal process produce rain and snow?

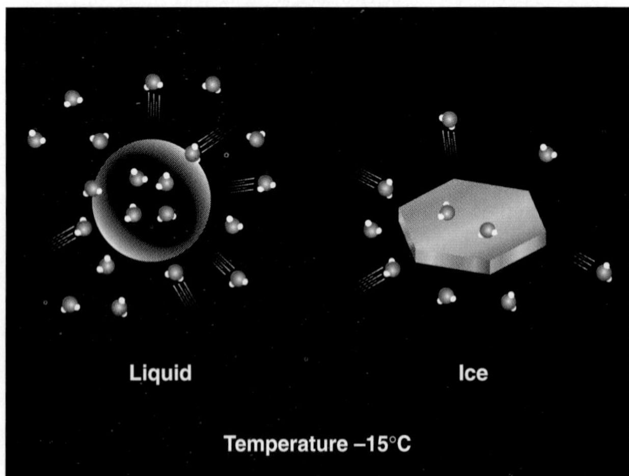

FIGURE 5.19 In a saturated environment, the water droplet and the ice crystal are in equilibrium, as the number of molecules leaving the surface of each droplet and ice crystal equals the number returning. The greater number of vapor molecules above the liquid indicates, however, that the saturation vapor pressure over water is greater than it is over ice.

In the subfreezing air of a cloud, many supercooled liquid droplets will surround each ice crystal. Suppose that the ice crystal and liquid droplet in Fig. 5.19 are part of a cold (−15°C), supercooled, saturated cloud. Since the air is saturated, both the liquid droplet and the ice crystal are in equilibrium, meaning that the number of molecules leaving the surface of both the droplet and the ice crystal must equal the number of molecules returning. Observe, however, that there are more vapor molecules above the liquid. The reason for this fact is that molecules escape the surface of water much easier than they escape the surface of ice. Consequently, more molecules escape the water surface at a given temperature, requiring more in the vapor phase to maintain saturation. Therefore, it takes more vapor molecules to saturate the air directly above the water droplet than it does to saturate the air directly above the ice crystal. Put another way, at the same subfreezing temperature, *the saturation vapor pressure just above the water surface is greater than the saturation vapor pressure above the ice surface.**

This difference in vapor pressure causes water vapor molecules to move (diffuse) from the droplet toward the ice crystal. The removal of vapor molecules reduces the vapor pressure above the droplet. Since the droplet is now out of equilibrium with its surroundings, it evaporates to replenish the diminished supply of water vapor above it. This process provides a continuous source of moisture for the ice crystal, which absorbs the water vapor and grows rapidly (see Fig. 5.20). Hence, during the **ice-crystal (Bergeron) process,** *ice crystals grow larger at the expense of the surrounding water droplets.*

The ice crystals may now grow even larger. For example, in some clouds, ice crystals might collide with supercooled liquid droplets. Upon contact, the liquid droplets freeze into ice and stick to the ice crystal—a process called **accretion,** or *riming.* The icy matter (rime) that forms is called *graupel* (or *snow pellets*). As the graupel falls, it may fracture or splinter into tiny ice particles when it collides with cloud droplets. These splinters may then go on themselves to become new graupel, which, in turn, may produce more splinters. In colder clouds, the delicate ice crystals may collide with other crystals and fracture into smaller ice particles, or tiny seeds, which freeze hundreds of supercooled droplets on contact. In both cases a chain reaction may develop, producing many

*This concept is illustrated in the insert in Fig. 4.5, p. 81.

ice crystals (see Fig. 5.21). As they fall, they may collide and stick to one another, forming an aggregate of ice crystals called a *snowflake.* If the snowflake melts before reaching the ground, it continues its fall as a raindrop. Therefore, much of the rain falling in middle and northern latitudes—even in summer—begins as snow.

Cloud Seeding and Precipitation The primary goal in many experiments concerning **cloud seeding** is to inject (or seed) a cloud with small particles that will act as nuclei, so that the cloud particles will grow large enough to fall to the surface as precipitation. The first ingredient in any seeding project is, of course, the presence of clouds, as seeding does not generate clouds. However, at least a portion of the cloud (preferably the upper part) must be supercooled because cloud seeding uses the ice-crystal process to cause the cloud particles to grow. The idea is to find clouds that have too low a ratio of ice crystals to droplets and then to add enough artificial ice nuclei so that the ratio of crystals to droplets is optimal (about 1:100,000) for producing precipitation.

Some of the first experiments in cloud seeding were conducted by Vincent Schaefer and Irving Langmuir during the late 1940s. To seed a cloud, they dropped crushed pellets of *dry ice* (solid carbon dioxide) from a plane. Because dry ice has a temperature of −78°C (−108°F), it acts as a cooling agent. As the extremely cold, dry ice pellets fall through the cloud, they quickly cool the air around them.

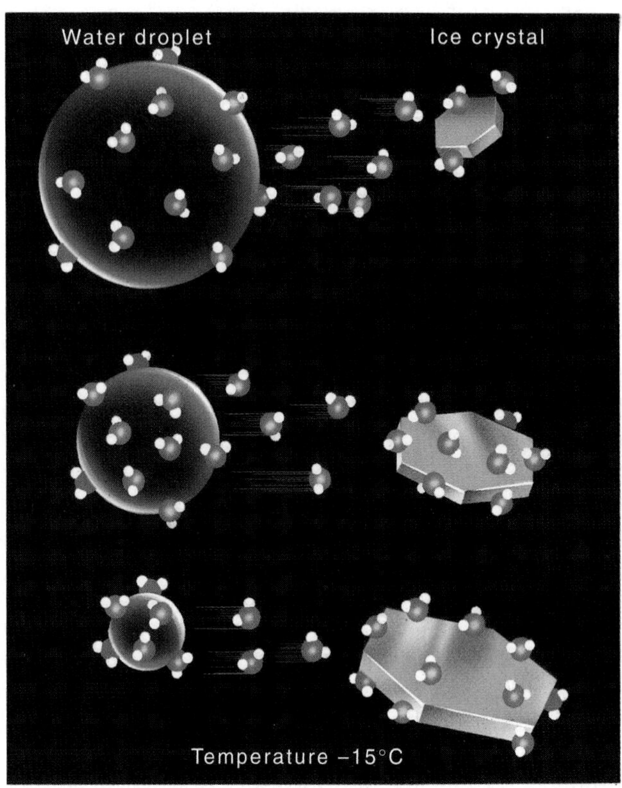

Meteorology Now™ ACTIVE FIGURE 5.20
The ice-crystal process. The greater number of water vapor molecules around the liquid droplets causes water molecules to diffuse from the liquid drops toward the ice crystals. The ice crystals absorb the water vapor and grow larger, while the water droplets grow smaller.
Watch this Active Figure at http://earthscience.brookscole.com/ahrens/ess4e.

(a) Falling ice crystals may freeze supercooled droplets on contact (accretion), producing larger ice particles.

(b) Falling ice particles may collide and fracture into many tiny (secondary) ice particles.

(c) Falling ice crystals may collide and stick to other ice crystals (aggregation), producing snowflakes.

FIGURE 5.21 Ice particles in clouds.

FOCUS ON AN ENVIRONMENTAL ISSUE

Does Cloud Seeding Enhance Precipitation?

Just how effective is artificial seeding with silver iodide in increasing precipitation? This is a much-debated question among meteorologists. First of all, it is difficult to evaluate the results of a cloud-seeding experiment. When a seeded cloud produces precipitation, the question always remains as to how much precipitation would have fallen had the cloud not been seeded.

Other factors must be considered when evaluating cloud-seeding experiments: the type of cloud, its temperature, moisture content, droplet size distribution, and updraft velocities in the cloud.

Although some experiments suggest that cloud seeding does not increase precipitation, others seem to indicate that seeding *under the right conditions* may enhance precipitation between 5 percent and 20 percent. And so the controversy continues.

Some cumulus clouds show an "explosive" growth after being seeded. The latent heat given off when the droplets freeze functions to warm the cloud, causing it to become more buoyant. It grows rapidly and becomes a longer-lasting cloud, which may produce more precipitation.

The business of cloud seeding can be a bit tricky, since overseeding can produce too many ice crystals. When this phenomenon occurs, the cloud becomes glaciated (all liquid droplets become ice) and the ice particles, being very small, do not fall as precipitation. Since few liquid droplets exist, the ice crystals cannot grow by the ice-crystal (Bergeron) process; rather, they evaporate, leaving a clear area in a thin, stratified cloud. Because dry ice can produce the most ice crystals in a supercooled cloud, it is the substance most suitable for deliberate overseeding. Hence, it is the substance most commonly used to dissipate cold fog at airports (see Chapter 4, p. 95).

Warm clouds with temperatures above freezing have also been seeded in an attempt to produce rain. Tiny water drops and particles of hygroscopic salt are injected into the base (or top) of the cloud. These particles (called *seed drops),* when carried into the cloud by updrafts, create large cloud droplets, which grow even larger by the collision-coalescence process. Apparently, the seed drop size plays a major role in determining the effectiveness of seeding with hygroscopic particles. To date, however, the results obtained using this method are inconclusive.

Cloud seeding may be inadvertent. Some industries emit large concentrations of condensation nuclei and ice nuclei into the air. Studies have shown that these particles are at least partly responsible for increasing precipitation in, and downwind of, cities. On the other hand, studies have also indicated that the burning of certain types of agricultural waste may produce smoke containing many condensation nuclei. These produce clouds that yield less precipitation because they contain numerous, but very small, droplets.

In summary, cloud seeding in certain instances may lead to more precipitation; in others, to less precipitation, and, in still others, to no change in precipitation amounts. Many of the questions about cloud seeding have yet to be resolved.

This cooling causes the air around the pellet to become supersaturated. In this supersaturated air, water vapor forms directly into many tiny cloud droplets. In the very cold air created by the falling pellets (below −40°C), the tiny droplets instantly freeze into tiny ice crystals. The newly formed ice crystals then grow larger by deposition at the expense of the nearby liquid droplets and, upon reaching a sufficiently large size, fall as precipitation.

In 1947, Bernard Vonnegut demonstrated that silver iodide (AgI) could be used as a cloud-seeding agent. Because silver iodide has a crystalline structure similar to an ice crystal, it acts as an effective ice nucleus at temperatures of −4°C (25°F) and lower. Silver iodide causes ice crystals to form in two primary ways:

1. Ice crystals form when silver iodide crystals come in contact with supercooled liquid droplets.
2. Ice crystals grow in size as water vapor deposits onto the silver iodide crystal.

Silver iodide is much easier to handle than dry ice, since it can be supplied to the cloud from burners located either on the ground or on the wing of a small aircraft. Although other substances, such as lead iodide and cupric sulfide, are also effective ice nuclei, silver iodide still remains the most commonly used substance in cloud-seeding projects. (Additional information on the controversial topic, the effectiveness of cloud seeding, is given in the Focus section above.)

Under certain conditions, clouds may be seeded naturally. For example, when cirriform clouds lie directly above a lower cloud deck, ice crystals may descend from the higher cloud and seed the cloud below. As the ice crystals mix into the lower cloud, supercooled droplets are converted to ice crystals, and the precipitation process is enhanced. Sometimes the ice crystals in the lower cloud may settle out, leaving a clear area or "hole" in the cloud. When the cirrus clouds form waves downwind from a mountain chain, bands of precipitation often form (see Fig. 5.22).

Precipitation in Clouds In cold, strongly convective clouds, precipitation may begin only minutes after the

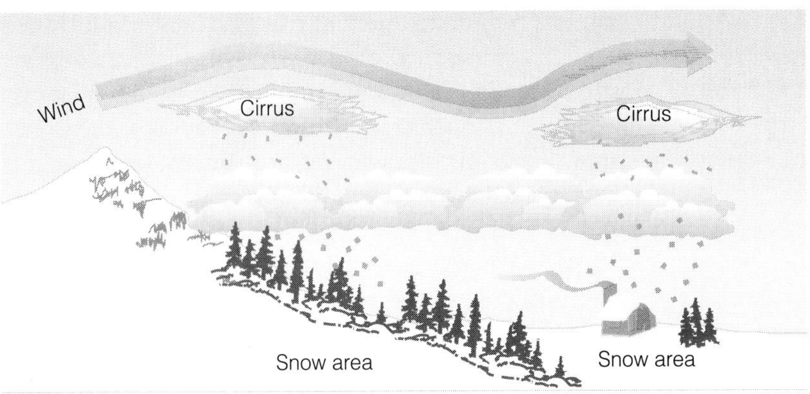

FIGURE 5.22 Natural seeding by cirrus clouds may form bands of precipitation downwind of a mountain chain.

cloud forms and may be initiated by either the collision-coalescence or the ice-crystal process. Once either process begins, most precipitation growth is by accretion. Although precipitation is commonly absent in warm-layered clouds, such as stratus, it is often associated with such cold-layered clouds as nimbostratus and altostratus. This precipitation is thought to form principally by the ice-crystal (Bergeron) process because the liquid water content of these clouds is generally lower than that in convective clouds, thus making the collision-coalescence process much less effective. Nimbostratus clouds are normally thick enough to extend to levels where air temperatures are quite low, and they usually last long enough for the ice-crystal process to initiate precipitation.

BRIEF REVIEW

In the last few sections we encountered a number of important concepts and ideas about how cloud droplets can grow large enough to fall as precipitation. Before examining the various types of precipitation, here is a summary of some of the important ideas presented so far:

- Cloud droplets are very small, much too small to fall as rain.
- Cloud droplets form on cloud condensation nuclei. Hygroscopic nuclei, such as salt, allow condensation to begin when the relative humidity is less than 100 percent.
- Cloud droplets, in above-freezing air, can grow larger as faster-falling, bigger droplets collide and coalesce with smaller droplets in their path.
- In the ice-crystal (Bergeron) process of rain formation, both ice crystals and liquid cloud droplets must coexist at below-freezing temperatures. The difference in saturation vapor pressure between liquid and ice causes water vapor to diffuse from the liquid droplets (which shrink) toward the ice crystals (which grow).
- Most of the rain that falls over middle latitudes results from melted snow that formed from the ice-crystal (Bergeron) process.
- Cloud seeding with silver iodide can only be effective in coaxing precipitation from clouds if the cloud is supercooled and the proper ratio of cloud droplets to ice crystals exists.

PRECIPITATION TYPES

Up to now, we have seen how cloud droplets are able to grow large enough to fall to the ground as rain or snow. While falling, raindrops and snowflakes may be altered by atmospheric conditions encountered beneath the cloud and transformed into other forms of precipitation that can profoundly influence our environment.

Rain Most people consider **rain** to be any falling drop of liquid water. To the meteorologist, however, that falling drop must have a diameter equal to, or greater than, 0.5 mm (0.02 in.) to be considered rain. Fine uniform drops of water whose diameters are smaller than 0.5 mm (which is a diameter about one-half the width of the letter "o" on this page) are called **drizzle.** Most drizzle falls from stratus clouds; however, small raindrops may fall through air that is unsaturated, partially evaporate, and reach the ground as drizzle. Occasionally, the rain falling from a cloud never reaches the surface because the low humidity causes rapid evaporation. As the drops become smaller, their rate of fall decreases, and they appear to hang in the air as a rain streamer. These evaporating streaks of precipitation are called **virga*** (see Fig. 5.23).

Raindrops may also fall from a cloud and not reach the ground if they encounter the rapidly rising air of an updraft. If the updraft weakens or changes direction and becomes a downdraft, the suspended drops will fall to the ground as a sudden rain **shower.** The showers falling from cumuliform clouds are usually brief and sporadic, as the cloud moves overhead and then drifts on by. If the shower is excessively heavy, it is termed a *cloudburst.* Beneath a cumulonimbus cloud, which normally contains

*Studies suggest that the "rain streamer" is actually caused by ice (which is more reflective) changing to water (which is less reflective). Apparently, most evaporation occurs below the virga line.

DID YOU KNOW?

Often before it rains the air has a distinctive (somewhat earthy) smell to it. This odor may originate from soil bacteria that produce aromatic gases. As rain falls onto the soil, it pushes these gases into the air, where winds carry them out ahead of the advancing rain shower.

large convection currents, it is entirely possible that one side of a street may be dry (updraft side), while a heavy shower is occurring across the street (downdraft side). Continuous rain, on the other hand, usually falls from a layered cloud that covers a large area and has smaller vertical air currents. These are the conditions normally associated with nimbostratus clouds.

Raindrops that reach the earth's surface are seldom larger than about 6 mm (0.2 in.), the reason being that the collisions (whether glancing or head-on) between raindrops tend to break them up into many smaller drops. Additionally, when raindrops grow too large they become unstable and break apart. (Earlier, we learned that for a falling drop of water to be called rain, the drop must have a diameter greater than 0.5 mm. Is the shape of this falling drop round or is it tear-shaped? If you are unsure, read the Focus section on p. 129.)

After a rainstorm, visibility usually improves primarily because precipitation removes (scavenges) many of the suspended particles. When rain combines with gaseous pollutants, such as oxides of sulfur and nitrogen, it becomes acidic. *Acid rain,* which has an adverse effect on plants and water resources, is becoming a major problem in many industrialized regions of the world. We will examine the acid rain problem more thoroughly in Chapter 12.

Snow We have learned that much of the precipitation reaching the ground actually begins as **snow.** In summer, the freezing level is usually high and the snowflakes falling from a cloud melt before reaching the surface. In winter, however, the freezing level is much lower, and falling snowflakes have a better chance of survival. In fact, snowflakes can generally fall about 300 m (or 1000 ft) below the freezing level before completely melting. When the warmer air beneath the cloud is relatively dry, the snowflakes partially melt. As the liquid water evaporates, it chills the snowflake, which retards its rate of melting. Consequently, in air that is relatively dry, snowflakes may reach the ground even when the air temperature is considerably above freezing.

Is it ever "too cold to snow"? Although many believe this expression, the fact remains that it is *never* too cold to snow. True, more water vapor will condense from warm saturated air than from cold saturated air. But, no matter how cold the air becomes, it always contains some water vapor that could produce snow. In fact, tiny ice crystals have been observed falling at temperatures as low as −47°C (−53°F). We usually associate ex-

FIGURE 5.23 The streaks of falling precipitation that evaporate before reaching the ground are called *virga.*

FOCUS ON A SPECIAL TOPIC

Are Raindrops Tear-Shaped?

As rain falls, the drops take on a characteristic shape. Choose the shape in Fig. 1 that you feel most accurately describes that of a falling raindrop. Did you pick number 1? The tear-shaped drop has been depicted by artists for many years. Unfortunately, *raindrops are not tear-shaped.* Actually, the shape depends on the drop size. Raindrops less than 2 mm (0.08 in.) in diameter are nearly spherical and look like raindrop number 2. The attraction among the molecules of the liquid (surface tension) tends to squeeze the drop into a shape that has the smallest surface area for its total volume—a sphere.

Large raindrops, with diameters exceeding 2 mm, take on a different shape as they fall. Believe it or not, they look like number 3, slightly elongated, flattened on the bottom, and rounded on top. As the larger drop falls, the air pressure against the drop is greatest on the bottom and least on the sides. The pressure of the air on the bottom flattens the drop, while the lower pressure on its sides allows it to expand a little. This shape has been described as everything from a falling parachute to a loaf of bread, or even a hamburger bun. You may call it what you wish, but remember: it is not tear-shaped.

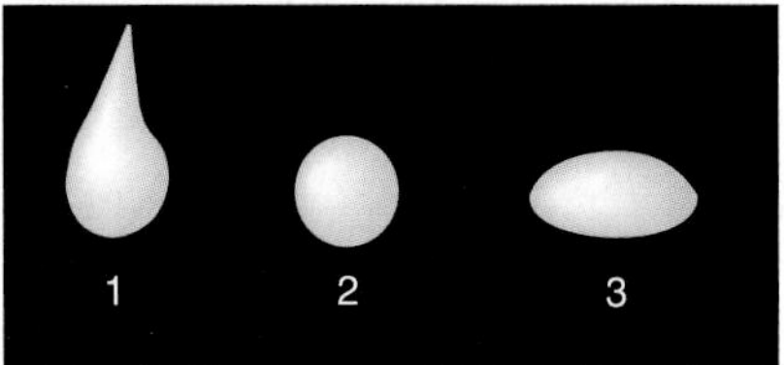

FIGURE 1 Which of the three drops drawn here represents the real shape of a falling raindrop?

tremely cold air with "no snow" because the coldest winter weather occurs on clear, calm nights—conditions that normally prevail with strong high pressure areas that have few if any clouds.

When ice crystals and snowflakes fall from high cirrus clouds they are called **fallstreaks.** Fallstreaks behave in much the same way as virga—as the ice particles fall into drier air, they usually disappear as they change from ice into vapor (called *sublimation*). Because the wind at higher levels moves the cloud and ice particles horizontally more quickly than do the slower winds at lower levels, fallstreaks often appear as dangling white streamers (see Fig. 5.24). Moreover, fallstreaks descending into lower, supercooled clouds may actually seed them.

Snowflakes falling through moist air that is slightly above freezing slowly melt as they descend. A thin film

Photo by author

FIGURE 5.24 The dangling white streamers of ice crystals beneath these cirrus clouds are known as *fallstreaks.* The bending of the streaks is due to the changing wind speed with height.

DID YOU KNOW?

Snow may not always be white. Blood-red snow actually fell over central France on March 10, 1869. Apparently, strong winds swept reddish dust from the Sahara Desert all the way into France, where it became mixed with the snow.

of water forms on the edge of the flakes, which acts like glue when other snowflakes come in contact with it. In this way, several flakes join to produce giant snowflakes that often measure an inch or more in diameter. These large, soggy snowflakes are associated with moist air and temperatures near freezing. However, when snowflakes fall through extremely cold air with a low moisture content, they do not readily stick together and small, powdery flakes of "dry" snow accumulate on the ground.

If you catch falling snowflakes on a dark object and examine them closely, you will see that the most common snowflake form is a fernlike branching shape called *dendrite* (see Fig. 5.25). As ice crystals fall through a cloud, they are constantly exposed to changing temperatures and moisture conditions. Since many ice crystals can join together *(aggregate)* to form a much larger snowflake, ice crystals may assume many complex patterns.

Snow falling from developing cumulus clouds is often in the form of **flurries.** These are usually light showers that fall intermittently for short durations and produce only light accumulations. A more intense snow shower is called a **snow squall.** These brief but heavy falls of snow are comparable to summer rain showers and, like snow flurries, usually fall from cumuliform clouds. A more continuous snowfall (sometimes steadily, for several hours) accompanies nimbostratus and altostratus clouds. The intensity of snow is based on its reduction of horizontal visibility at the time of observation (see Table 5.1).

When a strong wind is blowing at the surface, snow can be picked up and deposited into huge drifts. Drifting snow is usually accompanied by *blowing snow;* that is, snow lifted from the surface by the wind and blown about in such quantities that horizontal visibility is greatly restricted. The combination of drifting and

TABLE 5.1 Snowfall Intensity

SNOWFALL DESCRIPTION	VISIBILITY
Light	Greater than 1/2 mile*
Moderate	Greater than 1/4 mile, less than or equal to 1/2 mile
Heavy	Less than or equal to 1/4 mile

*In the United States, the National Weather Service determines visibility (the greatest distance you can see) in miles.

FIGURE 5.25 Computer color-enhanced image of dendrite snowflakes.

blowing snow, after falling snow has ended, is called a *ground blizzard.* A true **blizzard** is a weather condition characterized by low temperatures and strong winds (greater than 30 knots) bearing large amounts of fine, dry, powdery particles of snow, which can reduce visibility to only a few meters.

Sleet and Freezing Rain Consider the falling snowflake in Fig. 5.26. As it falls into warmer air, it begins to melt. When it falls through the deep subfreezing surface layer of air, the partially melted snowflake or cold raindrop turns back into ice, not as a snowflake, but as a tiny transparent (or translucent) *ice pellet* called **sleet.*** Generally, these ice pellets bounce when striking the ground and produce a tapping sound when they hit a window or piece of metal.

The cold surface layer beneath a cloud may be too shallow to freeze raindrops as they fall. In this case, they reach the surface as supercooled liquid drops. Upon striking a cold object, the drops spread out and almost immediately freeze, forming a thin veneer of ice. This form of precipitation is called **freezing rain,** or *glaze.* If the drops are quite small, the precipitation is called *freezing drizzle.* When small, supercooled cloud or fog droplets strike an object whose temperature is below freezing, the tiny droplets freeze, forming an accumulation of white or milky granular ice called **rime** (see Fig. 5.27).**

Freezing rain can create a beautiful winter wonderland by coating everything with silvery, glistening ice. At the same time, highways turn into skating rinks for automobiles, and the destructive weight of the ice—which can be many tons on a single tree—breaks tree branches, power lines, and telephone cables (see Fig. 5.28). A case in point is the huge ice storm of January, 1998, which left millions of people without power in northern New England and Canada, and caused over $1 billion in damages. The area most frequently hit by these storms extends over a broad region from Texas into Minnesota and eastward into the middle Atlantic states and New England. Such storms are extremely rare in southern California and Florida. (For additional information on freezing rain and its effect on aircraft, read the Focus section on p. 132.)

*Occasionally, the news media incorrectly use the term *sleet* to represent a mixture of rain and snow. The term used in this manner is, however, the British meaning.

**When a sheet of ice covering a road surface or pavement appears relatively dark, it is often referred to as *black ice.* Black ice commonly forms when light rain, drizzle, or supercooled fog droplets come in contact with surfaces (especially those of bridges and overpasses) that have cooled to a temperature below freezing.

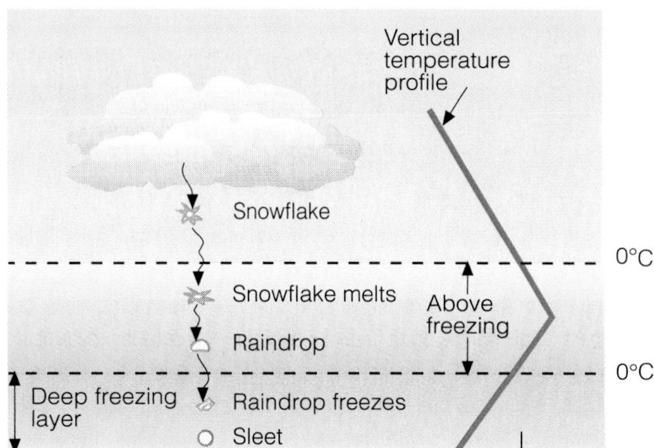

Meteorology Now™ ACTIVE FIGURE 5.26
Sleet forms when a partially melted snowflake or a cold raindrop freezes into a pellet of ice before reaching the ground.
Watch this Active Figure at http://earthscience.brookscole.com/ahrens/ess4e.

FIGURE 5.27 An accumulation of rime forms on tree branches as supercooled fog droplets freeze on contact in the below-freezing air.

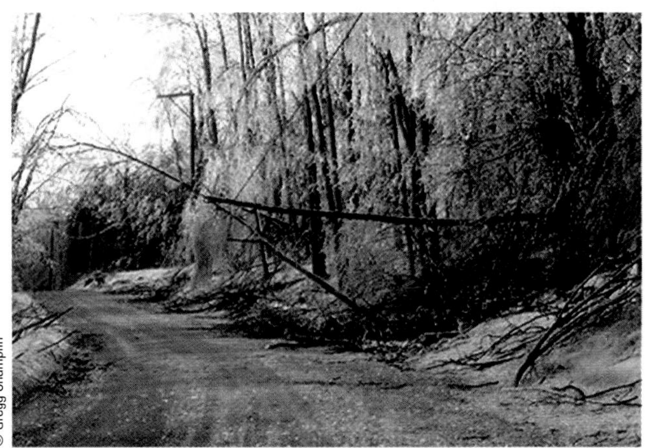

FIGURE 5.28 A heavy coating of freezing rain during this ice storm caused tree limbs to break and power lines to sag.

FOCUS ON AN OBSERVATION

Aircraft Icing

Consider an aircraft flying through an area of freezing rain or through a region of large supercooled droplets in a cumuliform cloud. As the large, supercooled drops strike the leading edge of the wing, they break apart and form a film of water, which quickly freezes into a solid sheet of ice. This smooth, transparent ice—called *clear ice*—is similar to the freezing rain or glaze that coats trees during ice storms. Clear ice can build up quickly; it is heavy and difficult to remove, even with modern de-icers.

When an aircraft flies through a cloud composed of tiny, supercooled liquid droplets, *rime ice* may form. Rime ice forms when some of the cloud droplets strike the wing and freeze before they have time to spread, thus leaving a rough and brittle coating of ice on the wing. Because the small, frozen droplets trap air between them, rime ice usually appears white (see Fig. 5.27). Even though rime ice redistributes the flow of air over the wing more than clear ice does, it is lighter in weight and is more easily removed with de-icers.

Because the raindrops and cloud droplets in most clouds vary in size, a mixture of clear and rime ice usually forms on aircraft. Also, because concentrations of liquid water tend to be greatest in warm air, icing is usually heaviest and most severe when the air temperature is between 0°C and −10°C (32°F and 14°F).

A major hazard to aviation, icing reduces aircraft efficiency by increasing weight. Icing has other adverse effects, depending on where it forms. On a wing or fuselage, ice can disrupt the air flow and decrease the plane's flying capability. When ice forms in the air intake of the engine, it robs the engine of air, causing a reduction in power. Icing may also affect the operation of brakes, landing gear, and instruments. Because of the hazards of ice on an aircraft, its wings are usually sprayed with a type of antifreeze before taking off during cold, inclement weather.

DID YOU KNOW?

The worst ice storm to hit Kansas and Missouri in 100 years deposited 5 cm (2 in.) of ice over sections of these states during January, 2002, causing over 300,000 people to lose power.

Snow Grains and Snow Pellets **Snow grains** are small, opaque grains of ice, the solid equivalent of drizzle. They fall in small quantities from stratus clouds, and never in the form of a shower. Upon striking a hard surface, they neither bounce nor shatter. **Snow pellets,** on the other hand, are white, opaque grains of ice about the size of an average raindrop. They are sometimes confused with snow grains. The distinction is easily made, however, by remembering that, unlike snow grains, snow pellets are brittle, crunchy, and bounce (or break apart) upon hitting a hard surface. They usually fall as showers, especially from cumulus congestus clouds. Snow pellets form as ice crystals collide with supercooled water droplets that freeze into a spherical aggregate of icy matter *(rime)* containing many air spaces. When the ice particle accumulates so much rime that it can no longer be recognized as an ice crystal (or snowflake), it is called *graupel.* During the winter, when the freezing level is at a low elevation, the graupel reaches the surface as a light, round clump of snowlike ice called a *snow pellet.* In a thunderstorm, when the freezing level is well above the surface, graupel that reaches the ground is sometimes called *soft hail.* During the summer, the graupel may melt and reach the surface as a large raindrop. In vigorously convective clouds, however, the graupel may develop into full-fledged hailstones.

Hail **Hailstones** are pieces of ice either transparent or partially opaque, ranging in size from that of small peas to that of golf balls or larger. (See Fig. 5.29). Some are round, others take on irregular shapes. The largest authenticated hailstone measured in the United States fell on Aurora, Nebraska, during June, 2003. This giant hailstone had a measured diameter of 17.8 cm (7 in.) and a circumference of 47.6 cm (18.7 in.) (See Fig. 5.30). Although an accurate weight was difficult to obtain, the hailstone (being almost as large as a soccer ball) probably weighed over 1.75 lbs. Canada's record hailstone fell on Cedoux, Saskatchewan, during August, 1973. It weighed 290 grams (0.6 lb) and measured about 10 cm (4 in.) in diameter. Needless to say, large hailstones are quite destructive as they can break windows, dent cars, batter roofs of homes, and cause extensive damage to livestock and crops. In fact, a single hailstorm can de-

Photo by author

FIGURE 5.29 The accumulation of small hail after a thunderstorm. The hail formed as supercooled cloud droplets collected on ice particles called *graupel* inside a cumulonimbus cloud.

NOAA

FIGURE 5.30 This giant hailstone — the largest ever reported in the United States with a diameter of 17.8 cm (7 in.) — fell on Aurora, Nebraska, during June, 2003.

stroy a farmer's crop in a matter of minutes. Estimates are that, in the United States alone, hail damage amounts to hundreds of millions of dollars annually. Although hailstones are potentially lethal, only two fatalities due to falling hail have been documented in the United States during the twentieth century.

Hail is produced in a cumulonimbus cloud when graupel, large frozen raindrops, or just about any particles (even insects) act as embryos that grow by accumulating supercooled liquid droplets—*accretion*. For a hailstone to grow to a golf ball size, it must remain in the cloud for between 5 and 10 minutes. Violent, upsurging air currents within the cloud carry small embryos high above the freezing level. When the updrafts are tilted, the embryos are swept laterally through the cloud. Studies reveal that the width and tilt of the main updraft are very important to hailstone growth, with the best trajectory being one that is nearly horizontal through the cloud (see Fig. 5.31).

As the embryos pass through regions of varying liquid water content, a coating of ice forms around them and they grow larger and larger. When the ice particles are of appreciable size, they become too large and heavy to be supported by the rising air, and they then begin to fall as hail. As they slowly descend, the hailstones may get caught in a violent updraft, only to be carried upward once again to repeat the cycle. Or, they may fall through the cloud and begin to melt in the warmer air below. Small hailstones often melt before reaching the ground, but, in the violent thunderstorms of summer, hailstones may grow large enough to reach the ground before completely melting. Strangely, then, the largest form of frozen precipitation occurs during the warmest time of the year.

As the cumulonimbus cloud moves along, it may deposit its hail in a long, narrow band known as a *hailstreak*. If the cloud should remain almost stationary for a period of time, substantial accumulation of hail is

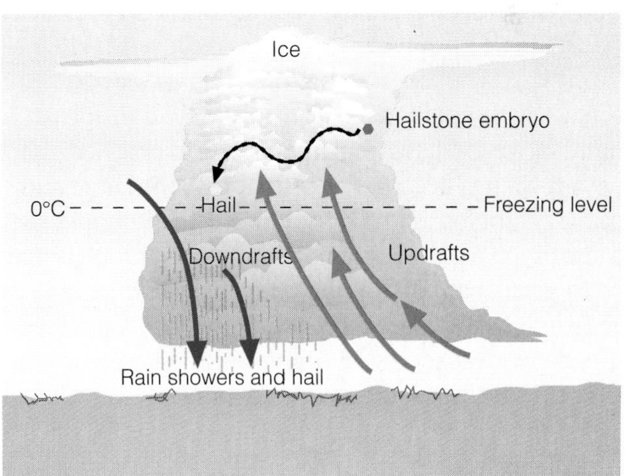

Meteorology Now™ ACTIVE FIGURE 5.31
Hailstones begin as embryos (usually ice particles) that remain suspended in the cloud by violent updrafts. When the updrafts are tilted, the ice particles are swept horizontally through the cloud, producing the optimal trajectory for hailstone growth. Along their path, the ice particles collide with supercooled liquid droplets, which freeze on contact. The ice particles eventually grow large enough and heavy enough to fall toward the ground as hailstones. Watch this Active Figure at http://earthscience.brookscole.com/ahrens/ess4e.

DID YOU KNOW?

Each year between 500 and 700 hailstorms produce hail large enough to cause damage or injury. However, in the United States, it is rare for a person to be killed by hail from one of these storms. But hail can be lethal—a hailstone the size of a baseball will whiz earthward at a speed close to 87 knots (100 mi/hr). During April, 1996, for instance, 92 people tragically perished in a hailstorm near Gopolganj, Bangladesh. Some of the hailstones in that storm reportedly weighed about 1 kg (2.2 lbs). In another example, 200 sheep caught in an open field during a violent hailstorm were killed in Montana during April, 1978, as baseball-size hailstones became lethal missiles.

possible. For example, in June, 1984, a devastating hailstorm lasting over an hour dumped knee-deep hail on the suburbs of Denver, Colorado. And during November, 2003, a rare hailstorm dumped more than 12.5 cm (5 in.) of hail over sections of Los Angeles, California, causing gutters to clog and floods to occur. In addition to its destructive effect, accumulation of hail on a roadway is a hazard to traffic as when, for example, four people lost their lives near Soda Springs, California, in a 15-vehicle pileup on a hail-covered freeway in September, 1989.

Because hailstones are so damaging, various methods have been tried to prevent them from forming in thunderstorms. One method employs the seeding of clouds with large quantities of silver iodide. These nuclei freeze supercooled water droplets and convert them into ice crystals. The ice crystals grow larger as they come in contact with additional supercooled cloud droplets. In time, the ice crystals grow large enough to be called graupel, which then becomes a hailstone embryo. Large numbers of embryos are produced by seeding in hopes that competition for the remaining supercooled droplets may be so great that none of the embryos would be able to grow into large and destructive hailstones. Russian scientists claim great success in suppressing hail using ice nuclei, such as silver iodide and lead iodide. In the United States, the results of most hail-suppression experiments are still inconclusive.

MEASURING PRECIPITATION

Instruments Any instrument that can be used to collect and measure rainfall is called a *rain gauge.* A **standard rain gauge** is commonly used to measure rainfall. This instrument consists of a funnel-shaped collector attached to a long measuring tube (see Fig. 5.32). The cross-sectional area of the collector is ten times that of the tube. Hence, rain falling into the collector is amplified tenfold in the tube, permitting measurements of great precision—to as low as one-hundredth (0.01) of an inch. An amount less than this is called a **trace.**

Another instrument that measures rainfall is the *tipping bucket rain gauge.* In Fig. 5.33, notice that this gauge has a receiving funnel leading to two small metal collectors (buckets). The bucket beneath the funnel collects the rain water. When it accumulates the equivalent of one-hundredth of an inch of rain, the weight of the water causes it to tip and empty itself. The second bucket immediately moves under the funnel to catch the water. When it fills, it also tips and empties itself, while the original bucket moves back beneath the funnel. Each time a bucket tips, an electric contact is made, causing a pen to register a mark on a remote recording chart. Adding up the total number of marks gives the rainfall for a certain time period. A problem with the tipping bucket rain gauge is that during each "tip" it loses some rainfall and, therefore, under-measures rainfall amounts, especially during heavy downpours. The tipping bucket is the rain gauge used in the automated (ASOS) weather stations.

Remote recording of precipitation can also be made with a *weighing-type rain gauge.* With this gauge,

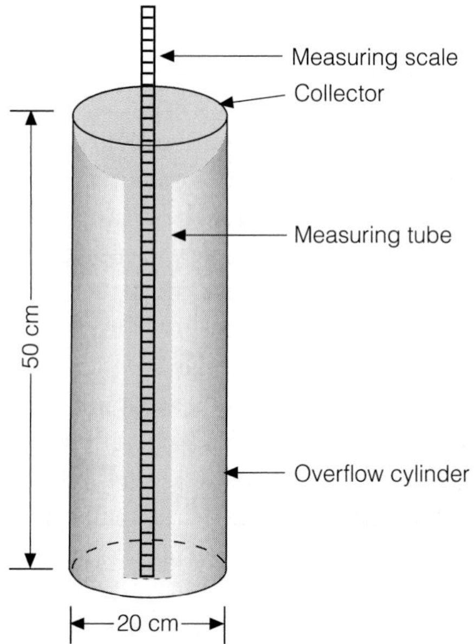

FIGURE 5.32 Components of the standard rain gauge.

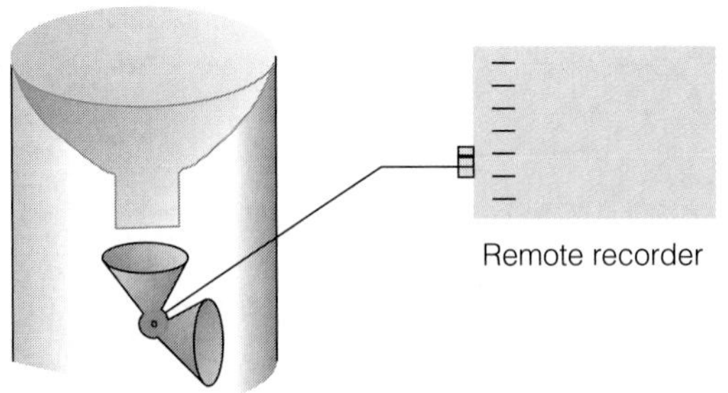

FIGURE 5.33 The tipping bucket rain gauge. Each time the bucket fills with one-hundredth of an inch of rain, it tips, sending an electric signal to the remote recorder.

precipitation is caught in a cylinder and accumulates in a bucket. The bucket sits on a sensitive weighing platform. Special gears translate the accumulated weight of rain or snow into millimeters or inches of precipitation. The precipitation totals are recorded by a pen on chart paper, which covers a clock-driven drum. By using special electronic equipment, this information can be transmitted from rain gauges in remote areas to satellites or land-based stations, thus providing precipitation totals from previously inaccessible regions.

The depth of snow in a region is determined by measuring its depth at three or more representative areas. The amount of snowfall is defined as the average of these measurements. Snow depth may also be measured by removing the collector and inner cylinder of a standard rain gauge and allowing snow to accumulate in the outer tube. Generally, about 10 inches of snow will melt down to about 1 inch of water, giving a typical fresh snowpack a **water equivalent*** of 10:1. This ratio, however, will vary greatly, depending on the texture and packing of the snow. Knowing the water equivalent of snow can provide valuable information about spring runoff and the potential for flooding, especially in mountain areas.

Doppler Radar and Precipitation **Radar** (*ra*dio *d*etection *a*nd *r*anging) has become an essential tool of the atmospheric scientist, for it gathers information about storms and precipitation in previously inaccessible regions. Atmospheric scientists use radar to examine the inside of a cloud much like physicians use X-rays to examine the inside of a human body. Essentially, the radar unit consists of a transmitter that sends out short, powerful microwave pulses. When this energy encounters a foreign object—called a *target*—a fraction of the energy is scattered back toward the transmitter and is detected by a receiver. The returning signal is amplified and displayed on a screen, producing an image or "echo" from the target. The elapsed time between transmission and reception indicates the target's distance.

The brightness of the echo is directly related to the amount (intensity) of rain falling in the cloud. So, the radar screen shows not only where precipitation is occurring, but also how intense it is. Typically the radar image is displayed using various colors to denote the intensity of precipitation within the range of the radar unit.

During the 1990s, **Doppler radar** replaced the conventional radar units that were put into service shortly after World War II. Doppler radar is like conventional radar in that it can detect areas of precipitation and measure rainfall intensity (see Fig. 5.34a). Using special computer programs called *algorithms*, the rainfall intensity, over a given area for a given time, can be computed and displayed as an estimate of total rainfall over that particular area (see Fig. 5.34b). But the Doppler radar can do more than conventional radar.

Because the Doppler radar uses the principle called *Doppler shift*,* it has the capacity to measure the speed at which falling rain is moving horizontally toward or away from the radar antenna. Falling rain moves with the wind. Consequently, Doppler radar allows scientists to peer into a tornado-generating thunderstorm and observe its wind. We will investigate these ideas further in Chapter 10, when we consider the formation of severe thunderstorms and tornadoes.

In some instances, radar displays indicate precipitation where there is none reaching the surface. This situation happens because the radar beam travels in a straight line and the earth curves away from it. Hence, the return echo is not necessarily that of precipitation reaching the ground, but is that of raindrops in the cloud.

**Water equivalent* is the depth of water that would result from the melting of a snow sample.

*The Doppler shift (or effect) is the change in the frequency of waves that occurs when the emitter or the observer is moving toward or away from the other. As an example, suppose a high-speed train is approaching you. The higher-pitched (higher frequency) whistle you hear as the train approaches will shift to a lower pitch (lower frequency) after the train passes.

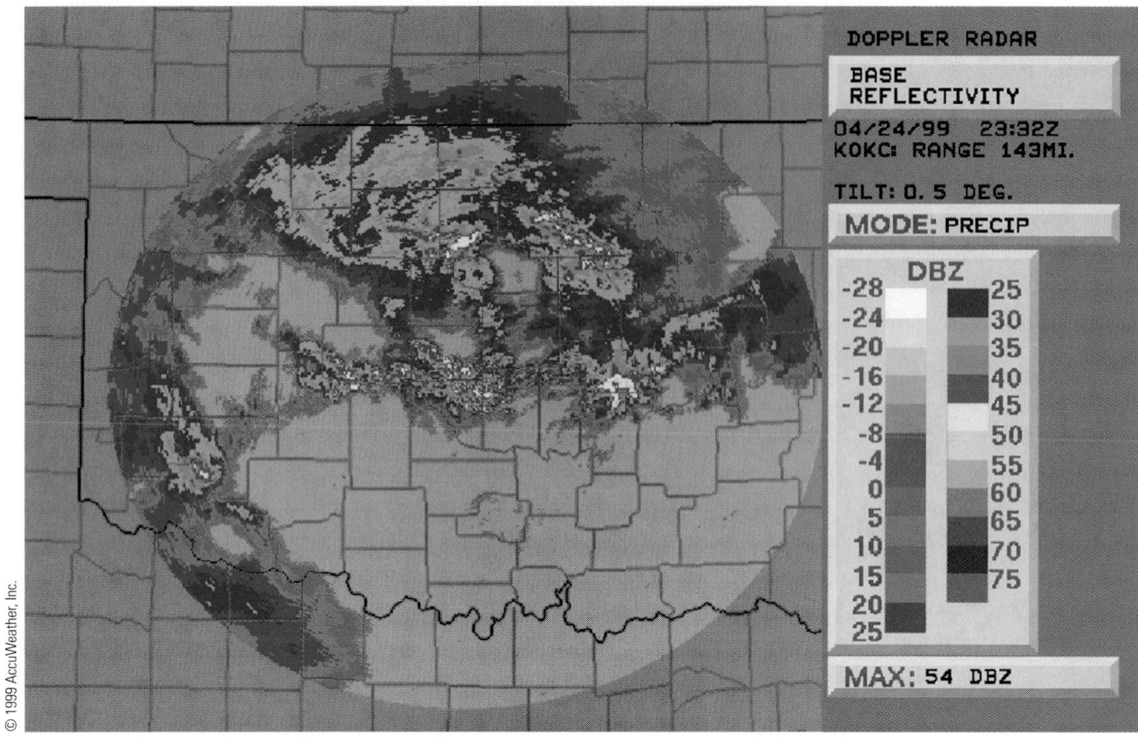

(a)

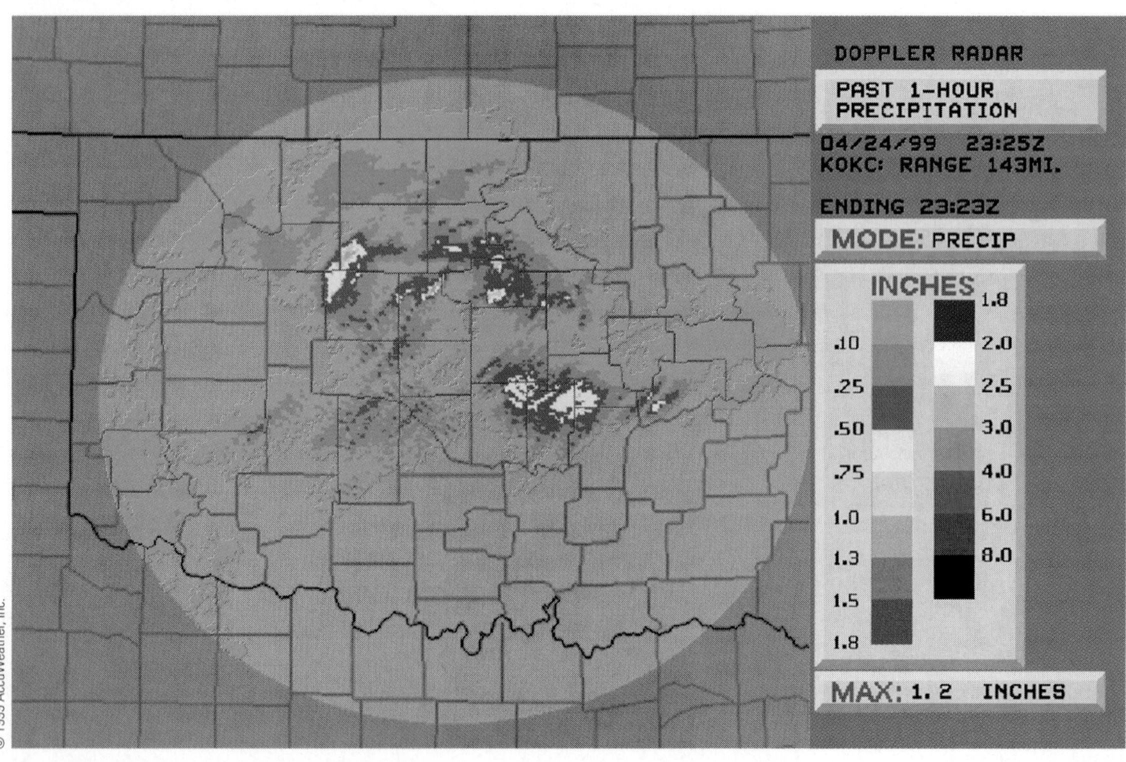

(b)

FIGURE 5.34 (a) Doppler radar display showing precipitation intensity over Oklahoma for April 24, 1999. The numbers under the letters DBZ represent the logarithmic scale for measuring the size and volume of precipitation particles. (b) Doppler radar display showing 1-hour rainfall amounts over Oklahoma for April 24, 1999.

SUMMARY

In this chapter, we tied together the concepts of stability, cloud formation, and precipitation. We learned that because stable air tends to resist upward vertical motions, clouds forming in a stable atmosphere often spread horizontally and have a stratified appearance. A stable atmosphere may be caused by either the surface air being cooled or the air aloft being warmed.

An unstable atmosphere tends to favor vertical air currents and produce cumuliform clouds. Instability may be brought on by either the surface air being warmed or the air aloft being cooled. In a conditionally unstable atmosphere, rising unsaturated air may be lifted to a level where condensation begins, latent heat is released, and instability results.

We looked at cloud droplets and found that, individually, they are too small and light to reach the ground as rain. They can grow in size as large cloud droplets, falling through a cloud, collide and merge with smaller droplets in their path. In clouds where the temperature is below freezing, ice crystals can grow larger at the expense of the surrounding liquid droplets. As an ice crystal begins to fall, it may grow larger by colliding with liquid droplets, which freeze on contact. In an attempt to coax more precipitation from them, some clouds are seeded with silver iodide.

We examined the various forms of precipitation, from raindrops that freeze on impact (producing freezing rain) to raindrops that freeze into tiny ice pellets called sleet. We learned that strong updrafts in a cumulonimbus cloud may carry ice particles high above the freezing level, where they acquire a further coating of ice and form destructive hailstones. We looked at instruments and found that although the rain gauge is still the most commonly used method of measuring precipitation, Doppler radar has become an important tool for determining precipitation intensity and estimating rainfall amount.

Meteorology Now™ Assess your understanding of this chapter's topics with additional quizzing and tutorials at http://earthscience.brookscole.com/ahrens/ess4e.

KEY TERMS

The following terms are listed in the order they appear in the text. Define each. Doing so will aid you in reviewing the material covered in this chapter.

adiabatic process
dry adiabatic rate
moist adiabatic rate
environmental lapse rate
absolutely stable atmosphere
absolutely unstable atmosphere
condensation level
conditionally unstable atmosphere
orographic uplift
rain shadow
coalescence
supercooled (water droplet)
ice nuclei
ice-crystal (Bergeron) process
accretion
cloud seeding
rain
drizzle
virga
shower (rain)
snow
fallstreaks
flurries (of snow)
snow squall
blizzard
sleet
freezing rain
rime
snow grains
snow pellets
hailstones
standard rain gauge
trace (of precipitation)
water equivalent
radar
Doppler radar

QUESTIONS FOR REVIEW

1. What is an adiabatic process?
2. How would one normally obtain the environmental lapse rate?
3. Why are the moist and dry adiabatic rates of cooling different?
4. How can the atmosphere be made more stable? More unstable?
5. If the atmosphere is conditionally unstable, what does this mean? What condition is necessary to bring on instability?
6. Explain why an inversion represents an extremely stable atmosphere.

7. What type of clouds would you most likely expect to see in a stable atmosphere? In an unstable atmosphere?
8. Why are cumulus clouds more frequently observed during the afternoon?
9. There are usually large spaces of blue sky between cumulus clouds. Explain why this is so.
10. Why do most thunderstorms have flat tops?
11. List the four primary ways in which clouds form.
12. Explain why rain shadows form on the leeward side of mountains.
13. On which side of a mountain (windward or leeward) would lenticular clouds most likely form?
14. What is the primary difference between a cloud droplet and a raindrop?
15. Why do typical cloud droplets seldom reach the ground as rain?
16. Describe how the process of collision and coalescence produces rain.
17. How does the ice-crystal (Bergeron) process produce precipitation? What is the *main* premise behind this process?
18. Explain the main principle behind cloud seeding.
19. How does rain differ from drizzle?
20. Why do heavy showers usually fall from cumuliform clouds? Why does steady precipitation normally fall from stratiform clouds?
21. Why is it *never* too cold to snow?
22. How would you be able to distinguish between virga and fallstreaks?
23. What is the difference between freezing rain and sleet?
24. How do the atmospheric conditions that produce sleet differ from those that produce hail?
25. Describe how a standard rain gauge measures precipitation.
26. (a) What is Doppler radar? (b) How does Doppler radar measure the intensity of precipitation?

QUESTIONS FOR THOUGHT AND EXPLORATION

1. Suppose a mountain climber is scaling the outside of a tall skyscraper. Two thermometers (shielded from the sun) hang from the climber's belt. One thermometer hangs freely, while the other is enclosed in a partially inflated balloon. As the climber scales the building, describe the change in temperature measured by each thermometer.
2. Where would you expect the moist adiabatic rate to be *greater:* in the tropics or near the North Pole? Explain why.
3. What changes in weather conditions near the earth's surface are needed to transform an absolutely stable atmosphere into an absolutely unstable atmosphere?
4. Under what circumstances can a rain shadow be formed on the western side of a mountain range?
5. A major snowstorm occurred in northern New Jersey. Three volunteer weather observers measured the snowfall. Observer #1 measured the depth of newly fallen snow every hour. At the end of the storm, Observer #1 added up the measurements and came up with a total of 12 inches of new snow. Observer #2 measured the depth of new snow twice: once in the middle of the storm and once at the end, and came up with a total snowfall of 10 inches. Observer #3 measured the new snowfall only once, after the storm had stopped, and reported 8.4 inches. Which of the three observers do you feel has the correct snowfall total? List *at least five* possible reasons why the snowfall totals were different.
6. Why is a warm, tropical cumulus cloud more likely to produce precipitation than a cold, stratus cloud?
7. Suppose a thick nimbostratus cloud contains ice crystals and cloud droplets all about the same size. Which precipitation process will be most important in producing rain from this cloud? Why?
8. Clouds that form over water are usually more efficient in producing precipitation than clouds that form over land. Why?

9. It is −12°C (10°F) in Albany, New York, and freezing rain is falling. Can you explain why? Draw a vertical profile of the air temperature (a sounding) that illustrates why freezing rain is occurring at the surface.

10. When falling snowflakes become mixed with sleet, why is this condition often followed by the snowflakes changing into rain?

11. Weather Radar Loop (http://www.intellicast.com/LocalWeather/World/UnitedStates/RadarLoop/): Examine current precipitation patterns as measured by Doppler radar.

Go to the Brooks/Cole Earth Sciences Resource Center (http://earthscience.brookscole.com) for critical thinking exercises, articles, and additional readings from InfoTrac College Edition, Brooks/Cole's online student library.

Photo: NASA

Wind-blown dust from the Sahara Desert during February, 2001, circulates into a storm off the coast of Spain.

Meteorology Now™ This icon, appearing throughout the book, indicates an opportunity to explore interactive tutorials, animations, or practice problems available on the MeteorologyNow Web site at http://earthscience.brookscole.com/ahrens/ess4e.

CHAPTER

6

Air Pressure and Winds

December 19, 1980, was a cool day in Lynn, Massachusetts, but not cool enough to dampen the spirits of more than 2000 people who gathered in Central Square—all hoping to catch at least one of the 1500 dollar bills that would be dropped from a small airplane at noon. Right on schedule, the aircraft circled the city and dumped the money onto the people below. However, to the dismay of the onlookers, a westerly wind caught the currency before it reached the ground and carried it out over the cold Atlantic Ocean. Had the pilot or the sponsoring leather manufacturer examined the weather charts beforehand, they might have been able to predict that the wind would ruin their advertising scheme.

CONTENTS

This opening scenario raises two questions: (1) Why does the wind blow? and (2) How can one tell its direction by looking at weather charts? Chapter 1 has already answered the first question: Air moves in response to horizontal differences in pressure. This happens when we open a vacuum-packed can—air rushes from the higher pressure region outside the can toward the region of lower pressure inside. In the atmosphere, the wind blows in an attempt to equalize imbalances in air pressure. Does this mean that the wind always blows directly from high to low pressure? Not really, because the movement of air is controlled not only by pressure differences but by other forces as well.

In this chapter, we will first consider how and why atmospheric pressure varies. Then we will look at the forces that influence atmospheric motions aloft and at the surface. Through studying these forces, we will be able to tell how the wind should blow in a particular region by examining surface and upper-air charts.

ATMOSPHERIC PRESSURE

In Chapter 1, we learned several important concepts about atmospheric pressure. One stated that **air pressure** is simply the mass of air above a given level. As we climb in elevation above the earth's surface, there are fewer air molecules above us; hence, atmospheric pressure always decreases with increasing height. Another concept we learned was that most of our atmosphere is crowded close to the earth's surface, which causes air pressure to decrease with height, rapidly at first, then more slowly at higher altitudes. But there are other important concepts to consider. For example, air pressure, air density (the mass of air in a given volume), and air temperature are all interrelated. If one of these variables changes, the other two usually change as well. The relationship among these three variables is expressed by the gas law, which is described in the Focus section on p. 144.

To help eliminate some of the complexities of the atmosphere, scientists construct *models*. Figure 6.1 shows a simple atmospheric model—a column of air, extending well up into the atmosphere. In the column, the dots represent air molecules. Our model assumes: (1) that the air molecules are not crowded close to the surface and, unlike the real atmosphere, the air density remains constant from the surface up to the top of the column, and (2) that the width of the column does not change with height.

Suppose we somehow force more air into the column in Fig. 6.1. What would happen? If the air temperature in the column does not change, the added air would make the column more dense, and the added mass of the air in the column would increase the surface air pressure. Likewise, if a great deal of air were removed from the column, the surface air pressure would decrease. With the same assumptions for our model in Fig. 6.1, let's look at Fig. 6.2.

Suppose the two air columns in Fig. 6.2a are located at the same elevation and have identical surface air pressures. This condition, of course, means that there must be the same number of molecules (same mass of air) in each column above both cities. Further suppose that the surface air pressure for both cities remains the same, while the air above city 1 cools and the air above city 2 warms (see Fig. 6.2b).

As the air in column 1 cools, the molecules move more slowly and crowd closer together—the air becomes more dense. In the warm air above city 2, the molecules move faster and spread farther apart—the air becomes less dense. Since the width of the columns does not change (and if we assume an invisible barrier exists between the columns), the surface pressure does not vary and the total number of molecules above each city must remain the same. Therefore, in the more dense cold air above city 1, the column shrinks, while the column rises in the less dense warm air above city 2.

We now have a cold shorter column of air above city 1 and a warm taller air column above city 2. From this situation, we can conclude that *it takes a shorter col-*

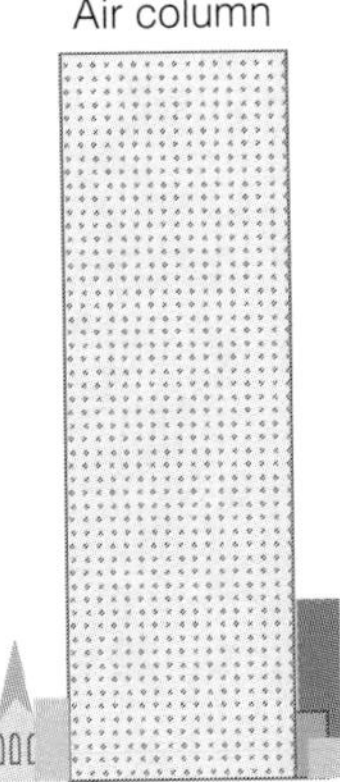

FIGURE 6.1 A model of the atmosphere where air density remains constant with height. The air pressure at the surface is related to the number of molecules above. When air of the same temperature is stuffed into the column, the surface air pressure rises. When air is removed from the column, the surface pressure falls.

umn of cold, more dense air to exert the same surface pressure as a taller column of warm, less dense air. This concept has a great deal of meteorological significance.

Atmospheric pressure decreases more rapidly with elevation in the cold column of air. In the cold air above city 1 (Fig. 6.2b), move up the column and observe how quickly you pass through the densely packed molecules. This activity indicates a rapid change in pressure. In the warmer, less dense air, the pressure does not decrease as rapidly with height, simply because you climb above fewer molecules in the same vertical distance.

In Fig. 6.2c, move up the warm, red column until you come to the letter *H*. Now move up the cold, blue column the same distance until you reach the letter *L*. Notice that there are more molecules above the letter *H* in the warm column than above the letter *L* in the cold column. The fact that the number of molecules above any level is a measure of the atmospheric pressure leads to an important concept: *Warm air aloft is normally associated with high atmospheric pressure, and cold air aloft is associated with low atmospheric pressure.*

In Fig. 6.2c, the horizontal difference in temperature creates a horizontal difference in pressure. The pressure difference establishes a force (called the *pressure gradient force*) that causes the air to move from higher pressure toward lower pressure. Consequently, if we remove the invisible barrier between the two columns and allow the air aloft to move horizontally, the air will move from column 2 toward column 1. As the air aloft leaves column 2, the mass of the air in the column decreases, and so does the surface air pressure. Meanwhile, the accumulation of air in column 1 causes the surface air pressure to increase.

In summary, heating or cooling a column of air can establish horizontal variations in pressure that cause the air to move. The net accumulation of air above the surface causes the surface air pressure to rise, whereas a net decrease in the amount of air above the surface causes the surface air pressure to fall.

MEASURING AIR PRESSURE

Up to this point, we have described air pressure as the mass of the atmosphere above any level. We can also define air pressure as the *force exerted by the air molecules over a given area.* Billions of air molecules constantly push on the human body. This force is exerted equally in all directions. We are not crushed by the force because billions of molecules inside the body push outward just as hard.

Even though we do not actually feel the constant bombardment of air, we can detect quick changes in it. For example, if we climb rapidly in elevation our ears may "pop." This experience happens because air collisions outside the eardrum lessen as the air pressure decreases. The popping comes about as air collisions between the inside and outside of the ear equalize.

Instruments that detect and measure pressure changes are called *barometers,* which literally means an instrument that measures bars. In meteorology, the *bar*

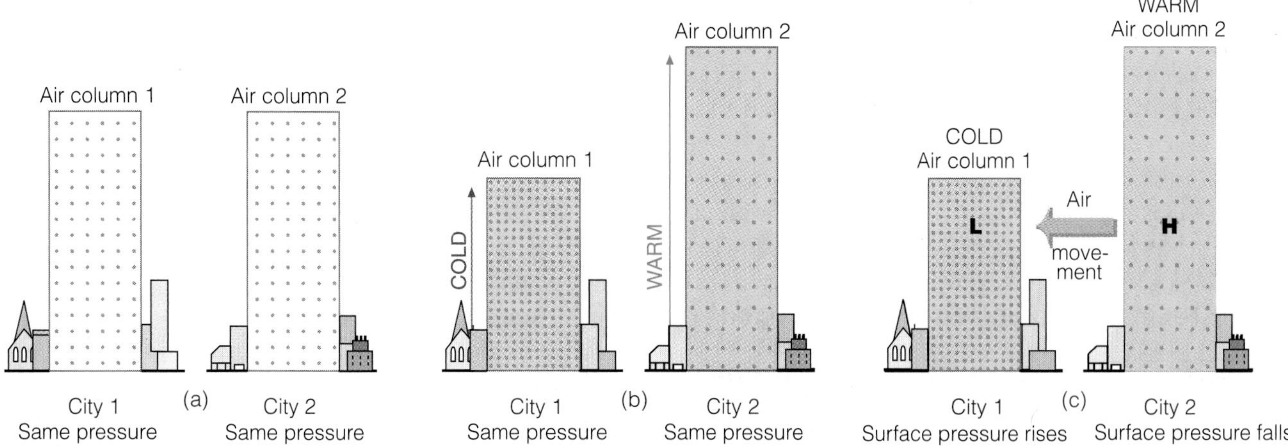

Meteorology Now™ ACTIVE FIGURE 6.2
It takes a shorter column of cold air to exert the same pressure as a taller column of warm air. Because of this fact, aloft, cold air is associated with low pressure and warm air with high pressure. The pressure differences aloft create a force that causes the air to move from a region of higher pressure toward a region of lower pressure. The removal of air from column 2 causes its surface pressure to drop, whereas the addition of air into column 1 causes its surface pressure to rise. (The difference in height between the two columns is greatly exaggerated.) Watch this Active Figure at http://earthscience.brookscole.com/ahrens/ess4e.

FOCUS ON A SPECIAL TOPIC

The Atmosphere Obeys the Gas Law

The relationship among the pressure, temperature, and density of air can be expressed by

$$\text{Pressure} = \text{temperature} \times \text{density} \times \text{constant.}$$

This simple relationship, often referred to as the *gas law* (or *equation of state*), tells us that the pressure of a gas is equal to its temperature times its density times a constant. When we ignore the constant and look at the gas law in symbolic form, it becomes

$$p \sim T \times \rho$$

where, of course, p is pressure, T is temperature, and ρ (the Greek letter rho, pronounced "row") represents air density. The line ~ is a symbol meaning "is proportional to." A change in one variable causes a corresponding change in the other two variables. Thus, it will be easier to understand the behavior of a gas if we keep one variable from changing and observe the behavior of the other two.

Suppose, for example, we hold the temperature constant. The relationship then becomes

$$p \sim \rho \text{ (temperature constant).}$$

This expression says that the pressure of the gas is proportional to its density, as long as its temperature does not change. Consequently, if the temperature of a gas (such as air) is held constant, as the pressure increases the density increases, and as the pressure decreases the density decreases. In other words, at *the same temperature, air at a higher pressure is more dense than air at a lower pressure.* If we apply this concept to the atmosphere, then with nearly the same temperature and elevation, air above a region of surface high pressure is more dense than air above a region of surface low pressure (see Fig. 1).

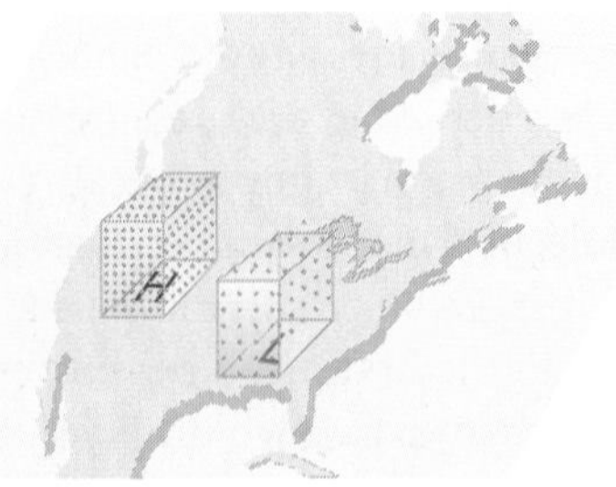

FIGURE 1 Air above a region of surface high pressure is more dense than air above a region of surface low pressure (at the same temperature). (The dots in each column represent air molecules.)

We can see, then, that for surface high-pressure areas (anticyclones) and surface low-pressure areas (mid-latitude storms) to form, the air density (mass of air) above these systems must change.

Earlier, we considered how pressure and density are related when the temperature is not changing. What happens to the gas law when the pressure of a gas remains constant? In shorthand notation, the law becomes

$$\text{(Constant pressure)} \times \text{constant} = T \times \rho.$$

This relationship tells us that when the pressure of a gas is held constant, the gas becomes less dense as the temperature goes up, and more dense as the temperature goes down. Therefore, *at a given atmospheric pressure, air that is cold is more dense than air that is warm.* Keep in mind that the idea that cold air is more dense than warm air applies only when we compare volumes of air at the same level, where pressure changes are small in any horizontal direction.

---

is a unit of pressure that describes a force over a given area.* Because the bar is a relatively large unit, and because surface pressure changes are normally small, the unit of pressure most commonly found on surface weather maps is the **millibar** (mb), where one millibar is equal to one-thousandth of a bar. Presently the *hectopascal* (hPa)** is gradually replacing the millibar as the preferred unit of pressure on surface maps. A common pressure unit used in aviation and on television and radio weather broadcasts is *inches of mercury* (Hg). At sea level, **standard atmospheric pressure*** is

$$1013.25 \text{ mb} = 1013.25 \text{ hPa} = 29.92 \text{ in. Hg.}$$

Figure 6.3 compares pressure readings in millibars and in inches of mercury. An understanding of how the unit "inches of mercury" is obtained is found in the following section on barometers.

Barometers Because we measure atmospheric pressure with an instrument called a **barometer,** atmospheric pressure is also referred to as *barometric pressure.*

*By definition, a bar is a force of 100,000 newtons acting on a surface area of 1 square meter. A *newton* is the amount of force required to move an object with a mass of 1 kilogram so that it increases its speed at a rate of 1 meter per second each second.

**The unit of pressure designed by the International System (SI) of measurement is the *pascal* (Pa), where 1 pascal is the force of 1 newton acting on a surface of 1 square meter. A more common unit is the *hectopascal* (hPa), as 1 hectopascal equals 1 millibar. (Additional pressure units and conversions are given in Appendix A.)

*Standard atmospheric pressure at sea level is the pressure extended by a column of mercury 29.92 in. (760 mm) high, having a density of 1.36×10^4 kg/m^3, and subject to an acceleration of gravity of 9.80 m/sec^2.

Evangelista Torricelli, a student of Galileo's, invented the **mercury barometer** in 1643. His barometer, similar to those used today, consisted of a long glass tube open at one end and closed at the other (see Fig. 6.4). Removing air from the tube and covering the open end, Torricelli immersed the lower portion into a dish of mercury. He removed the cover, and the mercury rose up the tube to nearly 30 inches above the level in the dish. Torricelli correctly concluded that the column of mercury in the tube was balancing the weight of the air above the dish, and, hence, its height was a measure of atmospheric pressure.

The most common type of home barometer—the **aneroid barometer**—contains no fluid. Inside this instrument is a small, flexible metal box called an *aneroid cell.* Before the cell is tightly sealed, air is partially removed, so that small changes in external air pressure cause the cell to expand or contract. The size of the cell is calibrated to represent different pressures, and any change in its size is amplified by levers and transmitted to an indicating arm, which points to the current atmospheric pressure (see Fig. 6.5).

Notice that the aneroid barometer often has descriptive weather-related words printed above specific pressure values. These descriptions indicate the most likely weather conditions when the needle is pointing to

DID YOU KNOW?

Although 1013.25 mb (29.92 in.) is the *standard atmospheric pressure* at sea level, it is *not* the average sea-level pressure. The earth's average sea-level pressure is 1011.0 mb (29.85 in.). Because much of the earth's surface is above sea level, the earth's annual average *surface pressure* is estimated to be 984.43 mb (29.07 in.).

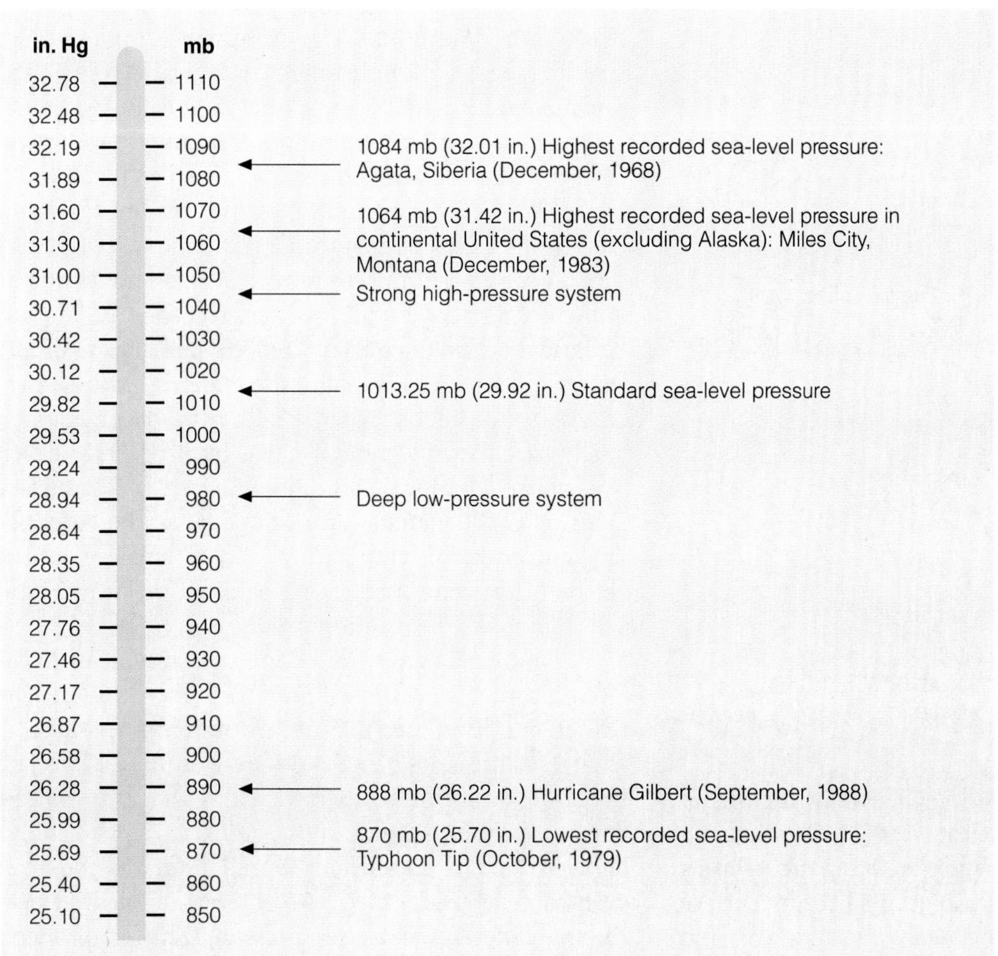

FIGURE 6.3 Atmospheric pressure in inches of mercury and in millibars.

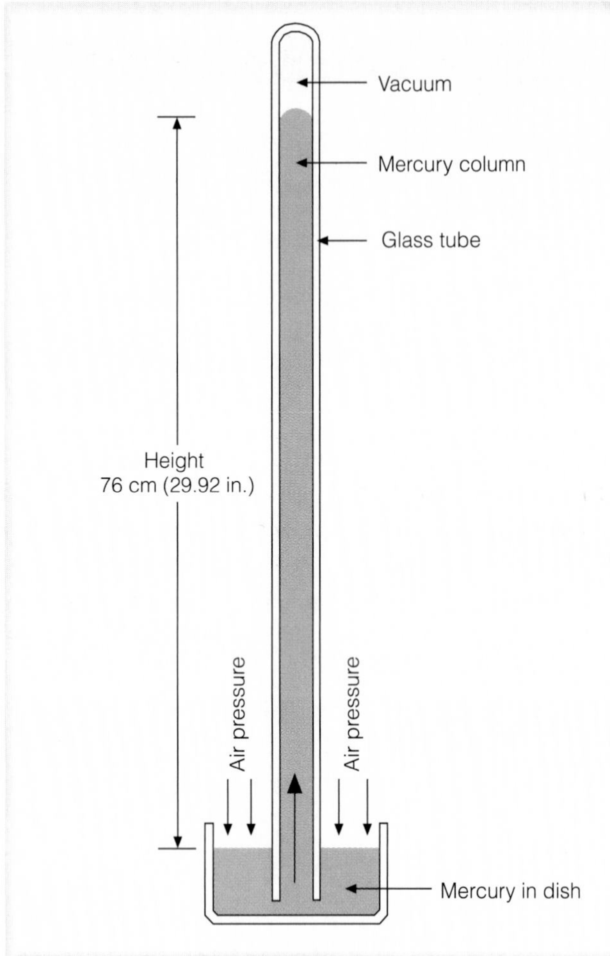

FIGURE 6.4 The mercury barometer. The height of the mercury column is a measure of atmospheric pressure.

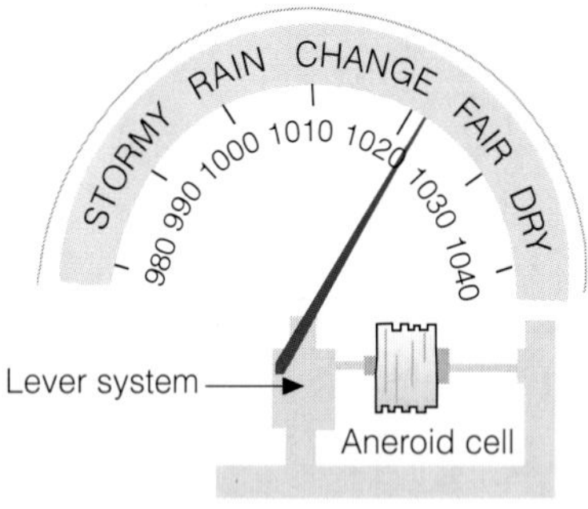

FIGURE 6.5 The aneroid barometer.

that particular pressure reading. Generally, the higher the reading, the more likely clear weather will occur, and the lower the reading, the better are the chances for inclement weather. This situation occurs because surface high-pressure areas are associated with sinking air and normally fair weather, whereas surface low-pressure areas are associated with rising air and usually cloudy, wet weather. A steady rise in atmospheric pressure (a rising barometer) usually indicates clearing weather or fair weather, whereas a steady drop in atmospheric pressure (a falling barometer) often signals the approach of a storm with inclement weather.

The *altimeter* and *barograph* are two types of aneroid barometers. Altimeters are aneroid barometers that measure pressure, but are calibrated to indicate altitude. Barographs are recording aneroid barometers. Basically, the barograph consists of a pen attached to an indicating arm that marks a continuous record of pressure on chart paper. The chart paper is attached to a drum rotated slowly by an internal mechanical clock (see Fig. 6.6).

> **DID YOU KNOW?**
>
> Mercury is 13.6 times more dense than water. Consequently, a water barometer resting on the ground near sea level would have to be read from a ladder over 10.3 m (34 ft) tall.

Pressure Readings The seemingly simple task of reading the height of the mercury column to obtain the air pressure is actually not all that simple. Being a fluid, mercury is sensitive to changes in temperature; it will expand when heated and contract when cooled. Consequently, to obtain accurate pressure readings without the influence of temperature, all mercury barometers are corrected as if they were read at the same temperature. Because the earth is not a perfect sphere, the force of gravity is not a constant. Since small gravity differences influence the height of the mercury column, they must be considered when reading the barometer. Finally, each barometer has its own "built-in" error, called *instrument error,* which is caused, in part, by the surface tension of the mercury against the glass tube. After being corrected for temperature, gravity, and instrument error, the barometer reading at a particular location and elevation is termed **station pressure.**

Figure 6.7a gives the station pressure measured at four locations only a few hundred kilometers apart. The different station pressures of the four cities are due primarily to the cities being at different altitudes. This fact becomes even clearer when we realize that atmospheric pressure changes much more quickly when we move upward than it does when we move sideways. A small vertical difference between two observation sites can yield a large difference in station pressure. Thus, to properly monitor horizontal changes in pressure, barometer readings must be corrected for altitude.

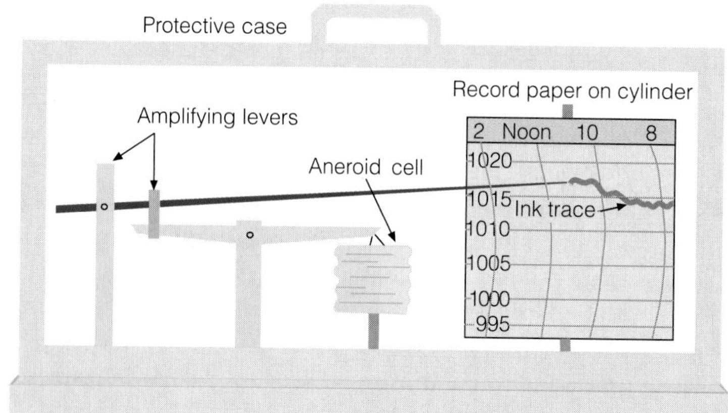

FIGURE 6.6 A recording barograph.

Altitude corrections are made so that a barometer reading taken at one elevation can be compared with a barometer reading taken at another. Station pressure observations are normally adjusted to a level of mean sea level—the level representing the average surface of the ocean. The adjusted reading is called **sea-level pressure.** The size of the correction depends primarily on how high the station is above sea level.

Near the earth's surface, atmospheric pressure decreases on the average by about 10 mb for every 100 m increase in elevation (about 1 in. of mercury for each 1000-ft rise).* Notice in Fig. 6.7a that city A has a station pressure of 952 mb. Notice also that city A is 600 m above sea level. Adding 10 mb per 100 m to its station pressure yields a sea-level pressure of 1012 mb (Fig. 6.7b). After all the station pressures are adjusted to sea level (Fig. 6.7b), we are able to see the horizontal variations in

*This decrease in atmospheric pressure with height (10 mb/100 m) occurs when the air temperature decreases at the standard lapse rate of 6.5°C/1000 m. Because atmospheric pressure decreases more rapidly with height in cold (more dense) air than it does in warm (less dense) air, the vertical rate of pressure change is typically greater than 10 mb per 100 m in cold air and less than that in warm air.

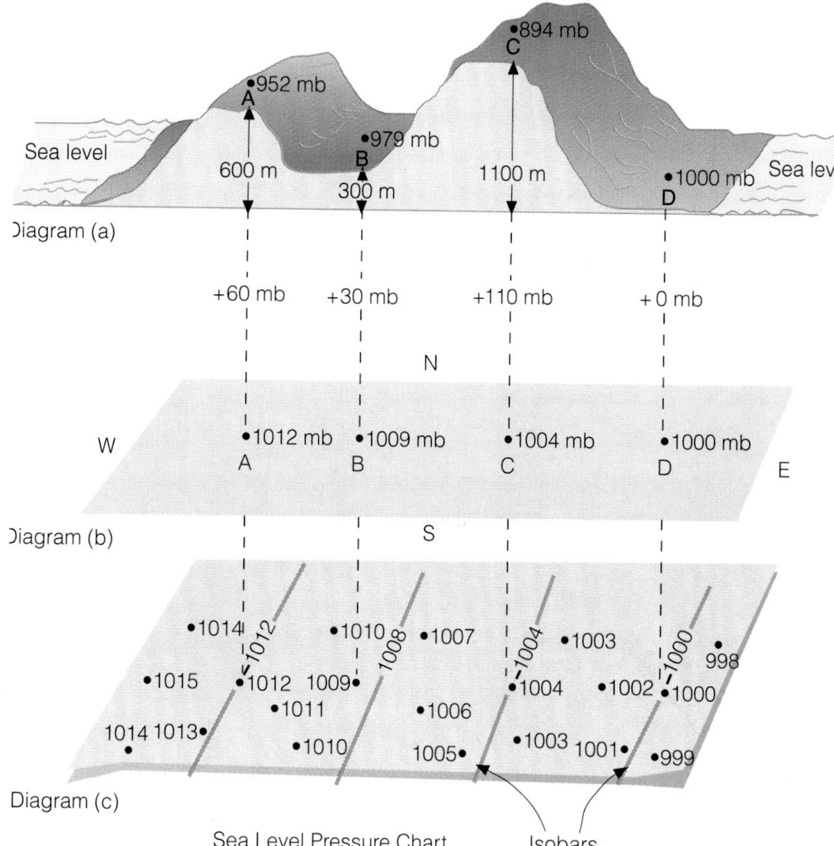

FIGURE 6.7 The top diagram (a) shows four cities (A, B, C, and D) at varying elevations above sea level, all with different station pressures. The middle diagram (b) represents sea-level pressures of the four cities plotted on a sea-level chart. The bottom diagram (c) shows isobars drawn on the chart (dark lines) at intervals of 4 millibars.

DID YOU KNOW?

The difference in air pressure from the base to the top of a New York City skyscraper about 0.5 km (1600 ft) tall is typically much greater than the horizontal difference in air pressure between New York City and Miami, Florida—a distance of more than 1800 km (1100 mi).

sea-level pressure—something we were not able to see from the station pressures alone in Fig. 6.7a.

When more pressure data are added (Fig. 6.7c), the chart can be analyzed and the pressure pattern visualized. **Isobars** (lines connecting points of equal pressure) are drawn as solid dark lines at intervals of 4 mb, with 1000 mb being the base value. Note that the isobars do not pass through each point, but, rather, between many of them, with the exact values being interpolated from the data given on the chart. For example, follow the 1008-mb line from the top of the chart southward and observe that there is no plotted pressure of 1008 mb. The 1008-mb isobar, however, comes closer to the station with a sea-level pressure of 1007 mb than it does to the station with a pressure of 1010 mb. With its isobars, the bottom chart (Fig. 6.7c) is now called a *sea-level pressure chart,* or simply a **surface map.** When weather data are plotted on the map, it becomes a *surface weather map.*

SURFACE AND UPPER-AIR CHARTS

Figure 6.8a is a simplified surface map that shows areas of high and low pressure and arrows that indicate *wind direction—the direction from which the wind is blowing.* The large blue *H*'s on the map indicate the centers of high pressure, which are also called **anticyclones.** The large red *L*'s represent centers of low pressure, also known as *depressions,* **mid-latitude cyclones,** or *extratropical cyclones* because they form in the middle latitudes, outside of the tropics. The solid dark lines are isobars with units in millibars. Notice that the surface winds tend to blow across the isobars toward regions of lower pressure. In fact, as we briefly observed in Chapter 1, in the Northern Hemisphere the winds blow counterclockwise and inward toward the center of the lows and clockwise and outward from the center of the highs.

Figure 6.8b shows an upper-air chart for the same day as the surface map in Fig. 6.8a. The upper-air map is a *constant pressure chart* because it is constructed to show height variations along a constant pressure *(isobaric)* surface, which is why these maps are also known as **isobaric maps.** This particular isobaric map shows height variations at a pressure level of 500 mb (which is about 5600 m or 18,000 ft above sea level). Hence, this map is called a *500-millibar map.* The solid dark lines on the map are **contour lines**—lines that connect points of equal eleva-

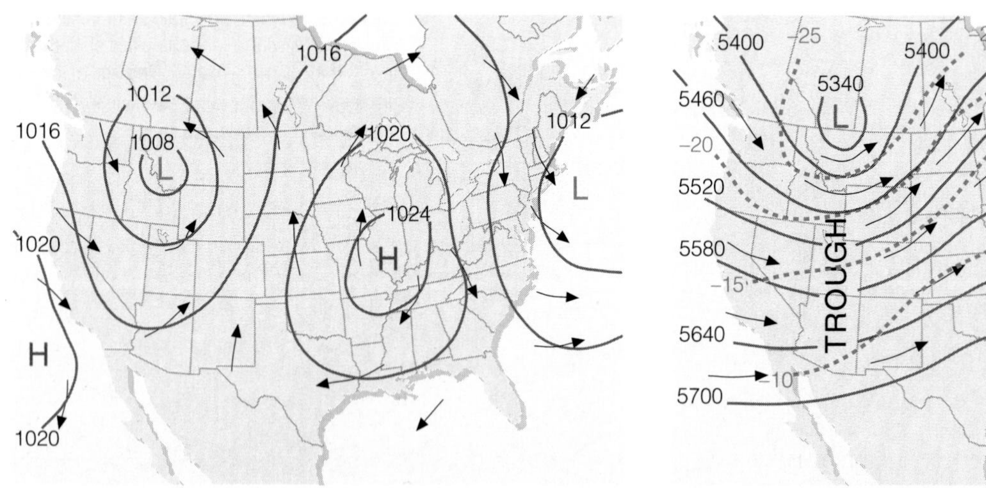

FIGURE 6.8 (a) Surface map showing areas of high and low pressure. The solid lines are isobars drawn at 4-mb intervals. The arrows represent wind direction—the direction from which the wind is blowing. Notice that the wind blows *across* the isobars. (b) The upper-level (500-mb) map for the same day as the surface map. Solid lines on the map are contour lines in meters above sea level. Dashed red lines are isotherms in °C. Arrows show wind direction. Notice that, on this upper-air map, the wind blows *parallel* to the contour lines.

FOCUS ON A SPECIAL TOPIC

Isobaric Maps

Figure 2 shows a column of air where warm, less-dense air lies to the south and cold, more-dense air lies to the north. The area shaded gray at the top of the column represents a constant pressure (isobaric) surface, where the atmospheric pressure at all points along the surface is 500 mb.

Notice that the height of the pressure surface varies. In the warmer air, a pressure reading of 500 mb is found at a higher level, while in the colder air, 500 mb is observed at a much lower level. The variations in height of the 500-mb constant pressure surface are shown as contour lines on the constant pressure (500-mb) map, situated at the bottom of the column. Each contour line tells us the elevation above sea level at which we would obtain a pressure reading of 500 mb. As we would expect, the elevations are higher in the warm air and lower in the cold air. Although contour lines are height lines, keep in mind that they illustrate pressure in the same manner as do isobars, as contour lines of high height (warm air aloft) represent regions of higher pressure, and contour lines of low height (cold air aloft) represent regions of low pressure.

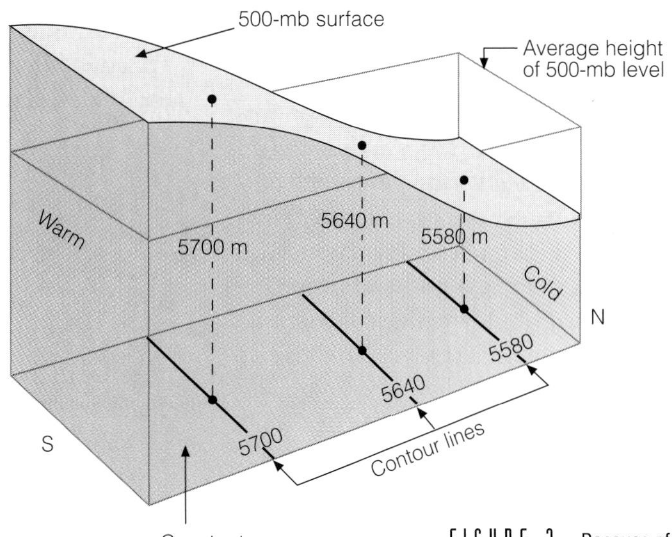

FIGURE 2 Because of the changes in air density, a surface of constant pressure (the shaded gray area) rises in warm, less-dense air and lowers in cold, more-dense air. These changes in elevation of a constant pressure (500-mb) surface show up as contour lines on a constant pressure (isobaric) 500-mb map.

tion above sea level. Although contour lines are height lines, they illustrate pressure much like isobars do. Consequently, *contour lines of low height represent a region of lower pressure, and contour lines of high height represent a region of higher pressure.* (Additional information on isobaric maps is given in the Focus section above.)

Notice on the 500-mb map (Fig. 6.8b) that the contour lines typically decrease in value from south to north. The reason for this fact is illustrated by the dashed red lines, which are *isotherms*—lines of equal temperature. Observe that colder air is generally to the north and warmer air to the south, and recall from our earlier discussion (see pp. 142–143) that cold air aloft is associated with low pressure, warm air aloft with high pressure. The contour lines are not straight, however, they bend and turn, indicating **ridges** *(elongated highs)* where the air is warmer and indicating depressions, or **troughs** *(elongated lows)* where the air is colder. The arrows on the 500-mb map show the wind direction. Notice that, unlike the surface winds that cross the isobars in Fig. 6.8a, the winds on the 500-mb chart tend to flow *parallel* to the contour lines in a wavy west-to-east direction.

Surface and upper-air charts are valuable tools for the meteorologist. Surface maps describe where the centers and high and low pressure are found, as well as the winds and weather associated with these systems. Upper-air charts, on the other hand, are extremely important in forecasting the weather. The upper-level winds not only determine the movement of surface pressure systems but, as we will see in Chapter 8, they determine whether these surface systems will intensify or weaken.

At this point, however, our interest lies mainly in the movement of air. Consequently, now that we have looked at surface and upper-air maps, we will use them to study why the wind blows the way it does, at both the surface and aloft.

WHY THE WIND BLOWS

Our understanding of why the wind blows stretches back through several centuries, with many scientists contributing to our knowledge. When we think of the movement of air, however, one great scholar stands out—Isaac Newton (1642–1727), who formulated several fundamental laws of motion.

Newton's Laws Of Motion Newton's first law of motion states that *an object at rest will remain at rest and an object in motion will remain in motion (and travel at a constant velocity along a straight line) as long as no force is exerted on the object.* For example, a baseball in a pitcher's hand will remain there until a force (a push) acts upon the ball. Once the ball is pushed (thrown), it would continue to move in that direction forever if it were not for the force of air friction (which slows it down), the force of gravity (which pulls it toward the ground), and the catcher's mitt (which exerts an equal but opposite force to bring it to a halt). Similarly, to start air moving, to speed it up, to slow it down, or even to change its direction requires the action of an external force. This brings us to Newton's second law.

Newton's second law states that *the force exerted on an object equals its mass times the acceleration produced.** In symbolic form, this law is written as

$$F = ma.$$

From this relationship we can see that, when the mass of an object is constant, the force acting on the object is directly related to the acceleration that is produced. A force in its simplest form is a push or a pull. *Acceleration is the speeding up, the slowing down, or the changing of direction of an object.* (More precisely, acceleration is the change of velocity** over a period of time.)

Because more than one force may act upon an object, Newton's second law always refers to the *net*, or total, force that results. An object will always accelerate in the direction of the total force acting on it. Therefore, to determine in which direction the wind will blow, we must identify and examine all of the forces that affect the horizontal movement of air. These forces include

1. pressure gradient force
2. Coriolis force
3. centripetal force
4. friction

*Newton's second law may also be stated in this way: The acceleration of an object (times its mass) is caused by all of the forces acting on it.

**Velocity specifies both the speed of an object and its direction of motion.

We will first study the forces that influence the flow of air aloft. Then we will see which forces modify winds near the ground.

Forces That Influence the Wind We have already learned that horizontal differences in atmospheric pressure cause air to move and, hence, the wind to blow. Since air is an invisible gas, it may be easier to see how pressure differences cause motion if we examine a visible fluid, such as water.

In Fig. 6.9, the two large tanks are connected by a pipe. Tank A is two-thirds full and tank B is only one-half full. Since the water pressure at the bottom of each tank is proportional to the weight of water above, the pressure at the bottom of tank A is greater than the pressure at the bottom of tank B. Moreover, since fluid pressure is exerted equally in all directions, there is a greater pressure in the pipe directed from tank A toward tank B than from B toward A.

Since pressure is force per unit area, there must also be a net force directed from tank A toward tank B. This force causes the water to flow from left to right, from higher pressure toward lower pressure. The greater the pressure difference, the stronger the force, and the faster the water moves. In a similar way, horizontal differences in atmospheric pressure cause air to move.

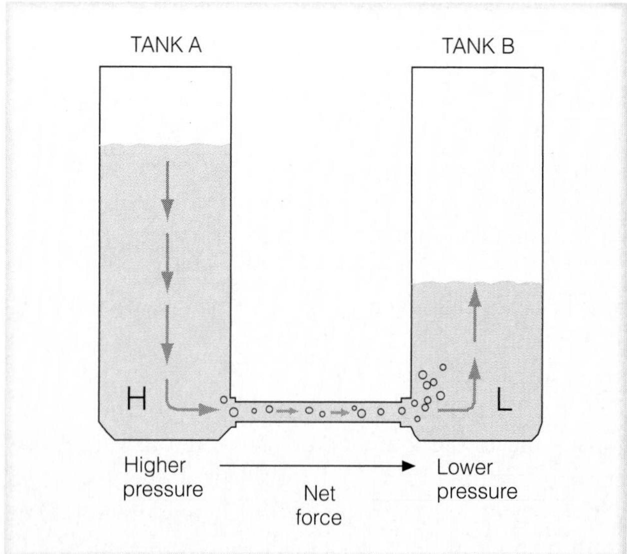

FIGURE 6.9 The higher water level creates higher fluid pressure at the bottom of tank A and a net force directed toward the lower fluid pressure at the bottom of tank B. This net force causes water to move from higher pressure toward lower pressure.

Pressure Gradient Force Figure 6.10 shows a region of higher pressure on the map's left side, lower pressure on the right. The isobars show how the horizontal pressure is changing. If we compute the amount of pressure change that occurs over a given distance, we have the **pressure gradient;** thus

$$\text{Pressure gradient} = \frac{\text{difference in pressure}}{\text{distance}}.$$

In Fig. 6.10, the pressure gradient between points 1 and 2 is 4 mb per 100 km.

Suppose the pressure in Fig. 6.10 were to change, and the isobars become closer together. This condition would produce a rapid change in pressure over a relatively short distance, or what is called a *steep* (or *strong*) *pressure gradient.* However, if the pressure were to change such that the isobars spread farther apart, then the difference in pressure would be small over a relatively large distance. This condition is called a *gentle* (or *weak*) *pressure gradient.*

Notice in Fig. 6.10 that when differences in horizontal air pressure exist there is a net force acting on the air. This force, called the **pressure gradient force** *(PGF), is directed from higher toward lower pressure at right angles to the isobars.* The magnitude of the force is directly related to the pressure gradient. Steep pressure gradients correspond to strong pressure gradient forces and vice versa. Figure 6.11 shows the relationship between pressure gradient and pressure gradient force.

The *pressure gradient force is the force that causes the wind to blow.* Because of this fact, closely spaced isobars on a weather chart indicate steep pressure gradients, strong forces, and high winds. On the other hand, widely spaced isobars indicate gentle pressure gradients, weak forces, and light winds. An example of a steep pressure gradient producing strong winds is illustrated on the surface weather map in Fig. 6.12. Notice that the tightly packed isobars along the green line are producing a steep pressure gradient of 32 mb per 500 km and strong surface winds of 40 knots.

If the pressure gradient force were the only force acting upon air, we would always find winds blowing directly from higher toward lower pressure. However, the moment air starts to move, it is deflected in its path by the *Coriolis force.*

Coriolis Force The **Coriolis force** describes an apparent force that is due to the rotation of the earth. To understand how it works, consider two people playing catch as they sit opposite one another on the rim of a merry-go-round (see Fig. 6.13, platform A). If the merry-go-round is not moving, each time the ball is thrown, it moves in a straight line to the other person.

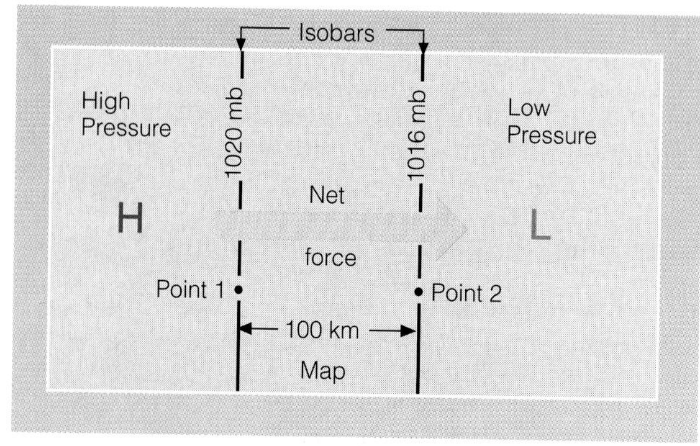

FIGURE 6.10 The pressure gradient between point 1 and point 2 is 4 mb per 100 km. The net force directed from higher toward lower pressure is the *pressure gradient force.*

Suppose the merry-go-round starts turning counterclockwise—the same direction the earth spins as viewed from above the North Pole. If we watch the game of catch from above, we see that the ball moves in a straight-line path just as before. However, to the people playing catch on the merry-go-round, the ball seems to veer to its right each time it is thrown, always landing to the right of the point intended by the thrower (see Fig. 6.13, platform B). This perspective is due to the fact that, while the ball moves in a straight-line path, the merry-go-round rotates beneath it; by the time the ball reaches the opposite side, the catcher has moved. To anyone on the merry-go-round, it seems as if there is

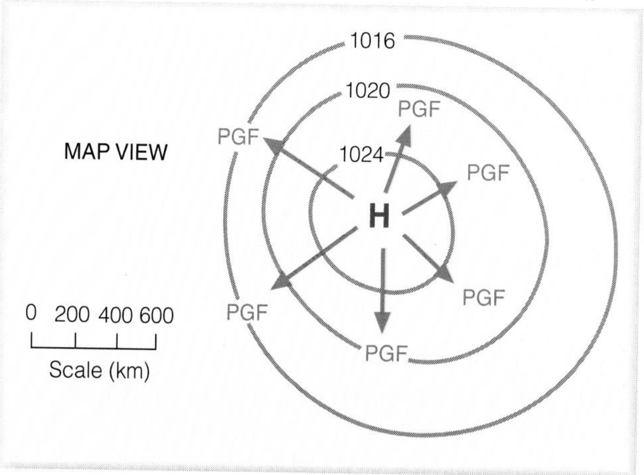

FIGURE 6.11 The closer the spacing of the isobars, the greater the pressure gradient. The greater the pressure gradient, the stronger the pressure gradient force *(PGF).* The stronger the *PGF,* the greater the wind speed. The red arrows represent the relative magnitude of the force, which is always directed from higher toward lower pressure.

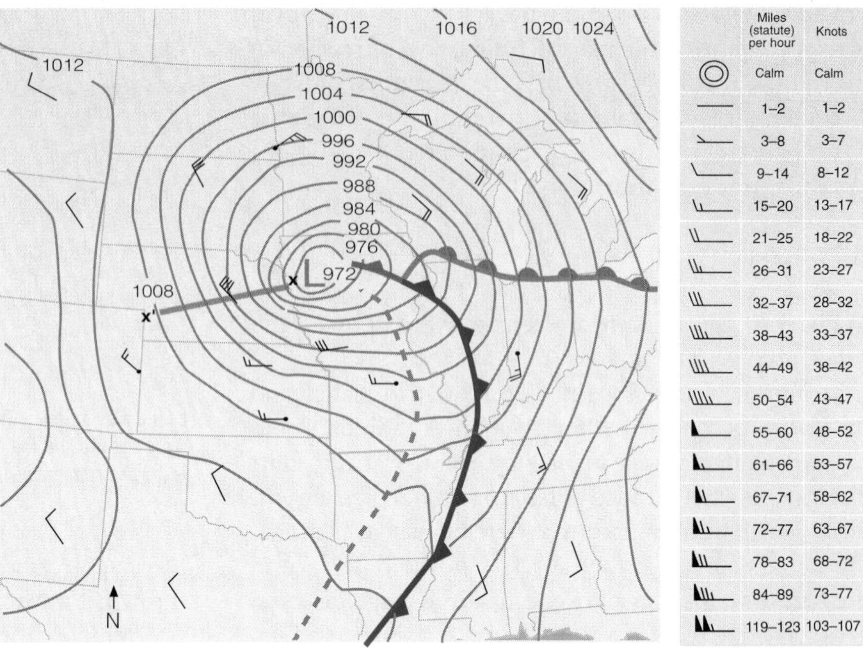

FIGURE 6.12 Surface weather map for 6 A.M. (CST), Tuesday, November 10, 1998. Dark gray lines are isobars with units in millibars. The interval between isobars is 4 mb. A deep low with a central pressure of 972 mb (28.70 in.) is moving over northwestern Iowa. The distance along the green line X-X' is 500 km. The difference in pressure between X and X' is 32 mb, producing a pressure gradient of 32 mb/500 km. The tightly packed isobars along the green line are associated with strong northwesterly winds of 40 knots, with gusts even higher. Wind directions are given by lines that parallel the wind. Wind speeds are indicated by barbs and flags. (A wind indicated by the symbol ⸌ would be a wind from the northwest at 10 knots. See blue insert.) The solid blue line is a cold front, the solid red line a warm front, and the solid purple line an occluded front. The heavy dashed line is a trough.

some force causing the ball to deflect to the right. This apparent force is called the *Coriolis force* after Gaspard Coriolis, a nineteenth-century French scientist who worked it out mathematically. (Because it is an *apparent force* due to the rotation of the earth, it is also called the *Coriolis effect.*) This effect occurs on the rotating earth, too. All free-moving objects, such as ocean currents, aircraft, artillery projectiles, and air molecules seem to deflect from a straight-line path because the earth rotates under them.

The Coriolis force *causes the wind to deflect to the right of its intended path in the Northern Hemisphere and to the left of its intended path in the Southern Hemisphere.* To illustrate, consider a satellite in polar circular orbit. If the earth were not rotating, the path of the satellite would be observed to move directly from north to south, parallel to the earth's meridian lines. However, the earth *does* rotate, carrying us and meridians eastward with it. Because of this rotation, in the Northern Hemisphere we see the satellite moving southwest instead of due south; it seems to veer off its path and move toward *its right.* In the Southern Hemisphere, the earth's direction of rotation is clockwise as viewed from above the South Pole. Consequently, a satellite moving northward from the South Pole would appear to move northwest and, hence, would veer to the *left* of its path.

As the wind speed increases, the Coriolis force increases; hence, *the stronger the wind, the greater the deflection.* Additionally, the Coriolis force increases for all wind speeds from a value of *zero at the equator to a max-*

DID YOU KNOW?

The deep, low-pressure area illustrated in Fig. 6.12 was quite a storm. The intense low with its tightly packed isobars and strong pressure gradient produced extremely high winds that gusted over 90 knots in Wisconsin and in Michigan's Mackinac Island. The extreme winds caused blizzard conditions over the Dakotas, closed many Interstate highways, shut down airports, and overturned trucks. The winds pushed a school bus off the road near Albert Lea, Minnesota, injuring two children, and blew the roofs off homes in Wisconsin. This notorious deep storm set an all-time record low pressure of 963 mb (28.43 in.) for Minnesota on November 10, 1998.

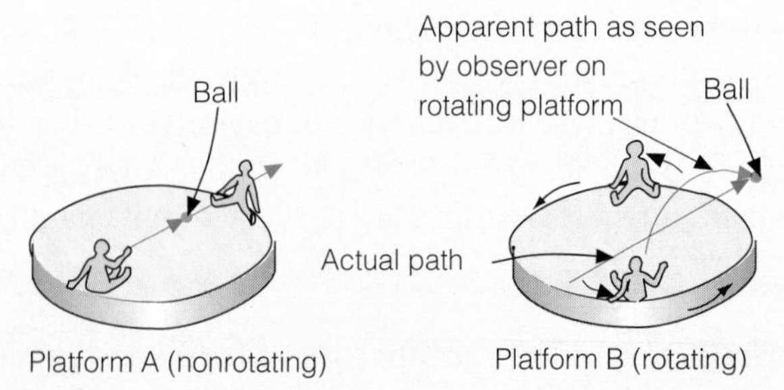

FIGURE 6.13 On nonrotating platform A, the thrown ball moves in a straight line. On platform B, which rotates counterclockwise, the ball continues to move in a straight line. However, platform B is rotating while the ball is in flight; thus, to anyone on platform B, the ball appears to deflect to the right of its intended path.

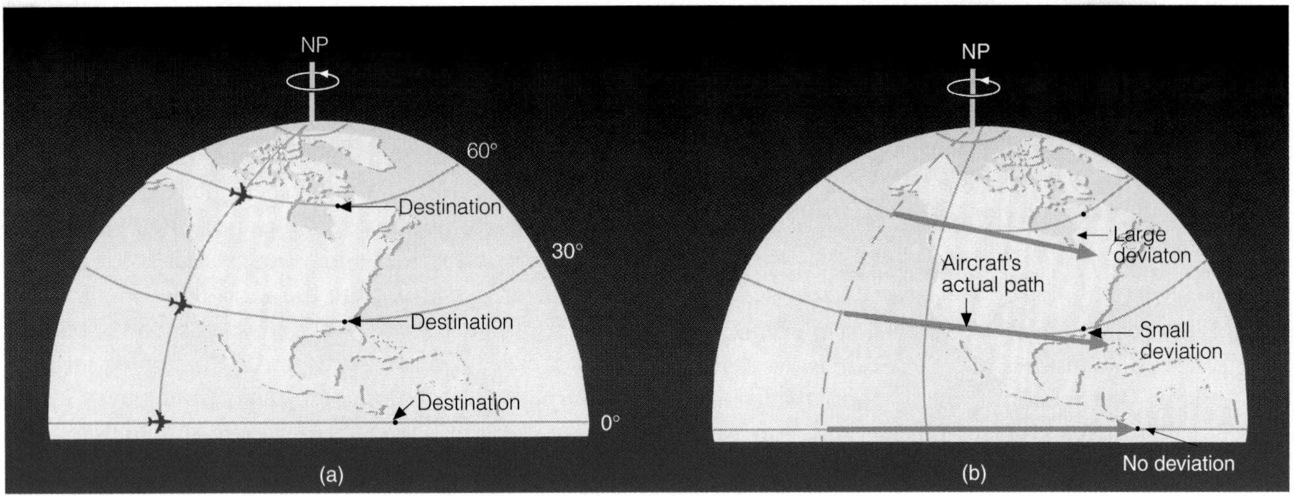

Meteorology Now™ ACTIVE FIGURE 6.14
Except at the equator, a free-moving object heading either east or west (or any other direction) will appear from the earth to deviate from its path as the earth rotates beneath it. The deviation (Coriolis force) is greatest at the poles and decreases to zero at the equator.
Watch this Active Figure at http://earthscience.brookscole.com/ahrens/ess4e.

imum at the poles. This phenomenon is illustrated in Fig. 6.14 where three aircraft, each at a different latitude, are flying along a straight-line path, with no external forces acting on them. The destination of each aircraft is due east and is marked on the illustration in Fig. 6.14a. Each plane travels in a straight path relative to an observer positioned at a fixed spot in space. The earth rotates beneath the moving planes, causing the destination points at latitudes 30° and 60° to change direction slightly—to the observer in space (see Fig. 6.14b). To an observer standing on the earth, however, it is the plane that appears to deviate. The amount of deviation is greatest toward the pole and nonexistent at the equator. Therefore, the Coriolis force has a far greater effect on the plane at high latitudes (large deviation) than on the plane at low latitudes (small deviation). On the equator, it has no effect at all. The same is true of its effect on winds.

In summary, to an observer on the earth, objects moving in *any direction* (north, south, east, or west) are deflected to the *right* of their intended path in the Northern Hemisphere and to the *left* of their intended path in the Southern Hemisphere. The amount of deflection depends upon

1. the rotation of the earth
2. the latitude
3. the object's speed

In addition, *the Coriolis force acts at right angles to the wind, only influencing wind direction and never wind speed.*

The Coriolis "force" behaves as a real force, constantly tending to "pull" the wind to its right in the Northern Hemisphere and to its left in the Southern Hemisphere. Moreover, this effect is present in all motions relative to the earth's surface. However, in most of our everyday experiences, the Coriolis force is so small (compared to other forces involved in those experiences) that it is negligible and, contrary to popular belief, does not cause water to turn clockwise or counterclockwise when draining from a sink.

The Coriolis force is also minimal on small-scale winds, such as those that blow inland along coasts in summer. Here, the Coriolis force might be strong because of high winds, but the force cannot produce much deflection over the relatively short distances. Only where winds blow over vast regions is the effect significant.

Meteorology Now™ Click "Coriolis Force" to visualize the effects of the earth's rotation on airplane flight.

BRIEF REVIEW

In summary, we know that:

- The pressure gradient force is always directed from higher pressure toward lower pressure and it is the pressure gradient force that causes the air to move.
- Steep pressure gradients (tightly packed isobars) indicate strong pressure gradient forces and high winds; gentle pressure gradients (widely spaced isobars) indicate weak pressure gradient forces and light winds.
- Once the wind starts to blow, the Coriolis force causes it to bend to the right of its intended path in the Northern Hemisphere and to the left of its intended path in the Southern Hemisphere.

DID YOU KNOW?

Does the Coriolis force influence a moving vehicle? When you drive along a highway (at the speed limit), the Coriolis force would "pull" your car about 460 m (1500 ft) to the right for every 160 km (100 mi) you travel, if it were not for the friction between your tires and the road surface.

With this information in mind, we will first examine how the pressure gradient force and the Coriolis force produce straight-line winds aloft (that is, above the friction layer). We will then see what influence the centripetal force has on winds that blow along a curved path.

Straight-Line Flow Aloft Earlier in this chapter, we saw that the winds aloft on an upper-level chart blow more or less parallel to the isobars or contour lines. We can see why this phenomenon happens by carefully looking at Fig. 6.15, which shows a map in the Northern Hemisphere, above the earth's frictional influence,* with horizontal pressure variations at an altitude of about 1 km above the earth's surface. The evenly spaced isobars indicate a constant pressure gradient force *(PGF)* directed from south toward north as indicated by the red arrow at the left. Why, then, does the map show a wind blowing from the west? We can answer this question by placing a parcel of air at position 1 in the diagram and watching its behavior.

At position 1, the *PGF* acts immediately upon the air parcel, accelerating it northward toward lower pressure.

*The friction layer (the layer where the wind is influenced by frictional interaction with objects on the earth's surface) usually extends from the surface up to about 1000 m (3300 ft) above the ground.

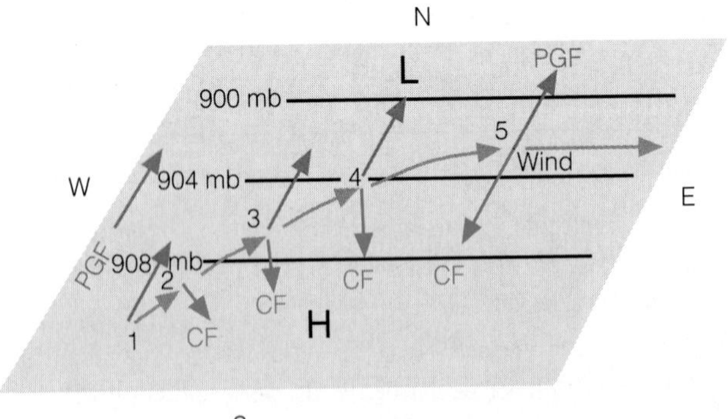

FIGURE 6.15 Above the level of friction, air initially at rest will accelerate until it flows parallel to the isobars at a steady speed with the pressure gradient force *(PGF)* balanced by the Coriolis force *(CF)*. Wind blowing under these conditions is called geostrophic.

However, the instant the air begins to move, the Coriolis force deflects the air toward its right, curving its path. As the parcel of air increases in speed (positions 2, 3, and 4), the magnitude of the Coriolis force increases (as shown by the blue arrows), bending the wind more and more to its right. Eventually, the wind speed increases to a point where the Coriolis force just balances the *PGF.* At this point (position 5), the wind no longer accelerates because the net force is zero. Here the wind flows in a straight path, parallel to the isobars at a constant speed.* This flow of air is called a **geostrophic** (*geo:* earth; *strophic:* turning) **wind.** Notice that the geostrophic wind blows in the Northern Hemisphere with lower pressure to its left and higher pressure to its right.

When the flow of air is purely geostrophic, the isobars (or contour lines) are straight and evenly spaced, and the wind speed is constant. In the atmosphere, isobars are rarely straight or evenly spaced, and the wind normally changes speed as it flows along. So, the geostrophic wind is usually only an approximation of the real wind. However, the approximation is generally close enough to help us more clearly understand the behavior of the winds aloft.

As we would expect from our previous discussion of winds, the speed of the geostrophic wind is directly related to the pressure gradient. In Fig. 6.16, we can see that a geostrophic wind flowing parallel to the isobars is similar to water in a stream flowing parallel to its banks. At position 1, the wind is blowing at a low speed; at position 2, the pressure gradient increases and the wind speed picks up. Notice also that at position 2, where the wind speed is greater, the Coriolis force is greater and balances the stronger pressure gradient force.

We know that the winds aloft do not always blow in a straight line; frequently, they curve and bend into meandering loops as they tend to follow the patterns of the isobars. In the Northern Hemisphere, winds blow counterclockwise around lows and clockwise around highs. The next section explains why.

Curved Winds Around Lows and Highs Aloft Because lows are also known as cyclones, the counterclockwise flow of air around them is often called *cyclonic flow.* Likewise, the clockwise flow of air around a high,

*At first, it may seem odd that the wind blows at a constant speed with no net force acting on it. But when we remember that the net force is necessary only to accelerate $(F = ma)$ the wind, it makes more sense. For example, it takes a considerable net force to push a car and get it rolling from rest. But once the car is moving, it only takes a force large enough to counterbalance friction to keep it going. There is no net force acting on the car, yet it rolls along at a constant speed.

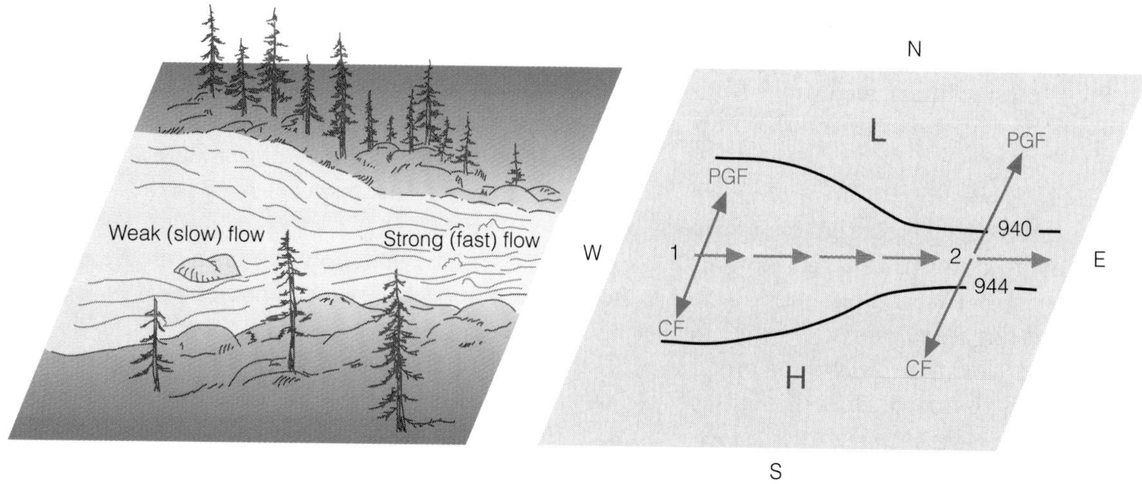

FIGURE 6.16 The isobars and contours on an upper-level chart are like the banks along a flowing stream. When they are widely spaced, the flow is weak; when they are narrowly spaced, the flow is stronger. The increase in winds on the chart results in a stronger Coriolis force *(CF)*, which balances a larger pressure gradient force *(PGF)*.

or anticyclone, is called *anticyclonic flow.* Look at the wind flow around the upper-level low (Northern Hemisphere) in Fig. 6.17a. At first, it appears as though the wind is defying the Coriolis force by bending to the left as it moves counterclockwise around the system. Let's see why the wind blows in this manner.

Suppose we consider a parcel of air initially at rest at position 1 in Fig. 6.17a. The pressure gradient force accelerates the air inward toward the center of the low and the Coriolis force deflects the moving air to its right, until the air is moving parallel to the isobars at position 2. If the wind were geostrophic, at position 3 the air would move northward parallel to straight-line isobars at a constant speed. The wind is blowing at a constant speed, but parallel to curved isobars. A wind that blows at a constant speed parallel to *curved isobars* above the level of frictional influence is termed a **gradient wind.**

Earlier in this chapter we learned that an object accelerates when there is a change in its speed or direction (or both). Therefore, the gradient wind blowing *around* the low-pressure center is constantly accelerating because it is constantly changing direction. This acceleration, called the *centripetal acceleration,* is directed at right angles to the wind, inward toward the low center.

Remember from Newton's second law that, if an object is accelerating, there must be a net force acting on it. In this case, the net force acting on the wind must be directed toward the center of the low, so that the air will

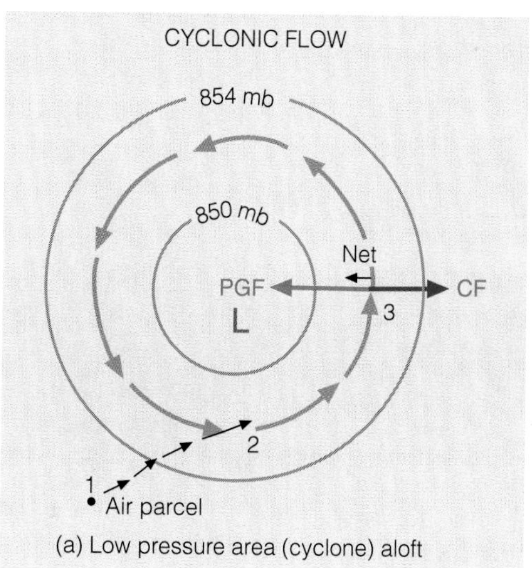

(a) Low pressure area (cyclone) aloft

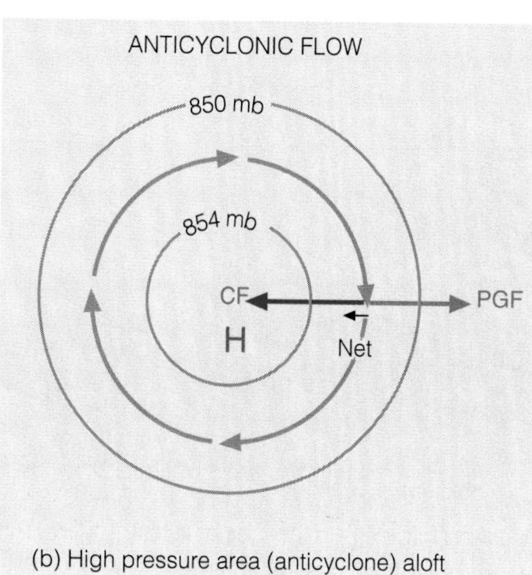

(b) High pressure area (anticyclone) aloft

FIGURE 6.17 Winds and related forces around areas of low and high pressure above the friction level in the Northern Hemisphere. Notice that the pressure gradient force *(PGF)* is in red, while the Coriolis force *(CF)* is in blue.

keep moving in a counterclockwise, circular path. This inward-directed force is called the **centripetal force** (*centri:* center; *petal:* to push toward), and results from an imbalance between the Coriolis force and the pressure gradient force.*

Again, look closely at position 3 (Fig. 6.17a) and observe that the inward-directed pressure gradient force *(PGF)* is greater than the outward-directed Coriolis force *(CF)*. The difference between these forces—the net force—is the inward-directed centripetal force. In Fig. 6.17b, the wind blows clockwise around the center of the high. The spacing of the isobars tells us that the magnitude of the *PGF* is the same as in Fig. 6.17a. However, to keep the wind blowing in a circle, the inward-directed Coriolis force must now be greater in magnitude than the outward-directed pressure gradient force, so that the centripetal force (again, the net force) is directed inward.

In the Southern Hemisphere, the pressure gradient force starts the air moving and the Coriolis force deflects it to the *left,* thereby causing the wind to blow *clockwise around lows* and *counterclockwise around highs.* We know that in the Northern Hemisphere the winds aloft tend to blow in a west-to-east direction. Does this mean that the winds aloft in the Southern Hemisphere blow from east to west? The answer is given in the Focus section on p. 157.

So far we have seen how winds blow in theory, but how do they appear on an actual map?

*In some cases, it is more convenient to express the centripetal force (and the centripetal acceleration) as the *centrifugal force,* an apparent force that is equal in magnitude to the centripetal force, but directed outward from the center of rotation. The gradient wind is then described as a balance of forces between the centrifugal force, the pressure gradient force, and the Coriolis force.

Winds On Upper-Level Charts On the upper-level 500-mb chart (Fig. 6.18), notice that, as we would expect, the winds tend to parallel the contour lines. The wind is geostrophic where it blows in a straight path parallel to evenly spaced lines; it is gradient where it blows parallel to curved contour lines. Moreover, where the lines are closer together, winds are stronger; where the lines are farther apart, the winds are weaker. Where the wind flows in large, looping meanders, following a more or less north-south trajectory (such as along the west coast of North America), the wind-flow pattern is called **meridional.** Where the winds are blowing in a west-to-east direction (such as over the eastern third of the United States), the flow is termed **zonal.**

Because the winds aloft in middle and high latitudes generally blow from west to east, planes flying in

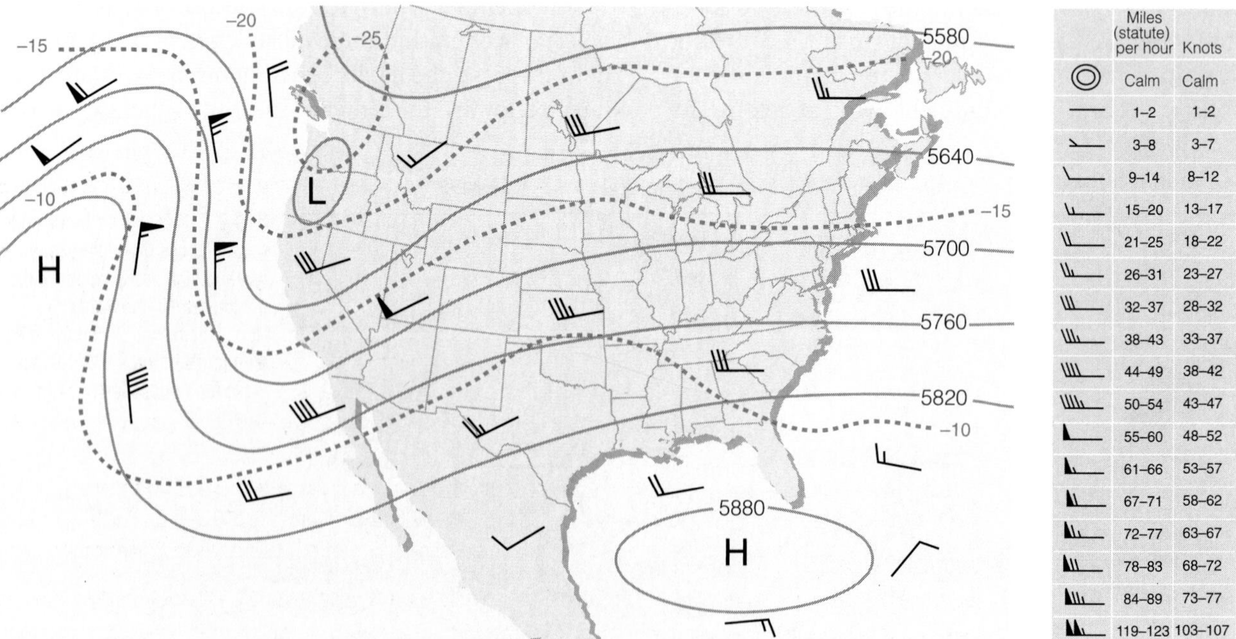

FIGURE 6.18 An upper-level 500-mb map showing wind direction, as indicated by lines that parallel the wind. Wind speeds are indicated by barbs and flags. (See the blue insert.) Solid gray lines are contours in meters above sea level. Dashed red lines are isotherms in °C.

FOCUS ON AN OBSERVATION

Winds Aloft in the Southern Hemisphere

In the Southern Hemisphere, just as in the Northern Hemisphere, the winds aloft blow because of horizontal differences in pressure. The pressure differences, in turn, are due to variations in temperature. Recall from an earlier discussion of pressure that warm air aloft is associated with high pressure and cold air aloft with low pressure. Look at Fig. 3. It shows an upper-level chart that extends from the Northern Hemisphere into the Southern Hemisphere. Over the equator where the air is warmer, the pressure aloft is high. North and south of the equator, where the air is colder, the pressure aloft is lower.

Let's assume, to begin with, that there is no wind on the chart. In the Northern Hemisphere, the pressure gradient force directed northward starts the air moving toward lower pressure. Once the air is set in motion, the Coriolis force bends it to the *right* until it is a *west wind,* blowing parallel to the isobars. In the Southern Hemisphere, the pressure gradient force directed southward starts the air moving south. But notice that the Coriolis force in the Southern Hemisphere bends the moving air to its *left,* until the wind is blowing parallel to the isobars *from the west.* Hence, in the middle and high latitudes of both hemispheres, we generally find westerly winds aloft.

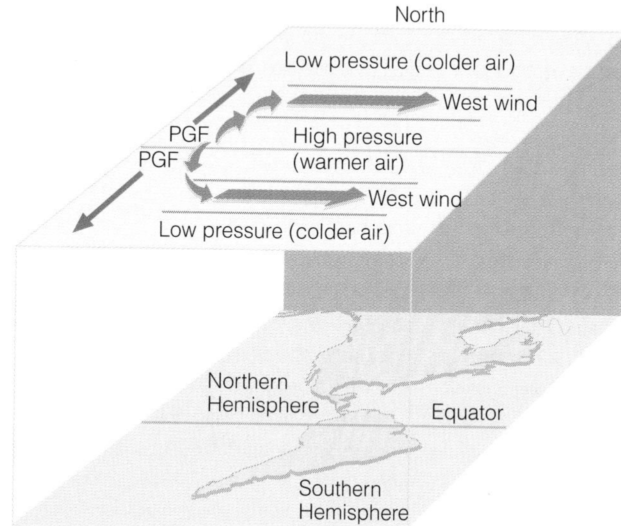

FIGURE 3 Upper-level chart that extends over the Northern and Southern hemispheres. Solid gray lines on the chart are isobars.

this direction have a beneficial tail wind, which explains why a flight from San Francisco to New York City takes about thirty minutes less than the return flight. If the flow aloft is zonal, clouds, storms, and surface anticyclones tend to move more rapidly from west to east. However, where the flow aloft is meridional, as we will see in Chapter 8, surface storms tend to move more slowly, often intensifying into major storm systems.

As we can see from Fig. 6.18, if we know the contour or isobar patterns on an upper-level chart, we also know the direction and relative speed of the wind, even for regions where no direct wind measurements have been made. Similarly, if we know the wind direction and speed, we can estimate the orientation and spacing of the contours or isobars, even if we do not have a current upper-air chart. (It is also possible to estimate the wind flow and pressure pattern aloft by watching the movement of clouds. The Focus section on p. 158 illustrates this further.)

Take a minute and look back at Fig. 6.12 on p. 152. Observe that the winds on this surface map tend to cross the isobars, blowing from higher pressure toward lower pressure. Observe also that along the green line, the tightly packed isobars are producing a steady surface wind of 40 knots. However, this same pressure gradient (with the same air temperature) would, on an upper-level chart, produce a much stronger wind. Why do surface winds normally cross the isobars and why do they blow more slowly than the winds aloft?

SURFACE WINDS

Winds on a surface weather map do not blow exactly parallel to the isobars; instead, they cross the isobars, moving from higher to lower pressure. The angle at which the wind crosses the isobars varies, but averages about 30°. The reason for this behavior is *friction.*

FOCUS ON AN OBSERVATION

Estimating Wind Direction and Pressure Patterns Aloft

Both the wind direction and the orientation of the isobars aloft can be estimated by observing middle- and high-level clouds from the earth's surface. Suppose, for example, we are in the Northern Hemisphere watching clouds directly above us move from southwest to northeast at an elevation of about 3000 m or 10,000 ft (see Fig. 4a). This indicates that the geostrophic wind at this level is southwesterly. Looking downwind, the geostrophic wind blows parallel to the isobars with lower pressure on the left and higher pressure on the right. Thus, if *we stand with our backs to the direction from which the clouds are moving, lower pressure aloft will always be to our left and higher pressure to our right.** From this observation, we can draw a rough upper-level chart (Fig. 4b), which shows isobars and wind direction for an elevation of approximately 10,000 ft.

The isobars aloft will not continue in a southwest-northeast direction indefinitely; rather, they will often bend into wavy patterns. We may carry our observation one step farther, then, by assuming a bending of the lines (Fig. 4c). Thus, with a southwesterly wind aloft, a trough of low pressure will be found to our west and a ridge of high pressure to our east. What would be the pressure pattern if the winds aloft were blowing *from* the northwest? Answer: A trough would be to the east and a ridge to the west.

*This statement for wind and pressure aloft in the Northern Hemisphere is often referred to as *Buys-Ballot's law,* after the Dutch meteorologist Christoph Buys-Ballot (1817–1890), who formulated it.

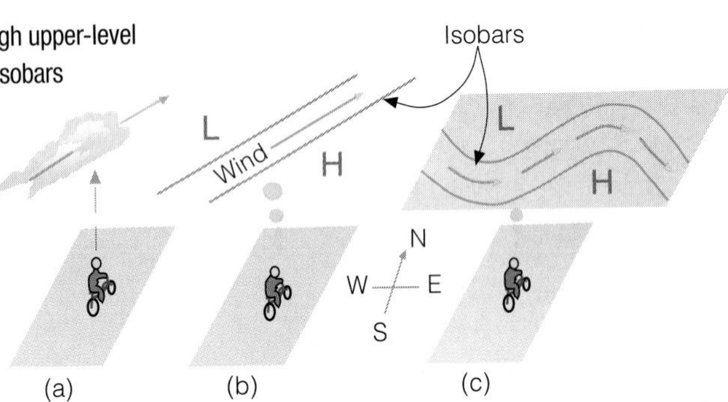

FIGURE 4 This drawing of a simplified upper-level chart is based on cloud observations. Upper-level clouds moving from the southwest (a) indicate isobars and winds aloft (b). When extended horizontally, the upper-level chart appears as in (c), where lower pressure is to the northwest and higher pressure is to the southeast.

The frictional drag of the ground slows the wind down. Because the effect of friction decreases as we move away from the earth's surface, wind speeds tend to increase with height above the ground. The atmospheric layer that is influenced by friction, called the **friction layer** (or *planetary boundary layer*), usually extends upward to an altitude near 1000 m or 3000 ft above the surface, but this altitude may vary somewhat since both strong winds and rough terrain can extend the region of frictional influence.

In Fig. 6.19a, the wind aloft is blowing at a level above the frictional influence of the ground. At this level, the wind is approximately geostrophic and blows parallel to the isobars with the pressure gradient force *(PGF)* on its left balanced by the Coriolis force *(CF)* on its right. Notice, however, that at the surface the wind speed is slower. Apparently, the same pressure gradient force aloft will not produce the same wind speed at the surface, and the wind at the surface will not blow in the same direction as it does aloft.

Near the surface, *friction reduces the wind speed, which in turn reduces the Coriolis force.* Consequently, the weaker Coriolis force no longer balances the pressure gradient force, and the wind blows across the isobars toward lower pressure. The pressure gradient force is now balanced by the sum of the frictional force and the Coriolis force. Therefore, in the Northern Hemisphere, we find surface winds blowing counterclockwise and *into* a low; they flow clockwise and *out* of a high (see Fig. 6.19b).

In the Southern Hemisphere, winds blow clockwise and inward around surface lows, counterclockwise and outward around surface highs. See Fig. 6.20, which shows a surface weather map and the general wind flow pattern for South America.

MeteorologyNow™ Click "Winds in the Two Hemispheres" to find examples of high and low pressure systems in different parts of the world.

WINDS AND VERTICAL AIR MOTIONS

Up to this point, we have seen that surface winds blow in toward the center of low pressure and outward away from the center of high pressure. As air moves inward toward the center of a low-pressure area (see Fig. 6.21), it must go somewhere. Since this converging air cannot go into the ground, it slowly rises. Above the surface low (at about 6 km or so), the air begins to spread apart (diverge) to compensate for the converging surface air.

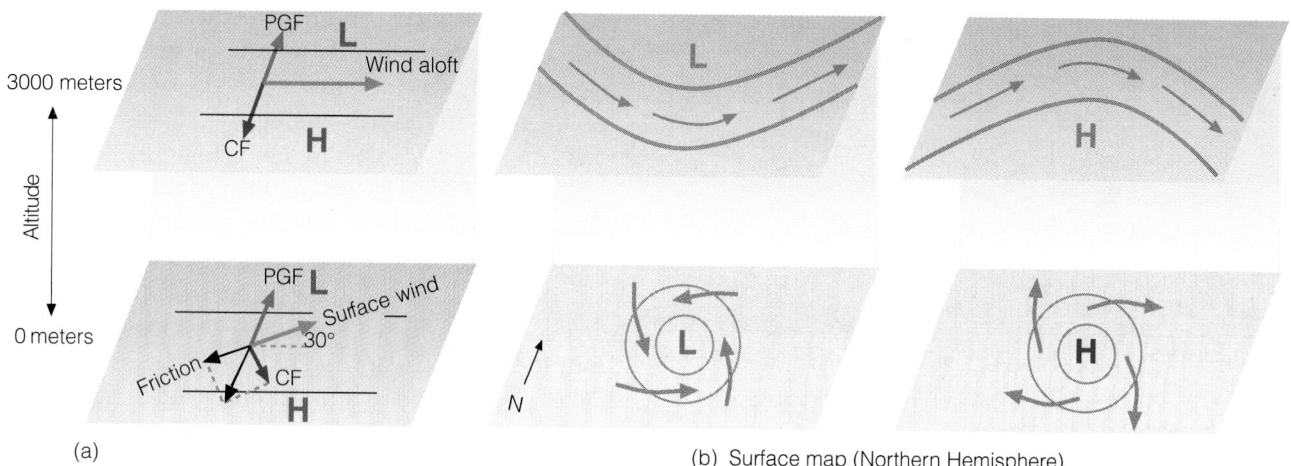

FIGURE 6.19 (a) The effect of surface friction is to slow down the wind so that, near the ground, the wind crosses the isobars and blows toward lower pressure. (b) This phenomenon at the surface produces an outflow of air around a high. Aloft, the winds blow parallel to the lines, usually in a wavy west-to-east pattern.

As long as the upper-level diverging air balances the converging surface air, the central pressure in the low does not change. However, the surface pressure *will change* if upper-level divergence and surface convergence are not in balance. For example, if upper-level divergence exceeds surface convergence (that is, more air is removed at the top than is taken in at the surface), the central pressure of the low will decrease, and isobars around the low will become more tightly packed. This situation increases the pressure gradient (and, hence, the pressure gradient force), which, in turn, increases the surface winds.

Surface winds move outward (diverge), away from the center of a high-pressure area. To replace this laterally spreading air, the air aloft converges and slowly descends (see Fig. 6.21). Again, as long as upper-level converging air balances surface diverging air, the central pressure in the high will not change. (Convergence and divergence of air are so important to the development or weakening of surface pressure systems that we will examine this topic again when we look more closely at the vertical structure of pressure systems in Chapter 8.)

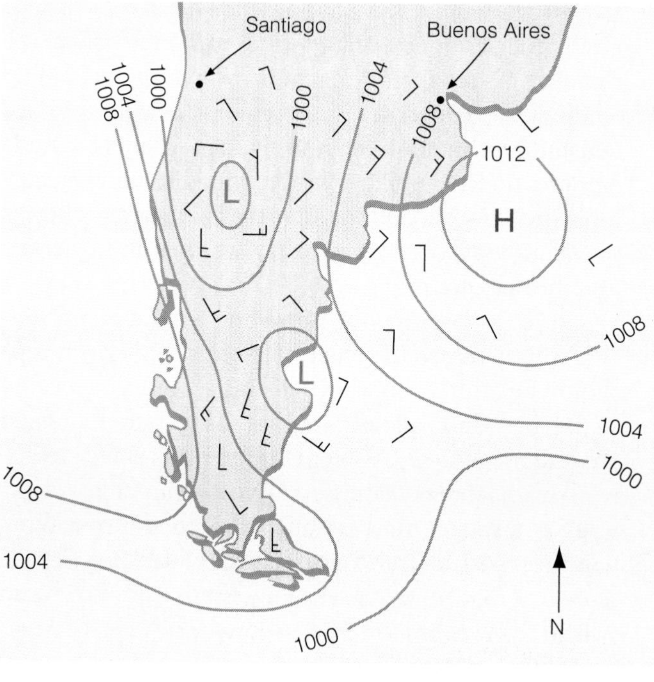

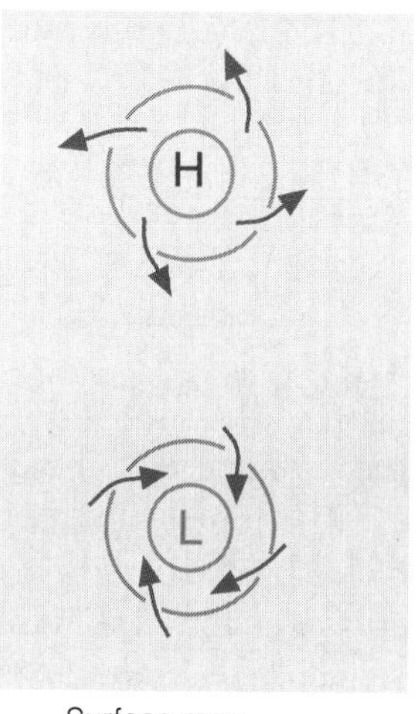

FIGURE 6.20 (a) Surface weather map showing isobars and winds on a day in December in South America. (b) The boxed area shows the idealized flow around surface-pressure systems in the Southern Hemisphere.

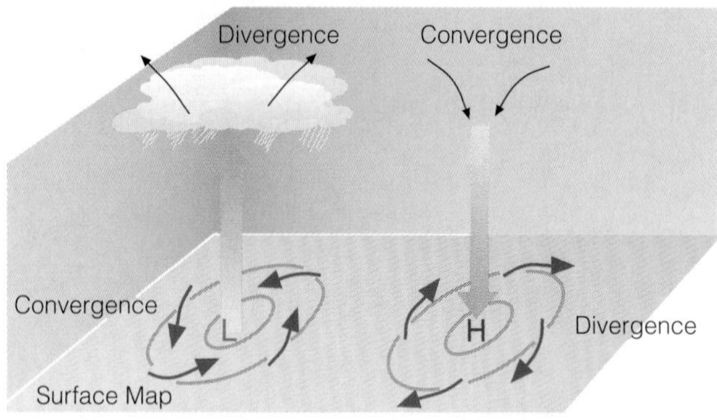

FIGURE 6.21 Winds and air motions associated with surface highs and lows in the Northern Hemisphere.

The rate at which air rises above a low or descends above a high is small compared to the horizontal winds that spiral about these systems. Generally, the vertical motions are usually only about several centimeters per second, or about 1.5 km (or 1 mi) per day.

Earlier in this chapter we learned that air moves in response to pressure differences. Because air pressure decreases rapidly with increasing height above the surface, there is always a strong pressure gradient force directed upward, much stronger than in the horizontal. Why, then, doesn't the air rush off into space?

Air does not rush off into space because the upward-directed pressure gradient force is nearly always exactly balanced by the downward force of gravity. When these two forces are in exact balance, the air is said to be in **hydrostatic equilibrium.** When air is in hydrostatic equilibrium, there is no net vertical force acting on it, and so there is no net vertical acceleration. Most of the time, the atmosphere approximates hydrostatic balance, even when air slowly rises or descends at a constant speed. However, this balance does not exist in violent thunderstorms and tornadoes, where the air shows appreciable vertical acceleration. But these occur over relatively small vertical distances, considering the total vertical extent of the atmosphere.

MEASURING AND DETERMINING WINDS

Wind is characterized by its direction, speed, and gustiness. Because air is invisible, we cannot really see it. Rather, we see things being moved by it. Therefore, we can determine wind direction by watching the movement of objects as air passes them. For example, the rustling of small leaves, smoke drifting near the ground, and flags waving on a pole all indicate wind direction. In a light breeze, a tried and true method of determining wind direction is to raise a wet finger into the air. The dampness quickly evaporates on the windward side, cooling the skin. Traffic sounds carried from nearby railroads or airports can be used to help figure out the direction of the wind. Even your nose can alert you to the wind direction as the smell of fried chicken or broiled hamburgers drifts with the wind from a local restaurant.

We already know that *wind direction* is given as the direction from which it is blowing—a north wind blows from the north toward the south. However, near large bodies of water and in hilly regions, wind direction may be expressed differently. For example, wind blowing from the water onto the land is referred to as an **onshore wind,** whereas wind blowing from land to water is called an **offshore wind** (see Fig. 6.22). Air moving uphill is an *upslope wind;* air moving downhill is a *downslope wind.* The wind direction may also be given as degrees about a 360° circle. These directions are expressed by the numbers shown in Fig. 6.23. For example: A wind direction of 360° is a north wind; an east wind is 90°; a south wind is 180°; and calm is expressed as zero. It is also common practice to express the wind direction in terms of compass points, such as N, NW, NE, and so on. (Helpful hints for estimating wind speed from surface observations may be found in the *Beaufort Wind Scale,* located in Appendix F, on p. 438.)

The Influence of Prevailing Winds At many locations, the wind blows more frequently from one direction than from any other. The **prevailing wind** is the name given to the wind direction most often observed during a given time period. Prevailing winds can greatly affect the climate of a region. For example, where the prevailing winds are upslope, the rising, cooling air makes clouds, fog, and precipitation more likely than where the winds are downslope. Prevailing onshore winds in summer carry moisture, cool air, and fog into coastal regions, whereas prevailing offshore breezes carry warmer and drier air into the same locations.

In city planning, the prevailing wind can help decide where industrial centers, factories, and city dumps should be built. All of these, of course, must be located so that the wind will not carry pollutants into populated areas. Sewage disposal plants must be situated downwind from large housing developments, and major runways at airports must be aligned with the prevailing wind to assist aircraft in taking off or landing. In the high country, strong prevailing winds can bend and twist tree branches toward the downwind side, producing *wind-sculptured "flag" trees* (see Fig. 6.24).

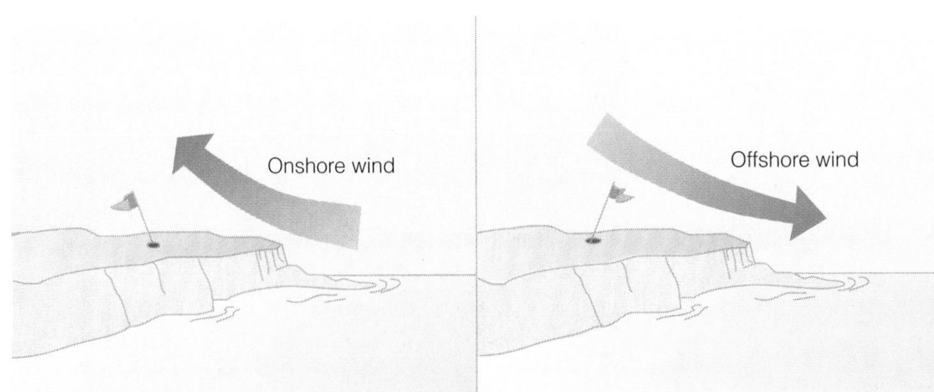

FIGURE 6.22 An onshore wind blows from water to land, whereas an offshore wind blows from land to water.

The prevailing wind can even be a significant factor in building an individual home. In the northeastern half of the United States, the prevailing wind in winter is northwest and in summer it is southwest. Thus, houses built in the northeastern United States should have windows facing southwest to provide summertime ventilation and few, if any, windows facing the cold winter winds from the northwest. The northwest side of the house should be thoroughly insulated and even protected by a windbreak.

The prevailing wind can be represented by a **wind rose** (Fig. 6.25), which indicates the percentage of time the wind blows from different directions. Extensions from the center of a circle point to the wind direction, and the length of each extension indicates the percentage of time the wind blew from that direction.

Wind Instruments A very old, yet reliable, weather instrument for determining wind direction is the **wind vane.** Most wind vanes consist of a long arrow with a tail, which is allowed to move freely about a vertical post (see Fig. 6.26). The arrow always points into the wind and, hence, always gives the wind direction. Wind vanes can be made of almost any material. At airports, a cone-shaped bag opened at both ends so that it extends horizontally as the wind blows through it sits near the runway. This form of wind vane, called a *wind sock,* enables pilots to tell the surface wind direction when landing.

The instrument that measures wind speed is the **anemometer.** Most anemometers consist of three (or more) hemispherical cups *(cup anemometer)* mounted on a vertical shaft as shown in Fig. 6.26. The difference in wind pressure from one side of a cup to the other causes the cups to spin about the shaft. The rate at which they rotate is directly proportional to the speed of the wind. The spinning of the cups is usually translated into

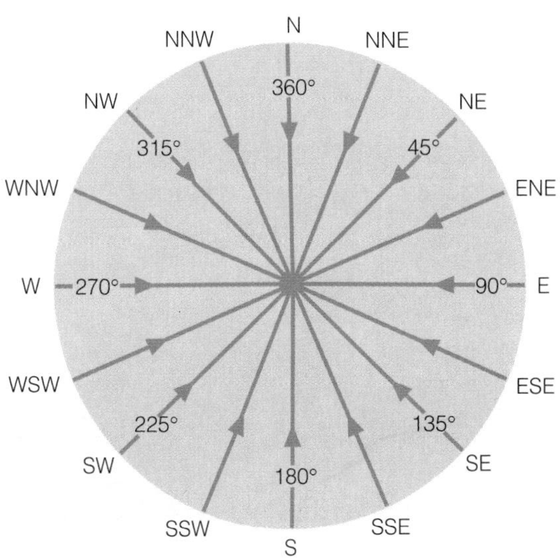

FIGURE 6.23 Wind direction can be expressed in degrees about a circle or as compass points.

Photo by author

FIGURE 6.24 In the high country, trees standing unprotected from the wind are often sculpted into "flag" trees.

DID YOU KNOW?

A challenge to the world wind record. On December 16, 1997, Typhoon Paka ravaged the island of Guam. On the north side of the island, the weather station at Anderson Air Force Base reported a wind gust of 205 knots (236 mi/hr) as the storm moved through. At first, the wind gust was declared a new world record, surpassing the record of 200 knots (230 mi/hr) set on Mt. Washington in 1934. However, after reviewing the wind record at the site, authorities concluded that the peak wind of 205 knots was unreliable, as the combination of high winds and heavy rain caused the anemometer to record an unrealistically high wind. However, on May 3, 1999, Doppler radar measured an incredible wind speed of 276 knots (318 mi/hr) inside a huge tornado over Oklahoma, making this the highest wind speed ever measured near the earth's surface.

wind speed through a system of gears, and may be read from a dial or transmitted to a recorder.

The **aerovane** *(skyvane)* is an instrument that indicates both wind speed and direction. It consists of a bladed propeller that rotates at a rate proportional to the wind speed. Its streamlined shape and a vertical fin keep the blades facing into the wind (see Fig. 6.27). When attached to a recorder, a continuous record of both wind speed and direction is obtained.

The wind-measuring instruments described thus far are "ground-based" and only give wind speed or direction at a particular fixed location. But the wind is influenced by local conditions, such as buildings, trees, and so on. Also, wind speed normally increases rapidly with height above the ground. Thus, wind instruments should be exposed to freely flowing air well above the roofs of buildings. In practice, unfortunately, anemometers are placed at various levels; the result, then, is often erratic wind observations.

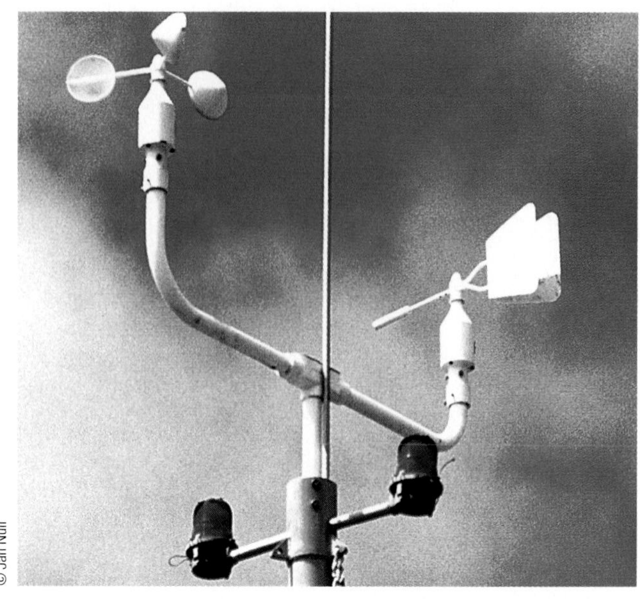
© Jan Null

FIGURE 6.26 A wind vane and a cup anemometer. These instruments are part of the ASOS system. (For a complete picture of the system, see Fig. 3.17, p. 74).

Wind information can be obtained during a radiosonde observation. A balloon rises from the surface carrying a *radiosonde* (an instrument package designed to measure the vertical profile of temperature, pressure, and humidity—see Chapter 1, p. 11). Equipment located on the ground constantly tracks the balloon, measuring its vertical and horizontal angles as well as its height above the ground. From this information, a computer determines and prints the vertical profile of wind from the surface up to where the balloon normally pops, typically in the stratosphere near 30 km or 100,000 ft. The observation of winds using a radiosonde balloon is called a *rawinsonde observation.*

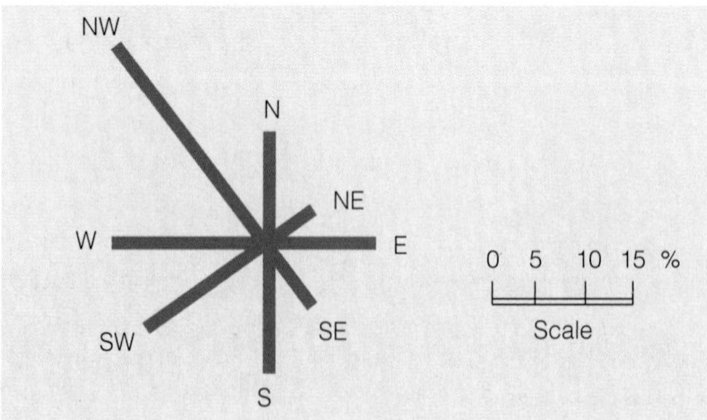

FIGURE 6.25 This wind rose represents the percent of time the wind blew from different directions at a given site during the month of January for the past ten years. The prevailing wind is NW and the wind direction of least frequency is NE.

Photo by author

FIGURE 6.27 The aerovane (skyvane).

FOCUS ON A SPECIAL TOPIC

Wind Power

For many decades thousands of small windmills—their arms spinning in a stiff breeze—have pumped water, sawed wood, and even supplemented the electrical needs of small farms. It was not until the energy crisis of the early 1970s, however, that we seriously considered wind-driven turbines, called *wind turbines,* to run generators that produce electricity.

Wind power seems an attractive way of producing energy—it is nonpolluting and, unlike solar power, is not restricted to daytime use. It does, however, pose some problems, as the cost of a single wind turbine can exceed $1 million. In addition, a region dotted with large wind machines is unaesthetic. (Probably, though, it is no more of an eyesore than the parades of huge electrical towers marching across many open areas.) And, unfortunately, each year the blades of spinning turbines kill countless birds. To help remedy this problem, many wind turbine companies hire avian specialists to study bird behavior, and some turbines are actually shut down during nesting time.

Photo by author

FIGURE 5 A portion of a wind farm near the summit of Altamont Pass, California. With over 7000 wind turbines, this is the world's largest wind energy development project.

If the wind turbine is to produce electricity, there must be wind, not just any wind, but a flow of air neither too weak nor too strong. A slight breeze will not turn the blades, and a powerful wind gust could severely damage the machine. Thus, regions with the greatest potential for wind-generated power would have moderate, steady winds.

Sophisticated advanced technology allows many modern turbines to sense meteorological data from their surroundings. Wind turbines actually produce energy in winds as low as 5 knots, and as high as 45 knots.

As fossil fuels diminish, the wind can help fill the gap by providing a pollution-free alternative form of energy. For example, in 1997, over 15,000 wind machines were generating over 3.5 billion kilowatt-hours of electricity in the United States, which is enough energy to supply the annual needs of more than 3 million people. In California alone, there are thousands of wind turbines, many of which are on *wind farms*—clusters of 50 or more wind turbines (see Fig. 5). In 1997, California's wind farms were producing a little less than 2 percent of the state's total electrical needs. Present estimates are that wind power may be able to furnish up to a few percent of the nation's total energy needs during the first half of this century.

In remote regions of the world where upper-air observations are lacking, wind speed and direction can be obtained from satellites. So far, the most reliable wind data have come from geostationary satellites positioned above a particular location. From this vantage point, the satellite shows the movement of clouds. The direction of cloud movement indicates wind direction, and the horizontal distance the cloud moves during a given time period indicates the wind speed.

Recently, Doppler radar has been employed to obtain a vertical profile of wind speed and direction up to an altitude of 16 km or so above the ground. Such a profile is called a *wind sounding,* and the radar, a **wind profiler** (or simply a *profiler*). Doppler radar, like conventional radar, emits pulses of microwave radiation that are returned (backscattered) from a target, in this case the irregularities in moisture and temperature created by turbulent, twisting eddies that move with the wind. Doppler radar works on the principle that, as these eddies move toward or away from the receiving antenna, the returning radar pulse will change in frequency. The Doppler radar wind profilers are so sensitive that they can translate the backscattered energy from these eddies into a vertical picture of wind speed and direction in a column of air 16 km (10 mi) thick. Presently, there is a network of wind profilers scattered across the central United States.

The wind is a strong weather element that can affect our environment in many ways. It can shape the landscape, transport material from one area to another, and generate ocean waves. It can also turn the blades of a windmill and blow down a row of trees. The reason the wind is capable of such feats is that, as the wind blows against an object, it exerts a force upon it. The amount of force exerted by the wind over an area increases as the square of the wind velocity. So when the wind velocity doubles, the force it exerts on an object goes up by a factor of four. In an attempt to harness some of the wind's energy and turn it into electricity, many countries are building wind generators. More information on this topic is given in the Focus section on p. 163.

(**Meteorology Now**™ Assess your understanding of this chapter's topics with additional quizzing and tutorials at http://earthscience.brookscole.com/ahrens/ess4e.

SUMMARY

This chapter gives us a broad view of how and why the wind blows. Aloft where horizontal variations in temperature exist, there is a corresponding horizontal change in pressure. The difference in pressure establishes a force, the pressure gradient force *(PGF)*, which starts the air moving from higher toward lower pressure.

Once the air is set in motion, the Coriolis force bends the moving air to the right of its intended path in the Northern Hemisphere and to the left in the Southern Hemisphere. Above the level of surface friction, the wind is bent enough so that it blows nearly parallel to the isobars, or contours. Where the wind blows in a straight-line path, and a balance exists between the pressure gradient force and the Coriolis force, the wind is termed geostrophic. Where the wind blows parallel to curved isobars (or contours), the wind is called a gradient wind. When the wind-flow pattern aloft is west-to-east, the flow is called *zonal*; when the wind-flow-pattern aloft is more north-south, the flow is called *meridional.*

The interaction of the forces causes the wind in the Northern Hemisphere to blow clockwise around regions of high pressure and counterclockwise around areas of low pressure. In the Southern Hemisphere, the wind blows counterclockwise around highs and clockwise around lows. The effect of surface friction is to slow down the wind. This causes the surface air to blow across the isobars from higher pressure toward lower pressure. Consequently, in both hemispheres, surface winds blow outward, away from the center of a high, and inward, toward the center of a low.

At the end of the chapter, we looked at various methods and instruments used to determine and measure wind speed and direction.

KEY TERMS

The following terms are listed in the order they appear in the text. Define each. Doing so will aid you in reviewing the material covered in this chapter.

air pressure
millibar
standard atmospheric pressure
barometer
mercury barometer
aneroid barometer
station pressure
sea-level pressure
isobar
surface map
anticyclone
mid-latitude cyclone
isobaric map
contour line
ridge
trough
pressure gradient
pressure gradient force
Coriolis force
geostrophic wind
gradient wind
centripetal force
meridional flow
zonal flow
friction layer
hydrostatic equilibrium
onshore wind
offshore wind
prevailing wind
wind rose
wind vane
anemometer
aerovane
wind profiler

QUESTIONS FOR REVIEW

1. Explain why atmospheric pressure always decreases with increasing altitude.
2. Why is the decrease of air pressure with increasing altitude more rapid when the air is cold?
3. What is considered standard sea-level atmospheric pressure in millibars? In inches of mercury? In hectopascals?
4. Would a sea-level pressure of 1040 mb be considered high or low pressure?
5. With the aid of a diagram, describe how a mercury barometer works.
6. How does an aneroid barometer measure atmospheric pressure? How does it differ from a mercury barometer?
7. How does sea-level pressure differ from station pressure? Can the two ever be the same? Explain.
8. Why will Denver, Colorado, always have a lower station pressure than Chicago, Illinois?

9. What are isobars?
10. On an upper-level map, is cold air aloft generally associated with low or high pressure? What about warm air aloft?
11. What do Newton's first and second laws of motion tell us?
12. What does a steep (or strong) pressure gradient mean?
13. How does a gentle (or weak) pressure gradient appear on a surface map?
14. What is the name of the force that initially sets the air in motion and, hence, causes the wind to blow?
15. Explain why, on a map, closely spaced isobars (or contours) indicate strong winds, and widely spaced isobars (or contours) indicate weak winds.
16. What does the Coriolis force do to moving air
    (a) in the Northern Hemisphere?
    (b) in the Southern Hemisphere?
17. Explain how each of the following influences the Coriolis force:
    (a) wind speed
    (b) latitude.
18. Why do upper-level winds in the middle latitudes of both hemispheres generally blow from west to east?
19. What is a geostrophic wind?
20. What are the forces that affect the horizontal movement of air?
21. Describe how the wind blows around high-pressure areas and low-pressure areas aloft and near the surface
    (a) in the Northern Hemisphere; and
    (b) in the Southern Hemisphere.
22. If the clouds overhead are moving from north to south, would the upper-level center of low pressure be to the east or west of you?
23. On a surface map, why do surface winds tend to cross the isobars and flow from higher pressure toward lower pressure?
24. Since there is always an upward-directed pressure gradient force, why doesn't air rush off into space?
25. List as many ways as you can of determining wind direction and wind speed.
26. Below is a list of instruments. Describe how each one is able to measure wind speed, wind direction, or both.
    (a) wind vane
    (b) cup anemometer
    (c) aerovane (skyvane)
    (d) radiosonde
    (e) satellite
    (f) wind profiler
27. An upper wind direction is reported as 225°. From what compass direction is the wind blowing?

QUESTIONS FOR THOUGHT AND EXPLORATION

1. The gas law states that pressure is proportional to temperature times density. Use the gas law to explain why a basketball seems to deflate when placed in a refrigerator.
2. Can the station pressure ever *exceed* the sea-level pressure? Explain.
3. Can two isobars drawn on a surface weather map ever intersect? Explain your reasoning.
4. The pressure gradient force causes air to move from higher pressures toward lower pressures (perpendicular to the isobars), yet actual winds rarely blow in this fashion. Explain why they don't.
5. The Coriolis force causes winds to deflect to the right of their intended path in the Northern Hemisphere, yet around a surface low-pressure area, winds blow counterclockwise, appearing to bend to their left. Explain why.
6. Explain why, on a sunny day, an aneroid barometer would indicate "stormy" weather when carried to the top of a hill or mountain.
7. Pilots often use the expression "high to low, look out below." In terms of upper-level temperature and pressure, explain what this can mean.
8. Suppose an aircraft using a pressure altimeter flies along a constant pressure surface from standard temperature into warmer-than-standard air without any corrections. Explain why the altimeter would indicate an altitude lower than the aircraft's true altitude.
9. If the earth were not rotating, how would the wind blow with respect to centers of high and low pressure?
10. Why are surface winds that blow over the ocean closer to being geostrophic than those that blow over the land?
11. In the Northern Hemisphere, you observe surface winds shift from N to NE to E, then to SE. From this observation, you determine that a west-to-east moving high-pressure area (anticyclone) has passed north of your location. Describe with the aid of a diagram how you were able to come to this conclusion.
12. As a cruise ship crosses the equator, the entertainment director exclaims that water in a tub will drain in the opposite direction now that the ship is in the Southern Hemisphere. Give *two* reasons to the entertainment director why this assertion is not so.

• Go to the Brooks/Cole Earth Sciences Resource Center (http://earthscience.brookscole.com) for critical thinking exercises, articles, and additional readings from InfoTrac College Edition, Brooks/Cole's online student library.

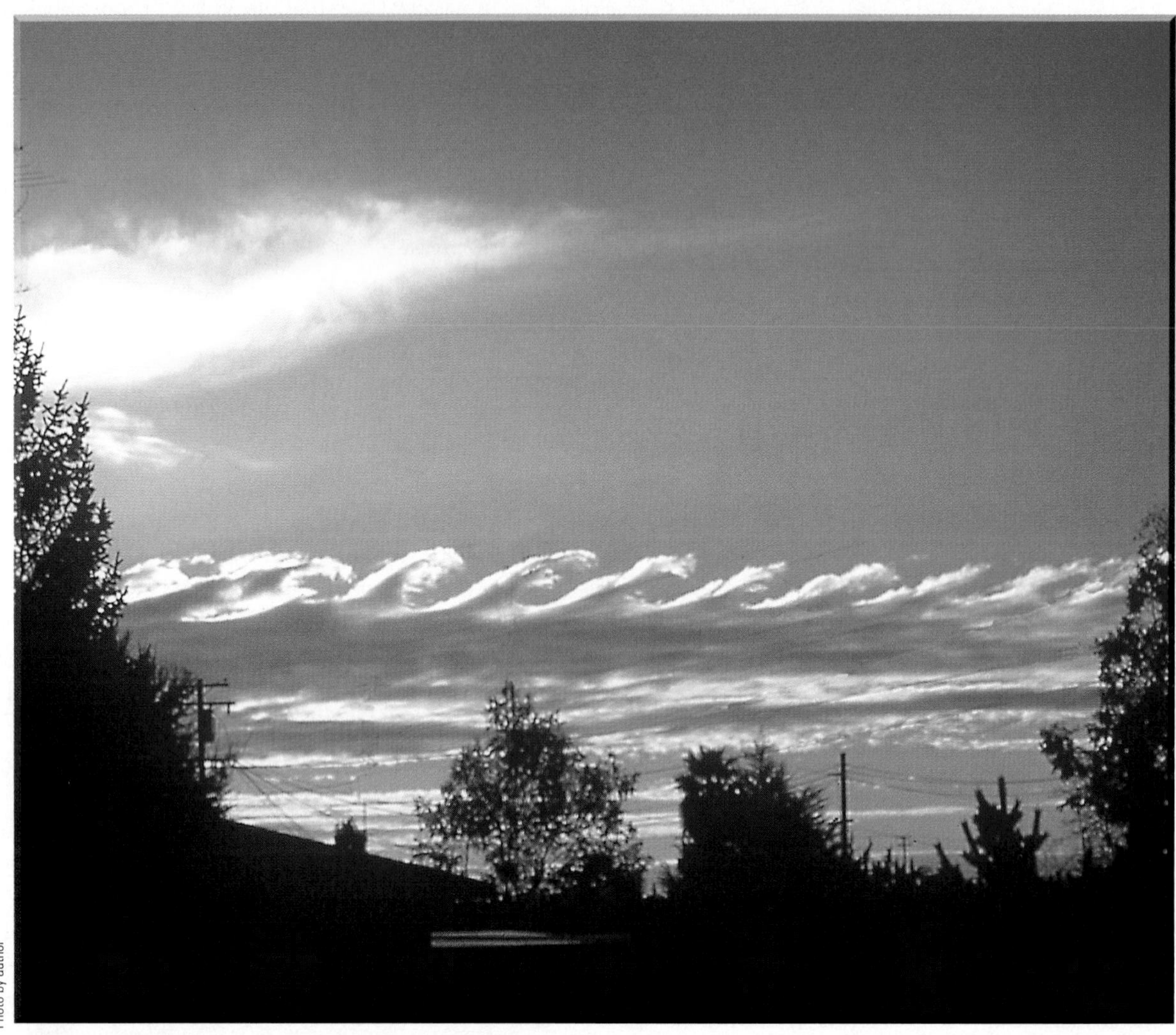

Photo by author

Wave clouds, called *billow clouds,* forming in a region of wind shear.

Meteorology Now™ This icon, appearing throughout the book, indicates an opportunity to explore interactive tutorials, animations, or practice problems available on the MeteorologyNow Web site at http://earthscience.brookscole.com/ahrens/ess4e.

Atmospheric Circulations

On December 30, 1997, a United Airlines' Boeing 747 carrying 374 passengers was en route to Hawaii from Japan. Dinner had just been served, and the aircraft had reached a cruising altitude of 33,000 feet. Suddenly, east of Tokyo and over the Pacific Ocean, this routine, uneventful flight turned tragic. Without warning, the aircraft entered a region of severe air turbulence and a vibration ran through the aircraft. The plane nosed upward, then plunged toward the earth for about 1000 feet before stabilizing. Screaming, terrified passengers not fastened to their seats were flung against the walls of the aircraft, then dropped. Bags, serving trays, and luggage that slipped out from under the seats were flying about inside the plane. Within seconds, the entire ordeal was over. At least 110 people were injured, 12 seriously. Tragically, there was one fatality: a 32-year-old woman, who had been hurled against the ceiling of the plane, died of severe head injuries. What sort of atmospheric phenomenon could cause such turbulence?

CONTENTS

The aircraft in our opener encountered a turbulent eddy—an "air pocket"—in perfectly clear weather. Such eddies are not uncommon, especially in the vicinity of jet streams. In this chapter, we will examine a variety of eddy circulations. First, we will look at the formation of small-scale winds. Then, we will examine slightly larger circulations—local winds —such as the sea breeze and the chinook, describing how they form and the type of weather they generally bring. Finally, we will look at the general wind flow pattern around the world.

SCALES OF ATMOSPHERIC MOTION

The air in motion—what we commonly call wind—is invisible, yet we see evidence of it nearly everywhere we look. It sculptures rocks, moves leaves, blows smoke, and lifts water vapor upward to where it can condense into clouds. The wind is with us wherever we go. On a hot day, it can cool us off; on a cold day, it can make us shiver. A breeze can sharpen our appetite when it blows the aroma from the local bakery in our direction. The wind is a powerful element. The workhorse of weather, it moves storms and large fair-weather systems around the globe. It transports heat, moisture, dust, insects, bacteria, and pollens from one area to another.

Circulations of all sizes exist within the atmosphere. Little whirls form inside bigger whirls, which encompass even larger whirls—one huge mass of turbulent, twisting *eddies.** For clarity, meteorologists arrange circulations according to their size. This hierarchy of motion from tiny gusts to giant storms is called the **scales of motion.**

Consider smoke rising into the otherwise clean air from a chimney in the industrial section of a large city (see Fig. 7.1a). Within the smoke, small chaotic motions—tiny eddies—cause it to tumble and turn. These eddies constitute the smallest scale of motion—the **microscale.** At the microscale, eddies with diameters of a few meters or less not only disperse smoke, they also sway branches and swirl dust and papers into the air. They form by convection or by the wind blowing past obstructions and are usually short-lived, lasting only a few minutes at best.

In Fig. 7.1b observe that, as the smoke rises, it drifts toward the center of town. Here the smoke rises even higher and is carried back toward the industrial section. This circulation of city air constitutes the next larger scale—the **mesoscale** (meaning middle scale). Typical mesoscale winds range from a few kilometers to about a hundred kilometers in diameter. Generally, they last longer than microscale motions, often many minutes, hours, or in some cases as long as a day. Mesoscale circulations include local winds (which form along shorelines and mountains), as well as thunderstorms, tornadoes, and small tropical storms.

When we look for the smokestack on a surface weather map (Fig. 7.1c), neither the smokestack nor the

*Eddies are spinning globs of air that have a life history of their own.

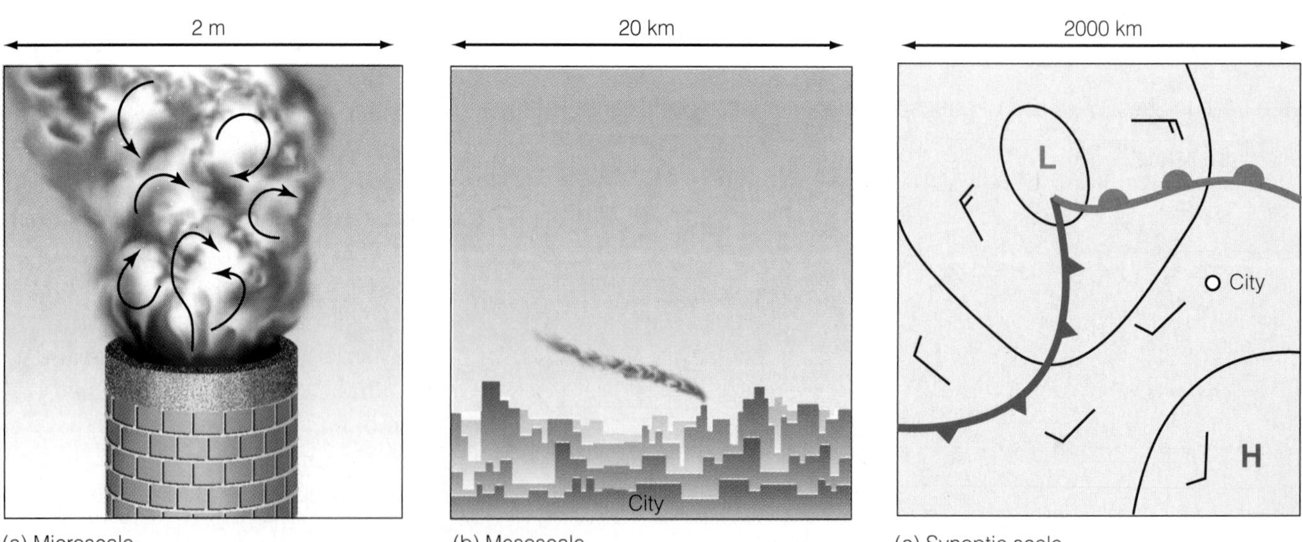

FIGURE 7.1 Scales of atmospheric motion. The tiny microscale motions constitute a part of the larger mesoscale motions, which, in turn, are part of the much larger synoptic scale. Notice that as the scale becomes larger, motions observed at the smaller scale are no longer visible.

TABLE 7.1 The Scales of Atmospheric Motion with the Phenomena's Average Size and Life Span*

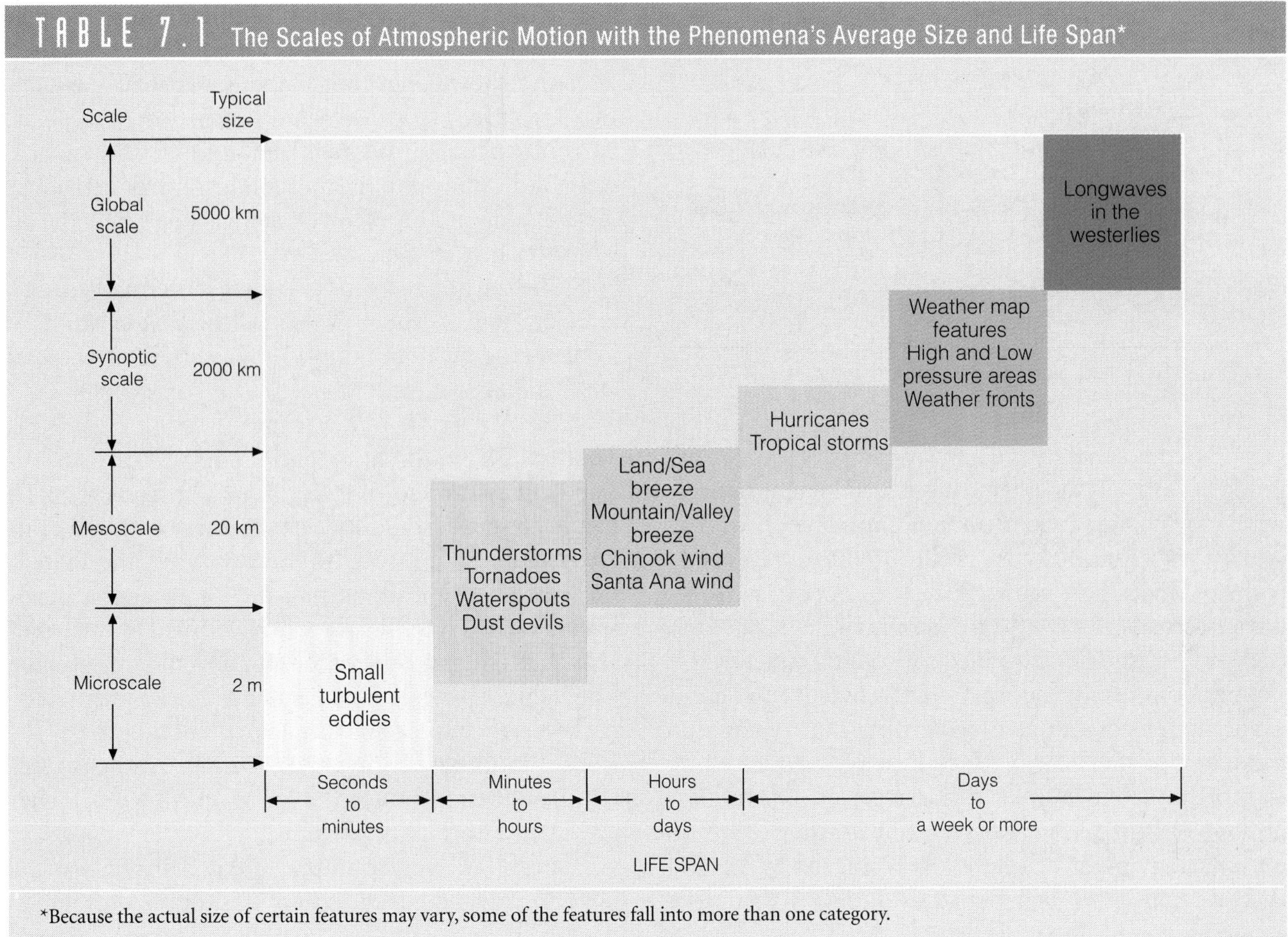

*Because the actual size of certain features may vary, some of the features fall into more than one category.

circulation of city air shows up. All that we see are the circulations around high- and low-pressure areas—the cyclones and anticyclones of the middle latitude. We are now looking at the **synoptic scale,** or weather map scale. Circulations of this magnitude dominate regions of hundreds to even thousands of square kilometers and, although the life spans of these features vary, they typically last for days and sometimes weeks. The largest wind patterns are seen at the **planetary** *(global)* **scale.** Here, we have wind patterns ranging over the entire earth. Sometimes, the synoptic and planetary scales are combined and referred to as the *macroscale.* (Table 7.1 summarizes the various scales of motion and their average life span.)

EDDIES—BIG AND SMALL

When the wind encounters a solid object, a whirl of air—or *eddy*—forms on the object's downwind side. The size and shape of the eddy often depend upon the size and shape of the obstacle and on the speed of the wind. Light winds produce small stationary eddies. Wind moving past trees, shrubs, and even your body produces small eddies. (You may have had the experience of dropping a piece of paper on a windy day only to have it carried away by a swirling eddy as you bend down to pick it up.) Air flowing over a building produces larger eddies that will, at best, be about the size of the building. Strong winds blowing past an open sports stadium can produce eddies that may rotate in such a way as to create surface winds on the playing field that move in a direction opposite to the wind flow above the stadium. Wind blowing over a fairly smooth surface produces few eddies, but when the surface is rough, many eddies form.

DID YOU KNOW?

During January, 1982, high winds in Boulder, Colorado, generated turbulent, twisting eddies that popped the windows out of 100 cars parked at a shopping center.

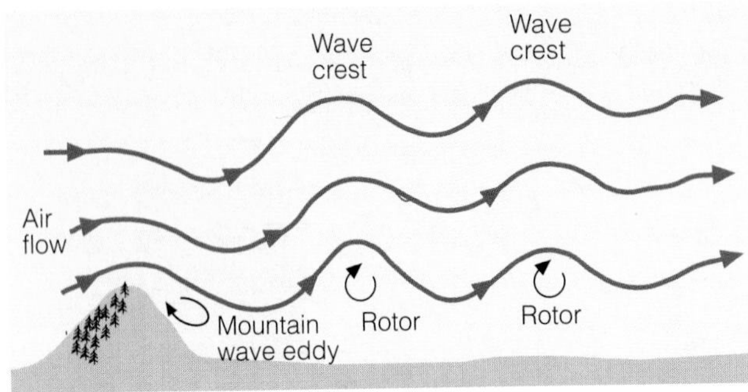

FIGURE 7.2 Under stable conditions, air flowing past a mountain range can create eddies many kilometers downwind from the mountain itself.

The eddies that form downwind from obstacles can produce a variety of interesting effects. For instance, wind moving over a mountain range in stable air with a speed greater than 40 knots usually produces waves and eddies, such as those shown in Fig. 7.2. We can see that eddies form both close to the mountain and beneath each wave crest. These so-called **rotors** have violent vertical motions that produce extreme turbulence and hazardous flying conditions. Strong winds blowing over a mountain in stable air may produce a *mountain wave eddy* on the downwind side, with a reverse flow near the ground. On a much smaller scale, the howling of wind on a blustery night is believed to be caused by eddies that are constantly being shed around obstructions, such as chimneys and roof corners.

Turbulent eddies form aloft as well as near the surface. Turbulence aloft can occur suddenly and unexpectedly, especially where the wind changes its speed or direction (or both) abruptly. Such a change is called **wind shear.** The shearing creates forces that produce eddies along a mixing zone. If the eddies form in clear air, this form of turbulence is called **clear air turbulence,** or **CAT.** (Additional information on this topic is given in the Focus section on p. 171.)

LOCAL WIND SYSTEMS

Every summer, millions of people flock to the New Jersey shore, hoping to escape the oppressive heat and humidity of the inland region. On hot, humid afternoons, these travelers often encounter thunderstorms about twenty miles or so from the ocean, thunderstorms that invariably last for only a few minutes. In fact, by the time the vacationers arrive at the beach, skies are generally clear and air temperatures are much lower, as cool ocean breezes greet them. If the travelers return home in the afternoon, these "mysterious" showers often occur at just about the same location as before.

The showers are not really mysterious. Actually, they are caused by a local wind system—the *sea breeze.* As cooler ocean air pours inland, it forces the warmer, unstable humid air to rise and condense, producing majestic clouds and rain showers along a line where the air of contrasting temperatures meet.

The sea breeze forms as part of a thermally driven circulation. Consequently, we will begin our study of local winds by examining the formation of thermal circulations.

Thermal Circulations Consider the vertical distribution of pressure shown in Fig. 7.3a. The isobars* all lie parallel to the earth's surface; thus, there is no horizontal variation in pressure (or temperature), and there is no pressure gradient and no wind. Suppose the atmosphere is cooled to the north and warmed to the south (Fig. 7.3b). In the cold, dense air above the surface, the isobars bunch closer together, while in the warm, less-dense air, they spread farther apart. This dipping of the isobars produces a horizontal pressure gradient force *(PGF)* aloft that causes the air to move from higher pressure toward lower pressure.

At the surface, the air pressure remains unchanged until the air aloft begins to move. As the air aloft moves from south to north, air leaves the southern area and "piles up" above the northern area. This redistribution of air reduces the surface air pressure to the south and raises it to the north. Consequently, a pressure gradient force is established at the earth's surface from north to south and, hence, surface winds begin to blow from north to south.

We now have a distribution of pressure and temperature and a circulation of air, as shown in Fig. 7.3c. As the cool surface air flows southward, it warms and becomes less dense. In the region of surface low pressure, the warm air slowly rises, expands, cools, and flows out the top at an elevation of about 1 km (3300 ft) above the surface. At this level, the air flows horizontally northward toward lower pressure, where it completes the circulation by slowly sinking and flowing out the bottom of the surface high. Circulations brought on by changes in air temperature, in which warmer air rises and colder air sinks, are termed **thermal circulations.**

*The isobars depicted here actually represent a surface of constant pressure (an *isobaric surface*), rather than a line, or isobar. Information on isobaric surfaces is given in the Focus section in Chapter 6 on p. 149.

FOCUS ON AN OBSERVATION

Eddies and "Air Pockets"

To better understand how eddies form along a zone of wind shear, imagine that, high in the atmosphere, there is a stable layer of air having vertical wind speed shear (changing wind speed with height) as depicted in Fig. 1. The top half of the layer slowly slides over the bottom half, and the relative speed of both halves is low. As long as the wind shear between the top and bottom of the layer is small, few if any eddies form. However, if the shear and the corresponding relative speed of these layers increase, wavelike undulations may form. When the shearing exceeds a certain value, the waves break into large swirls, with significant vertical movement. Eddies such as these often form in the upper troposphere near jet streams, where large wind speed shears exist. They also occur in conjunction with mountain waves, which may extend upward into the stratosphere (see Fig. 2). As we learned earlier, when these huge eddies develop in clear air, this form of turbulence is referred to as *clear air turbulence,* or *CAT.*

The eddies that form in clear air may have diameters ranging from a couple of meters to several hundred meters. An unsuspecting aircraft entering such a region may be in for more than just a bumpy ride. If the aircraft flies into a zone of descending air, it may drop suddenly, producing the sensation that there is no air to support the wings. Consequently, these regions have come to be known as *air pockets.*

Commercial aircraft entering an air pocket have dropped hundreds of meters, injuring passengers and flight attendants not strapped into their seats. For example, a DC-10 jetliner flying at 11,300 m (37,000 ft) over central Illinois during April, 1981, encountered a region of severe clear air turbulence and reportedly plunged about 600 m (2000 ft) toward the earth before stabilizing. Twenty-one of the 154 people aboard were injured; one person sustained a fractured hip and another person, after hitting the ceiling, jabbed himself in the nose with a fork, then landed in the seat in front of him. Clear air turbulence has occasionally caused structural damage to aircraft by breaking off vertical stabilizers and tail structures. Fortunately, the effects are usually not this dramatic.

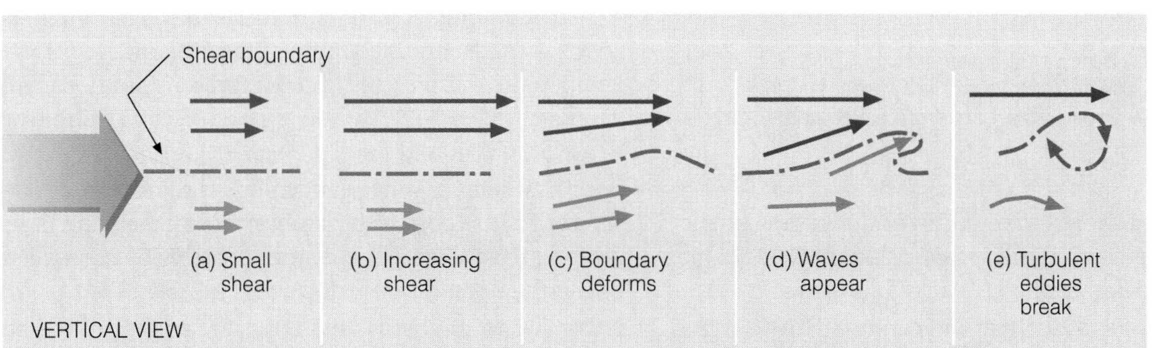

FIGURE 1 The formation of clear air turbulence (CAT) along a boundary of increasing wind speed shear. The view of wind is from the side, with a top layer of air moving over a layer below.

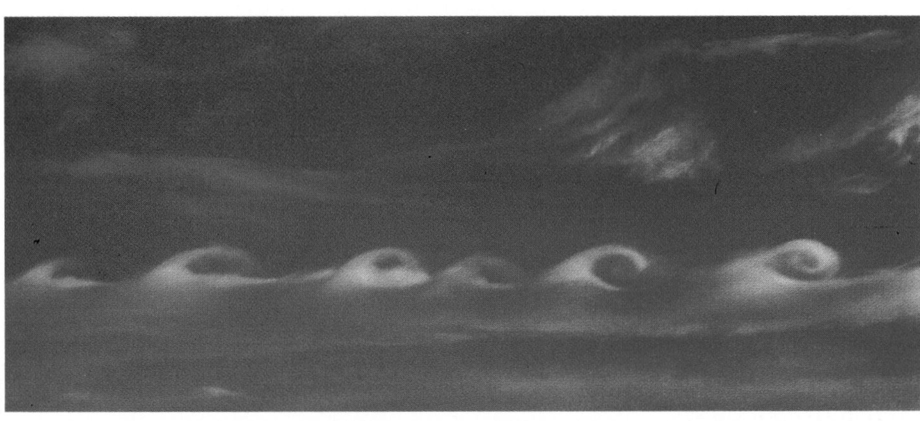

NCAR

FIGURE 2 Turbulent eddies forming downwind of a mountain chain in a wind shear zone produce these waves called *Kelvin Helmholtz* waves. The visible clouds that form are called *billow clouds.*

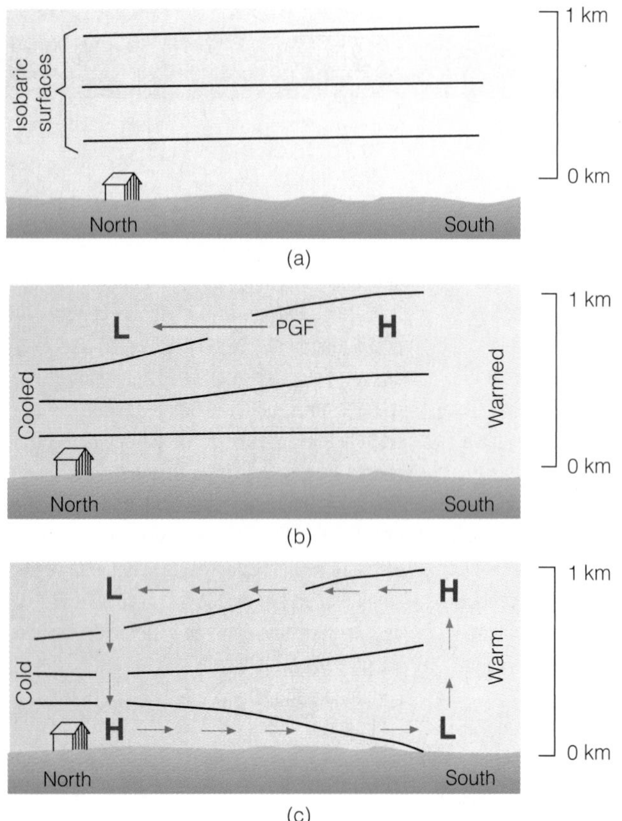

FIGURE 7.3 A thermal circulation produced by the heating and cooling of the atmosphere near the ground. The H's and L's refer to atmospheric pressure. The lines represent surfaces of constant pressure (isobaric surfaces).

The regions of surface high and low atmospheric pressure created as the atmosphere either cools or warms are called *thermal* (cold core) *highs* and *thermal* (warm core) *lows.* In general, they are shallow systems, usually extending no more than a few kilometers above the ground.

Sea and Land Breezes The sea breeze is a type of thermal circulation. The uneven heating rates of land and water (described in Chapter 3) cause these mesoscale coastal winds. During the day, the land heats more quickly than the adjacent water, and the intensive heating of the air above produces a shallow thermal low. The air over the water remains cooler than the air over the land; hence, a shallow thermal high exists above the water. The overall effect of this pressure distribution is a **sea breeze** that blows from the sea toward the land (see Fig. 7.4a). Since the strongest gradients of temperature and pressure occur near the land-water boundary, the strongest winds typically occur right near the beach and diminish inland. Further, since the greatest contrast in temperature between land and water usually occurs in the afternoon, sea breezes are strongest at this time. (The same type of breeze that develops along the shore of a large lake is called a *lake breeze.*)

At night, the land cools more quickly than the water. The air above the land becomes cooler than the air over the water, producing a distribution of pressure, such as the one shown in Fig. 7.4b. With higher surface pressure now over the land, the wind reverses itself and becomes a **land breeze**—a breeze that flows from the land toward the water. Temperature contrasts between land and water are generally much smaller at night, hence, land breezes are usually weaker than their daytime counterpart, the sea breeze. In regions where greater nighttime temperature contrasts exist, stronger land breezes occur over the water, off the coast. They are not usually noticed much on shore, but are frequently observed by ships in coastal waters.

Look at Fig. 7.4 again and observe that the rising air is over the land during the day and over the water during the night. Therefore, along the humid East Coast, daytime clouds tend to form over land and nighttime clouds over water. This explains why, at night, distant lightning flashes are sometimes seen over the ocean.

The leading edge of the sea breeze is called the *sea breeze front.* As the front moves inland, a rapid drop in temperature occurs just behind it. In some locations, this temperature change may be 5°C (9°F) or more during the first hours—a refreshing experience on a hot, sultry day. In regions where the water temperature is warm, the cooling effect of the sea breeze is hardly evident. Since cities near the ocean usually experience the sea breeze by noon, their highest temperature usually occurs much earlier than in inland cities. Along the East Coast, the passage of the sea breeze front is marked by a wind shift, usually from west to east. In the cool ocean air, the relative humidity rises as the temperature drops. If the relative humidity increases to above 70 percent, water vapor begins to condense upon particles of sea salt or industrial smoke, producing haze. When the ocean air is highly concentrated with pollutants, the sea breeze front may meet relatively clear air and thus appear as a *smoke front,* or a *smog front.* If the ocean air becomes saturated, a mass of low clouds and fog will mark the leading edge of the marine air.

When there is a sharp contrast in air temperature across the frontal boundary, the warmer, lighter air will converge and rise. In many regions, this makes for good sea breeze glider soaring. If this rising air is sufficiently moist, a line of cumulus clouds will form along the sea breeze front, and, if the air is also conditionally unstable, thunderstorms may form. As previously mentioned, on a

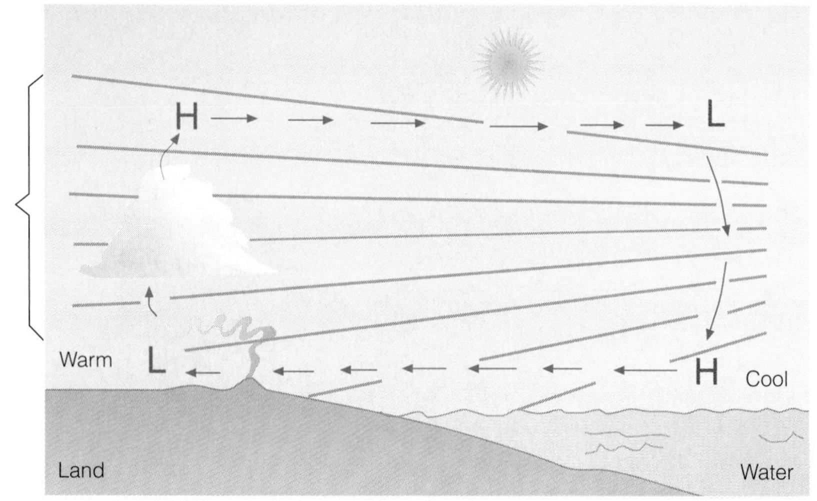

(a) Sea breeze

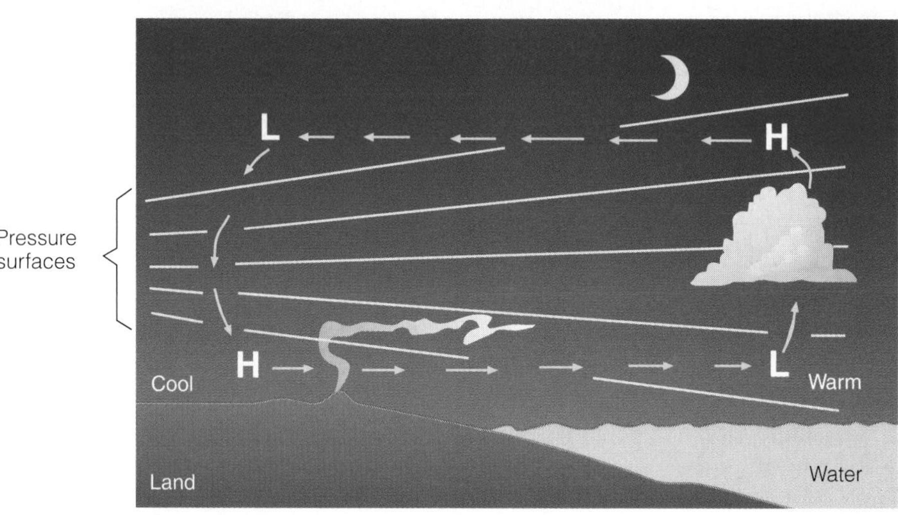

(b) Land breeze

FIGURE 7.4 Development of a sea breeze and a land breeze. (a) At the surface, a sea breeze blows from the water onto the land, whereas (b) the land breeze blows from the land out over the water. Notice that the pressure at the surface changes more rapidly with the sea breeze. This situation indicates a stronger pressure gradient force and higher winds with a sea breeze.

hot, humid day one can drive toward the shore, encounter heavy showers several miles from the ocean, and arrive at the beach to find it sunny with a steady onshore breeze.

Sea breezes in Florida help produce that state's abundant summertime rainfall. On the Atlantic side of the state, the sea breeze blows in from the east; on the Gulf shore, it moves in from the west. The convergence of these two moist wind systems, coupled with daytime convection, produces cloudy conditions and showery weather over the land (see Fig. 7.5). Over the water (where cooler, more stable air lies close to the surface), skies often remain cloud-free. On many days during June and July of 1998, however, Florida's converging wind system did not materialize. The lack of converging surface air and its accompanying showers left much of the state parched. Huge fires broke out over northern and central Florida, which left hundreds of people homeless and burned many thousands of acres of grass and woodlands.

Convergence of coastal breezes is not restricted to ocean areas. Both Lake Michigan and Lake Superior are capable of producing well-defined lake breezes. In upper Michigan, where these large bodies of water are separated by a narrow strip of land, the two breezes push inland and converge near the center of the peninsula, creating afternoon clouds and showers, while the lakeshore area remains sunny, pleasantly cool, and dry.

Seasonally Changing Winds—The Monsoon The word *monsoon* derives from the Arabic *mausim,* which means seasons. A **monsoon wind system** is one that *changes direction seasonally,* blowing from one direction in summer and from the opposite direction in winter.

FIGURE 7.5 Surface heating and lifting of air along a sea breeze combine to form thunderstorms almost daily during the summer in southern Florida.

This seasonal reversal of winds is especially well developed in eastern and southern Asia.

In some ways, the monsoon is similar to a large-scale sea breeze. During the winter, the air over the continent becomes much colder than the air over the ocean (see Fig. 3.8, p. 64). A large, shallow high-pressure area develops over continental Siberia, producing a *clockwise* circulation of air that flows out over the Indian Ocean and South China Sea (see Fig. 7.6a). Subsiding air of the anticyclone and the downslope movement of northeasterly winds from the inland plateau provide eastern and southern Asia with generally fair weather and the dry season. Hence, the *winter monsoon* means clear skies, with winds that blow from land to sea.

In summer, the wind flow pattern reverses itself as air over the continents becomes much warmer than air above the water (see Fig. 3.9, p. 64.) A shallow thermal low develops over the continental interior. The heated air within the low rises, and the surrounding air responds by flowing *counterclockwise* into the low center. This condition results in moisture-bearing winds sweeping into the continent from the ocean. The humid air

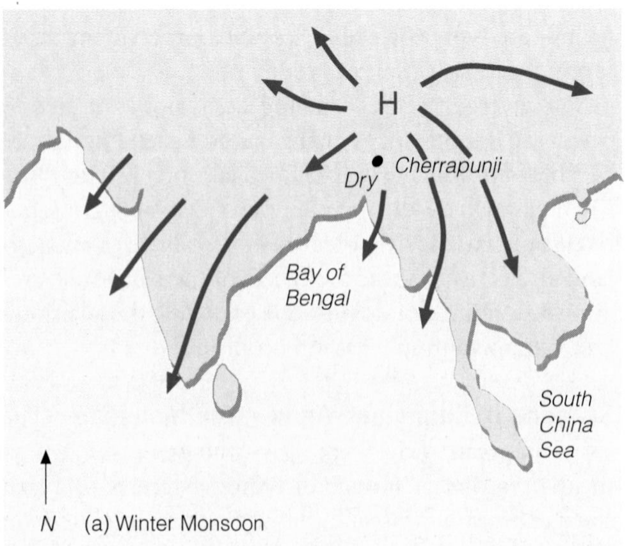

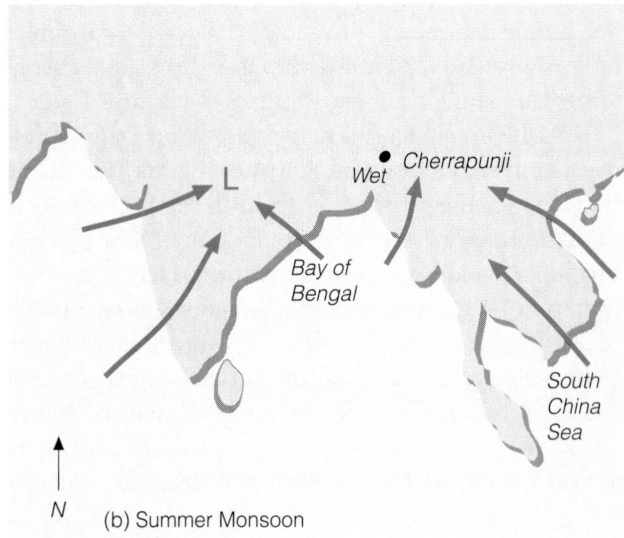

FIGURE 7.6 Changing annual wind flow patterns associated with the winter and summer Asian monsoon.

converges with a drier westerly flow, causing it to rise; further lifting is provided by hills and mountains. Lifting cools the air to its saturation point, resulting in heavy showers and thunderstorms. Thus, the *summer monsoon* of southeastern Asia, which lasts from about June through September, means wet, rainy weather (wet season) with winds that blow from sea to land (see Fig. 7.6b). Although the majority of rain falls during the wet season, it does not rain all the time. In fact, rainy periods of between 15 to 40 days are often followed by several weeks of hot, sunny weather.

The strength of the Indian monsoon appears to be related to the reversal of surface air pressure that occurs at irregular intervals about every two to seven years at opposite ends of the tropical South Pacific Ocean. As we will see later in this chapter, this reversal of pressure (which is known as the *Southern Oscillation*) is linked to an ocean warming phenomenon known as *El Niño.* During a major El Niño event, surface water near the equator becomes much warmer over the central and eastern Pacific. Over the region of warm water we find rising air, convection, and heavy rain. Meanwhile, to the west of the warm water (over the region influenced by the summer monsoon), sinking air inhibits cloud formation and convection. Hence, during El Niño years, monsoon rainfall is likely to be deficient.

Summer monsoon rains over southern Asia can reach record amounts. Located inland on the southern slopes of the Khasi Hills in northeastern India, Cherrapunji receives an average of 1080 cm (425 in.) of rainfall each year, most of it during the summer monsoon between April and October. The summer monsoon rains are essential to the agriculture of that part of the world. With a population of over 900 million people, India depends heavily on the summer rains so that food crops will grow. The people also depend on the rains for drinking water. Unfortunately, the monsoon can be unreliable in both duration and intensity. Since the monsoon is vital to the survival of so many people, it is no wonder that meteorologists have investigated it extensively. They have tried to develop methods of accurately forecasting the intensity and duration of the monsoon. With the aid of current research projects and the latest climate models (which tie in the interaction of ocean and atmosphere), there is hope that monsoon forecasts will begin to improve in accuracy.

Monsoon wind systems exist in other regions of the world, where large contrasts in temperature develop between oceans and continents. (Usually, however, these systems are not as pronounced as in southeast Asia.) For example, a monsoonlike circulation exists in the southwestern United States, especially in Arizona, New Mexico, Nevada, and the southern part of California, where spring and early summer are normally dry, as warm westerly winds sweep over the region. By mid-July, however, southerly or southeasterly winds are more common, and so are afternoon showers and thunderstorms (see Fig. 7.7).

DID YOU KNOW?

Cherrapunji, India, received 2647 cm (87 ft) of rain in 1861, most of which fell between April and October—the summer monsoon. In fact, during July, 1861, Cherrapunji recorded a whopping 930 cm (30.5 ft) of rain.

Mountain and Valley Breezes Mountain and valley breezes develop along mountain slopes. Observe in Fig. 7.8 that, during the day, sunlight warms the valley walls, which in turn warm the air in contact with them. The heated air, being less dense than the air of the same altitude above the valley, rises as a gentle upslope wind

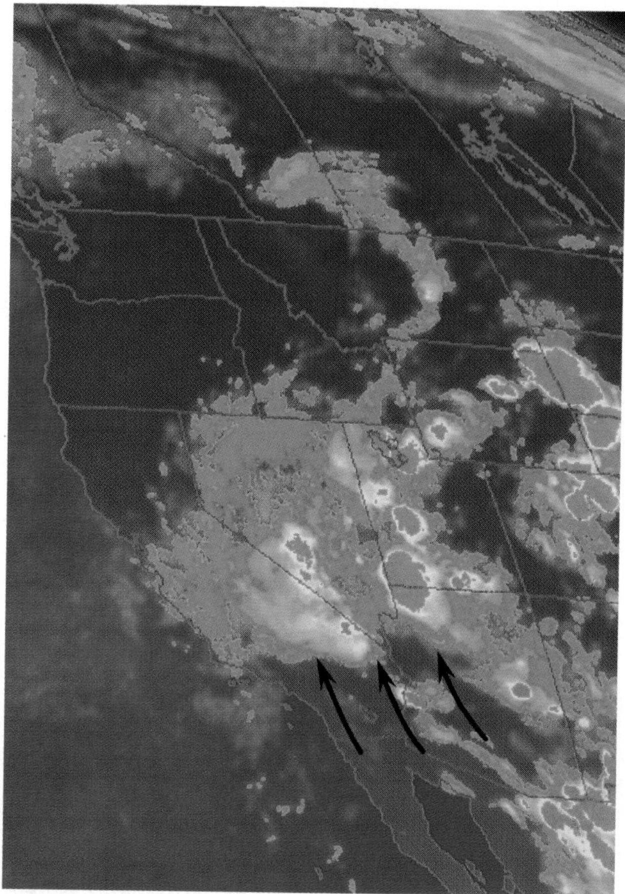

NOAA

FIGURE 7.7 Enhanced infrared satellite image with heavy arrows showing strong monsoonal circulation with showers and thunderstorms (yellow and red areas) forming over the southwestern section of the United States during July, 2001.

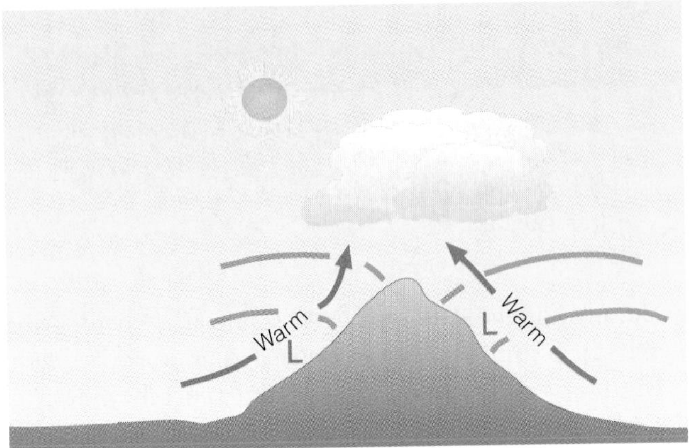

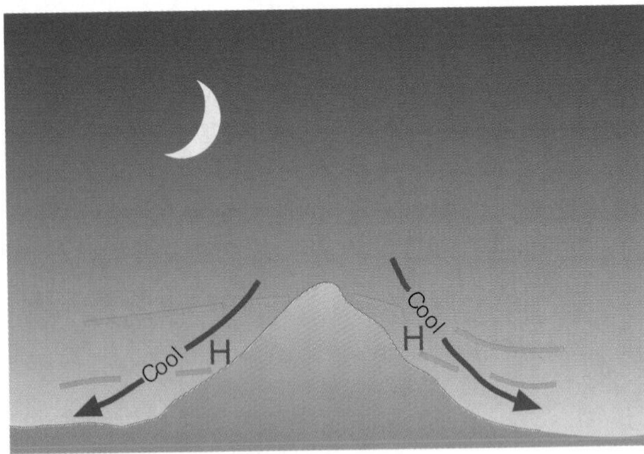

FIGURE 7.8 Valley breezes blow uphill during the day; mountain breezes blow downhill at night. (The L's and H's represent pressure, whereas the purple lines represent surfaces of constant pressure.)

known as a **valley breeze.** At night, the flow reverses. The mountain slopes cool quickly, chilling the air in contact with them. The cooler, more-dense air glides downslope into the valley, providing a **mountain breeze.** (Because gravity is the force that directs these winds downhill, they are also referred to as *gravity winds,* or *nocturnal drainage winds.*) This daily cycle of wind flow is best developed in clear, summer weather when prevailing winds are light.

When the upslope valley winds are well developed and have sufficient moisture, they can reveal themselves as building cumulus clouds above mountain summits (see Fig. 7.9). Since valley breezes usually reach their maximum strength in the early afternoon, cloudiness, showers, and even thunderstorms are common over mountains during the warmest part of the day—a fact well known to climbers, hikers, and seasoned mountain picnickers.

Katabatic Winds Although any downslope wind is technically a **katabatic wind,** the name is usually reserved for downslope winds that are much stronger than mountain breezes. Katabatic (or *fall*) winds can rush down elevated slopes at hurricane speeds, but most are not that intense and many are on the order of 10 knots or less.

Photo by author

FIGURE 7.9 As mountain slopes warm during the day, air rises and often condenses into cumuliform clouds, such as these.

The ideal setting for a katabatic wind is an elevated plateau surrounded by mountains, with an opening that slopes rapidly downhill. When winter snows accumulate on the plateau, the overlying air grows extremely cold. Along the edge of the plateau the cold, dense air begins to descend through gaps and saddles in the hills, usually as a gentle or moderate cold breeze. If the breeze, however, is confined to a narrow canyon or channel, the flow of air can increase, often destructively, as cold air rushes downslope like water flowing over a fall.

Katabatic winds are observed in various regions of the world. For example, along the northern Adriatic coast in the former Yugoslavia, a polar invasion of cold air from Russia descends the slopes from a high plateau and reaches the lowlands as the *bora*—a cold, gusty, northeasterly wind with speeds sometimes in excess of 100 knots. A similar, but often less violent, cold wind known as the *mistral* descends the western mountains into the Rhone Valley of France, and then out over the Mediterranean Sea. It frequently causes frost damage to exposed vineyards and makes people bundle up in the otherwise mild climate along the Riviera. Strong, cold katabatic winds also blow downslope off the icecaps in Greenland and Antarctica, occasionally with speeds greater than 100 knots.

In North America, when cold air accumulates over the Columbia plateau, it may flow westward through the Columbia River Gorge as a strong, gusty, and sometimes violent wind. Even though the sinking air warms by compression, it is so cold to begin with that it reaches the ocean side of the Cascade Mountains much colder than the marine air it replaces. The *Columbia Gorge wind* (called the *coho*) is often the harbinger of a prolonged cold spell.

Strong downslope katabatic-type winds funneled through a mountain canyon can do extensive damage. For example, during January, 1984, a ferocious downslope wind blew through Yosemite National Park at speeds estimated at 100 knots. The wind toppled trees and, unfortunately, caused a fatality when a tree fell on a park employee sleeping in a tent.

Chinook (Foehn) Winds The **chinook wind** is a warm, dry wind that descends the eastern slope of the Rocky Mountains. The region of the chinook is rather narrow and extends from northeastern New Mexico northward into Canada. Similar winds occur along the leeward slopes of mountains in other regions of the world. In the European Alps, for example, such a wind is called a *foehn.* When these winds move through an area, the temperature rises sharply, sometimes over 20°C (36°F) in one hour, and a corresponding sharp drop in the relative humidity occurs, occasionally to less than 5 percent. (More information on temperature changes associated with chinooks is given in the Focus section on p. 178.)

Chinooks occur when strong westerly winds aloft flow over a north-south-trending mountain range, such as the Rockies and Cascades. Such conditions can produce a trough of low pressure on the mountain's eastern side, a trough that tends to force the air downslope. As the air descends, it is compressed and warms. So the main source of warmth for a chinook is *compressional heating,* as potentially warmer (and drier) air is brought down from aloft.

When clouds and precipitation occur on the mountain's windward side, they can enhance the chinook. For example, as the cloud forms on the upwind side of the mountain in Fig. 7.10, the release of latent heat inside the cloud supplements the compressional heating on the downwind side. This phenomenon makes the descending air at the base of the mountain on the leeward side warmer than it was before it started its upward journey on the windward side. The air is also drier, since much of its moisture was removed as precipitation on the windward side.

Along the front range of the Rockies, a bank of clouds forming over the mountains is a telltale sign of an impending chinook. This *chinook wall cloud,* (which looks like a wall of clouds) usually remains stationary as air rises, condenses, and then rapidly descends the leeward slopes, often causing strong winds in foothill communities. Figure 7.11 shows how a chinook wall cloud appears as one looks west toward the Rockies from the Colorado plains. The photograph was taken on a winter afternoon with the air temperature about –7°C (20°F). That evening, the chinook moved downslope at high speeds through foothill valleys, picking up sand and pebbles (which dented cars and cracked windshields). The chinook spread out over the plains like a warm blanket, raising the air temperature the following day to a mild 15°C (59°F). The chinook and its

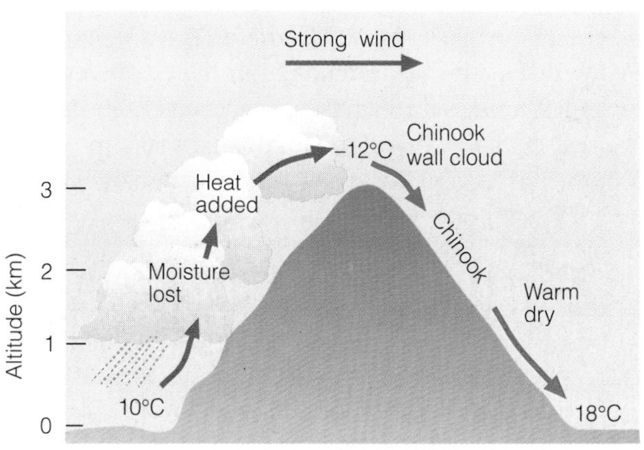

FIGURE 7.10 Conditions that may enhance a chinook.

FOCUS ON A SPECIAL TOPIC

Snow Eaters and Rapid Temperature Changes

Chinooks are thirsty winds. As they move over a heavy snow cover, they can melt and evaporate a foot of snow in less than a day. This situation has led to some tall tales about these so-called snow eaters. Canadian folklore has it that a sled-driving traveler once tried to outrun a chinook. During the entire ordeal his front runners were in snow while his back runners were on bare soil.

Actually, the chinook is important economically. It not only brings relief from the winter cold, but it uncovers prairie grass, so that livestock can graze on the open range. Also, these warm winds have kept railroad tracks clear of snow, so that trains can keep running. On the other hand, the drying effect of a chinook can create an extreme fire hazard. And when a chinook follows spring planting, the seeds may die in the parched soil. Along with the dry air comes a buildup of static electricity, making a simple handshake a shocking experience. These warm, dry winds have sometimes adversely affected human behavior. During periods of chinook winds some people feel irritable and depressed and others become ill. The exact reason for this phenomenon is not clearly understood.

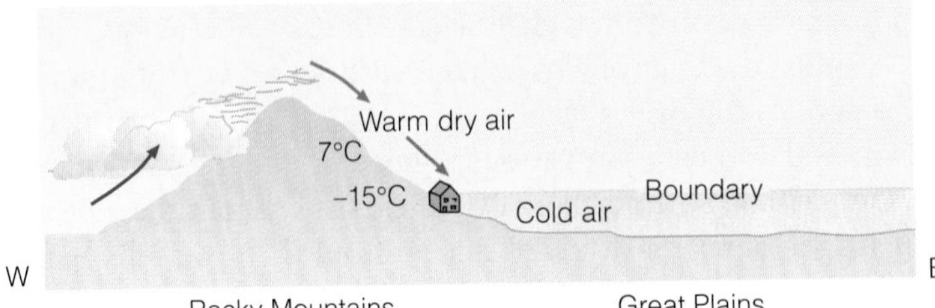

FIGURE 3 Cities near the warm air–cold air boundary can experience sharp temperature changes, if cold air should rock up and down like water in a bowl.

Chinook winds have been associated with rapid temperature changes. Figure 3 shows a shallow layer of extremely cold air that has moved southward out of Canada and is now resting against the Rocky Mountains. In the cold air, temperatures are near –15°C (5°F), while just a short distance up the mountain a warm chinook wind raises the air temperature to 7°C (45°F). The cold air behaves just as any fluid, and, in some cases, atmospheric conditions may cause the air to move up and down much like water does when a bowl is rocked back and forth. This rocking motion can cause extreme temperature variations for cities located at the base of the hills along the periphery of the cold air–warm air boundary, as they are alternately in and then out of the cold air. Such a situation is held to be responsible for the unbelievable two-minute temperature change of 49°F recorded at Spearfish, South Dakota, during the morning of January 22, 1943. On the same morning, in nearby Rapid City, the temperature fluctuated from –4°F at 5:30 A.M. to 54°F at 9:40 A.M., then down to 11°F at 10:30 A.M. and up to 55°F just 15 minutes later. At nearby cities, the undulating cold air produced similar temperature variations that lasted for several hours.

wall of clouds remained for several days, bringing with it a welcomed break from the cold grasp of winter.

Santa Ana Winds A warm, dry wind that blows from the east or northeast into southern California is the **Santa Ana wind.** As the air descends from the elevated desert plateau, it funnels through mountain canyons in the San Gabriel and San Bernardino Mountains, finally spreading over the Los Angeles Basin and San Fernando Valley. The wind often blows with exceptional speed in the Santa Ana Canyon (the canyon from which it derives its name).

These warm, dry winds develop as a region of high pressure builds over the Great Basin. The clockwise circulation around the anticyclone forces air downslope from the high plateau. Thus, *compressional heating* provides the primary source of warming. The air is dry, since it originated in the desert, and it dries out even more as it is heated. Figure 7.12 shows a typical wintertime Santa Ana situation.

As the wind rushes through canyon passes, it lifts dust and sand and dries out vegetation, which sets the stage for serious brush fires, especially in autumn, when chaparral-covered hills are already parched from the dry summer.* One such fire in November of 1961—the infa-

DID YOU KNOW?

Dressing for the weather in Granville, North Dakota, on February 21, 1918, would have been a difficult task, as a chinook wind caused the air temperature to jump 83°F—from a bone-chilling –33°F in the morning to a mild 50°F in the afternoon.

*Chaparral denotes a shrubby environment, in which many of the plant species contain highly flammable oils.

Photo by author

FIGURE 7.11 A chinook wall cloud forming over the Colorado Rockies (viewed from the plains).

mous Bel Air fire—burned for three days, destroying 484 homes and causing over $25 million in damage. During October, 1993, 15 wildfires driven by Santa Ana winds swept through wealthy suburbs and rural communities near Los Angeles. The fires charred over 200,000 acres, destroyed over 1000 homes, injured hundreds of people, and caused an estimated $1 billion in damages. Four hundred miles to the north in Oakland, California, a ferocious Santa Ana–type wind was responsible for the disastrous Oakland hills fire during October, 1991, that damaged or destroyed over 3000 dwellings and took 25 lives. With the protective vegetation cover removed, the land is ripe for erosion, as winter rains may wash away topsoil and, in some areas, create serious mudslides. The adverse effects of a wind-driven Santa Ana fire may be felt long after the fire itself has been put out.

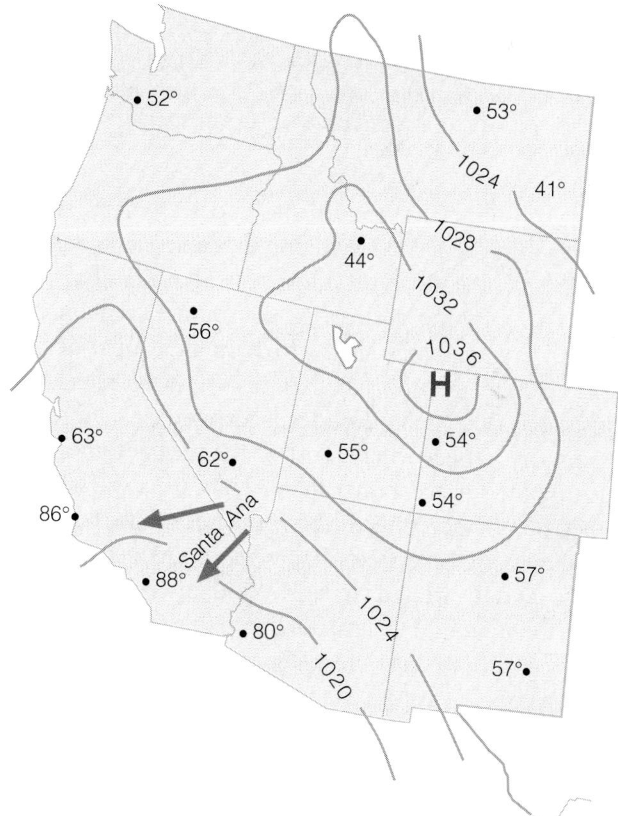

FIGURE 7.12 Surface weather map showing Santa Ana conditions in January. Maximum temperatures for this particular day are given in °F. Observe that the downslope winds blowing into southern California raised temperatures into the upper 80s, while elsewhere temperature readings were much lower.

Desert Winds Local winds form in deserts, too. *Dust storms* form in dry regions, where strong winds are able to lift and fill the air with particles of fine dust. A huge dust storm formed over the African Sahara during February, 2001. The storm—about the size of Spain—swept westward off the African coast, then northeastward as depicted in the opening photo of Chapter 6 (see p. 140). In desert areas where loose sand is more prevalent, *sandstorms* develop, as high winds enhanced by surface heating rapidly carry sand particles close to the ground. A spectacular example of a storm composed of dust or sand is the **haboob** (from Arabic *hebbe:* blown). The haboob forms as cold downdrafts along the leading edge of a thunderstorm lift dust or sand into a huge, tumbling dark cloud that may extend horizontally for over a hundred kilometers and rise vertically to the base of the thunderstorm. Spinning whirlwinds of dust frequently form along the turbulent cold air boundary, giving rise to sightings of huge *dust devils* and even tornadoes. Haboobs are

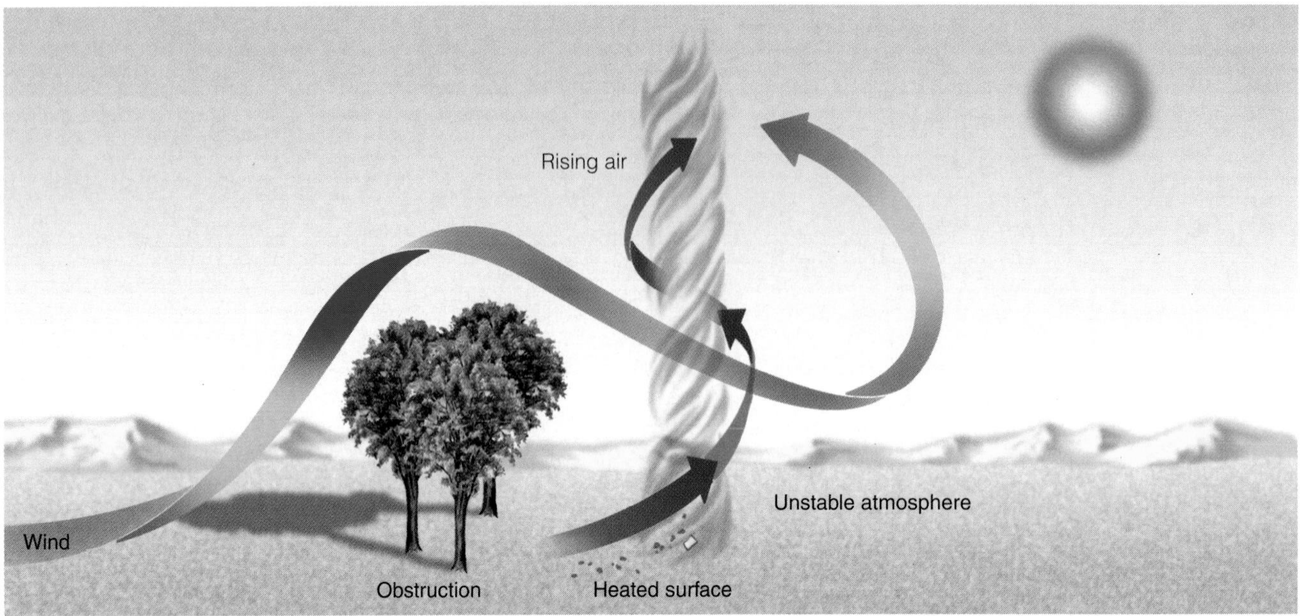

Meteorology Now™ ACTIVE FIGURE 7.13 The formation of a dust devil. On a hot, dry day, the atmosphere next to the ground becomes unstable. As the heated air rises, wind blowing past an obstruction twists the rising air, forming a rotating column, or *dust devil.* Air from the sides rushes into the rising column, lifting sand, dust, leaves, or any other loose material from the surface.
Watch this Active Figure at http://earthscience.brookscole.com/ahrens/ess4e.

most common in the African Sudan (where about twenty-four occur each year) and in the desert southwest of the United States, especially in southern Arizona.

The spinning vortices so commonly seen on hot days in dry areas are called **whirlwinds, or dust devils.** (In Australia, the Aboriginal word *willy-willy* refers to a dust devil.) Generally, dust devils form on clear, hot days over a dry surface where most of the sunlight goes into heating the surface, rather than evaporating water from vegetation. The air directly above the hot surface becomes absolutely unstable (see Chapter 5, p. 115), convection sets in, and the heated air rises. Wind, often deflected by small topographic barriers, flows into this region, rotating the rising air (see Fig. 7.13). Depending on the nature of the topographic feature, the spin of a dust devil around its central eye may be cyclonic or anticyclonic, and both directions occur with about equal frequency.

Having diameters of only a few meters and heights of less than a hundred meters (see Fig. 7.14), most dust

FIGURE 7.14 A dust devil forming on a clear, hot summer day just south of Phoenix, Arizona.

DID YOU KNOW?

Driven by fierce dry Santa Ana Winds, 11 wildfires raged through areas of Southern California during late October, 2003. All told, the fires claimed more than 740,000 acres, destroyed over 2800 homes, and took 20 lives.

devils are small and last only a short time. There are, however, some dust devils of sizable dimension, extending upward from the surface for many hundreds of meters. Such whirlwinds are capable of considerable damage; winds exceeding 75 knots may overturn mobile homes and tear the roofs off buildings. Fortunately, the majority of dust devils are small. Also keep in mind that dust devils *are not* tornadoes. The circulation of a tornado (as we will see in Chapter 10) usually descends downward from the base of a thunderstorm, whereas the circulation of a dust devil begins at the surface, normally in sunny weather, although some form beneath convective-type clouds.

GLOBAL WINDS

Up to now, we have seen that local winds vary considerably from day to day and from season to season. As you may suspect, these winds are part of a much larger circulation—the little whirls within larger whirls that we spoke of earlier in this chapter. Indeed, if the rotating high- and low-pressure areas are like spinning eddies in a huge river, then the flow of air around the globe is like the meandering river itself. When winds throughout the world are averaged over a long period, the local wind patterns vanish, and what we see is a picture of the winds on a global scale—what is commonly called the **general circulation of the atmosphere.**

General Circulation of the Atmosphere Before we study the general circulation, we must remember that it only represents the *average* air flow around the world. Actual winds at any one place and at any given time may vary considerably from this average. Nevertheless, the average can answer why and how the winds blow around the world the way they do—why, for example, prevailing surface winds are northeasterly in Honolulu and westerly in New York City. The average can also give a picture of the driving mechanism behind these winds, as well as a model of how heat is transported from equatorial regions poleward, keeping the climate in middle latitudes tolerable.

The underlying cause of the general circulation is the unequal heating of the earth's surface. We learned in Chapter 2 that, averaged over the entire earth, incoming solar radiation is roughly equal to outgoing earth radiation. However, we also know that this energy balance is not maintained for each latitude, since the tropics experience a net gain in energy, while polar regions suffer a net loss. To balance these inequities, the atmosphere transports warm air poleward and cool air equatorward. Although seemingly simple, the actual flow of air is complex; certainly not everything is known about it. In order to better understand it, we will first look at some models (that is, artificially constructed analogies) that eliminate some of the complexities of the general circulation.

Single-Cell Model The first model is the single-cell model, in which we assume that:

1. The earth's surface is uniformly covered with water (so that differential heating between land and water does not come into play).
2. The sun is always directly over the equator (so that the winds will not shift seasonally).
3. The earth does not rotate (so that the only force we need deal with is the pressure gradient force).

With these assumptions, the general circulation of the atmosphere would look much like the representation in Fig. 7.15a, a huge thermally driven convection cell in each hemisphere. (For reference, the names of the different regions of the world and their approximate latitudes are given in Figure 7.15b.)

The circulation of air described in Fig. 7.15a is the **Hadley cell** (named after the eighteenth-century English meteorologist George Hadley, who first proposed the idea). It is driven by energy from the sun. Excessive heating of the equatorial area produces a broad region of surface low pressure, while at the poles excessive cooling creates a region of surface high pressure. In response to the horizontal pressure gradient, cold surface polar air flows equatorward, while at higher levels air flows toward the poles. The entire circulation consists of a closed loop with rising air near the equator, sinking air over the poles, an equatorward flow of air near the surface, and a return flow aloft. In this manner, some of the excess energy of the tropics is transported as sensible and latent heat to the regions of energy deficit at the poles.

Such a simple cellular circulation as this does not actually exist on the earth. For one thing, the earth rotates, so the Coriolis force would deflect the southward-moving surface air in the Northern Hemisphere to the right, producing easterly surface winds at practically all latitudes. These winds would be moving in a direction opposite to that of the earth's rotation and, due to friction with the surface, would slow down the earth's spin. We know that this does not happen and that prevailing winds in middle latitudes actually blow from the west. Therefore, observations alone tell us that a closed circulation of air between the equator and the poles is not the proper model for a rotating earth. (Models that simulate air flow around the globe have also verified this.) How, then, does the wind blow on a rotating planet? To answer,

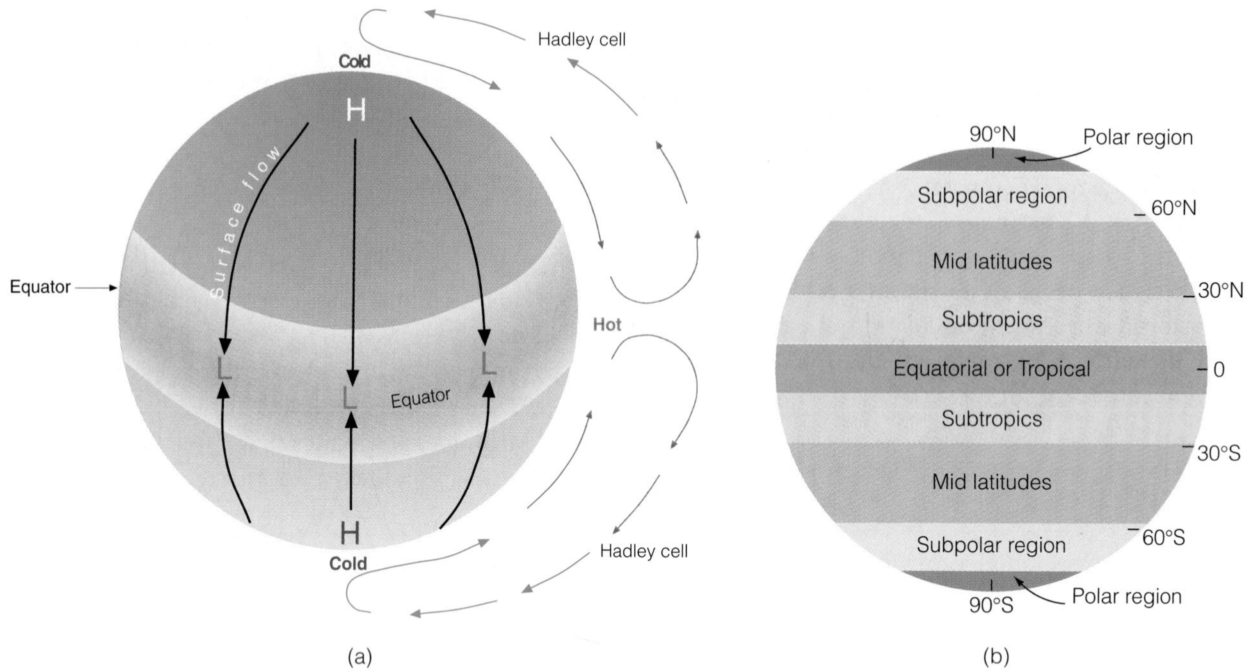

FIGURE 7.15 Diagram (a) shows the general circulation of air on a nonrotating earth uniformly covered with water and with the sun directly above the equator. (Vertical air motions are highly exaggerated in the vertical.) Diagram (b) shows the names that apply to the different regions of the world and their approximate latitudes.

we will keep our model simple by retaining our first two assumptions—that is, that the earth is covered with water and that the sun is always directly above the equator.

Three-Cell Model If we allow the earth to spin, the simple convection system breaks into a series of cells as shown in Fig. 7.16a. Although this model is considerably more complex than the single-cell model, there are some similarities. The tropical regions still receive an excess of heat and the poles a deficit. In each hemisphere, three cells instead of one have the task of energy redistribution. A surface high-pressure area is located at the poles, and a broad trough of surface low pressure still exists at the equator. From the equator to latitude 30°, the circulation closely resembles that of a Hadley cell, as does the circulation from the poles to about latitude 60°. Let's look at this model more closely by examining what happens to the air above the equator. (Refer to Fig. 7.16 as you read the following section.)

Over equatorial waters, the air is warm, horizontal pressure gradients are weak, and winds are light. This region is referred to as the **doldrums.** (The monotony of the weather in this area has given rise to the expression "down in the doldrums.") Here, warm air rises, often condensing into huge cumulus clouds and thunderstorms that liberate an enormous amount of latent heat. This heat makes the air more buoyant and provides energy to drive the Hadley cell. The rising air reaches the tropopause, which acts like a barrier, causing the air to move laterally toward the poles. The Coriolis force deflects this poleward flow toward the right in the Northern Hemisphere and to the left in the Southern Hemisphere, providing westerly winds aloft in both hemispheres. (We will see later that these westerly winds reach maximum velocity and produce jet streams near 30° and 60° latitudes.)

Air aloft moving poleward from the tropics constantly cools by giving up infrared radiation, and at the same time it also begins to converge, especially as it approaches the middle latitudes.* This convergence (piling up) of air aloft increases the mass of air above the surface, which in turn causes the air pressure at the surface to increase. Hence, at latitudes near 30°, the convergence of air aloft produces belts of high pressure called **subtropical highs** (or anticyclones). As the converging, relatively dry air above the highs slowly descends, it warms by compression. This subsiding air produces generally clear skies and warm surface temperatures; hence, it is here that we find the major deserts of the world, such as the Sahara. Over the ocean, the weak pressure gradients in the center of the high produce only weak winds. According to legend, sail-

*You can see why the air converges if you have a globe of the world. Put your fingers on meridian lines at the equator and then follow the meridians poleward. Notice how the lines and your fingers bunch together in the middle latitudes.

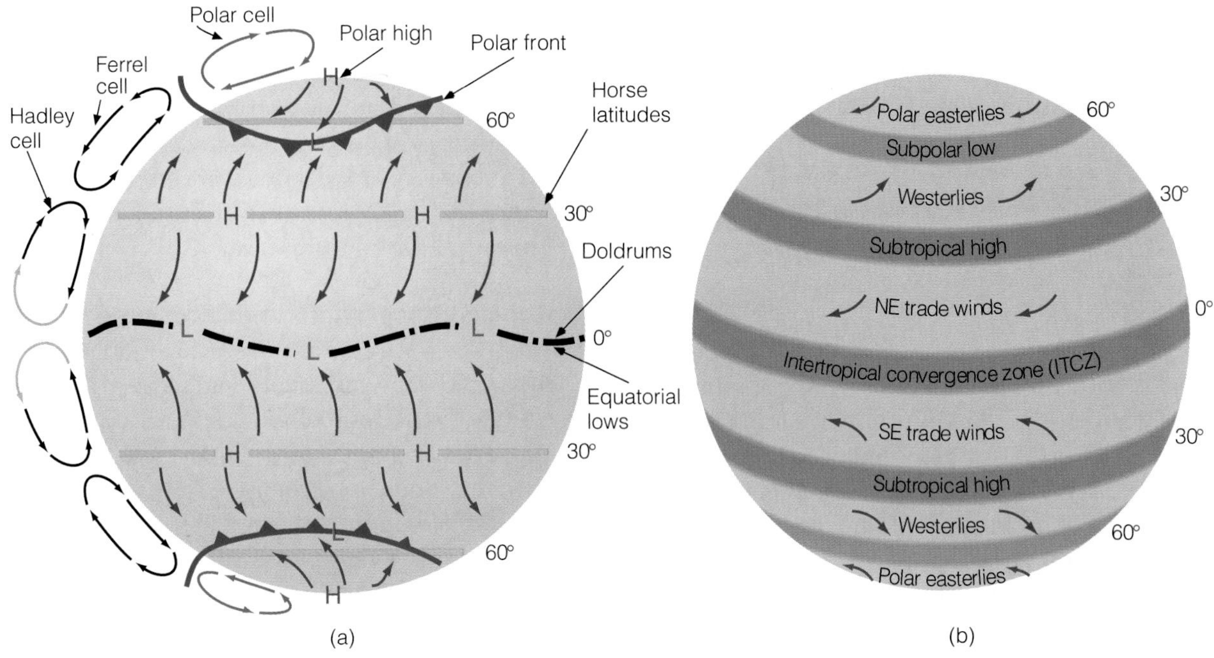

Meteorology Now™ ACTIVE FIGURE 7.16 Diagram (a) shows the idealized wind and surface-pressure distribution over a uniformly water-covered rotating earth. Diagram (b) gives the names of surface winds and pressure systems over a uniformly water-covered rotating earth. Watch this Active Figure at http://earthscience.brookscole.com/ahrens/ess4e.

ing ships traveling to the New World were frequently becalmed in this region; and, as food and supplies dwindled, horses were either thrown overboard or eaten. As a consequence, this region is sometimes called the *horse latitudes.*

From the horse latitudes, some of the surface air moves back toward the equator. It does not flow straight back, however, because the Coriolis force deflects the air, causing it to blow from the northeast in the Northern Hemisphere and from the southeast in the Southern Hemisphere. These steady winds provided sailing ships with an ocean route to the New World; hence, these winds are called the **trade winds.** Near the equator, the *northeast trades* converge with the *southeast trades* along a boundary called the **intertropical convergence zone (ITCZ).** In this region of surface convergence, air rises and continues its cellular journey.

Meanwhile, at latitude 30°, not all of the surface air moves equatorward. Some air moves toward the poles and deflects toward the east, resulting in a more or less westerly air flow—called the *prevailing westerlies,* or, simply, **westerlies**—in both hemispheres. Consequently, from Texas northward into Canada, it is much more common to experience winds blowing out of the west than from the east. The westerly flow in the real world is not constant because migrating areas of high and low pressure break up the surface flow pattern from time to time. In the middle latitudes of the Southern Hemisphere, where the surface is mostly water, winds blow more steadily from the west.

As this mild air travels poleward, it encounters cold air moving down from the poles. These two air masses of contrasting temperature do not readily mix. They are separated by a boundary called the **polar front,** a zone of low pressure—the **subpolar low**—where surface air converges and rises and storms develop. Some of the rising air returns at high levels to the horse latitudes, where it sinks back to the surface in the vicinity of the subtropical high. This middle cell (called the *Ferrel cell,* after the American meteorologist William Ferrel) is completed when surface air from the horse latitudes flows poleward toward the polar front.

Behind the polar front, the cold air from the poles is deflected by the Coriolis force, so that the general flow of air is from the northeast. Hence, this is the region of the **polar easterlies.** In winter, the polar front with its cold air can move into middle and subtropical latitudes, producing a cold polar outbreak. Along the front, a portion of the rising air moves poleward, and the Coriolis force deflects the air into a westerly wind at high levels. Air aloft eventually reaches the poles, slowly sinks to the surface, and flows back toward the polar front, completing the weak *polar cell.*

We can summarize all of this by referring back to Fig. 7.16 and noting that, at the surface, there are two major areas of high pressure and two major areas of low pressure. Areas of high pressure exist near latitude 30° and the poles; areas of low pressure exist over the equa-

DID YOU KNOW?

The African Sahara desert may have an influence on South America's Amazon basin. Annually, millions of tons of soil and trace elements from the Sahara drift westward with the northeast trades thousands of kilometers across the Atlantic, where they settle upon the Amazon basin. Some scientists suggest that this is one of the major sources of soil nutrients for the poor soil of the Amazon.

tor and near 60° latitude in the vicinity of the polar front. By knowing the way the winds blow around these systems, we have a generalized picture of surface winds throughout the world. The trade winds extend from the subtropical high to the equator, the westerlies from the subtropical high to the polar front, and the polar easterlies from the poles to the polar front.

How does this three-cell model compare with actual observations of winds and pressure? We know, for example, that upper-level winds at middle latitudes generally blow from the west. The middle cell, however, suggests an east wind aloft as air flows equatorward. Hence, discrepancies exist between this model and atmospheric observations. This model does, however, agree closely with the winds and pressure distribution at the *surface*, and so we will examine this next.

Average Surface Winds and Pressure: The Real World When we examine the real world with its continents and oceans, mountains and ice fields, we obtain an average distribution of sea-level pressure and winds for January and July, as shown in Figs. 7.17a and 7.17b. Even though these data are based on sparse observations, especially in unpopulated areas, we can see that there are regions where pressure systems appear to persist throughout the year. These systems are referred to as

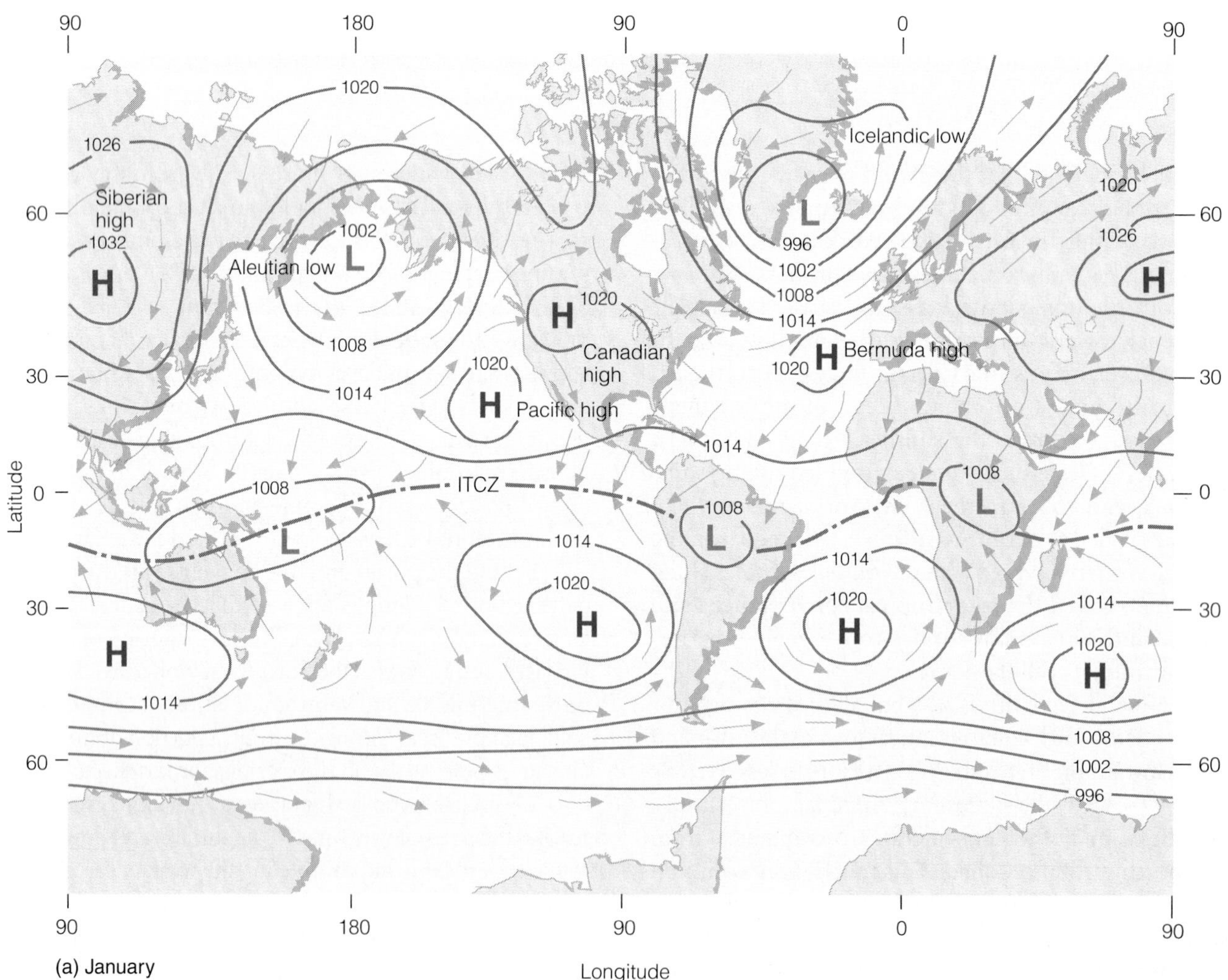

FIGURE 7.17 Average sea-level pressure distribution and surface wind-flow patterns for January (a) and for July (b). The heavy dashed line represents the position of the ITCZ.

semipermanent highs and lows because they move only slightly during the course of a year.

In Fig. 7.17a, we can see that there are four semipermanent pressure systems in the Northern Hemisphere during January. In the eastern Atlantic, between latitudes 25° and 35°N is the *Bermuda-Azores high,* often called the **Bermuda high,** and, in the Pacific Ocean, its counterpart, the **Pacific high.** These are the subtropical anticyclones that develop in response to the convergence of air aloft. Since surface winds blow clockwise around these systems, we find the trade winds to the south and the prevailing westerlies to the north. In the Southern Hemisphere, where there is relatively less land area, there is less contrast between land and water, and the subtropical highs show up as well-developed systems with a clearly defined circulation.

Where we would expect to observe the polar front (between latitudes 40° and 65°), there are two semipermanent subpolar lows. In the North Atlantic, there is the *Greenland-Icelandic low,* or simply **Icelandic low,** which covers Iceland and southern Greenland, while the **Aleutian low** sits over the Aleutian Islands in the North Pacific. These zones of cyclonic activity actually represent regions where numerous storms, having traveled eastward, tend to converge, especially in winter. In the Southern Hemisphere, the subpolar low forms a continuous trough that completely encircles the globe.

On the January map (Fig. 7.17a), there are other pressure systems, which are not semipermanent in nature. Over Asia, for example, there is a huge (but shallow) thermal anticyclone called the **Siberian high,** which forms because of the intense cooling of the land. South of this system, the winter monsoon shows up clearly, as air flows away from the high across Asia and out over the ocean. A similar (but less intense) anticyclone (called the *Canadian high*) is evident over North America.

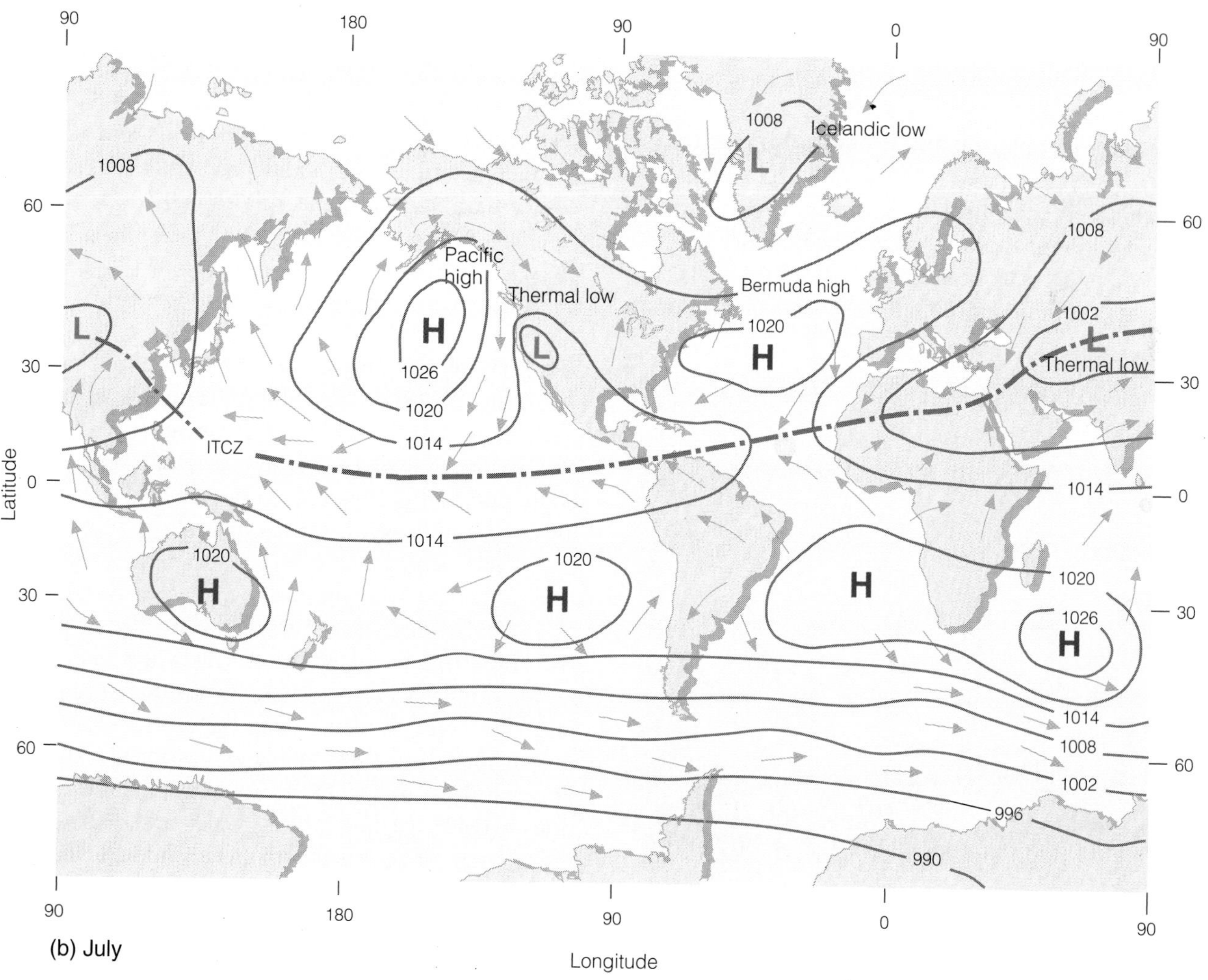

(b) July

As summer approaches, the land warms and the cold, shallow highs disappear. In some regions, areas of surface low pressure replace areas of high pressure. The lows that form over the warm land are *thermal lows.* On the July map (Fig. 7.17b), warm thermal lows are found over the desert southwest of the United States, over the plateau of Iran and north of India. As the thermal low over India intensifies, warm, moist air from the ocean is drawn into it, producing the wet summer monsoon so characteristic of India and Southeast Asia.

When we compare the January and July maps, we can see several changes in the semipermanent pressure systems. The strong subpolar lows so well developed in January over the Northern Hemisphere are hardly discernible on the July map. The subtropical highs, however, remain dominant in both seasons. Because the sun is overhead in the Northern Hemisphere in July and overhead in the Southern Hemisphere in January, the zone of maximum surface heating shifts seasonally. In response to this shift, the major pressure systems, wind belts, and ITCZ (heavy dashed line in Fig. 7.17) *shift toward the north in July and toward the south in January.*

The General Circulation and Precipitation Patterns

The position of the major features of the general circulation and their latitudinal displacement (which annually averages about 10° to 15°) strongly influence the climate of many areas. For example, on the global scale, we would expect abundant rainfall where the air rises and very little where the air sinks. Consequently, areas of high rainfall exist in the tropics, where humid air rises in conjunction with the ITCZ, and between 40° and 55° latitude, where middle-latitude storms and the polar front force air upward. Areas of low rainfall are found near 30° latitude in the vicinity of the subtropical highs and in polar regions where the air is cold and dry (see Fig. 7.18).

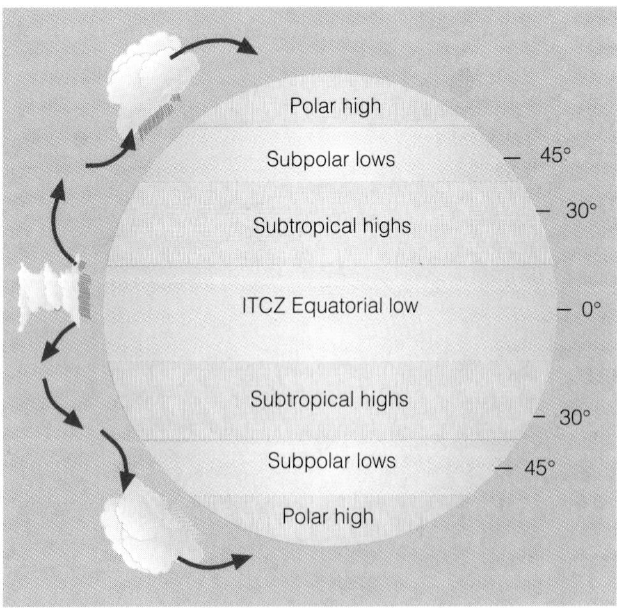

FIGURE 7.18 Major pressure systems and idealized air motions (heavy blue arrows) and precipitation patterns of the general circulation. (Areas shaded light blue represent abundant rainfall.)

During the summer, the Pacific high drifts northward to a position off the California coast (see Fig. 7.19). Sinking air on its eastern side produces a strong upper-level subsidence inversion, which tends to keep summer weather along the West Coast relatively dry. The rainy season typically occurs in winter when the high moves south and storms are able to penetrate the region. Observe in Fig. 7.19 that along the East Coast, the clockwise circulation of winds around the Bermuda high brings warm, tropical air northward into the United States and southern Canada from the Gulf of Mexico. Because subsiding air is not as well developed on this side of the high, the humid air can rise and condense into towering cumulus clouds and thunderstorms. So, in part, it is the air motions associated with the subtropical highs that keep summer weather dry in California and moist in Georgia. (Compare the rainfall patterns for Los Angeles, California, and Atlanta, Georgia, in Fig. 7.20.)

Meteorology Now™ Click "Global Atmosphere" to observe cloud movement in response to the atmosphere's general circulation.

Westerly Winds and the Jet Stream In Chapter 6, we learned that the winds above the middle latitudes in both hemispheres blow in a wavy west-to-east direction. The reason for these westerly winds is that, aloft, we generally find higher pressure over equatorial regions and lower pressures over polar regions. Where these upper-level winds tend to concentrate into narrow bands, we find rivers of fast-flowing air—what we call **jet streams.**

Atmospheric jet streams are swiftly flowing air currents hundreds of miles long, normally less than several hundred miles wide, and typically less than a mile thick (see Fig. 7.21). Wind speeds in the central core of a jet stream often exceed 100 knots and occasionally 200 knots. Jet streams are usually found at the tropopause at elevations between 10 and 14 km (33,000 and 46,000 ft) although they may occur at both higher and lower altitudes.

Jet streams were first encountered by high-flying military aircraft during World War II, but their existence was suspected before that time. Ground-based observations of fast-moving cirrus clouds had revealed that westerly winds aloft must be moving rapidly indeed.

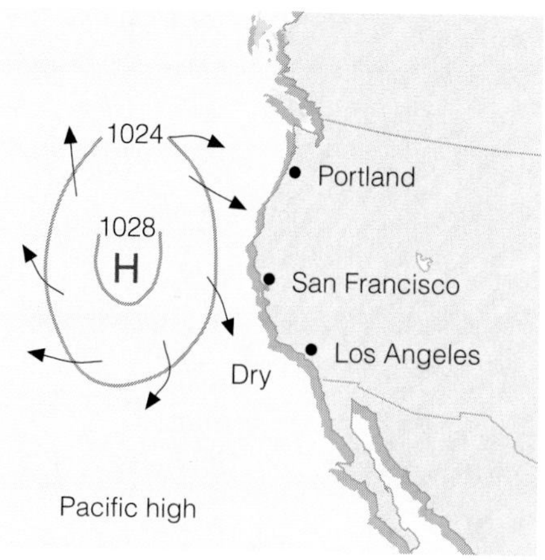

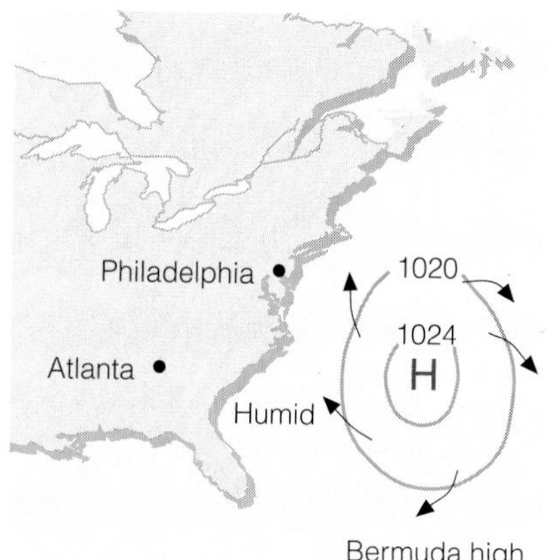

FIGURE 7.19 During the summer, the Pacific high moves northward. Sinking air along its eastern (over California) margin produces a strong subsidence inversion, which causes relatively dry weather to prevail. Along the western margin of the Bermuda high, southerly winds bring in humid air, which rises, condenses, and produces abundant rainfall.

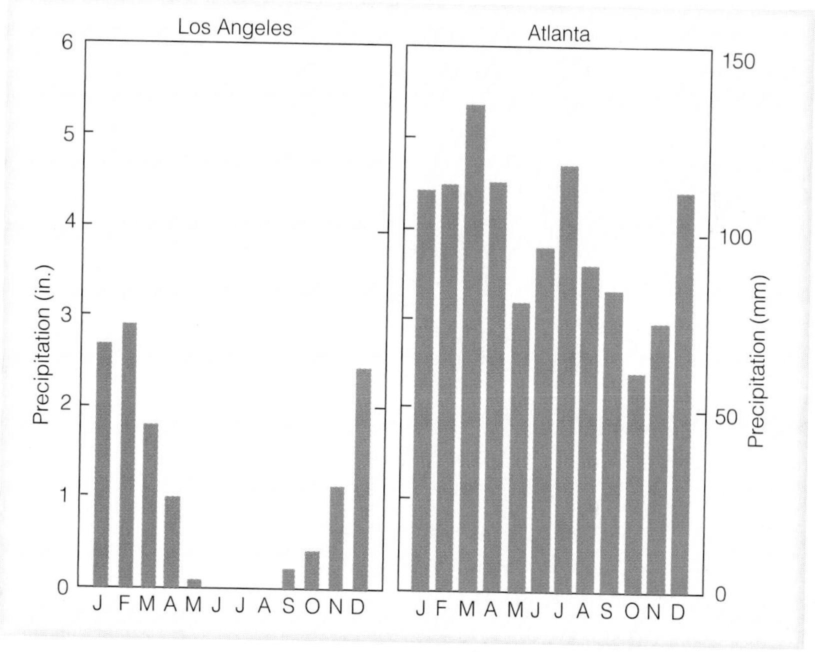

FIGURE 7.20 Average annual precipitation for Los Angeles, California, and Atlanta, Georgia.

Since jet streams are bands of strong winds, they form in the same manner as all winds—due to horizontal differences in pressure. In Fig. 7.21, notice that the Northern Hemisphere jet stream is situated along the boundary where cold, polar air lies to the north and milder, subtropical air lies to the south. Recall from our earlier discussion that this boundary is marked by the polar front (see Fig. 7.16, p. 183). Aloft, sharp contrast in temperatures along the front produces rapid horizontal pressure changes, which sets up a steep pressure gradient. This condition intensifies the wind speed along the front and causes the jet stream. The jet stream that forms in the vicinity of the polar front is called the **polar front jet stream** or, simply, the *polar jet stream.* Because

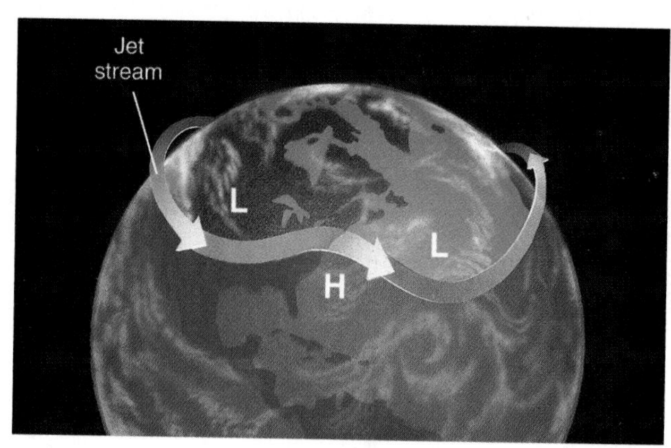

Meteorology Now™ ACTIVE FIGURE 7.21 [at right] A jet stream is a swiftly flowing current of air that moves in a wavy west-to-east direction. In the Northern Hemisphere, it forms along a boundary where colder air lies to the north and warmer air to the south. In the Southern Hemisphere, it forms where colder air lies to the south and warmer air to the north.
Watch this Active Figure at http://earthscience.brookscole.com/ahrens/ess4e.

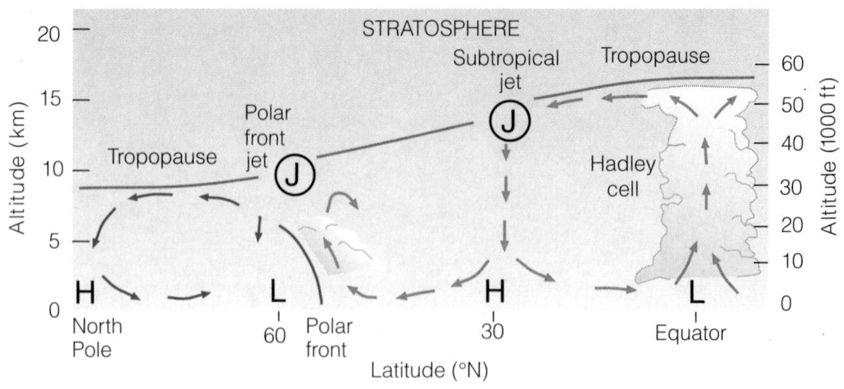

FIGURE 7.22 Average position of the polar front jet stream and the subtropical jet stream, with respect to a model of the general circulation in winter. Both jet streams are flowing into the page, away from the viewer, which would be from west to east.

DID YOU KNOW?

Time flies when traveling from west to east. Because of the jet stream, a plane flight from New York City to Europe takes about an hour less time than the return flight.

the north-south temperature contrast along the polar front is strongest in winter and weakest in summer, the polar jet stream shows seasonal variations. In winter, the winds blow stronger and the jet moves farther south, as the leading edge of the cold air may extend into southern California, south Texas, and even Florida. In summer, the polar jet stream is weaker and is usually found farther north, such as over southern Canada.

Figure 7.22 illustrates the average position of the jet streams, tropopause, and general air flow for the Northern Hemisphere in winter. From this diagram, we can see that there are two jet streams, both located in the tropopause gaps, where mixing between tropospheric and stratospheric air takes place. The jet stream situated at about 10 km (33,000 ft) near the polar front is the polar jet stream. The jet stream located near 13 km (43,000 ft) above the subtropical high is the **subtropical jet stream.***

In Fig. 7.22, the wind in the jet core would be flowing as a westerly wind away from the viewer. This direction, of course, is only an average, as jet streams often meander into broad loops that sweep north and south. When the polar jet stream develops this pattern, it may even merge with the subtropical jet. Occasionally, the polar jet splits into two jet streams. The jet stream to the north is often called the *northern branch* of the polar jet, whereas the one to the south is called the *southern branch.*

*The subtropical jet stream tends to form on the poleward side of the Hadley cell, as shown in Fig. 7.22. Here, warm air carried poleward by the Hadley cell produces sharp temperature contrasts, strong pressure gradients, and high winds.

We can see the looping pattern of the jet by studying Fig. 7.23. This diagram shows the position of the polar jet stream at the 300-mb level (near 9 km or 30,000 ft) on March 10, 1998. The air flow and jet core are represented by the colored arrows; the solid gray lines represent lines of equal wind speed *(isotachs)* in nautical miles per hour (knots). The map shows a strong subtropical jet stream over the Gulf states with a much weaker polar jet stream to the north. Notice how the subtropical jet swings northward, parallel to the East Coast. Observe also that the weaker polar jet displays a number of loops and actually merges with the subtropical jet over the central United States. Since the wind flow at the 300-mb level pretty much parallels the contour lines, a trough of low pressure exists off the northwest coast and over central Canada, while a ridge of high pressure extends northward over the Atlantic, just east of Canada.

The looping north-south (meridional) pattern of the jet stream has an important function. Observe in Fig. 7.23 that on the eastern side of the troughs, the red arrows indicate that swiftly moving air is directing warmer air poleward, while, on the trough's western side, the more northerly flow (indicated by blue arrows) directs cold air equatorward. Jet streams, therefore, play a major role in the global transfer of heat. Since jet streams tend to meander around the world, we can easily understand how pollutants or volcanic ash injected into the atmosphere in one part of the globe could eventually settle to the ground many thousands of kilometers to the west.

Although the polar and subtropical jets are the two most frequently in the news, there are other jet streams that deserve mentioning. For example, there is a *low-level jet stream* that forms just above the Central Plains of the United States. During the summer, this jet (which usually has peak winds of less than 60 knots) often contributes to the formation of nighttime thunderstorms by transporting moisture and warm air northward. Higher up in the

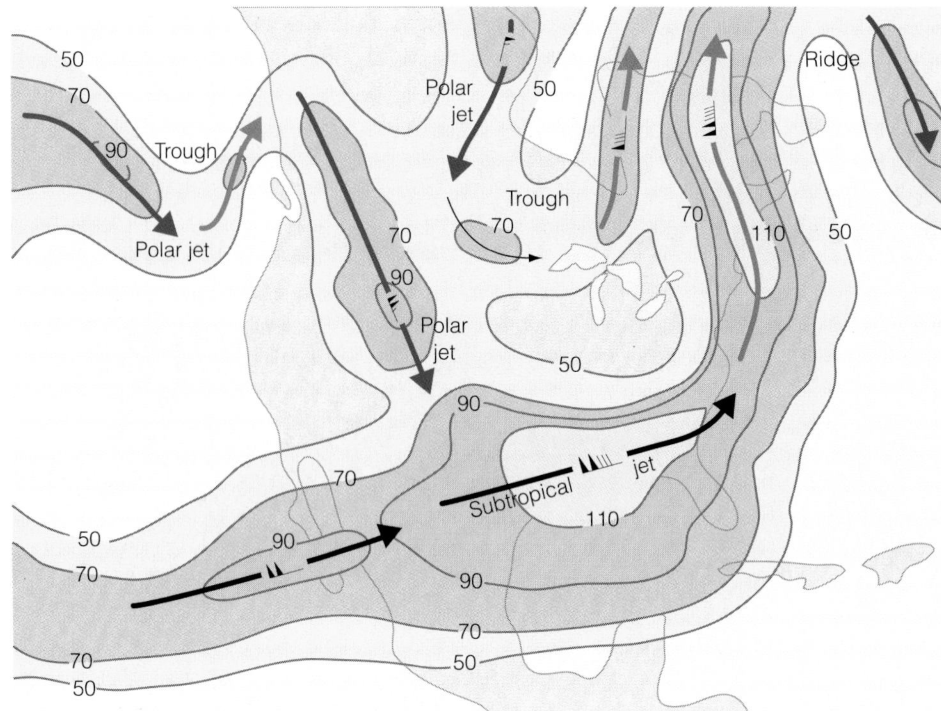

FIGURE 7.23 Position of the polar jet stream and the subtropical jet stream at the 300-mb level during the morning of March 10, 1998. Solid gray lines are lines of equal wind speed (isotachs) in knots. Heavy lines show the position of the jet streams. Heavy blue lines show where the jet stream directs cold air southward, while heavy red arrows show where the jet stream directs warm air northward.

atmosphere, over the subtropics, a summertime easterly jet called the *tropical easterly jet* forms at the base of the tropopause. And during the dark polar winter, a *stratospheric polar jet* forms near the top of the stratosphere.

BRIEF REVIEW

Before going on to the next section, which describes the many interactions between the atmosphere and the ocean, here is a review of some of the important concepts presented so far:

- The two major semipermanent subtropical highs that influence the weather of North America are the Pacific high situated off the west coast and the Bermuda high situated off the southeast coast.
- The polar front is a zone of low pressure where storms often form. It separates the mild westerlies of the middle latitudes from the cold, polar easterlies of the high latitudes.
- In equatorial regions, the intertropical convergence zone (ITCZ) is a boundary where air rises in response to the convergence of the northeast trades and the southeast trades.
- In the Northern Hemisphere, the major global pressure systems and wind belts shift northward in summer and southward in winter.
- The northward movement of the Pacific high in summer tends to keep summer weather along the west coast of North America relatively dry.
- Jet streams exist where strong winds become concentrated in narrow bands. The polar-front jet stream meanders in a wavy west-to-east pattern, becoming strongest in winter when the contrast in temperature between high and low latitudes is greatest.

GLOBAL WIND PATTERNS AND THE OCEANS

Although scientific understanding of all the interactions between the oceans and the atmosphere is far from complete, there are some relationships that deserve mentioning here.

As the wind blows over the oceans, it causes the surface water to drift along with it. The moving water gradually piles up, creating pressure differences within the water itself. This leads to further motion several hundreds of meters down into the water. In this manner, the general wind flow around the globe starts the major surface ocean currents moving. The relationship between the general circulation and ocean currents can be seen by comparing Figs. 7.17(a) and (b) (pp. 184-185) and Fig. 7.24 (p. 190).

Because of the larger frictional drag in water, ocean currents move more slowly than the prevailing wind. Typically, they range in speed from several kilometers per day to several kilometers per hour. In Fig. 7.24, we can see that ocean currents tend to spiral in semi-closed whirls. In the North Atlantic, flowing northward along the east coast of the United States, is a tremendous warm water current called the *Gulf Stream,* which carries vast quantities of warm, tropical water into higher latitudes. Off the coast of North Carolina, the Gulf Stream provides warmth and moisture for developing mid-latitude cyclones.

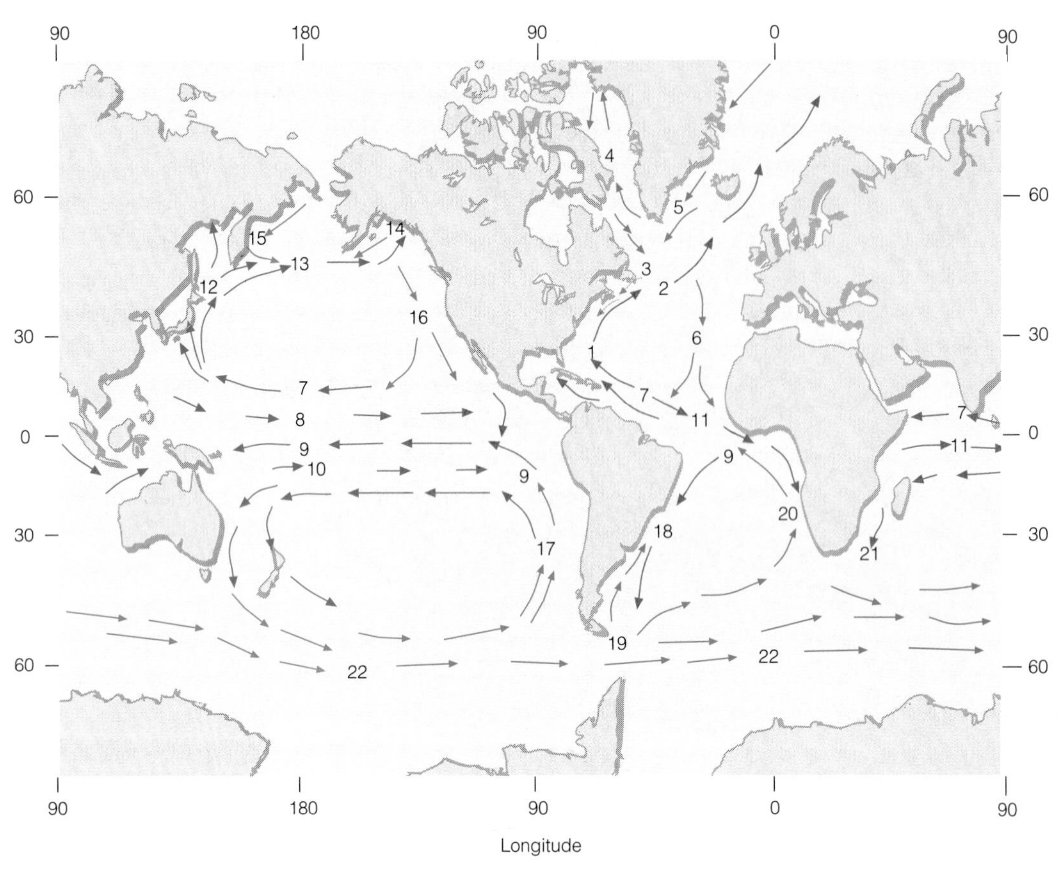

FIGURE 7.24 Average position and extent of the major surface ocean currents. Cold currents are shown in blue; warm currents are shown in red. Names of the ocean currents are given in Table 7.2.

TABLE 7.2 Major Ocean Currents

1. Gulf Stream	**9.** South Equatorial Current	**17.** Peru or Humbolt Current
2. North Atlantic Drift	**10.** South Equatorial Countercurrent	**18.** Brazil Current
3. Labrador Current	**11.** Equatorial Countercurrent	**19.** Falkland Current
4. West Greenland Drift	**12.** Kuroshio Current	**20.** Benguela Current
5. East Greenland Drift	**13.** North Pacific Drift	**21.** Agulhas Current
6. Canary Current	**14.** Alaska Current	**22.** West Wind Drift
7. North Equatorial Current	**15.** Oyashio Current	
8. North Equatorial Countercurrent	**16.** California Current	

Notice in Fig. 7.24 that as the Gulf Stream moves northward, the prevailing westerlies steer it away from the coast of North America and eastward toward Europe. Generally, it widens and slows as it merges into the broader *North Atlantic Drift.* As this current approaches Europe, part of it flows northward along the coasts of Great Britain and Norway, bringing with it warm water (which helps keep winter temperatures much warmer than one would expect this far north). The other part flows southward as the *Canary Current,* which transports cool, northern water equatorward. In the Pacific Ocean, the counterpart to the *Canary Current* is the *California Current* that carries cool water southward along the coastline of the western United States.

Up to now, we have seen that atmospheric circulations and ocean circulations are closely linked; wind blowing over the oceans produces surface ocean currents. The currents, along with the wind, transfer heat from tropical areas, where there is a surplus of energy, to polar regions, where there is a deficit. This helps to equalize the latitudinal energy imbalance with about 40 percent of the total heat transport in the Northern Hemisphere coming from surface ocean currents. The environmental implications of this heat transfer are tremendous. If the energy imbalance were to go unchecked, yearly temperature differences between low and high latitudes would increase greatly, and the climate would gradually change.

Meteorology Now™ Click "Global Ocean" and watch the patterns of ocean circulation over the course of a year.

Winds and Upwelling Earlier, we saw that the cool California Current flows roughly parallel to the west coast of North America. From this, we might conclude that summer surface water temperatures would be cool along the coast of Washington and gradually warm as we move south. A quick glance at the water temperatures along the west coast of the United States during August (Fig. 7.25) quickly alters that notion. The coldest water is observed along the northern California coast near Cape Mendocino. The reason for the cold, coastal water is **upwelling**—the rising of cold water from below.

For upwelling to occur, the wind must flow more or less parallel to the coastline. Notice in Fig. 7.26 that summer winds tend to parallel the coastline of California. As the wind blows over the ocean, the surface water beneath it is set in motion. As the surface water moves, it bends slightly to its right due to the Coriolis effect. (Remember, it would bend to the left in the Southern Hemisphere.) The water beneath the surface also moves, and it too bends slightly to its right. The net effect of this phenomenon is that a rather shallow layer of surface water moves at right angles to the surface wind and heads seaward. As the surface water drifts away from the coast, cold, nutrient-rich water from below rises (upwells) to replace it. Upwelling is strongest and surface water is coolest where the wind parallels the coast, such as it does in summer along the coast of northern California.

Because of the cold coastal water, summertime weather along the West Coast often consists of low clouds and fog, as the air over the water is chilled to its saturation point. On the brighter side, upwelling produces good fishing, as higher concentrations of nutrients are brought to the surface. But swimming is only for the hardiest of souls, since the average surface water temperature in summer is nearly 10°C (18°F) colder than the average coastal water temperature found at the same latitude along the Atlantic coast.

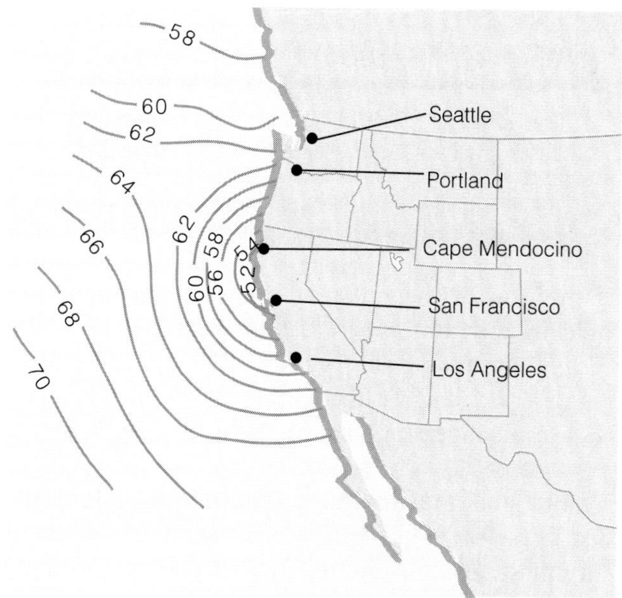

FIGURE 7.25 Average sea surface temperatures (°F) along the west coast of the United States during August.

Between the ocean surface and the atmosphere, there is an exchange of heat and moisture that depends, in part, on temperature differences between water and air. In winter, when air-water temperature contrasts are greatest, there is a substantial transfer of sensible and latent heat from the ocean surface into the atmosphere. This energy helps to maintain the global airflow. Consequently, even a

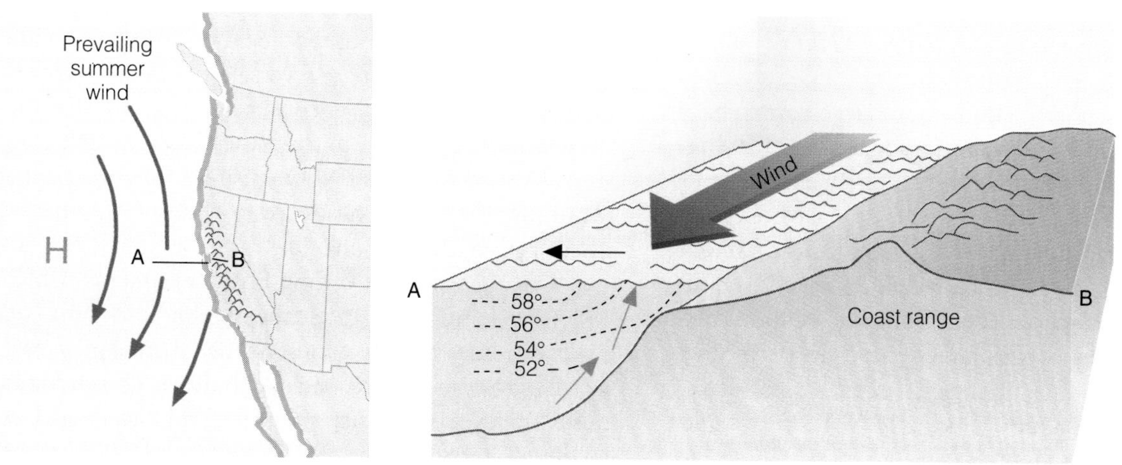

FIGURE 7.26 As winds blow parallel to the west coast of North America, surface water is transported to the right (out to sea). Cold water moves up from below (blue arrows) to replace the surface water.

DID YOU KNOW?

We have upwelling to thank for the famous quote of Mark Twain: "The coldest winter I ever experienced was a summer in San Francisco."

relatively small change in surface ocean temperatures could modify atmospheric circulations and have far-reaching effects on global weather patterns. The next section describes how weather events can be linked to surface ocean temperature changes in the tropical Pacific.

El Niño and the Southern Oscillation Along the west coast of South America, where the cool Peru Current sweeps northward, southerly winds promote upwelling of cold, nutrient-rich water that gives rise to large fish populations, especially anchovies. The abundance of fish supports a large population of sea birds whose droppings (called *guano*) produce huge phosphate-rich deposits, which support the fertilizer industry. Near the end of the calendar year, a warm current of nutrient-poor tropical water often moves southward, replacing the cold, nutrient-rich surface water. Because this condition frequently occurs around Christmas, local residents call it *El Niño* (Spanish for boy child), referring to the Christ child.

In most years, the warming lasts for only a few weeks to a month or more, after which weather patterns usually return to normal and fishing improves. However, when El Niño conditions last for many months, and a more extensive ocean warming occurs, the economic results can be catastrophic. This extremely warm episode, which occurs at irregular intervals of two to seven years and covers a large area of the tropical Pacific Ocean, is now referred to as a *major El Niño event,* or simply **El Niño.***

During a major El Niño event, large numbers of fish and marine plants may die. Dead fish and birds may litter the water and beaches of Peru; their decomposing carcasses deplete the water's oxygen supply, which leads to the bacterial production of huge amounts of smelly hydrogen sulfide. The El Niño of 1972–1973 reduced the annual Peruvian anchovy catch from 10.3 million metric tons in 1971 to 4.6 million metric tons in 1972. Since much of the harvest of this fish is converted into fishmeal and exported for use in feeding livestock and poultry, the world's fishmeal production in 1972 was greatly reduced. Countries such as the United States that rely on fishmeal for animal feed had to use soybeans as an alternative. This raised poultry prices in the United States by more than 40 percent.

Why does the ocean become so warm over the eastern tropical Pacific? Normally, in the tropical Pacific Ocean, the trades are persistent winds that blow westward from a region of higher pressure over the eastern Pacific toward a region of lower pressure centered near Indonesia (see Fig. 7.27a). The trades create upwelling that brings cold water to the surface. As this water moves westward, it is heated by sunlight and the atmosphere. Consequently, in the Pacific Ocean, surface water along the equator usually is cool in the east and warm in the west. In addition, the dragging of surface water by the trades raises sea level in the western Pacific and lowers it in the eastern Pacific, which produces a thick layer of warm water over the tropical western Pacific Ocean and a weak ocean current (called the *countercurrent*) that flows slowly eastward toward South America.

Every few years, the surface atmospheric pressure patterns break down, as air pressure rises over the region of the western Pacific and falls over the eastern Pacific (see Fig. 7.27b). This change in pressure weakens the trades, and, during strong pressure reversals, east winds are replaced by west winds. The west winds strengthen the countercurrent, causing warm water to head eastward toward South America over broad areas of the tropical Pacific. Toward the end of the warming period, which may last between one and two years, atmospheric pressure over the eastern Pacific reverses and begins to rise, whereas, over the western Pacific, it falls. This seesaw pattern of reversing surface air pressure at opposite ends of the Pacific Ocean is called the **Southern Oscillation.** Because the pressure reversals and ocean warming are more or less simultaneous, scientists call this phenomenon the *El Niño/Southern Oscillation* or **ENSO** for short. Although most ENSO episodes follow a similar evolution, each event has its own personality, differing in both strength and behavior.

During especially strong ENSO events (such as in 1982–1983 and 1997–1998) the easterly trades may actually become westerly winds. As these winds push eastward, they drag surface water with them. This dragging raises sea level in the eastern Pacific and lowers sea level in the western Pacific (see Fig. 7.27b). The eastward-moving water gradually warms under the tropical sun,

*It was thought that El Niño was a local event that occurs along the west coast of Peru and Ecuador. It is now known that the ocean-warming associated with a major El Niño can cover an area of the tropical Pacific much larger than the continental United States.

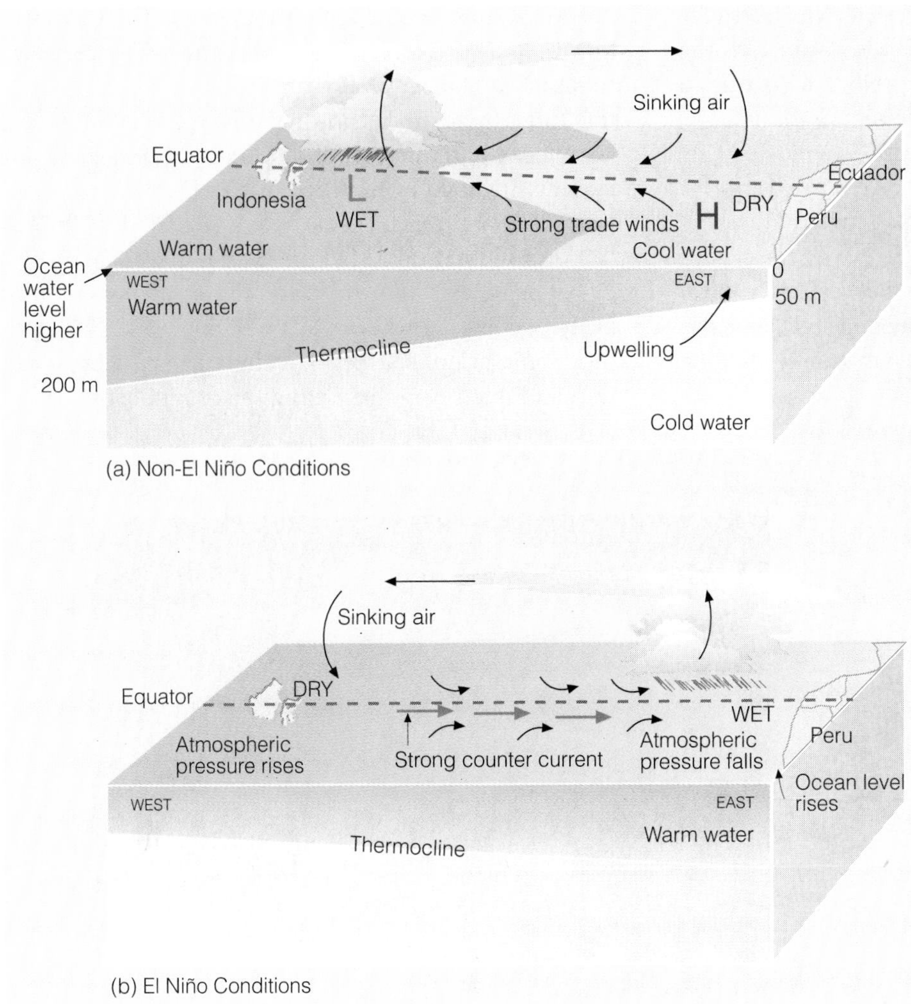

FIGURE 7.27 In diagram (a), under ordinary conditions higher pressure over the southeastern Pacific and lower pressure near Indonesia produce easterly trade winds along the equator. These winds promote upwelling and cooler ocean water in the eastern Pacific, while warmer water prevails in the western Pacific. The trades are part of a circulation that typically finds rising air and heavy rain over the western Pacific and sinking air and generally dry weather over the eastern Pacific. When the trades are exceptionally strong, water along the equator in the eastern Pacific becomes quite cool. This cool event is called *La Niña.* During El Niño conditions—diagram (b)—atmospheric pressure decreases over the eastern Pacific and rises over the western Pacific. This change in pressure causes the trades to weaken or reverse direction. This situation enhances the countercurrent that carries warm water from the west over a vast region of the eastern tropical Pacific. The thermocline, which separates the warm water of the upper ocean from the cold water below, changes as the ocean conditions change from non-El Niño to El Niño.

becoming as much as 6°C (11°F) warmer than normal in the eastern equatorial Pacific. Gradually, a thick layer of warm water pushes into coastal areas of Ecuador and Peru, choking off the upwelling that supplies cold, nutrient-rich water to South America's coastal region. The unusually warm water may extend from South America's coastal region for many thousands of kilometers westward along the equator (see Fig. 7.28). The warm tropical water may even spread northward along the west coast of North America.

Such a large area of abnormally warm water can have an effect on global wind patterns. The warm tropical water fuels the atmosphere with additional warmth and moisture, which the atmosphere turns into additional storminess and rainfall. The added warmth from the oceans and the release of latent heat during condensation apparently influence the westerly winds aloft in such a way that certain regions of the world experience too much rainfall, whereas others have too little. Meanwhile, over the warm tropical central Pacific, the frequency of typhoons usually increases. However, over the tropical Atlantic, between Africa and Central America, the winds aloft tend to disrupt the organization of thunderstorms that is necessary for hurricane development; hence, there are fewer hurricanes in this region during strong El Niño events. And, as we saw earlier in this chapter, during a strong El Niño, summer monsoon conditions tend to weaken over India, although this weakening did not happen during the strong El Niño of 1997.

Although the actual mechanism by which changes in surface ocean temperatures influence global wind patterns is not fully understood, the by-products are plain to see. For example, during exceptionally warm El Niños, drought is normally felt in Indonesia, southern Africa, and Australia, while heavy rains and flooding often occur in Ecuador and Peru. In the Northern Hemisphere, a strong subtropical westerly jet stream normally directs storms into California and heavy rain into the Gulf Coast states. The total damage worldwide due to flooding, winds, and drought may exceed $8 billion.

Following an ENSO event, the trade winds usually return to normal. However, if the trades are exceptionally strong, unusually cold surface water moves over the central and eastern Pacific, and the warm water and rainy weather is confined mainly to the western tropical Pacific (see Fig. 7.28b). This cold-water episode, which is the opposite of El Niño conditions, has been termed **La Niña** (the girl child).

As we have seen, El Niño and the Southern Oscillation are part of a large-scale ocean-atmosphere interaction that can take several years to run its course. During this time, there are certain regions in the world where significant climatic responses to an ENSO event are likely. Using data from previous ENSO episodes, scientists at the National Oceanic and Atmospheric Administration's Climatic Prediction Center have obtained a global picture of where climatic abnormalities are most likely (see Fig. 7.29).

Some scientists feel that the trigger necessary to start an ENSO event lies within the changing of the sea-

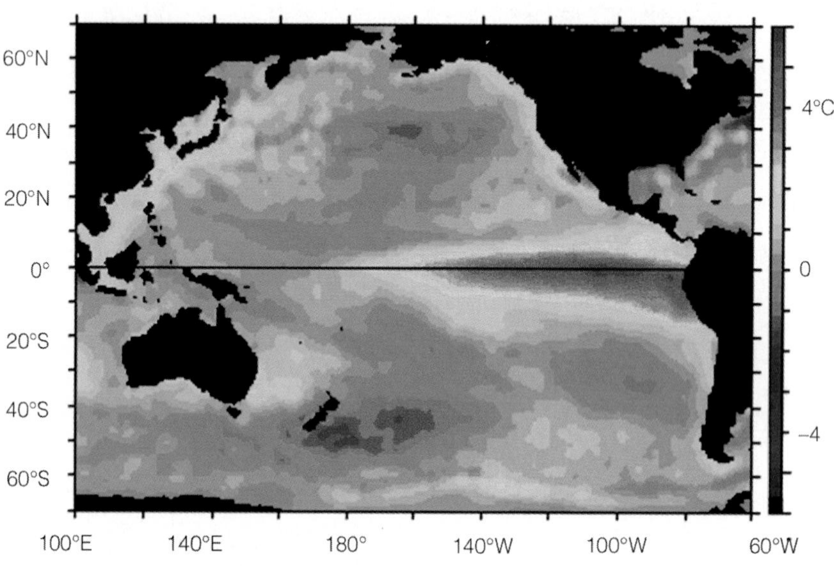

(a) El Niño Conditions, December, 1997

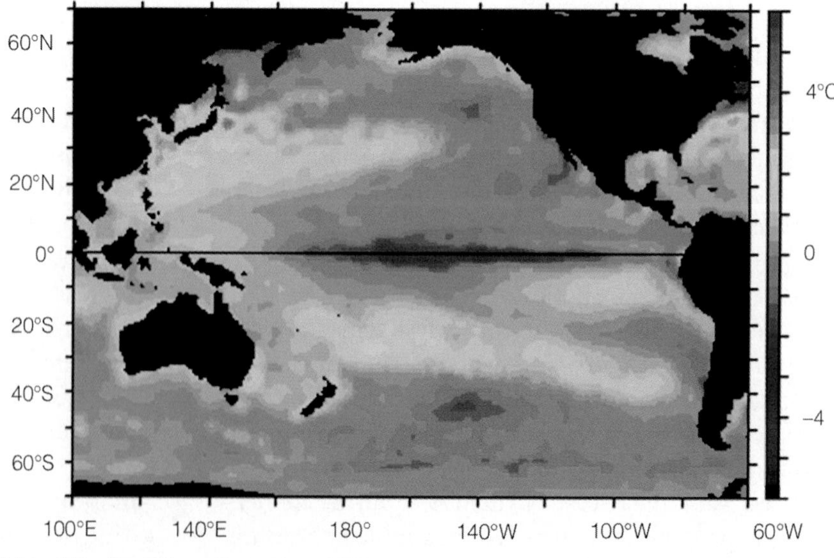

(b) La Niña Conditions, December, 1998

FIGURE 7.28 (a) Average sea surface temperature departures from normal as measured by satellite. During El Niño conditions upwelling is greatly diminished and warmer than normal water (deep red color), extends from the coast of South America westward, across the Pacific. (b) During La Niña conditions, strong trade winds promote upwelling, and cooler than normal water (dark blue color) extends over the eastern and central Pacific. (NOAA/PHEL/TAO)

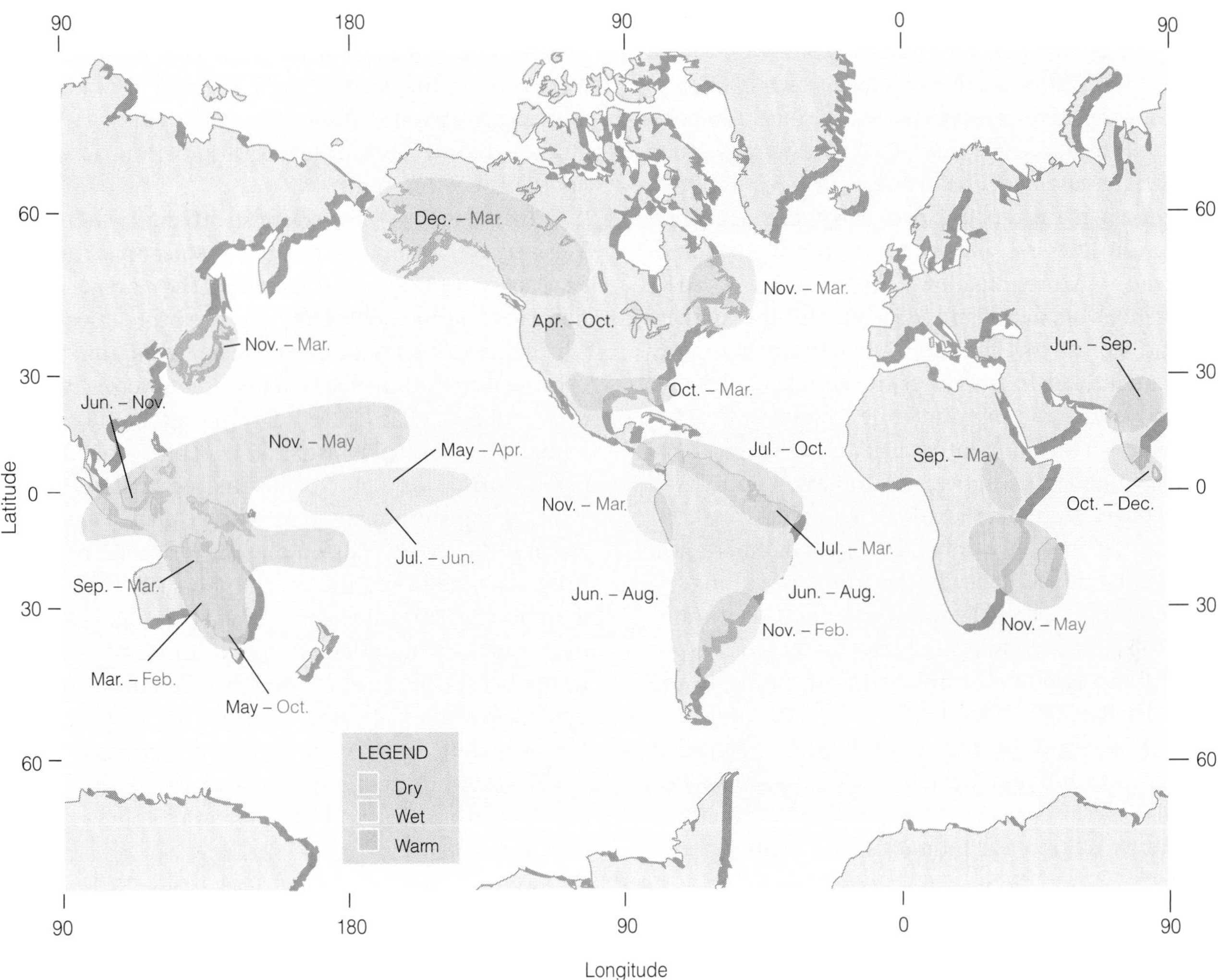

FIGURE 7.29 Regions of climatic abnormalities associated with El Niño–Southern Oscillation conditions. A strong ENSO event may trigger a response in nearly all indicated areas, whereas a weak event will likely play a role in only some areas. Note that the months in black type indicate months during the same years the major warming began; months in red type indicate the following year. (After NOAA Climatic Prediction Center.)

sons, especially the transition periods of spring and fall. Others feel that the winter monsoon plays a major role in triggering a major El Niño event. As noted earlier, it appears that an ENSO episode and the monsoon system are intricately linked, so that a change in one brings about a change in the other.

Presently, scientists (with the aid of coupled general circulation models) are trying to simulate atmospheric and oceanic conditions, so that El Niño and the Southern Oscillation can be anticipated. At this point, several models have been formulated that show promise in predicting the onset and life history of an ENSO event. In addition, an in-depth study known as TOGA (*T*ropical *O*cean and *G*lobal *A*tmosphere), which began in 1985 and ended in 1994, is providing scientists with valuable information about the interactions that occur between the ocean and the atmosphere. The primary aim of TOGA, a major component of the *W*orld *C*limate *R*esearch *P*rogram (WCRP), is to provide enough scientific information so that researchers can better predict climatic fluctuations (such as ENSO) that occur over periods of months and years. The hope is that a better understanding of El Niño and the Southern Oscillation will provide improved long-range forecasts of weather and climate.

Meteorology Now™ Click "Southern Oscillation" to explore multi-year changes in sea surface temperatures associated with El Niño and La Niña events.

Other Atmosphere-Ocean Interactions Is there a similar pattern in the Atlantic that compares to the Southern Oscillation in the Pacific? Over the Atlantic there is a reversal of pressure (called the *North Atlantic Oscillation*) that has an effect on the weather in Europe and along the east coast of North America. For example, in winter when atmospheric pressure in the vicinity of the Icelandic low rises, and the pressure in the region of the Bermuda-Azores high lowers (see Fig. 7.17a, p. 184 for reference), northern Europe's winter turns extremely cold, and storms that normally take a more northerly route slam into Spain. In the northeastern United States, extremely cold arctic air invades the region.

There is also an event similar to the tropical Pacific's El Niño that occurs over the tropical Atlantic. Typically in the eastern Atlantic (off the coast of Africa), the water is cooler and the weather drier than in the tropical western Atlantic. Periodically, the cool water along the African coast is replaced by warm water and heavy rainfall. This Atlantic warming, however, occurs more sporadically, and is not as strong as the warming in the tropical Pacific.

In recent years, scientists have unveiled a reversal of ocean temperatures in the Indian Ocean. In the equatorial Indian Ocean, warm water normally accumulates in the eastern section, near Indonesia, and relatively cool water exists in the western part, near the African coast. When this pattern reverses, cooler water in the east fosters dry conditions and even drought in Indonesia, while warmer water along the coast of Africa enhances storm activity, which brings devastating rains to Kenya and neighboring countries.

In the mid 1990s, scientists at the University of Washington, while researching connections between Alaskan salmon production and Pacific climate, identified a long-term Pacific Ocean temperature fluctuation, which they called the **Pacific Decadal Oscillation (PDO)** because the ocean surface temperature reverses every 20 to 30 years. The Pacific Decadal Oscillation is like ENSO in that it has a warm phase and a cool phase, but its temperature behavior is much different from that of El Niño.

During the warm (or positive) phase, unusually warm surface water exists along the west coast of North America, while over the central North Pacific, cooler than normal surface water prevails (see Fig. 7.30a). At the same time, the Aleutian low in the Gulf of Alaska strengthens, which causes more Pacific storms to move into Alaska and California. This situation causes winters, as a whole, to be warmer and drier over northwestern North America. Elsewhere, winters tend to be drier over the Great Lakes, and cooler and wetter in the southern United States.

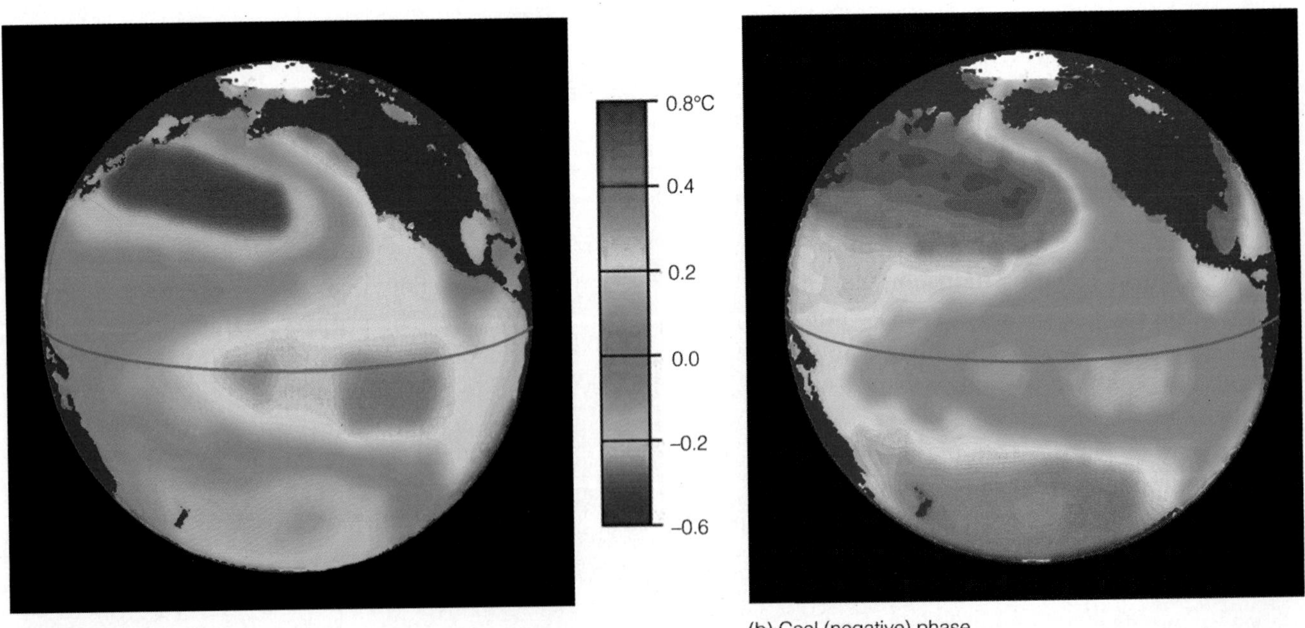

(a) Warm (positive) phase

(b) Cool (negative) phase

FIGURE 7.30 Typical winter sea surface temperature departure from normal in °C during the Pacific Decadal Oscillation's warm phase (a) and cool phase (b). (Source: JISAO, University of Washington, obtained via the www:http://tao.atmos.washington.edu/pdo. Used with permission of N. Mantua.)

The present cool (or negative) phase finds cooler-than-average surface water along the west coast of North America and an area of warmer-than-normal surface water extending from Japan into the central North Pacific (see Fig. 7.30b). Winters in the cool phase tend to be cooler and wetter than average over northwestern North America, wetter over the Great Lakes, and warmer and drier in the southern United States.

The climate patterns described so far only represent average conditions, as individual years within either phase may vary considerably. Hopefully, as our understanding of the Pacific Decadal Oscillation improves, climate forecasts across North America and elsewhere will improve as well.

SUMMARY

In this chapter, we examined a variety of atmospheric circulations. We looked at small-scale winds and found that eddies can form in a region of strong wind shear, especially in the vicinity of a jet stream. On a slightly larger scale, land and sea breezes blow in response to local pressure differences created by the uneven heating and cooling rates of land and water. Monsoon winds change direction seasonally, while mountain and valley winds change direction daily.

A warm, dry wind that descends the eastern side of the Rocky Mountains is the chinook. The same type of wind in the Alps is the foehn. A warm, dry downslope wind that blows into southern California is the Santa Ana wind. Local intense heating of the surface can produce small rotating winds, such as the dust devil, while downdrafts in a thunderstorm are responsible for the desert haboob.

The largest pattern of winds that persists around the globe is called the general circulation. At the surface in both hemispheres, winds tend to blow from the east in the tropics, from the west in the middle latitudes, and from the east in polar regions. Where upper-level westerly winds tend to concentrate into narrow bands, we find jet streams. The annual shifting of the major pressure systems and wind belts—northward in July and southward in January—strongly influences the annual precipitation of many regions.

Toward the end of the chapter we examined the interaction between the atmosphere and oceans. Here we found the interaction to be an ongoing process where everything, in one way or another, seems to influence everything else. On a large scale, winds blowing over the surface of the water drive the major ocean currents; the oceans, in turn, release energy to the atmosphere, which helps to maintain the general circulation.

When atmospheric circulation patterns change over the Tropical Pacific, and the trade winds weaken or reverse direction, warm tropical water is able to flow eastward toward South America where it chokes off upwelling and produces disastrous economic conditions. When the warm water extends over a vast area of the Tropical Pacific, the warming is called a major El Niño event, and the associated reversal of pressure over the Pacific Ocean is called the Southern Oscillation. The large-scale interaction between the atmosphere and the ocean during El Niño and the Southern Oscillation (ENSO) affects global atmospheric circulation patterns. The sweeping winds aloft provide too much rain in some areas and not enough in others.

The Tropical Atlantic and the Indian oceans experience weak temperature fluctuations similar to El Niño in the Tropical Pacific. Over the northern central Pacific and along the west coast of North America there is a reversal of surface water temperature that occurs every 20 to 30 years, called the *Pacific Decadal Oscillation* (PDO). Studies now in progress are designed to determine how the interchange between atmosphere and ocean can produce such events.

Meteorology Now™ Assess your understanding of this chapter's topics with additional quizzing and tutorials at http://earthscience.brookscole.com/ahrens/ess4e.

KEY TERMS

The following terms are listed in the order they appear in the text. Define each. Doing so will aid you in reviewing the material covered in this chapter.

scales of motion
microscale
mesoscale
synoptic scale
planetary scale
rotor
wind shear
clear air turbulence (CAT)
thermal circulation
sea breeze
land breeze
monsoon wind system
valley breeze
mountain breeze
katabatic wind
chinook wind
Santa Ana wind
haboob
dust devils (whirlwinds)
general circulation of the atmosphere
Hadley cell
doldrums
subtropical highs

trade winds
intertropical convergence zone (ITCZ)
westerlies
polar front
subpolar low
polar easterlies
Bermuda high
Pacific high
Icelandic low
Aleutian low
Siberian high
jet stream
polar front jet stream
subtropical jet stream
upwelling
El Niño
Southern Oscillation
ENSO
La Niña
Pacific Decadal Oscillation (PDO)

QUESTIONS FOR REVIEW

1. Describe the various scales of motion and give an example of each.
2. What is wind shear and how does it relate to clear air turbulence?
3. Using a diagram, explain how a thermal circulation develops.
4. Why does a sea breeze blow from sea to land and a land breeze from land to sea?
5. (a) Briefly explain how the monsoon wind system develops over eastern and southern Asia.
   (b) Why in India is the summer monsoon wet and the winter monsoon dry?
6. Which wind will produce clouds: a valley breeze or a mountain breeze? Why?
7. What are katabatic winds? How do they form?
8. Explain why chinook winds are warm and dry.
9. (a) What is the primary source of warmth for a Santa Ana wind?
   (b) What atmospheric conditions contribute to the development of a strong Santa Ana?
10. Describe how dust devils usually form.
11. Draw a large circle. Now, place the major surface semi-permanent pressure systems and the wind belts of the world at their appropriate latitudes.
12. According to Fig. 7.16 (p. 183), most of the United States is located in what wind belt?
13. Explain how and why the average surface pressure features shift from summer to winter.
14. How does the polar front influence the development of the polar-front jet stream?
15. Why is the polar jet stream more strongly developed in winter?
16. Explain the relationship between the general circulation of air and the circulation of ocean currents.
17. Describe how the winds along the west coast of North America produce upwelling.
18. (a) What is a major El Niño event?
    (b) What happens to the surface pressure at opposite ends of the Pacific Ocean during the Southern Oscillation?
    (c) Describe how an ENSO event may influence the weather in different parts of the world.
19. What are the conditions over the tropical eastern and central Pacific Ocean during the phenomenon known as La Niña?
20. Describe the ocean surface temperatures associated with the Pacific Decadal Oscillation.

QUESTIONS FOR THOUGHT AND EXPLORATION

1. Suppose you are fishing in a mountain stream during the early morning. Is the wind more likely to be blowing upstream or downstream? Explain why.
2. Why, in Antarctica, are winds on the high plateaus usually lighter than winds in steep, coastal valleys?
3. What atmospheric conditions must change so that the westerly flowing polar-front jet stream reverses direction and becomes an easterly flowing jet stream?
4. After a winter snowstorm, Cheyenne, Wyoming, reports a total snow accumulation of 48 cm (19 in.), while the maximum depth in the surrounding countryside is only 28 cm (11 in.). If the storm's intensity and duration were practically the same for a radius of 50 km around Cheyenne, explain why Cheyenne received so much more snow.
5. The prevailing winds in southern Florida are northeasterly. Knowing this, would you expect the strongest sea breezes to be along the east or west coast of southern Florida? What about the strongest land breezes?
6. Explain why icebergs tend to move at right angles to the direction of the wind.
7. Give *two* reasons why pilots would prefer to fly in the core of a jet stream rather than just above or below it.
8. Why do the major ocean currents in the North Indian Ocean reverse direction between summer and winter?
9. Explain why the surface water temperature along the northern California coast is warmer in winter than it is in summer.
10. The Coriolis force deflects moving water to the right in the Northern Hemisphere and to the left in the Southern Hemisphere. Why, then, does upwelling tend to occur along the western margin of continents in both hemispheres?
11. Pacific and Atlantic satellite images (http://www.earthwatch.com/WX_HDLINES/tropical.html): Examine current infrared satellite images of the Pacific and At-

lantic Ocean regions. Describe the types and sizes of the eddies that appear in the images.

12. Local winds (http://freespace.virgin.net/mike.ryding/local/local.htm): Look up several local wind circulations that affect specific localized areas around the globe.

- Go to the Brooks/Cole Earth Sciences Resource Center (http://earthscience.brookscole.com) for critical thinking exercises, articles, and additional readings from InfoTrac College Edition, Brooks/Cole's online student library.

A middle-latitude cyclonic storm spins counterclockwise over the eastern Atlantic.

Meteorology Now™ This icon, appearing throughout the book, indicates an opportunity to explore interactive tutorials, animations, or practice problems available on the MeteorologyNow Web site at http://earthscience.brookscole.com/ahrens/ess4e.

CHAPTER 8

Air Masses, Fronts, and Middle-Latitude Cyclones

About two o'clock in the afternoon it began to grow dark from a heavy, black cloud which was seen in the northwest. Almost instantly the strong wind, traveling at the rate of 70 miles an hour, accompanied by a deep bellowing sound, with its icy blast, swept over the land, and everything was frozen hard. The water in the little ponds in the roads froze in waves, sharp edged and pointed, as the gale had blown it. The chickens, pigs and other small animals were frozen in their tracks. Wagon wheels ceased to roll, froze to the ground. Men, going from their barns or fields a short distance from their homes, in slush and water, returned a few minutes later walking on the ice. Those caught out on horseback were frozen to their saddles, and had to be lifted off and carried to the fire to be thawed apart. Two young men were frozen to death near Rushville. One of them was found with his back against a tree, with his horse's bridle over his arm and his horse frozen in front of him. The other was partly in a kneeling position, with a tinder box in one hand and a flint in the other, with both eyes wide open as if intent on trying to strike a light. Many other casualties were reported. As to the exact temperature, however, no instrument has left any record; but the ice was frozen in the stream, as variously reported, from six inches to a foot in thickness in a few hours.

John Moses, *Illinois: Historical and Statistical*

CONTENTS

The opening details the passage of a spectacular cold front as it moved through Illinois on December 21, 1836. Although no reliable temperature records are available, estimates are that, as the front swept through, air temperatures dropped almost instantly from the balmy 40s (°F) to 0 degrees. Fortunately, temperature changes of this magnitude are quite rare with cold fronts.

In this chapter, we will examine the more typical weather associated with cold fronts and warm fronts. We will address questions such as: Why are cold fronts usually associated with showery weather? How can warm fronts cause freezing rain and sleet to form over a vast area during the winter? And how can one read the story of an approaching warm front by observing its clouds? We will also see how weather fronts are an integral part of a mid-latitude cyclonic storm. But, first, so that we may better understand fronts and storms, we will examine air masses. We will look at where and how they form and the type of weather usually associated with them.

AIR MASSES

An **air mass** is an extremely large body of air whose properties of temperature and humidity are fairly similar in any horizontal direction at any given altitude. Air masses may cover many thousands of square kilometers. In Fig. 8.1, a large winter air mass, associated with a high pressure area, covers over half of the United States. Note that, although the surface air temperature and dew point vary somewhat, everywhere the air is cold and dry, with the exception of the zone of snow showers on the eastern shores of the Great Lakes. This cold, shallow anticyclone will drift eastward, carrying with it the temperature and moisture characteristic of the region where the air mass formed; hence, in a day or two, cold air will be located over the central Atlantic Ocean. Part of weather forecasting is, then, a matter of determining air mass characteristics, predicting how and why they change, and in what direction the systems will move.

Source Regions Regions where air masses originate are known as **source regions.** In order for a huge mass of air to develop uniform characteristics, its source region should be generally flat and of uniform composition, with light surface winds. The longer the air remains stagnant over its source region, or the longer the path over which the air moves, the more likely it will acquire properties of the surface below. Consequently, ideal source regions are usually those areas dominated by surface high pressure. They include the ice- and snow-covered arctic plains in winter and subtropical oceans in summer. The

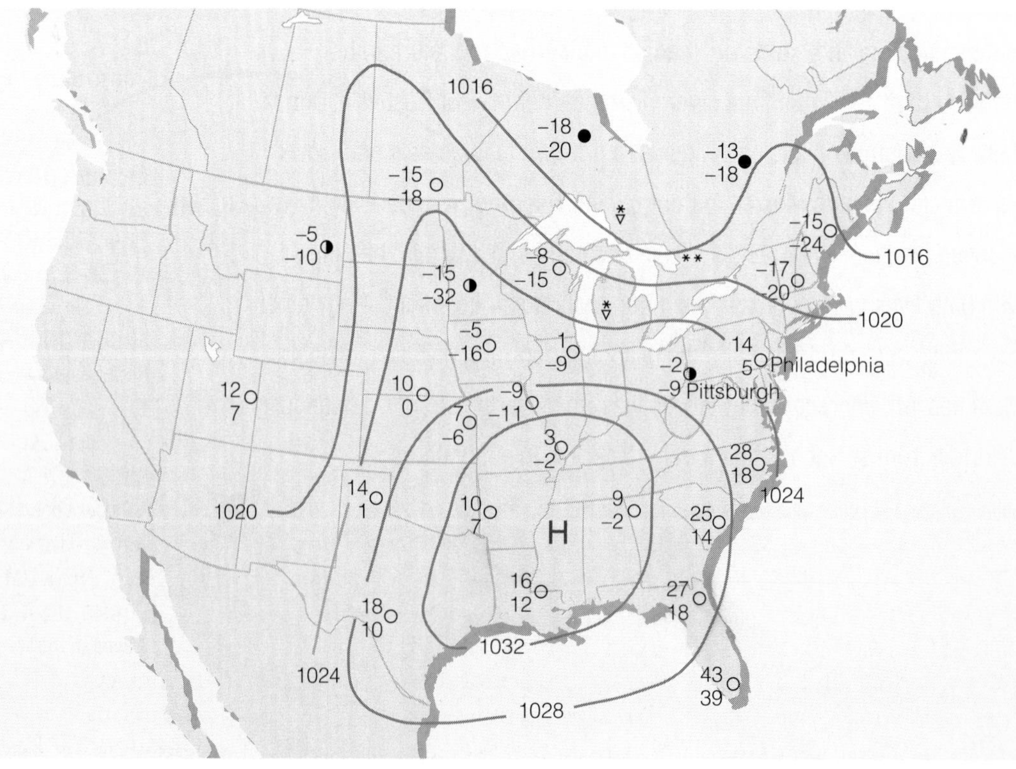

FIGURE 8.1 Here, a large, extremely cold winter air mass is dominating the weather over much of the United States. At almost all cities, the air is cold and dry. Upper number is air temperature (°F); bottom number is dew point (°F).

middle latitudes, where surface temperatures and moisture characteristics vary considerably, are not good source regions. Instead, this region is a transition zone where air masses with different physical properties move in, clash, and produce an exciting array of weather activity.

Classification Air masses are usually classified according to their temperature and humidity, both of which usually remain fairly uniform in any horizontal direction. There are cold and warm air masses, humid and dry air masses. Air masses are grouped into four general categories according to their source region. Air masses that originate in polar latitudes are designated by the capital letter "P" (for *polar*); those that form in warm tropical regions are designated by the capital letter "T" (for *tropical*). If the source region is land, the air mass will be dry and the lowercase letter "c" (for *continental*) precedes the P or T. If the air mass originates over water, it will be moist—at least in the lower layers—and the lowercase letter "m" (for *maritime*) precedes the P or T. We can now see that polar air originating over land will be classified cP on a surface weather map, whereas tropical air originating over water will be marked as mT. In winter, an extremely cold air mass is designated as cA, *continental arctic.* Often, however, it is difficult to distinguish between arctic and polar air masses, especially when the arctic air mass has traveled over warmer terrain. Table 8.1 lists the four basic air masses.

After the air mass spends some time over its source region, it usually begins to move in response to the winds aloft. As it moves away from its source region, it encounters surfaces that may be warmer or colder than itself. When the air mass is colder than the underlying surface, it is warmed from below, which produces instability at low levels. In this case, increased convection and turbulent mixing near the surface usually result in good visibility, cumuliform clouds, and showers of rain or snow. On the other hand, when the air mass is warmer than the surface below, the lower layers are chilled by contact with the cold earth. Warm air above cooler air produces stable air with little vertical mixing. This causes the accumulation of dust, smoke, and pollutants, which restricts surface visibilities. In moist air, stratiform clouds accompanied by drizzle or fog may form.

Air Masses of North America The principal air masses (with their source regions) that enter the United States are shown in Fig. 8.2. We are now in a position to study the formation and modification of each of these air masses and the variety of weather that accompanies them.

TABLE 8.1
Air Mass Classification and Characteristics

SOURCE REGION	POLAR (P)	TROPICAL (T)
Land continental (c)	cP Cold, dry, stable	cT Hot, dry, stable air aloft; unstable surface air
Water maritime (m)	mP Cool, moist, unstable	mT Warm, moist; usually unstable

cP (Continental Polar) and cA (Continental Arctic) Air Masses The bitterly cold weather that invades southern Canada and the United States in winter is associated with **continental polar** and **continental arctic** air masses. These air masses originate over the ice- and snow-covered regions of northern Canada and Alaska where long, clear nights allow for strong radiational cooling of the surface. Air in contact with the surface becomes quite cold and stable. Since little moisture is added to the air, it is also quite dry. Eventually a portion of this cold air breaks away and, under the influence of the airflow aloft, moves southward as an enormous shallow high pressure area.

As the cold air moves into the interior plains, there are no topographic barriers to restrain it, so it continues southward, bringing with it cold wave warnings and frigid temperatures. As the air mass moves over warmer land to the south, the air temperature moderates slightly.

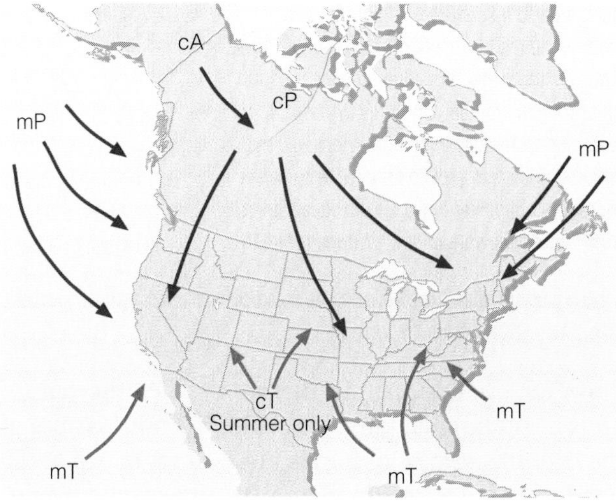

FIGURE 8.2 Air mass source regions and their paths.

FOCUS ON A SPECIAL TOPIC

Lake-Effect (Enhanced) Snows

During the winter, when the weather in the Midwest is dominated by clear and cold cP (or cA) air, people living on the eastern shores of the Great Lakes brace themselves for heavy snow showers. Snowstorms that form on the downwind side of one of these lakes are known as *lake-effect snows.* (Since the lakes are responsible for enhancing the amount of snow that falls, these snowstorms are also called *lake-enhanced snows,* especially when the snow is associated with a cold front or mid-latitude cyclone.) Such storms are highly localized, extending from just a few kilometers to more than 50 km inland. The snow usually falls as a heavy shower or squall in a concentrated zone. So centralized is the region of snowfall, that one part of a city may accumulate many centimeters of snow, while, in another part, the ground is bare.

Lake-effect snows are most numerous from November to January. During these months, cold air moves over the lakes when they are relatively warm and not quite frozen. The contrast in temperature between water and air can be as much as 25°C (45°F). Studies show that the greater the contrast in temperature, the greater the potential for snow showers. In Fig. 1, we can see that, as the cold air moves over the warmer water, the air mass is quickly warmed from below, making it more buoyant and less stable. Rapidly, the air sweeps up moisture, soon becoming saturated. Out over the water, the vapor condenses into steam fog. As the air continues to warm, it rises and forms billowing cumuliform clouds, which continue to grow as the air becomes more unstable. Eventually, these clouds produce heavy showers of snow, which make the lake seem like a snow factory. Once the air and clouds reach the downwind side of the lake, additional lifting is provided by low hills and the convergence of air as it slows down over the rougher terrain. In late winter, the frequency and intensity of lake-effect snows often taper off as the temperature contrast between water and air diminishes and larger portions of the lakes freeze.

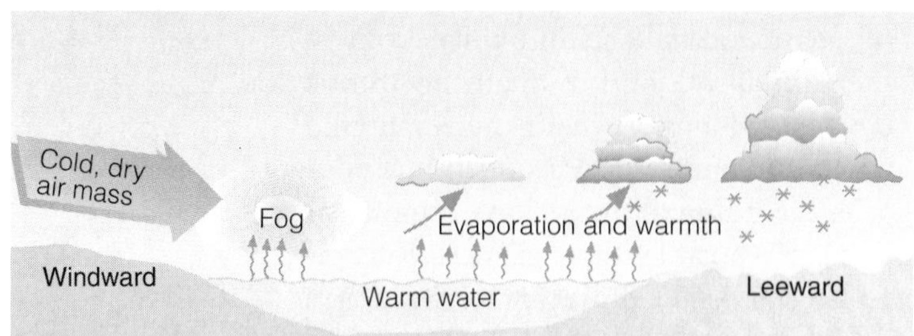

FIGURE 1 The formation of lake-effect snows. Cold, dry air crossing the lake gains moisture and warmth from the water. The more buoyant air now rises, forming clouds that deposit large quantities of snow on the lake's leeward shores.

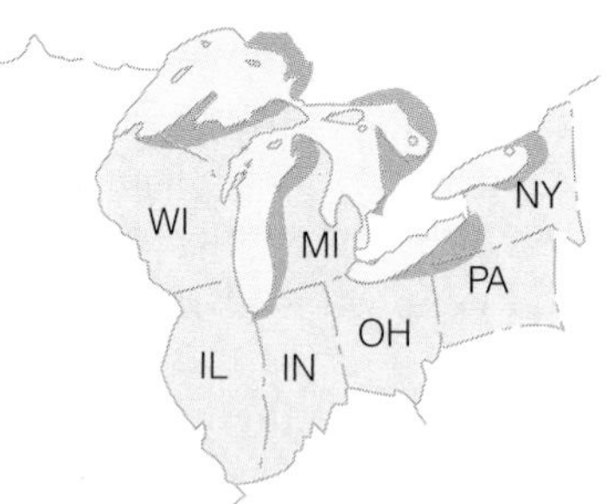

FIGURE 2 Areas shaded purple show regions that experience heavy lake-effect snows.

Generally, the longer the stretch of water over which the air mass travels (the longer the fetch), the greater the amount of warmth and moisture derived from the lake, and the greater the potential for heavy snow showers. Consequently, forecasting lake-effect snowfalls depends to a large degree on determining the trajectory of the air as it flows over the lake. Regions that experience heavy lake-effect snowfalls are shown in Fig. 2.

As the cold air moves farther east, the heavy snow showers usually taper off; however, the western slope of the Appalachian Mountains produces further lifting, enhancing the possibility of more and heavier showers. The heat given off during condensation warms the air and, as the air descends the eastern slope, compressional heating warms it even more. Snowfall ceases, and by the time the air arrives at Philadelphia, New York, or Boston, the only remaining trace of the snow showers occurring on the other side of the mountains are the puffy cumulus clouds drifting overhead.

Lake-effect (or enhanced) snows are not confined to the Great Lakes. In fact, any large unfrozen lake (such as the Great Salt Lake) can enhance snowfall when cold, relatively dry air sweeps over it.

However, even during the afternoon, when the surface air is most unstable, cumulus clouds are rare because of the extreme dryness of the air mass. At night, when the winds die down, rapid radiational surface cooling and clear skies combine to produce low minimum temperatures. If the cold air moves as far south as central or southern Florida, the winter vegetable crop may be severely damaged. When the cold, dry air mass moves over a relatively warm body of water, such as the Great Lakes, heavy snow showers—called **lake-effect snows**—often form on the eastern shores. (More information on lake-effect snows is provided in the Focus section above.)

In winter, the generally fair weather accompanying cP air is due to the stable nature of the atmosphere aloft. Sinking air develops above the large dome of high pressure. The subsiding air warms by compression and creates warmer air, which lies above colder surface air. Therefore, a strong upper-level temperature inversion often forms. Should the anticyclone stagnate over a region for several days, the visibility gradually drops as pollutants become trapped in the cold air near the ground. Usually, however, winds aloft move the cold air mass either eastward or southeastward.

The Rockies, Sierra Nevada, and Cascades normally protect the Pacific Northwest from the onslaught of cP air, but, occasionally, cP air masses do invade these regions. When the upper-level winds over Washington and Oregon blow from the north or northeast on a trajectory beginning over northern Canada or Alaska, cold cP (and cA) air can slip over the mountains and extend its icy fingers all the way to the Pacific Ocean. As the air moves off the high plateau, over the mountains, and on into the lower valleys, compressional heating of the sinking air causes its temperature to rise, so that by the time it reaches the lowlands, it is considerably warmer than it was originally. However, in no way would this air be considered warm. In some cases, the subfreezing temperatures slip over the Cascades and extend southward into the coastal areas of southern California.

DID YOU KNOW?

Heavy lake-effect snows can bury a city in a matter of days. For example, in less than four days Buffalo, New York, received nearly seven feet of snow during December, 2001. And Montague, New York (which lies on the eastern side of Lake Ontario), received over seven feet of snow in less than forty-eight hours during January, 1997.

A similar but less dramatic warming of cP and cA air occurs along the east coast of the United States. Air rides up and over the lower Appalachian Mountains. Turbulent mixing and compressional heating increase the air temperatures on the downwind side. Consequently, cities located to the east of the Appalachian Mountains usually do not experience temperatures as low as those on the west side. In Fig. 8.1, notice that for the same time of day—in this case 7 A.M. EST—Philadelphia, with an air temperature of 14°F, is 16°F warmer than Pittsburgh, at –2°F.

Figure 8.3 shows two upper-air patterns that led to extremely cold outbreaks of arctic air during December 1989 and 1990. Upper-level winds typically blow from west to east, but, in both of these cases, the flow, as given by the heavy, dark arrows, had a strong north-south (meridional) trajectory. The *H* represents the positions of

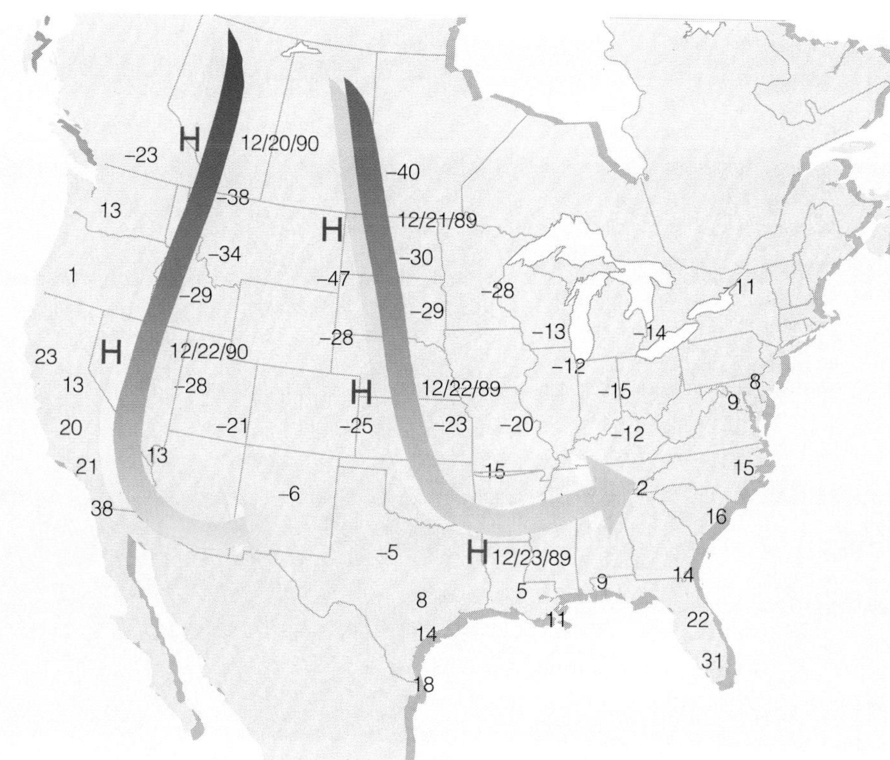

FIGURE 8.3 Average upper-level wind flow (heavy arrows) and surface position of anticyclones (H) associated with two extremely cold outbreaks of arctic air during December. Numbers on the map represent minimum temperatures (°F) measured during each cold snap.

the cold surface anticyclones. Numbers on the map represent minimum temperatures (°F) recorded during the cold spells. East of the Rocky Mountains, over 350 record low temperatures were set between December 21 and 24, 1989, with the arctic outbreak causing an estimated $480 million in damage to the fruit and vegetable crops in Texas and Florida. Along the West Coast, the frigid air during December, 1990, caused over $300 million in damage to the vegetable and citrus crops, as temperatures over parts of California plummeted to their lowest readings in more than fifty years. Notice in both cases how the upper-level wind directs the paths of the air masses.

The cP air that moves into the United States in summer has properties much different from its winter counterpart. The source region remains the same but is now characterized by long summer days that melt snow and warm the land. The air is only moderately cool, and surface evaporation adds water vapor to the air. A summertime cP air mass usually brings relief from the oppressive heat in the central and eastern states, as cooler air lowers the air temperature to more comfortable levels. Daytime heating warms the lower layers, producing surface instability. With its added moisture, the rising air may condense and create a sky dotted with fair weather cumulus clouds.

When an air mass moves over a large body of water, its original properties may change considerably. For instance, cold, dry cP air moving over the Gulf of Mexico warms rapidly and gains moisture. The air quickly assumes the qualities of a maritime air mass. Notice in Fig. 8.4 that rows of cumulus clouds are forming over the Gulf of Mexico parallel to northerly surface winds as cP air is being warmed by the water beneath it. As the air continues its journey southward into Mexico and Central America, strong, moist northerly winds build into heavy clouds (bright area) and showers along the northern coast. Hence, a once cold, dry, and stable air mass can be modified to such an extent that its original characteristics are no longer discernible. When this happens, the air mass is given a new designation.

In summary, polar and arctic air masses are responsible for the bitter cold winter weather that can cover wide sections of North America. When the air mass originates over the Canadian Northwest Territories, frigid air can bring record-breaking low temperatures. Such was the case on Christmas Eve, 1983, when arctic air covered most of North America. (A detailed look at this air mass and its accompanying record-setting low temperatures is given in the Focus section on p. 207.)

mP (Maritime Polar) Air Masses During the winter, cP and cA air originating over Asia and frozen polar regions is carried eastward and southward over the Pacific Ocean by the circulation around the Aleutian low. The ocean water modifies these cold air masses by adding warmth and moisture to them. Since this air travels over water many hundreds or even thousands of kilometers, it gradually changes into a **maritime polar air mass.**

By the time this air mass reaches the Pacific coast it is cool, moist, and conditionally unstable. The ocean's effect is to keep air near the surface warmer than the air aloft. Temperature readings in the 40s and 50s (°F) are common near the surface, while air at an altitude of about a

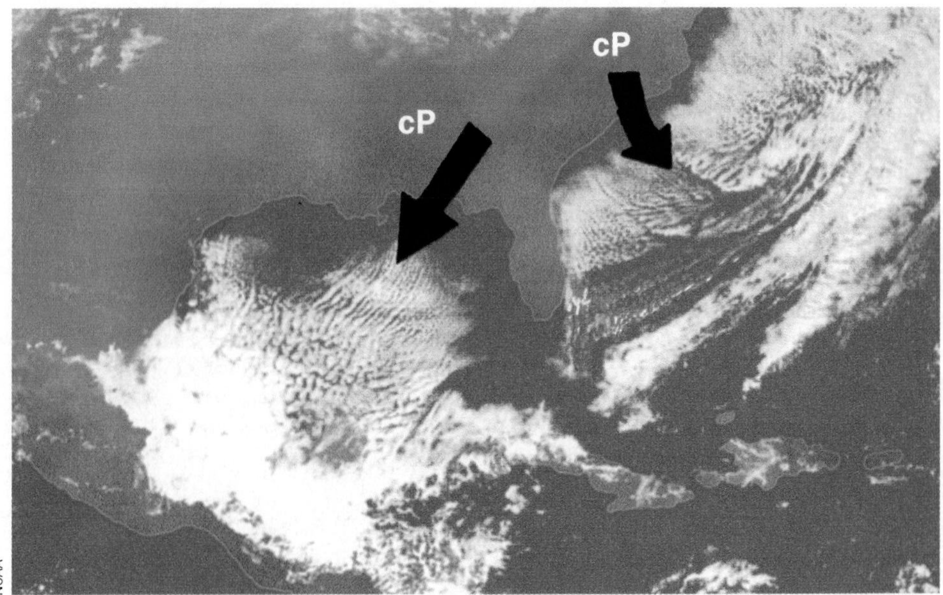

FIGURE 8.4 Visible satellite image showing the modification of cP air as it moves over the warmer Gulf of Mexico and the Atlantic Ocean.

FOCUS ON A SPECIAL TOPIC

The Return of the Siberian Express

The winter of 1983–1984 was one of the coldest on record across North America. Unseasonably cold weather arrived in December, which, for much of the country, was one of the coldest Decembers since records have been kept. During the first part of the month, continental polar air covered most of the northern and central plains. As the cold air moderated slightly, far to the north a huge mass of bitter cold arctic air was forming over the frozen reaches of the Canadian Northwest Territories.

By mid-month, the frigid air, associated with a massive high pressure area, covered all of northwest Canada. Meanwhile, aloft, strong northerly winds directed the leading edge of the frigid air southward over the prairie provinces of Canada and southward into the United States. Because the extraordinarily cold air was accompanied in some regions by winds gusting to 45 knots, at least one news reporter dubbed the onslaught of this arctic blast, "the Siberian Express."

The Express dropped temperatures to some of the lowest readings ever recorded during the month of December. On December 22, Elk Park, Montana, recorded an unofficial low of –53°C (–64°F), only 4°C higher than the all-time low of –57°C (–70°F) for the nation (excluding Alaska) recorded at Rogers Pass, Montana, on January 20, 1954.

The center of the massive anticyclone gradually pushed southward out of Canada. By December 24, its center was over eastern Montana (Fig. 3), where the sea level pressure at Miles City reached an incredible 1064 mb (31.42 in.)—a new United States record that topped the old mark of 1063 mb set in Helena, Montana, on January 10, 1962. An enormous ridge of high pressure stretched from the Canadian arctic coast to the Gulf of Mexico. On the east side of the ridge, cold westerly winds brought lake-effect snows to the eastern shores of the Great Lakes. To the south of the high pressure center, cold easterly winds, rising along the elevated plains, brought light amounts of *upslope snow**

*Upslope snow forms as cold air moving from east to west gradually rises (and cools even more) as it approaches the Rocky Mountains.

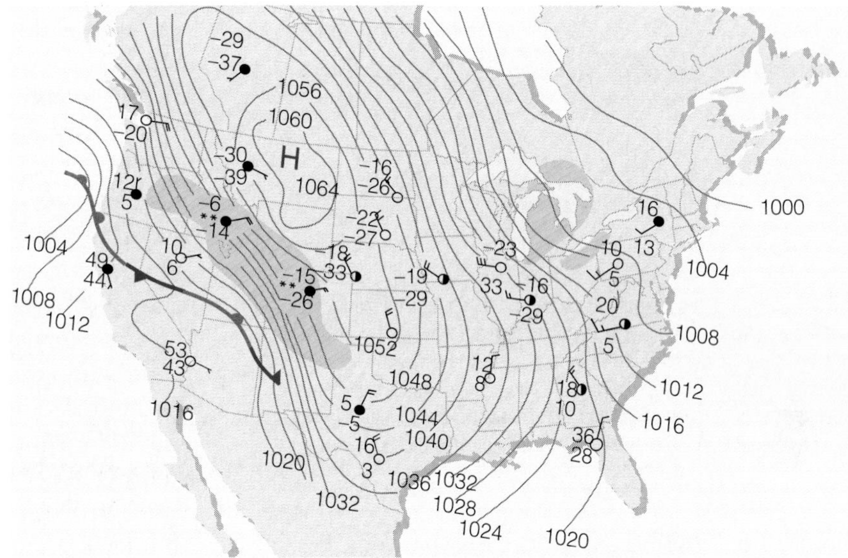

FIGURE 3 Surface weather map for 7 A.M., EST, December 24, 1983. Solid lines are isobars. Areas shaded green represent precipitation. An extremely cold arctic air mass covers nearly 90 percent of the United States. (Weather symbols for the surface map are given in Appendix B.)

to sections of the Rocky Mountain states. Notice in Fig. 3 that, on Christmas Eve, arctic air covered almost 90 percent of the United States. As the cold air swept eastward and southward, a hard freeze caused hundreds of millions of dollars in damage to the fruit and vegetable crops in Texas, Louisiana, and Florida. On Christmas Day, 125 record low temperature readings were set in twenty-four states. That afternoon, at 1:00 P.M., it was actually colder in Atlanta, Georgia, than it was in Fairbanks, Alaska. One of the worst cold waves to occur in December during the twentieth century continued through the week, as many new record lows were established in the Deep South from Texas to Louisiana.

By January 1, the extreme cold had moderated, as the upper-level winds became more westerly. These winds brought milder Pacific air eastward into the Great Plains. The warmer pattern continued until about January 10, when the Siberian Express decided to make a return visit. Driven by strong upper-level northerly winds, impulse after impulse of arctic air from Canada swept across the United States. On January 18, an all-time record low of –54°C (–65°F) was recorded for the state of Utah at Middle Sinks. On January 19, temperatures plummeted to a new low of –22°C (–7°F) for the airports in Philadelphia and Baltimore. Toward the end of the month, the upper-level winds once again became more westerly. Over much of the nation, the cold air moderated. But the Express was to return at least one more time.

The beginning of February saw relatively warm air covering much of the nation from California to the Atlantic coast. On February 4, an arctic outbreak spread southward and eastward across the nation. Although freezing air extended southward into central Florida, the Express ran out of steam, and a February heat wave soon engulfed most of the United States east of the Rocky Mountains as warm, humid air from the Gulf of Mexico spread northward.

Even though February was one of the warmest months on record over parts of the United States, the winter of 1983–1984 (December, January, and February) will go down in the record books as one of the coldest winters for the United States as a whole since reliable record keeping began in 1931.

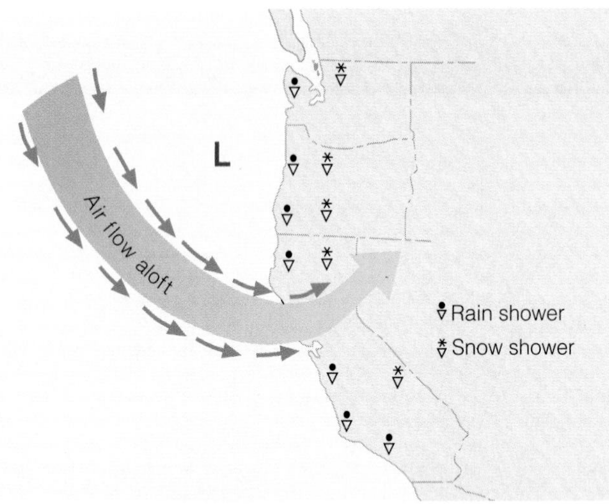

FIGURE 8.5 A winter upper-air pattern that brings mP air into the west coast of North America. The large arrow represents the upper-level flow. Note the trough of low pressure along the coast. The small arrows show the trajectory of the mP air at the surface. Regions that normally experience precipitation under these conditions are also shown on the map. Showers are most prevalent along the coastal mountains and in the Sierra Nevada.

kilometer or so above the surface may be at the freezing point. Within this colder air, characteristics of the original cold, dry air mass may still prevail. As the air moves inland, coastal mountains force it to rise, and much of its water vapor condenses into rain-producing clouds. In the colder air aloft, the rain changes to snow, with heavy amounts accumulating in mountain regions. A typical upper-level wind flow pattern that brings mP air onto the west coast of North America is shown in Fig. 8.5.

When the maritime polar air moves inland, it loses much of its moisture as it crosses a series of mountain ranges. Beyond these mountains, it travels over a cold, elevated plateau that chills the surface air and slowly transforms the lower level into dry, stable cP air. East of the Rockies this air mass is referred to as *Pacific air* (see Fig. 8.6). Here, it often brings fair weather and temperatures that are cool but not nearly as cold as the continental polar and arctic air that invades this region from northern Canada. In fact, when Pacific air from the west replaces retreating cold air from the north, Chinook winds often develop. Furthermore, when the modified maritime polar air replaces moist tropical air, storms can form along the boundary separating the two air masses.

Along the East Coast, mP air originates in the North Atlantic as cP air moves southward some distance off the Atlantic coast. Steered by northeasterly winds, mP air then swings southwestward toward the northeastern states. Because the water of the North Atlantic is very cold and the air mass travels only a short distance over water, wintertime Atlantic mP air masses are usually much colder than their Pacific counterparts. Because the prevailing winds aloft are westerly, Atlantic mP air masses are also much less common.

Figure 8.7 illustrates a typical late winter or early spring surface weather pattern that carries mP air from the Atlantic into the New England and middle Atlantic states. A slow-moving, cold anticyclone drifting to the east (north of New England) causes a northeasterly flow of mP air to the south. The boundary separating this invading colder air from warmer air even farther south is marked by a stationary front. North of this front, northeasterly winds provide generally undesirable weather, consisting of damp air and low, thick clouds from which light precipitation falls in the form of rain, drizzle, or snow. As we will see later in this chapter, when upper atmospheric conditions are right, storms may develop along the stationary front, move eastward, and intensify near the shores of Cape Hatteras. Such storms, called *Hatteras lows* (see Fig. 8.21, p. 222), sometimes swing northeastward along the coast, where they become *northeasters* (or *nor'easters*) bringing with them strong northeasterly winds, heavy rain or snow, and coastal flooding. (We will examine northeasters later in this chapter when we examine mid-latitude cyclonic storms.)

FIGURE 8.6 After crossing several mountain ranges, cool moist mP air from off the Pacific ocean descends the eastern side of the Rockies as modified, relatively dry Pacific air.

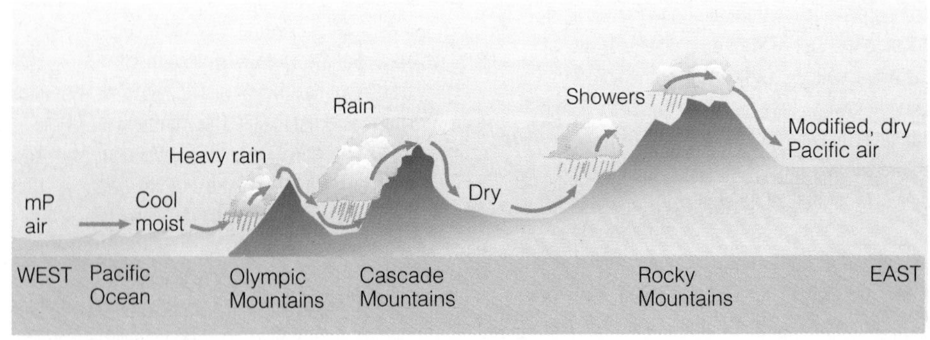

mT (Maritime Tropical) Air Masses The wintertime source region for Pacific **maritime tropical** air masses is the subtropical east Pacific Ocean (Look back at Fig. 8.2, p. 203). Air from this region must travel over many kilometers of water before it reaches the California coast. Consequently, these air masses are very warm and moist by the time they arrive along the West Coast. In winter, the warm air produces heavy precipitation usually in the form of rain, even at high elevations. Melting snow and rain quickly fill rivers, which overflow into the low-lying valleys. The rapid snowmelt leaves local ski slopes barren, and the heavy rain can cause disastrous mud slides in the steep canyons.

Figure 8.8 shows maritime tropical air (usually referred to as *subtropical air*) streaming into northern California on January 1, 1997. The humid, subtropical air, which originated near the Hawaiian Islands, was termed by at least one weathercaster as "*the pineapple express.*" After battering the Pacific Northwest with heavy rain, the pineapple express roared into northern and central California, causing catastrophic floods that sent over 100,000 people fleeing from their homes, mud slides that closed roads, property damage (including crop losses) that amounted to more than $1.5 billion, and eight fatalities. Yosemite National Park, which sustained

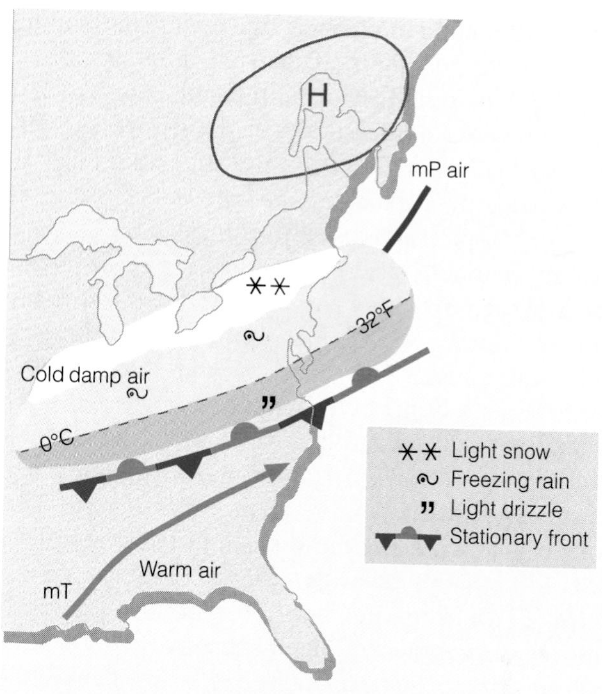

FIGURE 8.7 Winter and early spring surface weather patterns that usually prevail during the invasion of cold, moist mP air into the mid-Atlantic and New England states. (Green-shaded area represents light rain and drizzle; pink-shaded region represents freezing rain and sleet; white-shaded area is experiencing snow.)

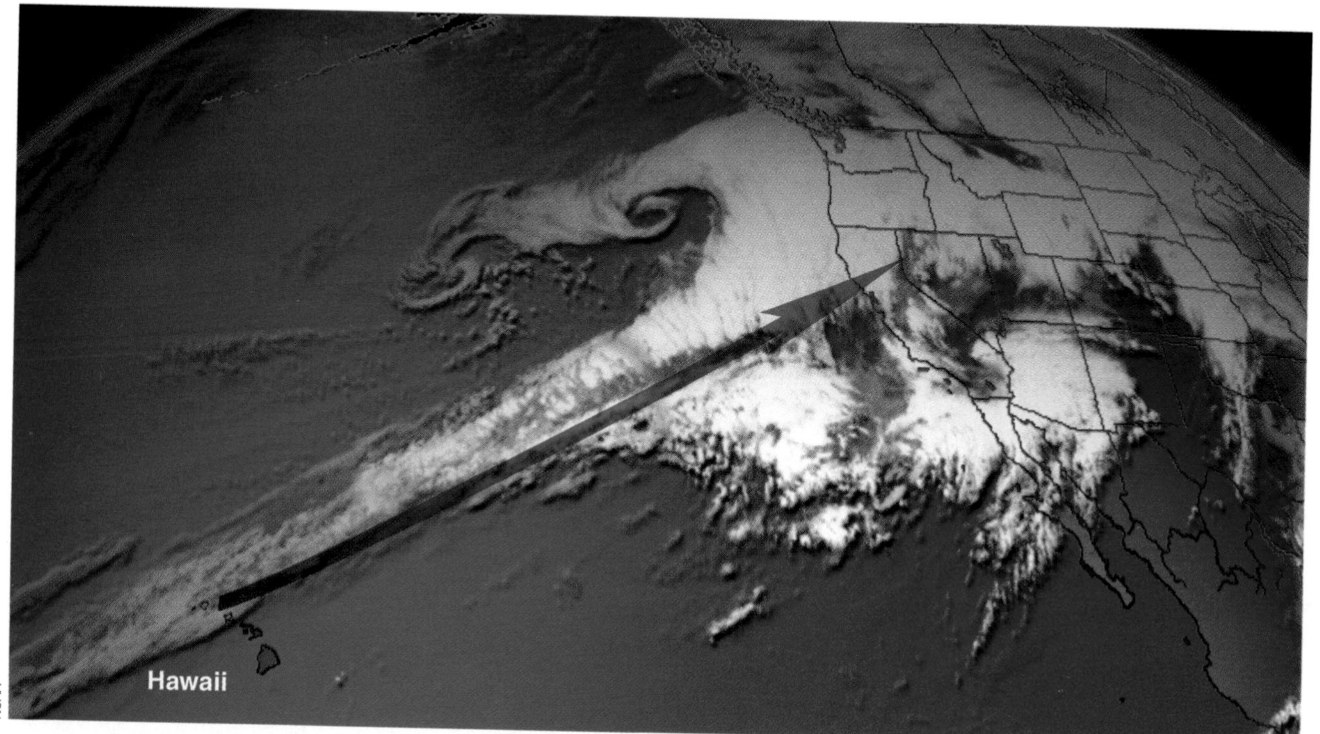

FIGURE 8.8 An infrared satellite image that shows maritime tropical air (heavy red arrow) moving into northern California on January 1, 1997. The warm, humid airflow (sometimes called "the pineapple connection") produced heavy rain and extensive flooding in northern and central California.

over $170 million in damages due mainly to flooding, was forced to close for more than two months.

The humid subtropical air that influences much of the weather east of the Rockies originates over the Gulf of Mexico and Caribbean Sea. In winter, cold polar air tends to dominate the continental weather scene, so maritime tropical air is usually confined to the Gulf and extreme southern states. Occasionally, a slow-moving storm system over the Central Plains draws warm, humid air northward. Gentle, south or southwesterly winds carry this air into the central and eastern parts of the nation in advance of the system. Since the land is still extremely cold, air near the surface is chilled to its dew point. Fog and low clouds form in the early morning, dissipate by midday, and re-form in the evening. This mild winter weather in the Mississippi and Ohio valleys lasts, at best, only a few days. Soon cold polar air will move down from the north behind the eastward-moving storm system. Along the boundary between the two air masses, the warm, humid air is lifted above the more dense cold, polar air, which often leads to heavy and widespread precipitation and storminess.

When a large, mid-latitude storm system stalls over the Central Plains, a constant supply of mT air from the Gulf of Mexico can bring record-breaking maximum temperatures to the eastern half of the country. Sometimes the air temperatures are higher in the mid-Atlantic states than they are in the Deep South, as compressional heating warms the air even more as it moves downslope after crossing the Appalachian Mountains.

Figure 8.9 shows a surface weather map and the associated upper airflow (heavy arrow) that brought unseasonably warm mT air into the central and eastern states during April, 1976. A large high centered off the southeast coast coupled with a strong southwesterly flow aloft carried warm, moist air into the Midwest and East, causing a record-breaking April heat wave. The flow aloft prevented the surface low and the cold, polar (cP) air behind it from making much eastward progress, so that the warm spell lasted for five days. Note that, on the west side of the surface low, the winds aloft funneled cold air from the north into the western states, creating unseasonably cold weather from California to the Rockies. Hence, while people in the Southwest were huddled around heaters, others several thousand kilometers away in the Northeast were turning on air conditioners. We can see that it is the upper-level flow, directing cP air southward and mT air northward, that makes these contrasts in temperature possible.

As maritime air moves inland over the hot continent, it warms, rises, and frequently causes cumuliform clouds, which produce afternoon showers and thunderstorms. You can almost count on thunderstorms developing along the Gulf Coast each afternoon in summer. As evening approaches, thunderstorm activity typically dies off. Nighttime cooling lowers the air temperature and, if the air becomes saturated, fog or low clouds form. These, of course, dissipate by late morning as surface heating warms the air again.

A weak, but often persistent, flow around an upper-level anticyclone in summer will spread humid, maritime

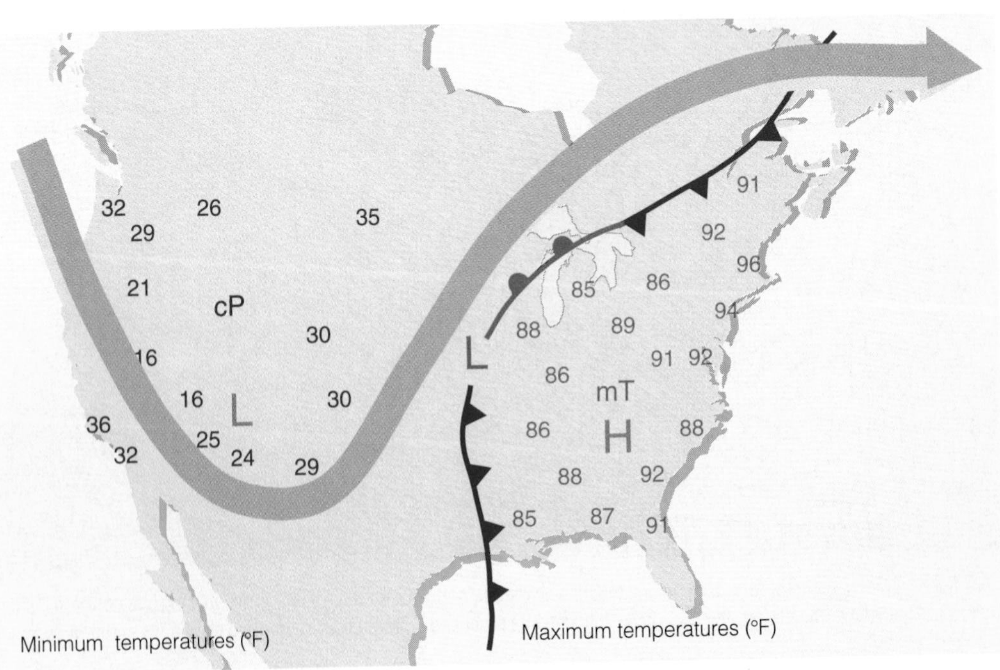

FIGURE 8.9 Weather conditions during an unseasonably hot spell in the eastern portion of the United States that occurred between the 15th and 20th of April, 1976. The surface low-pressure area and fronts are shown for April 17. Numbers to the east of the surface low (in red) are maximum temperatures recorded during the hot spell, while those to the west of the low (in blue) are minimums reached during the same time period. The heavy arrow is the average upper-level flow during the period. The faint L and H show average positions of the upper-level trough and ridge.

tropical air from the Gulf of Mexico or from the Gulf of California into the southern and central Rockies, where it causes afternoon thunderstorms. Occasionally, this easterly flow may work its way even farther west, producing shower activity in the otherwise dry southwestern desert.

During the summer, humid subtropical air originating over the southeastern Pacific and Gulf of California normally remains south of California. Occasionally, a weak upper-level southerly flow will spread this humid air northward into the southwestern United States, most often Arizona, Nevada, and the southern part of California. In many places, the moist, conditionally unstable air aloft only shows up as middle and high cloudiness. However, where the moist flow meets a mountain barrier, it usually rises and condenses into towering shower-producing clouds. (For an exceptionally strong flow of subtropical air into this region, look at Fig. 7.7 on p. 175.)

cT (Continental Tropical) Air Masses The only real source region for hot, dry **continental tropical** air masses in North America is found during the summer in northern Mexico and the adjacent arid southwestern United States. Here, the air mass is hot, dry, and conditionally unstable at low levels, with frequent dust devils forming during the day. Because of the low relative humidity (typically less than 10 percent during the afternoon), air must rise to great heights before condensation begins. Furthermore, an upper-level ridge usually produces weak subsidence over the region, tending to make the air aloft rather stable and the surface air even warmer. Consequently, skies are generally clear, the weather is hot, and rainfall is practically nonexistent where continental tropical air masses prevail. If this air mass moves outside its source region and into the Great Plains and stagnates over that region for any length of time, a severe drought may result. Figure 8.10 shows a weather map situation where continental tropical air covers a large portion of the western United States and produces hot, dry weather northward to Canada.

So far, we have examined the various air masses that enter North America annually. The characteristics of each depend upon the air mass source region and the type of surface over which the air mass moves. The winds aloft determine the trajectories of these air masses. Occasionally, an air mass will control the weather in a region for some time. These persistent weather conditions are sometimes referred to as **air mass weather.**

Air mass weather is especially common in the southeastern United States during summer as, day after

> **DID YOU KNOW?**
>
> A continental tropical air mass, stretching from southern California to the heart of Texas, brought record warmth to the desert southwest during the last week of June, 1990. The temperature, which on June 26 soared to a sweltering peak of 50°C (122°F) in Phoenix, Arizona, caused officials to suspend aircraft takeoffs at Sky Harbor Airport. The extreme heat had lowered air density to the point where it reduced aircraft lift.

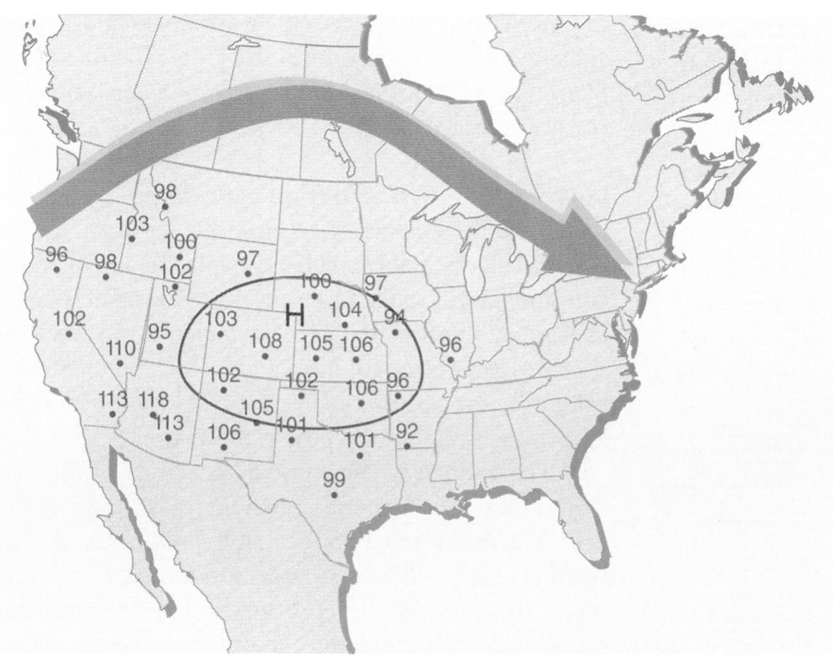

FIGURE 8.10 During June 29 and 30, 1990, continental tropical air covered a large area of the central and western United States. Numbers on the map represent maximum temperatures (°F) during this period. The large H with the isobar shows the upper-level position of the subtropical high. Sinking air associated with the high contributed to the hot weather. Winds aloft were weak, with the main flow shown by the heavy arrow.

day, humid subtropical air from the Gulf brings sultry conditions and afternoon thunderstorms. It is also common in the Pacific Northwest in winter when conditionally unstable, cool maritime air accompanied by widely scattered showers dominates the weather for several days or more. The real weather action, however, usually occurs not within air masses but at their margins, where air masses with sharply contrasting properties meet—in the zone marked by weather fronts.*

BRIEF REVIEW

Before we examine fronts, here is a review of some of the important facts about air masses:

- An air mass is a large body of air whose properties of temperature and humidity are fairly similar in any horizontal direction.
- Source regions for air masses tend to be generally flat, of uniform composition, and in an area of light winds dominated by surface high pressure.
- Continental air masses form over land. Maritime air masses form over water. Polar air masses originate in cold, polar regions, and extremely cold air masses form over arctic regions. Tropical air masses originate in warm, tropical regions.
- Continental polar (cP) air masses are cold and dry; continental arctic (cA) air masses are extremely cold and dry; continental tropical (cT) air masses are hot and dry; maritime tropical (mT) air masses are warm and moist; maritime polar (mP) air masses are cold and moist.

FRONTS

Although we briefly looked at fronts in Chapter 1, we are now in a position to study them in depth, which will aid us in forecasting the weather. We will now learn about the general nature of fronts—how they move and what weather patterns are associated with them.

A **front** is the transition zone between two air masses of different densities. Since density differences are most often caused by temperature differences, fronts usually separate air masses with contrasting temperatures. Often, they separate air masses with different humidities as well. Remember that air masses have both horizontal and vertical extent; consequently, the upward extension of a front is referred to as a *frontal surface*, or a *frontal zone.*

Figure 8.11 shows a weather map illustrating four different fronts. Notice that the fronts are associated with lower pressure and that the fronts separate differing air masses. As we move from west to east across the map, the fronts appear in the following order: a stationary front between points A and B; a cold front between points B and C; a warm front between points C and D; and an occluded front between points C and L. Let's examine the properties of each of these fronts.

*The word *front* is used to denote the clashing or meeting of two air masses, probably because it resembles the fighting in Western Europe during World War I, when the term originated.

Stationary Fronts

A **stationary front** has essentially no movement. On a colored weather map, it is drawn as an alternating red and blue line. Semicircles face toward colder air on the red line and triangles point toward warmer air on the blue line. The stationary front between points A and B in Fig. 8.11 marks the boundary where cold, dense cP air from Canada butts up against the north-south trending Rocky Mountains. Unable to cross the barrier, the cold air shows little or no westward movement. The stationary front is drawn along a line separating the cP from the milder more humid mP air to the west. Notice that the surface winds tend to blow parallel to the front, but in opposite directions on either side of it. Moreover, upper-level winds often blow parallel to a stationary front.

The weather along the front is clear to partly cloudy, with much colder air lying on its eastern side. Because both air masses are relatively dry, there is no precipitation. This is not, however, always the case. When warm, moist air rides up and over the cold air, widespread cloudiness with light precipitation can cover a vast area. These are the conditions that prevail north of the east-west running stationary front depicted in Fig. 8.7, p. 209.

If the warmer air to the west begins to move and replace the colder air to the east, the front in Fig. 8.11 will no longer remain stationary; it will become a warm front. If, on the other hand, the colder air slides up over the mountain and replaces the warmer air on the other side, the front will become a cold front. If either a cold front or a warm front should stop moving, it would become a stationary front.

Cold Fronts

The **cold front** between points B and C on the surface weather map (Fig. 8.11) represents a zone where cold, dry, stable polar air is replacing warm, moist, conditionally unstable subtropical air. The front is drawn as a solid blue line with the triangles along the front showing its direction of movement. How did the meteorologist know to draw the front at that location? A closer look at the situation will give us the answer.

The weather in the immediate vicinity of this cold front in the southeastern United States is shown in Fig. 8.12. The data plotted on the map represent the current weather at selected cities. The station model used to represent the data at each reporting station is a simplified one that shows temperature, dew point, present weather,

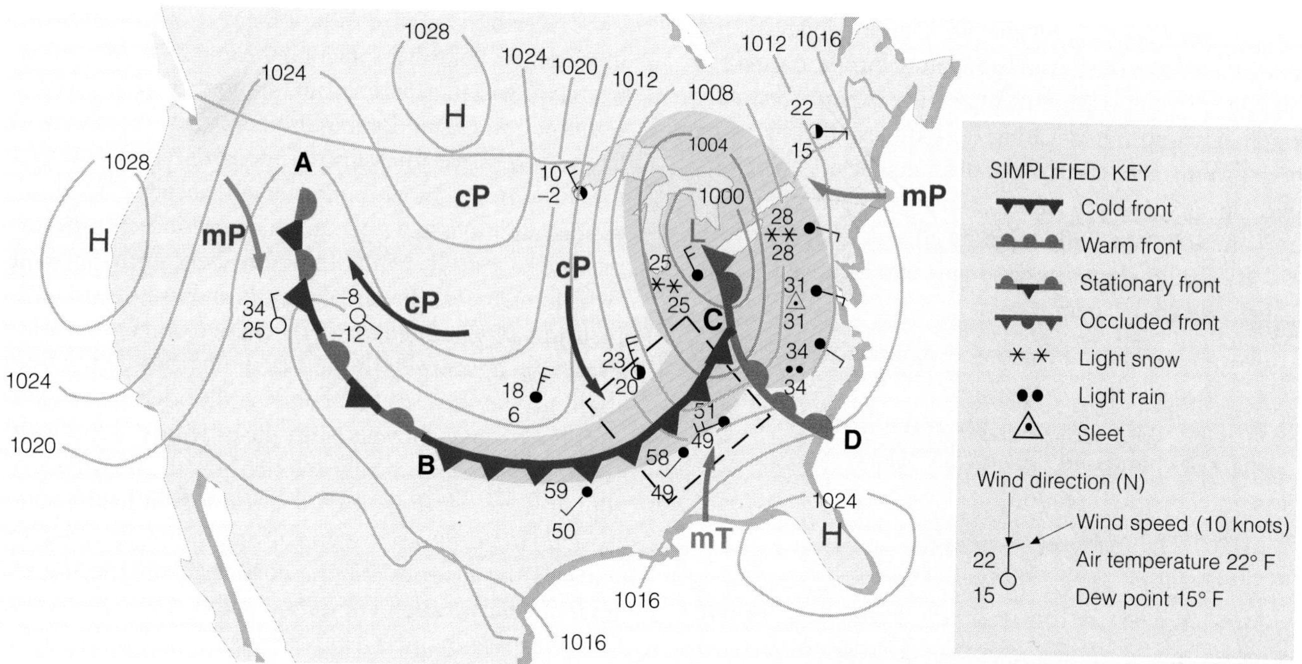

FIGURE 8.11 A weather map showing surface-pressure systems, air masses, fronts, and isobars (in millibars) as solid gray lines. Large arrows in color show air flow. (Green-shaded area represents precipitation.)

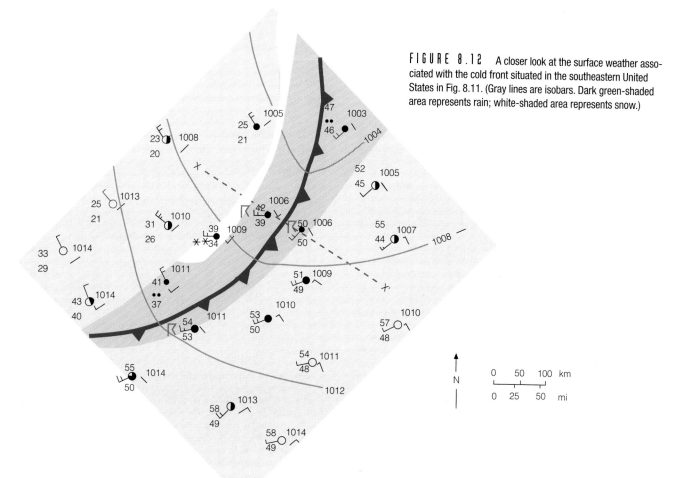

FIGURE 8.12 A closer look at the surface weather associated with the cold front situated in the southeastern United States in Fig. 8.11. (Gray lines are isobars. Dark green-shaded area represents rain; white-shaded area represents snow.)

cloud cover, sea level pressure, wind direction and speed. The little line in the lower right-hand corner of each station shows the pressure change—the *pressure tendency*, whether rising (/) or falling (\)—during the last three hours. With all of this information, the front can be properly located.* (Appendix C explains the weather symbols and the station model more completely.)

The following criteria are used to locate a front on a surface weather map:

1. sharp temperature changes over a relatively short distance
2. changes in the air's moisture content (as shown by marked changes in the dew point)
3. shifts in wind direction
4. pressure and pressure changes
5. clouds and precipitation patterns

In Fig. 8.12, we can see a large contrast in air temperature and dew point on either side of the front. There is also a wind shift from southwesterly ahead of the front, to northwesterly behind it. Notice that each isobar kinks as it crosses the front, forming an elongated area of low pressure—a *trough*—which accounts for the wind shift. Since surface winds normally blow across the isobars toward lower pressure, we find winds with a southerly component ahead of the front and winds with a northerly component behind it.

Since the cold front is a trough of low pressure, sharp changes in pressure can be significant in locating the front's position. One important fact to remember is that the lowest pressure usually occurs just as the front passes a station. Notice that, as you move toward the front, the pressure drops, and, as you move away from it, the pressure rises.

*Locating any front on a weather map is not always a clear-cut process. Even meteorologists can disagree on an exact position.

The cloud and precipitation patterns are better seen in a side view of the front along the line *X–X'*. We can see from Fig. 8.13 that, at the front, the cold, dense air wedges under the warm air, forcing the warm air upward, much like a snow shovel forces snow upward as it glides through the snow. As the moist, conditionally unstable air rises, it condenses into a series of cumuliform clouds. Strong, upper-level westerly winds blow the delicate ice crystals (which form near the top of the cumulonimbus) into cirrostratus (Cs) and cirrus (Ci). These clouds usually appear far in advance of the approaching front. At the front itself, a relatively narrow band of thunderstorms (Cb) produces heavy showers with gusty winds. Behind the front, the air cools quickly. (Notice how the freezing level dips as it crosses the front.) The winds shift from southwesterly to northwesterly, pressure rises, and precipitation ends. As the air dries out, the skies clear, except for a few lingering fair weather cumulus clouds.

Observe that the leading edge of the front is steep. The steepness is due to friction, which slows the airflow near the ground. The air aloft pushes forward, blunting the frontal surface. If we could walk from where the front touches the surface back into the cold air, a distance of 50 km, the front would be about 1 km above us. Thus, the slope of the front—the ratio of vertical rise to horizontal distance—is 1:50. This is typical for a fast-moving cold front—those that move about 25 knots. In a slower-moving cold front, the slope is much more gentle.

With slow-moving cold fronts, clouds and precipitation usually cover a broad area behind the front. When the ascending warm air is stable, stratiform clouds, such as nimbostratus, become the predominate cloud type and fog may even develop in the rainy area. Occasionally, along a fast-moving front, a line of active showers and thunderstorms, called a *squall line*, develops parallel to and often ahead of the advancing front.

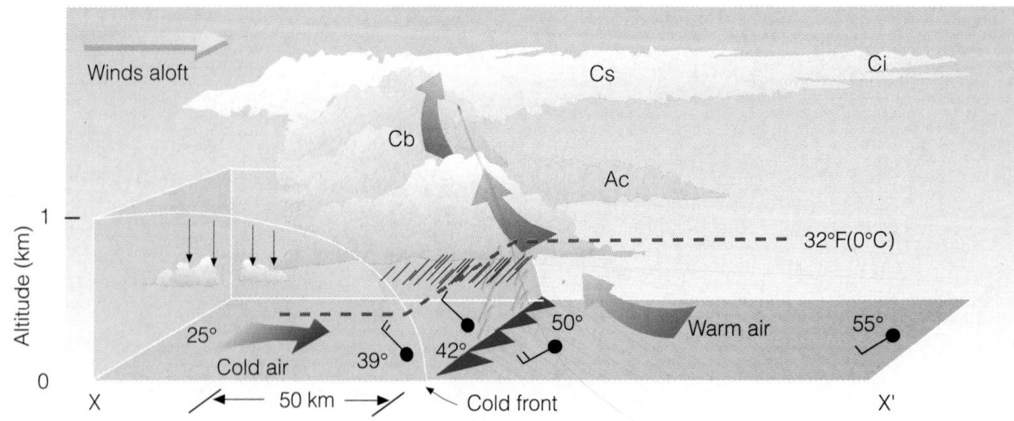

Meteorology Now™

ACTIVE FIGURE 8.13
A vertical view of the weather across the cold front in Fig. 8.12 along the line *X–X'*.
Watch this Active Figure at http://earthscience.brookscole.com/ahrens/ess4e.

DID YOU KNOW?

Taking a walk to the store in Portsmouth, New Hampshire, on the evening of January 19, 1810, could have been life-threatening. In the early evening the temperature measured a relatively mild 5°C (41°F). But, within a few hours, after the passage of a strong cold front, the temperature plummeted to −25°C (−13°F).

So far, we have considered the general weather patterns of "typical" cold fronts. There are, of course, many exceptions. In fact, no two fronts are exactly alike. In some, the cold air is shallow; in others, it is much deeper. If the rising warm air is dry and stable, scattered clouds are all that form, and there is no precipitation. In extremely dry weather, a marked change in the dew point, accompanied by a slight wind shift, may be the only clue to a passing front.

During the winter, a series of cold polar outbreaks may travel across the United States so quickly that warm air is unable to develop ahead of the front. In this case, frigid arctic air usually replaces cold polar air, and a drop in temperature is the only indication that a cold front has moved through your area. Along the West Coast, the Pacific Ocean modifies the air so much that cold fronts, such as those described in the previous section, are never seen. In fact, as a cold front moves inland from the Pacific Ocean, the surface temperature contrast across the front may be quite small. Topographic features usually distort the wind pattern so much that locating the position of the front and the time of its passage are exceedingly difficult. In this case, the pressure tendency is the most reliable indication of a frontal passage.

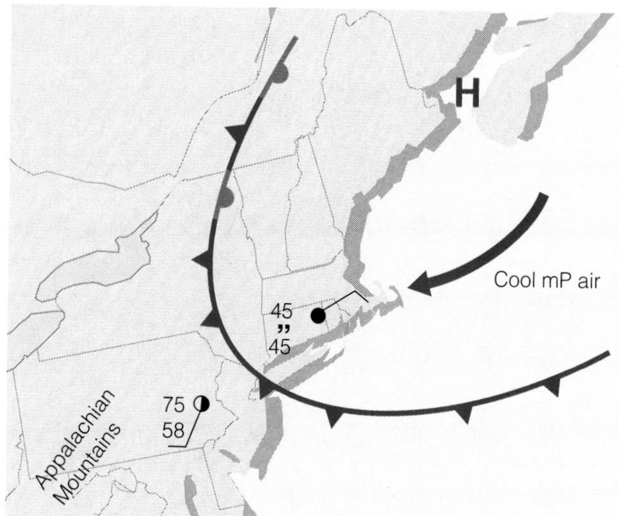

FIGURE 8.14 A "back door" cold front moving into New England during the spring. Notice that, behind the front, the weather is cold and damp with drizzle, while to the south, ahead of the front, the weather is partly cloudy and warm.

Most cold fronts move toward the south, southeast, or east. But sometimes they will move southwestward out of Canada. Cold fronts that move in from the east, or northeast, are called **back door cold fronts.** Typically, as the front passes, westerly surface winds shift to easterly or northeasterly, and temperatures drop (see Fig. 8.14).

Even though cold-front weather patterns have many exceptions, learning these patterns can be to your advantage if you live in an area that experiences well-defined cold fronts. Knowing them improves your own ability to make short-range weather forecasts. For your reference, Table 8.2 summarizes idealized cold-front weather in the Northern Hemisphere.

TABLE 8.2 Typical Weather Conditions Associated with a Cold Front in the Northern Hemisphere

WEATHER ELEMENT	BEFORE PASSING	WHILE PASSING	AFTER PASSING
Winds	South or southwest	Gusty, shifting	West or northwest
Temperature	Warm	Sudden drop	Steadily dropping
Pressure	Falling steadily	Minimum, then sharp rise	Rising steadily
Clouds	Increasing Ci, Cs, then either Tcu* or Cb	Tcu or Cb*	Often Cu, Sc* when ground is warm
Precipitation	Short period of showers	Heavy showers of rain or snow, sometimes with hail, thunder, and lightning	Decreasing intensity of showers, then clearing
Visibility	Fair to poor in haze	Poor, followed by improving	Good except in showers
Dew point	High; remains steady	Sharp drop	Lowering

*Tcu stands for towering cumulus, such as cumulus congestus; whereas Cb stands for cumulonimbus. Sc stands for stratocumulus.

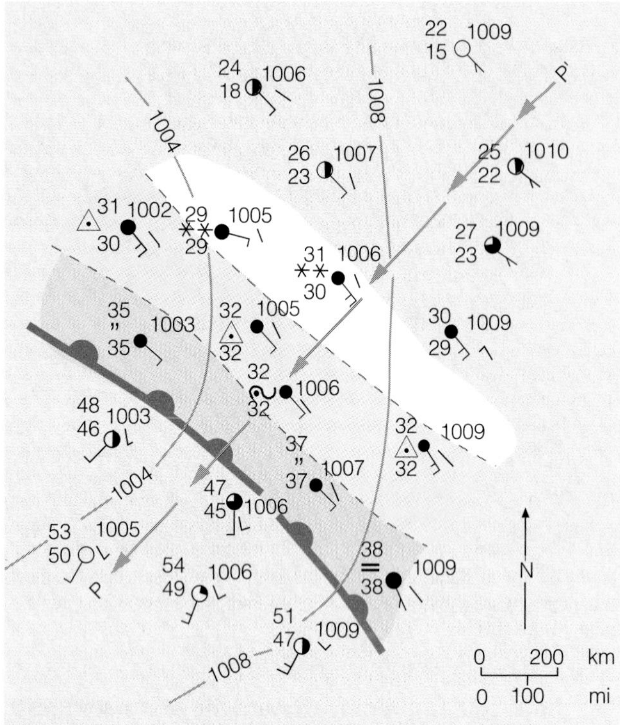

Meteorology Now™ ACTIVE FIGURE 8.15
Surface weather associated with a typical warm front. (Green-shaded area represents rain; pink-shaded area represents freezing rain and sleet; white-shaded area represents snow.)
Watch this Active Figure at http://earthscience.brookscole.com/ahrens/ess4e.

Warm Fronts In Fig. 8.11, p. 213, a **warm front** is drawn along the solid red line running from points C to D. Here, the leading edge of advancing warm, moist subtropical (mT) air from the Gulf of Mexico replaces the retreating cold maritime polar air from the North Atlantic. The direction of frontal movement is given by the half circles, which point into the cold air; this front is heading toward the northeast. As the cold air recedes, the warm front slowly advances. The average speed of a warm front is about 10 knots, or about half that of an average cold front. During the day, as mixing occurs on both sides of the front, its movement may be much faster. Warm fronts often move in a series of rapid jumps, which show up on successive weather maps. At night, however, radiational cooling creates cool, dense surface air behind the front. This inhibits both lifting and the front's forward progress. When the forward surface edge of the warm front passes a station, the wind shifts, the temperature rises, and the overall weather conditions improve. To see why, we will examine the weather commonly associated with the warm front both at the surface and aloft.

Look at Figs. 8.15 and 8.16 closely and observe that the warmer, less-dense air rides up and over the colder, more-dense surface air. This rising of warm air over cold, called **overrunning,** produces clouds and precipitation well in advance of the front's surface boundary. The warm front that separates the two air masses has an average slope of about 1:300—a much more gentle or inclined shape than that of a typical cold front.*

Suppose we are standing at the position marked P′ in Figs. 8.15 and 8.16. Note that we are over 1200 km (750 mi) ahead of where the warm front is touching the surface. Here, the surface winds are light and variable. The air is cold and about the only indication of an approaching warm front is the high cirrus clouds overhead. We know the front is moving slowly toward us and that within a day or so it will pass our area. Suppose that, instead of waiting for the front to pass us, we drive toward it, observing the weather as we go.

*This slope of 1:300 is a more gentle slope than that of most warm fronts. Typically, the slope of a warm front is on the order of 1:150 to 1:200.

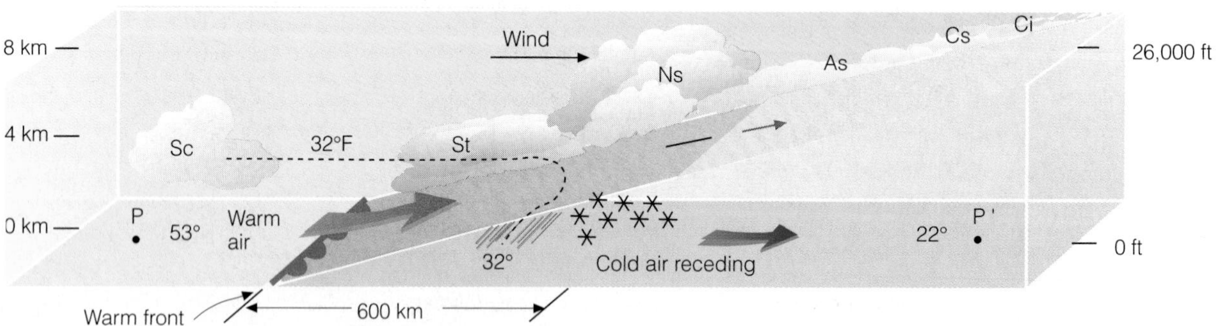

Meteorology Now™ ACTIVE FIGURE 8.16
Vertical view of clouds, precipitation, and winds across the warm front in Fig. 8.15 along the line P–P′.
Watch this Active Figure at http://earthscience.brookscole.com/ahrens/ess4e.

Heading toward the front, we notice that the cirrus (Ci) clouds gradually thicken into a thin, white veil of cirrostratus (Cs) whose ice crystals cast a halo around the sun.* Almost imperceptibly, the clouds thicken and lower, becoming altocumulus (Ac) and altostratus (As) through which the sun shows only as a faint spot against an overcast gray sky. Snowflakes begin to fall, and we are still over 600 km (370 mi) from the surface front. The snow increases, and the clouds thicken into a sheetlike covering of nimbostratus (Ns). The winds become brisk and out of the southeast, while the atmospheric pressure slowly falls. Within 400 km (250 mi) of the front, the cold surface air mass is now quite shallow. The surface air temperature moderates and, as we approach the front, the light snow changes first into sleet. It then becomes freezing rain and finally rain and drizzle as the air temperature climbs above freezing. Overall, the precipitation remains light or moderate but covers a broad area. Moving still closer to the front, warm, moist air mixes with cold, moist air producing ragged windblown stratus (St) and fog. (Thus, flying in the vicinity of a warm front can be quite hazardous.)

Finally, after a trip of over 1200 km, we reach the warm front's surface boundary. As we cross the front, the weather changes are noticeable, but much less pronounced than those experienced with the cold front; they show up more as a gradual transition rather than a sharp change. On the warm side of the front, the air temperature and dew point rise, the wind shifts from southeast to south or southwest, and the atmospheric pressure stops falling. The light rain ends and, except for a few stratocumulus, the fog and low clouds vanish.

This scenario of an approaching warm front represents average, if not idealized, warm-front weather. In some instances, the weather can differ from this dramatically. For example, if the overrunning warm air is relatively dry and stable, only high and middle clouds will form, and no precipitation will occur. On the other hand, if the warm air is relatively moist and conditionally unstable (as is often the case during the summer), heavy showers can develop as thunderstorms become embedded in the cloud mass. In the southern Great Plains, warm, humid air may be separated from warm, dry air along a boundary called a *dryline.* We will look more closely at drylines and their effect on developing thunderstorms in Chapter 10.

Along the west coast, the Pacific Ocean significantly modifies the surface air so that warm fronts are difficult to locate on a surface weather map. Also, not all warm fronts move northward or northeastward. On rare occasions, a front will move into the eastern seaboard from the Atlantic Ocean as the front spins all the way around a deep storm positioned off the coast. Cold northeasterly winds ahead of the front usually become warm northeasterly winds behind it. Even with these exceptions, knowing the normal sequence of warm-front weather will be useful, especially if you live where warm fronts become well developed. You can look for certain cloud and weather patterns and make reasonably accurate short-range forecasts of your own. Table 8.3 summarizes typical warm-front weather.

*If the warm air is relatively unstable, ripples or waves of cirrocumulus clouds will appear as a "mackerel sky."

TABLE 8.3 Typical Weather Conditions Associated with a Warm Front in the Northern Hemisphere

WEATHER ELEMENT	BEFORE PASSING	WHILE PASSING	AFTER PASSING
Winds	South or southeast	Variable	South or southwest
Temperature	Cool to cold, slow warming	Steady rise	Warmer, then steady
Pressure	Usually falling	Leveling off	Slight rise, followed by fall
Clouds	In this order: Ci, Cs, As, Ns, St, and fog; occasionally Cb in summer	Stratus-type	Clearing with scattered Sc, especially in summer; occasionally Cb in summer
Precipitation	Light-to-moderate rain, snow, sleet, or drizzle; showers in summer	Drizzle or none	Usually none; sometimes light rain or showers
Visibility	Poor	Poor, but improving	Fair in haze
Dew point	Steady rise	Steady	Rise, then steady

Occluded Fronts If a cold front catches up to and overtakes a warm front, the frontal boundary created between the two air masses is called an **occluded front,** or, simply, an **occlusion** (meaning "closed off"). On the surface weather map, it is represented as a purple line with alternating cold-front triangles and warm-front half circles; both symbols point in the direction toward which the front is moving. In Fig. 8.11 (p. 213), notice that the air behind the occluded front is colder than the air ahead of it. This is known as a *cold-type occluded front,* or *cold occlusion.* Let's see how this front develops.

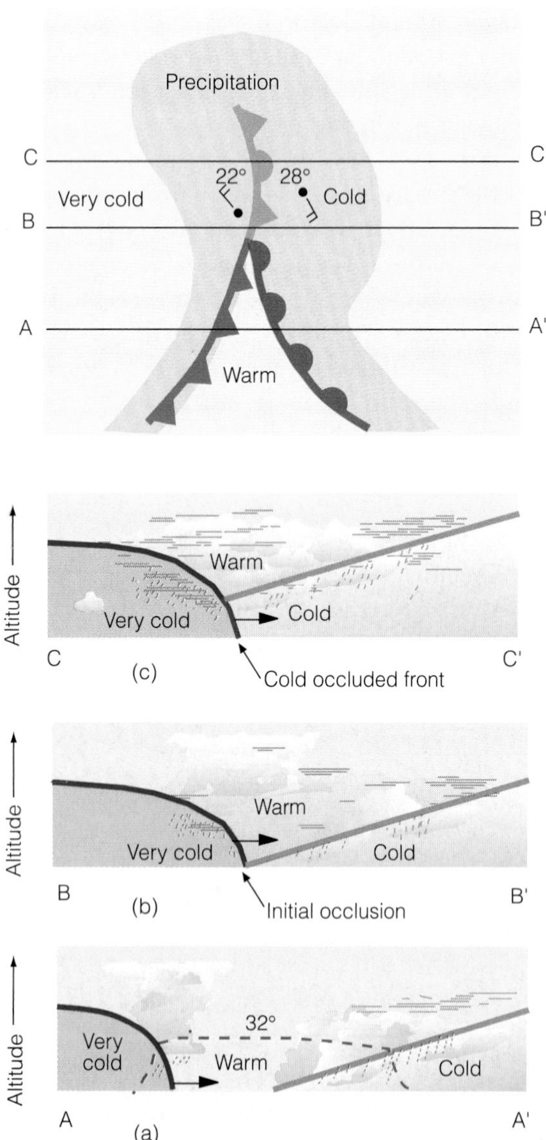

FIGURE 8.17 The formation of a cold occluded front. The faster-moving cold front in (a) catches up to the slower-moving warm front in (b) and forces it to rise off the ground (c). (Green-shaded area represents precipitation.)

The development of a cold occlusion is shown in Fig. 8.17. Along line *A–A′*, the cold front is rapidly approaching the slower-moving warm front. Along line *B–B′*, the cold front overtakes the warm front, and, as we can see in the vertical view across *C–C′*, it underrides and lifts both the warm front and the warm air mass off the ground. As a cold-occluded front approaches, the weather sequence is similar to that of a warm front with high clouds lowering and thickening into middle and low clouds, with precipitation forming well in advance of the surface front. Since the front represents a trough of low pressure, southeasterly winds and falling atmospheric pressure occurs ahead of it. The frontal passage, however, brings weather similar to that of a cold front: heavy, often showery precipitation with winds shifting to west or northwest. After a period of wet weather, the sky begins to clear, atmospheric pressure rises, and the air turns colder. The most violent weather usually occurs where the cold front is just overtaking the warm front, at the point of occlusion, where the greatest contrast in temperature occurs. Cold occlusions are the most prevalent type of front that moves into the Pacific coastal states and into the interior of North America. Occluded fronts frequently form over the North Pacific and North Atlantic, as well as in the vicinity of the Great Lakes.

Continental polar air over eastern Washington and Oregon may be much colder than milder maritime polar air moving inland from the Pacific Ocean. Figure 8.18 illustrates this situation. Observe that the air ahead of the warm front is colder than the air behind the cold front. Consequently, when the cold front catches up to and overtakes the warm front, the milder, lighter air behind the cold front is unable to lift the colder, heavier air off the ground. As a result, the cold front rides "piggyback" along the sloping warm front. This produces a *warm-type occluded front,* or a *warm occlusion.* The surface weather associated with a warm occlusion is similar to that of a warm front.*

Contrast Fig. 8.17 and Fig. 8.18. Note that the primary difference between the warm- and cold-type occluded front is the location of the upper-level front. In a warm occlusion, the upper-level cold front *precedes* the surface occluded front, whereas in a cold occlusion the upper warm front *follows* the surface occluded front.

In the world of weather fronts, occluded fronts are the mavericks. In our discussion, we treated occluded

*The relatively mild winter air that moves into Europe from the north Atlantic causes many of the occlusions that move into this region in winter to be of the warm variety.

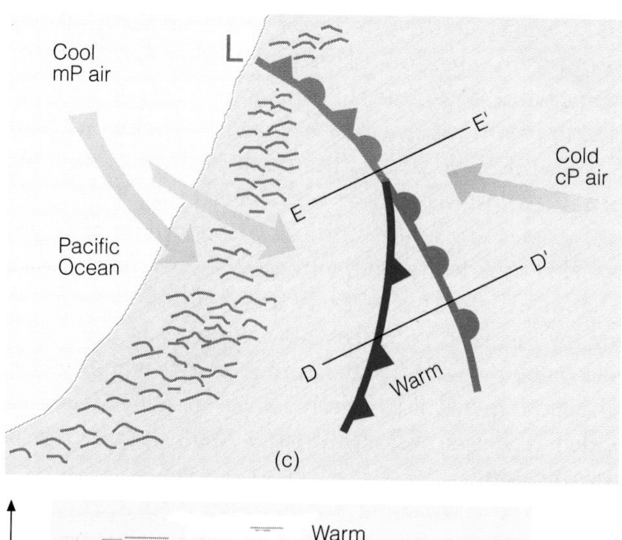

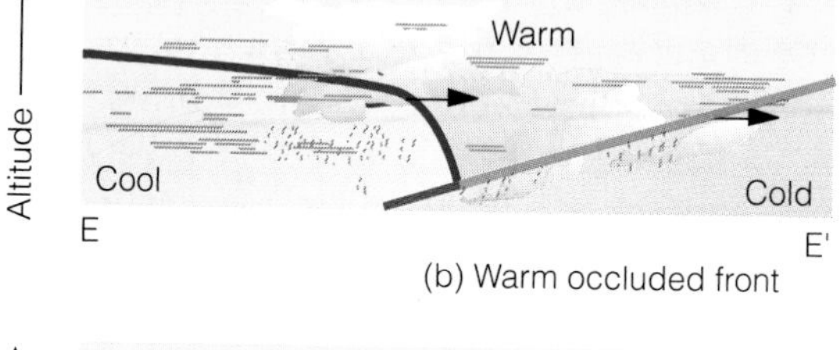

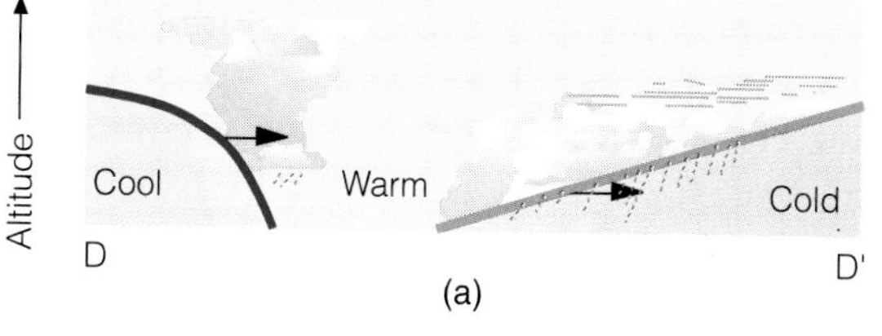

FIGURE 8.18 The formation of a warm-type occluded front. The faster-moving cold front in (a) overtakes the slower-moving warm front in (b). The lighter air behind the cold front rises up and over the denser air ahead of the warm front. Diagram (c) shows a surface map of the situation.

fronts as forming when a cold front overtakes a warm front. Some may form in this manner, but others apparently form as new fronts, which develop when a surface mid-latitude cyclone intensifies in a region of cold air after its trailing cold and warm fronts have broken away and moved eastward. The new occluded front shows up on a surface chart as a trough of low pressure separating two cold air masses. Because of this, locating and defining occluded fronts at the surface is often difficult for the meteorologist. Similarly, you too may find it hard to recognize an occlusion. In spite of this, we will assume that the weather associated with occluded fronts in North America behaves in a similar way to that shown in Table 8.4.

The frontal systems described so far are actually part of a much larger storm system—the middle-latitude cyclone. The following section details these storms, explaining where, why, and how they form.

MIDDLE-LATITUDE CYCLONES

Early weather forecasters were aware that precipitation generally accompanied falling barometers and areas of low pressure. However, it was not until the early part of the twentieth century that scientists began to piece together the information that yielded the ideas of modern meteorology and cyclonic storm development.

Working largely from surface observations, a group of scientists in Bergen, Norway, developed a model explaining the life cycle of an *extratropical,* or *middle-latitude cyclonic storm;* that is, a storm that forms at middle and high latitudes outside of the tropics. This

TABLE 8.4 Typical Weather Most Often Associated with Occluded Fronts in North America

WEATHER ELEMENT	BEFORE PASSING	WHILE PASSING	AFTER PASSING
Winds	East, southeast, or south	Variable	West or northwest
Temperature			
Cold type	Cold or cool	Dropping	Colder
Warm type	Cold	Rising	Milder
Pressure	Usually falling	Low point	Usually rising
Clouds	In this order: Ci, Cs, As, Ns	Ns, sometimes Tcu and Cb	Ns, As, or scattered Cu
Precipitation	Light, moderate, or heavy precipitation	Light, moderate, or heavy continuous precipitation or showers	Light-to-moderate precipitation followed by general clearing
Visibility	Poor in precipitation	Poor in precipitation	Improving
Dew point	Steady	Usually slight drop, especially if cold-occluded	Slight drop, although may rise a bit if warm-occluded

extraordinary group of meteorologists included Vilhelm Bjerknes, his son Jakob, Halvor Solberg, and Tor Bergeron. They published their theory shortly after World War I. It was widely acclaimed and became known as the "polar front theory of a developing wave cyclone" or, simply, the **polar front theory.** What these meteorologists gave to the world was a working model of how a mid-latitude cyclone progresses through the stages of birth, growth, and decay. An important part of the model involved the development of weather along the polar front. As new information became available, the original work was modified, so that, today, it serves as a convenient way to describe the structure and weather associated with a migratory middle-latitude storm system.

Polar Front Theory The development of a wave cyclone, according to the Norwegian model, begins along the polar front. You will remember (from our discussion of the general circulation in Chapter 7) that the polar front is a semicontinuous global boundary separating cold polar air from warm subtropical air. The stages of a developing wave cyclone are illustrated in the sequence of surface weather maps shown in Fig. 8.19.

Figure 8.19a shows a segment of the polar front as a stationary front. It represents a trough of lower pressure with higher pressure on both sides. Cold air to the north and warm air to the south flow parallel to the front, but in opposite directions. This type of flow sets up a cyclonic wind shear. You can conceptualize the shear more clearly if you place a pen between the palms of your hands and move your left hand toward your body; the pen turns counterclockwise, cyclonically.

Under the right conditions (described on p. 225), a wavelike kink forms on the front, as shown in Fig. 8.19b. The wave that forms is known as a **frontal wave.** Watching the formation of a frontal wave on a weather map is like watching a water wave from its side as it approaches a beach: It first builds, then breaks, and finally dissipates, which is why a cyclonic storm system is known as a **wave cyclone.**

Figure 8.19b shows the newly formed wave with a cold front pushing southward and a warm front moving northward. The region of lowest pressure is at the junction of the two fronts. As the cold air displaces the warm air upward along the cold front, and as *overrunning* occurs ahead of the warm front, a narrow band of precipitation forms (shaded green area). Steered by the winds aloft, the system typically moves east or northeastward and gradually becomes a fully developed **open wave** in 12 to 24 hours (Fig. 8.19c). The central pressure is now much lower, and several isobars encircle the wave's apex. These more tightly packed isobars create a stronger cyclonic

Meteorology Now™

ACTIVE FIGURE 8.19
The idealized life cycle of a wave cyclone (a through f) in the Northern Hemisphere based on the polar front theory. As the life cycle progresses, the system moves eastward in a dynamic fashion. The small arrow next to each L shows the direction of storm movement.
Watch this Active Figure at http://earthscience.brookscole.com/ahrens/ess4e.

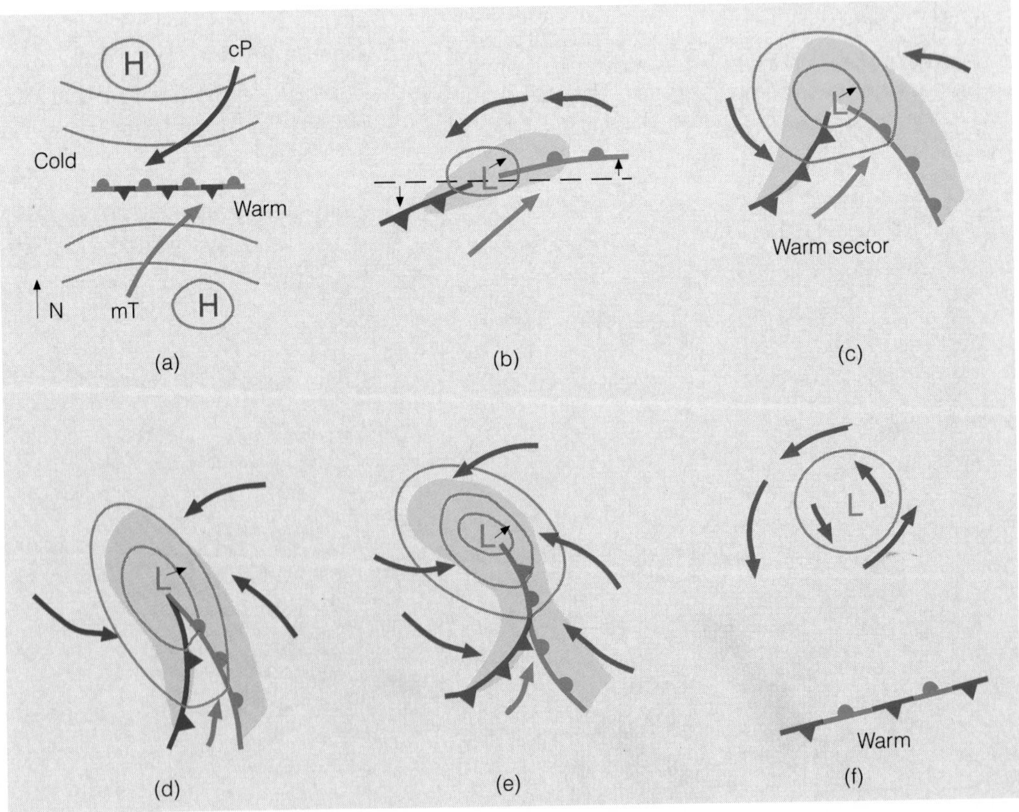

flow, as the winds swirl counterclockwise and inward toward the low's center. Precipitation forms in a wide band *ahead* of the warm front and along a narrow band of the cold front. The region of warm air between the cold and warm fronts is known as the *warm sector.* Here, the weather tends to be partly cloudy, although scattered showers may develop if the air is conditionally unstable.

Energy for the storm is derived from several sources. As the air masses try to attain equilibrium, warm air rises and cold air sinks, transforming potential energy into kinetic energy (that is, energy of motion). Condensation supplies energy to the system in the form of latent heat. And, as the surface air converges toward the low center, wind speeds may increase, producing an increase in kinetic energy.

As the open wave moves eastward, central pressures continue to decrease, and the winds blow more vigorously. The faster-moving cold front constantly inches closer to the warm front, squeezing the warm sector into a smaller area, as shown in Fig. 8.19d. In this model, the cold front eventually overtakes the warm front and the system becomes occluded. At this point, the storm is usually most intense, with clouds and precipitation covering a large area. The intense storm system shown in Fig. 8.19e gradually dissipates, because cold air now lies on both sides of the occluded front. Without the supply of energy provided by the rising warm, moist air, the old storm system dies out and gradually disappears (Fig. 8.19f). Occasionally, however, a new wave will form on the westward end of the trailing cold front. We can think of the sequence of a developing wave cyclone as a whirling eddy in a stream of water that forms behind an obstacle, moves with the flow, and gradually vanishes downstream. The entire life cycle of a wave cyclone can last from a few days to over a week.

Figure 8.20 shows a series of wave cyclones at various stages of development along the polar front in winter. Such a succession of storms is known as a *"family" of cyclones.* Observe that to the north of the front are cold anticyclones; to the south over the Atlantic Ocean is the warm, semipermanent Bermuda high. The polar front

DID YOU KNOW?

Mid-latitude storms have been known to change the course of history. One such storm, in March of 1776, forced the British to cancel a massive amphibious assault at Dorchester Heights, south of Boston, where General Washington held a well-fortified position. The storm kept Dorchester Heights from becoming a bloody, historic battlefield, such as Lexington, Concord, and Bunker Hill.

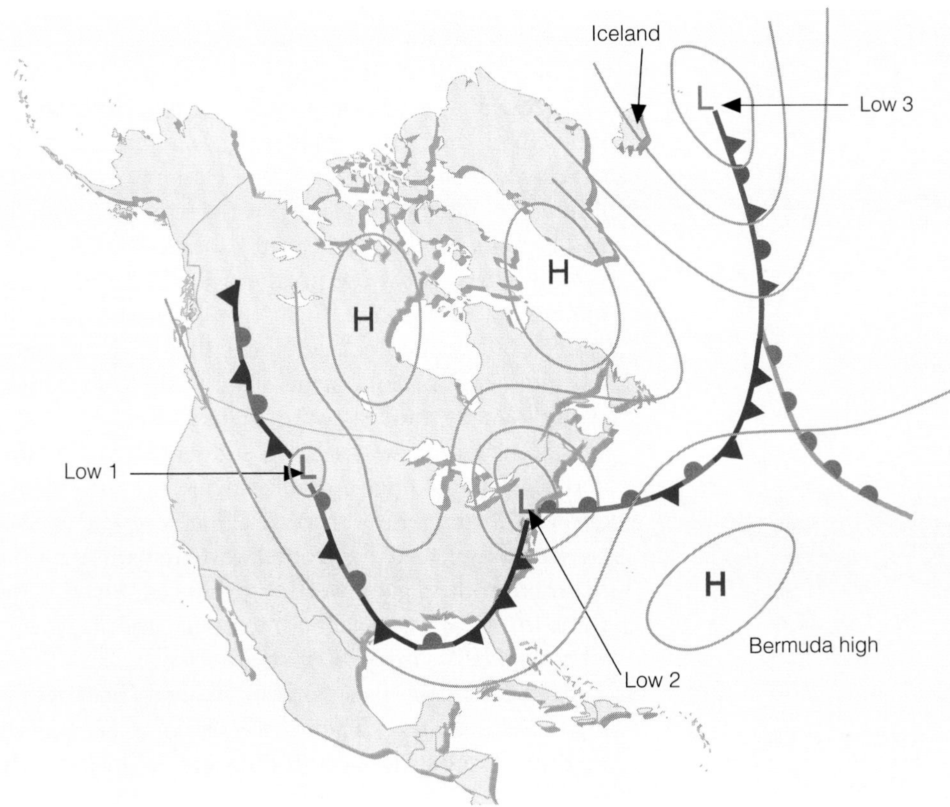

FIGURE 8.20 A series of wave cyclones (a "family" of cyclones) forming along the polar front.

itself has developed into a series of loops, and at the apex of each loop is a cyclone. The cyclone over the northern plains (Low 1) is just forming; the one along the east coast (Low 2) is an open wave; and the system near Iceland (Low 3) is dying out. If the average rate of movement of a wave cyclone from birth to decay is 25 knots, then it is entirely possible for a storm to develop over the central part of the United States, intensify into a large storm over New England, become occluded over the ocean, and reach the coast of England in its dissipating stage less than a week after it formed.

Up to now, we have considered the polar front model of a developing wave cyclone, which represents a rather simplified version of the stages that a mid-latitude storm system must go through. In fact, few (if any) storms adhere to the model exactly. Nevertheless, it serves as a good foundation for understanding the structure of cyclonic storms. So keep the model in mind as you read the following sections.

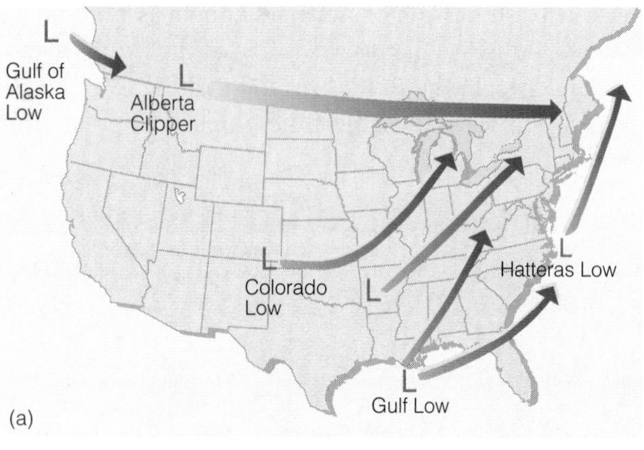

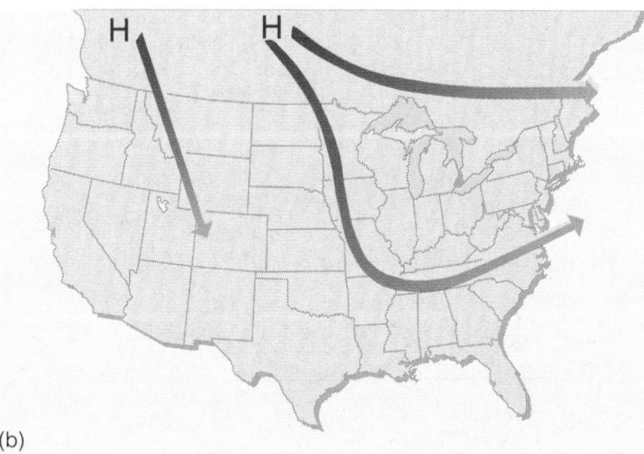

FIGURE 8.21 (a) Typical paths of winter mid-latitude cyclones. The lows are named after the region where they form. (b) Typical paths of winter anticyclones.

Where Do Mid-Latitude Cyclones Tend to Form?

Any development or strengthening of a mid-latitude cyclone is called **cyclogenesis.** There are regions of North America that show a propensity for cyclogenesis, including the eastern slopes of the Rockies, where a strengthening or developing storm is called a **lee-side low** because it is forming on the leeward (downwind) side of the mountain. Additional areas that exhibit cyclogenesis are the Great Basin, the Gulf of Mexico, and the Atlantic Ocean east of the Carolinas. Near Cape Hatteras, North Carolina, for example, warm Gulf Stream water can supply moisture and warmth to the region south of a stationary front, thus increasing the contrast between air masses to a point where storms may suddenly spring up along the front. As noted earlier, these cyclones normally move northeastward along the Atlantic coast, bringing high winds and heavy snow or rain to coastal areas. Before the age of modern satellite imagery and weather prediction such coastal storms would often go undetected during their formative stages; and sometimes an evening weather forecast of "fair and colder" along the eastern seaboard would have to be changed to "heavy snowfall" by morning. Fortunately, with today's weather information gathering and forecasting techniques, these storms rarely strike by surprise. (Storms that form along the eastern seaboard of the United States and then move northeastward are called **northeasters** or *nor'easters.* Additional information on northeasters is given in the Focus section on p. 223.)

Figure 8.21 shows the typical paths taken in winter by mid-latitude cyclones and anticyclones. Notice in Fig. 8.21a that some of the lows are named after the region where they form, such as the *Hatteras Low* which develops off the coast near Cape Hatteras, North Carolina. The *Alberta Clipper* forms (or redevelops) on the eastern side of the Rockies in Alberta, Canada, then rapidly skirts across the northern tier states. The *Colorado Low,* in contrast, forms (or redevelops) on the eastern side of the Rockies. Notice that the lows generally move eastward or northeastward, whereas the highs typically move southeastward, then eastward.

Some frontal waves form suddenly, grow in size, and develop into huge cyclonic storms. They slowly dissipate with the entire process taking several days to a week to complete. Other frontal waves remain small and never grow into a giant weather-producer. Why is it that some frontal waves develop into huge cyclonic storms, whereas others simply dissipate in a day or so?

This question poses one of the real challenges in weather forecasting. The answer is complex. Indeed, there are many surface conditions that do influence the

FOCUS ON A SPECIAL TOPIC

Northeasters

Northeasters (commonly called *nor'easters*) are mid-latitude cyclonic storms that develop or intensify off the eastern seaboard of North America then move northeastward along the coast. They often bring gale force northeasterly winds to coastal areas, along with heavy rain, snow or sleet. They usually deepen and become most intense off the coast of New England. The ferocious northeaster of December, 1992, (shown in Fig. 4) produced strong northeasterly winds from Maryland to Massachusetts. Huge waves accompanied by hurricane-force winds that reached 78 knots (90 mi/hr) in Wildwood, New Jersey, pounded the shoreline, causing extensive damage to beaches, beach front homes, sea walls, and boardwalks. Heavy snow and rain, which lasted for several days, coupled with high winds and high tides, put many coastal areas and highways under water, including parts of the New York City subway. Another strong northeaster dumped between one and three feet of snow over portions of the northeast during late March, 1997.

Studies suggest that some of the northeasters, which batter the coastline in winter, may actually possess some of the characteristics of a tropical hurricane. For example, the northeaster shown in Fig. 4 actually developed something like a hurricane's "eye" as the winds at its center went calm when it moved over Atlantic City, New Jersey. (We will examine hurricanes and their characteristics in more detail in Chapter 11.)

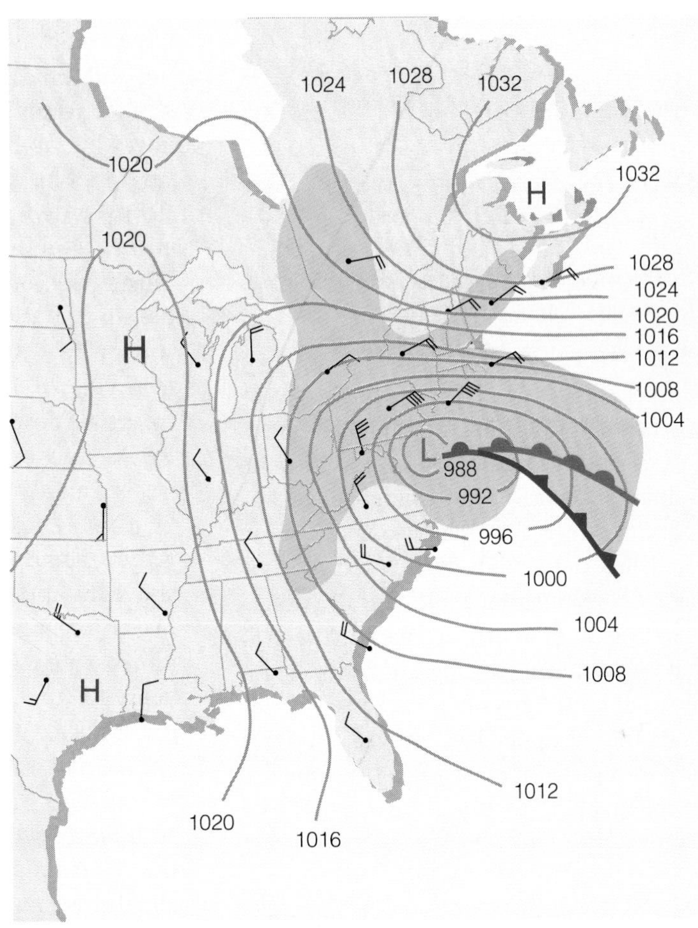

FIGURE 4 The surface weather map for 7:00 A.M. (EST) December 11, 1992, shows an intense low-pressure area (central pressure 988 mb, or 29.18 in.), which is generating strong northeasterly winds and heavy precipitation (area shaded green) from the mid-Atlantic states into New England. This northeaster devastated a wide area of the eastern seaboard, causing damage in the hundreds of millions of dollars.

formation of a wave, including mountain ranges and land-ocean temperature contrasts. However, the real key to the development of a wave cyclone is found in the *upper-wind flow,* in the region of the high-level westerlies. Therefore, before we can arrive at a reasonable answer to our question, we need to see how the winds aloft influence surface pressure systems.

Developing Mid-Latitude Cyclones and Anticyclones In Chapter 7, we learned that thermal pressure systems are shallow and weaken with increasing elevation. On the other hand, developing surface cyclonic storm systems are deep lows that usually intensify with height. This means that a surface low-pressure area will appear on an upper-level chart as either a closed low or a trough.

Suppose the upper-level low is directly above the surface low, as illustrated in Fig. 8.22. Notice that only at the surface (because of friction) do the winds blow inward toward the low's center. As these winds converge (flow together), the air "piles up." This piling up of air,

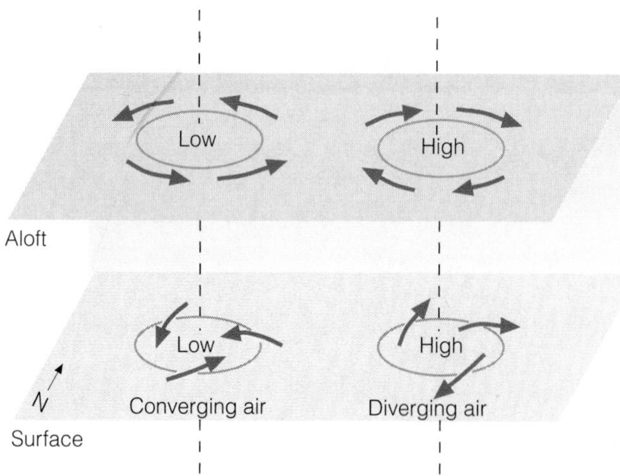

FIGURE 8.22 If lows and highs aloft were always directly above lows and highs at the surface, the surface systems would quickly dissipate.

called **convergence,** causes air density to increase directly above the surface low. This increase in mass causes surface pressures to rise; gradually, the low fills and the surface low dissipates. The same reasoning can be applied to surface anticyclones. Winds blow outward away from the center of a surface high. If a closed high or ridge lies directly over the surface anticyclone, **divergence** (the spreading out of air) at the surface will remove air from the column directly above the high. Surface pressures fall and the system weakens. Consequently, it appears that, if upper-level pressure systems were always located directly above those at the surface (such as shown in Fig. 8.22), cyclones and anticyclones

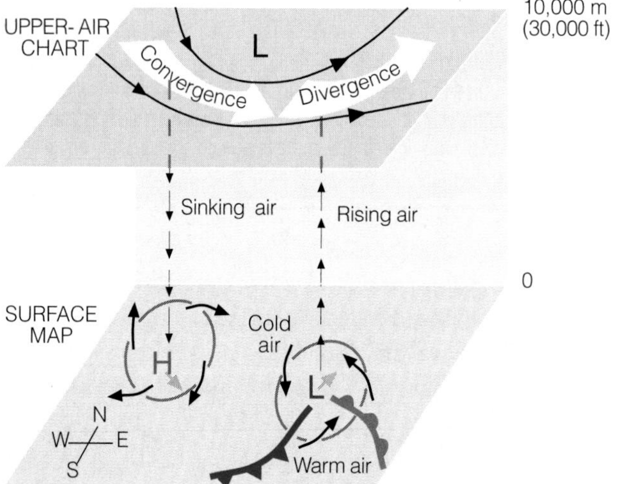

Meteorology Now™ ACTIVE FIGURE 8.23
Convergence, divergence, and vertical motions associated with surface pressure systems. Notice that for the surface storm to intensify, the upper trough of low pressure must be located to the left (or west) of the surface low. Watch this Active Figure at http://earthscience.brookscole.com/ahrens/ess4e.

would die out soon after they form (if they could form at all). What, then, is it that allows these systems to develop and intensify? (Before reading on, you may wish to review the additional information on convergence and divergence given in the Focus section on p. 225.)

For mid-latitude cyclones and anticyclones to maintain themselves or intensify, the winds aloft must blow in such a way that zones of converging and diverging air form. For example, notice in Fig. 8.23 that the surface winds are converging about the center of the low; while aloft, directly above the low, the winds are diverging. For the surface low to develop into a major storm system, *upper-level divergence of air must be greater than surface convergence of air;* that is, more air must be removed above the storm than is brought in at the surface. When this event happens, surface air pressure decreases, and we say that the storm system is *intensifying* or *deepening.* If the reverse should occur (more air flows in at the surface than is removed at the top), surface pressure will rise, and the storm system will weaken and gradually dissipate in a process called *filling.*

Notice also in Fig. 8.23 that surface winds are diverging about the center of the high, while aloft, directly above the anticyclone, they are converging. In order for the surface high to strengthen, *upper-level convergence of air must exceed low-level divergence of air* (more air must be brought in above the anticyclone than is removed at the surface). When this occurs, surface air pressure increases, and we say that the high pressure area is *building.*

In Fig. 8.23, the convergence of air aloft causes an accumulation of air above the surface high, which allows the air to sink slowly and replace the diverging surface air. Above the surface low, divergence allows the converging surface air to rise and flow out the top of the column.

We can see from Fig. 8.23 that, when an upper-level trough is as sufficiently deep as is illustrated here, a region of converging air usually forms on the west side of the trough and a region of diverging air forms on the east side. (For reference, compare Fig. 8.23 with Fig. 5 on p. 225.) Aloft, the area of diverging air is directly above the surface low, and the area of convergence is directly above the surface high. This configuration means that, for a surface storm to intensify, the upper-level trough of low pressure must be located behind (or to the *west*) of the surface low. When the upper-level trough is in this position, the atmosphere is able to redistribute its mass, as regions of low-level convergence are compensated for by regions of upper-level divergence, and vice versa. (Notice that the upper-level trough in Fig. 8.23 is in the form of a wave. More information on these upper-level waves are given in the Focus section on p. 226.)

FOCUS ON A SPECIAL TOPIC

A Closer Look at Convergence and Divergence

We know that *convergence* is the piling up of air above a region, while *divergence* is the spreading out of air above some region. Convergence and divergence of air may result from changes in wind direction and wind speed. For example, convergence occurs when moving air is funneled into an area, much in the way cars converge when they enter a crowded freeway. Divergence occurs when moving air spreads apart, much as cars spread out when a congested two-lane freeway becomes three lanes. On an upper-level chart, this type of convergence (also called *confluence*) occurs when contour lines move closer together, as a steady wind flows parallel to them (see the upper-level chart in Fig. 5). On the same chart, this type of divergence (also called *diffluence*) occurs when the contour lines move apart as a steady wind flows parallel to them. Notice that below the area of divergence lies the surface middle-latitude cyclonic storm.

Convergence and divergence may also result from changes in wind speed. *Speed convergence* occurs when the wind slows down as it moves along, whereas *speed divergence* occurs when the wind speeds up. We can grasp these relationships more clearly if we imagine air molecules to be marching in a band. When the marchers in front slow down, the rest of the band members squeeze together, causing convergence; when the marchers in front start to run, the band members spread apart, or diverge.

In summary, *speed convergence* takes place when the wind speed decreases downwind, and *speed divergence* takes place when the wind speed increases downwind.

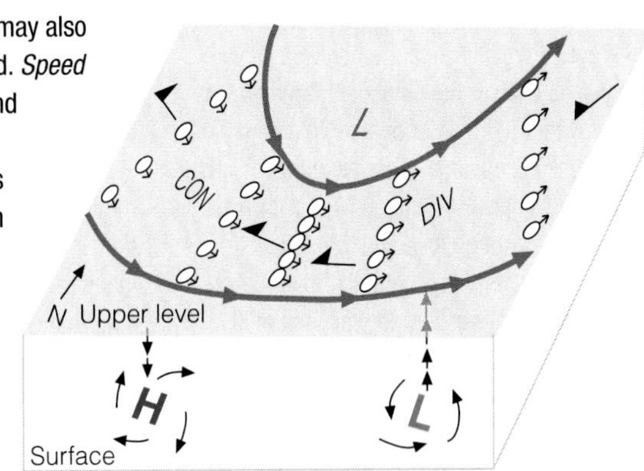

FIGURE 5 The formation of convergence (CON) and divergence (DIV) of air with a constant wind speed (indicated by flags) in the upper troposphere. Circles represent air parcels that are moving parallel to the contour lines on a constant pressure chart. Below the area of convergence the air is sinking, and we find the surface high (H). Below the area of divergence the air is rising, and we find the surface low (L).

Winds aloft steer the movement of the surface pressure systems. Since the winds above the storm in Fig. 8.23 are blowing from the southwest, the surface low should move northeastward. The northwesterly winds above the surface high should direct it toward the southeast. These paths are typical of the average movement of surface pressure systems in the eastern two-thirds of the United States (see Fig. 8.21, p. 222).

Jet Streams and Developing Mid-Latitude Cyclones

Jet streams play an additional part in the formation of surface mid-latitude cyclones and anticyclones. When the polar jet stream flows in a wavy west-to-east pattern (as illustrated in Fig. 8.24a), deep troughs and ridges exist in the flow aloft. Notice that, in the trough, the area shaded orange represents a strong core of winds called a *jet maximum*, or **jet streak.** The curving of the jet stream coupled with the changing wind speeds around the jet streak produce regions of strong convergence and divergence along the flanks of the jet. The region of diverging air above the storm (marked D in Fig. 8.24a) draws warm surface air upward to the jet stream, which quickly sweeps the air downstream. Since the air above the storm is being removed more quickly than converging surface winds can supply air to the storm's center, the central pressure of the storm drops rapidly. As surface pressure gradients increase, the wind speed increases. Above the high-pressure area, a region of converging air (marked C in Fig. 8.24a) feeds cold air

DID YOU KNOW?

The great March storm of 1993 that hit the East Coast of the United States was one whopper of a mid-latitude cyclonic storm. It blanketed 50 billion tons of snow from Alabama to Canada. Fierce winds piled the snow into huge drifts that closed roads, leaving hundreds of motorists stranded. The storm shut down every major airport along the East Coast, and more than 3 million people lost electric power. This great cyclonic storm, sometimes called "the storm of the century," damaged or destroyed hundreds of homes, produced 27 tornadoes, caused an estimated $800 million in damage, and claimed the lives of at least 270 people.

FOCUS ON A SPECIAL TOPIC

Waves in the Westerlies

Regions of strong upper-level divergence and convergence typically occur when well-developed waves exist in the flow aloft. Recall from Chapter 6 that the flow above the middle latitudes usually consists of a series of waves in the form of troughs and ridges. The distance from trough to trough (or ridge to ridge) is known as the *wavelength.* When the wavelength is on the order of many thousands of kilometers, the wave is called a *longwave.* Observe in Fig. 6a that the length of the longwave is greater than the width of North America. Typically, at any given time, there are between three and six longwaves looping around the earth. These longwaves are also known as *Rossby waves,* after C. G. Rossby, a famous meteorologist who carefully studied their motion. In Fig. 6a, we can see that embedded in longwaves are *shortwaves,* which are small disturbances or ripples.

By comparing Fig. 6a with Fig. 6b, we can see that while the longwaves move eastward very slowly, the shortwaves move fairly quickly around the longwaves. Generally, shortwaves deepen when they approach a longwave trough and weaken when they approach a ridge. Also, notice in Fig. 6b that, when a shortwave moves into a longwave trough, the trough tends to deepen. The upper flow is now capable of providing the proper ingredients for the development or intensification of surface low and high pressure areas as illustrated in Fig. 8.23, p. 224.

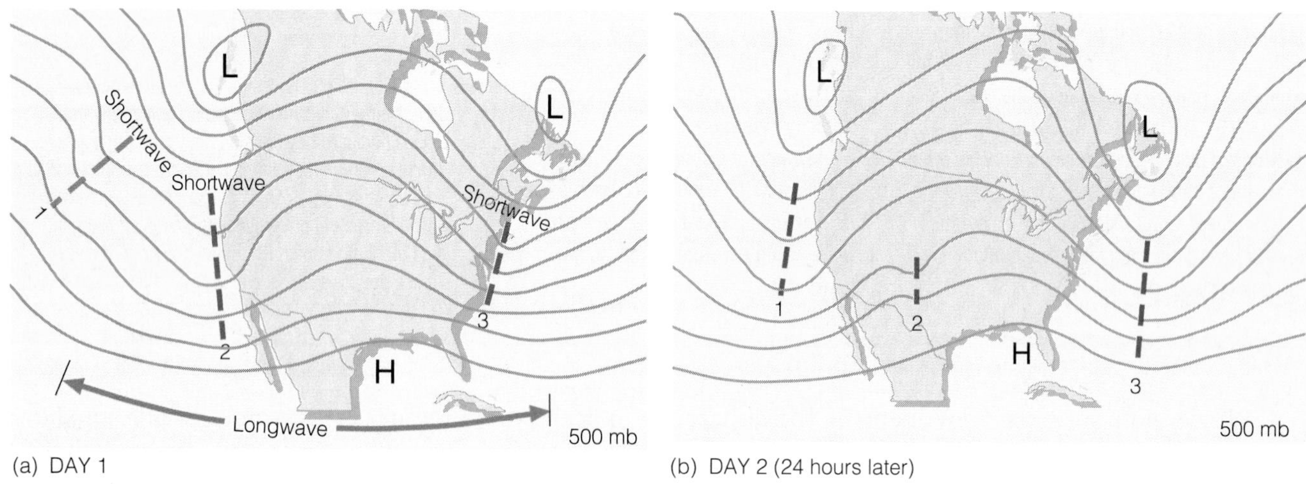

Meteorology Now™ ACTIVE FIGURE 6 (a) Upper-air chart showing a longwave with three shortwaves (heavy dashed lines) embedded in the flow.(b) Twenty-four hours later the shortwaves have moved rapidly around the longwave. Notice that the shortwaves labeled 1 and 3 tend to deepen the longwave trough, while shortwave 2 has weakened as it moves into a ridge. Watch this Active Figure at http://earthscience.brookscole.com/ahrens/ess4e.

downward into the anticyclone to replace the diverging surface air. Hence, we find the jet stream *removing air above the surface cyclone and supplying air to the surface anticyclone.* Additionally, the sinking of cold air and the rising of warm air provide energy for the developing cyclone as potential energy is transformed into energy of motion (kinetic energy).

As the jet stream steers the storm along (toward the northeast, in this case), the surface storm occludes, and cold air surrounds the surface low (see Fig. 8.24b). Since the surface low has moved out from under the pocket of diverging air aloft, the occluded storm gradually fills as the surface air flows into the system.

Since the polar jet stream is strongest and moves farther south in winter, we can see why mid-latitude cyclonic storms are better developed and move more quickly during the coldest months. During the summer when the polar jet shifts northward, developing mid-latitude storm activity shifts northward and occurs principally over the Canadian provinces of Alberta and the Northwest Territories.

In general, we now have a fairly good picture as to why some surface lows intensify into huge storms while others do not. For a storm to intensify, there must be an upper-level counterpart—a trough of low pressure—that lies to the *west* of the surface low. At the same time,

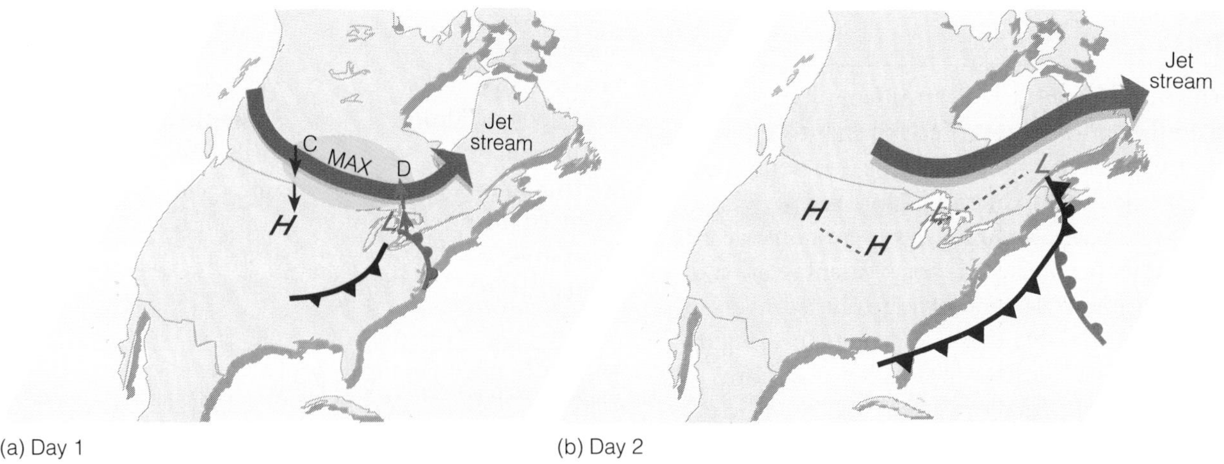

FIGURE 8.24 (a) As the polar jet stream and its area of maximum winds (the jet streak, or MAX), swings over a developing mid-latitude cyclone, an area of divergence *(D)* draws warm surface air upward, and an area of convergence *(C)* allows cold air to sink. The jet stream removes air above the surface storm, which causes surface pressures to drop and the storm to intensify. (b) When the surface storm moves northeastward and occludes, it no longer has the upper-level support of diverging air, and the surface storm gradually dies out.

the polar jet stream must form into waves and swing slightly south of the developing storm. When these conditions exist, zones of converging and diverging air, along with rising and sinking air, provide energy conversions for the storm's growth. When these conditions do not exist, we say that the surface storm does not have the proper *upper-air support* for its development. The horizontal and vertical motions, cloud patterns, and weather that typically occur with a developing open-wave mid-latitude cyclone are summarized in Fig. 8.25.

Meteorology Now™ Click "Cyclogenesis" to examine the complete life cycle of a mid-latitude cyclone.

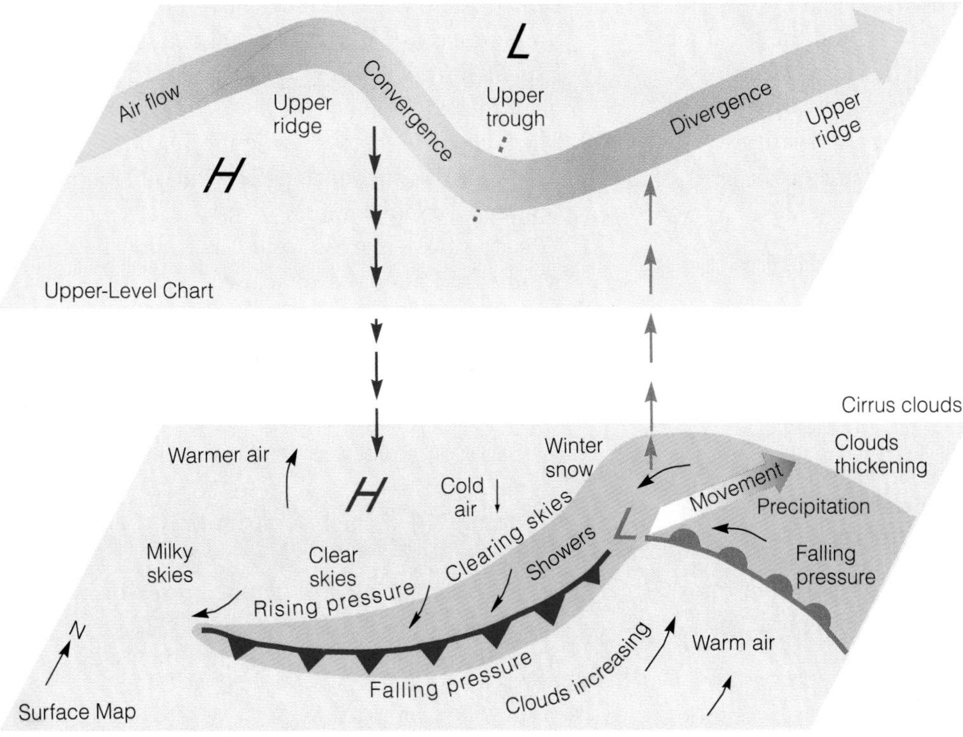

FIGURE 8.25 Summary of clouds, weather, vertical motions, and upper-air support associated with a developing mid-latitude cyclone.

SUMMARY

In this chapter, we considered the different types of air masses and the various weather each brings to a particular region. Continental arctic air masses are responsible for the extremely cold (arctic) outbreaks of winter, whereas continental polar air masses are responsible for cold, dry weather in winter and cool, pleasant weather in summer. Maritime polar air, having traveled over an ocean for a considerable distance, brings to a region cool, moist weather. The hot, dry weather of summer is associated with continental tropical air masses, whereas warm, humid conditions are due to maritime tropical air masses. Where air masses with sharply contrasting properties meet, we find weather fronts.

Along the leading edge of a cold front, where colder air replaces warmer air, showers are prevalent, especially if the warmer air is moist and conditionally unstable. Along a warm front, warmer air rides up and over colder surface air, producing widespread cloudiness and precipitation. Occluded fronts, which are often difficult to locate and define on a surface weather map, may have characteristics of both cold and warm fronts.

We learned that fronts are actually part of the mid-latitude cyclone. We examined where, why, and how these storms form and found that an important influence on their development is the upper-air flow, including the jet stream. We learned that when an upper-level low lies to the west of the surface low, and the polar jet stream bends and then dips south of the surface storm, an area of divergence above the surface low provides the necessary ingredients for the surface mid-latitude cyclone to develop into a deep low-pressure area.

Meteorology Now™ Assess your understanding of this chapter's topics with additional quizzing and tutorials at http://earthscience.brookscole.com/ahrens/ess4e.

KEY TERMS

The following terms are listed in the order they appear in the text. Define each. Doing so will aid you in reviewing the material covered in this chapter.

air mass
source region (for air masses)
continental polar (air mass)
continental arctic (air mass)
lake-effect snows
maritime polar (air mass)
maritime tropical (air mass)
continental tropical (air mass)
air mass weather
front
stationary front
cold front
back door cold front
warm front
overrunning
occluded front (occlusion)
polar front theory
frontal wave
wave cyclone
open wave
cyclogenesis
lee-side low
northeaster
convergence
divergence
jet streak

QUESTIONS FOR REVIEW

1. (a) What is an air mass?
   (b) If an area is described as a "good air-mass source region," what information can you give about it?
2. How does a cA air mass differ from a cP air mass?
3. Why is cP air not welcome to the Central Plains in winter and yet very welcome in summer?
4. What are lake-effect snows and how do they form? On which side of a lake do they typically occur?
5. Explain why the central United States is not a good air-mass source region.
6. List the temperature and moisture characteristics of each of the major air mass types.
7. Which air mass only forms in summer over the United States?
8. Why are mP air masses along the east coast of the United States usually colder than those along the nation's west coast? Why are they also *less* prevalent?
9. Explain how the airflow aloft regulates the movement of air masses.
10. The boundaries between neighboring air masses tend to be more distinct during the winter than during the summer. Explain why.
11. What type of air mass would be responsible for the weather conditions listed below?
   (a) hot, muggy summer weather in the Midwest and the East
   (b) refreshing, cool, dry breezes after a long summer hot spell on the Central Plains
   (c) persistent cold, damp weather with drizzle along the East Coast
   (d) drought with high temperatures over the Great Plains
   (e) record-breaking low temperatures over a large portion of North America
   (f) cool weather with showers over the Pacific Northwest
   (g) daily afternoon thunderstorms along the Gulf Coast

12. Describe the typical characteristics of:
    (a) a warm front
    (b) a cold front
    (c) an occluded front
13. Sketch side views of a typical cold front, warm front, and cold-occluded front. Include in each diagram cloud types and patterns, areas of precipitation, and relative temperature on each side of the front.
14. Describe the stages of a developing wave cyclone using the polar front theory.
15. Why do cyclones tend to develop along the polar front?
16. List four regions in North America where wave cyclones tend to develop.
17. Why is it important that for a surface low to develop or intensify, its upper-level counterpart must be to the left (or west) of the surface storm?
18. Describe some of the necessary ingredients (upper-air support) for a wave cyclone to develop into a huge mid-latitude cyclonic storm system.
19. Explain the role that upper-level divergence plays in the development of a wave cyclone.
20. How does the polar jet stream influence the formation of a wave cyclone?
21. Explain why, in the eastern half of the United States, a mid-latitude cyclonic storm often moves eastward or northeastward.

QUESTIONS FOR THOUGHT AND EXPLORATION

1. If Lake Erie freezes over in January, do you feel it is still possible to have lake-effect snow on its eastern shores in February? Explain your answer.
2. Explain how an autumn anticyclone can bring record low temperatures and cP air to the southeastern United States and, only a day or so later, record-high temperatures and mT air to the same region.
3. During the winter, cold-front weather is typically more violent than warm-front weather. Why is this so? Explain why this is not necessarily true during the summer.
4. You are in upstate New York and observe the wind shifting from the east to the south. This wind shift is accompanied by a sudden rise in both air temperature and dew-point temperature. What type of front is passing?
5. Why does the same cold front typically produce more rain over Kentucky than over western Kansas?
6. Explain why the boundaries between neighboring air masses tend to be more distinct during the winter than during the summer?
7. Sketch a Southern Hemisphere mid-latitude cyclonic storm, complete with isobars and at least two types of fronts. Compare and contrast this Southern Hemisphere cyclone with its Northern Hemisphere counterpart.
8. Why are mid-latitude cyclones described as waves?
9. Explain how this can happen: At the same time a mid-latitude cyclonic storm over the eastern United States is moving northeastward, a large surface high-pressure area over the northern plains is moving southeastward.
10. Would a wave cyclone intensify or dissipate if the upper trough was located to the *east* of the surface low pressure area? Explain your answer with the aid of a diagram.
11. Current surface temperatures (http://weather.unisys.com/surface/sfc_con_temp_na.html): Using a current map showing surface temperatures over North America, identify any air masses that are present over the continent.
12. Upper-air 300-mb map (http://weather.unisys.com/upper_air/ua_300.html) and current surface map (http://weather.unisys.com/surface/sfc_map.html): Using the current 300-mb map, find areas of upper-level convergence and divergence. Where can current surface cyclones and anticyclones be found relative to these upper-level features? Is this consistent with the information in Fig. 8.23?

- Go to the Brooks/Cole Earth Sciences Resource Center (http://earthscience.brookscole.com) for critical thinking exercises, articles, and additional readings from InfoTrac College Edition, Brooks/Cole's online student library.

Photo by author

A watchful eye and a little weather information are important when making a local short-range weather forecast. Increasing southwesterly surface winds, and a falling barometer all suggest that a storm is approaching from the west.

Meteorology Now™ This icon, appearing throughout the book, indicates an opportunity to explore interactive tutorials, animations, or practice problems available on the MeteorologyNow Web site at http://earthscience.brookscole.com/ahrens/ess4e.

Weather Forecasting

Sometimes there is no job security in weather forecasting. In fact, a weather forecaster actually lost his job for not altering his forecast. On April 15, 2001, a function honoring a well-known conservative radio talk show host was scheduled outdoors at the Madera, California, fairgrounds. The story goes that a local forecaster at the radio station that sponsored the event had called for a "chance of rain" on April 15th. Upset that such a forecast might discourage people from attending the function, the station manager told the forecaster to alter his forecast and predict a greater possibility of sunshine. The forecaster refused and was promptly fired. Apparently, retribution reigned supreme—it poured on the event.

CONTENTS

Weather forecasts are issued to save lives, to save property and crops, and to tell us what to expect in our atmospheric environment. In addition, knowing what the weather will be like in the future is vital to many human activities. For example, a summer forecast of extended heavy rain and cool weather would have construction supervisors planning work under protective cover, department stores advertising umbrellas instead of bathing suits, and ice cream vendors vacationing as their business dropped off. The forecast would alert farmers to harvest their crops before their fields became too soggy to support the heavy machinery needed for the job. And the commuter? Well, the commuter knows that prolonged rain could mean clogged gutters, flooded highways, stalled traffic, blocked railway lines, and late dinners.

On the other side of the coin, a forecast calling for extended high temperatures with low humidity has an entirely different effect. As ice cream vendors prepare for record sales, the dairy farmer anticipates a decrease in milk and egg production. The forest ranger prepares warnings of fire danger in parched timber and grasslands. The construction worker is on the job outside once again, but the workday begins in the early morning and ends by early afternoon to avoid the oppressive heat. And the commuter prepares for increased traffic stalls due to overheated car engines.

Put yourself in the shoes of a weather forecaster: It is your responsibility to predict the weather accurately so that thousands (possibly millions) of people will know whether to carry an umbrella, wear an overcoat, or prepare for a winter storm. Since weather forecasting is not an exact science, your predictions will occasionally be incorrect. If your erroneous forecast misleads many people, you may become the target of jokes, insults, and even anger. There are even people who expect you to be able to predict the unpredictable. For example, on Monday you may be asked whether two Mondays from now will be a nice day for a picnic. And, of course, what about next winter? Will it be bitterly cold?

Unfortunately, accurate answers to such questions are beyond meteorology's present technical capabilities. Will forecasters ever be able to answer such questions confidently? If so, what steps are being taken to improve the forecasting art? How are forecasts made, and why do they sometimes go wrong? These are just a few of the questions we will address in this chapter.

ACQUISITION OF WEATHER INFORMATION

Weather forecasting basically entails predicting how the present state of the atmosphere will change. Consequently, if we wish to make a weather forecast, present weather conditions over a large area must be known. To obtain this information, a network of observing stations is located throughout the world. Over 10,000 land-based stations and hundreds of ships and buoys provide surface weather information four times a day. Most airports observe conditions hourly. Additional information, especially upper-air data, is supplied by radiosondes, aircraft, and satellites.

A United Nations agency—the World Meteorological Organization (WMO)—consists of over 175 nations. The WMO is responsible for the international exchange of weather data and certifies that the observation procedures do not vary among nations, an extremely important task, since the observations must be comparable.

Weather information from all over the world is transmitted electronically to the National Center for Environmental Prediction (NCEP), located in Camp Springs, Maryland, just outside Washington, D.C. Here, the massive job of analyzing the data, preparing weather maps and charts, and predicting the weather on a global and national scale begins. From NCEP, weather information is transmitted to private and public agencies, such as weather forecast offices that use the information to issue local and regional weather forecasts.

The public hears weather forecasts over radio or television. Many stations hire private meteorological companies or professional meteorologists to make their own forecasts aided by NCEP material or to modify a weather service forecast. Other stations hire meteorologically untrained announcers who paraphrase or read the forecasts of the National Weather Service word for word.

Today, the forecaster has access to many hundreds of maps and charts, as well as vertical profiles (called *soundings*) of temperature, dew point, and winds. Also available are visible and infrared satellite images, as well as Doppler radar information that can detect and monitor the severity of precipitation and thunderstorms.

When hazardous weather is likely, the National Weather Service issues advisories in the form of weather watches and warnings. A **watch** indicates that atmospheric conditions favor hazardous weather occurring over a particular region during a specified time period. A **warning,** on the other hand, indicates that hazardous weather is either imminent or actually occurring within the specified forecast area. *Advisories* are issued to in-

FOCUS ON A SPECIAL TOPIC

Watches, Warnings, and Advisories

As we have seen, where severe or hazardous weather is either occurring or possible, the National Weather Service issues a forecast in the form of a watch or warning. The public, however, is not always certain as to what this forecast actually means. For example, a *high wind warning* indicates that there will be high winds—but how high and for how long? The following describes a few of the various watches, warnings, and advisories issued by the National Weather Service and the necessary precautions that should be taken during the event.

Wind advisory Issued when sustained winds reach 25 to 39 mi/hr or when wind gusts are up to 57 mi/hr.

High wind warning Issued when sustained winds are at least 40 mi/hr, or when wind gusts exceed 57 mi/hr. Caution should be taken when driving high-profile vehicles, such as trucks, trailers, and motor homes.

Wind-chill advisory Issued for wind-chill temperatures of –30° to –35°F or below.

Heat advisory/warning Advisory issued when the daytime Heat Index is expected to reach 105°F for 3 hours or more and nighttime lows do not drop below 80°F. Warning issued when Heat Index reaches 115°F or above.

Flash-flood watch Heavy rains may result in flash flooding in the specified area. Be alert and prepared for the possibility of a flood emergency that will require immediate action.

Flash-flood warning Flash flooding is occurring or is imminent in the specified area. Move to safe ground immediately.

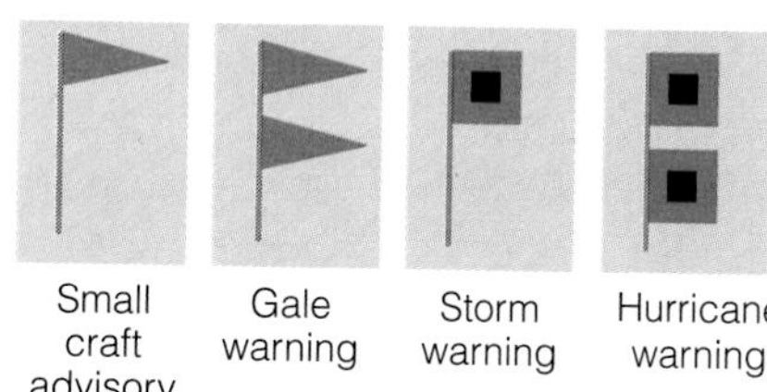

FIGURE 1 Flags indicating advisories and warnings in maritime areas.

Urban and small stream advisory Issued when flooding is occurring in small streams, streets, or in low-lying areas, such as railroad underpasses and urban storm drains.

Severe thunderstorm watch Thunderstorms (with winds exceeding 57 mi/hr and/or hail three-fourths of an inch or more in diameter) are possible.

Severe thunderstorm warning Severe thunderstorms have been visually sighted or indicated by Doppler radar. Be prepared for lightning, heavy rains, strong winds, and large hail. (Tornadoes can form with severe thunderstorms.)

Tornado watch Issued to alert people that tornadoes may develop within a specified area during a certain time period.

Tornado warning Issued to alert people that a tornado has been spotted either visually or by Doppler radar. Take shelter immediately.

Snow advisory In nonmountainous areas, expect a snowfall of 2 in. or more in 12 hours, or 3 in. or more in 24 hours.

Winter storm warning (formerly heavy snow warning) In nonmountainous areas, expect a snowfall of 4 in. or more in 12 hours or 6 in. or more in 24 hours. (Where heavy snow is infrequent, a snow fall of several inches may justify a warning.)

Blizzard warning Issued when falling or blowing snow and winds of at least 35 mi/hr frequently restrict visibility to less than 1/4 mile for several hours.

Dense fog advisory Issued when fog limits visibility to less than 1/4 mile, or in some parts of the country to less than 1/8 mile.

WARNINGS OVER THE WATER

Small craft advisories Issued to alert mariners that weather or sea conditions might be hazardous to small boats. Expect winds of 18 to 34 knots (21 to 39 mi/hr). Figure 1 displays the posted advisory and warning flags.

Gale warning Winds will range between 34 and 47 knots (39 to 54 mi/hr) in the forecast area.

Storm warning Winds in excess of 47 knots (54 mi/hr) are to be expected in the forecast area.

Hurricane watch Issued when a tropical storm or hurricane becomes a threat to a coastal area. Be prepared to take precautionary action in case hurricane warnings are issued.

Hurricane warning Issued when it appears that the storm will strike an area within 24 hours. Expect wind speeds in excess of 64 knots (74 mi/hr).

form the public of less hazardous conditions caused by wind, dust, fog, snow, sleet, or freezing rain. (Additional information on watches, warnings, and advisories is given in the Focus section above.)

WEATHER FORECASTING METHODS

As late as the mid-1950s, all weather maps and charts were plotted by hand and analyzed by individuals. Meteorologists predicted the weather using certain rules that related to the particular weather system in question. For short-range forecasts of six hours or less, surface weather systems were moved along at a steady rate. Upper-air charts were used to predict where surface storms would develop and where pressure systems aloft would intensify or weaken. The predicted positions of these systems were extrapolated into the future using linear graphical techniques and current maps. Experi-

DID YOU KNOW?

Each day a weather forecaster has access to hundreds of computer-drawn progs. To produce a single prog that predicts the weather a few days into the future may require as many as one trillion calculations.

ence played a major role in making the forecast. In many cases, these forecasts turned out to be amazingly accurate. They were good but, with the advent of the modern computers, along with our present observing techniques, today's forecasts are even better.

The Computer and Weather Forecasting: Numerical Weather Prediction Modern electronic computers can analyze large quantities of data extremely fast. Each day the many thousands of observations transmitted to NCEP are fed into a high-speed computer, which plots and draws lines on surface and upper-air charts. Meteorologists interpret the weather patterns and then correct any errors that may be present. The final chart is referred to as an **analysis.**

The computer not only plots and analyzes data, it also predicts the weather. The routine daily forecasting of weather by the computer has come to be known as **numerical weather prediction.**

Because the many weather variables are constantly changing, meteorologists have devised **atmospheric models** that describe the present state of the atmosphere. These are not physical models that paint a picture of a developing storm; they are, rather, mathematical models consisting of dozens of mathematical equations that describe how atmospheric temperature, pressure, winds, and moisture will change with time. Actually, the models do not fully represent the real atmosphere but are approximations formulated to retain the most important aspects of the atmosphere's behavior.

The models are programmed into the computer, and surface and upper-air observations of temperature, pressure, moisture, winds, and air density are fed into the equations. To determine how each of these variables will change, each equation is solved for a small increment of future time—say, five minutes—for a large number of locations called *grid points,* each situated a given distance apart.* In addition, each equation is solved for as many as 50 levels in the atmosphere. The results of these computations are then fed back into the original equations. The computer again solves the equations with the new "data," thus predicting weather over the following five minutes. This procedure is done repeatedly until it reaches some desired time in the future, usually 12, 24, 36, or 48 hours. The computer then analyzes the data and draws the projected positions of pressure systems with their isobars or contour lines. The final forecast chart representing the atmosphere at a specified future time is called a **prognostic chart,** or, simply, a **prog.** Computer-drawn progs have come to be known as "machine-made" forecasts.

The forecaster uses the progs as a guide to predicting the weather. At present, there are a variety of models (and, hence, progs) from which to choose, each producing a slightly different interpretation of the weather for the same projected time and atmospheric level (see Fig. 9.1). The differences between progs may result from the way the models use the equations, or the distance between grid points. Some models predict some features better than others: One model may work best in predicting the position of troughs on upper-level charts, whereas another forecasts the position of surface lows quite well. Some models even forecast the state of the atmosphere 384 hours (16 days) into the future.

A good forecaster knows the idiosyncrasies of each model and carefully scrutinizes all the progs. The forecaster then makes a prediction based on the *guidance* from the computer, a personalized practical interpretation of the weather situation, and any local geographic features that influence the weather within the specific forecast area.

Forecasting Tools To help forecasters handle all the available charts and maps, high-speed data modeling systems using computers are employed. The communication system in use today is known as **AWIPS** (*A*dvanced *W*eather *I*nteractive *P*rocessing *S*ystem). The AWIPS system is shown in Fig. 9.2.

The AWIPS system has data communications, storage, processing, and display capabilities (including graphical overlays) to better help the individual forecaster extract and assimilate information from the mass of available data. In addition, AWIPS is able to process information received from the Doppler radar system (the WSR-88D) and the new *A*utomated *S*urface *O*bserving *S*ystems (*ASOS*) that are operational at selected airports and other sites throughout the United States. The ASOS system is designed to provide nearly continuous information about wind, temperature, pressure, cloud-base height, and runway visibility at various airports. Meteorologists are hopeful that information from

*Some models have a grid spacing as small as 10 km, whereas the spacing in others exceeds 180 km. There are models that actually describe the atmosphere using a set of mathematical equations with wavelike characteristics rather than a set of discrete numbers associated with grid points.

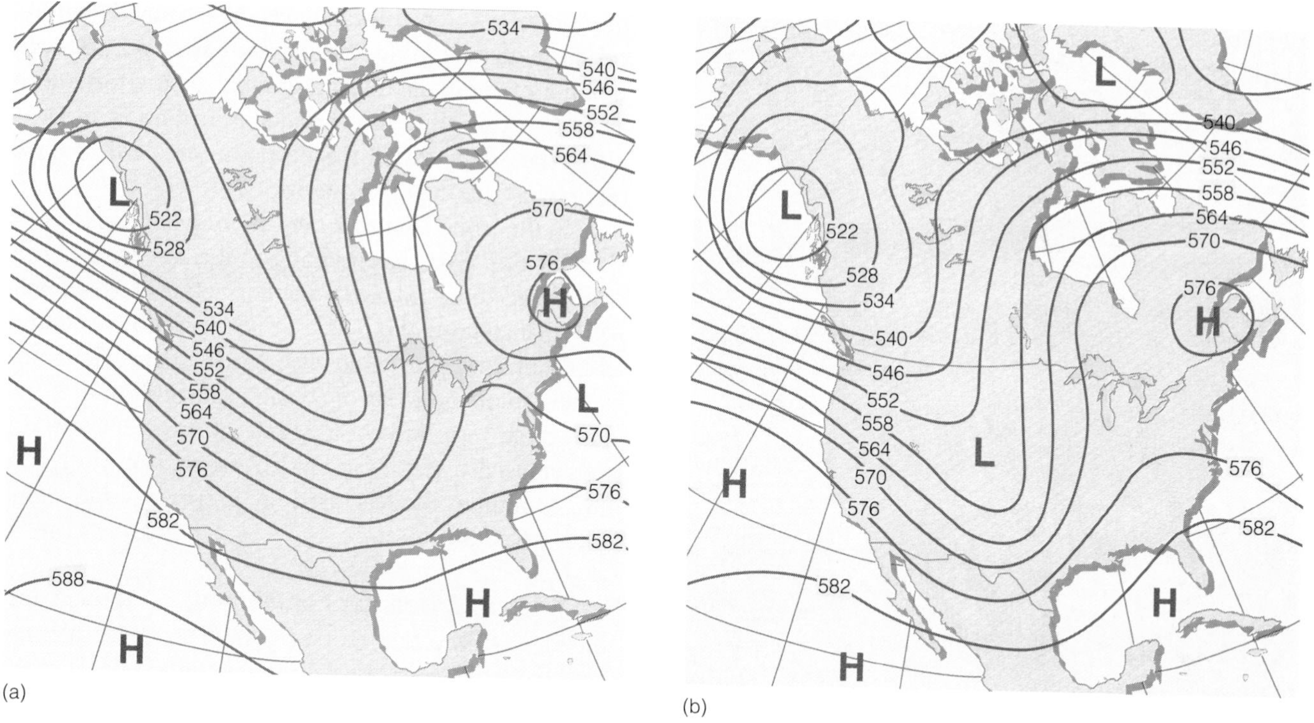

FIGURE 9.1 Two 500-mb progs for 7 P.M. EST, May 4, 1999—48 hours into the future. Prog (a) is the MRF model, whereas Prog (b) is the ETA model. Solid lines on the map are height contours, where 552 equals 5520 meters. Notice how the two progs (models) agree on the atmosphere's large-scale circulation.

all of these sources will improve the accuracy of weather forecasts by providing previously unobtainable data for integration into numerical models. Moreover, much of the information from ASOS and Doppler radar is processed by software according to predetermined formulas, or *algorithms,* before it goes to the forecaster. Certain criteria or combinations of measurements can alert the forecaster to an impending weather situation, such as the severe weather illustrated in Fig. 9.3.

With so much information at the forecaster's disposal, it is essential that the data be easily accessible and in a format that allows several weather variables to be viewed at one time. The **meteogram** is a chart that shows how one or more weather variables has changed at a station over a given period of time. As an example, the chart may represent how air temperature, dew point, and sea-level pressure have changed over the past five days, or it may illustrate how these same

© Jan Null

FIGURE 9.2 The AWIPS computer work station provides various weather maps and overlays on different screens.

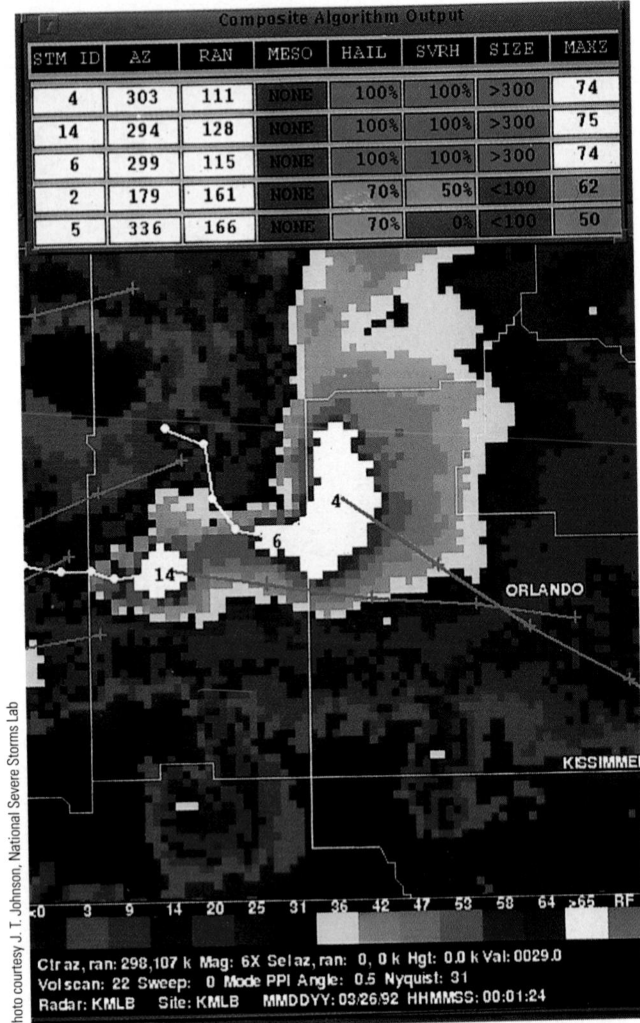

Composite Algorithm Output
STM ID AZ RAN MESO HAIL SVRH SIZE MAXZ
4 303 111 NONE 100% 100% >300 74
14 294 128 NONE 100% 100% >300 75
6 299 115 NONE 100% 100% >300 74
2 179 161 NONE 70% 50% <100 62
5 336 166 NONE 70% 0% <100 50
4
6
14
ORLANDO
KISSIMMEE
<0 3 9 14 20 25 31 36 42 47 53 58 64 >65 RF
Ctr az, ran: 298,107 k Mag: 6X Sel az, ran: 0, 0 k Hgt: 0.0 k Val: 0029.0
Vol scan: 22 Sweep: 0 Mode PPI Angle: 0.5 Nyquist: 31
Radar: KMLB Site: KMLB MMDDYY: 03/26/92 HHMMSS: 00:01:24

Photo courtesy J. T. Johnson, National Severe Storms Lab

FIGURE 9.3 Doppler radar data from Melbourne, Florida, on March 25, 1992, during the time of a severe hailstorm that caused $60 million in damages in the Orlando area. In the table near the top of the display, the hail algorithm determined that there was 100 percent probability that the storm was producing hail and severe hail. The algorithm also estimated the maximum size of the hailstones to be greater than 3 inches. A forecaster can project the movement of the storm and adequately warn those areas in the immediate path of severe weather.

variables are projected to change over the next five days (see Fig. 9.4).

Another aid in weather forecasting is the use of *soundings*—a two-dimensional vertical profile of temperature, dew point, and winds (see Fig. 9.5). The analysis of a sounding can be especially helpful when making a short-range forecast that covers a relatively small area, such as the mesoscale. The forecaster examines the sounding of the immediate area (or closest proximity), as well as the soundings of those sites upwind, to see how the atmosphere might be changing. Computer programs then automatically calculate from the sounding a number of meteorological *indexes* that can aid the forecaster in determining the likelihood of smaller-scale weather phenomena, such as thunderstorms, tornadoes, and hail. Soundings also provide information that can aid in the prediction of fog, air pollution alerts, and the downwind mixing of strong winds.

In the United States, a network of *wind profilers* (see Chapter 6, p. 163) is providing forecasters with hourly wind speed and wind direction information at 72 different levels in a column of air 16 km thick. The almost continuous monitoring of winds is especially beneficial when briefing pilots on areas of strong headwinds and on regions of strong wind shear. Wind information from the profilers is also sent to the NCEP, where it is integrated into computer models used in preparing mid-range weather forecasts.

Satellites and Weather Forecasting Another valuable tool utilized to forecast the weather is the satellite. Satellites provide extremely valuable cloud photographs of areas where there are no ground-based observations. Because water covers over 70 percent of the earth's surface, there are vast regions where few (if any) surface cloud observations are made. Before weather satellites were in use, tropical storms, such as hurricanes and typhoons, often went undetected until they moved dangerously near inhabited areas. Residents of the regions affected had little advance warning. Today, satellites spot these storms while they are still far out in the ocean and track them accurately.

There are two primary types of weather satellites in use for viewing clouds. The first are called **geostationary satellites** (or *geosynchronous satellites*) because they orbit the equator at the same rate the earth spins and, hence, remain at nearly 36,000 km (22,300 mi) above a fixed spot on the earth's surface (see Fig. 9.6). This positioning allows continuous monitoring of a specific region.

Geostationary satellites are also important because they use a "real time" data system, meaning that the satellites transmit photographs to the receiving system on the ground as soon as the camera takes the picture. Successive cloud photographs from these satellites can be put into a time-lapse movie sequence to show the

DID YOU KNOW?

America's first *official* weather forecaster, Professor Increase A. Lapham, began predictions for the United States Weather (Signal) Service on November 8, 1870.

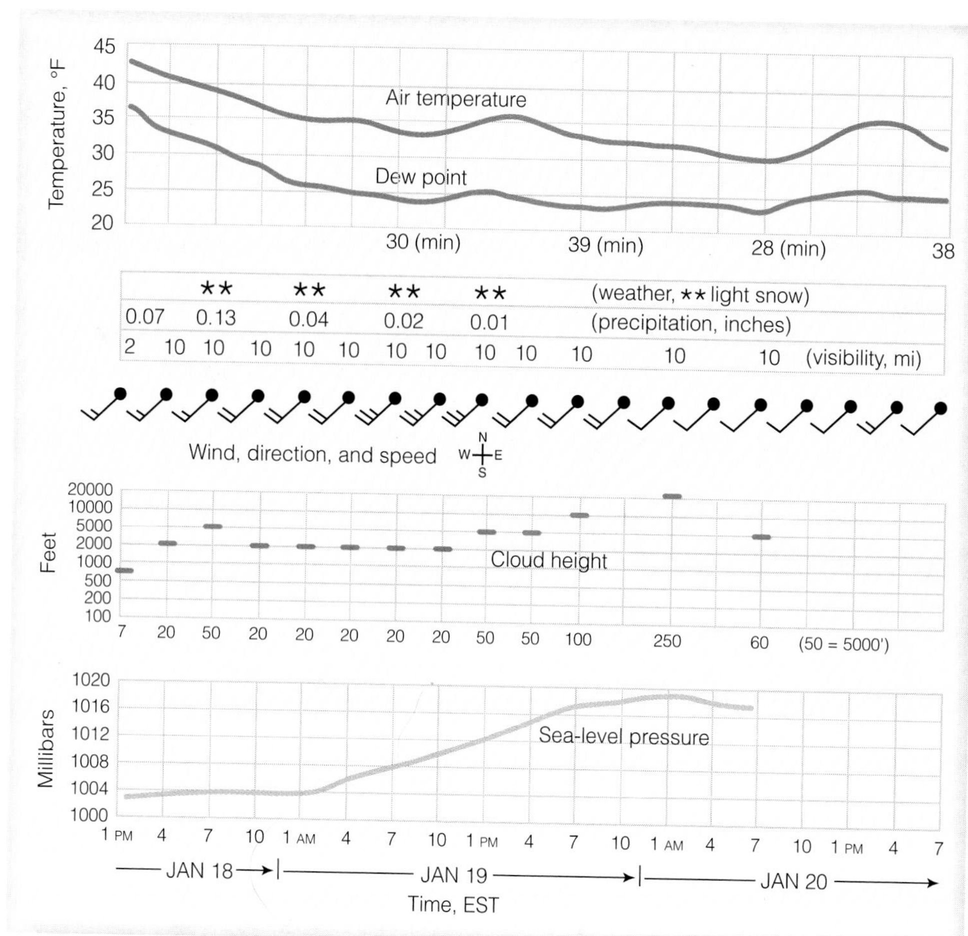

FIGURE 9.4 Meteogram illustrating predicted weather elements at Buffalo, New York, from 1 P.M. January 18, 1999, to 7 P.M. January 20, 1999. Notice that at 7 A.M. on January 19, the weather is projected to be: sea-level pressure 1007 mb; cloud height 2000 ft.; southwest winds at 15 knots; visibility 10 miles; light snow; air temperature 33°F; dew point 24°F; and the minimum temperature for the day should be 30°F. The forecast is derived from the model output statistics (MOS) of the Nested Grid Model (NGM).

cloud movement, dissipation, or development associated with weather fronts and storms. This information is a great help in forecasting the progress of large weather systems. Wind directions and speeds at various levels may also be approximated by monitoring cloud movement with the geostationary satellite.

To complement the geostationary satellites, there are **polar-orbiting satellites,** which closely parallel the earth's meridian lines. These satellites pass over the north and south polar regions on each revolution. As the earth rotates to the east beneath the satellite, each pass monitors an area to the west of the previous pass (see Fig. 9.7). Eventually, the satellite covers the entire earth.

Polar-orbiting satellites have the advantage of photographing clouds directly beneath them. Thus, they provide sharp pictures in polar regions, where photographs from a geostationary satellite are distorted because of the low angle at which the satellite "sees" this region. Polar orbiters also circle the earth at a much lower altitude (about 850 km, or 530 mi) than geostationary satellites and provide detailed photographic information about objects, such as violent storms and cloud systems.

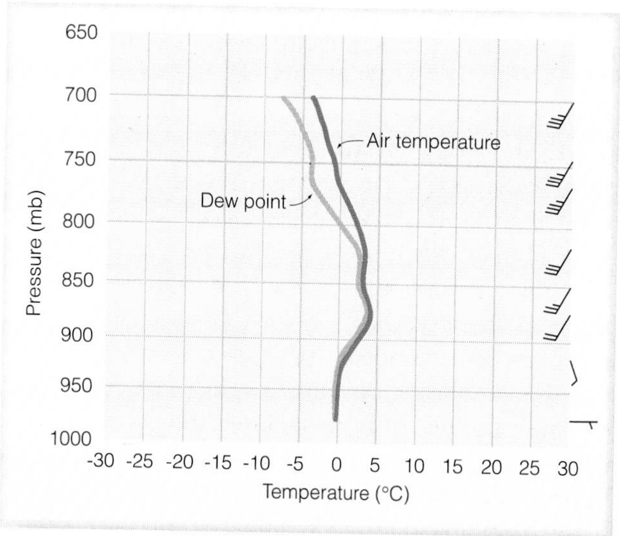

FIGURE 9.5 A sounding of air temperature, dew point, and winds at Pittsburgh, PA, on January 14, 1999. Looking at this sounding, a forecaster would see that saturated air extends up to about 820 mb. The forecaster would also observe that below-freezing temperatures only exist in a shallow layer near the surface and that the freezing rain presently falling over the Pittsburgh area would continue or possibly change to rain, as cold easterly surface winds are swinging around to warmer southwesterly winds aloft.

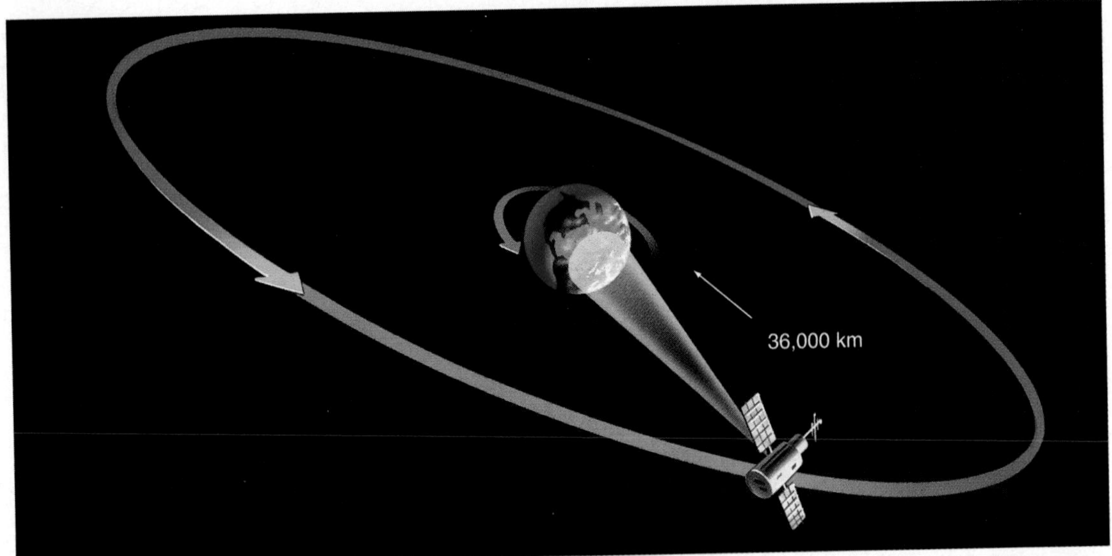

FIGURE 9.6 The geostationary satellite moves through space at the same rate that the earth rotates, so it remains above a fixed spot on the equator and monitors one area constantly.

Continuously improved detection devices make weather observation by satellites more versatile than ever. Early satellites, such as *TIROS I*, launched on April 1, 1960, used television cameras to photograph clouds. Contemporary satellites use radiometers, which can observe clouds during both day and night by detecting radiation that emanates from the top of the clouds. Additionally, the new generation *Geostationary Operational Environmental Satellite (GOES)* series has the capacity to obtain cloud images and, at the same time, provide vertical profiles of atmospheric temperature and moisture by detecting emitted radiation from atmospheric gases, such as water vapor. In modern satellites, a special type of advanced radiometer (called an *imager*) provides satellite pictures with much better resolution than did previous imagers. Moreover, another type of special radiometer (called a *sounder*) gives a more accurate profile of temperature and moisture at different levels in the atmosphere than did earlier instruments. In the latest *GOES* series, the imager and sounder are able to operate independently of each other.

The forecaster can obtain information on cloud thickness and height from satellite photographs. Visible photographs show the sunlight reflected from a cloud's upper surface. Because thick clouds have a higher reflectivity than thin clouds, they appear brighter on a visible satellite photograph. However, high, middle, and low clouds have just about the same reflectivity, so it is difficult to distinguish among them simply by using visible light photographs. To make this distinction, *infrared cloud pictures* are used. Such pictures produce a better image of the actual radiating surface because they do not show the strong visible reflected light. Since warm objects radiate more energy than cold objects, high temperature regions can be artificially made to appear darker on an infrared photograph. Because the tops of low clouds are warmer than those of high clouds, cloud observations made in the infrared can distinguish between warm low clouds (dark) and cold high clouds (light)—see Fig. 9.8. Moreover, cloud temperatures can be converted by a computer into a three-dimensional image of the cloud. These are the 3-D cloud photos presented on television by many weathercasters.

Figure 9.9a shows a visible satellite image (from a geostationary satellite) of an occluded storm system in the eastern Pacific. Notice that all of the clouds in the photo appear white. However, in the infrared photograph (Fig. 9.9b), taken on the same day (and just about the same time), the clouds appear to have many shades of gray. In the visible photograph, the clouds covering part of Oregon and northern California appear relatively thin compared to the thicker, bright clouds to the west. Furthermore, these thin clouds must be high because they also appear bright in the infrared picture.

Along the elongated band of clouds associated with the occluded front, the clouds appear white and bright in both pictures, indicating a zone of thick, heavy clouds. Behind the front, the forecaster knows that the lumpy clouds are probably cumulus because they appear gray in the infrared photo, suggesting that their tops are low and relatively warm.

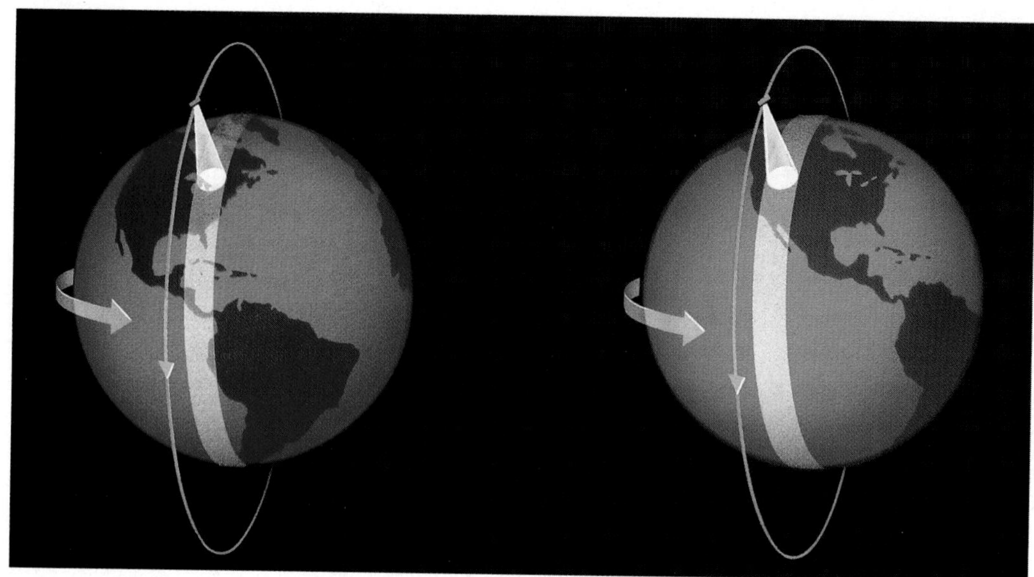

FIGURE 9.7 Polar-orbiting satellites scan from north to south, and on each successive orbit the satellite scans an area farther to the west.

When temperature differences are small, it is difficult to directly identify significant cloud and surface features on an infrared picture. Some way must be found to increase the contrast between features and their backgrounds. This can be done by a process called *computer enhancement.* Certain temperature ranges in the infrared photograph are assigned specific shades of gray—grading from black to white. Figure 9.10 is an infrared-enhanced picture for the same day and area as shown in Fig. 9.9. Note the dark and light contouring in the picture. Clouds with cold tops, and those with tops near freezing, are assigned the darkest gray color.

To make these types of features more obvious, often dark blue, red, or purple is assigned to clouds with the coldest (highest) tops. Hence, the dark red areas embedded along the front represent the region where the coldest and, therefore, highest and thickest clouds are found. It is here where the stormiest weather is probably occurring. Also notice that, near the southern tip of the picture, the dark red blotches surrounded by areas of white are thunderstorms that have developed over warm tropical waters. They show up clearly as white, thick clouds in both the visible and infrared photographs. By examining the movement of these clouds on successive satellite photographs, the forecaster can predict the arrival of clouds and storms, and the passage of weather fronts.

In regions where there are no clouds, it is difficult to observe the movement of the air. To help with this situation, the latest geostationary satellites are equipped with water-vapor sensors that can profile the distribution of atmospheric water vapor in the middle and upper troposphere (see Fig. 9.11). In time-lapse films, the swirling patterns of moisture clearly show wet regions and dry regions, as well as middle tropospheric swirling wind patterns and jet streams.

Up to this point, we have examined some of the tools that a forecaster might use in making a weather

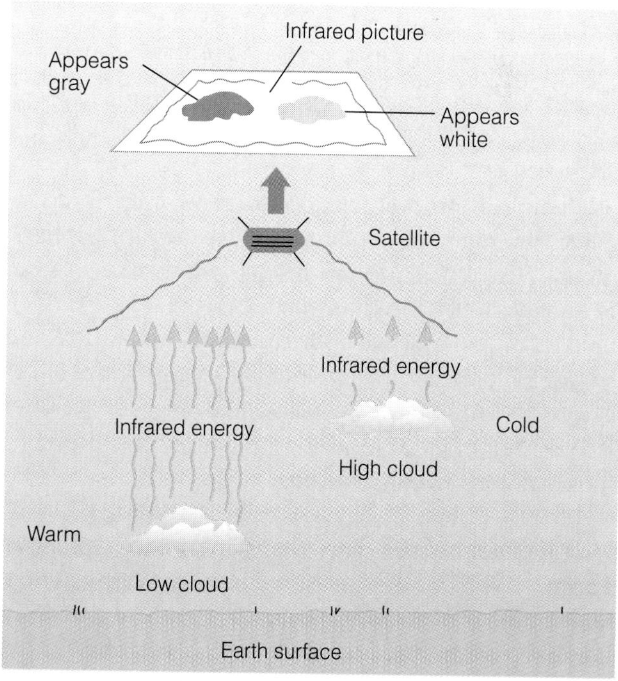

FIGURE 9.8 Generally, the lower the cloud, the warmer its top. Warm objects emit more infrared energy than do cold objects. Thus, an infrared satellite picture can distinguish warm, low (gray) clouds from cold, high (white) clouds.

(a)

(b)

Meteorology Now™ ACTIVE FIGURE 9.9
A visible image (a) and the infrared image (b) of the eastern Pacific were taken at just about the same time on the same day. Notice that the clouds in the visible image appear white. In the infrared image, the clouds appear in various shades of gray.
Watch this Active Figure at http://earthscience.brookscole.com/ahrens/ess4e.

prediction. Moreover, we looked at weather forecasts made by high-speed computers using atmospheric models. Currently, forecast models predict the weather reasonably well 4 to 6 days into the future. These models tend to do a better job of predicting temperature and jet-stream patterns than predicting precipitation. However, even with all of the modern advances in weather forecasting provided by ever more powerful computers, forecasts are sometimes wrong.

Why Forecasts Go Awry and Steps to Improve Them Why do forecasts sometimes go wrong? There are a number of reasons for this unfortunate situation. For one, computer models have inherent flaws that limit the accuracy of weather forecasts. For example, computer-forecast models idealize the real atmosphere, meaning that each model makes certain assumptions about the atmosphere. These assumptions may be on target for some weather situations and be way off for others. Consequently, the computer may produce a prog that on one day comes quite close to describing the actual state of the atmosphere, and not so close on another. A forecaster who bases a prediction on an "off day" computer prog may find a forecast of "rain and windy" turning out to be a day of "clear and colder."

Another forecasting problem arises because the majority of models are not global in their coverage, and errors are able to creep in along the model's boundaries. For example, a model that predicts the weather for North America may not accurately treat weather systems that move in along its boundary from the western Pacific. Obviously, a global model would be preferred. But a global model of similar sophistication would require an incredible number of computations.

Even though many thousands of weather observations are taken worldwide each day, there are still re-

NOAA

FIGURE 9.10 An enhanced infrared image of the eastern Pacific taken on the same day as the images shown in Fig. 9.9(a) and (b).

gions where observations are sparse, particularly over the oceans and at higher latitudes. As we saw in the previous section, to help alleviate this problem, the newest *GOES* satellite, with advanced atmospheric sounders, is providing a more accurate profile of temperature and humidity for the computer models. Wind information now comes from a variety of sources, such as Doppler radar, commercial aircraft, and satellites that translate ocean surface roughness into surface wind speed.

Earlier, we saw that the computer solves the equations that represent the atmosphere at many locations called grid points, each about 4 to 200 km apart. As a consequence, on computer models with large spacing between grid points (say 80 km), weather systems, such as extensive mid-latitude cyclones and anticyclones, show up on computer progs, whereas much smaller systems, such as thunderstorms, do not. The computer models that forecast for a large area such as North America are, therefore, better at predicting the widespread precipitation associated with a large cyclonic storm than local showers and thunderstorms. In summer, when much of the precipitation falls as local showers, a computer prog may have indicated fair weather, while outside it is pouring rain.

To capture the smaller-scale weather features as well as the terrain of the region, the distance between grid points on some models is being reduced. For example, the forecast model known as MM5 has a grid spacing as low as 4 km. This model predicts mesoscale atmospheric condi-

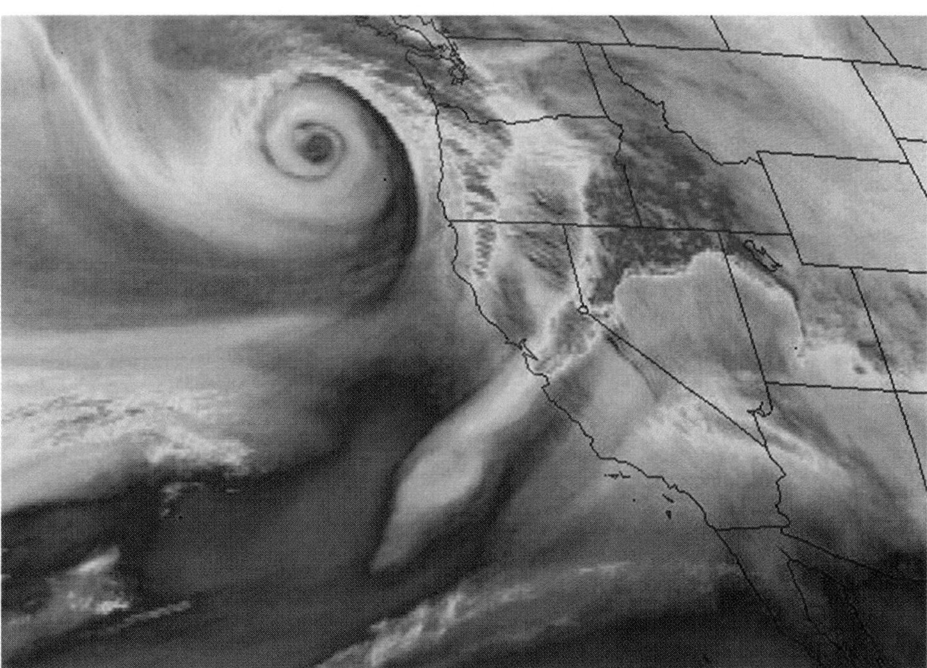

NOAA

FIGURE 9.11 Infrared water vapor image. The darker areas represent dry air aloft; the brighter the gray, the more moist the air in the middle or upper troposphere. Bright white areas represent dense cirrus clouds or the tops of thunderstorms. The area in color represents the coldest cloud tops. The swirl of moisture off the West Coast represents a well-developed mid-latitude cyclonic storm.

DID YOU KNOW?

Nightly news weather presentations have come a long way since the early days of television. The "weather girl," a fad that became popular during the 1960s, employed crazy gimmicks to attract viewers. Females gave weather forecasts in various attire (from bathing suits to bunny outfits), sometimes with the aid of hand puppets that resembled odd-looking turtles. One West Coast television station actually hired a woman to do the nightly news weather segment with the requirement that she be able to write backwards on a clear, Plexiglass screen.

tions over a limited region, such as a coastal area where terrain might greatly impact the local weather. The problem with models that have a small grid spacing (high resolution) is that, as the horizontal spacing between grid points decreases, the number of computations increases. When the distance is halved, there are 8 times as many computations to perform, and the time required to run the model goes up by a factor of 16.

Another forecasting problem is that many computer models cannot adequately interpret many of the factors that influence surface weather, such as the interactions of water, ice, surface friction, and local terrain on weather systems. Many large-scale models now take mountain regions and oceans into account. Some models (such as the MM5) take even smaller factors into account—features that large-scale computers miss due to their longer grid spacing. Given the effect of local terrain, as well as the impact of some of the other problems previously mentioned, computer models that forecast the weather over a vast area do an inadequate job of predicting local weather conditions, such as surface temperatures, winds, and precipitation.

Even with better observing techniques and near perfect computer models, there are countless small, unpredictable atmospheric fluctuations that fall under the heading of **chaos.** For example, tiny eddies are much smaller than the grid spacing on the computer model and, therefore, go unmeasured. These small disturbances, as well as small errors (uncertainties) in the data, generally amplify with time as the computer tries to project the weather farther and farther into the future. After a number of days, these initial imperfections tend to dominate, and the forecast shows little or no accuracy in predicting the behavior of the real atmosphere. In essence, what happens is that the small uncertainty in the initial atmospheric conditions eventually leads to a huge uncertainty in the forecast.

Because of the atmosphere's chaotic nature, meteorologists are turning to a technique called **ensemble forecasting** to improve medium-range forecasts. The ensemble approach is based on running several forecast models—or different versions (simulations) of a single model—each beginning with slightly different weather information to reflect the errors inherent in the measurements. Suppose, for example, a forecast model predicts the state of the atmosphere 24 hours into the future. For the ensemble forecast, the entire model simulation is repeated, but only after the initial conditions are "tweaked" just a little. The "tweaking," of course, represents the degree of uncertainty in the observations. Repeating this process several times creates an ensemble of forecasts for a range of small initial changes.

If, at the end of a specific time, the progs match each other fairly well, the forecast is considered *robust.* This situation allows the forecaster to issue a prediction with a high degree of confidence. If the progs disagree, the forecaster, with little faith in the computer model prediction, issues a forecast with limited confidence, perhaps by giving a number ranging from 0 (no confidence) to 5 (great confidence). In essence, *the less agreement among the progs, the less predictable the weather.* Consequently, it would not be wise to make outdoor plans for Saturday when on Monday the weekend forecast calls for "sunny and warm" with a low degree of confidence.

In summary, imperfect numerical weather predictions may result from flaws in the computer models, from errors that creep in along the models' boundaries, from the sparseness of data, and/or from inadequate representation of many pertinent processes, interactions, and inherently chaotic behavior that occurs within the atmosphere.

Up to this point, we have looked primarily at weather forecasts made by high-speed computers using atmospheric models. There are, however, other forecasting methods, many of which have stood the test of time and are based mainly on the experience of the forecaster. Many of these techniques are of value, but often they give more of a general overview of what the weather should be like, rather than a specific forecast. (Before going on, you may wish to read the Focus section on p. 243 that describes how TV weathercasters present weather visuals.)

Other Forecasting Methods Probably the easiest weather forecast to make is a **persistence forecast,** which is simply a prediction that future weather will be the same as present weather. If it is snowing today, a persistence forecast would call for snow through tomorrow.

FOCUS ON AN OBSERVATION

TV Weathercasters—How Do They Do It?

As you watch the TV weathercaster, you typically see a person describing and pointing to specific weather information, such as satellite photos, radar images, and weather maps, as illustrated in Fig. 2. What you may not know is that the weathercaster is actually pointing to a blank board (usually green or blue) on which there is nothing (Fig. 3). This process of electronically superimposing weather information in the TV camera against a blank wall is called color-separation overlay, or *chroma key.*

The chroma key process works because the studio camera is constructed to pick up all colors except (in this case) blue. The various maps, charts, satellite photos, and other graphics are electronically inserted from a computer into this blue area of the color spectrum. The person in the TV studio should not wear blue clothes because such clothing would not be picked up by the camera—what you would see on your home screen would be a head and hands moving about the weather graphics!

How, then, does a TV weathercaster know where to point on the blank wall? Positioned on each side of the blue wall are TV monitors (look carefully at Fig. 3) that weathercasters watch so that they know where to point.

Photo by author

FIGURE 2 On your home television, the weather forecaster Tom Loffman appears to be pointing to weather information directly behind him.

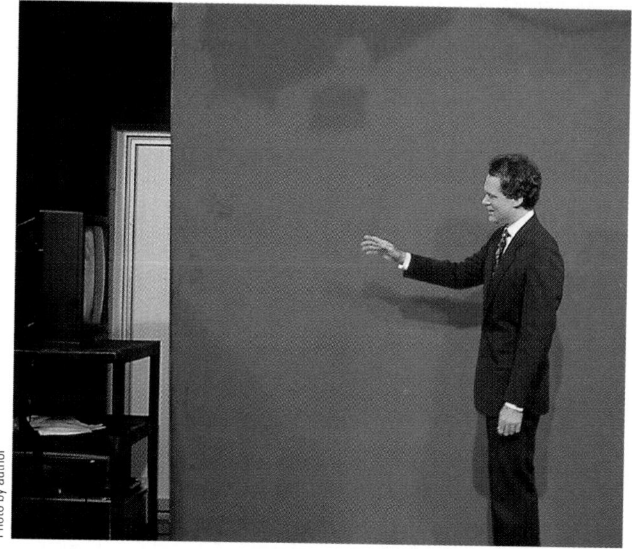

Photo by author

FIGURE 3 In the studio, however, he is actually standing in front of a blank board.

Such forecasts are most accurate for time periods of several hours and become less and less accurate after that.

Another method of forecasting is the **steady-state,** or **trend method.** The principle involved here is that surface weather systems tend to move in the same direction and at approximately the same speed as they have been moving, providing no evidence exists to indicate otherwise. Suppose, for example, that a cold front is moving eastward at an average speed of 30 mi/hr and it is 90 mi west of your home. Using the steady-state method, we might extrapolate and predict that the front should pass through your area in three hours.

The **analogue method** is yet another form of weather forecasting. Basically, this method relies on the fact that existing features on a weather chart (or a series of charts) may strongly resemble features that produced certain weather conditions sometime in the past. To the forecaster, the weather map "looks familiar," and for this reason the analogue method is often referred to as *pattern recognition.* A forecaster might look at a prog and say "I've seen this weather situation before, and this happened." Prior weather events can then be utilized as a guide to the future. The problem here is that, even though weather situations may appear similar, they are never *exactly* the

TABLE 9.1
Forecast wording used by the National Weather Service to describe the percentage probability of measurable precipitation (0.01 inch or greater) for steady precipitation and for convective, showery precipitation.

PERCENT PROBABILITY OF PRECIPITATION	FORECAST WORDING FOR STEADY PRECIPITATION	FORECAST WORDING FOR SHOWERY PRECIPITATION
20 percent	*Slight chance* of precipitation	*Widely scattered* showers
30 to 50 percent	*Chance* of precipitation	*Scattered* showers
60 to 70 percent	Precipitation *likely*	*Numerous* showers
≥ 80 percent	Precipitation,* *rain, snow*	*Showers**

*A forecast that calls for an 80 percent chance of rain in the afternoon might read like this: "... cloudy today with rain this afternoon...." For an 80 percent chance of rain showers, the forecast might read "... cloudy today with rain showers this afternoon...."

same. There are always sufficient differences in the variables to make applying this method a challenge.

The analogue method can be used to predict a number of weather elements, such as maximum temperature. Suppose that in New York City the average maximum temperature on a particular date for the past 30 years is 10°C (50°F). By statistically relating the maximum temperatures on this date to other weather elements—such as the wind, cloud cover, and humidity—a relationship between these variables and maximum temperature can be drawn. By comparing these relationships with current weather information, the forecaster can predict the maximum temperature for the day.

Presently, **statistical forecasts** are made routinely of weather elements based on the past performance of computer models. Known as *Model Output Statistics*, or MOS, these predictions, in effect, are statistically weighted analogue forecast corrections incorporated into the computer model output. For example, a forecast of tomorrow's maximum temperature might be derived from a statistical equation that uses a numerical model's forecast of relative humidity, cloud cover, wind direction, and air temperature.

When the Weather Service issues a forecast calling for rain, it is usually followed by a probability. For example: "The chance of rain is 60 percent." Does this mean (a) that it will rain on 60 percent of the forecast area or (b) that there is a 60 percent chance that it will rain within the forecast area? Neither one! The expression means that there is a 60 percent chance that any random place in the forecast area, such as your home, will receive measurable rainfall.* Looking at the forecast in another way, if the forecast for 10 days calls for a 60 percent chance of rain, it should rain where you live on 6 of those days. The verification of the forecast (as to whether it actually rained or not) is usually made at the Weather Service office, but remember that the computer models forecast for a given region, not for an individual location. When the National Weather Service issues a forecast calling for a "slight chance of rain," what is the probability (percentage) that it will rain? Table 9.1 provides this information.

An example of a **probability forecast** using climatological data is given in Fig. 9.12. The map shows the probability of a "White Christmas"—1 inch or more of snow on the ground—across the United States. The map is based on the average of 30 years of data and gives the likelihood of snow in terms of a probability. For instance, the chances are greater than 90 percent (9 Christmases out of 10) that portions of northern Minnesota, Michigan, and Maine will experience a White Christmas. In Chicago, it is close to 50 percent; and in Washington, D.C., about 20 percent. Many places in the far west and south have probabilities less than 5 percent, but nowhere is the probability exactly 0, for there is always some chance (no matter how small) that a mantle of white will cover the ground on Christmas day.

Predicting the weather by **weather types** employs the analogue method. In general, weather patterns are categorized into similar groups or "types," using such criteria as the position of the subtropical highs, the upper-level flow, and the prevailing storm track. As an example, when the Pacific high is weak or depressed southward and the flow aloft is zonal (west-to-east), surface storms tend to travel rapidly eastward across the Pacific Ocean and into the United States without developing into deep systems. But when the Pacific high is to the north of its normal position and the upper airflow is meridional (north-south), looping waves form in the flow with surface lows usually developing into huge storms. Since upper-level longwaves move slowly, usually remaining almost stationary for perhaps a few days to a week or more, the particular surface weather at dif-

*The 60 percent chance of rain does not apply to a situation that involves rain showers. In the case of showers, the percentage refers to the expected area over which the showers will fall.

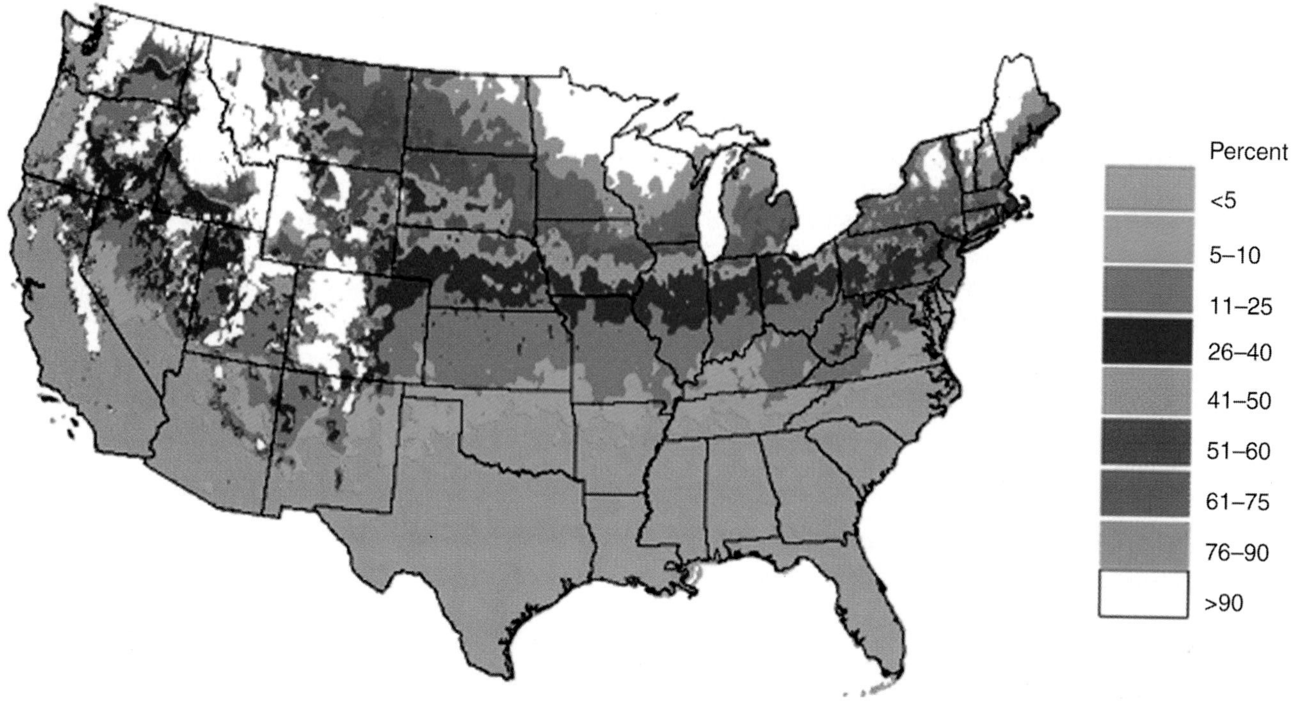

FIGURE 9.12 Probability of a "White Christmas"—one inch or more of snow on the ground—based on a 30-year average. The probabilities do not include the mountainous areas in the western United States.

ferent positions around the wave is likely to persist for some time. Figure 9.13 presents an example of weather conditions most likely to prevail with a winter meridional weather type.

A forecast based on the climatology (average weather) of a particular region is known as a **climatological forecast.** Anyone who has lived in Los Angeles for a while knows that July and August are practically rain-free. In fact, rainfall data for the summer months taken over many years reveal that rainfall amounts of more than a trace occur in Los Angeles about 1 day in every 90, or only about 1 percent of the time. Therefore, if we predict that it will not rain on some day next year during July or August in Los Angeles, our chances are nearly 99 percent that the forecast will be correct based on past records. Since it is unlikely that this pattern will significantly change in the near future, we can confidently make the same forecast for the year 2020.

Types of Forecasts Weather forecasts are normally grouped according to how far into the future the forecast extends. For example, a weather forecast for up to a few hours (usually not more than 6 hours) is called a **very short-range forecast,** or *nowcast.* The techniques used in making such a forecast normally involve subjective interpretations of surface observations, satellite imagery, and Doppler radar information. Often weather systems are moved along by the steady state or trend method of forecasting, with human experience and pattern recognition coming into play.

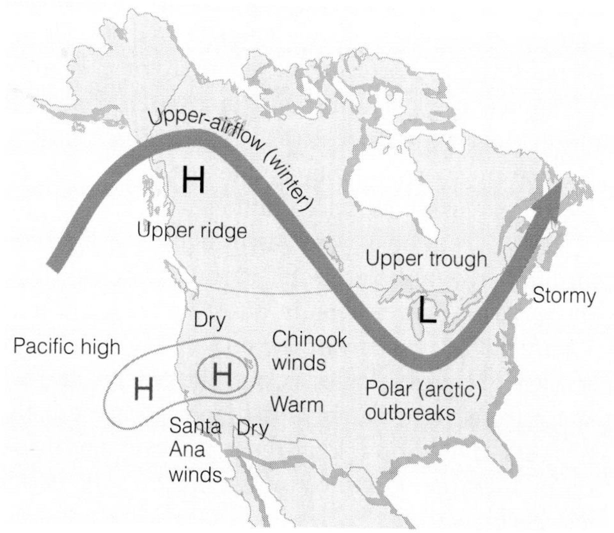

FIGURE 9.13 Winter weather type showing upper-airflow (heavy arrow), surface position of Pacific high, and general weather conditions that should prevail.

DID YOU KNOW?

When a weather forecast calls for "Fair tonight," does the "Fair" mean that the weather is better than "poor" but not up to being "good?" According to the National Weather Service, the subjective term "Fair" implies a rather pleasant weather situation where there is no precipitation, no extremes in temperature, good visibility, and less than 40 percent of the sky is covered by opaque clouds such as stratus.

Weather forecasts that range from about 6 hours to a few days (generally 2.5 days or 60 hours) are called **short-range forecasts.** The forecaster may incorporate a variety of techniques in making a short-range forecast, such as satellite imagery, Doppler radar, surface weather maps, upper-air winds, and pattern recognition. As the forecast period extends beyond about 12 hours, the forecaster tends to weight the forecast heavily on computer-drawn progs and statistical information, such as Model Output Statistics (MOS).

A **medium-range forecast** is one that extends from about 3 to 8.5 days (200 hours) into the future. Medium-range forecasts are almost entirely based on computer-derived products, such as forecast progs and statistical forecasts (MOS). A forecast that extends beyond 3 days is often called an *extended forecast.*

A forecast that extends beyond about 8.5 days (200 hours) is called a **long-range forecast.** Although computer progs are available for up to 16 days into the future, they are not accurate in predicting temperature and precipitation, and at best only show the broad-scale weather features. Presently, the Climate Prediction Center issues forecasts, called *outlooks,* of average weather conditions for a particular month or a season. These are not forecasts in the strict sense, but rather an overview of how average precipitation and temperature patterns may compare with normal conditions.

Initially, outlooks were based mainly on the relationship between the projected average upper-air flow and the surface weather conditions that the type of flow will create. Today, many of the outlooks are based on persistence statistics that carry over the general weather pattern from immediately preceding months, seasons, and years. In addition, long-range forecasts are made from models that link the atmosphere with the ocean surface temperature.

In Chapter 7, we saw how a vast warming (El Niño) or cooling (La Niña) of the equatorial tropical Pacific can affect the weather in different regions of the world. These interactions, where a warmer or cooler tropical Pacific can influence rainfall in California, are called **teleconnections.** These types of interactions between widely separated regions are identified through statistical correlations. For example, over regions of North America, where temperature and precipitation patterns tend to depart from normal during El Niño and La Niña events, the Climate Prediction Center can issue a *seasonal outlook* of an impending wetter or drier winter, months in advance. Forecasts using teleconnections have shown promise. For example, as the tropical equatorial Pacific became much warmer than normal during the spring and early summer of 1997, forecasters predicted a wet rainfall season over central and southern California. Although the heavy rains didn't begin until December, the weather during the winter of 1997–1998 was wet and wild: Storm after storm pounded the region, producing heavy rains, mud slides, road closures, and millions of dollars in damages.

In most locations throughout North America, the weather is fair more often than rainy. Consequently, there is a forecasting bias toward fair weather, which means that, if you made a forecast of no-rain where you live for each day of the year, your forecast would be correct more than 50 percent of the time. But did you show any *skill* in making your correct forecast? What constitutes skill, anyway? And how accurate are the forecasts issued by the National Weather Service?

Accuracy and Skill in Forecasting In spite of the complexity and ever-changing nature of the atmosphere, forecasts made for between 12 and 24 hours are usually quite accurate. Those made for between 2 and 5 days are fairly good. Beyond about 7 days, due to the chaotic nature of the atmosphere, forecast accuracy falls off rapidly. Although weather predictions made for up to 3 days are by no means perfect, they are far better than simply flipping a coin. But how accurate are they?

One problem with determining forecast accuracy is deciding what constitutes a right or wrong forecast. Suppose tomorrow's forecast calls for a minimum temperature of 5°C. If the official minimum turns out to be 6°C, is the forecast incorrect? Is it as incorrect as one 10 degrees off? By the same token, what about a forecast for snow over a large city, and the snow line cuts the city in half with the southern portion receiving heavy amounts and the northern portion none? Is the forecast right or wrong? At present, there is no clear-cut answer to the question of determining forecast accuracy.

How does forecast accuracy compare with forecast skill? Suppose you are forecasting the daily summertime weather in Los Angeles. It is not raining today and your forecast for tomorrow calls for "no rain." Suppose that tomorrow it doesn't rain. You made an accurate forecast, but did you show any skill in so doing? Earlier, we saw that the chance of measurable rain in Los Angeles on any summer day is very small indeed; chances are good that day after day it will not rain. For a forecast to show skill, it should be better than one based solely on the current weather *(persistence)* or on the "normal" weather *(climatology)* for a given region. Therefore, during the summer in Los Angeles, a forecaster will have many accurate forecasts calling for "no measurable rain," but will need skill to predict correctly on which summer days it will rain.

Meteorological forecasts, then, show skill when they are more accurate than a forecast utilizing only persistence or climatology. Persistence forecasts are usually difficult to improve upon for a period of time of several hours or less. Weather forecasts ranging from 12 hours to a few days generally show much more skill than those of persistence. However, as the range of the forecast period increases, because of chaos the skill drops quickly. The 6- to 14-day mean outlooks both show some skill (which has been increasing over the last several decades) in predicting temperature and precipitation. However, the accuracy of precipitation forecasts is less than that for temperature. Presently, 7-day forecasts now show about as much skill as 3-day forecasts did a decade ago. Beyond 15 days, specific forecasts are only slightly better than climatology.

Forecasting large-scale weather events several days in advance (such as the blizzard of 1996 along the eastern seaboard of the United States) is far more accurate than forecasting the precise evolution and movement of small-scale, short-lived weather systems, such as tornadoes and severe thunderstorms. In fact, 3-day forecasts of the development and movement of a major low-pressure system show more skill today than 36-hour forecasts did 15 years ago.

Even though the *precise* location where a tornado will form is presently beyond modern forecasting techniques, the general area where the storm is *likely* to form can often be predicted up to 3 days in advance. With improved observing systems, such as Doppler radar and advanced satellite imagery, the lead time of watches and warnings for severe storms has increased. In fact, the lead time* for tornado warnings has more than doubled over the last decade.

Although scientists may never be able to skillfully predict the weather beyond about 15 days using available observations, the prediction of *climatic trends* appears to be more promising. Whereas individual weather systems vary greatly and are difficult to forecast very far in advance, global-scale patterns of winds and pressure frequently show a high degree of persistence and predictable change over periods of a few weeks to a month or more. With the latest generation of high-speed supercomputers, general circulation models (GCMs) are doing a far better job at predicting large-scale atmospheric behavior than did the earlier models.

As new knowledge and methods of modeling are fed into the GCMs, it is hoped that they will become a reliable tool in the forecasting of weather and climate. (In Chapter 14, we will examine in more detail the climatic predictions based on numerical models.)

**Lead time* is the interval of time between the issue of the warning and actual observance of the tornado.

BRIEF REVIEW

Up to this point, we have looked at the various methods of weather forecasting. Before going on, here is a review of some of the important ideas presented so far:

- The forecasting of weather by high-speed computers is known as *numerical weather prediction.* Mathematical models that describe how atmospheric temperature, pressure, winds, and moisture will change with time are programmed into the computer. The computer then draws surface and upper-air charts, and produces a variety of forecast charts called *progs.*
- After a number of days, flaws in the computer models—atmospheric chaos and small errors in the data—greatly limit the accuracy of weather forecasts.
- Available to the forecaster are a number of tools that can be used when making a forecast, including surface and upper-air maps, computer progs, meteograms, soundings, and satellite information.
- Geostationary satellites remain fixed nearly 37,000 kilometers above a spot on the equator and orbit the earth at the same rate the earth spins. Polar-orbiting satellites, at about 850 kilometers above the earth, closely parallel meridian lines and scan the earth from north to south.
- Ensemble forecasting is a technique based on running several forecast models (or different versions of a single model), each beginning with slightly different weather information to reflect errors in the measurements.
- A *persistence forecast* is a prediction that future weather will be the same as the present weather, whereas a *climatological forecast* is based on the climatology of a particular region.
- Weather forecasts for up to a few hours are called *very short-range forecasts;* those that range from about 6 hours to a few days are called *short-range forecasts; medium-range forecasts* extend from about 3 to 5 days into the future, whereas *long-range forecasts* extend beyond, to about 8.5 days.
- For a forecast to show skill, it must be better than a persistence forecast or a climatological forecast.

Predicting the Weather from Local Signs Because the weather affects every aspect of our daily lives, attempts to predict it accurately have been made for centuries. One of the earliest attempts was undertaken by Theophrastus, a pupil of Aristotle, who in 300 B.C. compiled all sorts of weather indicators in his *Book of Signs.* A dominant influence in the field of weather forecasting for 2000 years, this work consists of ways to foretell the weather by examining natural signs, such as the color and shape of clouds, and the intensity at which a fly bites. Some of these signs have validity and are a part of our own weather folklore—"a halo around the moon portends rain" is one of these. Today, we realize that the halo is caused by the bending of light as it passes through ice crystals and that ice crystal-type clouds (cirrostratus) are often the forerunners of an approaching storm.

Weather predictions can be made by observing the sky and using a little weather wisdom. If you keep your eyes open and your senses keenly tuned to your environment, you should, with a little practice, be able to make fairly good short-range local weather forecasts by interpreting the messages written in the weather elements. Table 9.2 is designed to help you with this endeavor.

TABLE 9.2

Forecast at a Glance–Forecasting the Weather from Local Weather Signs. Listed below are a few forecasting rules that may be applied when making a short-range local weather forecast.

OBSERVATION	INDICATION	LOCAL WEATHER FORECAST
Surface winds from the S or from the SW; clouds building to the west; warm (hot) and humid (pressure falling)	Possible cool front and thunderstorms approaching from the west	Possible showers; possibly turning cooler; windy
Surface winds from the E or from the SE, cool or cold; high clouds thickening and lowering; halo around the sun or moon (pressure falling)	Possible approach of a warm front	Possibility of precipitation within 12–24 hours; windy (rain with possible thunderstorms during the summer; snow changing to sleet or rain in winter)
Strong surface winds from the NW or W; cumulus clouds moving overhead (pressure rising)	A low-pressure area may be moving to the east, away from you; and an area of high pressure is moving toward you from the west	Continued clear to partly cloudy, cold nights in winter; cool nights with low humidity in summer
Winter night		
(a) If clear, relatively calm with low humidity (low dew-point temperature)	(a) Rapid radiational cooling will occur	(a) A very cold night
(b) If clear, relatively calm with low humidity and snow covering the ground	(b) Rapid radiational cooling will occur	(b) A very cold night with minimum temperatures lower than in (a)
(c) If cloudy, relatively calm with low humidity	(c) Clouds will absorb and radiate infrared (IR) energy back to surface	(c) Minimum temperature will not be as low as in (a) or (b)
Summer night		
(a) Clear, hot, humid (high dew points)	(a) Strong absorption and emission of IR energy back to surface by water vapor	(a) High minimum temperatures
(b) Clear and relatively dry	(b) More rapid radiational cooling	(b) Lower minimum temperatures
Summer afternoon		
Scattered cumulus clouds that show extensive vertical growth by mid-morning	Atmosphere is relatively unstable	Possible showers or thunderstorms by afternoon with gusty winds
Afternoon cumulus clouds with limited vertical growth and with tops at just about the same level	Stable layer above clouds (region dominated by high pressure)	Continued partly cloudy with no precipitation; probably clearing by nightfall

WEATHER FORECASTING USING SURFACE CHARTS

We are now in a position to forecast the weather, utilizing more sophisticated techniques. Suppose, for example, that we wish to make a short-range weather prediction and the only information available is a surface weather map. Can we make a forecast from such a chart? Most definitely. And our chances of that forecast being correct improve markedly if we have maps available from several days back. We can use these past maps to locate the previous position of surface features and predict their movement.

A simplified surface weather map is shown in Fig. 9.14. The map portrays early winter weather conditions on Tuesday morning at 6:00 A.M. A single isobar is drawn around the pressure centers to show their positions without cluttering the map. Note that an open wave cyclone is developing over the Central Plains with showers forming along a cold front and light rain and snow ahead of a warm front. The dashed lines on the map represent the position of the weather systems six hours ago. Our first question is: How will these systems move?

Determining the Movement of Weather Systems

There are several methods we can use in forecasting the movement of surface pressure systems and fronts. The following are a few of these forecasting rules of thumb:

1. For short-time intervals, mid-latitude cyclonic storms and fronts tend to move in the same direction and at approximately the same speed as they did during the previous six hours (providing, of course, there is no evidence to indicate otherwise).
2. Low-pressure areas tend to move in a direction that parallels the isobars in the warm air ahead of the cold front.
3. Lows tend to move toward the region of greatest surface pressure drop, whereas highs tend to move toward the region of greatest surface pressure rise.
4. Surface pressure systems tend to move in the same direction as the wind at 5500 m (18,000 ft)—the

> **DID YOU KNOW?**
>
> Groundhog Day, February 2, derives from certain religious ceremonies performed before the birth of Christ. Somehow, this date came to be considered the midpoint of winter, and people, in an attempt to forecast what the remaining half would be like, placed the burden of weather prognosticator on the backs (or rather, the shadows) of animals such as the groundhog.

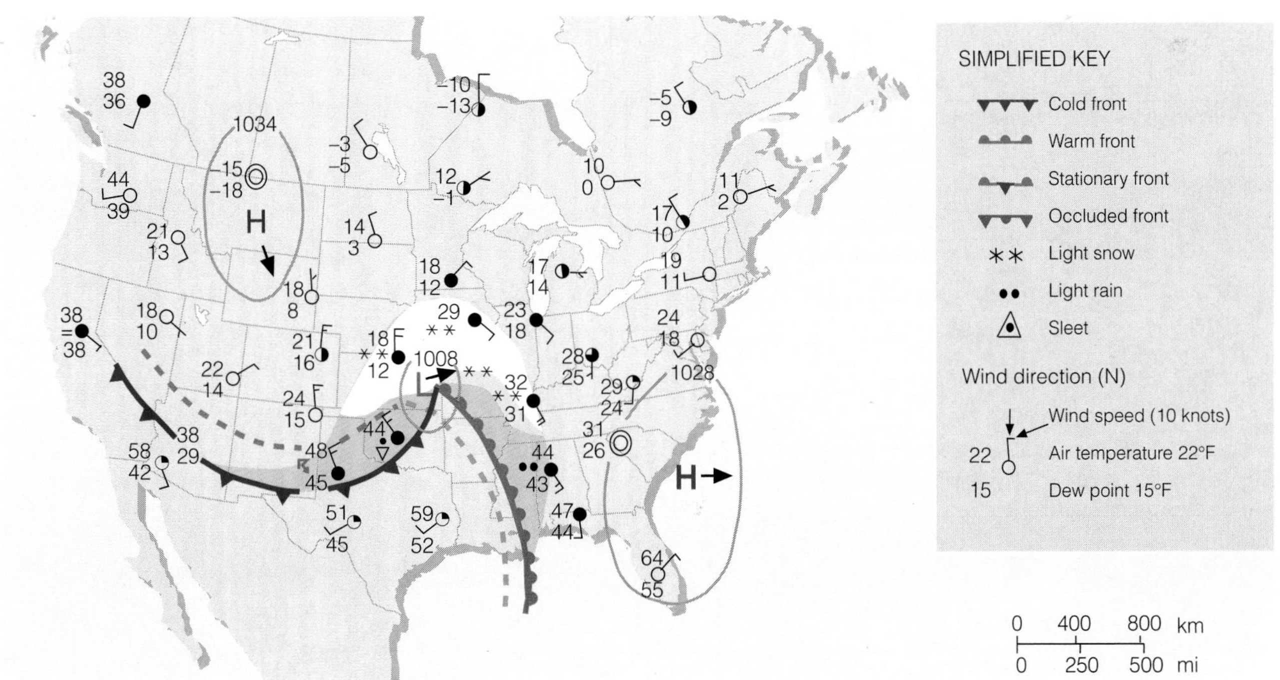

FIGURE 9.14 Surface weather map for 6:00 A.M. Tuesday. Dashed lines indicate positions of weather features six hours ago. Areas shaded green are receiving rain, while areas shaded white are receiving snow, and those shaded pink, freezing rain or sleet.

500-mb level. The speed at which surface systems move is about half the speed of the winds at this level.

When the surface map (Fig. 9.14) is examined carefully and when rules of thumb 1 and 2 are applied, it appears that—based on present trends—the storm center over the Central Plains should move northeast. When we observe the 500-mb upper-air chart (Fig. 9.15), it too suggests that the surface low should move northeast at a speed of about 25 knots.

A Forecast for Six Cities We are now in a position to make a weather forecast for six cities. To do this, we will project the pressure systems, fronts, and current weather into the future by assuming steady-state conditions. Figure 9.16 gives the 12- and 24-hour projected positions of these features.

A word of caution before we make our forecasts. We are assuming that the pressure systems and fronts are moving at a constant rate, which may or may not occur. Low-pressure areas, for example, tend to accelerate until they occlude, after which their rate of movement slows. Furthermore, the direction of moving systems may change due to "blocking" highs and lows that exist in their path or because of shifting upper-level wind patterns. We will assume a constant rate of movement and forecast accordingly, always keeping in mind that the longer our forecasts extend into the future, the more susceptible they are to error.

If we move the low- and high-pressure areas eastward, as illustrated in Fig. 9.16, we can make a basic weather forecast for a variety of regions. For example, the cold front moving into north Texas on Tuesday morning is projected to pass Dallas by that evening, so a forecast for the Dallas area would be "warm with showers, then turning colder." But we can do much better than this. Knowing the weather conditions that accompany advancing pressure areas and fronts, we can make more detailed weather forecasts that will take into account changes in temperature, pressure, humidity, cloud cover, precipitation, and winds. Our forecast will include the 24-hour period from Tuesday morning to Wednesday morning for the cities of Augusta, Georgia; Washington, D.C.; Chicago, Illinois; Memphis, Tennessee; Dallas, Texas; and Denver, Colorado. We will begin with Augusta.

Weather Forecast for Augusta, Georgia On Tuesday morning, continental polar air associated with a high pressure area brought freezing temperatures and fair weather to the Augusta area (see Fig. 9.14). Clear skies, light winds, and low humidities allowed rapid nighttime cooling so that, by morning, temperatures were in the low thirties. Now look

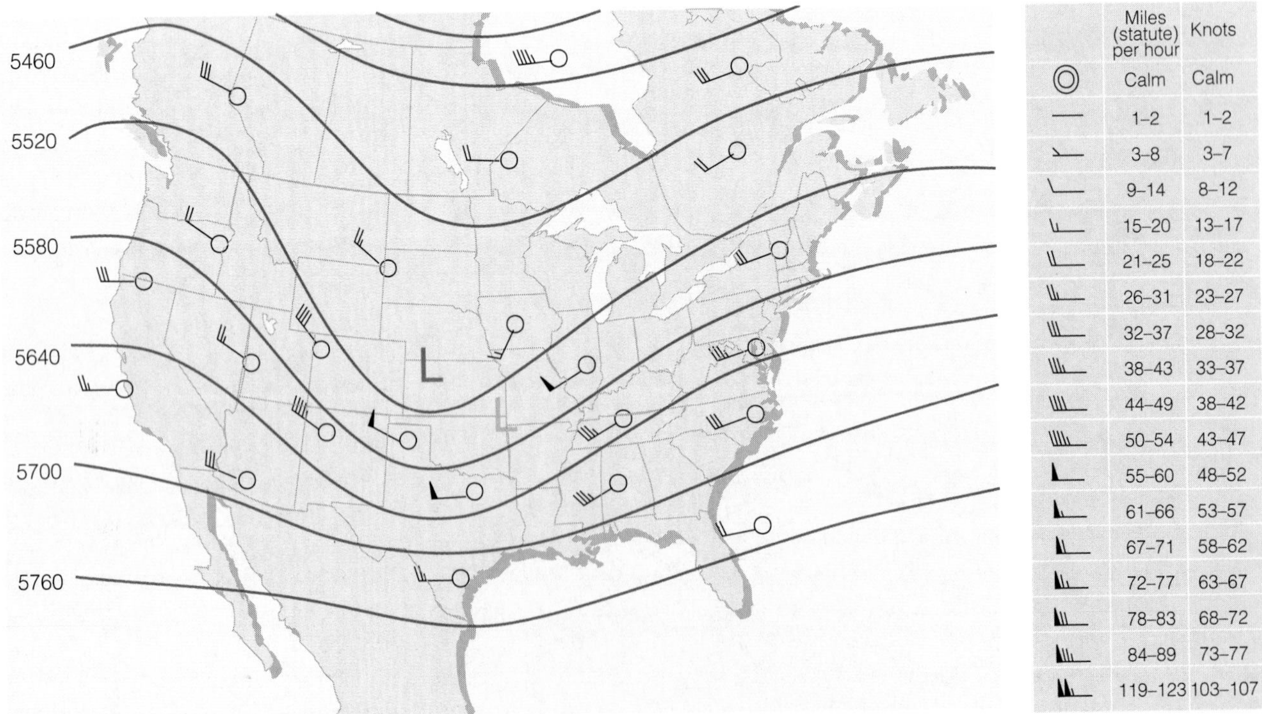

FIGURE 9.15 A 500-mb chart for 6:00 A.M. Tuesday, showing wind flow. The light red L represents the position of the surface low. The winds aloft tend to steer surface pressure systems along and, therefore, indicate that the surface low should move northeastward at about half the speed of the winds at this level, or 25 knots. Solid lines are contours in meters above sea level.

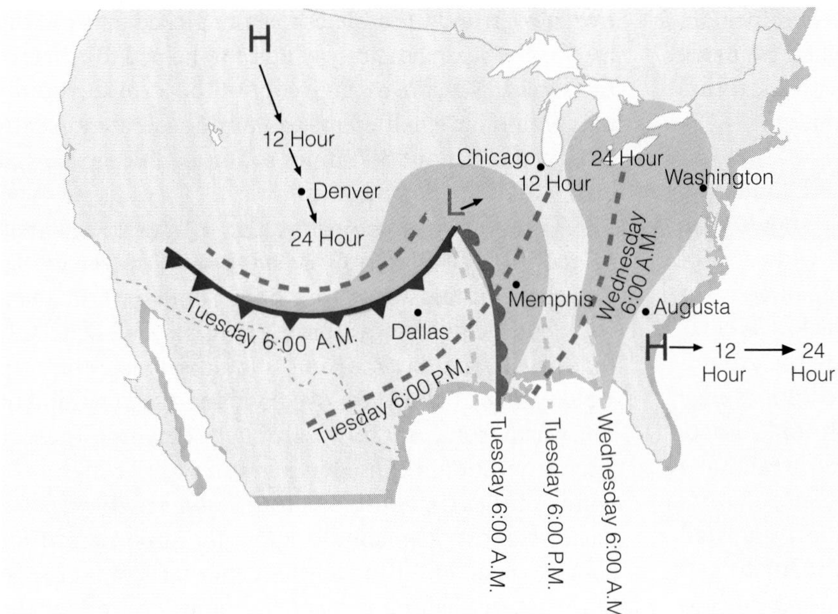

FIGURE 9.16 Projected 12- and 24-hour movement of fronts, pressure systems, and precipitation (shaded green area) from 6:00 A.M. Tuesday until 6:00 A.M. Wednesday.

closely at Fig. 9.16 and observe that the anticyclone is moving slowly eastward. Southerly winds on the western side of this system will bring warmer and more moist air to the region. Therefore, afternoon temperatures will be warmer than those of the day before. As the warm front approaches from the west, clouds will increase, appearing first as cirrus, then thickening and lowering into the normal sequence of warm-front clouds. Barometric pressure should fall. Clouds and high humidity should keep minimum temperatures well above freezing on Tuesday night. Note that the projected area of precipitation (green-shaded region) does not quite reach Augusta. With all of this in mind, our forecast might sound something like this:

> Clear and cold this morning with moderating temperatures by afternoon. Increasing high clouds with skies becoming overcast by evening. Cloudy and not nearly as cold tonight and tomorrow morning. Winds will be light and out of the south or southeast. Barometric pressure will fall slowly.

Wednesday morning we discover that the weather in Augusta is foggy with temperatures in the upper 40s (°F). But fog was not in the forecast. What went wrong? We forgot to consider that the ground was still cold from the recent cold snap. The warm, moist air moving over the cold surface was chilled below its dew point, resulting in fog. Above the fog were the low clouds we predicted. The minimum temperatures remained higher than anticipated because of the release of latent heat during fog formation and the absorption of infrared energy by the fog droplets. Not bad for a start. Now we will forecast the weather for Washington, D.C.

Rain or Snow for Washington, D.C.? Look at Fig. 9.16 and observe that the low-pressure area is slowly approaching Washington, D.C., from the west. Hence, the clear weather, light southwesterly winds, and low temperatures on Tuesday morning (Fig. 9.14) will gradually give way to increasing cloudiness, winds becoming southeasterly, and slightly higher temperatures. By Wednesday morning, the projected band of precipitation will be over the city. Will it be in the form of rain or snow? Without a vertical profile of temperature (a sounding), this question is difficult to answer. We can see in Fig. 9.16, however, that cities south of Washington, D.C.'s latitude are receiving snow. So a reasonable forecast would call for snow, possibly changing to rain as warm air moves in aloft in advance of the approaching fronts. A 24-hour forecast for Washington, D.C., might sound like this:

> Increasing clouds today and continued cold. Snow beginning by early Wednesday morning, possibly changing to rain. Winds will be out of the southeast. Pressures will fall.

Wednesday morning a friend in Washington, D.C., calls to tell us that the sleet began to fall but has since changed to rain. Sleet? Another fractured forecast! Well, almost. What we forgot to account for this time was the intensification of the storm. As the storm moved eastward, it deepened; central pressure lowered, pressure gradients tightened, and southeasterly winds blew stronger than anticipated. As air moved inland off the warmer Atlantic, it rode up and over the colder surface air. Snow falling into this warm layer at least partially melted; it then refroze as it entered the colder air near ground level. The influx of warmer air from the ocean slowly raised the surface temperatures, and the sleet soon became rain. Although we did not see this possibility when we made our forecast, a forecaster more familiar with local surroundings would have. Let's move on to Chicago.

Big Snowstorm for Chicago From Figs. 9.14 and 9.16, it appears that Chicago is in for a major snowstorm. Overrunning of warm air has produced a wide area of snow which, from all indications, is heading directly for the Chicago area. Since cold air north of the low's center will be over Chicago, precipitation reaching the ground should be frozen. On Tuesday morning the leading edge of precipitation is less than six hours away from Chicago. Based on the projected path of the storm, light snow should begin to fall around noon.

By evening, as the storm intensifies, snowfall should become heavy. It should taper off and finally end around midnight as the center of the low moves on east. If it snows for a total of twelve hours—six hours as light snow (around one inch every three hours) and six hours as heavy snow (around one inch per hour)—then the total expected accumulation will be between six and ten inches. As the low moves eastward, passing south of Chicago, winds on Tuesday will gradually shift from southeasterly to easterly, then northeasterly by evening. Since the system is intensifying, it should produce strong winds that will swirl the snow into huge drifts, which may bring traffic to a crawl.

The winds will continue to shift to the north and finally become northwesterly by Wednesday morning. By then the storm center will probably be far enough east so that skies should begin to clear. Cold air moving in from the northwest behind the storm will cause temperatures to drop further. Barometer readings during the storm will fall as the low's center approaches and reach a low value sometime Tuesday night, after which they will begin to rise. A weather forecast for Chicago might be:

> Cloudy and cold with light snow beginning by noon, becoming heavy by evening and ending by Wednesday morning. Total accumulations will range between six and ten inches. Winds will be strong and gusty out of the east or northeast today, becoming northerly tonight and northwesterly by Wednesday morning. Barometric pressure will fall sharply today and rise tomorrow.

A call Wednesday morning to a friend in Chicago reveals that our forecast was correct except that the total snow accumulation so far is 13 inches. We were off in our forecast because the storm system slowed as it became occluded. We did not consider this because we moved the system by the steady-state method. At this time of year (early winter), Lake Michigan is not quite frozen over, and the added moisture picked up from the lake by the strong easterly winds also helped to enhance the snowfall. Again, a knowledge of the local surroundings would have helped make a more accurate forecast. The weather about 500 miles south of Chicago should be much different from this.

Mixed Bag of Weather for Memphis Observe in Fig. 9.16 that, within twenty-four hours, both a warm and a cold front should move past Memphis. The light rain that began Tuesday morning should saturate the cool air, creating a blanket of low clouds and fog by midday. The warm front, as it moves through sometime Tuesday afternoon, should cause temperatures to rise slightly as winds shift to the south or southwest. At night, clear to partly cloudy skies should allow the ground and air above to cool, offsetting any tendency for a rapid rise in temperature. Falling pressures should level off in the warm air, then fall once again as the cold front approaches. According to the projection in Fig. 9.16, the cold front should arrive sometime before midnight on Tuesday, bringing with it gusty northwesterly winds, showers, the possibility of thunderstorms, rising pressures, and colder air. Taking all of this into account, our weather forecast for Memphis will be:

> Cloudy and cool with light rain, low clouds, and fog early today, becoming partly cloudy and warmer by late this afternoon. Clouds increasing with possible showers and thunderstorms later tonight or early Wednesday morning and turning colder. Winds southeasterly this morning, becoming southerly or southwesterly this evening and shifting to northwesterly by Wednesday morning. Pressures falling this morning, leveling off this evening, then falling again tonight and rising by Wednesday morning.

A friend who lives near Memphis calls Wednesday to inform us that our forecast was correct except that the thunderstorms did not materialize and that Tuesday night dense fog formed in low-lying valleys, but by Wednesday morning it had dissipated. Apparently, in the warm air, winds were not strong enough to mix the cold, moist air that had settled in the valleys with the warm air above. It's on to Dallas.

Cold Wave for Dallas From Fig. 9.16, it appears that our weather forecast for Dallas should be straightforward, since a cold front is expected to pass the area around noon. Weather along the front is showery with a few thunderstorms developing; behind the front the air is clear but cold. By Wednesday morning it looks as if the cold front will be far to the east and south of Dallas and an area of high pressure will be centered over Colorado.

North or northwesterly winds on the east side of the high will bring cold continental polar air into Texas, dropping temperatures as much as 40°F within a 24-hour period. With minimum temperatures well below freezing, Dallas will be in the grip of a cold wave. Our weather forecast should therefore sound something like this:

> Increasing cloudiness and mild this morning with the possibility of showers and thunderstorms this afternoon. Clearing and turning much colder tonight and tomorrow. Winds will be southwesterly today, becoming gusty north or northwesterly this afternoon and tonight. Pressures falling this morning, then rising later today.

How did our forecast turn out? A quick call to Dallas on Wednesday morning reveals that the weather there is cold but not as cold as expected, and the sky is overcast. Cloudy weather? How can this be?

The cold front moved through on schedule Tuesday afternoon, bringing showers, gusty winds, and cold weather with it. Moving southward, the front gradually slowed and became stationary along a line stretching from the Gulf of Mexico westward through southern Texas and northern Mexico. (From the surface map alone, we had no way of knowing this would happen.) Along the stationary front a wave of low pressure formed. This wave caused warm, moist Gulf air to slide northward up and over the cold surface air. Clouds formed, minimum temperatures did not go as low as expected, and we are left with a fractured forecast. Let's try Denver.

Clear but Cold for Denver In Fig. 9.16, we can see that, based on our projections, the cold high-pressure area will be almost directly over Denver by Wednesday morning. Sinking air aloft should keep the sky relatively free of clouds. Weak pressure gradients will produce only weak winds and this, coupled with dry air, will allow for intense radiational cooling. Minimum temperatures will probably drop to well below 0°F. Our forecast should therefore read:

> Clear and cold through tomorrow. Northerly winds today becoming light and variable by tonight. Low temperatures tomorrow morning will be below zero. Barometric pressure will continue to rise.

Almost reluctantly Wednesday morning, we inquire about the weather conditions at Denver. "Clear and very cold" is the reply. A successful forecast at last! We are told, however, that the minimum temperature did not go below zero; in fact, 13°F was as cold as it got. A downslope wind coming off the mountains to the west of Denver kept the air mixed and the minimum temperature higher than expected. Again, a forecaster familiar with the local topography of the Denver area would have foreseen the conditions that lead to such downslope winds and would have taken this into account when making the forecast.

A complete picture of the surface weather systems for 6:00 A.M. Wednesday morning is given in Fig. 9.17. By comparing this chart with Fig. 9.16, we can summarize why our forecasts did not turn out exactly as we had predicted. For one thing, the center of the low-pressure area over the Central Plains moved slower than expected. This slow movement allowed a southeasterly flow of mild Atlantic air to overrun cooler surface air ahead of the storm while, behind the low, cities remained in the snow area for a longer time. The weak wave that developed along

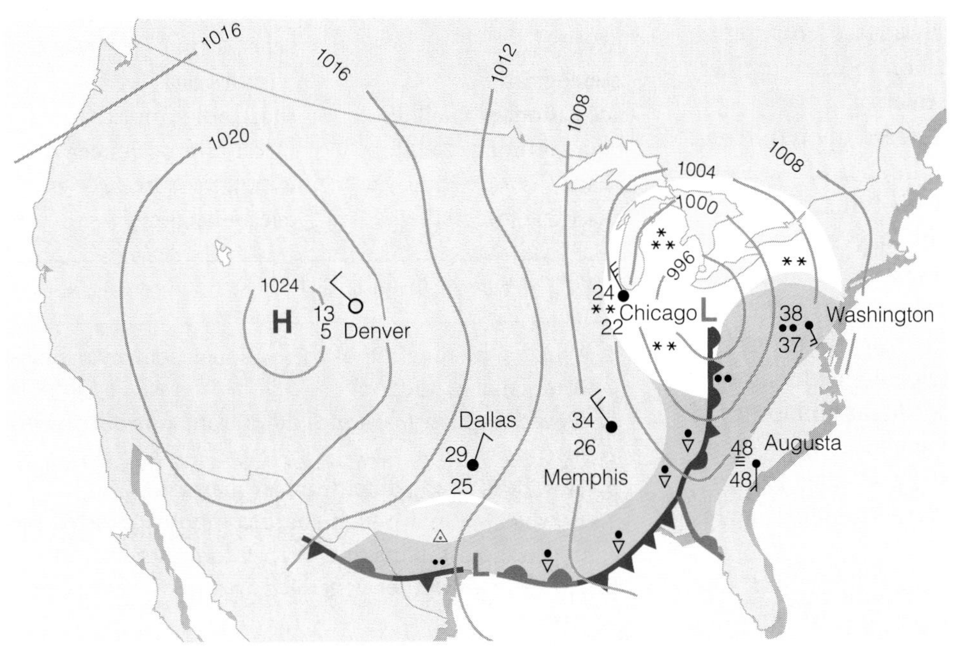

FIGURE 9.17 Surface weather map for 6:00 A.M. Wednesday.

the trailing cold front brought cloudiness and precipitation to Texas and prevented the really cold air from penetrating deep into the south. Further west, the high-pressure area originally over Montana moved more southerly than southeasterly, which set up a pressure gradient that brought westerly downslope winds to eastern Colorado.

In summary, the forecasting techniques discussed in this section are those you can use when making a short-range weather prediction. Keep in mind, however, that this chapter was not intended to make you an expert weather forecaster, nor was it designed to show you all the methods of weather prediction. It is hoped that you now have a better understanding of some of the problems confronting anyone who attempts to predict the behavior of this churning mass of air we call our atmosphere.

Meteorology Now™ Click "Forecasting" to examine current weather charts and to see how your own forecasts compare with those made by meteorological computer models.

SUMMARY

Forecasting tomorrow's weather entails a variety of techniques and methods. Persistence, surface maps, satellite imagery, and Doppler radar are all useful when making a very short-range (0–6 hour) prediction. For short- and medium-range forecasts, the current analysis, satellite data, pattern recognition, meteorologist intuition, and experience, along with statistical information and guidance from the many computer progs supplied by the National Weather Service, all go into making a prediction. For monthly and seasonal long-range forecasts, meteorologists incorporate changes in sea-surface temperature in the tropical Pacific into seasonal outlooks of temperature and precipitation in North America.

Different computer progs are based upon different atmospheric models that describe the state of the atmosphere and how it will change with time. The atmosphere's chaotic behavior, along with flaws in the models and tiny errors (uncertainties) in the data, generally amplify as the computer tries to project weather farther and farther into the future. At present, computer progs that predict the weather over a vast region are better at forecasting the position of mid-latitude highs and lows and their development than at forecasting local showers and thunderstorms. To skillfully forecast smaller features, the grid spacing on some models is reduced to as low as 4 kilometers.

Satellites aid the forecaster by providing a bird's-eye view of clouds and storms. Polar-orbiting satellites obtain data covering the earth from pole to pole, whereas geostationary satellites situated above the equator supply the forecaster with dynamic images of cloud and storm development and movement. To show where the highest and thickest clouds are located in a particular storm, infrared pictures are often enhanced by computer.

In the latter part of this chapter, we learned how people, by observing the weather around them, and by watching the weather systems on surface weather maps, can make fairly good short-range weather predictions.

Most of the forecasting methods in this chapter apply mainly to skill in predicting events associated with large-scale weather systems, such as fronts and mid-latitude cyclones. The next chapter on severe weather deals with the formation and forecasting of smaller-scale (mesoscale) systems, such as thunderstorms, squall lines, and tornadoes.

Meteorology Now™ Assess your understanding of this chapter's topics with additional quizzing and tutorials at http://earthscience.brookscole.com/ahrens/ess4e.

KEY TERMS

The following terms are listed in the order they appear in the text. Define each. Doing so will aid you in reviewing the material covered in this chapter.

watch (weather)
warning (weather)
analysis
numerical weather prediction
atmospheric models
prognostic chart (prog)
AWIPS
meteogram
geostationary satellites
polar-orbiting satellites
chaos
ensemble forecasting
persistence forecast
steady-state (trend)
analogue method
statistical forecast
probability forecast
weather types
climatological forecast
very short range forecast (nowcast)
short-range forecast
medium-range forecast
long-range forecast
teleconnections

QUESTIONS FOR REVIEW

1. What is the function of the National Center for Environmental Prediction?
2. How does a *weather watch* differ from a *weather warning*?
3. How does a prog differ from an analysis?
4. In what ways have high-speed computers assisted the meteorologist in making weather forecasts?
5. How are computer-generated weather forecasts prepared?

6. What are some of the problems associated with computer-model forecasts?
7. List some of the tools a weather forecaster might use when making a short-range forecast.
8. How do geostationary satellites differ from polar-orbiting satellites?
9. What are some of the problems associated with computer-model forecasts?
10. (a) Explain how satellites aid in forecasting the weather.
    (b) Using infrared satellite information, how can a forecaster distinguish high clouds from low clouds?
    (c) Why is it often necessary to enhance infrared satellite images?
11. Describe four methods of forecasting the weather and give an example for each one.
12. How does pattern recognition aid a forecaster in making a prediction?
13. Suppose that where you live, the middle of January is typically several degrees warmer than the rest of the month. If you forecast this "January thaw" for the middle of next January, what type of weather forecast will you have made?
14. (a) Look out the window and make a persistence forecast for tomorrow at this time.
    (b) Did you use any skill in making this prediction?
15. How can ensemble forecasts improve medium-range forecasts?
16. Explain how teleconnections are used in making a long-range seasonal outlook.
17. If today's weather forecast calls for a *"chance of snow,"* what is the percentage probability that it will snow today? (Hint: See Table 9.1, p. 248.)
18. Do all accurate forecasts show skill on the part of the forecaster? Explain.
19. List three methods that you would use to predict the movement of a surface mid-latitude cyclonic storm.
20. Do monthly and seasonal forecasts make specific predictions of rain or snow? Explain.

QUESTIONS FOR THOUGHT AND EXPLORATION

1. What types of watches and warnings are most commonly issued for your area?
2. Since computer models have difficulty in adequately considering the effects of small-scale geographic features on a weather map, why don't numerical weather forecasters simply reduce the grid spacing to, say, 1 kilometer on all models?
3. Suppose it's warm and raining outside. A cold front will pass your area in 3 hours. Behind the front, it is cold and snowing. Make a persistence forecast for your area 6 hours from now. Would you expect this forecast to be correct? Explain. Now, make a forecast for your area using the steady-state or trend method.
4. Why isn't the steady-state method very accurate when forecasting the weather more than a few hours into the future? What considerations can be taken into account to improve a steady-state forecast?
5. Go outside and observe the weather. Make a weather forecast using the weather signs you observe. Explain the rationale for your forecast.
6. Explain how the phrase "sensitive dependence on initial conditions" relates to the final outcome of a computer-based weather forecast.
7. Suppose the chance for a "White Christmas" at your home is 10 percent. Last Christmas was a white one. If for next year you forecast a "nonwhite" Christmas, will you have shown any skill if your forecast turns out to be correct? Explain.
8. Compare the visible satellite picture (Fig. 9.9a) with the infrared image (Fig. 9.9b). With the aid of the infrared picture, label on the visible picture the regions of middle, high, and low clouds. On the enhanced infrared picture (Fig. 9.10), label where the highest and thickest clouds appear to be located.
9. Computer model forecasts (http://cirrus.sprl.umich.edu/wxnet/model/model.html): Look at 12-, 24-, 36- and 48-hour forecast maps from a numerical weather prediction model. Can you observe the life cycle of a mid-latitude cyclone in the forecasts? Describe the major weather conditions that are affecting the forecast area.
10. Satellite Water Vapor Images (http://www.ssec.wisc.edu/data/g8/latest_g8wv.gif and http://www.ssec.wisc.edu/data/g9/latest_g9wv.gif): Examine current water-vapor patterns as measured by satellites. How do high areas of water vapor appear in the image? What do dry areas look like?

• Go to the Brooks/Cole Earth Sciences Resource Center (http://earthscience.brookscole.com) for critical thinking exercises, articles, and additional readings from InfoTrac College Edition, Brooks/Cole's online student library.

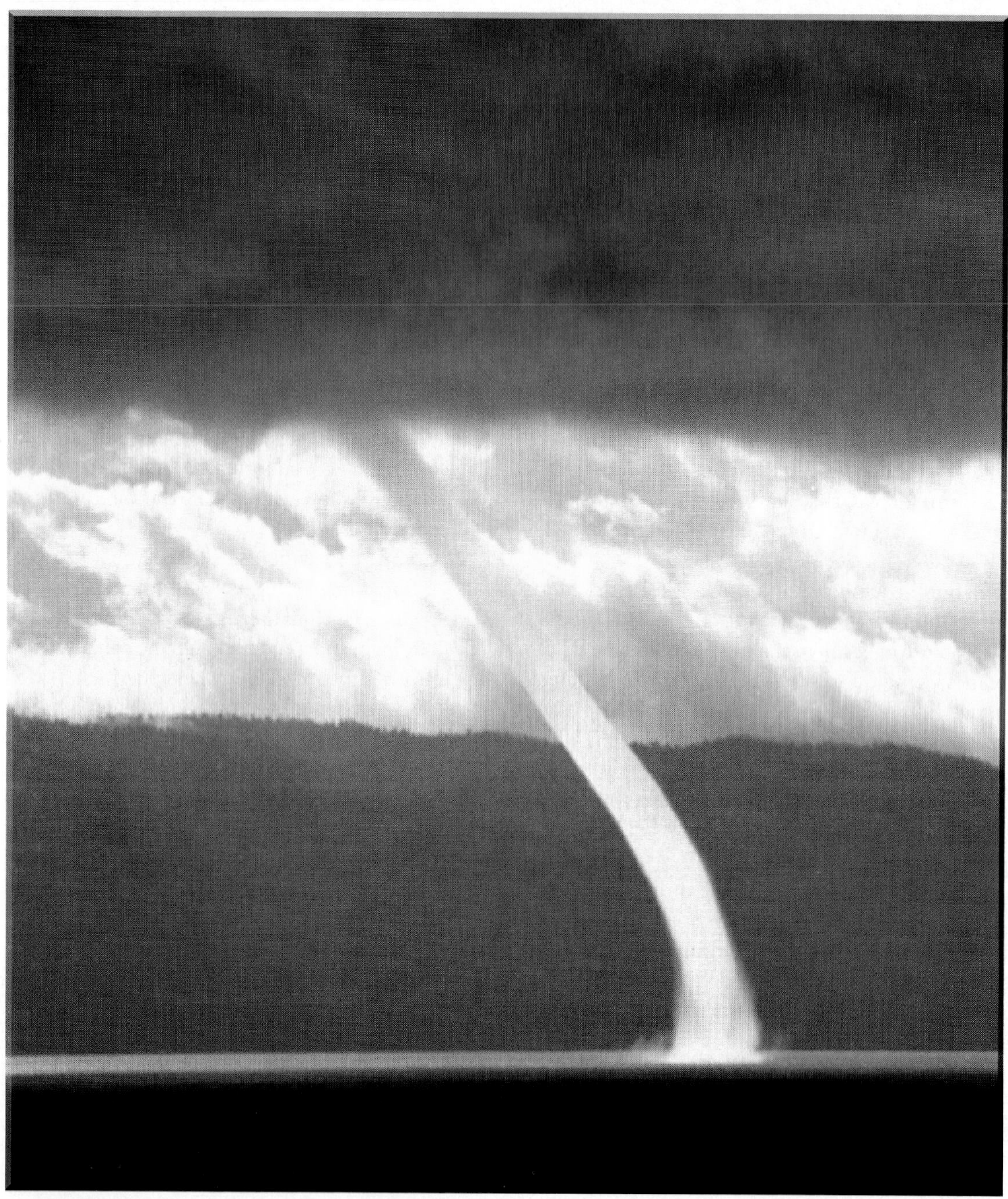

© G. Kaufman

A powerful waterspout moves across Lake Tahoe, California.

Meteorology Now™ This icon, appearing throughout the book, indicates an opportunity to explore interactive tutorials, animations, or practice problems available on the MeteorologyNow Web site at http://earthscience.brookscole.com/ahrens/ess4e.

CHAPTER 10

Thunderstorms and Tornadoes

Wednesday, March 18, 1925, was a day that began uneventfully, but within hours turned into a day that changed the lives of thousands of people and made meteorological history. Shortly after 1:00 P.M., the sky turned a dark greenish-black and the wind began whipping around the small town of Murphysboro, Illinois. Arthur and Ella Flatt lived on the outskirts of town with their only son, Art, who would be four years old in two weeks. Arthur was working in the garage when he heard the roar of the wind and saw the threatening dark clouds whirling overhead. Instantly concerned for the safety of his family, he ran toward the house as the tornado began its deadly pass over the area. With debris from the house flying in his path and the deafening thunder of destruction all around him, Arthur reached the front door. As he struggled in vain to get to his family, whose screams he could hear inside, the porch and its massive support pillars caved in on him. Inside the house, Ella had scooped up young Art in her arms and was making a panicked dash down the front hallway towards the door when the walls collapsed, knocking her to the floor, with Art cradled beneath her. Within seconds, the rest of the house fell down upon them. Both Arthur and Ella were killed instantly, but Art was spared, nestled safely under his mother's body.

As the dead and survivors were pulled from the devastation that remained, the death toll mounted. Few families escaped the grief of lost loved ones. The infamous tri-state tornado killed 234 people in Murphysboro and leveled 40 percent of the town.

CONTENTS

The devastating tornado described in our opening cut a mile-wide path for a distance of more than 200 miles through the states of Missouri, Illinois, and Indiana. The tornado (which was most likely a series of tornadoes) totally obliterated 4 towns, killed an estimated 695 persons, and left over 2000 injured. Tornadoes such as these, as well as much smaller ones, are associated with severe thunderstorms. Consequently, we will first examine the different types of thunderstorms. Later, we will focus on tornadoes, examining how and where they form, and why they are so destructive.

THUNDERSTORMS

It probably comes as no surprise that a *thunderstorm* is merely a storm containing lightning and thunder. Sometimes a thunderstorm produces gusty surface winds with heavy rain and hail. The storm itself may be a single cumulonimbus cloud, or several thunderstorms may form into a cluster. In some cases, a line of thunderstorms will form that may extend for hundreds of kilometers.

Thunderstorms are *convective storms* that form with rising air. So the birth of a thunderstorm often begins when warm, humid air rises in a conditionally unstable environment.* The rising air may be a parcel of air ranging in size from a large balloon to a city block, or an entire layer, or slab of air, may be lifted. As long as a rising air parcel is warmer (less dense) than the air surrounding it, there is an upward-directed *buoyant force* acting on it. The warmer the parcel is compared to the air surrounding it, the greater the buoyant force and the stronger the convection. The trigger (or "forcing mechanism") needed to start air moving upward may be unequal heating at the surface, the effect of terrain, or the lifting of air along shallow boundaries of converging surface winds. Diverging upper-level winds, coupled with converging surface winds and rising air, also provide a favorable condition for thunderstorm development. Moreover, thunderstorms often form when warm air rises along a frontal zone. Usually, several of these mechanisms work together to generate severe thunderstorms.

Scattered thunderstorms that form in summer are often referred to as *ordinary thunderstorms,* or *air-mass thunderstorms,* because they tend to develop in warm, humid air masses away from weather fronts. These storms are usually short-lived and rarely produce strong winds or large hail. On the other hand, *severe thunderstorms* may produce high winds, flash floods, damaging hail, and even tornadoes. Let's examine the ordinary thunderstorms first.

Ordinary Thunderstorms **Ordinary thunderstorms** tend to form in a region where there is limited wind shear—that is, where the wind speed and wind direction do not abruptly change with increasing height above the surface. Many ordinary thunderstorms appear to form as parcels of air are lifted from the surface by turbulent overturning in the presence of wind. Moreover, ordinary storms often form along shallow zones where surface winds converge. Such zones may be due to any number of things, such as topographic irregularities, sea-breeze fronts, or the cold outflow of air from inside a thunderstorm that reaches the ground and spreads horizontally. These converging wind boundaries are normally zones of contrasting air temperature and humidity and, hence, air density. Because insects and insect-eating birds often concentrate along these boundaries, Doppler radar can detect these zones, which are displayed as *fine lines* on the radar screen.

Extensive studies indicate that ordinary thunderstorms go through a cycle of development from birth to maturity to decay. The first stage is known as the **cumulus stage,** or *growth stage.* As a parcel of warm, humid air rises, it cools and condenses into a single cumulus cloud or a cluster of clouds (see Fig. 10.1a). If you have ever watched a thunderstorm develop, you may have noticed that at first the cumulus cloud grows upward only a short distance, then it dissipates. The top of the cloud dissipates because the cloud droplets evaporate as the drier air surrounding the cloud mixes with it. However, after the water drops evaporate, the air is more moist than before. So, the rising air is now able to condense at successively higher levels, and the cumulus cloud grows taller, often appearing as a rising dome or tower.

As the cloud builds, the transformation of water vapor into liquid or solid cloud particles releases large quantities of latent heat, a process that keeps the rising air inside the cloud warmer than the air surrounding it. The cloud continues to grow in the unstable atmosphere as long as it is constantly fed by rising air from below. In this manner, a cumulus cloud may show extensive vertical development and grow into a towering cumulus cloud (cumulus congestus) in just a few minutes. During the cumulus stage, there normally is insufficient time for precipitation to form, and the updrafts keep water drop-

*A conditionally unstable atmosphere exists when cold, dry air aloft overlies warm, moist surface air. However, thunderstorms may form when a cold "pool" of air moves over a region where the surface air temperature is no more than 10°C (50°F). This situation often occurs during the winter along the west coast of North America. Additionally, thunderstorms occasionally form in wintertime snowstorms. In both of these cases, the air aloft is considerably colder than the surface air, which generates instability. More information on atmospheric instability is given in Chapter 5, beginning on p. 114.)

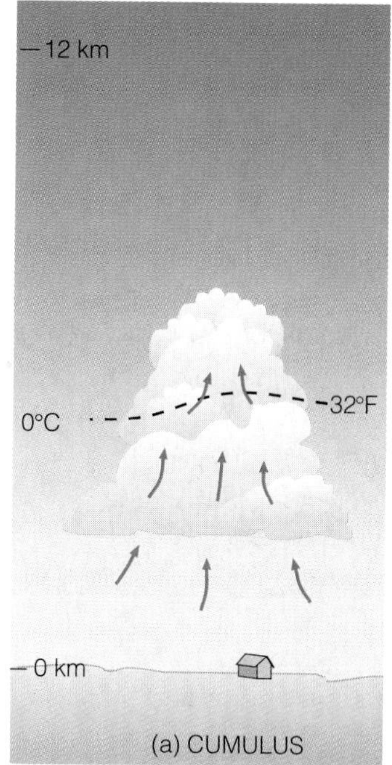

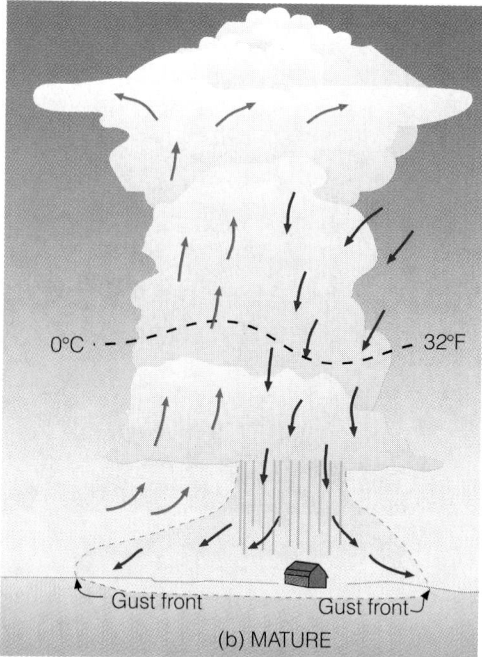

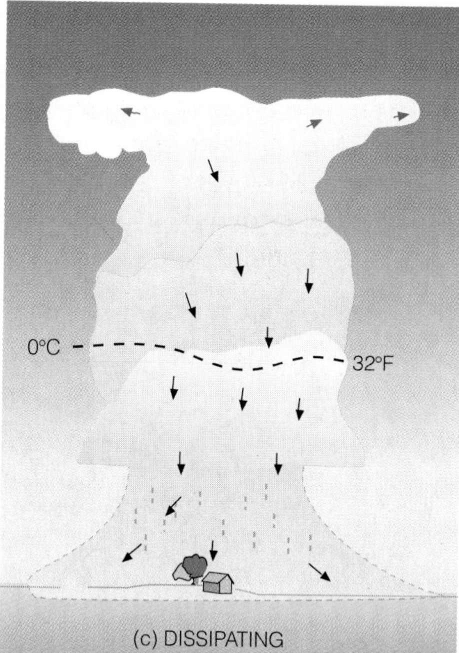

Meteorology Now™ ACTIVE FIGURE 10.1 Simplified model depicting the life cycle of an ordinary thunderstorm that is nearly stationary. (Arrows show vertical air currents. Dashed line represents freezing level, 0°C isotherm.) Watch this Active Figure at http://earthscience.brookscole.com/ahrens/ess4e.

lets and ice crystals suspended within the cloud. Also, there is no lightning or thunder during this stage.

As the cloud builds well above the freezing level, the cloud particles grow larger. They also become heavier. Eventually, the rising air is no longer able to keep them suspended, and they begin to fall. While this phenomenon is taking place, drier air from around the cloud is being drawn into it in a process called *entrainment.* The entrainment of drier air causes some of the raindrops to evaporate, which chills the air. The air, now colder and heavier than the air around it, begins to descend as a *downdraft.* The downdraft may be enhanced as falling precipitation drags some of the air along with it.

The appearance of the downdraft marks the beginning of the **mature stage.** The downdraft and updraft within the mature thunderstorm constitute a *cell.** In some storms, there are several cells, each of which may last for less than 30 minutes.

During its mature stage, the thunderstorm is most intense. The top of the cloud, having reached a stable region of the atmosphere (which may be the stratosphere), begins to take on the familiar anvil shape, as upper-level winds spread the cloud's ice crystals horizontally (see Fig. 10.2). The cloud itself may extend upward to an altitude of over 12 km (40,000 ft) and be several kilometers in diameter near its base. Updrafts and downdrafts reach their greatest strength in the middle of the cloud, creating severe turbulence. In some storms, the updrafts may intrude above the cloud top into the stable atmosphere, a condition known as *overshooting.* Lightning and thunder are also present in the mature stage. Heavy rain (and occasionally small hail) falls from the cloud. And, at the surface, there is often a downrush of cold air with the onset of precipitation.

Where the cold downdraft reaches the surface, the air spreads out horizontally in all directions. The surface boundary that separates the advancing cooler air from the surrounding warmer air is called a **gust front.** Along the gust front, winds rapidly change both direction and speed. Look at Fig. 10.lb and notice that the gust front forces warm, humid air up into the storm, which enhances the cloud's updraft. In the region of the downdraft, rainfall may or may not reach the surface, depending on the relative humidity beneath the storm. In the dry air of the desert Southwest, for example, a mature thunderstorm may look ominous and contain all of the ingredients of any other storm, except that the raindrops evaporate before reaching the ground. However, intense downdrafts from the storm may reach the surface, producing strong, gusty winds and a gust front.

After the storm enters the mature stage, it begins to dissipate in about 15 to 30 minutes. The **dissipating**

*In convection, the cell may be a single updraft or a single downdraft, or a combination of the two.

FIGURE 10.2 An ordinary thunderstorm in its mature stage. Note the distinctive anvil top.

stage occurs when the updrafts weaken as the gust front moves away from the storm and no longer enhances the updrafts. At this stage, as illustrated in Fig. 10.1c, downdrafts tend to dominate throughout much of the cloud. The reason the storm does not normally last very long is that the downdrafts inside the cloud tend to cut off the storm's fuel supply by destroying the humid updrafts. Deprived of the rich supply of warm, humid air, cloud droplets no longer form. Light precipitation now falls from the cloud, accompanied by only weak downdrafts. As the storm dies, the lower-level cloud particles evaporate rapidly, sometimes leaving only the cirrus anvil as the reminder of the once mighty presence (see Fig. 10.3). A single ordinary thunderstorm may go through its three stages in one hour or less.

Not only do thunderstorms produce summer rainfall for a large portion of the United States but they also bring with them momentary cooling after an oppressively hot day. The cooling comes during the mature stage, as the downdraft reaches the surface in the form of a blast of welcome relief. Sometimes, the air temperature may lower as much as 10°C (18°F) in just a few minutes. Unfortunately, the cooling effect is short-lived, as the downdraft diminishes or the thunderstorm moves on. In fact, after the storm has ended, the air temperature usually rises; and as the moisture from the rainfall evaporates into the air, the humidity increases, sometimes to a level where it actually feels more oppressive after the storm than it did before.

The cold downdraft, upon reaching the surface, has yet another effect. It may force warm, moist surface air upward along its advancing edge. This rising air then condenses and gradually builds into a new thunderstorm. This process may repeat over and over as old cells die out and new ones form. Thus, it is entirely possible for a series of thunderstorms to grow in a line, one next to the other, each in a different stage of development (see Fig. 10.4). Thunderstorms that form in this manner are termed **multicell storms.** Most ordinary thunderstorms are multicell storms, as are many severe thunderstorms.

Earlier we saw that for a thunderstorm to develop there must be rising, moist air in a conditionally unstable atmosphere. The ingredient necessary to start the air rising may be any one of a number of factors, such as the unequal heating of the surface, topographical irregularities, or even the leading edge of a sea breeze. Rising air may also occur along frontal boundaries and mountain ranges. Most of the thunderstorms that form in this manner are not severe, and their life cycle usually follows the pattern described for ordinary thunderstorms.

Severe thunderstorms also form as air rises. But the rising air is forced to spin as the wind rapidly changes direction with increasing height above the surface.

Severe Thunderstorms By definition, a **severe thunderstorm** is any thunderstorm that produces hail at least three-quarters of an inch in diameter and/or surface wind gusts of 50 knots or greater and/or produces a tornado. The likelihood that a thunderstorm will become severe increases with the length of time the storm sur-

DID YOU KNOW?

On July 13, 1999, in Sattley, California, a strong downdraft from a mature thunderstorm dropped the air temperature from 97°F at 4:00 P.M. to a chilly 57°F one hour later.

FIGURE 10.3 A dissipating thunderstorm near Naples, Florida. Most of the cloud particles in the lower half of the storm have evaporated.

vives. The longer the storm remains in existence, the greater the chance of producing large hail, strong downdrafts, and tornadoes. Just as the ordinary thunderstorm, severe thunderstorms form as moist air is forced to rise into a conditionally unstable atmosphere. However, severe thunderstorms usually form in areas with a strong vertical wind shear.

In the previous section we learned that ordinary thunderstorms tend to form in regions of low wind shear. Because of this fact, the storm's precipitation (along with the downdraft) can fall into the updraft. The downdraft then cuts off the storm's fuel supply, which causes the storm to dissipate in a relatively short time. If the winds aloft, however, increase in strength (moderate shear), the precipitation is pushed downwind so that it does not fall into the updraft. Hence, the updraft is not suppressed. If the outflow of cold air in the downdraft undercuts the updraft, new cells can form, producing a long-lasting *multicell storm.* If convection is strong and the updraft is intense, the storm can become severe. In fact, the updrafts in a severe thunderstorm may be so strong that the cloud top is able to intrude well into the stable stratosphere. In

FIGURE 10.4 A multicell storm. This storm is composed of a series of cells in successive stages of growth. The thunderstorm in the middle is in its mature stage, with a well-defined anvil. Heavy rain is falling from its base. To the right of this cell, a thunderstorm is in its cumulus stage. To the left, a well-developed cumulus congestus cloud is about ready to become a mature thunderstorm.

DID YOU KNOW?

In a thunderstorm, not all downdrafts are cool. On the evening of May 22, 1996, in the town of Chickasha, Oklahoma, a blast of hot air from a mature thunderstorm caused the air temperature to jump from 88°F to 102°F in twenty-five minutes. Apparently, the hot downburst (called a *"heat burst"*) originated high up in the drier air of the cloud and was warmed by compressional heating as it plunged toward the earth.

some cases, the top of the cloud may extend to as high as 18 km (60,000 ft) above the surface. The violent updrafts keep hailstones suspended in the cloud long enough for them to grow to considerable size. Once they are large enough, they either fall out the bottom of the cloud with the downdraft, or a strong updraft may toss them out the side of the cloud, or even from the base of the anvil. Aircraft have actually encountered hail in clear air several kilometers from a storm. Also, downdrafts within the anvil produce beautiful mammatus clouds.

If the winds aloft become even stronger (strong shear) and change direction with height (from more southerly at the surface to more westerly aloft), the storm may move along in such a way that the outflow of cold air from the downdraft never undercuts the updraft. The wind shear may be strong enough to create horizontal spin, which when tilted into the updraft causes it to rotate. In this situation, the thunderstorm may grow into a larger, long-lasting (longer than an hour) severe storm called a **supercell** that has a violent updraft and a single cell. As we will see later in this chapter, it is the rotational aspect of supercells that can lead to the formation of tornadoes.

The *supercell storm* is an enormous thunderstorm that consists primarily of a single rotating updraft. The internal structure of the storm is organized in such a way that the storm may maintain itself as a single entity for hours on end (see Fig. 10.5). Storms of this type are capable of producing updrafts that can exceed 90 knots, hail the size of grapefruit, damaging surface winds, and large, long-lasting tornadoes. Normally, precipitation does not form in the region of the strong updraft. If pre-

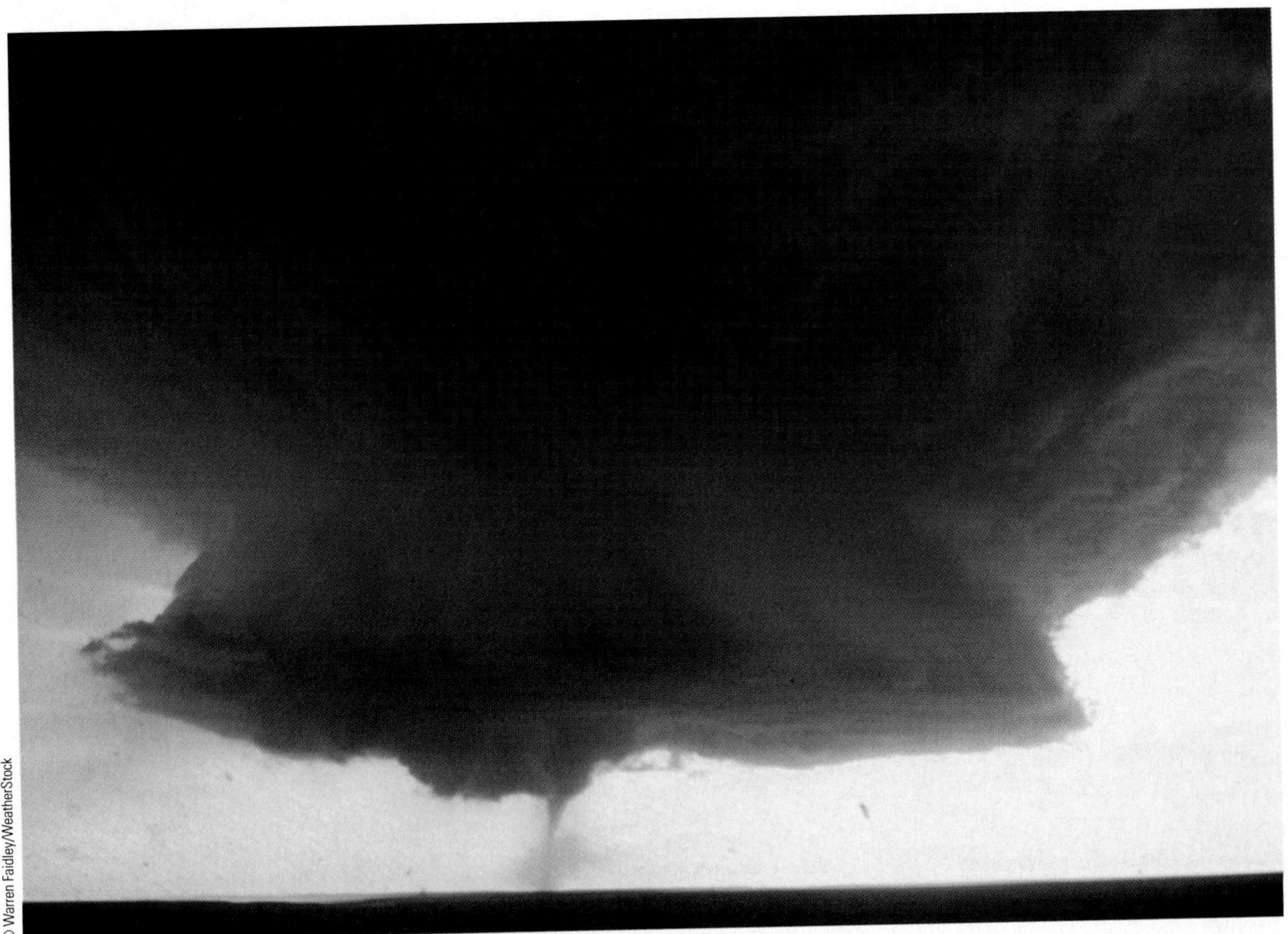

FIGURE 10.5 A supercell thunderstorm with a tornado sweeps over Texas.

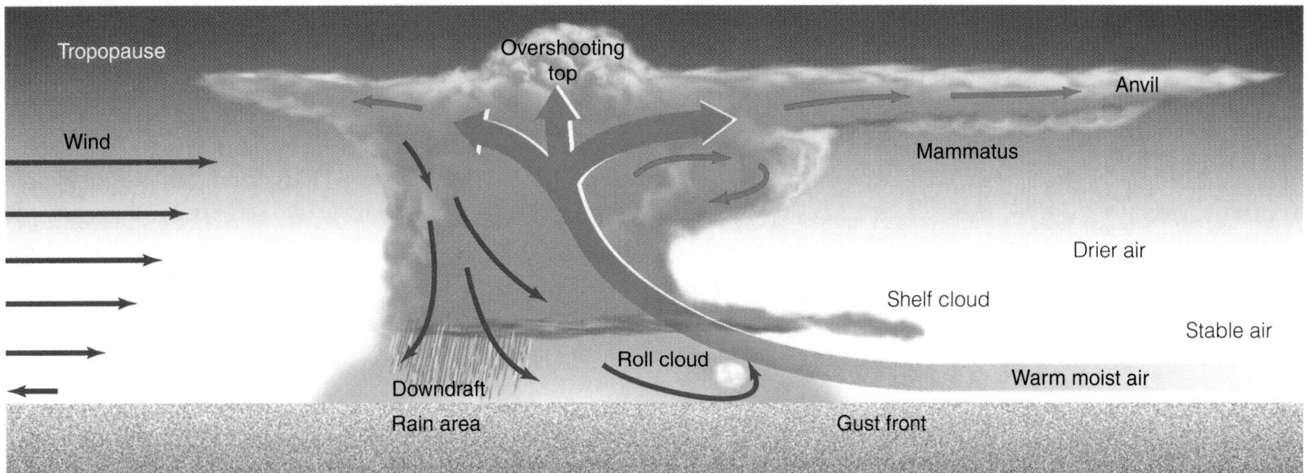

Meteorology Now™ ACTIVE FIGURE 10.6 A simplified model describing air motions and other features associated with an intense thunderstorm that has a tilted updraft. The severity depends on the intensity of the storm's circulation pattern. Watch this Active Figure at http://earthscience.brookscole.com/ahrens/ess4e.

cipitation does form, it may be swept laterally out of the region by the rapidly rotating air.

The Gust Front and Microburst Earlier we learned that a *gust front* represents the leading edge of the cold downdraft originating inside a thunderstorm. Gust fronts occur with both ordinary and severe thunderstorms. Figure 10.6 illustrates a strong gust front extending outward from an intense thunderstorm. To an observer on the ground, the passage of the gust front resembles that of a cold front. During its passage, the wind shifts and becomes strong and gusty, with speeds occasionally exceeding 55 knots; temperatures drop sharply and, in the cold heavy air of the downdraft, the surface pressure often rises—sometimes several millibars. Near the surface, the cold, dense air extending outward from the thunderstorm may produce a small shallow area of high pressure called a *mesohigh* (meaning "mesoscale high"). The cold air may even linger close to the ground for several hours, well after the thunderstorm activity has ceased.

Along the leading edge of the gust front, the air is quite turbulent. Here, strong winds can pick up loose dust and soil and lift them into a huge tumbling cloud—the haboob that we described in Chapter 7. As warm, moist air rises along the forward edge of the gust front, a **shelf cloud** (also called an *arcus cloud*) may form, such as the one shown in Fig. 10.7. These clouds are especially prevalent when the atmosphere is very stable near the base of the thunderstorm. Look again at Fig. 10.6 and notice that the shelf cloud is attached to the base of the thunderstorm. Occasionally, an elongated ominous-looking cloud forms just behind the gust front. These clouds, which appear to slowly spin about a horizontal axis, are called **roll clouds** (see Fig. 10.8).

Downdrafts in a thunderstorm are not only instrumental in creating a gust front, they may also produce downbursts. A **downburst** forms when the downdraft becomes localized so that it hits the ground and spreads horizontally in a radial burst of wind, much like water pouring from a tap and striking the sink below. A downburst with winds extending only 4 km or less is termed a **microburst.** In spite of its small size, an intense microburst can induce damaging winds as high as 146 knots. (A larger downburst with winds extending more than 4 kilometers is termed a *macroburst.*) Figure 10.9 shows the dust clouds generated from a microburst north of Denver, Colorado. Since a microburst is an intense downdraft, its leading edge can evolve into a gust front.

Microbursts are capable of blowing down trees and inflicting heavy damage upon poorly built structures as well as upon sailing vessels that encounter microbursts over open water. In fact, microbursts may be responsible for some damage once attributed to tornadoes. Moreover, microbursts and their accompanying *wind shear* (that is, rapid changes in wind speed or wind direction) appear to be responsible for several airline crashes. When an aircraft flies through a microburst at a relatively low altitude, say 300 m (1000 ft) above the ground, it first encounters a headwind that generates extra lift. This is position (a) in Fig. 10.10. At this point, the aircraft tends to climb (it gains lift), and if the pilot noses the aircraft downward there could be grave consequences, for in a matter of seconds the aircraft encounters the powerful downdraft (position b), and the headwind is replaced by a tail wind (po-

© Richard F. Piconso

FIGURE 10.7 A dramatic example of a shelf cloud (or arcus cloud) associated with an intense thunderstorm. The photograph was taken in the Philippines as the thunderstorm approached from the northwest.

© Howard B. Bluestein

FIGURE 10.8 A roll cloud forming behind a gust front.

Photo by author

FIGURE 10.9 Dust clouds rising in response to the outburst winds of a microburst north of Denver, Colorado.

sition c). This situation causes a sudden loss of lift and a subsequent decrease in the performance of the aircraft, which is now accelerating toward the ground.

One accident attributed to a microburst occurred north of Dallas–Fort Worth Regional Airport during August, 1985. Just as an aircraft was making its final approach, it encountered severe wind shear beneath a small but intense thunderstorm. The aircraft then dropped to the ground and crashed, killing over 100 passengers. To detect the hazardous wind shear associated with microbursts, a ground-based wind-shear detection system called LLWSAS (for *L*ow-*L*evel *W*ind *S*hear *A*lert *S*ystem) is in operation at several airports. In addition, the installation of Doppler radars throughout the United States is providing a better system of detecting microbursts and other severe weather phenomena.

Microbursts can be associated with severe thunderstorms, producing strong, damaging winds. But studies show that they can also occur with ordinary thunderstorms and with clouds that produce only isolated showers—clouds that may or may not contain thunder and lightning.

Meteorology Now™
Click "Microbursts" to try your hand at landing an aircraft in a microburst.

The strong downburst winds associated with a cluster of severe thunderstorms can produce damaging *straight-line winds** that may exceed 90 knots. If the wind damage extends for several hundred kilometers along the storm's path, the winds are called a **derecho** (day-ray-sho), after the Spanish word for "straight ahead." Most derechoes are caused by a cluster of downbursts from intense thunderstorms. Often the thunderstorms on a radar screen appear in the shape of a *bow,* and are referred to as

**Straight-line winds* are thunderstorm-generated winds that are not associated with rotation.

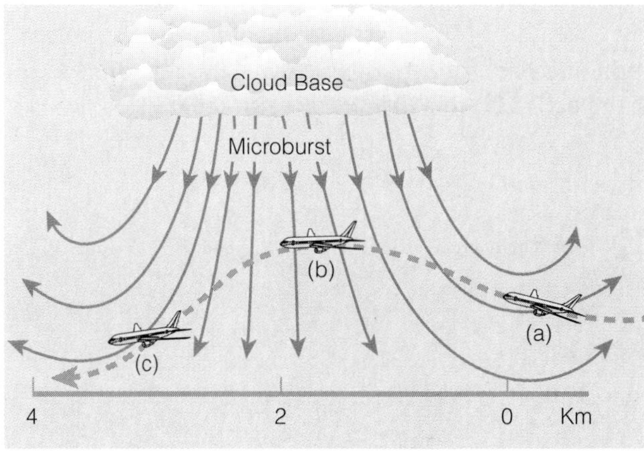

FIGURE 10.10 Flying into a microburst. At position (a), the pilot encounters a headwind; at position (b), a strong downdraft; and at position (c), a tailwind that reduces lift and causes the aircraft to lose altitude.

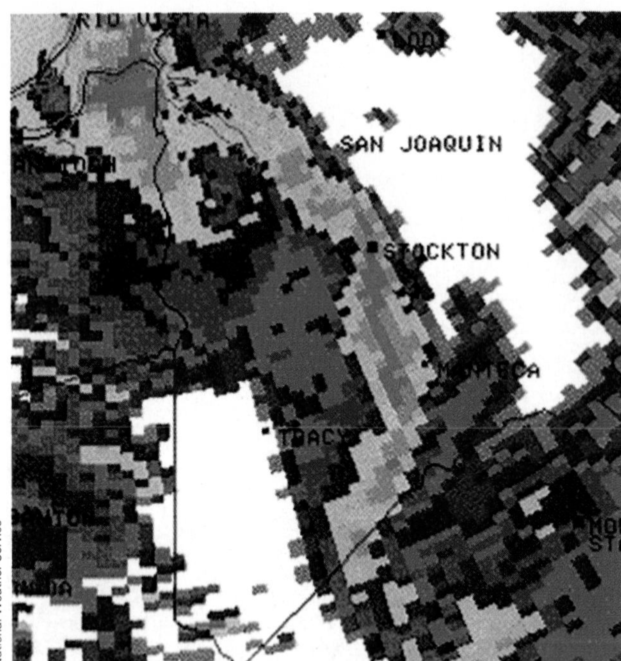

FIGURE 10.11 Doppler radar display showing a line of thunderstorms (a squall line) bent in the shape of a bow (colors red, orange, and yellow) as they move eastward across the San Joaquin Valley of California. Such *bow echos* often produce damaging surface winds near the center of the bow. Sometimes the left (usually northern) end of the bow will develop cyclonic rotation and produce a tornado.

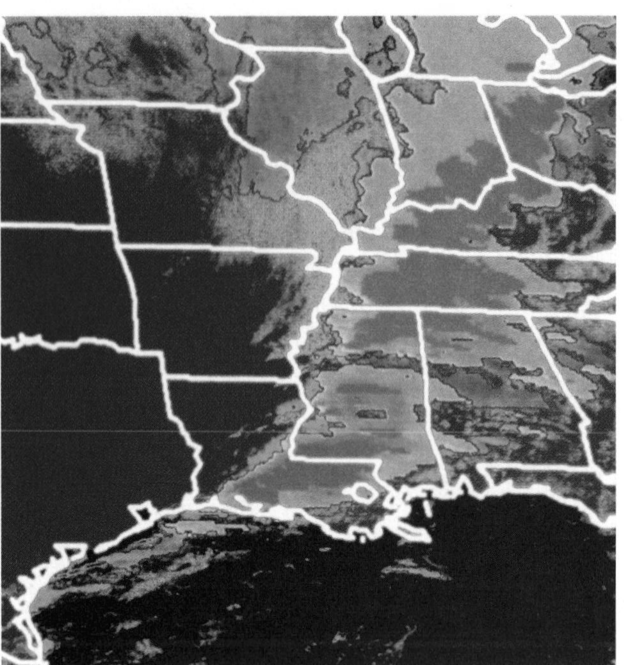

FIGURE 10.12 An enhanced infrared satellite image showing a pre-frontal squall-line (in dark red) that stretches from the Ohio Valley into the southern United States.

a *bow echo* (see Fig. 10.11). Typically, derechoes form in the early evening and last throughout the night. An especially powerful derecho roared through New York State during the early morning of July 15, 1995, where it blew down millions of trees in Adirondack State Park.

Mesoscale Convective Systems Favorable atmospheric conditions may enable thunderstorms to form either as a line of storms, called a *squall line,* or as a cluster of storms, called a *mesoscale convective complex.* Such an ensemble of thunderstorms that form from convection and have an extensive width is called a **Mesoscale Convective System (MCS).**

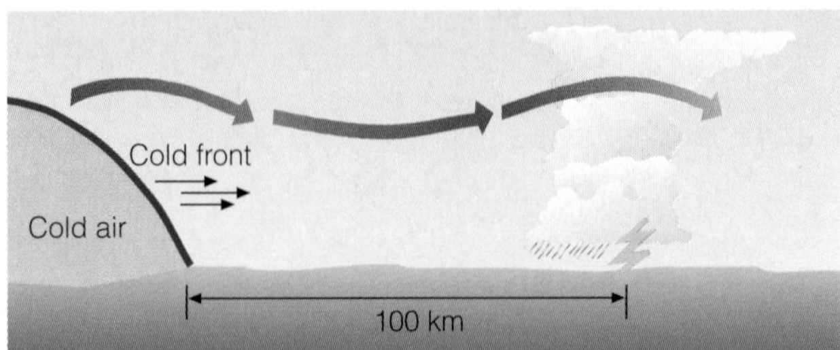

FIGURE 10.13 Pre-frontal squall-line thunderstorms may form ahead of an advancing cold front as the upper-air flow develops waves downwind from the cold front.

The **squall line** forms as a line of thunderstorms (see Fig. 10.11). Sometimes they are right along a cold front but they also form in the warm air 100 to 300 km out ahead of it. These *pre-frontal squall-line thunderstorms* of the middle latitudes represent the largest and most severe type of squall line. The line of storms may extend for over 1000 km (600 mi), with huge thunderstorms (some of which may be supercells) causing severe weather over much of its length (see Fig. 10.12).

There is still debate as to exactly how pre-frontal squall lines form. Models that simulate their formation suggest that, initially, convection begins along the cold front then re-forms further away. Moreover, the surging nature of the main cold front itself, or developing cumulus clouds along the front, may cause the air aloft to develop into waves (called *gravity waves*), much like the waves that form downwind of a mountain chain (see Fig. 10.13). Out ahead of the cold front, the rising motion of the wave may be the trigger that initiates the development of cumulus clouds and a pre-frontal squall line.

Where conditions are favorable for convection, a number of individual thunderstorms may occasionally grow in size and organize into a large convective weather system. These convectively driven systems, called **Mesoscale Convective Complexes (MCCs),** are quite large—they can be as much as 1000 times larger than an individual ordinary thunderstorm. In fact, they are often large enough to cover an entire state, an area in excess of 100,000 square kilometers (see Fig. 10.14).

Within the MCC, the individual thunderstorms apparently work together to generate a long-lasting weather system that moves slowly (normally less than 20 knots) and often exists for periods exceeding 12 hours. The circulation of the MCC supports the growth of new thunderstorms as well as a region of widespread precipitation. These systems are beneficial, as they provide a significant portion of the growing season rainfall over much of the corn and wheat belts of the United States. However, MCCs can also produce a wide variety of severe weather, including hail, high winds, destructive flash floods, and tornadoes. In fact, about one-fourth of all MCCs produce damaging straight-line winds, or derechoes.

Mesoscale convective complexes tend to form during the summer in regions where the upper-level winds are weak, which is often beneath a ridge of high pressure. If a weak cold front should stall beneath the ridge, surface heating and moisture may be sufficient to generate thunderstorms on the cool side of the front. Often moisture from the south is brought into the system by a low-level jet stream.

FIGURE 10.14 An enhanced infrared satellite image showing a Mesoscale Convective Complex (dark red color) extending from central Kansas across western Missouri. This organized mass of thunderstorms brought hail, heavy rain, and flooding to this area.

Dryline Thunderstorms Severe thunderstorms may form along or just east of a boundary called a dryline. The **dryline** represents a narrow zone where there is a sharp horizontal change in moisture. Because dew-point temperatures may drop along this boundary by as much as 9°C (16°F) per km, drylines have been referred to as *dew-point fronts.* Although drylines can occur as far north as the Dakotas, and as far east as the Texas-Louisiana border, they are most frequently observed in the western half of Texas, Oklahoma, and Kansas, especially during spring and early summer.

Figure 10.15 shows springtime weather conditions that can lead to the development of a dryline and intense thunderstorms. The map shows a developing mid-latitude cyclone with a cold front, a warm front, and three distinct air masses. Behind the cold front, cold, dry continental polar (cP) air or modified cool dry Pacific air pushes in from the northwest. In the warm air, ahead of the cold front, warm, dry continental tropical (cT) air moves in from the southwest. Further east, warm but very moist maritime tropical (mT) air sweeps northward from the Gulf of Mexico. The dryline is the north–south oriented boundary that separates the warm, dry air and the warm, moist air.

Along the cold front—where cold, dry air replaces warm, dry air—there is insufficient moisture for thunderstorm development. The moisture boundary lies along the dryline. Because the Central Plains of North America are elevated to the west, some of the hot, dry air from the southwest is able to ride over the slightly cooler, more humid air from the Gulf. This condition sets up a potentially unstable atmosphere just east of the dryline. Converging surface winds in the vicinity of the dryline, coupled with upper-level outflow, may result in rising air and the development of thunderstorms. As thunderstorms form, the cold downdraft from inside the storm may produce an outflow boundary that moves along the ground and initiates the uplift necessary for generating new (possibly more severe) thunderstorms.

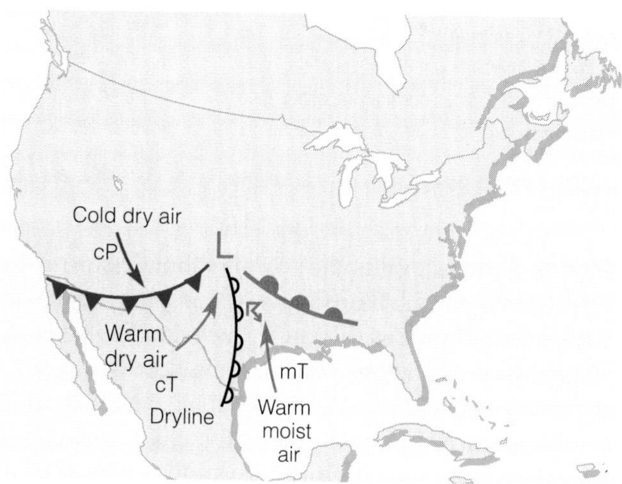

FIGURE 10.15 Surface conditions that can produce a dryline with severe thunderstorms.

FOCUS ON A SPECIAL TOPIC

The Terrifying Flash Flood in the Big Thompson Canyon

July 31, 1976, was like any other summer day in the Colorado Rockies, as small cumulus clouds with flat bases and dome-shaped tops began to develop over the eastern slopes near the Big Thompson and Cache La Poudre rivers. At first glance, there was nothing unusual about these clouds, as almost every summer afternoon they form along the warm mountain slopes. Normally, strong upper-level winds push them over the plains, causing rainshowers of short duration. But the cumulus clouds on this day were different. For one thing, they were much lower than usual, indicating that the southeasterly surface winds were bringing in a great deal of moisture. Also, their tops were somewhat flattened, suggesting that an inversion aloft was stunting their growth. But these harmless-looking clouds gave no clue that later that evening in the Big Thompson Canyon more than 135 people would lose their lives in a terrible flash flood.

By late afternoon, a few of the cumulus clouds were able to puncture the inversion. Fed by moist southeasterly winds, these clouds soon developed into gigantic multicell thunderstorms with tops exceeding 18 km (60,000 ft). By early evening, these same clouds were producing incredible downpours in the mountains.

In the narrow canyon of the Big Thompson River, some places received as much as 30.5 cm (12 in.) of rain in the four hours between 6:30 P.M. and 10:30 P.M. local time. This is an incredible amount of precipitation, considering that the area normally receives about 40.5 cm (16 in.) for an entire year. The heavy downpours turned small creeks into raging torrents, and the Big Thompson River was quickly filled to capacity. Where the canyon narrowed, the river overflowed its banks and water covered the road. The relentless pounding of water caused the road to give way.

Soon cars, tents, mobile homes, resort homes, and campgrounds were being claimed by the river. Where the debris entered a narrow constriction, it became a dam. Water backed up behind it, then broke through, causing a wall of water to rush downstream.

Figure 1 shows the weather conditions during the evening of July 31, 1976. A cool front moved through earlier in the day and is now south of Denver. The weak inversion layer associated with the front kept the cumulus clouds from building to great heights earlier in the afternoon. However, the strong southeasterly flow behind the cool front pushed unusually moist air upslope along the mountain range. Heated from below, the conditionally unstable air eventually punctured the inversion and developed into a huge multicell thunderstorm complex that remained nearly stationary for several hours due to the weak southerly winds aloft. The deluge may have deposited 19 cm (7.5 in.) of rain on the main fork of the Big Thompson River in about one hour. Of the approximately 2000 people in the canyon that evening, over 135 lost their lives and property damage exceeded $35.5 million.

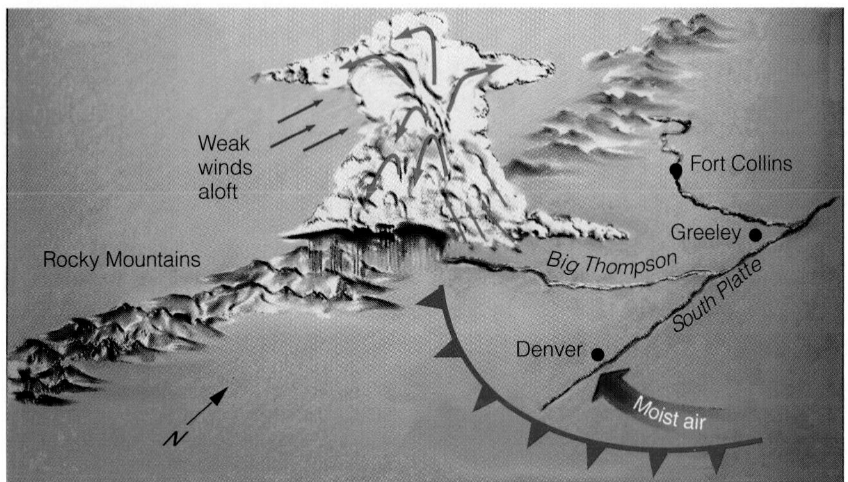

FIGURE 1 Weather conditions that led to the development of severe thunderstorms, which remained nearly stationary over the Big Thompson Canyon in the Colorado Rockies. The arrows within the thunderstorm represent air motions.

Floods and Flash Floods Intense thunderstorms are often associated with **flash floods**—floods that rise rapidly with little or no advance warning. Such flooding often results when thunderstorms stall or move very slowly, causing heavy rainfall over a relatively small area, such as in southwestern West Virginia during July, 2001, when flood waters swept through narrow valleys and damaged up to 3000 homes. Or, they can occur when thunderstorms move quickly, but keep passing over the same area, a phenomenon called *training*. (Like railroad cars, one after another, passing over the same tracks.) In recent years, flash floods in the United States have claimed an average of more than 100 lives a year, and have accounted for untold property and crop damage. (An example of a terrible flash flood that took the lives of more than 135 people is given in the Focus section above.)

In some areas, flooding occurs primarily in the spring when heavy rain and melting snow cause rivers to overflow their banks. During March, 1997, heavy downpours over the Ohio River Valley caused extensive flooding that forced thousands from their homes along rivers and smaller streams in Ohio, Kentucky, Tennessee, and West Virginia. One month later, heavy rain coupled with melting snow caused the Red River to overflow its banks, inundating 75 percent of the city of Grand Forks, North Dakota. Flooding also occurs with tropical storms that deposit torrential rains over an extensive area.

During the summer of 1993, thunderstorm after thunderstorm rumbled across the upper Midwest, causing the worst flood ever in that part of the United States (see Fig. 10.16). Estimates are that $6.5 billion in crops were lost as millions of acres of valuable farmland were inundated by flood waters. The worst flooding this area had ever seen took 45 human lives, damaged or destroyed 45,000 homes, and forced the evacuation of 74,000 people.

Distribution of Thunderstorms It is estimated that more than 50,000 thunderstorms occur each day throughout the world. Hence, over 18 million occur annually. The combination of warmth and moisture make equatorial land masses especially conducive to thunderstorm formation. Here, thunderstorms occur on about one out of every three days. Thunderstorms are also prevalent over water along the intertropical convergence zone, where the low-level convergence of air helps to initiate uplift. The heat energy liberated in these storms helps the earth maintain its heat balance by distributing heat poleward (see Chapter 7). Thunderstorms are much less prevalent in dry climates, such as the polar regions and the desert areas of the subtropical highs.

Figure 10.17 shows the average annual number of days having thunderstorms in various parts of the United States. Notice that they occur most frequently in the southeastern states along the Gulf Coast with a maximum in Florida. A secondary maximum exists over the central Rockies. The region with the fewest thunderstorms is the Pacific coastal and interior valleys.

In many areas, thunderstorms form primarily in summer during the warmest part of the day when the surface air is most unstable. There are some exceptions,

DID YOU KNOW?

The Great Flood of 1993 in the Mississippi and Missouri river basins left 56 small river towns completely inundated, and caused more than $10 billion in property and crop losses. In addition, the Great Flood took a huge toll on riverside cemeteries. Just as an example, the Hardin Cemetery in Missouri had more than 700 graves opened. Some caskets were swept away by raging flood waters and deposited many kilometers downstream, and some were never found.

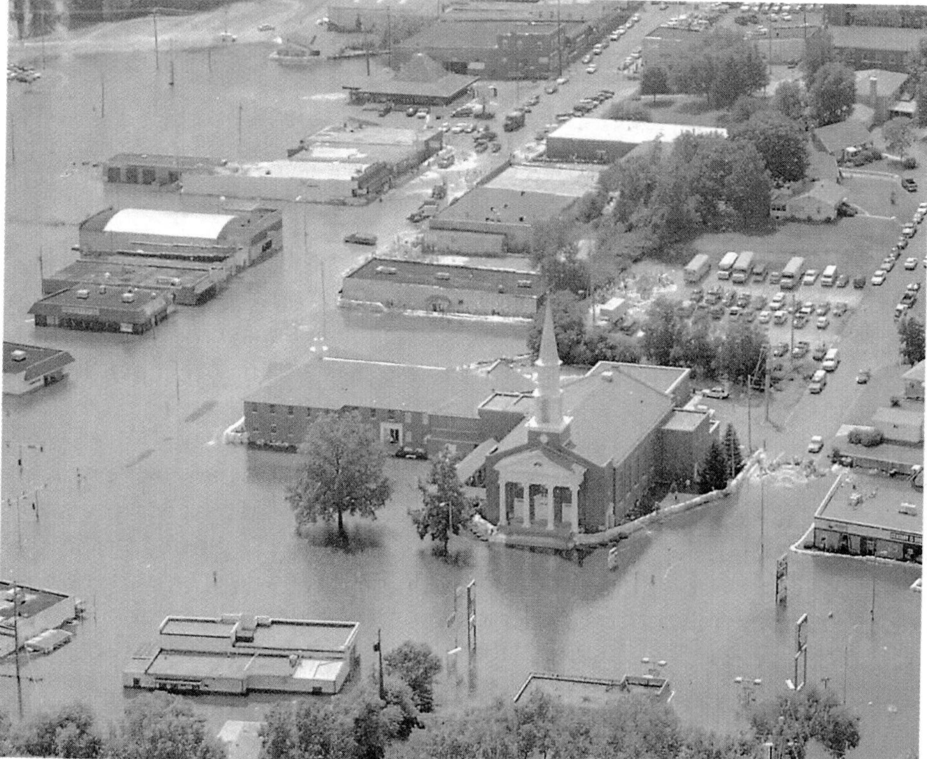

FIGURE 10.16 Flooding during the summer of 1993 covered a vast area of the upper Midwest. Here, floodwaters near downtown Des Moines, Iowa, during July, 1993, inundate buildings of the Des Moines waterworks facility. Flood-contaminated water left 250,000 people without drinking water.

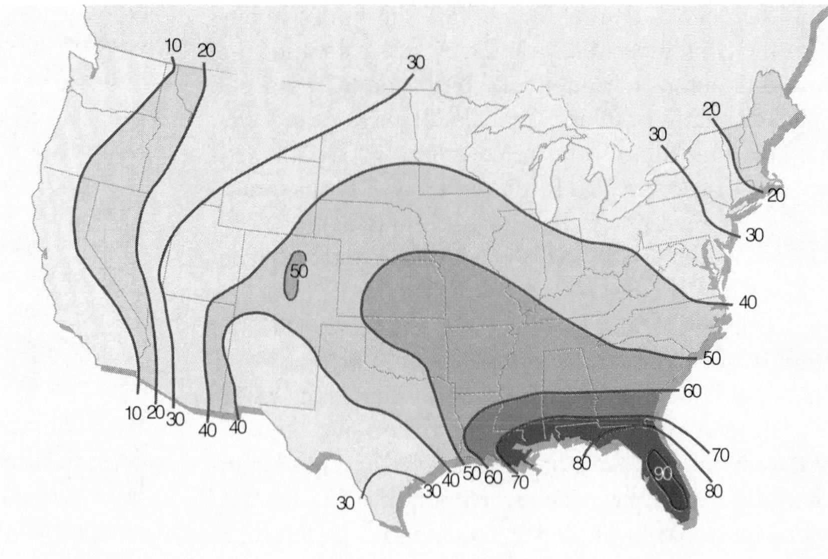

FIGURE 10.17 The average number of days each year on which thunderstorms are observed throughout the United States. (Due to the scarcity of data, the number of thunderstorms is underestimated in the mountainous west.)

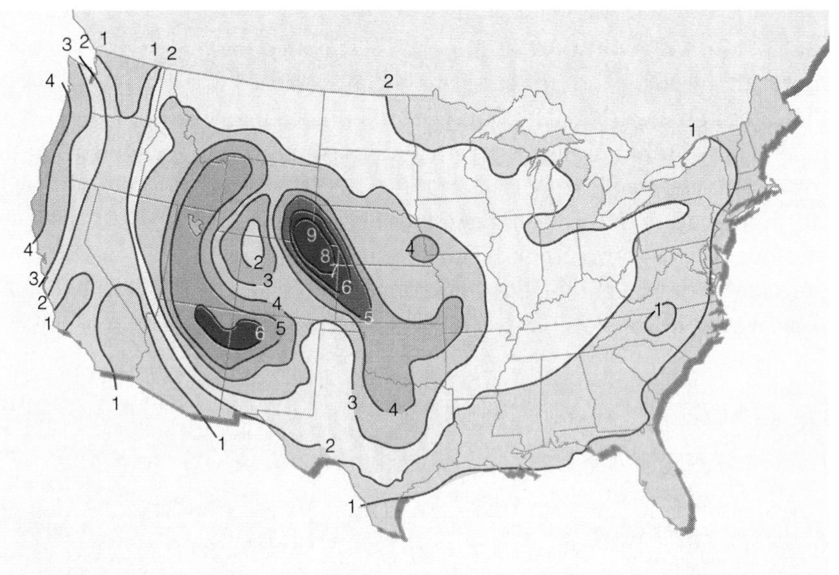

FIGURE 10.18 The average number of days each year on which hail is observed throughout the United States.

however. During the summer in the valleys of central and southern California, dry, sinking air produces an inversion that inhibits the development of towering cumulus clouds. In these regions, thunderstorms are most frequent in winter and spring, particularly when cold, moist, conditionally unstable air aloft moves over moist, mild surface air. The surface air remains relatively warm because of its proximity to the ocean. Over the Central Plains, thunderstorms tend to form more frequently at night. These storms may be caused by a low-level southerly jet stream that forms at night, and not only carries humid air northward but also initiates areas of converging surface air, which helps to trigger uplift. As the thunderstorms build, their tops cool by radiating infrared energy to space. This cooling process tends to destabilize the atmosphere, making it more suitable for nighttime thunderstorm development.

At this point, it is interesting to compare Fig. 10.17 and Fig. 10.18. Notice that, even though the greatest frequency of thunderstorms is near the Gulf Coast, the greatest frequency of hailstorms is over the western Great Plains. One reason for this situation is that conditions over the Great Plains are more favorable for the development of severe thunderstorms. We also find that, in summer along the Gulf Coast, a thick layer of warm, moist air extends upward from the surface. Most hailstones falling into this layer will melt before reaching the ground. Over the plains, the warm surface layer is much shallower and drier. Falling hailstones do begin to melt, but the water around their periphery quickly evaporates in the dry air.

This cools the hailstones and slows the melting rate so that many survive as ice all the way to the surface.

Now that we have looked at the development and distribution of thunderstorms, we are ready to examine an interesting, though yet not fully understood, aspect of all thunderstorms—lightning.

Lightning and Thunder **Lightning** is simply a discharge of electricity, a giant spark, which occurs in mature thunderstorms.* Lightning may take place within a cloud, from one cloud to another, from a cloud to the surrounding air, or from a cloud to the ground. (The majority of lightning strikes occur within the cloud, while only about 20 percent or so occur between cloud and ground.) The lightning stroke can heat the air through which it travels to an incredible 30,000°C (54,000°F), which is 5 times hotter than the surface of the sun. This extreme heating causes the air to expand explosively, thus initiating a shock wave that becomes a booming sound wave—called **thunder**—that travels outward in all directions from the flash.

Light travels so fast that we see light instantly after a lightning flash. But the sound of thunder, traveling at only about 330 m/sec (1100 ft/sec), takes much longer to reach the ear. If we start counting seconds from the moment we see the lightning until we hear the thunder, we can determine how far away the stroke is. Because it takes sound about 3 seconds to travel 1 kilometer (5 seconds for each mile), if we see lightning and hear the thunder 15 seconds later, the lightning stroke occurred 5 km (3 mi) away.

When the lightning stroke is very close (several hundred feet or less) thunder sounds like a clap or a crack followed immediately by a loud bang. When it is farther away, it often rumbles. The rumbling can be due to the sound emanating from different areas of the stroke. Moreover, the rumbling is accentuated when the sound wave reaches an observer after having bounced off obstructions, such as hills and buildings.

In some instances, lightning is seen but no thunder is heard. Does this mean that thunder was not produced by the lightning? Actually, there is thunder, but the atmosphere refracts (bends) and attenuates the sound waves, making the thunder inaudible. Sound travels faster in warm air than in cold air. Because thunderstorms form in a conditionally unstable atmosphere, where the temperature normally drops rapidly with height, a sound wave moving outward away from a lightning stroke will often bend upward, away from an observer at the surface. Consequently, an observer closer than about 5 km (3 mi) to a lightning stroke will usually hear thunder, while an observer 15 km (about 9 mi) away will not.

A sound occasionally mistaken for thunder is the **sonic boom.** Sonic booms are produced when an aircraft exceeds the speed of sound at the altitude at which it is flying. The aircraft compresses the air, forming a shock wave that trails out as a cone behind the aircraft. Along the shock wave, the air pressure changes rapidly over a short distance. The rapid pressure change causes the distinct boom. (Exploding fireworks generate a similar shock wave and a loud bang.)

As for lightning, what causes it? The normal fair weather electric field of the atmosphere is characterized by a negatively charged surface and a positively charged upper atmosphere. For lightning to occur, separate regions containing opposite electrical charges must exist within a cumulonimbus cloud. Exactly how this charge separation comes about is not totally comprehended; however, there are many theories to account for it.

Electrification of Clouds One theory proposes that clouds become electrified when graupel (small ice particles called *soft hail*) and hail fall through a region of supercooled droplets and ice crystals. As liquid droplets collide with a hailstone, they freeze on contact and release latent heat. This process keeps the surface of the hailstone warmer than that of the surrounding ice crystals. When the warmer hailstone comes in contact with a colder ice crystal, an important phenomenon occurs: *There is a net transfer of positive ions (charged molecules) from the warmer object to the colder object.* Hence, the hailstone (larger particle) becomes negatively charged and the ice crystal (smaller particle) positively charged, as the positive ions are incorporated into the ice crystal. The same effect occurs when colder supercooled droplets freeze on contact with a warmer hailstone and tiny splinters of positively charged ice break off. These lighter, positively charged particles are then carried to the upper part of the cloud by updrafts. The larger hailstones (or graupel), left with a negative charge, fall toward the bottom of the cloud. By this mechanism, the cold upper part of the cloud becomes positively charged, while the middle of the cloud becomes negatively charged. The lower part of the cloud is generally of negative and mixed charge except for an occasional positive region located in the falling precipitation near the melting level (see Fig. 10.19).

Another school of thought proposes that during the formation of precipitation, regions of separate charge exist within tiny cloud droplets and larger precipitation particles. In the upper part of these particles we find negative charge, while in the lower part we find positive charge.

*Lightning may also occur in snowstorms, in dust storms, on rare occasions in nimbostratus clouds, and in the gas cloud of an erupting volcano.

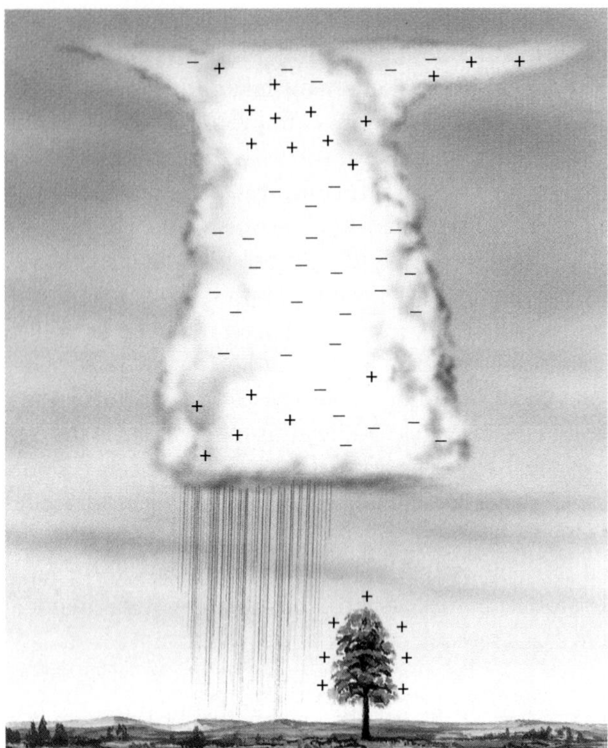

FIGURE 10.19 The generalized charge distribution in a mature thunderstorm.

When falling precipitation collides with smaller particles, the larger precipitation particles become negatively charged and the smaller particles, positively charged. Updrafts within the cloud then sweep the smaller positively charged particles into the upper reaches of the cloud, while the larger negatively charged particles either settle toward the lower part of the cloud or updrafts keep them suspended near the middle of the cloud.

The Lightning Stroke Because unlike charges attract one another, the negative charge at the bottom of the cloud causes a region of the ground beneath it to become positively charged. As the thunderstorm moves along, this region of positive charge follows the cloud like a shadow. The positive charge is most dense on protruding objects, such as trees, poles, and buildings. The difference in charges causes an electric potential between the cloud and ground. In dry air, however, a flow of current does not occur because the air is a good electrical insulator. Gradually, the electrical potential gradient builds, and when it becomes sufficiently large (on the order of one million volts per meter), the insulating properties of the air break down, a current flows, and lightning occurs.

Cloud-to-ground lightning begins within the cloud when the localized electric potential gradient exceeds 3 million volts per meter along a path perhaps 50 meters long. This situation causes a discharge of electrons to rush toward the cloud base and then toward the ground in a series of steps (see Fig. 10.20a). Each discharge covers about 50 to 100 meters, then stops for about 50-millionths of a second, then occurs again over another 50 meters or so. This **stepped leader** is very faint and is usually invisible to the human eye. As the tip of the stepped leader approaches the ground, the potential gradient (the voltage per meter) increases, and a current of positive charge starts upward from the ground (usually along elevated objects) to meet it (see Fig. 10.20b). After they meet, large numbers of electrons flow to the ground and a much larger, more luminous **return stroke** several centimeters in diameter surges upward to the cloud along the path followed by the stepped leader (Fig. 10.20c). Hence, the downward flow of electrons establishes the bright channel of upward propagating current. Even though the bright return stroke travels from the ground up to the cloud, it happens so quickly—in one ten-thousandth of a second—that our eyes cannot resolve the motion, and we see what appears to be a continuous bright flash of light (see Fig. 10.21).

Sometimes there is only one lightning stroke, but more often the leader-and-stroke process is repeated in the same ionized channel at intervals of about four-hundredths of a second. The subsequent leader, called a **dart leader,** proceeds from the cloud along the same channel as the original stepped leader; however, it proceeds downward more quickly because the electrical resistance of the path is now lower. As the leader approaches the ground, normally a less energetic return stroke than the first one travels from the ground to the cloud. Typically, a lightning flash will have three or four leaders, each followed by a return stroke. A lightning flash consisting of many strokes (one photographed flash had 26 strokes) usually lasts less than a second. During this short period of time, our eyes may barely be able to perceive the individual strokes, and the flash appears to flicker.

The lightning described so far (where the base of the cloud is negatively charged and the ground positively

DID YOU KNOW?

The folks of Elgin, Manitoba, literally had their "goose cooked" during April, 1932, when a lightning bolt killed 52 geese that were flying overhead in formation. As the birds fell to the ground, they were reportedly gathered up and distributed to the townspeople for dinner.

Meteorology Now™ ACTIVE FIGURE 10.20 The development of a lightning stroke. (a) When the negative charge near the bottom of the cloud becomes large enough to overcome the air's resistance, a flow of electrons—the stepped leader—rushes toward the earth. (b) As the electrons approach the ground, a region of positive charge moves up into the air through any conducting object, such as trees, buildings, and even humans. (c) When the downward flow of electrons meets the upward surge of positive charge, a strong electric current—a bright return stroke—carries positive charge upward into the cloud. Watch this Active Figure at http://earthscience.brookscole.com/ahrens/ess4e.

FIGURE 10.21 Time exposure of an evening thunderstorm with an intense lightning display near Denver, Colorado. The bright flashes are return strokes. The lighter forked flashes are probably stepped leaders that did not make it to the ground.

charged) is called *negative cloud-to-ground-lightning*, because the stroke carries negative charges from the cloud to the ground. About 90 percent of all cloud-to-ground lightning is negative. However, when the base of the cloud is positively charged and the ground negatively charged, a *positive cloud-to-ground lightning* flash may result. Positive lightning, most common with severe thunderstorms, has the potential to cause more damage because it generates a much higher current level and its flash lasts for a longer duration than negative lightning.

Types of Lightning Notice in Fig. 10.21 that lightning may take on a variety of shapes and forms. When a dart leader moving toward the ground deviates from the original path taken by the stepped leader, the lightning appears crooked or forked, and it is called *forked lightning*. An interesting type of lightning is *ribbon lightning* that forms when the wind moves the ionized channel between each return stroke, causing the lightning to appear as a ribbon hanging from the cloud. If the lightning channel breaks up, or appears to break up, the lightning (called *bead lightning*) looks like a series of beads tied to a string. *Ball lightning* looks like a luminous sphere that appears to float in the air or slowly dart about for several seconds. Although many theories have been proposed, the actual cause of ball lightning remains an enigma. *Sheet lightning* forms when either the lightning flash occurs inside a cloud or intervening clouds obscure the flash, such that a portion of the cloud (or clouds) appears as a luminous white sheet.

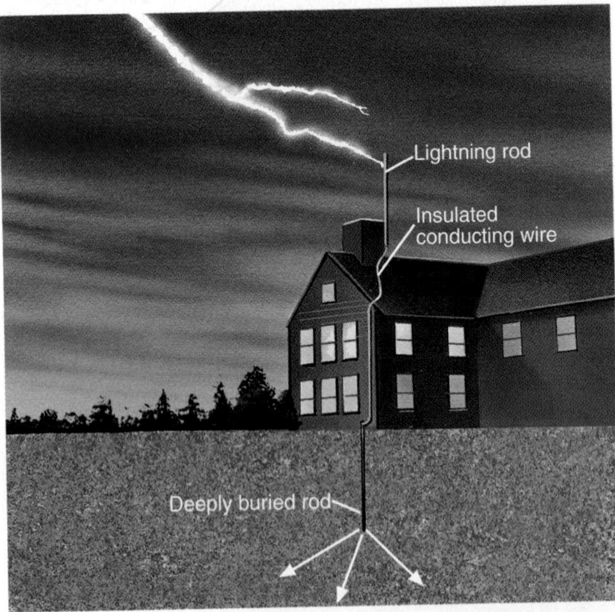

FIGURE 10.22 The lightning rod extends above the building, increasing the likelihood that lightning will strike the rod rather than some other part of the structure. After lightning strikes the metal rod, it follows an insulated conducting wire harmlessly into the ground.

Distant lightning from thunderstorms that is seen but not heard is commonly called **heat lightning** because it frequently occurs on hot summer nights when the overhead sky is clear. As the light from distant electrical storms is refracted through the atmosphere, air molecules and fine dust scatter the shorter wavelengths of visible light, often causing heat lightning to appear orange to a distant observer. Lightning may also shoot upward from the tops of thunderstorms into the upper atmosphere as a dim red flash called a *red sprite*, or as a narrow blue cone called a *blue jet*.

As the electric potential near the ground increases, a current of positive charge moves up pointed objects, such as antennas and masts of ships. However, instead of a lightning stroke, a luminous greenish or bluish halo may appear above them, as a continuous supply of sparks—a *corona discharge*—is sent into the air. This electric discharge, which can cause the top of a ship's mast to glow, is known as **St. Elmo's Fire,** named after the patron saint of sailors. St. Elmo's Fire is also seen around power lines and the wings of aircraft. When St. Elmo's Fire is visible and a thunderstorm is nearby, a lightning flash may occur in the near future, especially if the electric field of the atmosphere is increasing.

Lightning rods are placed on buildings to protect them from lightning damage. The rod is made of metal and has a pointed tip, which extends well above the structure (see Fig. 10.22). The positive charge concentration will be maximum on the tip of the rod, thus increasing the probability that the lightning will strike the tip and follow the metal rod harmlessly down into the ground, where the other end is deeply buried.

When lightning strikes an object such as a car, lightning normally leaves the passengers unharmed because it usually takes the quickest path to the ground along the outside metal casing of the vehicle. The lightning then jumps to the road through the air, or it enters the roadway through the tires (see Fig. 10.23). If you should be caught in the open in a thunderstorm, what should you do? Of course, seek shelter immediately, but under a tree? If you are not sure, please read the Focus section on p. 276.

Meteorology Now™
Click "Lightning" to observe the anatomy and evolution of a lightning strike.

Lightning Detection and Suppression For many years, lightning strokes were detected primarily by visual observation. Today, cloud-to-ground lightning is located by means of an instrument called a *lightning direction-*

Photo by author

FIGURE 10.23 The four marks on the road surface represent areas where lightning, after striking a car traveling along south Florida's Sunshine State Parkway, entered the roadway through the tires. Lightning flattened three of the car's tires and slightly damaged the radio antenna. The driver and a six-year-old passenger were taken to a nearby hospital, treated for shock, and released.

finder, which works by detecting the radio waves produced by lightning. A web of these magnetic devices is a valuable tool in pinpointing lightning strokes throughout the United States, Canada, and Alaska. Lightning detection devices allow scientists to examine in detail the lightning activity inside a storm as it intensifies and moves (see Fig. 10.24). This gives forecasters a better idea where intense lightning strokes might be expected.* In addition, when this information is correlated with satellite images, a more complete and precise structure of a thunderstorm is obtained.

Each year, approximately 10,000 fires are started by lightning in the United States alone and around $50 million worth of timber is destroyed. For this reason, tests have been conducted to see whether the number of cloud-to-ground lightning discharges can be reduced. One technique that has shown some success in suppressing lightning involves seeding a cumulonimbus cloud with hair-thin pieces of aluminum about 10 cm long. The idea is that these pieces of metal will produce many tiny sparks, or corona discharges, and prevent the electrical potential in the cloud from building to a point where lightning occurs. While the results of this experiment are inconclusive, many forestry specialists point out that nature itself may use a similar mechanism to prevent excessive lightning damage. The long, pointed needles of pine trees may act as tiny lightning rods, diffusing the concentration of electric charges and preventing massive lightning strokes.

*In fact, with the aid of these instruments and computer models of the atmosphere, the National Weather Service currently issues lightning probability forecasts for the western United States.

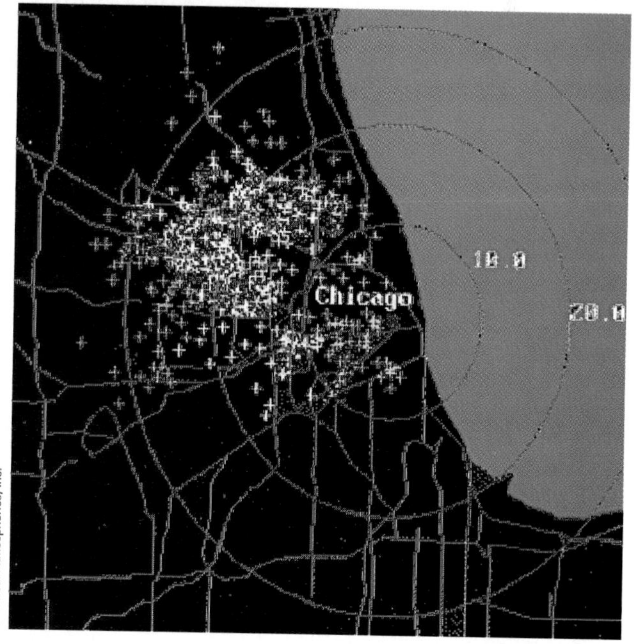

© Global Atmospherics, Inc.

FIGURE 10.24 Cloud-to-ground lightning strikes in the vicinity of Chicago, Illinois, as detected by the National Lightning Detection Network.

DID YOU KNOW?

Lightning is extremely dangerous. Just because you are not beneath a thunderstorm does not mean you are entirely safe. Lightning can travel horizontally outward from a storm, then turn downward, hitting the ground miles from the cloud. Even if lightning doesn't hit you directly, it can still kill you, as a lightning stroke hitting close by may fatally interfere with the rhythm of your heart.

FOCUS ON AN OBSERVATION

Don't Sit Under the Apple Tree

Because a single lightning stroke may involve a current as great as 100,000 amperes, animals and humans can be electrocuted when struck by lightning. The average yearly death toll in the United States attributed to lightning is nearly 100, with Florida accounting for the most fatalities. Many victims are struck in open places, riding on farm equipment, playing golf, or sailing in a small boat. Some live to tell about it, as did the champion golfer Lee Trevino. Others are less fortunate. When you see someone struck by lightning, immediately give CPR (cardiopulmonary resuscitation), as lightning normally leaves its victims unconscious without heartbeat and without respiration.

Most lightning fatalities occur in the vicinity of relatively isolated trees (see Fig. 2). Because a positive charge tends to concentrate in upward projecting objects, the upward return stroke that meets the stepped leader is most likely to originate from such objects. Clearly, sitting under a tree during an electrical storm is not wise. What *should* you do?

When caught in a thunderstorm, the best protection, of course, is to get inside a building. Automobiles and trucks (but not golf carts) may also provide protection. If no such shelter exists, be sure to avoid elevated places and isolated trees. If you are on level ground, try to keep your head as low as possible, but do not lie down. Because lightning channels usually emanate outward through the ground at the point of a lightning strike, a surface current may travel through your body and injure or kill you. Therefore, crouch down as low as possible and minimize the contact area you have with the ground. There are some warning signs to alert you to a strike. If your hair begins to stand on end or your skin begins to tingle and you hear clicking sounds, beware—lightning may be about to strike. And if you are standing upright, you may be acting as a lightning rod.

FIGURE 2 A cloud-to-ground lightning flash hitting a 65-foot sycamore tree. It should be apparent why one should *not* seek shelter under a tree during a thunderstorm.

BRIEF REVIEW

In the last several sections, we examined thunderstorms with their associated lightning and thunder. Listed below for your review are several important concepts we considered:

- Thunderstorms tend to form when warm, humid air rises in a conditionally unstable atmosphere.
- While ordinary (air-mass) thunderstorms are usually short-lived (typically less than an hour), and rarely produce strong winds and hail, severe thunderstorms such as the supercell and pre-frontal squall-line thunderstorms may exist for hours and are capable of producing large hail, strong surface winds (derechoes), flash floods, and tornadoes.
- Strong downdrafts of a thunderstorm, called downbursts (or if the downdrafts are smaller than 4 km, they are called microbursts) have been responsible for several airline crashes, because upon striking the surface, these winds produce extreme wind shear—rapid changes in wind speed and wind direction.
- While lightning is a visible electrical discharge (a giant spark that usually occurs in mature thunderstorms), thunder is the sound that results from the rapidly expanding heated air along the channel of the lightning stroke.

Now that we have looked at thunderstorms, we are ready to explore a product of a thunderstorm that is one of nature's most awesome phenomena: the tornado, a rapidly spiraling column of air that usually extends down from the base of a cumulonimbus cloud and can strike sporadically and violently.

TORNADOES

A **tornado** is a rapidly rotating column of air that blows around a small area of intense low pressure with a circulation that reaches the ground. A tornado's circulation is present on the ground either as a funnel-shaped cloud or as a swirling cloud of dust and debris. Sometimes called *twisters* or *cyclones,* tornadoes can assume a variety of shapes and forms that range from twisting rope-like funnels, to cylindrical-shaped funnels, to massive black funnels, to funnels that resemble an elephant's trunk hanging from a large cumulonimbus cloud. A **funnel cloud** is a tornado whose circulation has not reached the ground. When viewed from above, the majority of North American tornadoes rotate counterclockwise about their central core of low pressure. A few have been seen rotating clockwise, but those are rare.

The diameter of most tornadoes is between 100 and 600 m (about 300 to 2000 ft), although some are just a few meters wide and others have diameters exceeding 1600 m (1 mi). Tornadoes that form ahead of an advancing cold front are often steered by southwesterly winds and, therefore, tend to move from the southwest toward the northeast at speeds usually between 20 and 40 knots. However, some have been clocked at speeds greater than 70 knots. Most tornadoes last only a few minutes and have an average path length of about 7 km (4 mi). There are cases where they have reportedly traveled for hundreds of kilometers and have existed for many hours, such as the one that lasted over 7 hours and cut a path 470 km (292 mi) long through portions of Illinois and Indiana on May 26, 1917.*

*Actually, this situation may have been several tornadoes (a family) that were generated by a single thunderstorm as it moved along.

Tornado Life Cycle Major tornadoes usually evolve through a series of stages. The first stage is the *dust-whirl stage,* where dust swirling upward from the surface marks the tornado's circulation on the ground and a short funnel often extends downward from the thunderstorm's base. Damage during this stage is normally light. The next stage, called the *organizing stage,* finds the tornado increasing in intensity with an overall downward extent of the funnel. During the tornado's *mature stage,* damage normally is most severe as the funnel reaches its greatest width and is almost vertical (see Fig. 10.25). The *shrinking stage* is characterized by an overall decrease in the funnel's width, an increase in the funnel's tilt, and a narrowing of the damage swath at the surface, although the tornado may still be capable of intense and sometimes violent damage. The final stage, called the *decay stage,* usually finds the tornado stretched into the shape of a rope. Normally, the tornado becomes greatly contorted before it finally dissipates. Although these are the typical stages of a major tornado, minor tornadoes may evolve only through

FIGURE 10.25 A mature tornado with winds exceeding 150 knots rips through southern Illinois.

the organizing stage. Some even skip the mature stage and go directly into the decay stage. However, when a tornado reaches its mature stage, its circulation usually stays in contact with the ground until it dissipates.

Tornado Outbreaks Each year, tornadoes take the lives of many people. The yearly average is less than 100, although over 100 may die in a single day. In recent years, an alarming statistic is that 45 percent of all fatalities occurred in mobile homes. The deadliest tornadoes are those that occur in *families;* that is, different tornadoes spawned by the same thunderstorm. (Some thunderstorms produce a sequence of several tornadoes over 2 or more hours and over distances of 100 km or more.) Tornado families often are the result of a single, long-lived supercell thunderstorm. When a large number of tornadoes (typically 6 or more) forms over a particular region, this constitutes what is termed a **tornado outbreak.**

A particularly devastating outbreak occurred on May 3, 1999, when 78 tornadoes marched across parts of Texas, Kansas, and Oklahoma. One tornado, whose width at times reached one mile and whose wind speed was measured by Doppler radar at 276 knots (318 mi/hr), moved through the southwestern section of Oklahoma City. Within its 40-mile path, it damaged or destroyed thousands of homes, injured nearly 600 people, claimed 38 lives, and caused over $1 billion in property damage.

One of the most violent outbreaks ever recorded occurred on April 3 and 4, 1974. During a 16-hour period, 148 tornadoes cut through parts of 13 states, killing 307 people, injuring more than 6000, and causing an estimated $600 million in damage. Some of these tornadoes were among the most powerful ever witnessed. The combined path of all the tornadoes during this *super outbreak* amounted to 4181 km (2598 mi), well over half of the total path for an average year. The greatest loss of life attributed to tornadoes occurred during the tri-state outbreak of March 18, 1925, when an estimated 695 people died as at least 7 tornadoes traveled a total of 703 km (437 mi) across portions of Missouri, Illinois, and Indiana.

DID YOU KNOW?

The tornado outbreak on May 3, 1999, produced 78 tornadoes over parts of Texas, Kansas, and Oklahoma. One tornado, whose width at times reached one mile, moved through the southwestern section of Oklahoma City. Within its 40-mile path, it damaged or destroyed thousands of homes, injured nearly 600 people, claimed 38 lives, and caused over $1 billion in property damage.

DID YOU KNOW?

During May, 2003, a record 516 tornadoes touched down in the United States—the most in any month ever.

Tornado Occurrence Tornadoes occur in many parts of the world, but no country experiences more tornadoes than the United States, which averages more than 1000 annually and experienced a record 1424 tornadoes during 1998. Although tornadoes have occurred in every state, including Alaska and Hawaii, the greatest number occur in the tornado belt or *tornado alley* of the Central Plains, which stretches from central Texas to Nebraska* (see Fig. 10.26).

The Central Plains region is most susceptible to tornadoes because it provides the proper atmospheric setting for the development of the severe thunderstorms that spawn tornadoes. Here (especially in spring) warm, humid surface air is overlain by cooler, drier air aloft, producing a conditionally unstable atmosphere. When a strong vertical wind shear exists and the surface air is forced upward, large thunderstorms capable of spawning tornadoes may form. Therefore, tornado frequency is highest during the spring and lowest during the winter when the warm surface air is normally absent.

About three-fourths of all tornadoes in the United States develop from March to July. The month of May normally has the greatest number of tornadoes (the average is about 6 per day) while the most violent tornadoes seem to occur in April when vertical wind shear tends to be present as well as when horizontal and vertical temperature and moisture contrasts are greatest. Although tornadoes have occurred at all times of the day and night, they are most frequent in the late afternoon (between 4:00 P.M. and 6:00 P.M.), when the surface air is most unstable; they are least frequent in the early morning before sunrise, when the atmosphere is most stable.

Although large, destructive tornadoes are most common in the Central Plains, they can develop anywhere if conditions are right. For example, a series of at least 36 tornadoes, more typical of those that form over the plains, marched through North and South Carolina on March 28, 1984, claiming 59 lives and causing hundreds of millions of dollars in damage. One tornado was enormous, with a diameter of at least 4000 m (2.5 mi) and winds that exceeded 200 knots. No place is totally immune to a tornado's destructive force. On March 1, 1983,

*Many of the tornadoes that form along the Gulf Coast are generated by thunderstorms embedded within the circulation of hurricanes.

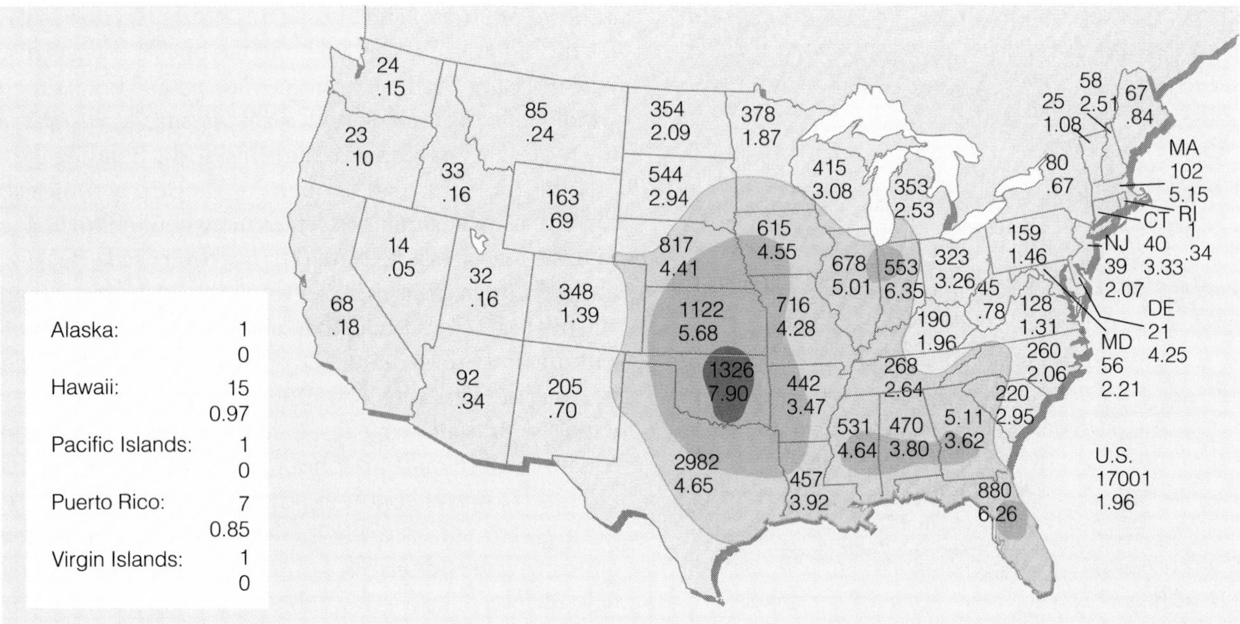

FIGURE 10.26 Tornado incidence by state. The upper figure shows the number of tornadoes reported by each state during a 25-year period. The lower figure is the average annual number of tornadoes per 10,000 square miles. The darker the shading, the greater the frequency of tornadoes.

a rare tornado cut a 5-km swath of destruction through downtown Los Angeles, California, damaging more than 100 homes and businesses and injuring 33 people.

Even in the central part of the United States, the statistical chance that a tornado will strike a particular place this year is quite small. However, tornadoes can provide many exceptions to statistics. Oklahoma City, for example, has been struck by tornadoes at least 33 times in the past 90 years. And the little town of Codell, Kansas, was hit by tornadoes in 3 consecutive years—1916, 1917, and 1918—and each time on the same date: May 20! Considering the many millions of tornadoes that must have formed during the geological past, it is likely that at least one actually moved across the land where your home is located, especially if it is in the Central Plains.

Tornado Winds The strong winds of a tornado can destroy buildings, uproot trees, and hurl all sorts of lethal missiles into the air. People, animals, and home appliances all have been picked up, carried several kilometers, then deposited. Tornadoes have accomplished some astonishing feats, such as lifting a railroad coach with its 117 passengers and dumping it in a ditch 25 meters away. Showers of toads and frogs have poured out of a cloud after tornadic winds sucked them up from a nearby pond. Other oddities include chickens losing all of their feathers, pieces of straw being driven into metal pipes, and frozen hot dogs being driven into concrete walls. Miraculous events have occurred, too. In one instance, a schoolhouse was demolished and the 85 students inside were carried over 100 meters without one of them being killed.

Our earlier knowledge of the furious winds of a tornado came mainly from observations of the damage done and the analysis of motion pictures. Today more accurate wind measurements are made with Doppler radar. Because of the destructive nature of the tornado, it was once thought that it packed winds greater than 500 knots. However, studies conducted after 1973 reveal that even the most powerful twisters seldom have winds exceeding 220 knots, and most tornadoes probably have winds of less than 125 knots. Nevertheless, being confronted with even a small tornado can be terrifying.

When a tornado is approaching from the southwest, its strongest winds are on its southeast side. We can see why in Fig. 10.27. The tornado is heading northeast at 50 knots. If its rotational speed is 100 knots, then its forward speed will add 50 knots to its southwestern side (position D) and subtract 50 knots from its northwestern side (position A). Hence, the most destructive and extreme winds will be on the tornado's southeastern side.

Many violent tornadoes (with winds exceeding 180 knots) contain smaller whirls that rotate within them. Such tornadoes are called *multi-vortex tornadoes* and the smaller whirls are called **suction vortices** (see Fig. 10.28). Suction vortices are only about 10 m (30 ft) in diameter, but they rotate very fast and apparently do a great deal of damage.

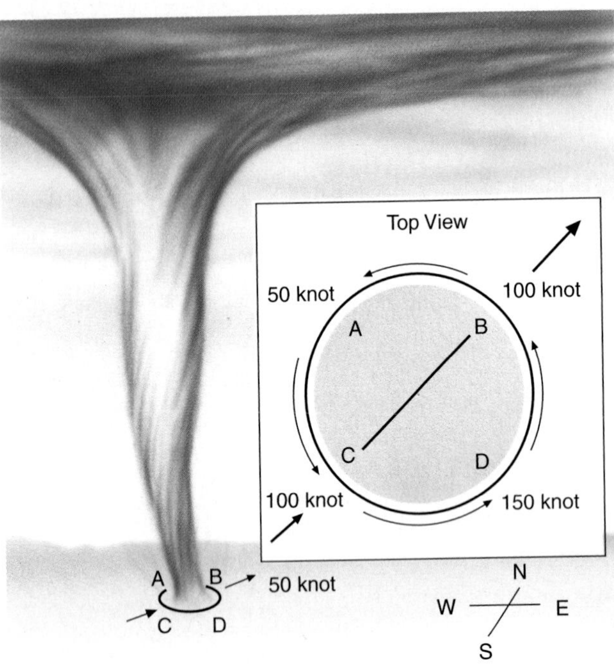

FIGURE 10.27 The total wind speed of a tornado is greater on one side than on the other. When facing an on-rushing tornado, the strongest winds will be on your left side.

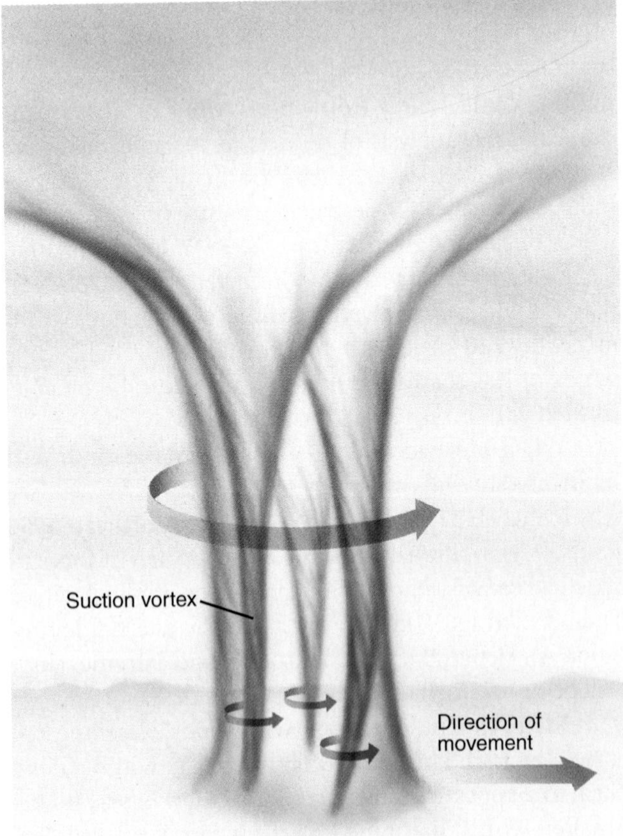

FIGURE 10.28 A powerful multi-vortex tornado with three suction vortices.

Seeking Shelter The high winds of the tornado cause the most damage as walls of buildings buckle and collapse when blasted by the extreme wind force. Also, as high winds blow over a roof, lower air pressure forms above the roof. The greater air pressure inside the building then lifts the roof just high enough for the strong winds to carry it away. A similar effect occurs when the tornado's intense low-pressure center passes overhead. Because the pressure in the center of a tornado may be more than 100 mb (3 in.) lower than that of its surroundings, there is a momentary drop in outside pressure when the tornado is above the structure. It was once thought that opening windows and allowing inside and outside pressures to equalize would minimize the chances of the building exploding. However, it now appears that opening windows during a tornado actually increases the pressure on the opposite wall and *increases* the chances that the building will collapse. (The windows are usually shattered by flying debris anyway.) So stay away from windows. Damage from tornadoes may also be inflicted on people and structures by flying debris. Hence, the wisest course to take when confronted with an approaching tornado is to *seek shelter immediately.*

At home, take shelter in a basement and stay away from windows. In a large building without a basement, the safest place is usually in a small room, such as a bathroom, closet, or interior hallway, preferably on the lowest floor and near the middle of the edifice. Pull a mattress around you as the handles on the side make it easy to hang onto. Wear a bike or football helmet to protect your head from flying debris. At school, move to the hallway and lie flat with your head covered. In a mobile home, leave immediately and seek substantial shelter. If none exists, lie flat on the ground in a depression or ravine. Don't try to outrun an oncoming tornado in a car or truck, as tornadoes often cover erratic paths with speeds sometimes exceeding 70 knots (80 mi/hr). Stop your car and let the tornado go by or turn around on the road's shoulder

DID YOU KNOW?

It may be almost impossible to survive the powerful winds of a violent tornado if you are inside the wrong type of structure, such as a mobile home. During the May 3, 1999, tornado outbreak many people who abandoned their unprotected homes in favor of muddy ditches survived largely because the ditches were below ground level and out of the path of wind-blown objects. Many who stayed in the confines of their inadequate homes perished when tornado winds blew their homes away, leaving only the foundations.

and drive in the opposite direction. And do not take shelter under a freeway overpass, as the tornado's winds are actually funneled (strengthened) by the overpass structure. If caught outdoors in an open field, look for a ditch, streambed, or ravine, and lie flat with your head covered.

When tornadoes are likely to form during the next few hours, a **tornado watch** is issued by the Storms Prediction Center in Norman, Oklahoma, to alert the public that tornadoes may develop within a specific area during a certain time period. Many communities have trained volunteer spotters, who look for tornadoes after the watch is issued. Once a tornado is spotted—either visually or on a radar screen—a **tornado warning** is issued by the local National Weather Service Office. In some communities, sirens are sounded to alert people of the approaching storm. Radio and television stations interrupt regular programming to broadcast the warning. Although not completely effective, this warning system is apparently saving many lives. Despite the large increase in population in the tornado belt during the past 30 years, tornado-related deaths have actually shown a decrease (see Table 10.1).

The Fujita Scale In the 1960s, the late Dr. T. Theodore Fujita, a noted authority on tornadoes at the University of Chicago, proposed a scale (called the **Fujita scale**) for classifying tornadoes according to their rotational wind speed based on the damage done by the storm. (A tornado's winds are actually estimated based on the assessment of the damage caused by the storm.) Table 10.2 presents this scale.

Statistics reveal that the majority of tornadoes are F0 and F1 (weak tornadoes) and only a few percent each year are above the F3 classification (violent) with perhaps one or two F5 tornadoes reported annually. However, it is the violent tornadoes that account for the majority of tornado-related deaths. As an example, a powerful F5 tornado marched through the southern part of Andover, Kansas, on the evening of April 26, 1991. The tornado, which stayed on the ground for nearly 110 km (68 mi), destroyed more than 100 homes and businesses, injured several hundred people, and out of the 39 tornado fatalities in 1991, this F5 tornado alone took the lives of 17. A powerful F5 tornado is shown in Fig. 10.29.

TABLE 10.1
Average Annual Number of Tornadoes and Tornado Deaths by Decade

DECADE	TORNADOES/YEAR	DEATHS/YEAR
1950–59	480	148
1960–69	681	94
1970–79	858	100
1980–89	819	52
1990–99	1,220*	56

*More tornadoes are being reported as populations increase and tornado-spotting technology improves.

Tornadic Thunderstorms Although everything is not known about the formation of a tornado, we do know that tornadoes tend to form with intense thunderstorms and that a conditionally unstable atmosphere is essential for their development.

Favorable Atmospheric Conditions One atmospheric situation that frequently leads to severe thunderstorms with tornadoes in the spring is shown in Fig. 10.30. At the surface, we find an open-wave middle-latitude cyclone with cold, dry air moving in behind a cold front, and warm, humid air pushing northward from the Gulf of Mexico behind a warm front. Above the warm surface air a wedge of warm, moist air is streaming northward. Directly above the moist layer is a wedge of colder, drier air moving in from the southwest. Higher up, at the

TABLE 10.2 **Fujita Scale for Damaging Wind**

SCALE	CATEGORY	MI/HR	KNOTS	EXPECTED DAMAGE
F0	Weak	40-72	35-62	Light: tree branches broken, sign boards damaged
F1		73-112	63-97	Moderate: trees snapped, windows broken
F2	Strong	113-157	98-136	Considerable: large trees uprooted, weak structures destroyed
F3		158-206	137-179	Severe: trees leveled, cars overturned, walls removed from buildings
F4	Violent	207-260	180-226	Devastating: frame houses destroyed
F5*		261-318	227-276	Incredible: structures the size of autos moved over 100 meters, steel-reinforced structures highly damaged

*The scale continues up to a theoretical F12. Very few (if any) tornadoes have wind speeds in excess of 318 mi/hr.

FIGURE 10.29 A devastating F5 tornado about 200 meters wide plows through Hesston, Kansas, on March 13, 1990, leaving almost 300 people homeless and 13 injured.

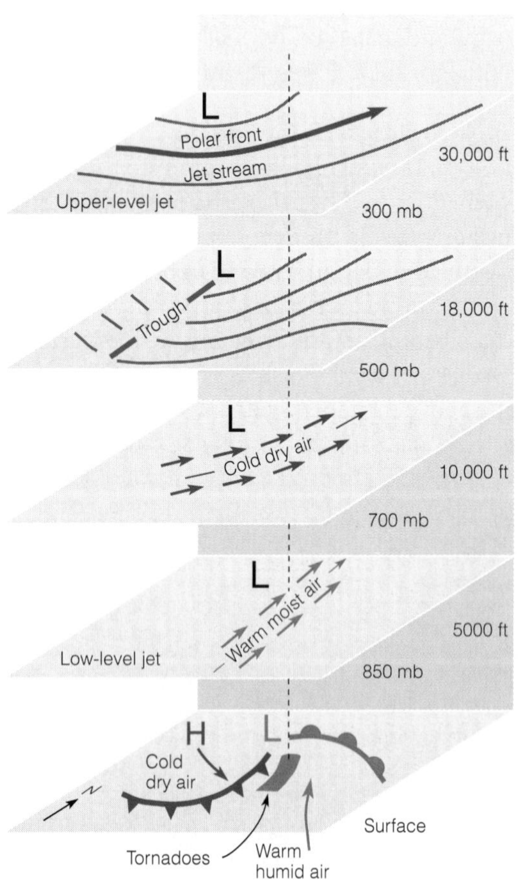

FIGURE 10.30 Conditions leading to the formation of severe thunderstorms that can spawn tornadoes.

500-mb level, a trough of low pressure exists to the west of the surface low and, at the 300-mb level, the polar-front jet stream swings over the region. At this level, the jet stream provides an area of divergence that initiates surface convergence and rising air. The stage is now set for the development of severe storms.

The boxed-off red area on the surface map (Fig. 10.30) shows where tornadoes are most likely to form. They tend to form in this region because the position of cold air above warm air produces a conditionally unstable atmosphere. However, there is more to it than this. Above the warm, humid surface air, there often exists a shallow temperature inversion that acts like a lid on the moist air below. During the morning, the inversion caps the moist air, and only small cumulus clouds form. As the day progresses, and the surface becomes warmer, rising blobs of air are able to break through the inversion at isolated places, and clouds build rapidly, sometimes explosively, as the humid air is vented upward through the opening. (The inversion is important because it prevents many small thunderstorms from forming.) Divergence at the jet stream level then draws this humid air upward into the cold, conditionally unstable air aloft, and a huge thunderstorm quickly develops.

While it is difficult to tell which thunderstorm will spawn a tornado, it is easier to predict where tornado-generating storms are most likely to form. Notice in Fig. 10.30 that this area (the boxed-off red area on the surface map) is situated where the polar front jet stream and the cold, dry air cross the warm, humid surface air. Knowing this helps to explain why the region of greatest tornado activity shifts northward from winter to summer.

During the winter, tornadoes are most likely to form over the southern Gulf states when the polar front jet is above this region, and the contrast between warm and cold air masses is greatest. In spring, humid Gulf air surges northward; contrasting air masses and the jet stream also move northward and tornadoes become more prevalent from the southern Atlantic states westward into the southern Great Plains. In summer, the contrast between air masses lessens, and the jet stream is normally near the Canadian border; hence, tornado activity tends to be concentrated from the northern plains eastward to New York State.

Supercell Tornadoes Tornadoes that form with supercell thunderstorms are called **supercell tornadoes.** Earlier we saw that a supercell is a severe thunderstorm that has a single rotating updraft that can exist for hours. Recall also that supercells form in a region of strong vertical wind shear. In Fig. 10.30, notice that the wind speed increases rapidly with height (vertical wind speed shear), and that the wind changes direction with height—from southerly at low levels to westerly at high levels (vertical wind direction shear). This type of shear can cause the updraft inside the storm to rotate.

For example, in Fig. 10.31 notice that there is wind direction shear as the surface winds are southeasterly, and aloft they are westerly. There is also wind speed shear as the wind speed increases as you move upward. This wind shear causes the air near the surface to rotate about a horizontal axis, much like a pencil rotates around its long axis. Such spiraling tubes of spinning air are called *vortex tubes.* Now suppose the strong updraft of a developing thunderstorm tilts the rotating tube and draws it into the storm, as illustrated in Fig. 10.32. This rising, spinning column on the south side of the storm, perhaps 5 to 10 kilometers across, is called a **mesocyclone** (see Fig. 10.33).

The updraft is so strong in a supercell (sometimes 90 knots) that precipitation cannot fall through it. South-

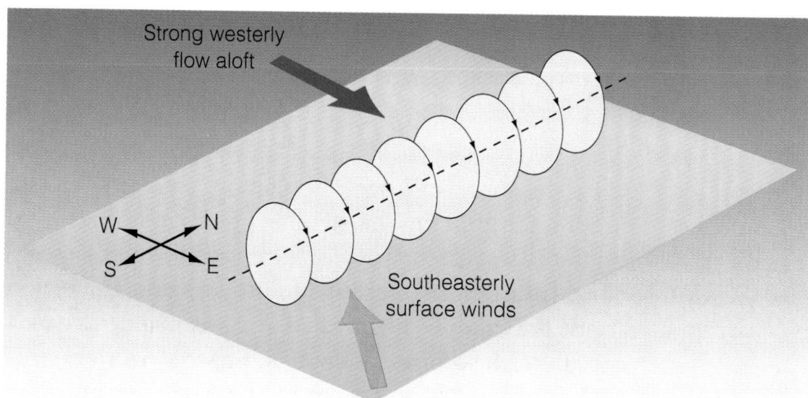

FIGURE 10.31 Spinning vortex tubes created by wind shear.

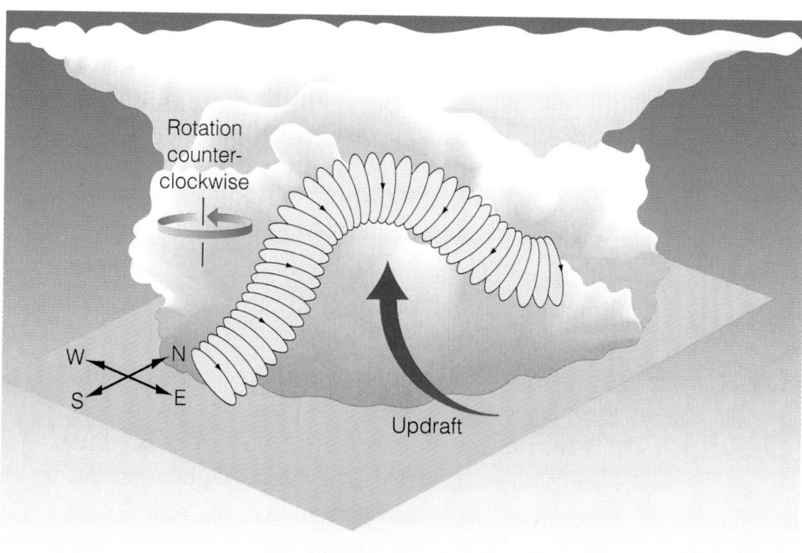

FIGURE 10.32 The strong updraft in the thunderstorm carries the vortex tube into the thunderstorm, producing a rotating air column that is oriented in the vertical plane.

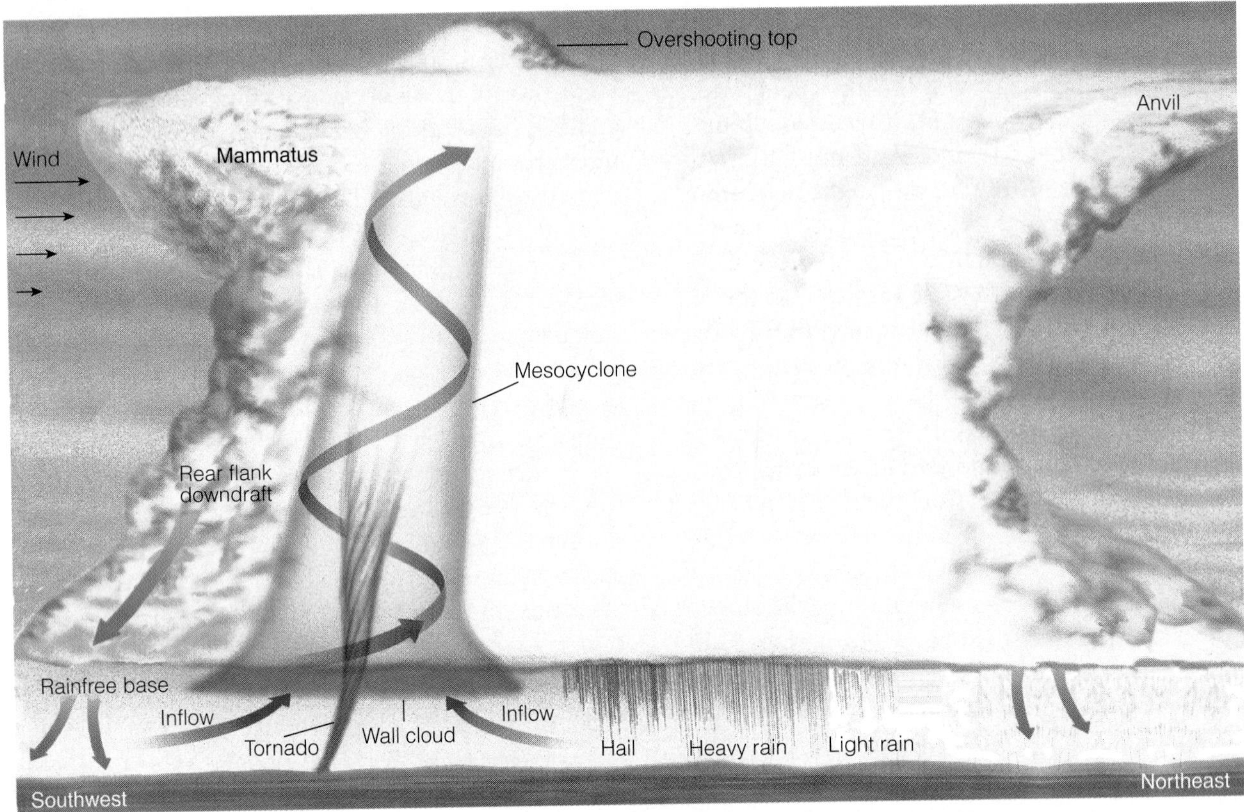

FIGURE 10.33 Some of the features associated with a tornado-breeding supercell thunderstorm as viewed from the southeast. The storm is moving to the northeast.

westerly winds aloft usually blow the precipitation northeastward. Notice in Fig. 10.33, that the largest hailstones, having remained in the cloud for some time, usually fall just north of the updraft, and the heaviest rain occurs just north of the falling hail. If the mesocyclone persists, it can circulate some of the precipitation counterclockwise around the updraft. This swirling precipitation shows up on the radar screen, whereas the area inside the mesocyclone (nearly void of precipitation) does not. The region inside the thunderstorm where radar is unable to detect precipitation is known as the *bounded weak echo region,* or *BWER.* Meanwhile, as the precipitation is drawn into a cyclonic spiral around the mesocyclone, the rotating precipitation may, on the radar screen, unveil itself in the shape of a hook, called a **hook echo,** as shown in Fig. 10.34.

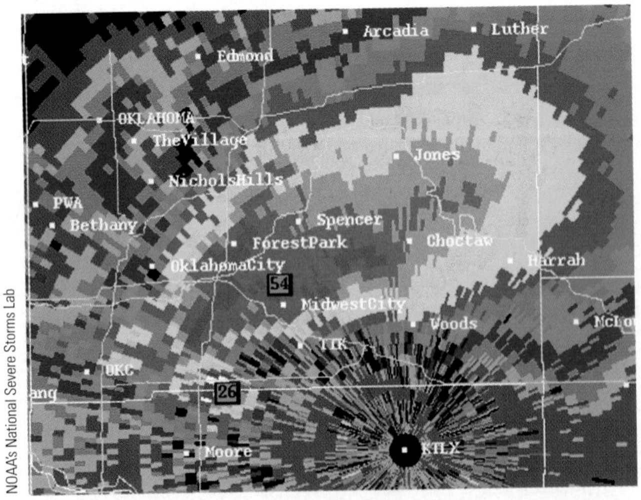

At this point in the storm's development, the updraft, the counterclockwise swirling precipitation, and the surrounding air may all interact to produce the *rear flank downdraft* (to the south of the updraft), as shown in Figure 10.33. When this downdraft strikes the ground, it may (under favorable shear conditions) interact with the region of surface inflow (beneath the mesocyclone) to produce a supercell tornado.

As air rushes upward into the low-pressure core of the mesocyclone, the air expands, cools, and, if sufficiently moist, condenses into a visible cloud—the *funnel cloud.* At the same time, the mesocyclone stretches vertically and shrinks horizontally (to between 2 and 4 km wide), and the spinning air is accelerated upward. As the

NOAA's National Severe Storms Lab

FIGURE 10.34 A tornado-spawning thunderstorm over Oklahoma City on May 3, 1999, shows a hook echo in its rainfall pattern on a Doppler radar screen. The colors red and orange represent the heaviest precipitation.

FIGURE 10.35 A wall cloud photographed southwest of Norman, Oklahoma.

air beneath the funnel cloud is drawn into its core, the air cools rapidly and condenses, and the funnel cloud descends toward the surface. Upon reaching the ground, the tornado usually picks up dirt and debris, making it appear both dark and ominous. Observations reveal that supercell tornadoes develop near the right rear sector of the storm, on the southwestern side of a northeastward-moving storm, as shown in Fig. 10.33.

Not all supercells produce tornadoes; in fact, perhaps less than 15 percent do. Many atmospheric situations may suppress tornado formation. For example, if the precipitation in the cloud is swept too far away from the updraft, or if too much precipitation wraps around the mesocyclone, the necessary interactions that produce the rear flank downdraft are disrupted, and a tornado is not likely to form.

As we have seen, the first sign that a supercell is about to give birth to a tornado is the sight of *rotating* clouds at the base of the storm.* If the area of rotating clouds lowers, it becomes a **wall cloud** (compare the wall cloud depicted in Fig. 10.33 with the photograph of a wall cloud in Fig. 10.35). Usually within the wall cloud, the rapidly rotating funnel extends toward the surface. Sometimes the air is so dry that the swirling wind remains invisible until it reaches the ground and begins to pick up dust. Unfortunately, people have mistaken these "invisible tornadoes" for dust devils, only to find out (often too late) that they were not. Occasionally, the funnel cannot be seen due to falling rain, clouds of dust, or darkness. Even when not clearly visible, many tornadoes have a distinctive roar that can be heard for several kilometers. This sound, which has been described as "a roar like a thousand freight trains," appears to be loudest when the tornado is touching the surface. However, not all tornadoes make this sound and, when these storms strike, they become silent killers.

*Occasionally, people will call a sky dotted with mammatus clouds "a tornado sky." Mammatus clouds may appear with both severe and nonsevere thunderstorms as well as with a variety of other cloud types (see Chapter 6). Mammatus clouds are not funnel clouds, do not rotate, and their appearance has no relationship to tornadoes.

DID YOU KNOW?

Although tornadoes are rare in Utah, with only about two per year being reported, a tornado rampaged through downtown Salt Lake City during August, 1999. While only on the ground for about 5 miles, the tornado damaged over 120 homes, injured a dozen people, produced one fatality, and caused over $50 million in damage.

Certainly, the likelihood of a thunderstorm producing a tornado increases when the storm becomes a supercell, but not all supercells produce tornadoes. And not all tornadoes come from rotating thunderstorms (supercells).

Nonsupercell Tornadoes Tornadoes that do not occur in association with a pre-existing wall cloud of a supercell are called **nonsupercell tornadoes.** These tornadoes may occur with intense multicell storms as well as with ordinary thunderstorms, even relatively weak ones. Some nonsupercell tornadoes extend from the base of a thunderstorm as a visible funnel cloud, as shown in Fig. 10.36, whereas others may begin on the ground and build upwards in the absence of a condensation funnel.

Nonsupercell tornadoes may form along a gust front where the cool downdraft of the thunderstorm forces warm, humid air upwards. Tornadoes that form along a gust front are commonly called **gustnadoes.** These relatively weak tornadoes normally are short-lived and rarely inflict significant damage. Gustnadoes are often seen as a rotating cloud of dust or debris rising above the surface.

Occasionally, rather weak, short-lived tornadoes will occur with rapidly building cumulus congestus clouds. Tornadoes such as these commonly form over east-central Colorado. Because they look similar to waterspouts that form over water, they are sometimes called **landspouts.**

Figure 10.37 illustrates how a landspout can form. Suppose, for example, that the winds at the surface converge along a boundary, as illustrated in Fig. 10.37a. (The wind may converge due to topographic irregularities or any number of other factors, including temperature and moisture variations.) Notice that along the boundary, the air is rising, condensing, and forming into a cumulus congestus cloud. Notice also that along the surface at the boundary there is horizontal rotation (spin) created by the wind blowing in opposite directions along the boundary. If the developing cloud should move over the region of rotating air, the spinning air may be drawn up into the cloud by the storm's updraft. As the spinning, rising air shrinks in diameter, it produces a tornado-like structure called a *landspout.* As with a tornado, the landspout increases in rotational speed much like a spinning skater increases in speed when the arms are brought in close to the body. Landspouts usually dissipate when rain falls through the cloud and destroys the updraft. Tornadoes may form in this manner along many types of converging wind boundaries, including sea breezes and gust fronts. Nonsupercell tornadoes and funnel clouds may also form with thunderstorms when cold air aloft (associated with an upper-level trough) moves over a region. Common along the west coast of North America, these short-lived tornadoes are sometimes called *cold-air funnels.*

FIGURE 10.36 A funnel cloud extends downward from the base of a nonsupercell thunderstorm over central California.

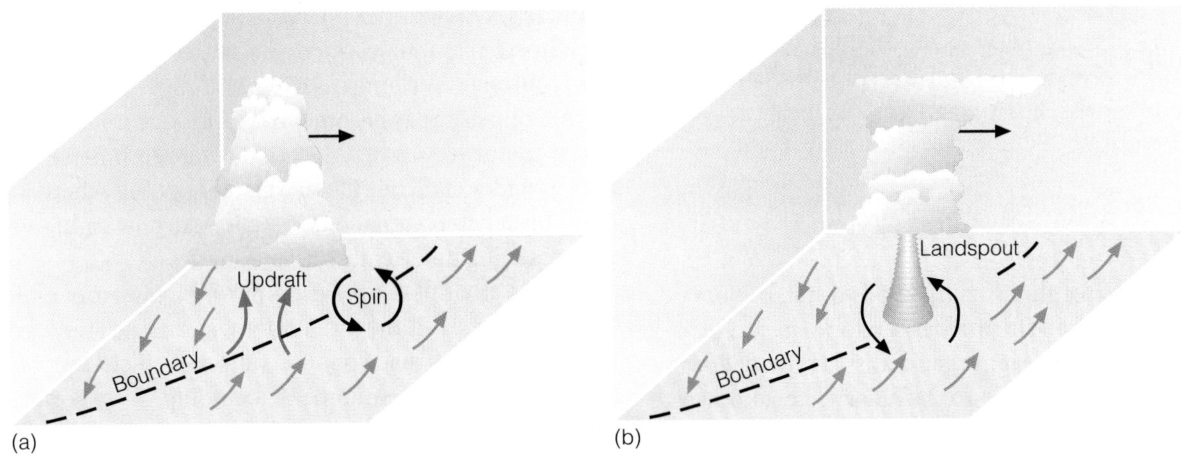

FIGURE 10.37 (a) Along the boundary of converging winds, the air rises and condenses into a cumulus congestus cloud. At the surface the converging winds along the boundary create a region of counterclockwise spin. (b) As the cloud moves over the area of rotation, the updraft draws the spinning air up into the cloud, producing a nonsupercell tornado, or landspout. (Modified after Wakimoto and Wilson)

SEVERE WEATHER AND DOPPLER RADAR

Most of our knowledge about what goes on inside a tornado-generating thunderstorm has been gathered through the use of *Doppler radar.* Remember from Chapter 5 that a radar transmitter sends out microwave pulses and that, when this energy strikes an object, a small fraction is scattered back to the antenna. Precipitation particles are large enough to bounce microwaves back to the antenna. As a consequence, the colorful area on the radar screen in Fig. 10.34 on p. 284 represents precipitation inside a severe thunderstorm.

Doppler radar can do more than measure rainfall intensity: It can actually measure the speed at which precipitation is moving horizontally toward or away from the radar antenna. Because precipitation particles are carried by the wind, Doppler radar can peer into a severe storm and reveal its winds.

Doppler radar works on the principle that, as precipitation moves toward or away from the antenna, the returning radar pulse will change in frequency. A similar change occurs when the high-pitched sound (high frequency) of an approaching noise source, such as a siren or train whistle, becomes lower in pitch (lower frequency) after it passes by the person hearing it. This change in frequency is called the *Doppler shift* and this, of course, is where the Doppler radar gets its name.

A single Doppler radar cannot detect winds that blow parallel to the antenna. Consequently, two or more units probing the same thunderstorm are needed to give a complete three-dimensional picture of the winds within the storm. To help distinguish the storm's air motions, wind velocities can be displayed in color. Color contouring the wind field gives a good picture of the storm (see Fig. 10.38).

Even a single Doppler radar can uncover many of the features of a severe thunderstorm. For example, studies conducted in the 1970s revealed, for the first

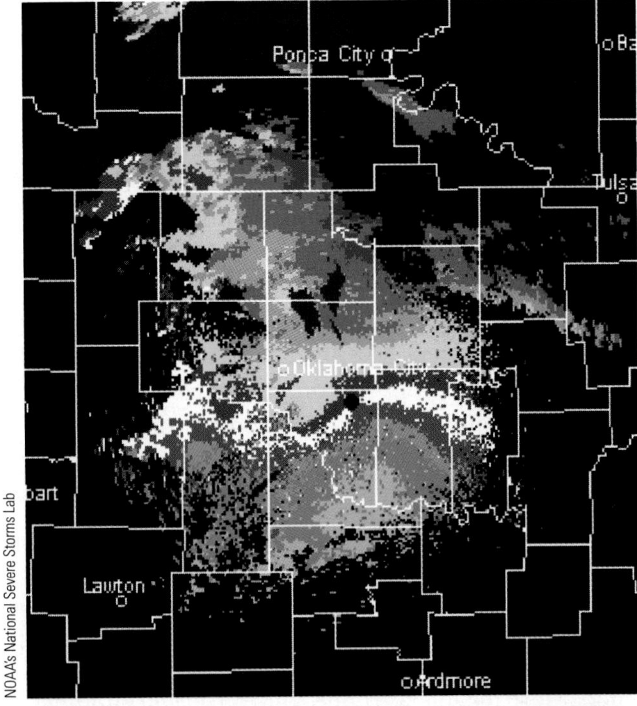

FIGURE 10.38 Doppler radar display of winds associated with the supercell storm that moved through parts of Oklahoma City during the afternoon of May 3, 1999. The close packing of the horizontal winds blowing toward the radar (green and blue shades), and those blowing away from the radar (yellow and red shades), indicate strong cyclonic rotation and the presence of a tornado.

DID YOU KNOW?

Storm-chasing scientists on May 3, 1999, using two Doppler radars mounted on separate vehicles, measured a record wind speed of 318 mi/hr inside a violent F5 tornado that ultimately ravaged sections of Oklahoma City.

time, the existence of the swirling winds of the mesocyclone inside a supercell storm. Mesocyclones have a distinct image (signature) on the radar display. Tornadoes also have a distinct signature on the radar screen, known as the *tornado vortex signature (TVS),* which shows up as a region of rapidly changing wind directions within the mesocyclone (see Fig. 10.38).

Unfortunately, the resolution of the Doppler radar is not high enough to measure actual wind speeds of most tornadoes, whose diameters are only a few hundred meters or less. However, a new and experimental Doppler system—called **Doppler lidar**—uses a light beam (instead of microwaves) to measure the change in frequency of falling precipitation, cloud particles, and dust. Because it uses a shorter wavelength of radiation, it has a narrower beam and a higher resolution than does Doppler radar. In an attempt to obtain tornado wind information at fairly close range (less than 10 km), smaller portable Doppler radar units (Doppler on wheels) are peering into tornado-generating storms.

The new network of more than 150 Doppler radar units deployed at selected weather stations within the continental United States is referred to as **NEXRAD** (an acronym for *NEX*t Generation Weather *RAD*ar). The NEXRAD system consists of the WSR-88D* Doppler radar and a set of computers that perform a variety of functions.

The computers take in data, display them on a monitor, and run computer programs called *algorithms,* which, in conjunction with other meteorological data, detect severe weather phenomena, such as storm cells, hail, mesocyclones, and tornadoes. Algorithms provide a great deal of information to the forecasters that allows them to make better decisions as to which thunderstorms are most likely to produce severe weather and possible flash flooding. In addition, the algorithms give advanced and improved warning of an approaching tornado. More reliable warnings, of course, will cut down on the number of false alarms.

Because the Doppler radar shows horizontal air motions within a storm, it can help to identify the magnitude of other severe weather phenomena, such as gust fronts, derechoes, microbursts, and wind shears that are dangerous to aircraft. Certainly, as more and more information from Doppler radar becomes available, our understanding of the processes that generate severe thunderstorms and tornadoes will be enhanced, and hopefully there will be an even better tornado and severe storm warning system, resulting in fewer deaths and injuries.

In an attempt to unravel some of the mysteries of the tornado, several studies are under way. In one study, scientists using an armada of observational vehicles, aircraft, and state-of-the-art equipment, including Doppler radar, pursued tornado-generating thunderstorms over portions of the Central Plains during the spring and summer. These observations are providing information on the inner workings of severe thunderstorms. At the same time, laboratory models of tornadoes in chambers (called *vortex chambers*), along with mathematical computer models, are offering new insights into the formation and development of these fascinating storms (see Fig. 10.39).

WATERSPOUTS

A **waterspout** is a rotating column of air over a large body of water. The waterspout may be a tornado that formed over land and then traveled over water. In such a case, the waterspout is sometimes referred to as a *tornadic waterspout.* Such tornadoes can inflict major damage to ocean-going vessels, especially when the tornadoes are of the supercell variety. Waterspouts that form over water, especially above the warm, shallow coastal waters of the Florida Keys, where almost 100 occur each month during the summer, are referred to as *"fair weather" waterspouts.** These waterspouts are generally much smaller than an average tornado, as they have diameters usually between 3 and 100 meters. Fair weather waterspouts are also less intense, as their rotating winds are typically less than 45 knots. (On the Fujita scale this speed is in the F0 range.) In addition, they tend to move more slowly than tornadoes and they only last for about 10 to 15 minutes, although some have existed for up to one hour.

Fair weather waterspouts tend to form in much the same way that landspouts do—when the air is conditionally unstable and cumulus clouds are developing. Some form with small thunderstorms, but most form with developing cumulus congestus clouds whose tops are frequently no higher than 3600 m (12,000 ft) and do not extend to the freezing level. Apparently, the warm,

*The name WSR-88D stands for *Weather Surveillance Radar, 1988 Doppler.*

*"Fair weather" waterspouts may form over any large body of warm water. Hence, they occur frequently over the Great Lakes in summer.

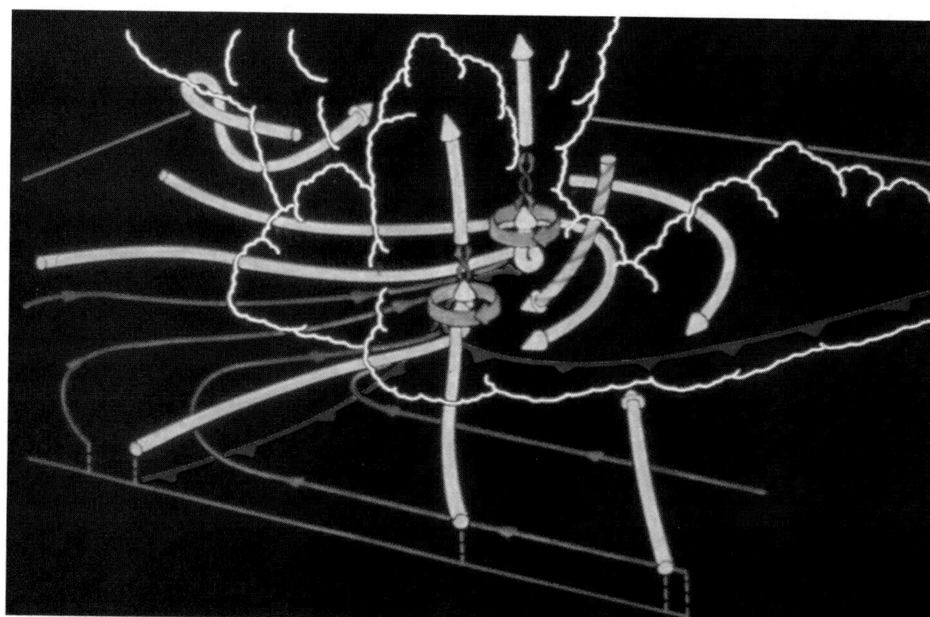

FIGURE 10.39 A computer model illustrating air motions inside a severe tornado-generating thunderstorm.

humid air near the water helps to create atmospheric instability, and the updraft beneath the resulting cloud helps initiate uplift of the surface air. Studies even suggest that gust fronts and converging sea breezes may play a role in the formation of some of the waterspouts that form over the Florida Keys.

The waterspout funnel is similar to the tornado funnel in that both are clouds of condensed water vapor with converging winds that rise about a central core. Contrary to popular belief, the waterspout does not draw water up into its core; however, swirling spray may be lifted several meters when the waterspout funnel touches the water. Apparently, the most destructive waterspouts are those that begin as tornadoes over land, then move over water. A photograph of a particularly well-developed and intense waterspout is shown in Fig. 10.40.

Meteorology Now™

ACTIVE FIGURE 10.40
A waterspout over the warm waters of the Florida Keys. Watch this Active Figure at http://earthscience.brookscole.com/ahrens/ess4e.

SUMMARY

In this chapter, we examined thunderstorms and the atmospheric conditions that produce them. The ingredients for an isolated ordinary thunderstorm are humid surface air, plenty of sunlight to heat the ground, and a conditionally unstable atmosphere. When these conditions prevail, and the air begins to rise, small cumulus clouds may grow into towering clouds and thunderstorms within 30 minutes.

When conditions are ripe for thunderstorm development, and a strong vertical wind shear exists, the stage is set for the generation of severe thunderstorms. Supercell storms are large, rotating thunderstorms that may exist for many hours, as their updrafts and downdrafts are nearly in balance and remain separated so they do not compete for the same air space. Thunderstorms that form in a line, along or ahead of an advancing cold front, are called a squall line, while those that form in clusters are called Mesoscale Convective Complexes.

Lightning is a discharge of electricity that occurs in mature thunderstorms. The lightning stroke momentarily heats the air to an incredibly high temperature. The rapidly expanding air produces a sound called thunder. Along with lightning and thunder, severe thunderstorms produce violent weather, such as destructive hail, strong downdrafts, and the most feared of all atmospheric storms—the tornado.

Tornadoes are rapidly rotating columns of air with a circulation that reaches the ground. The rotating air of the tornado may begin within the thunderstorm or it may begin at the surface and extend upwards. Tornadoes can form with supercells, as well as with less intense thunderstorms. Most tornadoes are less than a few hundred meters wide with wind speeds less than 100 knots, although violent tornadoes may have wind speeds that exceed 250 knots. A violent tornado may actually have smaller whirls (suction vortices) rotating within it. With the aid of Doppler radar, scientists are probing tornado-spawning thunderstorms, hoping to better predict tornadoes and to better understand where, when, and how they form.

A normally small and less destructive cousin of the tornado is the "fair weather" waterspout that commonly forms above the warm waters of the Florida Keys and the Great Lakes in summer.

Meteorology Now™
Assess your understanding of this chapter's topics with additional quizzing and tutorials at http://earthscience.brookscole.com/ahrens/ess4e.

KEY TERMS

The following terms are listed in the order they appear in the text. Define each. Doing so will aid you in reviewing the material covered in this chapter.

ordinary thunderstorms
cumulus stage
mature stage
gust front
dissipating stage
multicell storms
severe thunderstorm
supercell
shelf cloud
roll cloud
downburst
microburst
derecho
Mesoscale Convective System (MCS)
squall line
Mesoscale Convective Complex (MCC)
dryline
flash flood
lightning
thunder
sonic boom
stepped leader
return stroke
dart leader
heat lightning
St. Elmo's Fire
tornadoes
funnel cloud
tornado outbreak
suction vortices
tornado watch
tornado warning
Fujita scale
supercell tornadoes
mesocyclone
hook echo
wall cloud
nonsupercell tornadoes
gustnadoes
landspout
Doppler lidar
NEXRAD
waterspout

QUESTIONS FOR REVIEW

1. What is a thunderstorm?
2. Describe the stages of development of an ordinary (air-mass) thunderstorm.
3. How do downdrafts form in thunderstorms?
4. Why do ordinary thunderstorms most frequently form in the afternoon?
5. What atmospheric conditions are necessary for the development of an ordinary thunderstorm?
6. (a) What are gust fronts and how do they form?
   (b) If a gust front passes, what kind of weather will you experience?
7. (a) Describe how a microburst forms.
   (b) Why is the term *wind shear* often used in conjunction with a microburst?
8. How do severe thunderstorms differ from ordinary thunderstorms?

9. Explain why ordinary thunderstorms tend to dissipate much sooner than supercell thunderstorms.
10. Why are severe thunderstorms not very common in polar latitudes?
11. Give a possible explanation for the generation of pre-frontal squall-line thunderstorms.
12. In what region of the United States do dryline thunderstorms most frequently form?
13. When thunderstorms are *training*, what are they doing?
14. What is a Mesoscale Convective Complex (MCC)?
15. Where does the highest frequency of thunderstorms occur in the United States? Why there?
16. Why is large hail more common in Kansas than in Florida?
17. Explain how a cloud-to-ground lightning stroke develops.
18. How is thunder produced?
19. If you see lightning and ten seconds later you hear thunder, how far away is the lightning stroke?
20. Why is it unwise to seek shelter under a tree during a thunderstorm?
21. What is a tornado?
22. Give some average statistics about tornadoes, including their size, winds, and direction of movement.
23. How does a tornado *watch* differ from a tornado *warning?*
24. Why is it suggested that one *not* open windows when a tornado is approaching?
25. Explain why the central part of the United States is more susceptible to tornadoes than any other region of the world.
26. Describe the atmospheric conditions at the surface and aloft that are necessary for the development of the majority of supercell tornadic thunderstorms. Be sure to include how wind shear plays a role in the storm's formation.
27. Explain how a nonsupercell tornado, such as a landspout, might form.
28. How has Doppler radar helped in the prediction of severe weather?
29. Explain both how and why there is a shift in tornado activity from winter to summer within the continental United States.
30. What atmospheric conditions lead to the formation of "fair weather" waterspouts?

QUESTIONS FOR THOUGHT AND EXPLORATION

1. Why does the bottom half of a dissipating thunderstorm usually "disappear" before the top?
2. Sinking air warms, yet thunderstorm downdrafts are usually cold. Why?
3. If you are confronted by a large tornado in an open field and there is no way that you can outrun it, your only recourse might be to run and lie down in a depression. If given the choice, when facing the tornado, would you run toward your left or toward your right as the tornado approaches? Explain your reasoning.
4. Suppose while you are on a high mountain ridge a thundercloud passes overhead. What would be the wisest thing to do—stand upright? lie down? or crouch? Explain.
5. Tornadoes apparently form in the region of a strong updraft, yet they descend from the base of a cloud. Why?
6. On a map of the United States, place the surface weather conditions (air masses, fronts, and so on) that are necessary for the formation of most tornadoes.
7. Suppose several of your friends went on a storm-chasing adventure in the central United States. To help guide their chase, you stay behind, with an internet-connected computer and a cellular phone. Which current weather and forecast maps would you use to guide their storm chase? Explain why you choose those maps.
8. A multi-vortex tornado with a rotational wind speed of 125 knots is moving from southwest to northeast at 30 knots. Assume the suction vortices within this tornado have rotational winds of 100 knots:
   (a) What is the maximum wind speed of this multi-vortex tornado?
   (b) If you are facing the approaching tornado, on which side (northeast, northwest, southwest, or southeast) would the strongest winds be found? the weakest winds? Explain both of your answers.
   (c) According to the Fujita scale (Table 10.2, p. 281), how would this tornado be classified?

• Go to the Brooks/Cole Earth Sciences Resource Center (http://earthscience.brookscole.com) for critical thinking exercises, articles, and additional readings from InfoTrac College Edition, Brooks/Cole's online student library.

Photo: NOAA.

Hurricane Carlotta moving northeastward parallel to the coast of Mexico during June, 2000. With sustained winds of 100 knots and a central pressure near 950 mb, Carlotta ranks as a Category 3 hurricane on the Saffir-Simpson hurricane scale.

Meteorology Now™ This icon, appearing throughout the book, indicates an opportunity to explore interactive tutorials, animations, or practice problems available on the MeteorologyNow Web site at http://earthscience.brookscole.com/ahrens/ess4e.

Hurricanes

On September 18, 1926, as a hurricane approached Miami, Florida, people braced themselves for the devastating high winds and storm surge. Just before dawn the hurricane struck with full force—torrential rains, flooding, and easterly winds that gusted to over 100 miles per hour. Then, all of a sudden, it grew calm and a beautiful sunrise appeared. People wandered outside to inspect their property for damage. Some headed for work, and scores of adventurous young people crossed the long causeway to Miami Beach for the thrill of swimming in the huge surf. But the lull lasted for less than an hour. And from the south, ominous black clouds quickly moved overhead. In what seemed like an instant, hurricane force winds from the west were pounding the area and pushing water from Biscayne Bay over the causeway. Many astonished bathers, unable to swim against the great surge of water, were swept to their deaths. Hundreds more drowned as Miami Beach virtually disappeared under the rising wind-driven tide.

CONTENTS

Born over warm tropical waters and nurtured by a rich supply of water vapor, the *hurricane* can indeed grow into a ferocious storm that generates enormous waves, heavy rains, and winds that may exceed 150 knots. What exactly are hurricanes? How do they form? And why do they strike the east coast of the United States more frequently than the west coast? These are some of the questions we will consider in this chapter.

DID YOU KNOW?

The word *hurricane* derives from the Taino language of Central America. The literal translation of the Taino word *hurucan* is "god of evil." The word *typhoon* comes from the Chinese word *taifung,* meaning "big wind."

TROPICAL WEATHER

In the broad belt around the earth known as the tropics—the region 23½° north and south of the equator—the weather is much different from that of the middle latitudes. In the tropics, the noon sun is always high in the sky, and so diurnal and seasonal changes in temperature are small. The daily heating of the surface and high humidity favor the development of cumulus clouds and afternoon thunderstorms. Most of these are individual thunderstorms that are not severe. Sometimes, however, they grow together into loosely organized systems called *non-squall clusters.* On other occasions, the thunderstorms will align into a row of vigorous convective cells or a *squall line.* The passage of a squall line is usually noted by a sudden wind gust followed immediately by a heavy downpour. This deluge is then followed by several hours of relatively steady rainfall. Many of these tropical squall lines are similar to the middle-latitude squall lines described in Chapter 10.

As it is warm all year long in the tropics, the weather is not characterized by four seasons which, for the most part, are determined by temperature variations. Rather, most of the tropics are marked by seasonal differences in precipitation. The greatest cloudiness and precipitation occur during the high-sun period, when the intertropical convergence zone moves into the region. Even during the dry season, precipitation can be irregular, as periods of heavy rain, lasting for several days, may follow an extremely dry spell.

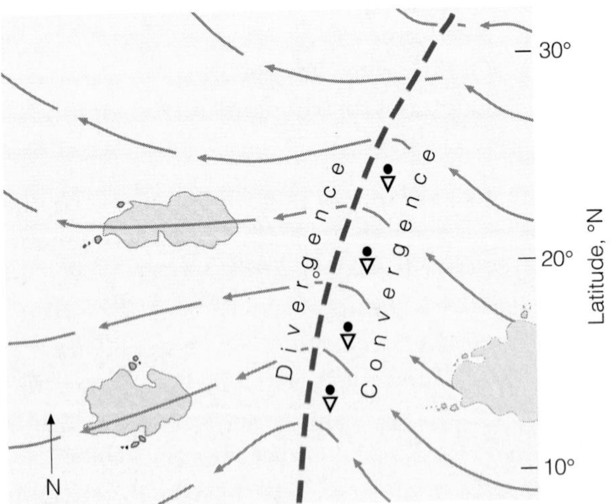

FIGURE 11.1 A tropical wave (also called an easterly wave) as shown by the bending of streamlines—lines that show wind flow patterns. (The heavy dashed line is the axis of the trough.) The wave moves slowly westward, bringing fair weather on its western side and showers on its eastern side.

The winds in the tropics generally blow from the east, northeast, or southeast—the trade winds. Because the variation of sea-level pressure is normally quite small, drawing isobars on a weather map provides little useful information. Instead of isobars, **streamlines** that depict wind flow are drawn. Streamlines are useful because they show where surface air converges and diverges. Occasionally, the streamlines will be disturbed by a weak trough of low pressure called a **tropical wave,** or **easterly wave** (see Fig. 11.1).

Tropical waves have wavelengths on the order of 2500 km (1550 mi) and travel from east to west at speeds between 10 and 20 knots. Look at Fig. 11.1 and observe that, on the western side of the trough (heavy dashed line), where easterly and northeasterly surface winds diverge, sinking air produces generally fair weather. On its eastern side, where the southeasterly winds converge, rising air generates showers and thunderstorms. Consequently, the main area of showers forms *behind* the trough. Occasionally, a tropical wave will intensify and grow into a hurricane.

ANATOMY OF A HURRICANE

A **hurricane** is an intense storm of tropical origin, with sustained winds exceeding 64 knots (74 mi/hr), which forms over the warm northern Atlantic and eastern North Pacific oceans. This same type of storm is given different names in different regions of the world. In the western North Pacific, it is called a **typhoon,** in India a *cyclone,* and in Australia a *tropical cyclone.* By international agreement, **tropical cyclone** is the general term for all hurricane-type storms that originate over tropical waters. For simplicity, we will refer to all of these storms as hurricanes.

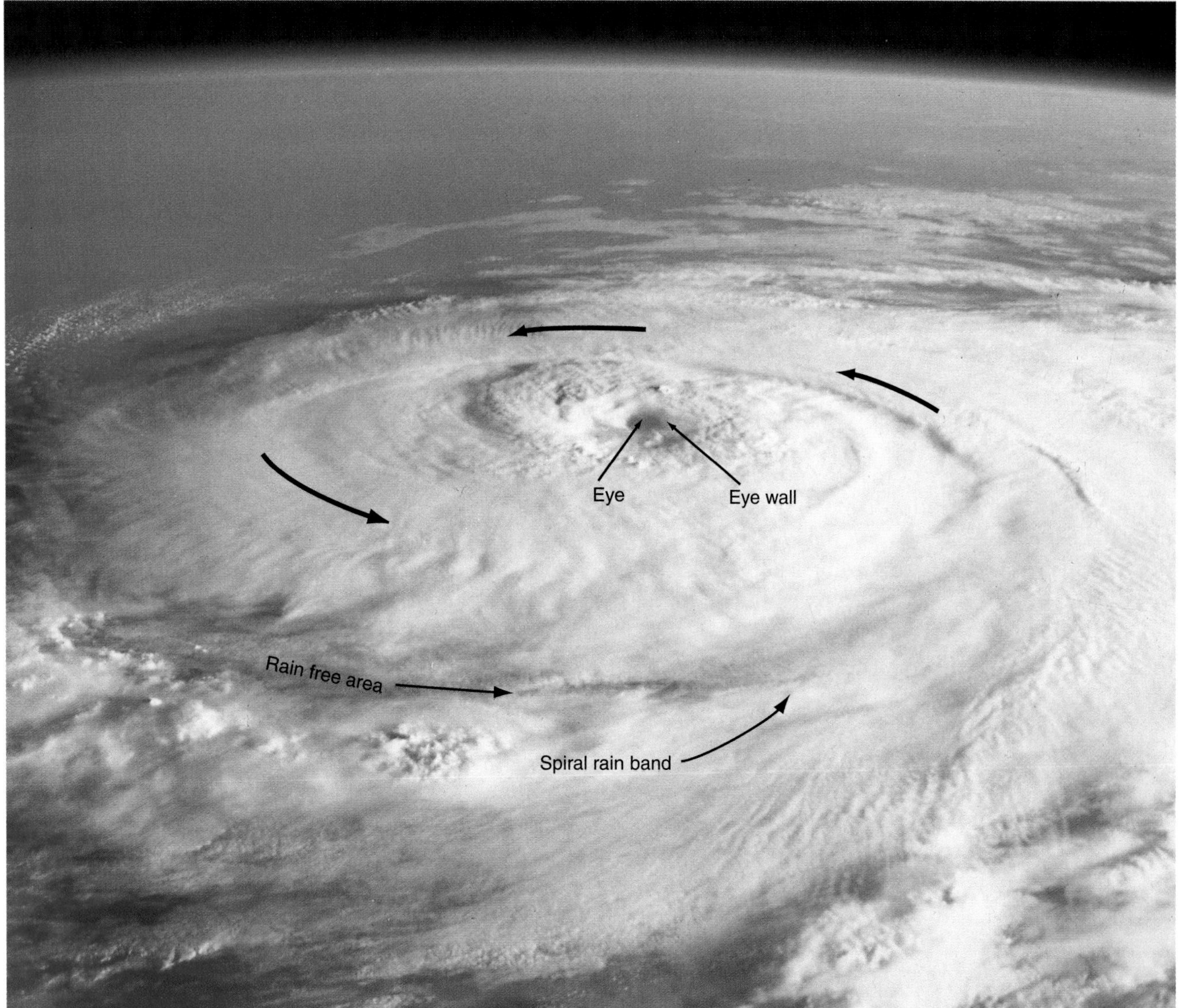

FIGURE 11.2 Hurricane Elena over the Gulf of Mexico, about 130 km (80 mi) southwest of Apalachicola, Florida, as photographed from the space shuttle *Discovery* during September, 1985. Because this storm is situated north of the equator, surface winds are blowing counterclockwise about its center (eye). The central pressure of the storm is 955 mb, with sustained winds of 105 knots near its eye.

Figure 11.2 is a photo of Hurricane Elena situated over the Gulf of Mexico. The storm is approximately 500 km (310 mi) in diameter, which is about average for hurricanes. The area of broken clouds at the center is its **eye.** Elena's eye is almost 40 km (25 mi) wide. Within the eye, winds are light and clouds are mainly broken. The surface air pressure is very low, nearly 955 mb (28.20 in.).* Notice that the clouds align themselves into spiraling bands (called *spiral rain bands*) that swirl in toward the storm's center, where they wrap themselves around the eye. Surface winds increase in speed as they blow counterclockwise and inward toward this center. (In the Southern Hemisphere, the winds blow clockwise around the center.) Adjacent to the eye is the **eye wall,** a ring of intense thunderstorms that whirl around the storm's center and extend upward to almost 15 km (49,000 ft) above sea level. Within the eye wall, we find the heaviest precipitation and the strongest winds, which, in this storm, are 105 knots, with peak gusts of 120 knots.

If we were to venture from west to east (left to right) at the surface through the storm in Fig. 11.2, what might we experience? As we approach the hurricane, the sky becomes overcast with cirrostratus clouds; barometric pressure drops slowly at first, then more rapidly as we move closer to the center. Winds blow from the north

*An extreme low pressure of 870 mb (25.70 in.) was recorded in Typhoon Tip during October, 1979, and Hurricane Gilbert had a pressure reading of 888 mb (26.22 in.) during September, 1988.

and northwest with ever-increasing speed as we near the eye. The high winds, which generate huge waves over 10 m (33 ft) high, are accompanied by heavy rain showers. As we move into the eye, the winds slacken, rainfall ceases, and the sky brightens, as middle and high clouds appear overhead. The atmospheric pressure is now at its lowest point (955 mb), some 50 mb lower than the pressure measured on the outskirts of the storm. The brief respite ends as we enter the eastern region of the eye wall. Here, we are greeted by heavy rain and strong southerly winds. As we move away from the eye wall, the pressure rises, the winds diminish, the heavy rain lets up, and eventually the sky begins to clear.

This brief, imaginary venture raises many unanswered questions. Why, for example, is the surface pressure lowest at the center of the storm? And why is the weather clear almost immediately outside the storm area? To help us answer such questions, we need to look at a vertical view, a profile of the hurricane along a slice that runs directly through its center. A model that describes such a profile is given in Fig. 11.3.

The model shows that the hurricane is composed of an organized mass of thunderstorms that are an integral part of the storm's circulation. Near the surface, moist tropical air flows in toward the hurricane's center. Adjacent to the eye, this air rises and condenses into huge cumulonimbus clouds that produce heavy rainfall, as much as 25 cm (10 in.) per hour. Near the top of the clouds, the relatively dry air, having lost much of its moisture, begins to flow outward away from the center. This diverging air aloft actually produces a clockwise (anticyclonic in the Northern Hemisphere) flow of air several hundred kilometers from the eye. As this outflow reaches the storm's periphery, it begins to sink and warm, inducing clear skies. In the vigorous thunderstorms of the eye wall, the air warms due to the release of large quantities of latent heat. This warming produces slightly higher pressures aloft, which initiate downward air motion within the eye. As the air subsides, it warms by compression. This process helps to account for the warm air and the absence of thunderstorms in the center of the storm.

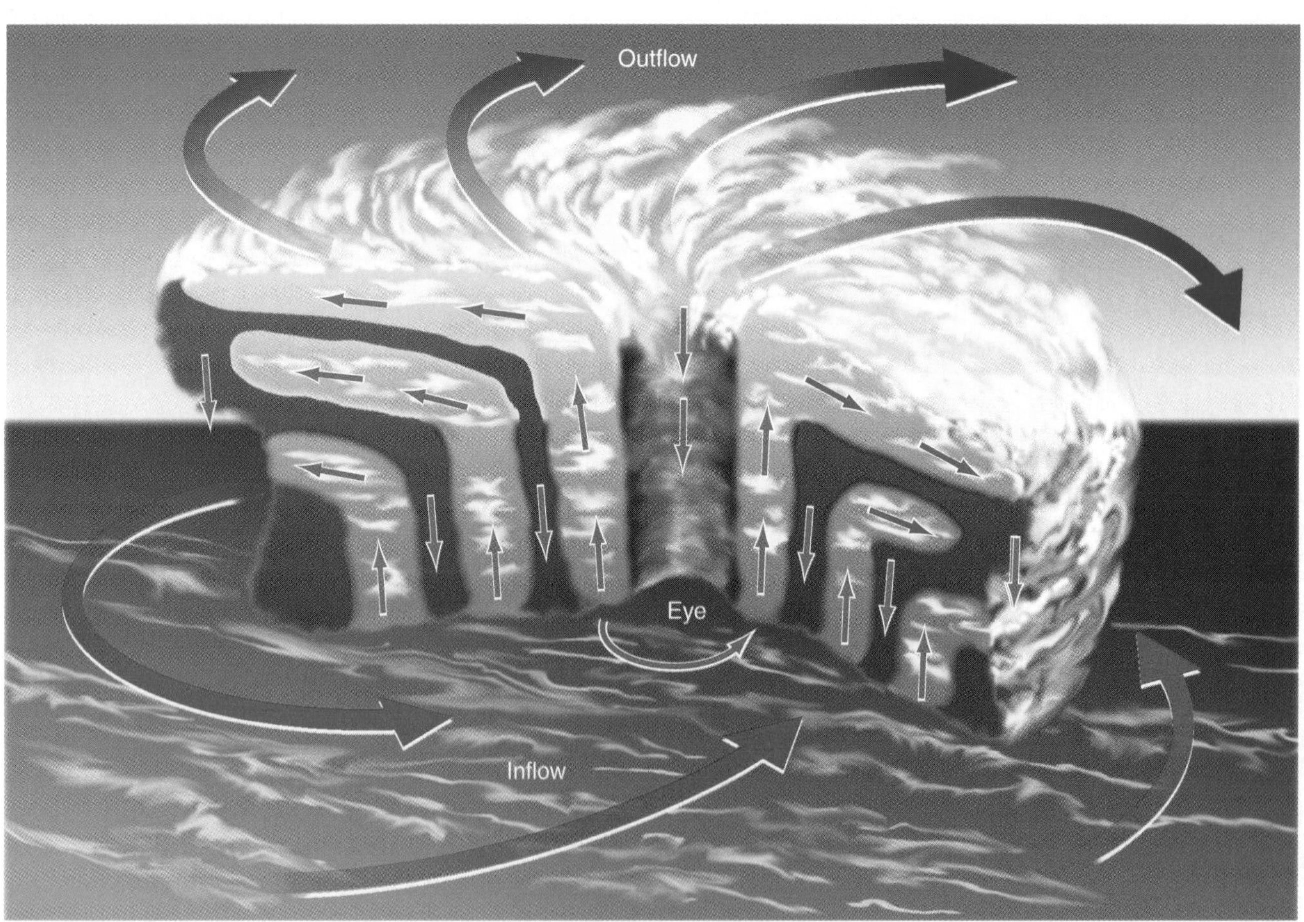

Meteorology Now™ ACTIVE FIGURE 11.3 A model that shows a vertical view of air motions and clouds in a typical hurricane. The diagram is exaggerated in the vertical. Watch this Active Figure at http://earthscience.brookscole.com/ahrens/ess4e.

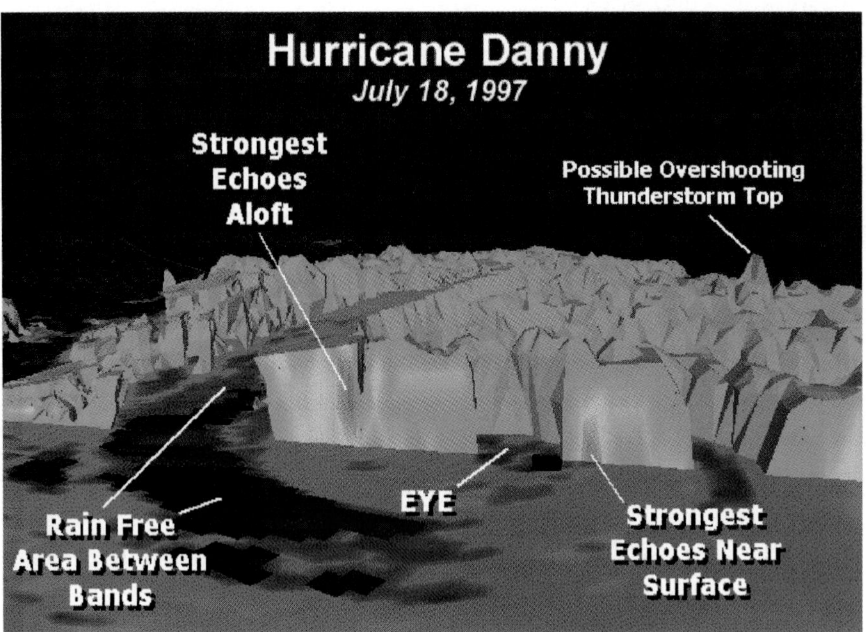

FIGURE 11.4 A radar composite of Hurricane Danny showing several features associated with the storm. The echoes in the composite are radar echoes that illustrate, in red and yellow, where the heaviest rain is falling.

As surface air rushes in toward the region of much lower surface pressure, it should expand and cool, and we might expect to observe cooler air around the eye, with warmer air further away. But, apparently, so much heat is added to the air from the warm ocean surface that the surface air temperature remains fairly uniform throughout the hurricane.

Figure 11.4 is a three-dimensional radar composite of Hurricane Danny as it sits near the mouth of the Mississippi River on July 18, 1997. Although Danny is a weak hurricane, compare its features with those of typical hurricanes illustrated in Fig. 11.2 and Fig. 11.3. Notice that the strongest radar echoes (heaviest rain) near the surface are located in the eye wall, adjacent to the eye.

We are now left with an important question: Where and how do hurricanes form? Although not everything is known about their formation, it is known that certain necessary ingredients are required before a weak tropical disturbance will develop into a full-fledged hurricane.

Meteorology Now™ Click "Virtual Hurricane" to explore changes in pressure, winds, clouds, and precipitation in different areas of a hurricane.

HURRICANE FORMATION AND DISSIPATION

Hurricanes form over tropical waters where the winds are light, the humidity is high in a deep layer extending up through the troposphere, and the surface water temperature is warm, typically 26.5°C (80°F) or greater, over a vast area* (see Fig. 11.5). Moreover, the warm surface water must extend downward to a depth of about 200 m (600 ft) before hurricane formation is possible. These conditions usually prevail over the tropical and subtropical North Atlantic and North Pacific oceans during the summer and early fall; hence, the hurricane season normally runs from June through November.

For a mass of unorganized thunderstorms to develop into a hurricane, the surface winds must converge. In the Northern Hemisphere, converging air spins counterclockwise about an area of surface low pressure. Because this type of rotation will not develop on the equator where the Coriolis force is zero (see Chapter 6), hurricanes form in tropical regions, usually between 5° and 20° latitude. (In fact, about two-thirds of all tropical cyclones form between 10° and 20° of the equator.)

Hurricanes do not form spontaneously, but require some kind of "trigger" to start the air converging. We know, for example, from Chapter 7 that surface winds converge along the intertropical convergence zone (ITCZ). Occasionally, when a wave forms along the ITCZ, an area of low pressure develops, convection becomes organized, and the system grows into a hurricane. Weak convergence also occurs on the eastern side of a tropical wave, where hurricanes have been known to form. In fact, many if not most Atlantic hurricanes can

*It was once thought that for hurricane formation, the ocean must be sufficiently warm through a depth of about 200 meters. It is now known that hurricanes can form in the eastern North Pacific when the warm layer of ocean water is only about 20 m (65 ft) deep.

FIGURE 11.5 Hurricanes form over warm, tropical waters. This image shows where sea surface temperatures in the tropical Atlantic exceed 28°C (82°F) — warm enough for tropical storm development — during May, 2002.

be traced to tropical waves that form over Africa. However, only a small fraction of all of the tropical disturbances that form over the course of a year ever grow into hurricanes. Studies suggest that major Atlantic hurricanes are more numerous when the western part of Africa is relatively wet. Apparently, during the wet years, tropical waves are stronger, better organized, and more likely to develop into strong Atlantic hurricanes.

Convergence of surface winds may also occur along a pre-existing atmospheric disturbance, such as a front that has moved into the tropics from middle latitudes. Although the temperature contrast between the air on both sides of the front is gone, developing thunderstorms and converging surface winds may form, especially when the front is accompanied by an upper-level trough.

Even when all of the surface conditions appear near perfect for the formation of a hurricane (for example, warm water, humid air, converging winds, and so forth), the storm may not develop if the weather conditions aloft are not just right. For instance, in the region of the trade winds, and especially near latitude 20°, the air is often sinking in association with the subtropical high. The sinking air warms and creates an inversion, known as the **trade wind inversion.** When the inversion is strong, it can inhibit the formation of intense thunderstorms and hurricanes. Also, hurricanes do not form where the upper-level winds are strong, creating strong wind shear. Strong wind shear tends to disrupt the organized pattern of convection and disperses heat and moisture, which are necessary for the growth of the storm. Hurricanes do form where the upper-level winds are diverging, spreading outward, and the air aloft is leaving a column of air more quickly than the air at the surface is entering it.

The situation of strong winds aloft typically occurs over the tropical Atlantic during a major El Niño event. As a consequence, during El Niño there are usually fewer Atlantic hurricanes than normal. However, the warmer water of El Niño in the northern tropical Pacific favors the development of hurricanes in that region. During the cold water episode in the tropical Pacific (known as La Niña), winds aloft over the tropical Atlantic usually weaken and become easterly—a condition that favors hurricane development.*

The energy for a hurricane comes from the direct transfer of sensible heat from the warm water into the atmosphere and from the transfer of latent heat from the ocean surface. One idea (known as the *organized convection theory*) proposes that for hurricanes to form, the

DID YOU KNOW?

In a warmer world, would hurricanes reach greater intensity? Studies conducted by meteorologists using mathematical models concluded that if the sea surface temperatures in the northwest Pacific Ocean increased by about 2.2°C (4°F), maximum winds in a strong hurricane would increase between 5 and 12 percent.

*El Niño and La Niña are covered in Chapter 7 beginning on p. 192.

thunderstorms must become organized so that the latent heat that drives the system can be confined to a limited area. If thunderstorms start to organize along the ITCZ or along a tropical wave, and if the trade wind inversion is weak, the stage may be set for the birth of a hurricane. The likelihood of hurricane development is enhanced if the air aloft is conditionally unstable. Such instability can be brought on when a cold upper-level trough from middle latitudes moves over the storm area. When this situation occurs, the cumulonimbus clouds are able to build rapidly and grow into enormous thunderstorms (see Fig. 11.6).

Although the upper air is initially cold, it warms rapidly due to the huge amount of latent heat released during condensation. As this cold air is transformed into much warmer air, the air pressure in the upper troposphere above the developing storm rises, producing an area of high pressure (see Chapter 6, p. 142 for more information on this topic). Now the air aloft begins to move outward, away from the region of developing thunderstorms. This diverging air aloft, coupled with warming of the air layer, causes the surface pressure to drop, and a small area of surface low pressure forms. The surface air begins to spin counterclockwise and in toward the region of low pressure. As it moves inward, its speed increases, just as ice skaters spin faster as their arms are brought in close to their bodies. The winds then generate rough seas, which increase the friction on the moving air. This increased friction causes the winds to converge and ascend about the center of the storm.

We now have a chain reaction or *feedback mechanism* in progress. The rising air, having picked up added moisture and warmth from the choppy sea, fuels more thunderstorms and releases more heat, which causes the surface pressure to lower even more. The lower pressure near the center creates a greater friction, more convergence, more rising air, more thunderstorms, more heat, lower surface pressure, stronger winds, and so on until a full-blown hurricane is born.

As long as the upper-level outflow of air is greater than the surface inflow, the storm will intensify and the surface pressure will drop. Because the air pressure within the system is controlled to a large extent by the warmth of the air, the storm will intensify only up to a point. The controlling factors are the temperature of the water and the release of latent heat. Consequently, when the storm is literally full of thunderstorms, it will use up just about all of the available energy, so that air temperature will no longer rise and pressure will level off. Because there is a limit to how intense the storm can become, peak wind gusts seldom exceed 200 knots. When the converging surface air near the center exceeds the outflow at the top, surface pressure begins to increase, and the storm dies out.

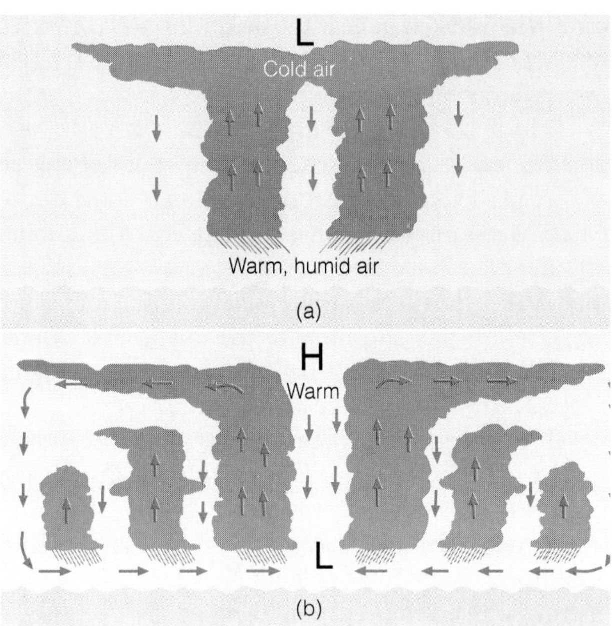

FIGURE 11.6 Development of a hurricane by the organized convection theory. (a) Cold air above an organized mass of tropical thunderstorms generates conditionally unstable air and large cumulonimbus clouds. (b) The release of latent heat warms the upper troposphere, creating an area of high pressure. Upper-level winds move outward away from the high. This movement, coupled with the warming of the air layer, causes surface pressures to drop. As air near the surface moves toward the lower pressure, it converges, rises, and fuels more thunderstorms. Soon a chain reaction develops, and a hurricane forms.

An alternative to the organized convection theory is the *heat engine theory,* which proposes that a hurricane is like a heat engine. In a heat engine, heat is taken in at a high temperature, converted into work, then ejected at a low temperature. In a hurricane, small swirling eddies transfer sensible and latent heat from the ocean surface into the overlying air. The warmer the water and the greater the wind speed, the greater the transfer of sensible and latent heat. As the air sweeps in toward the center of the storm, the rate of heat transfer increases because the wind speed increases toward the eye wall. Similarly, the higher wind speeds cause greater evaporation rates, and the overlying air becomes nearly saturated.

Near the eye wall, turbulent eddies transfer the warm moist air upward, where the water vapor condenses to form clouds. The release of latent heat inside the clouds causes the air temperature in the region of the eye wall to be much higher than the air temperature at the same altitude further out, away from the storm center. This situation causes a horizontal pressure gradient aloft that induces the air to move outward, away from the storm center in the anvils of the cumulonimbus clouds. At the top of the storm, heat is lost by clouds radiating infrared energy to space. Hence, in a hurricane, heat is taken in near the ocean surface, converted to kinetic energy (energy of motion) or wind, and lost at its top through radiational cooling.

The maximum strength a hurricane can achieve is determined by the difference in temperature between the ocean surface and the top of its clouds at the tropopause. The storm's strength is also related to the amount of water evaporated from the surface. As a consequence, the warmer the ocean surface, the lower the minimum pressure of the storm, and the higher its winds. Presently, most scientists believe that the heat engine process is the driving force behind a hurricane.

If the hurricane remains over warm water, it may survive for several weeks. However, most hurricanes last for less than a week; they weaken rapidly when they travel over colder water and lose their heat source. They also dissipate rapidly over land. Here, not only is their energy source removed, but their winds decrease in strength (due to the added friction) and blow more directly into the center, causing the central pressure to rise.

As a hurricane approaches land, will it intensify, maintain its strength, or weaken? This question has plagued meteorologists for some time. To help with the answer, forecasters have been using a statistical model that compares the behavior of the present storm with that of similar tropical storms in the past. However, the results using this model have not been encouraging. Another more recent model uses the depth of warm ocean water in front of the storm's path to predict the storm's behavior. If the reservoir of warm water ahead of the storm is relatively shallow, ocean waves generated by the hurricane's wind turbulently bring deeper, cooler water to the surface. Studies show that if the water beneath the eye wall (the region of thunderstorms adjacent to the eye) cools by 2.5°C (4.5°F), the storm's energy source is cut off, and the hurricane tends to dissipate. Whereas, if a deep layer of warm ocean water exists, the storm tends to maintain its strength or intensify, as long as other factors remain the same. So, knowing the depth of warm surface water is important in predicting whether a hurricane will intensify or weaken. Moreover, as new hurricane-prediction models are implemented, and as our understanding of the nature of hurricanes increases, improved forecasts of hurricane movement and intensification should become available.

Hurricane Stages of Development Hurricanes go through a set of stages from birth to death. Initially, the mass of thunderstorms with only a slight wind circulation is known as a **tropical disturbance,** or *tropical wave.* The tropical disturbance becomes a **tropical depression** when the winds increase to between 20 and 34 knots and several closed isobars appear about its center on a surface weather map. When the isobars are packed together and the winds are between 35 and 64 knots, the tropical depression becomes a **tropical storm.** The tropical storm is classified as a *hurricane* only when its winds exceed 64 knots (74 miles per hour).

Figure 11.7 shows four tropical systems in various stages of development. Moving from east to west, we see a weak tropical disturbance (a tropical wave) crossing over Panama. Further west, a tropical depression is organizing around a developing center with winds less than 25 knots. In a few days, this system will develop into Hurricane Gilma. Further west is a full-fledged hurricane with peak winds in excess of 110 knots. The swirling band of clouds to the northwest is Emilia; once a hurricane (but now with winds less than 40 knots), it is rapidly weakening over colder water.

BRIEF REVIEW

Before reading the next several sections, here is a review of some of the important points about hurricanes.

- Hurricanes are tropical cyclones, comprised of an organized mass of thunderstorms.
- Hurricanes have peak winds about a central core (eye) that exceed 64 knots (74 mi/hr).
- Hurricanes form over warm tropical waters, where light surface winds converge, the humidity is high in a deep layer, and the winds aloft are weak.
- For a mass of thunderstorms to organize into a hurricane there must be some mechanism that triggers the formation, such as converging surface winds along the ITCZ, a pre-existing atmospheric disturbance, or a tropical wave.
- Hurricanes derive their energy from the warm, tropical oceans and by evaporating water from the ocean's surface. Heat energy is converted to wind energy when the water vapor condenses inside deep convective clouds.
- Hurricanes grow stronger as long as the air aloft moves outward, away from the storm's center more quickly than the surface air moves in toward the center.
- Hurricanes dissipate rapidly when they move over colder water or over a large landmass.

Up to this point, it is probably apparent that tropical cyclones called hurricanes are similar to middle-latitude cyclones in that, at the surface, both have central cores of low pressure and winds that spiral counterclockwise about their respective centers (Northern Hemisphere). However, there are many differences between the two systems, which are described in the Focus section on p. 302.

FIGURE 11.7
Visible satellite image showing four tropical systems, each in a different stage of its life cycle.

Hurricane Movement Figure 11.8 shows where most hurricanes are born and the general direction in which they move. Notice that they form over tropical oceans, except in the South Atlantic and in the eastern South Pacific. The surface water temperatures are too cold in these areas for their development. It is also possible that the unfavorable location of the ITCZ during the Southern Hemisphere's warm season discourages their development.

Hurricanes that form over the North Pacific and North Atlantic are steered by easterly winds and move

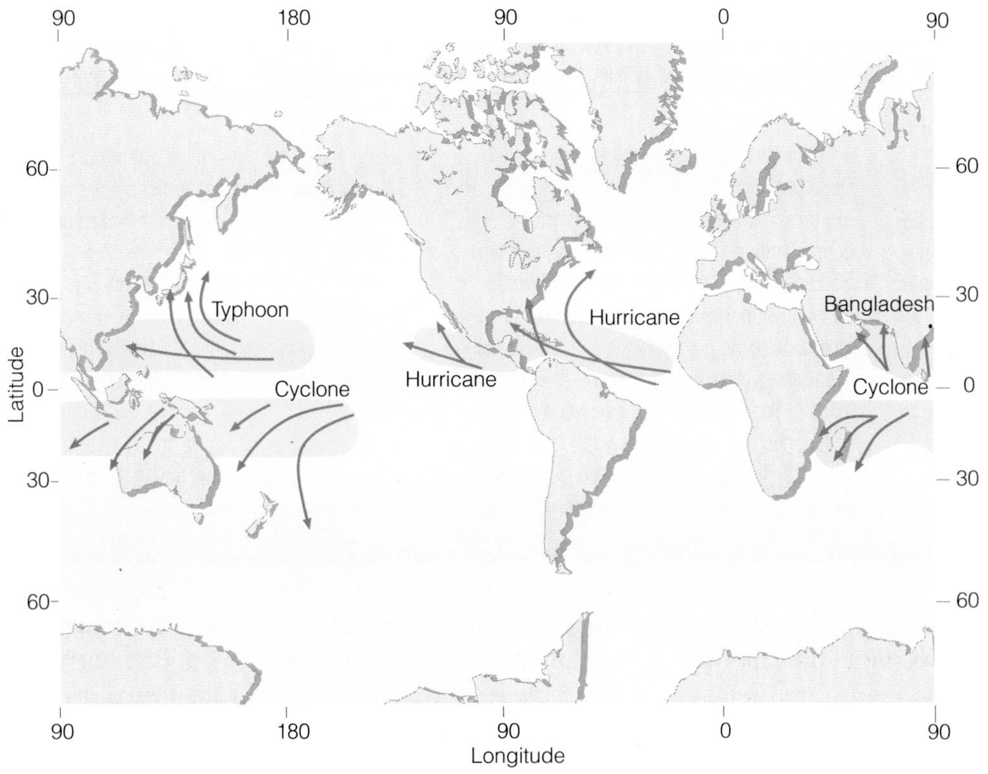

FIGURE 11.8
Regions where tropical storms form (red shading), the names given to storms, and the typical paths they take (red arrows).

FOCUS ON A SPECIAL TOPIC

How Do Hurricanes Compare with Middle-Latitude Storms?

By now, it should be apparent that a hurricane is much different from the mid-latitude cyclone that we discussed in Chapter 8. A hurricane derives its energy from the warm water and the latent heat of condensation, whereas the mid-latitude storm derives its energy from horizontal temperature contrasts. The vertical structure of a hurricane is such that its central column of air is warm from the surface upward; consequently, hurricanes are called *warm-core lows.* A hurricane weakens with height, and the area of low pressure at the surface may actually become an area of high pressure above 12 km (40,000 ft). Mid-latitude cyclones, on the other hand, are *cold-core lows* that usually intensify with increasing height, with a cold upper-level low or trough often existing above, or to the west of the surface low.

A hurricane usually contains an eye where the air is sinking, while mid-latitude cyclones are characterized by centers of rising air. Hurricane winds are strongest near the surface, whereas the strongest winds of the mid-latitude storm are found aloft in the jet stream.

Further contrasts can be seen on a surface weather map. Figure 1 shows Hurricane Allen over the Gulf of Mexico and a mid-latitude storm north of New England. Around the hurricane, the isobars are more circular, the pressure gradient is much steeper, and the winds are stronger. The hurricane has no fronts and is smaller (although Allen is larger than most hurricanes). There are similarities between the two systems: Both are areas of surface low pressure, with winds moving counterclockwise about their respective centers.

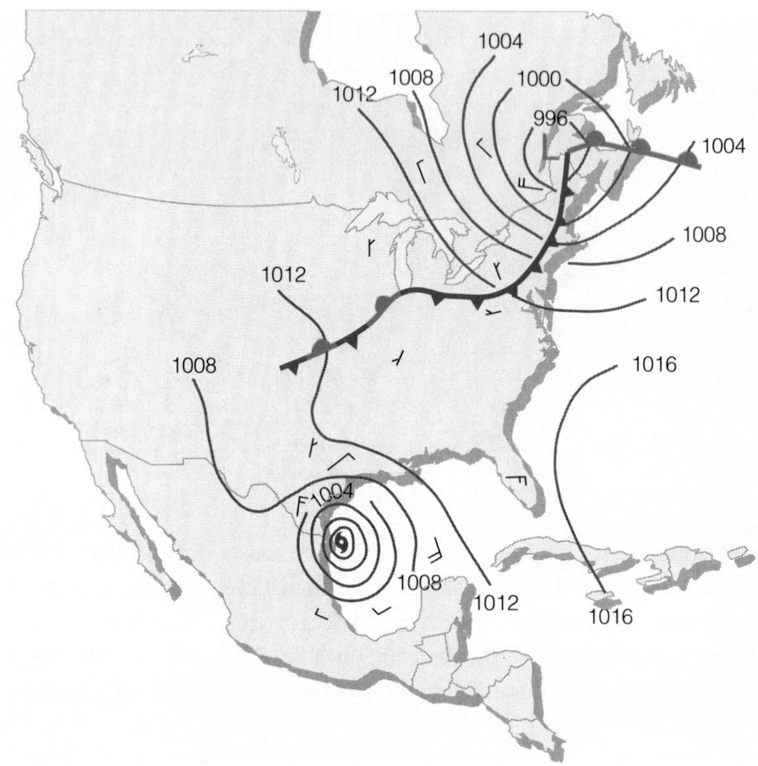

FIGURE 1 Surface weather map for the morning of August 9, 1980, showing Hurricane Allen over the Gulf of Mexico and a middle-latitude storm system north of New England.

It is interesting to note that some northeasters (winter storms that move northeastward along the coastline of North America, bringing with them heavy precipitation, high surf, and strong winds) may actually possess some of the characteristics of a hurricane. For example, a particularly powerful northeaster during January, 1989, was observed to have a cloud-free eye, with surface winds in excess of 85 knots spinning about a warm inner core. Moreover, some *polar lows*—lows that develop over polar waters during winter—may exhibit many of the observed characteristics of a hurricane, such as a symmetric band of thunderstorms spiraling inward around a cloud-free eye, a warm-core area of low pressure, and strong winds near the storm's center. In fact, when surface winds within these polar storms reach 58 knots, they are sometimes referred to as *Arctic hurricanes.*

Even though hurricanes weaken rapidly as they move inland, their circulation may draw in air with contrasting properties. If the hurricane links with an upper-level trough, it may actually become a mid-latitude cyclone. Swept eastward by upper-level winds, the remnants of an Atlantic hurricane can become a severe mid-latitude autumn storm in Europe.

west or northwestward at about 10 knots for a week or so. Gradually, they swing poleward around the subtropical high, and when they move far enough north, they become caught in the westerly flow, which curves them to the north or northeast. In the middle latitudes, the hurricane's forward speed normally increases, sometimes to more than 50 knots. The actual path of a hurricane (which appears to be determined by the structure of the storm and the storm's interaction with the environment) may vary considerably. Some take erratic paths and make odd turns that occasionally catch weather forecasters by surprise (see Fig. 11.9). There have been many instances where a storm heading directly for land suddenly veered away and spared the region from almost certain disaster. As an example, Hurricane Elena, with peak winds of 90 knots, moved northwestward into the Gulf of Mexico on August 29, 1985. It then veered eastward toward the west coast of Florida. After stalling offshore, it headed northwest. After weakening, it then moved onshore near Biloxi, Mississippi, on the morning of September 2.

Eastern Pacific Hurricanes As we saw in an earlier section, many hurricanes form off the coast of Mexico over the North Pacific. In fact, this area usually spawns about eight hurricanes each year, which is slightly more than the yearly average of six storms born over the tropical North Atlantic. Eastern North Pacific hurricanes normally move westward, away from the coast; hence, little is heard about them. When one does move northwestward, it normally weakens rapidly over the cool water of the North Pacific. Occasionally, however, one will curve northward or even northeastward and slam into Mexico, causing destructive flooding. Hurricane Tico left 25,000 people homeless and caused an estimated $66 million in property damage after passing over Mazatlán, Mexico, in October, 1983. The remains of Tico even produced record rains and flooding in Texas and Oklahoma. Even less frequently, a hurricane will stray far enough north to bring summer rains to southern California and Arizona, as did the remains of Hurricane Nora during September, 1997. (Nora's path is shown in Fig. 11.9.)

DID YOU KNOW?

Hurricane Tina in 1992 traveled for thousands of miles over warm, tropical waters and maintained hurricane force winds for 24 days, making it one of the longest-lasting North Pacific hurricanes on record.

The Hawaiian Islands, which are situated in the central North Pacific between about 20° and 23°N, appear to be in the direct path of many eastern Pacific hurricanes and tropical storms. By the time most of these storms have reached the islands, however, they have weakened considerably, and pass harmlessly to the south or northeast. The exceptions were Hurricane Iwa during November, 1982, and Hurricane Iniki during September, 1992. Iwa lashed part of Hawaii with 100-knot winds and huge surf, causing an estimated $312 million in damages. Iniki, the worst hurricane to hit Hawaii in the twentieth century, battered the island of Kauai with torrential rain, sustained winds of 114 knots that gusted to 140 knots, and 20-foot waves that crashed over coastal highways. Major damage was sustained by most of the hotels and about 50 percent of the homes on the island. Iniki (the costliest hurricane in Hawaiian history with damage estimates of $1.8 billion) flattened sugar cane fields, destroyed the

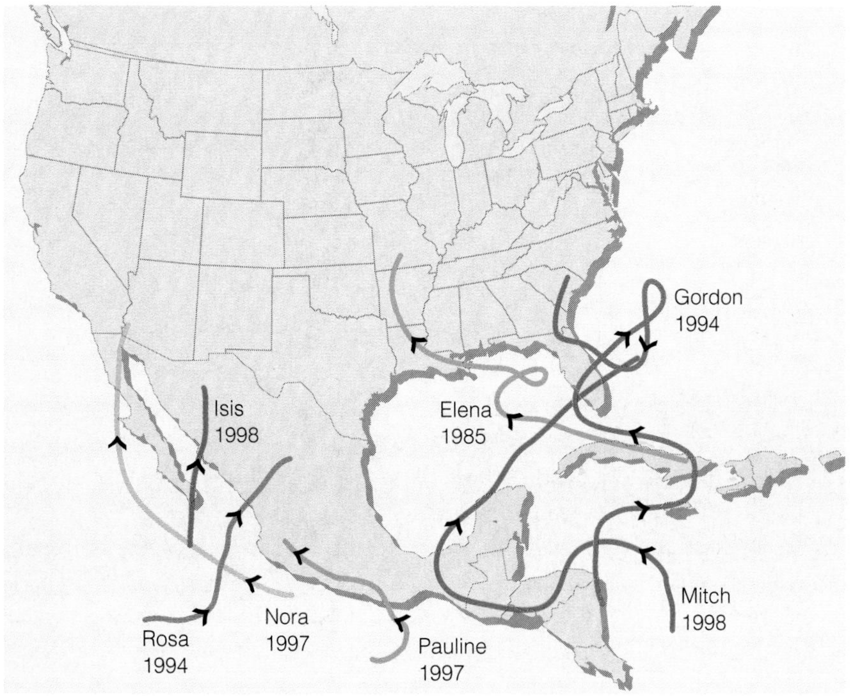

FIGURE 11.9 Some erratic paths taken by hurricanes.

macadamia nut crop, injured about 100 people, and caused at least 7 deaths.

North Atlantic Hurricanes Hurricanes that form over the tropical North Atlantic also move westward or northwestward on a collision course with Central or North America. Most hurricanes, however, swing away from land and move northward, parallel to the coastline of the United States. A few storms, perhaps three per year, move inland, bringing with them high winds, huge waves, and torrential rain that may last for days. Figure 11.10 is a collection of infrared satellite images of Hurricane Georges, showing its path from September 18 to September 28, 1998. As Georges moved westward it ravaged the large Caribbean Islands, causing extensive damage and taking the lives of more than 350 people. After raking the Florida Keys with high winds and heavy rain, its path curved toward the northwest, where it eventually slammed into Mississippi with torrential rains and winds exceeding 100 knots. Four people in the United States died due to Hurricane Georges.

A hurricane moving northward over the Atlantic will normally survive as a hurricane for a longer time than will its counterpart at the same latitude over the eastern Pacific. The reason is, of course, that the surface water of the Atlantic is much warmer.

Meteorology Now™ Click "Hurricane Forecasting" to track and forecast the movement of tropical storms.

Destruction and Warning When a hurricane is approaching from the east, its highest winds are usually on its north (poleward) side. The reason for this phenomenon is that the winds that push the storm along add to the winds on the north side and subtract from the winds on the south (equator) side. Hence, a hurricane with 110-knot winds moving westward at 10 knots will have 120-knot winds on its north side and 100-knot winds on its south side.

The same type of reasoning can be applied to a northward-moving hurricane. For example, as Hurricane Gloria moved northward along the coast of Virginia on the morning of September 27, 1985 (see Fig. 11.11), winds of 75 knots were swirling counterclockwise about its center. Because the storm was moving northward at about 25 knots, sustained winds on its eastern (right) side were about 100 knots, while on its western (left) side—on the coast—the winds were only about 50 knots. Even so, these winds were strong enough to cause significant beach erosion along the coasts of Maryland, Delaware, and New Jersey.

Even though Hurricane Gloria is moving northward in Fig. 11.11, there is a net transport of water directed eastward toward the coast. To understand this behavior, recall from Chapter 7 that as the wind blows over open water, the water beneath is set in motion. If we imagine the top layer of water to be broken into a series of layers, then we find each layer moving to the *right* of the layer above. This type of movement (bending) of water with depth (called the *Ekman Spiral*) causes a net transport of

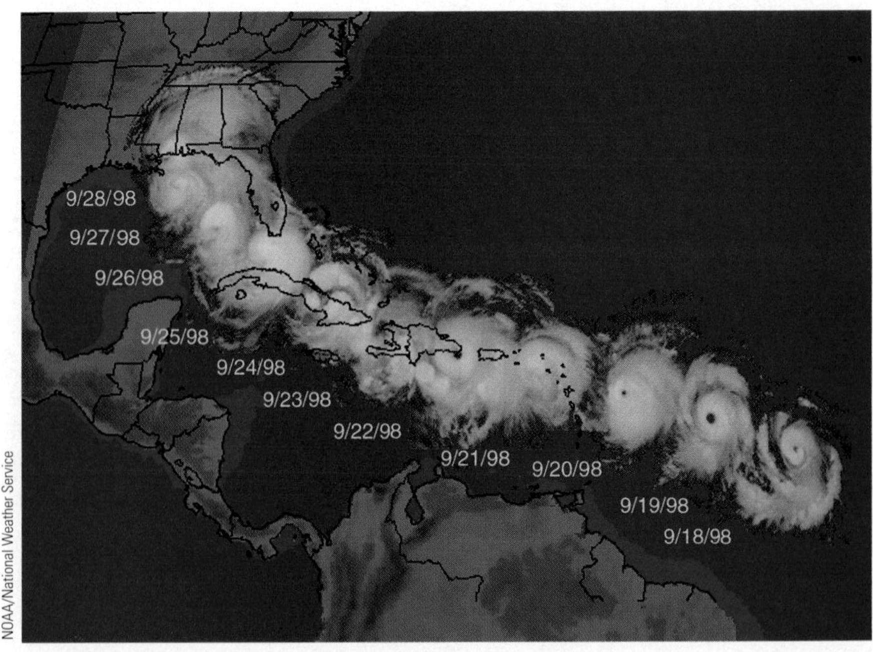

FIGURE 11.10 A composite of infrared satellite images of Hurricane Georges from September 18 to September 28, 1998, that shows its westward trek across the Caribbean, then northward into the United States.

water (known as **Ekman transport**) to the right of the surface wind in the Northern Hemisphere. Hence, the north wind on Hurricane Gloria's left (western) side causes a net transport of water toward the shore. Here, the water piles up and rapidly inundates the region.

The high winds of a hurricane also generate large waves, sometimes 10 to 15 m (33 to 49 ft) high. These waves move outward, away from the storm, in the form of *swells* that carry the storm's energy to distant beaches. Consequently, the effects of the storm may be felt days before the hurricane arrives.

Although the hurricane's high winds inflict a great deal of damage, it is the huge waves, high seas, and *flooding** that normally cause most of the destruction. The flooding is due, in part, to winds pushing water onto the shore and to the heavy rains, which may exceed 25 inches in 24 hours. Flooding is also aided by the low pressure of the storm. The region of low pressure allows the ocean level to rise (perhaps half a meter), much like a soft drink rises up a straw as air is withdrawn. (A drop of one millibar in air pressure produces a rise of one centimeter in ocean level.) The combined effect of high water (which is usually well above the high-tide level), high winds, and the net transport of water toward the coast, produces the **storm surge**—an abnormal rise of several meters in the ocean level—which inundates low-lying areas and turns beachfront homes into piles of splinters (see Fig. 11.12). The storm surge is particularly damaging when it coincides with normal high tides. Flooding, however, is not just associated with storms that reach hurricane strength, as destructive floods can occur with tropical storms that never become hurricanes. An example of such a storm is given in the Focus section on p. 306.

Considerable damage may also occur from hurricane-spawned tornadoes. About one-fourth of the hurricanes that strike the United States produce tornadoes. The exact mechanism by which these tornadoes form is not yet known; however, studies suggest that surface topography may play a role by initiating the convergence

*Hurricanes may sometimes have a beneficial aspect, in the sense that they can provide much needed rainfall in drought-stricken areas.

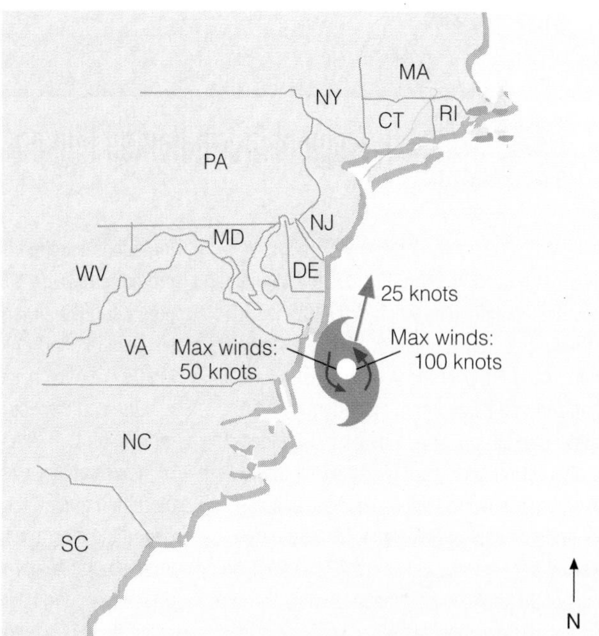

FIGURE 11.11 Hurricane Gloria on the morning of September 27, 1985. Moving northward at 25 knots, Gloria sustained winds of 100 knots on its right side and 50 knots on its left side. The central pressure of the storm is about 945 mb (27.91 in.).

FIGURE 11.12 When a storm surge moves in at high tide it can inundate and destroy a wide swath of coastal lowlands.

FOCUS ON A SPECIAL TOPIC

A Tropical Storm Named Allison

In late May, 2001, Allison began as a tropical wave that moved westward across the Atlantic. The wave continued its westward journey, and by the first of June it had moved across Central America and out over the Pacific Ocean. Here, it organized into a band of thunderstorms and a tropical depression. Upper-level winds guided the depression northward over the Gulf of Mexico, where the warm water fueled the circulation, and just east of Galveston, Texas, the depression became tropical storm Allison. Packing winds of 53 knots, Allison made landfall over the east end of Galveston Island on June 5. It drifted inland and weakened (see Fig. 2).

On the eastern side of the storm, heavy rain fell over parts of Texas and Louisiana. Some areas of southeast Texas received as much as 25 cm (10 in.) of rain in less than five hours. Homes, streets, and highways flooded as heavy rain continued to pound the area. But the worst was yet to come.

On June 7, as the upper-level winds began to change, the remnants of Allison drifted southwestward toward Houston. Heavy rain fell over southeast Texas and Louisiana, where several tornadoes touched down. Over the Houston area, more than 50 cm (20 in.) of rain fell within a 12-hour period, submerging a vast part of the city. In six days the Port of Houston received a staggering 94 cm (which is over 3 ft) of rain.

The center of circulation drifted southward, moving off the Texas coast and out over the Gulf of Mexico on the evening of June 9. The flow aloft then guided the storm northeastward, where the storm made landfall again, but this time in southeastern Louisiana. Heavy rain continued to pound Louisiana, creating one of the worst floods on record—a station in southern Louisiana reported a rainfall total of 76 cm (30 in.). On June 11, a zone of maximum winds aloft (a jet streak) associated with the subtropical jet stream enhanced the outflow above the surface storm, and the remains of tropical storm Allison actually began to intensify over land. As the storm entered Mississippi, its central pressure lowered, wind gusts reached 52 knots, and the center of circulation developed a weak-looking eye (see Fig. 3). As the system trekked eastward, it weakened and lost its eye, but continued to dump heavy rain over the southern Gulf States. Eventually, on June 14, the storm reached the Carolina coast.

Unfortunately, the storm slowed, then turned northward over North Carolina. Flooding became a major problem—Doppler radar estimated that up to 53 cm (21 in.) of rain had fallen over parts of the state. Severe weather broke out in Georgia and in the Carolinas, where some areas reported hail and downed trees due to gusty winds. The storm moved northeastward, parallel to the coast. A cold front moving in from the west eventually hooked up with the moisture from Allison. This situation caused heavy rain to fall over the mid-Atlantic states and southern New England. The storm finally accelerated to the northeast, away from the coast on June 18.

Allison, which never developed hurricane strength winds, claimed the lives of 43 people, whose deaths were mainly due to flooding. The total damage from the storm totaled in the billions of dollars, with the Houston area alone sustaining over $2 billion in damage. If all the rain that fell from Allison could be placed in Texas, it would cover two-thirds of the state with water a foot deep.

FIGURE 2 Visible satellite image showing the remains of tropical storm Allison centered over Texas on the morning of June 6, 2001. Heavy rain is falling from the thick clouds over Louisiana and eastern Texas.

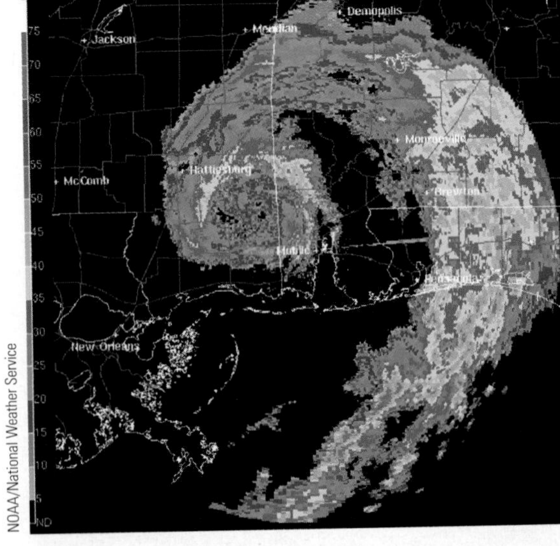

FIGURE 3 Doppler radar display on June 11, 2001, showing bands of heavy rain swirling counterclockwise into the center of once tropical storm Allison. The center of the storm, which is over Mississippi, has actually deepened and formed somewhat of an eye.

(and, hence, rising) of surface air. Moreover, tornadoes tend to form in the right front quadrant of an advancing hurricane, where vertical wind speed shear is greatest. Studies also suggest that swathlike areas of extreme damage once attributed to tornadoes may actually be due to downbursts associated with large thunderstorms around the eye wall.

In examining the extensive damage wrought by Hurricane Andrew during August, 1992, researchers initially theorized that the areas of most severe damage might have been caused by *spin-up vortices (mini-swirls)*—small whirling eddies perhaps 30 to 100 meters in diameter that occur in narrow bands. Many scientists today believe those rapidly rotating eddies were, in fact, small tornadoes. Lasting for about 10 seconds, the vortices appeared to have formed in a region of strong wind speed shear in the hurricane's eye wall, where the air was rapidly rising. As intense updrafts stretched the vortices vertically, they shrank horizontally, which induced them to spin faster (perhaps as fast as 70 knots), much like skaters spin faster as their arms are pulled inward. When the rotational winds of a vortice are added to the hurricane's steady wind, the total wind speed over a relatively small area may increase substantially. In the case of Hurricane Andrew, isolated wind speeds may have reached 174 knots (200 mi/hr) over narrow stretches of south Florida.

With the aid of ship reports, satellites, radar, buoys, and reconnaissance aircraft, the location and intensity of hurricanes are pinpointed and their movements carefully monitored. When a hurricane poses a direct threat to an area, a **hurricane watch** is issued, typically 24 to 48 hours before the storm arrives, by the National Hurricane Center in Miami, Florida, or by the Pacific Hurricane Center in Honolulu, Hawaii. When it appears that the storm will strike an area within 24 hours, a **hurricane warning** is issued. Along the east coast of North America, the warning is accompanied by a probability. The probability gives the percent chance of the hurricane's center passing within 105 km (65 mi) of a particular community. The warning is designed to give residents ample time to secure property and, if necessary, to evacuate the area.

A hurricane warning is issued for a rather large coastal area, usually about 550 km (342 mi) in length. Since the average swath of hurricane damage is normally about one-third this length, much of the area is "overwarned." As a consequence, many people in a warning area feel that they are needlessly forced to evacuate. The evacuation order is given by local authorities* and typically only for those low-lying coastal areas directly affected by the storm surge. People at higher elevations or further from the coast are not usually requested to leave, in part because of the added traffic problems this would create. This issue has engendered some controversy in the wake of Hurricane Andrew, since its winds were so devastating over inland south Florida during August, 1992. The time it takes to complete an evacuation puts a special emphasis on the timing and accuracy of the warning. As new hurricane-prediction models are implemented, and as our understanding of the nature of hurricanes increases, improved forecasts of hurricane movement and intensification should become available.

In recent years, the annual hurricane death toll in the United States has averaged less than 50 persons, although over 200 people died in Mexico when Hurricane Gilbert slammed the Gulf Coast of Mexico during September, 1988. This relatively low total is partly due to the advance warning provided by the National Weather Service and to the fact that only a few really intense storms have made *landfall** during the past 30 years. However, there is concern that as the population density continues to increase in vulnerable coastal areas, the potential for a hurricane-caused disaster continues to increase also.

Hurricane Camille (1969) stands out as one of the most intense hurricanes to reach the coastline of the United States during the twentieth century. With a central pressure of 909 mb, tempestuous winds reaching 160 knots (184 mi/hr), and a storm surge more than 7 m (23 ft) above the normal high-tide level, Camille unleashed its fury on Mississippi, destroying thousands of buildings. During its rampage, it caused an estimated $1.5 billion in property damage and took more than 200 lives.

During September, 1989, Hurricane Hugo was born as a cluster of thunderstorms became a tropical depression off the coast of Africa, southeast of the Cape Verde Islands. The storm grew in intensity, tracked westward for several days, then turned northwestward, striking the island of St. Croix with sustained winds of 125 knots.

DID YOU KNOW?

Even the aftermath of a hurricane can be deadly. In the Gulf states one of the greatest fears in the cleanup effort is encountering one of the poisonous snakes that seem to find their way into various nooks and crannies of debris.

*In the state of New Jersey, the Board of Casinos and the Governor must be consulted before an evacuation can be ordered.

*Landfall is the position along a coast where the center of a hurricane passes from ocean to land.

After passing over the eastern tip of Puerto Rico, this large, powerful hurricane took aim at the coastline of South Carolina. With maximum winds estimated at 120 knots (138 mi/hr), and a central pressure near 934 mb, Hugo made landfall near Charleston, South Carolina, about midnight on September 21 (see Fig. 11.13). The high winds and storm surge, which ranged between 2.5 and 6 m (8 and 20 ft), hurled a thundering wall of water against the shore. This knocked out power, flooded streets, and, as can be seen in Fig. 11.14, caused widespread destruction to coastal communities. The total damage in the United States attributed to Hugo was over $7 billion, with a death toll of 21 in the United States and 49 overall. But Hugo does not even come close to the costliest hurricane on record—that dubious distinction goes to Hurricane Andrew.

On August 21, 1992, as tropical storm Andrew churned westward across the Atlantic, it began to weaken, prompting some forecasters to surmise that this tropical storm would never grow to hurricane strength. But Andrew moved into a region favorable for hurricane development. Even though it was outside the tropics near latitude 25°N, warm surface water and weak winds aloft allowed Andrew to intensify rapidly. And in just two days Andrew's winds increased from 45 knots to 122 knots, turning an average tropical storm into one of the most intense hurricanes to strike Florida the last century (see Table 11.1).

With steady winds of 126 knots (145 mi/hr) and a powerful storm surge, Andrew made landfall south of Miami on the morning of August 24 (see Fig. 11.15). The eye of the storm moved over Homestead, Florida. Andrew's fierce winds completely devastated the area (see Fig. 11.16), as 50,000 homes were destroyed, trees were leveled, and steel-reinforced tie beams weighing tons were torn free of townhouses and hurled as far as several blocks. Swaths of severe damage led scientists to postulate that peak winds may have approached 174 knots (200 mi/hr). Such winds may have occurred with small tornadoes that added substantially to the storm's wind speed. In an instant, a wind gust of 142 knots (164 mi/hr) blew down a radar dome and inactivated several satellite dishes on the roof of the National Hurricane Center in Coral Gables. Observations reveal that some of Andrew's destruction may have been caused by microbursts in the severe thunderstorms of the eyewall. The hurricane roared westward across Southern Florida, weakened slightly, then regained strength over the warm Gulf of Mexico. Surging northwestward, Andrew slammed into Louisiana with 120-knot winds on the evening of August 25.

All told, Hurricane Andrew was the costliest natural disaster ever to hit the United States. It destroyed or damaged over 200,000 homes and businesses, left more than 160,000 people homeless, caused over $30 billion in damages, and took 53 lives, including 41 in Florida. Although Andrew may well be the most expensive hurricane on record, it is far from the deadliest.

Before the era of satellites and radar, catastrophic losses of life had occurred. In 1900, more than 6000 people lost their lives when a hurricane slammed into Galveston, Texas, with a huge storm surge (see Table 11.1, p. 309). Most of the deaths occurred in the low-lying

FIGURE 11.13 A color-enhanced infrared satellite image of Hurricane Hugo with its eye over the coast near Charleston, South Carolina.

TABLE 11.1 The Twelve Most Intense Hurricanes (at Landfall) to Strike the United States from 1900 to 2003

RANK	HURRICANE	YEAR	CENTRAL PRESSURE (millibars/inches)	CATEGORY	DEATH TOLL
1	Florida (Keys)	1935	892/26.35	5	408
2	Camille	1969	909/26.85	5	256
3	Andrew	1992	922/27.23	5	53
4	Florida (Keys)/South Texas	1919	927/27.37	4	>600*
5	Florida (Lake Okeechobee)	1928	929/27.43	4	1836
6	Donna	1960	930/27.46	4	50
7	Texas (Galveston)	1900	931/27.49	4	>6000
8	Louisiana (Grand Isle)	1909	931/27.49	4	350
9	Louisiana (New Orleans)	1915	931/27.49	4	275
10	Carla	1961	931/27.49	4	46
11	Hugo	1989	934/27.58	4	49
12	Florida (Miami)	1926	935/27.61	4	243

*More than 500 of this total were lost at sea on ships. (The > symbol means "greater than.")

coastal regions as flood waters pushed inland. In October, 1893, nearly 2000 people perished on the Gulf Coast of Louisiana as a giant storm surge swept that region. Spectacular losses are not confined to the Gulf Coast. Nearly 1000 people lost their lives in Charleston, South Carolina, during August of the same year. But these statistics are small compared to the more than 300,000 lives taken as a killer cyclone and storm surge ravaged the coast of Bangladesh with flood waters in 1970. Again in April, 1991, a similar cyclone devastated the area with reported winds of 127 knots and a storm surge of 7 m (23 ft). In all, the storm destroyed 1.4 million houses and killed 140,000 people and 1 million cattle. Unfortunately, the potential for a repeat of this type of disaster remains high in Bangladesh, as many people live along the relatively low, wide flood plain that slopes outward to the bay. And, historically, this region is in a path frequently taken by tropical cyclones (look back at Fig. 11.8, p. 301).

Even with modern satellite observation techniques, hurricane disasters can reach epic proportions. For example, Hurricane Mitch during late October, 1998, be-

DID YOU KNOW?

Hurricane Mitch brought a tragic end to the Ghost. On October 24, 1998, high winds and huge seas generated by Mitch pounded the majestic 234-foot sailing vessel *Fantome,* which in French means "ghost." The captain and a crew of 31 tried in vain to outmaneuver the monstrous storm, but all were lost as the $50 million ship sank in a maelstrom of winds and waves off the coast of Honduras.

NOAA/National Weather Service

(a)

NOAA/National Weather Service

(b)

FIGURE 11.14 Beach homes at Folly Beach, South Carolina, (a) before and (b) after Hurricane Hugo.

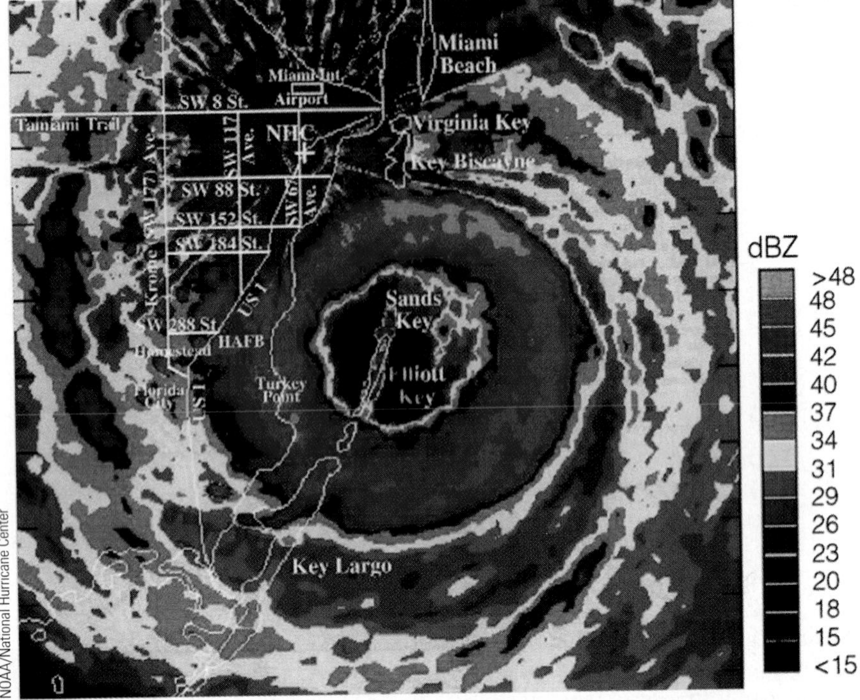

NOAA/National Hurricane Center

FIGURE 11.15 Color radar image of Hurricane Andrew as it moves on shore over south Florida on the morning of August 24, 1992. The National Hurricane Center (NHC) is located about 30 km (19 mi) from the center of the eye.

© Weather Catalog/Media Services

FIGURE 11.16 A community in Homestead, Florida, devastated by Hurricane Andrew on August 26, 1992.

TABLE 11.2 Saffir-Simpson Hurricane Damage-Potential Scale

SCALE NUMBER (CATEGORY)	CENTRAL PRESSURE mb	CENTRAL PRESSURE in.	WINDS mi/hr	WINDS knots	STORM SURGE ft	STORM SURGE m	DAMAGE
1	≥980*	≥28.94	74–95	64–82	4–5	~1.5	Damage mainly to trees, shrubbery, and unanchored mobile homes
2	965–979	28.50–28.91	96–110	83–95	6–8	~2.0–2.5	Some trees blown down; major damage to exposed mobile homes; some damage to roofs of buildings
3	945–964	27.91–28.47	111–130	96–113	9–12	~2.5–4.0	Foliage removed from trees; large trees blown down; mobile homes destroyed; some structural damage to small buildings
4	920–944	27.17–27.88	131–155	114–135	13–18	~4.0–5.5	All signs blown down; extensive damage to roofs, windows, and doors; complete destruction of mobile homes; flooding inland as far as 10 km (6 mi); major damage to lower floors of structures near shore
5	<920	<27.17	>155	>135	>18	>5.5	Severe damage to windows and doors; extensive damage to roofs of homes and industrial buildings; small buildings overturned and blown away; major damage to lower floors of all structures less than 4.5 m (15 ft) above sea level within 500 m of shore

*Symbol > means "greater than"; < means "less than"; ≥ means "equal to or greater than"; ~ means "approximately equal to."

came the most deadly hurricane to strike the Western Hemisphere since the Great Hurricane of 1780, which claimed approximately 22,000 lives in the eastern Caribbean. Mitch's high winds, huge waves (estimated maximum height 44 ft), and torrential rains destroyed vast regions of coastal Central America (for Mitch's path, see Fig. 11.9, p. 303). In the mountainous regions of Honduras and Nicaragua, rainfall totals from the storm may have reached 190 cm (75 in.). The heavy rains produced floods and deep mudslides that swept away entire villages, including the inhabitants. Mitch caused over $5 billion in damages, destroyed hundreds of thousands of homes, and killed over 11,000 people. More than 3 million people were left homeless or were otherwise severely affected by this deadly storm.

In an effort to estimate the possible damage a hurricane's sustained winds and storm surge could do to a coastal area, the **Saffir-Simpson scale** was developed (see Table 11.2). The scale numbers (which range from 1 to 5) are based on actual conditions at some time during the life of the storm. As the hurricane intensifies or weakens, the category, or scale number, is reassessed accordingly. Major hurricanes are classified as Category 3 and above. In the western Pacific, a typhoon with sustained winds of at least 130 knots (150 mi/hr)—at the upper end of the wind speed range in category 4 on the Saffir-Simpson scale—is called a **super-typhoon.**

Figure 11.17 shows the number of hurricanes that have made landfall along the coastline of the United States from 1900 through 1999. Out of a total of 167 hurricanes striking the American coastline, 66 (40 percent) were major hurricanes. Hence, along the Gulf and Atlantic coasts, on the average, about five hurricanes make landfall every three years, two of which are major hurricanes with winds in excess of 95 knots (110 mi/hr) and a storm surge exceeding 2.5 m (8 ft).

Modifying Hurricanes Because of the potential destruction and loss of lives that hurricanes can inflict, attempts have been made to reduce their winds by seeding them with silver iodide. The idea is to seed the clouds just outside the eye wall with just enough artificial ice nuclei so that the latent heat given off will stimulate cloud growth in this area of the storm. These clouds, which grow at the expense of the eye wall thunderstorms, actually form a new eye wall farther away from

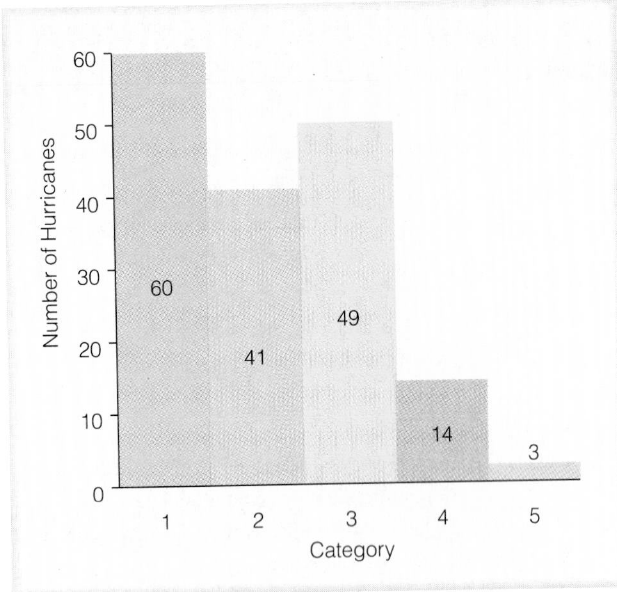

FIGURE 11.17 The number of hurricanes (by each category) that made landfall along the coastline of the United States from 1900 through 1999. All of the hurricanes struck the Gulf or Atlantic coasts. Categories 3, 4, and 5 are considered major hurricanes.

the hurricane's center. As the storm center widens, its pressure gradient should weaken, which may cause its spiraling winds to decrease in speed. During project STORMFURY, a joint effort of the National Oceanic and Atmospheric Administration (NOAA) and the U.S. Navy, several hurricanes were seeded by aircraft. In 1963, shortly after Hurricane Beulah was seeded with silver iodide, surface pressure in the eye began to rise and the region of maximum winds moved away from the storm's center. Even more encouraging results were obtained from the multiple seeding of Hurricane Debbie in 1969. After one day of seeding, Debbie showed a 30 percent reduction in maximum winds. However, the question remains: Would the winds have lowered naturally had the storm not been seeded? One study even casts doubt upon the theoretical basis for this kind of hurricane modification because hurricanes appear to contain too little supercooled water and too much natural ice. Consequently, there are many uncertainties about the effectiveness of seeding hurricanes in an attempt to reduce their winds, and all endeavors to modify hurricanes have been discontinued since the 1970s.

DID YOU KNOW?

Under the direction of Professor William Gray, scientists at Colorado State University issue hurricane forecasts. Their forecasts include the number and intensity of tropical storms and hurricanes that will develop each hurricane season. Their predictions are based upon such factors as seasonal rainfall in Africa, upper-level winds, and sea-level pressure over the tropical Atlantic and the Caribbean Sea. During the 1990s, they predicted for the North Atlantic region a total of 104 tropical storms, 63 hurricanes, and 22 intense hurricanes. The actual numbers were: 108, 64, and 25.

NAMING HURRICANES

In reading the previous sections, you noticed that hurricanes were assigned names. (The name is assigned when the storm reaches tropical storm strength.) Before this practice was started, hurricanes were identified according to their latitude and longitude. This method was confusing, especially when two or more storms were present over the same ocean. To reduce the confusion, hurricanes were identified by letters of the alphabet. During World War II, names like Able and Baker were used. (These names correspond to the radio code words associated with each letter of the alphabet.) This method also seemed cumbersome so, beginning in 1953, the National Weather Service began using female names to identify hurricanes. The list of names for each year was in alphabetical order, so that the name of the season's first storm began with the letter *A*, the second with *B*, and so on.

From 1953 to 1977, only female names were used. However, beginning in 1978, hurricanes in the eastern Pacific were alternately assigned female and male names, but not just English names, as Spanish and French ones were used too. This practice was started for North Atlantic hurricanes in 1979. (Remember that a storm only gets a name when it reaches tropical storm strength.) Table 11.3 gives the proposed list of names for both North Atlantic and eastern Pacific hurricanes. Once a storm has caused great damage and becomes infamous as a Category 3 or higher, its name is retired for at least ten years.

SUMMARY

Hurricanes are tropical cyclones with winds that exceed 64 knots (74 mi/hr) and blow counterclockwise about their centers in the Northern Hemisphere. A hurricane consists of a mass of organized thunderstorms that spiral in toward the extreme low pressure of the storm's eye. The most intense thunderstorms, the heaviest rain, and the highest winds occur outside the eye, in the region

TABLE 11.3 Names for Hurricanes and Tropical Storms

NORTH ATLANTIC HURRICANE NAMES				EASTERN NORTH PACIFIC HURRICANE NAMES			
2005	**2006**	**2007**	**2008**	**2005**	**2006**	**2007**	**2008**
Arlene	Alberto	Andrea	Arthur	Adrian	Aletta	Alvin	Alma
Bret	Beryl	Barry	Bertha	Beatriz	Bud	Barbara	Boris
Cindy	Chris	Chantal	Cristobal	Calvin	Carlotta	Cosme	Cristina
Dennis	Debby	Dean	Dolly	Dora	Daniel	Dalila	Douglas
Emily	Ernesto	Erin	Edouard	Eugene	Emilia	Erick	Elida
Franklin	Florence	Felix	Fay	Fernanda	Fabio	Flossie	Fausto
Gert	Gordon	Gabrielle	Gustav	Greg	Gilma	Gil	Genevieve
Harvey	Helene	Humberto	Hanna	Hilary	Hector	Henriette	Heman
Irene	Isaac	Ingrid	Ike	Irwin	Ileana	Ivo	Iselle
Jose	Joyce	Jerry	Josephine	Jova	John	Juliette	Julio
Katrina	Kirk	Karen	Kyle	Kenneth	Kristy	Kiko	Karina
Lee	Leslie	Lorenzo	Lili*	Lidia	Lane	Lorena	Lowell
Maria	Michael	Melissa	Marco	Max	Miriam	Manuel	Marie
Nate	Nadine	Noel	Nana	Norma	Norman	Narda	Norbert
Ophelia	Oscar	Olga	Omar	Otis	Olivia	Octave	Odile
Philippe	Patty	Pablo	Paloma	Pilar	Paul	Priscilla	Polo
Rita	Rafael	Rebekah	Rene	Ramon	Rosa	Raymond	Rachel
Stan	Sandy	Sebastien	Sally	Selma	Sergio	Sonia	Simon
Tammy	Tony	Tanya	Teddy	Todd	Tara	Tico	Trudy
Vince	Valerie	Van	Vicky	Veronica	Vicente	Velma	Vance
Wilma	William	Wendy	Wilfred	Wiley	Willa	Wallis	Winnie
				Xina	Xavier	Xina	Xavier
				York	Yolanda	York	Yolanda
				Zelda	Zeke	Zelda	Zeke

*Lili was retired after 2002 season, replacement name to be determined.

known as the eye wall. In the eye itself, the air is warm, winds are light, and skies may be broken or overcast.

Hurricanes (and all tropical cyclones) are born over warm tropical waters where the air is humid, surface winds converge, and thunderstorms become organized in a region of weak upper-level winds. Surface convergence may occur along the ITCZ, on the eastern side of a tropical wave, or along a front that has moved into the tropics from higher latitudes. If the disturbance becomes more organized, it becomes a tropical depression. If central pressures drop and surface winds increase, the depression becomes a tropical storm. At this point, the storm is given a name. Some tropical storms continue to intensify into full-fledged hurricanes, as long as they remain over warm water and are not disrupted by strong vertical wind shear.

In the organized convection theory, the energy source that drives the hurricane comes primarily from the release of latent heat. As vast quantities of latent heat are released in thunderstorms near the eye, surface pressures lower, inducing stronger winds, more rising air, and the development of even stronger thunderstorms. In the heat engine model, energy for the storm's growth is taken in at the surface in the form of sensible and latent heat, converted to kinetic energy in the form of winds, then lost at the cloud tops through radiational cooling.

The easterly winds in the tropics usually steer hurricanes westward. In the Northern Hemisphere, most storms then gradually swing northwestward around the subtropical high. If the storm moves into middle latitudes, the prevailing westerlies steer it northeastward. Because hurricanes derive their energy from the warm surface water and from the latent heat of condensation, they tend to dissipate rapidly when they move over cold water or over a large mass of land, where surface friction causes their winds to decrease and flow into their centers.

Although the high winds of a hurricane can inflict a great deal of damage, it is usually the huge waves and the flooding associated with the storm surge that cause the most destruction. The Saffir-Simpson hurricane scale was developed to estimate the potential destruction that a hurricane can cause.

Meteorology Now™ Assess your understanding of this chapter's topics with additional quizzing and tutorials at http://earthscience.brookscole.com/ahrens/ess4e.

KEY TERMS

The following terms are listed in the order they appear in the text. Define each. Doing so will aid you in reviewing the material covered in this chapter.

streamlines
tropical wave (easterly wave)
hurricane
typhoon
tropical cyclone
eye (of hurricane)
eye wall
trade wind inversion
tropical disturbance
tropical depression
tropical storm
Ekman transport
storm surge
hurricane watch
hurricane warning
Saffir-Simpson scale
super-typhoon

QUESTIONS FOR REVIEW

1. What is a tropical (easterly) wave? How do these waves generally move in the Northern Hemisphere? Are showers found on the eastern or western side of the wave?
2. Why are streamlines, rather than isobars, used on surface weather maps in the tropics?
3. What is the name given to a hurricane-like storm that forms over the western North Pacific Ocean?
4. Describe the horizontal and vertical structure of a hurricane.
5. Why are skies often clear or partly cloudy in a hurricane's eye?
6. What conditions at the surface and aloft are necessary for hurricane development?
7. (a) Describe the formation of a hurricane using the organized convection theory.
   (b) In this theory, where do hurricanes derive their energy?
8. (a) Hurricanes are sometimes described as a heat engine. According to this model, what is the "fuel" that drives the hurricane?
   (b) In this model, what determines the maximum strength (the highest winds) that the storm can achieve?
9. If a hurricane is moving westward at 10 knots, will the strongest winds be on its northern or southern side? Explain. If the same hurricane turns northward, will the strongest winds be on its eastern or western side?
10. What factors tend to weaken hurricanes?
11. Distinguish among a tropical disturbance, a tropical depression, a tropical storm, and a hurricane.
12. In what ways is a hurricane different from a mid-latitude cyclone? In what ways are these two systems similar?
13. Why do most hurricanes move westward over tropical waters?
14. If the high winds of a hurricane are not responsible for inflicting the most damage, what is?
15. Explain how a storm surge forms. How does it inflict damage in hurricane-prone areas?
16. Hurricanes are given names when the storm is in what stage of development?
17. How does a hurricane watch differ from a hurricane warning?
18. Why have hurricanes been seeded with silver iodide?
19. Give two reasons why hurricanes are more likely to strike New Jersey than Oregon.

QUESTIONS FOR THOUGHT AND EXPLORATION

1. A hurricane just off the coast of northern Florida is moving northeastward, parallel to the eastern seaboard. Suppose that you live in North Carolina along the coast.
   (a) How will the surface winds in your area change direction as the hurricane's center passes due *east* of you? Illustrate your answer by making a sketch of the hurricane's movement and the wind flow around it.
   (b) If the hurricane passes east of you, the strongest winds would most likely be blowing from which direction? Explain your answer. (Assume that the storm does not weaken as it moves northeastward.)
   (c) The lowest sea-level pressure would most likely occur with which wind direction? Explain.
2. Use the Saffir-Simpson Hurricane Damage-Potential Scale (Table 11.2, p. 311), and the associated text material to determine the category of each of the following hurricanes:
   (a) Hurricane Elena in Fig. 11.2, p. 295.
   (b) Hurricane Gloria in Fig. 11.11, p. 305.
   (c) Hurricane Hugo in Fig. 11.13, p. 308.
3. Give several reasons how a hurricane that once began to weaken can strengthen again.
4. Why are North Atlantic hurricanes more apt to form in October than in May?
5. Retired hurricane names (http://www.publicaffairs.noaa.gov/grounders/retirednames.html): Pick three hurricanes from the list of retired hurricane names

and explain why their names were retired. What were the particular features (strength, damage, movement, etc.) of these hurricanes?

6. Hurricane and tropical storm archives (http://www.earthwatch.com/hurricane.html): Pick three hurricanes or tropical storms from the list of archived images. Compare and contrast these storms. How large were they? Where did they occur? How large was the eye in each?
7. Online Guide to Hurricanes (http://ww2010.atmos.uiuc.edu/(Gh)/guides/mtr/hurr/home.rxml): How many tropical storms and hurricanes occur annually around the globe? Describe the convection processes that lead to the formation of a hurricane.

Go to the Brooks/Cole Earth Sciences Resource Center (http://earthscience.brookscole.com) for critical thinking exercises, articles, and additional readings from InfoTrac College Edition, Brooks/Cole's online student library.

Photo: © David Nunjk/Science Photo Library, Photo Researchers.

Dark smoke and sulfur oxides rise into the atmosphere from an industrial chimney.

Meteorology Now™ This icon, appearing throughout the book, indicates an opportunity to explore interactive tutorials, animations, or practice problems available on the MeteorologyNow Web site at http://earthscience.brookscole.com/ahrens/ess4e.

Air Pollution

Air pollution makes the earth a less pleasant place to live. It reduces the beauty of nature. This blight is particularly noticed in mountain areas. Views that once made the pulse beat faster because of the spectacular panorama of mountains and valleys are more often becoming shrouded in smoke. When once you almost always could see giant boulders sharply etched in the sky and the tapered arrowheads of spired pines, you now often see a fuzzy picture of brown and green. The polluted air acts like a translucent screen pulled down by an unhappy God.

Louis J. Battan, *The Unclean Sky*

CONTENTS

Every deep breath fills our lungs mostly with gaseous nitrogen and oxygen. Also inhaled, in minute quantities, may be other gases and particles, some of which could be considered pollutants. These contaminants come from car exhaust, chimneys, forest fires, factories, power plants, and other sources related to human activities.

Virtually every large city has to contend in some way with air pollution, which clouds the sky, injures plants, and damages property. Some pollutants merely have a noxious odor, whereas others can cause severe health problems. The cost is high. In the United States, for example, outdoor air pollution takes its toll in health care and lost work productivity at an annual expense that runs into *billions* of dollars. Estimates are that, worldwide, nearly 1 billion people in urban environments are continuously being exposed to health hazards from air pollutants.

This chapter takes a look at this serious contemporary concern. We begin by briefly examining the history of problems in this area, and then go on to explore the types and sources of air pollution, as well as the weather that can produce an unhealthful accumulation of pollutants. Finally, we investigate how air pollution influences the urban environment and also how it brings about unwanted acid precipitation.

A BRIEF HISTORY OF AIR POLLUTION

Strictly speaking, air pollution is not a new problem. More than likely it began when humans invented fire whose smoke choked the inhabitants of poorly ventilated caves. In fact, very early accounts of air pollution characterized the phenomenon as "smoke problems," the major cause being people burning wood and coal to keep warm.

To alleviate the smoke problem in old England, King Edward I issued a proclamation in 1273 forbidding the use of sea coal, an impure form of coal that produced a great deal of soot and sulfur dioxide when burned. One person was reputedly executed for violating this decree. In spite of such restrictions, the use of coal as a heating fuel grew during the fifteenth and sixteenth centuries.

As industrialization increased, the smoke problem worsened. In 1661, the prominent scientist John Evelyn wrote an essay deploring London's filthy air. And by the 1850s, London had become notorious for its "pea-soup" fog, a thick mixture of smoke and fog that hung over the city. These fogs could be dangerous. In 1873, one was responsible for as many as 700 deaths. Another, in 1911, claimed the lives of 1150 Londoners. To describe this chronic atmospheric event, a physician, Harold Des Voeux, coined (around 1911) the word *smog,* meaning a combination of smoke and fog.

Little was done to control the burning of coal as time went by, primarily because it was extremely difficult to counter the basic attitude of the powerful industrialists: "Where there's muck, there's money." London's acute smog problem intensified. Then, during the first week of December, 1952, a major disaster struck. The winds died down over London and the fog and smoke became so thick that people walking along the street literally could not see where they were going. This particular disastrous smog lasted 5 days and took nearly 4000 lives, prompting Parliament to pass a Clean Air Act in 1956. Additional air pollution incidents occurred in England during 1956, 1957, and 1962, but due to the strong legislative measures taken against air pollution, London's air today is much cleaner, and "pea soup" fogs are a thing of the past.

Air pollution episodes were by no means limited to Great Britain. During the winter of 1930, for instance, Belgium's highly industrialized Meuse Valley experienced an air pollution tragedy when smoke and other contaminants accumulated in a narrow steep-sided valley. The tremendous buildup of pollutants caused about 600 people to become ill, and ultimately 63 died. Not only did humans suffer, but cattle, birds, and rats fell victim to the deplorable conditions.

The industrial revolution brought air pollution to the United States, as homes and coal-burning industries belched smoke, soot, and other undesirable emissions into the air. Soon, large industrial cities, such as St. Louis and Pittsburgh (which became known as the "Smoky City"), began to feel the effects of the ever-increasing use of coal. As early as 1911, studies documented the irritating effect of smoke particles on the human respiratory system and the "depressing and devitalizing" effects of the constant darkness brought on by giant, black clouds of smoke. By 1940, the air over some cities had become so polluted that automobile headlights had to be turned on during the day.

The first major documented air pollution disaster in the United States occurred at Donora, Pennsylvania, during October, 1948, when industrial pollution became trapped in the Monongahela River Valley. During the ordeal, which lasted 5 days, more than 20 people died

DID YOU KNOW?

In London, England, during the severe smog episode of 1952, people wore masks over their mouths and found their way along the sidewalks by feeling the walls of buildings.

and thousands became ill.* Several times during the 1960s, air pollution levels became dangerously high over New York City. Meanwhile, on the West Coast, in cities such as Los Angeles, the ever-rising automobile population, coupled with the large petroleum processing plants, were instrumental in generating a different type of pollutant—one that forms in sunny weather and irritates the eyes. Toward the end of World War II, Los Angeles had its first (of many) smog alerts.

Air pollution episodes in Los Angeles, New York, and other large American cities led to the establishment of much stronger emission standards for industry and automobiles. The Clean Air Act of 1970, for example, empowered the federal government to set emission standards that each state was required to enforce. The Clean Air Act was revised in 1977 and updated by Congress in 1990 to include even stricter emission requirements for autos and industry. The new version of the Act also includes incentives to encourage companies to lower emissions of those pollutants contributing to the current problem of acid rain. Moreover, amendments to the Act have identified 189 toxic air pollutants for regulation. In 2001, the United States Supreme Court, in a unanimous ruling, made it clear that cost need not be taken into account when setting clean air standards.

*Additional information about the Donora air pollution disaster is given in the Focus section on p. 335.

TYPES AND SOURCES OF AIR POLLUTANTS

Air pollutants are airborne substances (either solids, liquids, or gases) that occur in concentrations high enough to threaten the health of people and animals, to harm vegetation and structures, or to toxify a given environment. Air pollutants come from both natural sources and human activities. Examples of natural sources include wind picking up dust and soot from the earth's surface and carrying it aloft, volcanoes belching tons of ash and dust into our atmosphere, and forest fires producing vast quantities of drifting smoke (see Fig. 12.1).

Human-induced pollution enters the atmosphere from both *fixed sources* and *mobile sources.* Fixed sources encompass industrial complexes, power plants, homes, office buildings, and so forth; mobile sources include motor

NASA

FIGURE 12.1 Strong northeasterly Santa Ana winds on October 28, 2003, blew the smoke from massive wild fires across southern California out over the Pacific Ocean.

TABLE 12.1 Some of the Sources of Primary Air Pollutants

SOURCES			POLLUTANTS
Natural			
	Volcanic eruptions		Particles (dust, ash), gases (SO_2, CO_2)
	Forest fires		Smoke, unburned hydrocarbons, CO_2, nitrogen oxides, ash
	Dust storms		Suspended particulate matter
	Ocean waves		Salt particles
	Vegetation		Hydrocarbons (VOCs),* pollens
	Hot springs		Sulfurous gases
Human caused			
Industrial	Paper mills		Particulate matter, sulfur oxides
	Power Plants	Coal	Ash, sulfur oxides, nitrogen oxides
		Oil	Sulfur oxides, nitrogen oxides, CO
	Refineries		Hydrocarbons, sulfur oxides, CO
	Manufacturing	Sulfuric acid	SO_2, SO_3, and H_2SO_4
		Phosphate fertilizer	Particulate matter, gaseous fluoride
		Iron and steel mills	Metal oxides, smoke, fumes, dust, organic and inorganic gases
		Plastics	Gaseous resin
		Varnish/paint	Acrolein, sulfur compounds
Personal	Automobiles		CO, nitrogen oxides, hydrocarbons (VOCs), particulate matter
	Home furnaces/fireplaces		CO, particulate matter
	Open burning of refuse		CO, particulate matter

*VOCs are volatile organic compounds; they represent a class of organic compounds, most of which are hydrocarbons.

vehicles, ships, and jet aircraft. Certain pollutants are called **primary air pollutants** because they enter the atmosphere directly—from smokestacks and tail pipes, for example. Other pollutants, known as **secondary air pollutants,** form only when a chemical reaction occurs between a primary pollutant and some other component of air, such as water vapor or another pollutant. Table 12.1 summarizes some of the sources of primary air pollutants.

We can see in Fig. 12.2 that transportation (motor vehicles, and so on) accounts for nearly 50 percent (by weight) of the pollution across the United States, with fuel combustion from stationary (fixed) sources coming in a distant second. Although hundreds of pollutants are found in our atmosphere, most fall into five groups, which are summarized in the following section.

Principal Air Pollutants The term **particulate matter** represents a group of solid particles and liquid droplets that are small enough to remain suspended in the air. Collectively known as *aerosols,* this grouping includes solid particles that may irritate people but are usually not poisonous, such as soot (tiny solid carbon particles), dust, smoke, and pollen. Some of the more dangerous substances include asbestos fibers and ar-

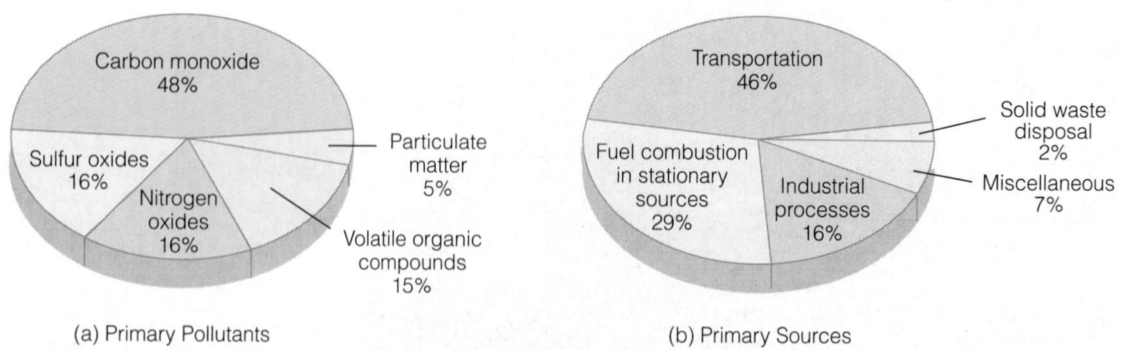

FIGURE 12.2 (a) Estimates of emissions of the primary air pollutants in the United States on a per weight basis; (b) the primary sources for the pollutants. (Data courtesy of United States Environmental Protection Agency.)

senic. Tiny liquid droplets of sulfuric acid, PCBs, oil, and various pesticides are also placed into this category.

Because it often dramatically reduces visibility in urban environments, particulate matter pollution is the most noticeable (see Fig. 12.3). Some particulate matter collected in cities includes iron, copper, nickel, and lead. This type of pollution can immediately influence the human respiratory system. Once inside the lungs, it can make breathing difficult, particularly for those suffering from chronic respiratory disorders. Lead particles especially are dangerous as they tend to fall out of the atmosphere and become absorbed into the body through ingestion of contaminated food and water supplies. Lead accumulates in bone and soft tissues, and in high concentrations can cause brain damage, convulsions, and death. Even at low doses, lead can be particularly dangerous to fetuses, infants, and children, who, when exposed, may suffer central nervous system damage.

Particulate pollution not only adversely affects the lungs, but recent studies suggest that particulate matter can interfere with the normal rhythm of the human heart. Apparently, as this type of pollution increases, there is a subtle change in a person's heart rate. For a person with an existing cardiac problem, a change in

Courtesy John Day

(a)

Courtesy John Day

(b)

Meteorology Now™

ACTIVE FIGURE 12.3

(a) Denver, Colorado, on a clear day, and (b) on a day when particulate matter and other pollutants greatly reduce visibility.
Watch this Active Figure at http://earthscience.brookscole.com/ahrens/ess4e.

DID YOU KNOW?

On any given day, estimates are that about 10 million tons of solid particulate matter are suspended in our atmosphere. And in a polluted environment, a volume of air about the size of a sugar cube can contain as many as 200,000 tiny particles.

heart rate can produce serious consequences. In fact, one study estimated that, each year, particulate pollution may be responsible for as many as 10,000 heart disease fatalities in the United States.

Industrial processes account for nearly 40 percent of the estimated 6.6 million metric tons of particulate matter emitted over the United States during the course of a year, whereas highway vehicles account for about 17 percent. One main problem is that particulate pollution may remain in the atmosphere for some time depending on the size and the amount of precipitation that occurs. For example, larger, heavier particles with diameters greater than about 10 μm* (0.01 mm) tend to settle to the ground in about a day or so after being emitted; whereas fine, lighter particles with diameters less than 1 μm (0.001 mm) can remain suspended in the lower atmosphere for several weeks.

Finer particles with diameters smaller than 10 μm are referred to as *PM-10.* These particles pose the greatest health risk, as they are small enough to penetrate the lung's natural defense mechanisms. Moreover, winds can carry these fine particles great distances before they finally reach the surface. In fact, suspended particles from sources in Europe and the former Soviet Union are believed responsible for the brownish cloud layer called *Arctic haze* that forms over the Arctic each spring. And strong winds over northern China can pick up dust particles and sweep them eastward, where they may settle on North America. This *Asian dust* can reduce visibility, produce spectacular sunrises and sunsets, and coat everything with a thin veneer of particles (see Fig. 12.4).

Studies show that particulate matter with diameters less than 2.5 μm, called *PM-2.5,* are especially dangerous. For one thing, they can penetrate farther into the lungs. Moreover, these tiny particles frequently consist of toxic or carcinogenic (cancer-causing) combustion products. Of recent concern are the PM-2.5 particles found in diesel soot. Relatively high amounts of these particles have been measured inside school busses with higher amounts observed downwind of traffic corridors and truck terminals.

Rain and snow remove many of these particles from the air; even the minute particles are removed by ice crystals and cloud droplets. In fact, numerical simulations of air pollution suggest that the predominant removal mechanism occurs when these particles act as nuclei for cloud droplets and ice crystals. As we will see in Chapter 14, a long-lasting accumulation of suspended particles (especially those rich in sulfur) is not only aesthetically unappealing but has the potential for affecting the climate.

*Recall that one micrometer (μm) is one-millionth of a meter. (The thickness of this page is about 100 micrometers.)

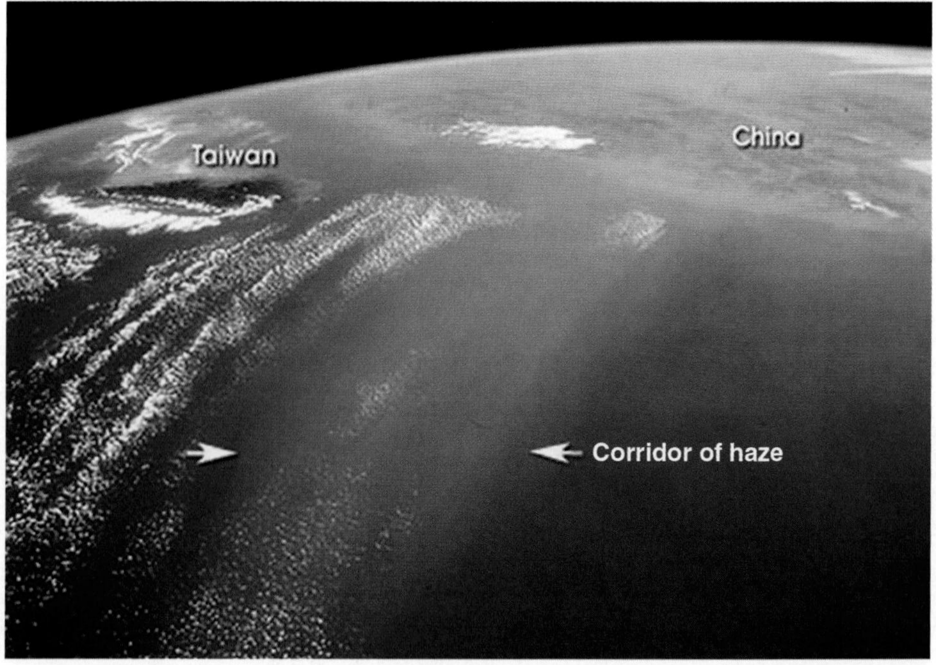

FIGURE 12.4 A thick haze about 200 km wide and about 600 km long covers a portion of the East China Sea on March 4, 1996. The haze is probably a mixture of industrial air pollution, dust, and smoke.

Photo by author

FIGURE 12.5 Cumulus clouds and a thunderstorm rise above the thick layer of haze that frequently covers the eastern half of the United States on humid summer days.

Many of the suspended particles are hygroscopic, as water vapor readily condenses onto them. As a thin film of water forms on the particles, they grow in size. When they reach a diameter between 0.1 and 1.0 μm these *wet haze* particles effectively scatter incoming sunlight to give the sky a milky white appearance. The particles are usually sulfate or nitrate particulate matter from combustion processes, such as those produced by diesel engines and power plants. The hazy air mass may become quite thick, and on humid summer days it often becomes well defined, as illustrated in Figure 12.5.

Carbon monoxide (CO), a major pollutant of city air, is a colorless, odorless, poisonous gas that forms during the incomplete combustion of carbon-containing fuels. From Fig. 12.2, we can see that carbon monoxide is the most plentiful of the primary pollutants.

The Environmental Protection Agency (EPA) estimates that over 60 million metric tons of carbon monoxide enter the air annually over the United States alone—about half from highway vehicles. However, due to stricter air quality standards and the use of emission-control devices, carbon monoxide levels have decreased by about 40 percent since the early 1970s.

Fortunately, carbon monoxide is quickly removed from the atmosphere by microorganisms in the soil, for even in small amounts this gas is dangerous. Hence, it poses a serious problem in poorly ventilated areas, such as highway tunnels and underground parking garages. Because carbon monoxide cannot be seen or smelled, it can kill without warning. Here's how: Normally, your cells obtain oxygen through a blood pigment called *hemoglobin,* which picks up oxygen from the lungs, combines with it, and carries it throughout your body. Unfortunately, human hemoglobin prefers carbon monoxide to oxygen, so if there is too much carbon monoxide in the air you breathe, your brain will soon be starved of oxygen, and headache, fatigue, drowsiness, and even death may result.*

Sulfur dioxide (SO_2) is a colorless gas that comes primarily from the burning of sulfur-containing fossil fuels (such as coal and oil). Its primary source includes power plants, heating devices, smelters, petroleum refineries, and paper mills. However, it can enter the atmosphere naturally during volcanic eruptions and as sulfate particles from ocean spray.

Sulfur dioxide readily oxidizes to form the secondary pollutants *sulfur trioxide* (SO_3) and, in moist air, highly corrosive *sulfuric acid* (H_2SO_4). Winds can carry these particles great distances before they reach the earth as undesirable contaminants. When inhaled into the lungs, high concentrations of sulfur dioxide aggravate respiratory problems, such as asthma, bronchitis, and emphysema. Sulfur dioxide in large quantities can cause injury to certain plants, such as lettuce and spinach, sometimes producing bleached marks on their leaves and reducing their yield.

Volatile organic compounds (VOCs) represent a class of organic compounds that are mainly **hydrocarbons**—individual organic compounds composed of hydrogen and carbon. At room temperature they occur as solids, liquids, and gases. Even though thousands of such compounds are known to exist, methane (which occurs naturally and poses no known dangers to health) is the most abundant. Other volatile organic compounds include benzene, formaldehyde, and some chlorofluorocarbons. The Environmental Protection Agency estimates that over 18 million metric tons of VOCs are emitted into

*Should you become trapped in your car during a snowstorm, and you have your engine and heater running to keep warm, roll down the window just a little. This action will allow the escape of any carbon monoxide that may have entered the car through leaks in the exhaust system.

the air over the United States each year, with about 34 percent of the total coming from vehicles used for transportation and about 50 percent from industrial processes.

Certain VOCs, such as benzene (an industrial solvent) and benzo-a-pyrene (a product of burning wood, tobacco and barbecuing), are known to be *carcinogens*—cancer-causing agents. Although many VOCs are not intrinsically harmful, some will react with nitrogen oxides in the presence of sunlight to produce secondary pollutants, which are harmful to human health.

Nitrogen oxides are gases that form when some of the nitrogen in the air reacts with oxygen during the high-temperature combustion of fuel. The two primary nitrogen pollutants are **nitrogen dioxide (NO_2)** and **nitric oxide (NO)**, which, together, are commonly referred to as NO_x—or simply, *oxides of nitrogen.*

Although both nitric oxide and nitrogen dioxide are produced by natural bacterial action, their concentration in urban environments is between 10 and 100 times greater than in nonurban areas. In moist air, nitrogen dioxide reacts with water vapor to form corrosive nitric acid (HNO_3), a substance that adds to the problem of acid rain, which we will address later.

The primary sources of nitrogen oxides are motor vehicles, power plants, and waste disposal systems. High concentrations are believed to contribute to heart and lung problems, as well as lowering the body's resistance to respiratory infections. Studies on test animals suggest that nitrogen oxides may encourage the spread of cancer. Moreover, nitrogen oxides are highly reactive gases that play a key role in producing ozone and other ingredients of photochemical smog.

Ozone in the Troposphere As mentioned earlier, the word **smog** originally meant the combining of smoke and fog. Today, however, the word mainly refers to the type of smog that forms in large cities, such as Los Angeles. Because this type of smog forms when chemical reactions take place in the presence of sunlight (called *photochemical reactions*), it is termed **photochemical smog,** or *Los Angeles-type smog.* When the smog is composed of sulfurous smoke and foggy air, it is usually called *London-type smog.*

The main component of photochemical smog is the gas **ozone (O_3).** Ozone is a noxious substance with an unpleasant odor that irritates eyes and the mucous membranes of the respiratory system, aggravating chronic diseases, such as asthma and bronchitis. Even in healthy people, exposure to relatively low concentrations of ozone for six or seven hours during periods of moderate exercise can significantly reduce lung function. This situation often is accompanied by symptoms such as chest pain, nausea, coughing, and pulmonary congestion. Ozone also attacks rubber, retards tree growth, and damages crops. Each year, in the United States alone, ozone is responsible for crop yield losses of several billion dollars.

We will see later that ozone forms naturally in the stratosphere through the combining of molecular oxygen and atomic oxygen. There, *stratospheric ozone* provides a protective shield against the sun's harmful ultraviolet rays. However, near the surface, in polluted air, ozone—often referred to as *tropospheric* (or *ground-level*) *ozone*—is a secondary pollutant that is not emitted directly into the air. Rather, it forms from a complex series of chemical reactions involving other pollutants, such as nitrogen oxides and volatile organic compounds (hydrocarbons). Because sunlight is required to produce ozone, concentrations of tropospheric ozone are normally higher during the afternoons (see Fig. 12.6) and during the summer months, when sunlight is more intense.

In polluted air, ozone production occurs along the following lines. Sunlight (with wavelengths shorter than about 0.41 μm) dissociates nitrogen dioxide into nitric oxide and atomic oxygen, which may be expressed by

$$NO_2 + \text{solar radiation} \rightarrow NO + O.$$

The atomic oxygen then combines with molecular oxygen (in the presence of a third molecule, M), to form ozone, as

$$O_2 + O + M \rightarrow O_3 + M.$$

The ozone is then destroyed by combining with nitric oxide; thus

$$O_3 + NO \rightarrow NO_2 + O_2.$$

If sunlight is present, however, the newly formed nitrogen dioxide will break down into nitric oxide and atomic oxygen. The atomic oxygen then combines with molecular oxygen to form ozone again.

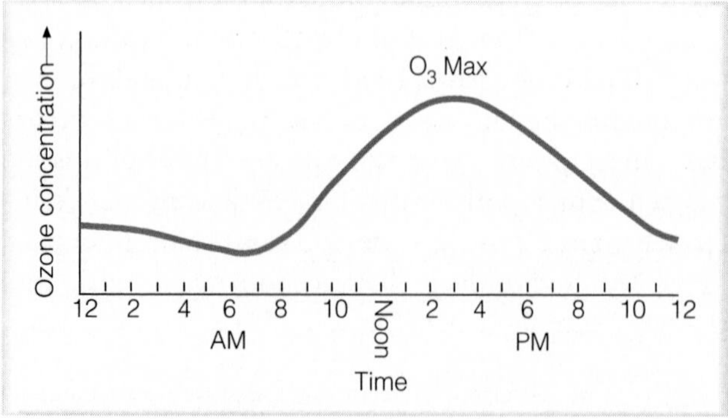

FIGURE 12.6 Average hourly concentrations of ozone measured at six major cities over a two-year period.

As a result of these reactions, large concentrations of ozone can form in polluted air only if some of the nitric oxide (NO) reacts with other gases *without removing ozone in the process.* This situation can take place in polluted air as unburned or partially burned hydrocarbons (released into the air by automobiles and industry) react with a variety of gases to form reactive molecules. These molecules then combine with nitric oxide (NO) to produce nitrogen dioxide (NO_2) and other products. In this manner, nitric oxide can react with hydrocarbons to form nitrogen dioxide *without removing ozone.* Hence, certain hydrocarbons in polluted air allow ozone concentrations to increase by preventing nitric oxide from destroying the ozone as rapidly as it forms.

The hydrocarbons (VOCs) also react with oxygen and nitrogen dioxide to produce other undesirable contaminants, such as *PAN* (peroxyacetyl nitrate)—a pollutant that irritates eyes and is extremely harmful to vegetation—and organic compounds. Ozone, PAN, and small amounts of other oxidating pollutants are the ingredients of photochemical smog. Instead of being specified individually, these pollutants are sometimes grouped under a single heading called *photochemical oxidants.**

Hydrocarbons (VOCs) do occur naturally in the atmosphere, as they are given off by vegetation. Oxides of nitrogen drifting downwind from urban areas can react with these natural hydrocarbons and produce smog in relatively uninhabited areas. This phenomenon has been observed downwind of cities such as Los Angeles, London, and New York. Some regions have so much natural (background) hydrocarbon that it may be difficult to reduce ozone levels as much as desired.

In spite of vast efforts to control ozone levels in major metropolitan areas, results have been generally disappointing because ozone, as we have seen, is a secondary pollutant that forms from chemical reactions involving other pollutants. Ozone production should decrease in most areas when emissions of *both* nitrogen oxides and hydrocarbons (VOCs) are reduced. However, the reduction of only one of these pollutants will not necessarily diminish ozone production because the oxides of nitrogen act as a catalyst for producing ozone in the presence of hydrocarbons.

Meteorology Now™ Click "Smog" to explore how reducing emissions of nitrogen oxides and VOCs can reduce ozone levels.

Ozone in the Stratosphere Recall from Chapter 1 that the stratosphere is a region of the atmosphere that lies above the troposphere between about 10 and 50 km (6 and 31 mi) above the earth's surface. The atmosphere is stable in the stratosphere, as there exists a strong temperature inversion—the air temperature increases rapidly with height (look back at Fig. 1.7, p. 10). The inversion is due, in part, to the gas ozone that absorbs ultraviolet radiation at wavelengths below about 0.3 micrometers.

In the stratosphere, above middle latitudes, notice in Fig. 12.7 that ozone is most dense at an altitude near 25 km (16 mi). Even at this altitude, its concentration is quite small, as there are only about 12 ozone molecules for every million air molecules (12 ppm).* Although thin, this layer of ozone is significant, for it shields earth's inhabitants from harmful amounts of ultraviolet solar radiation. This protection is fortunate because ultraviolet radiation at wavelengths below 0.3 micrometers has enough energy to cause skin cancer in humans.

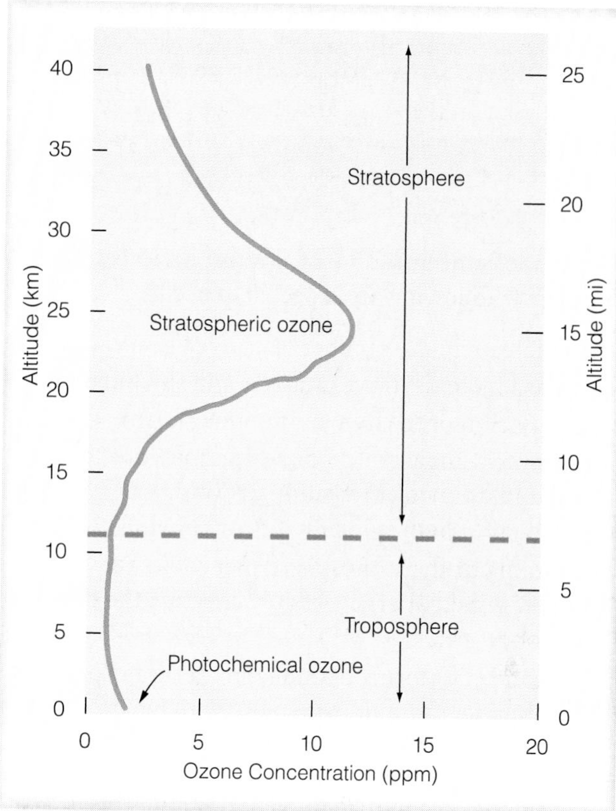

Meteorology Now™ ACTIVE FIGURE 12.7
The average distribution of ozone above the earth's surface in the middle latitudes. Watch this Active Figure at http://earthscience.brookscole.com/ahrens/ess4e.

*An oxidant is a substance (such as ozone) whose oxygen combines chemically with another substance.

*With a concentration of ozone of only 12 parts per million in the stratosphere, the composition of air here is about the same as it is near the earth's surface—mainly 78 percent nitrogen and 21 percent oxygen.

DID YOU KNOW?

Due to a thinning of the ozone layer above New Zealand (latitude 45°S), ultraviolet radiation reaching the surface there has increased by about 12 percent from 1990 to 1999.

If the concentration of stratospheric ozone decreases, the following are expected to occur:

- An increase in the number of cases of skin cancer.
- A sharp increase in eye cataracts and sun burning.
- Suppression of the human immune system.
- An adverse impact on crops and animals due to an increase in ultraviolet radiation.
- A reduction in the growth of ocean phytoplankton.
- A cooling of the stratosphere that could alter stratospheric wind patterns, possibly affecting the destruction of ozone.

Ozone (O_3) forms naturally in the stratosphere by the combining of atomic oxygen (O) with molecular oxygen (O_2) in the presence of another molecule. Although it forms mainly above 25 kilometers, ozone gradually drifts downward by mixing processes, producing a peak concentration in middle latitudes near 25 kilometers. (In polar regions, its maximum concentration is found at lower levels.) Ozone is broken down into molecular and atomic oxygen by absorbing ultraviolet (UV) solar radiation with wavelengths between 0.2 and 0.3 micrometers (see Fig. 12.8). Thus

$$O_3 + UV \rightarrow O_2 + O.$$

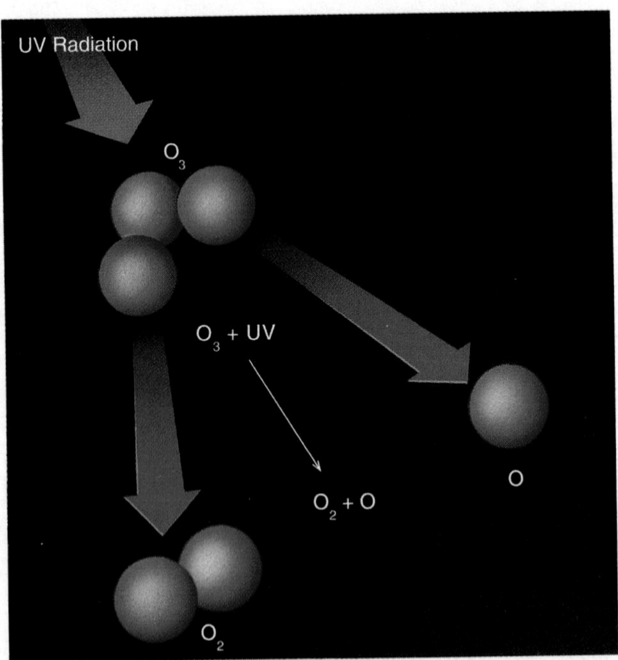

FIGURE 12.8 An ozone molecule absorbing ultraviolet radiation can become molecular oxygen (O_2) and atomic oxygen (O).

It is now apparent that human activities are altering the amount of stratospheric ozone. This possibility was first brought to light in the early 1970s as Congress pondered over whether or not the United States should build a supersonic jet transport. One of the gases emitted from the engines of this aircraft is nitric oxide. Although the aircraft was designed to fly in the stratosphere below the level of maximum ozone, it was feared that the nitric oxide would eventually have an adverse effect on the ozone. This factor was one of many considered when Congress decided to halt the development of the United States' version of the supersonic transport.

More recently, concerns involve emissions of chemicals at the earth's surface, such as nitrous oxide emitted from nitrogen fertilizers (which may drift into the stratosphere, where it could destroy ozone) and *chlorofluorocarbons* (CFCs). Until the late 1970s, when the United States banned all nonessential uses of chlorofluorocarbons, they were the most widely used propellants in spray cans, such as deodorants and hairsprays. In the troposphere, these gases are quite safe, being nonflammable, nontoxic, and unable to chemically combine with other substances.* Hence, these gases slowly diffuse upward without being destroyed. They apparently enter the stratosphere

1. near breaks in the tropopause; especially in the vicinity of jet streams
2. in building thunderstorms, especially those that develop in the tropics along the intertropical convergence zone and penetrate the lower stratosphere.

Once CFC molecules reach the middle stratosphere, ultraviolet energy that is normally absorbed by ozone breaks them up, releasing atomic *chlorine* in the process (and chlorine rapidly destroys ozone). In fact, estimates are that a single chlorine atom removes as many as 100,000 ozone molecules before it is taken out of action by combining with other substances.

Since the average lifetime of a CFC molecule is between 50 and 100 years, any increase in the concentration of CFCs is long lasting and a genuine threat to the concentration of ozone. Given this fact and the additional knowledge that CFCs contribute to the earth's greenhouse effect, an international agreement called the *Montreal Protocol* was signed in 1987. This agreement

*Recall from Chapter 3, however, that CFCs do act as strong greenhouse gases in the troposphere.

established a timetable for diminishing CFC emissions and the use of bromine compounds (halons), which destroy ozone at a rate more than 50 times greater than do chlorine compounds.*

During November, 1992, representatives of more than half the world's nations met in Copenhagen to update and revise the treaty. Provisions of the meeting called for a quicker phase-out of the previously targeted ozone-destroying chemicals and the establishment of a permanent fund to help Third World nations find the technology to develop ozone-friendly chemicals. The phase-out appears to be working, as the National Oceanic and Atmospheric Administration (NOAA) reported in 1996 that ground-based stations around the world had detected for the first time an overall slight decline (about 1 percent) in the total concentration of ozone-destroying chlorine and bromine compounds. Moreover, both satellite measurements and surface observations confirm that chlorine concentrations in the stratosphere peaked about 1997.

The question of just how much ozone is being depleted by chlorine is now under investigation. More than 5 billion kilograms of CFCs have already been released into the troposphere and will diffuse upward during the next few decades. In a 1991 study, an international panel of over 80 scientists concluded that the ozone layer thinned by about 3 percent during the summer from 1979 to 1991 over heavily populated areas of the Northern Hemisphere. More recent studies show an ozone reduction in the stratosphere of between 5 and 6 percent since 1978.

Satellite measurements in 1992 and 1993 revealed that ozone concentrations had dropped to record low levels over much of the globe. The decrease appears to stem from ozone-destroying chemicals and from the 1991 volcanic eruption of Mt. Pinatubo that sent tons of sulfur dioxide gas into the stratosphere, where it formed tiny droplets of sulfuric acid. These droplets not only enhance the ozone destructiveness of the chlorine chemicals but also alter the circulation of air in the stratosphere, making it more favorable for ozone depletion. During the mid-1990s, wintertime ozone levels dropped well below normal over much of the Northern Hemisphere. This decrease apparently was due to ozone-destroying pollution along with natural cold weather patterns that favored ozone reduction.

Presently, there are two major substitutes for CFCs, *hydrochlorofluorocarbons* (HCFCs) and *hydrofluorocarbons* (HFCs). The HCFCs contain fewer chlorine atoms per molecule than CFCs and, therefore, pose much less danger to the ozone layer, whereas HFCs contain no chlorine. These gases may have to be phased out, however, as both are greenhouse gases that can enhance global warming.

Ozone levels appear to be dropping more quickly above specific regions of the planet. Scientists point to the fact that ozone concentrations over springtime Antarctica have plummeted at an alarming rate. This sharp drop in ozone is known as the **ozone hole.** (More information on the ozone hole is provided in the Focus section on p. 330.)

Air Pollution: Trends and Patterns Over the past decades, strides have been made in the United States to improve the quality of the air we breathe. Figure 12.9 shows the estimated emission trends over the United States for the primary pollutants. Notice that since the Clean Air Act of 1970, emissions of most pollutants have fallen off substantially, with lead showing the greatest reduction, primarily due to the gradual elimination of leaded gasoline.

Although the situation has improved, we can see from Fig. 12.9 that much more needs to be done, as large quantities of pollutants still spew into our air. In fact, many areas of the country do not conform to the standards for air quality set by the Clean Air Act of 1990. A large part of the problem of pollution control lies in the fact that, even with stricter emission laws, increasing numbers of autos (estimates are that more than 198 million are on the road today) and other sources can overwhelm control efforts.

Clean air standards are established by the Environmental Protection Agency. *Primary ambient air quality standards* are set to protect human health, whereas *secondary standards* protect human welfare, as measured by the effects of air pollution on visibility, crops, and buildings. Areas that do not meet air quality standards are called *nonattainment areas.* Even with stronger emission laws, estimates are that presently more than 80 million Americans are breathing air that does not meet at least one of the standards.

To indicate the air quality in a particular region, the EPA developed the **air quality index (AQI).*** The index includes the pollutants carbon monoxide, sulfur dioxide, nitrogen dioxide, particulate matter, and ozone. On any given day, the pollutant measuring the highest value is the one used in the index. The pollutant's measurement is then converted to a number that ranges from 0 to 500 (see Table 12.2). When the pollutant's value is

*There are many chemical reactions that involve chlorine and bromine and the destruction of ozone in the stratosphere.

*In June, 2000, the EPA updated the pollutant standard index (PSI) and renamed it the air quality index (AQI).

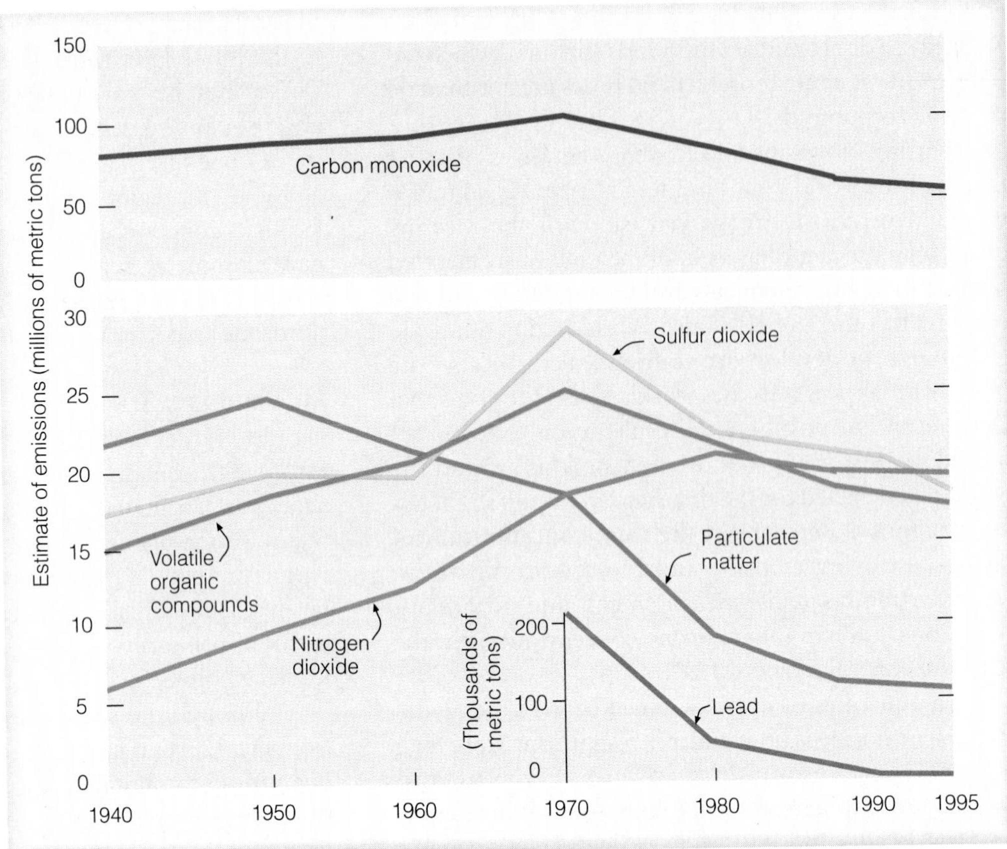

FIGURE 12.9 Emission estimates of six pollutants in the United States from 1940–1995. (Data courtesy of United States Environmental Protection Agency.)

TABLE 12.2 The Air Quality Index (AQI)

AQI VALUE	AIR QUALITY	GENERAL HEALTH EFFECTS	RECOMMENDED ACTIONS
0-50	Good	None	
51-100	Moderate	There may be a moderate health concern for a very small number of individuals. People unusually sensitive to ozone may experience respiratory symptoms.	When O_3 AQI values are in this range, unusually sensitive people should consider limiting prolonged outdoor exposure.
101-150	Unhealthy for sensitive groups	Mild aggravation of symptoms in susceptible persons.	Active people with respiratory or heart disease should limit prolonged outdoor exertion.
151-200	Unhealthy	Aggravation of symptoms in susceptible persons, with irritation symptoms in the healthy population.	Active children and adults with respiratory or heart disease should avoid extended outdoor activities; everyone else, especially children, should limit prolonged outdoor exertion.
201-300	Very Unhealthy	Significant aggravation of symptoms and decreased exercise tolerance in persons with heart or lung disease, with widespread symptoms in the healthy population.	Active children and adults with existing heart or lung disease should avoid outdoor activities and exertion. Everyone else, especially children, should limit outdoor exertion.
301-500	Hazardous	Significant aggravation of symptoms. Premature onset of certain diseases. Premature death may occur in ill or elderly people. Healthy people may experience a decrease in exercise tolerance.	Everyone should avoid all outdoor exertion and minimize physical outdoor activities. Elderly and persons with exisitng heart or lung disease should stay indoors.

the same as the primary ambient air quality standard, the pollutant is assigned an AQI number of 100. A pollutant is considered unhealthful when its AQI value exceeds 100. Figure 12.10 shows the number of unhealthful days across the United States during 1990. When the AQI value is between 51 and 100, the air quality is described as "moderate." Although these levels may not be harmful to humans during a 24-hour period, they may exceed long-term standards. Notice that the AQI is color-coded, with each color corresponding to an AQI level of health concern. The color green indicates "good" air quality; the color red, "unhealthy" air; and maroon, "hazardous" air quality. Table 12.2 also shows the health effects and the precautions that should be taken when the AQI value reaches a certain level.

Higher emission standards, along with cleaner fuels (such as natural gas), have made the air over our large cities cleaner today than it was years ago. In fact, total emissions of toxic chemicals spewed into the skies over the United States have been declining steadily since the EPA began its inventory of these chemicals in 1987. But the control of ozone in polluted air is still a pervasive problem. Because ozone is a secondary pollutant, its formation is controlled by the concentrations of other pollutants, namely nitrogen oxides and hydrocarbons (VOCs). Moreover, weather conditions play a vital role in ozone formation, as ozone reaches its highest concentrations in hot sunny weather when surface winds are light and a stagnant high-pressure area covers the region. As a result of these factors, year-to-year ozone trends are quite variable.

DID YOU KNOW?

Air pollution became so intolerable in Mexico City during January, 1989, that school children were given the entire month off.

BRIEF REVIEW

Before going on to the next several sections, here is a brief review of some of the important points presented so far.

- Near the surface, primary air pollutants (such as particulate matter, CO, SO_2, NO, NO_2, and VOCs) enter the atmosphere directly, whereas secondary air pollutants (such as O_3) form when a chemical reaction takes place between a primary pollutant and some other component of air.
- The word "smog" (coined in London in the early 1900s) originally meant the combining of smoke and fog. Today, the word mainly refers to photochemical smog—pollutants that form in the presence of sunlight.
- Stratospheric ozone forms naturally in the stratosphere and provides a protective shield against the sun's harmful ultraviolet rays. Tropospheric (ground-level) ozone that forms in polluted air is a health hazard and is the primary ingredient of photochemical smog.
- Human-induced chemicals, such as chlorofluorocarbons (CFCs), appear to be altering the amount of ozone in the stratosphere by releasing chlorine, which rapidly destroys ozone.
- Even though the emissions of most pollutants have declined across the United States since 1970, millions of Americans are breathing air that does not meet air quality standards.

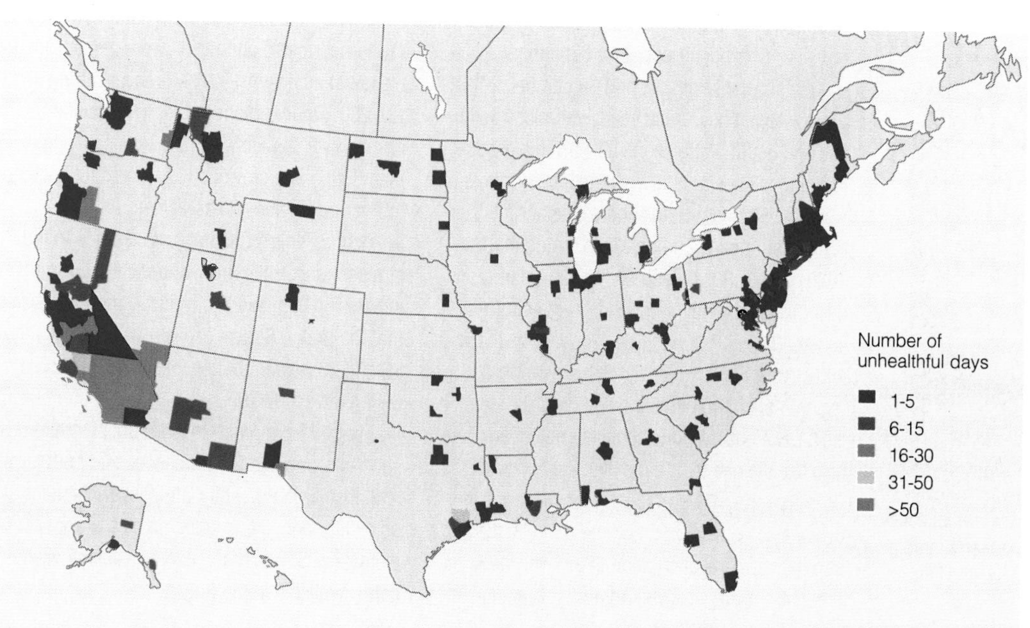

FIGURE 12.10 The number of unhealthful days (by county) across the United States for any one of the five pollutants (CO, SO_2, NO_2, O_3, and particulate matter) during 1990. (Data courtesy of United States Environmental Protection Agency.)

FOCUS ON AN ENVIRONMENTAL ISSUE

The Ozone Hole

In 1974, two chemists from the University of California at Irvine—F. Sherwood Rowland and Mario J. Molina—warned that increasing levels of CFCs would eventually deplete stratospheric ozone on a global scale. Their studies suggested that ozone depletion would occur gradually and would perhaps not be detectable for many years to come. It was surprising, then, when British researchers identified a year-to-year decline in stratospheric ozone over Antarctica. Their findings, corroborated later by satellites and balloon-borne instruments, showed that since the late 1970s ozone concentrations have diminished each year during the months of September and October. This decrease in stratospheric ozone over springtime Antarctica is known as the *ozone hole* (see Fig. 1). In years of severe depletion, such as in 1998, the ozone hole covers almost twice the area of the Antarctic continent.

To understand the causes behind the ozone hole, scientists in 1986 organized the first *National Ozone Expedition,* NOZE-1, which set up a fully instrumented observing station near McMurdo Sound, Antarctica. During 1987, with the aid of instrumented aircraft, NOZE-2 got under way. The findings from these research programs helped scientists put together the pieces of the ozone puzzle.

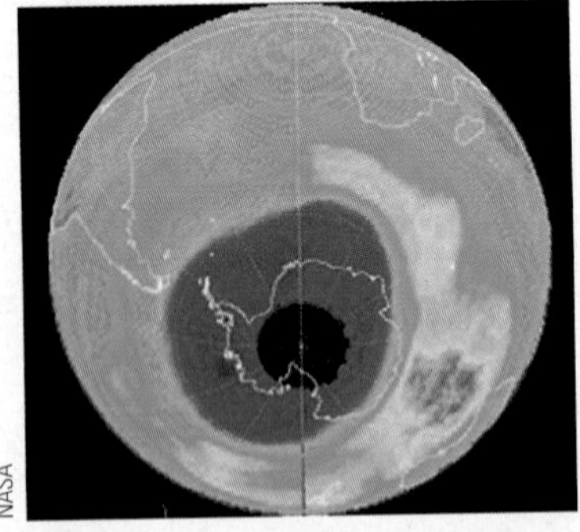

FIGURE 1 Ozone distribution over the Southern Hemisphere on September 11, 2003, as measured by the Total Ozone Mapping Spectrometer (TOMS). Notice that the area of lowest ozone concentration or ozone hole (purple shades) is larger than Antarctica and represents the second largest ozone hole ever observed.

The stratosphere above Antarctica has one of the world's highest ozone concentrations. Most of this ozone forms over the tropics and is brought to the Antarctic by stratospheric winds. During September and October (spring in the Southern Hemisphere), a belt of stratospheric winds called the *polar vortex* encircles the Antarctic region near 66°S latitude, essentially isolating the cold Antarctic stratospheric air from the warmer air of the middle latitudes. During the long dark Antarctic winter, temperatures inside the vortex can drop to −85°C (−121°F). This frigid air allows for the formation of *polar stratospheric clouds.* These ice clouds are critical in facilitating chemical interactions among nitrogen, hydrogen, and chlorine atoms, the end product of which is the destruction of ozone.

In 1986, the NOZE-1 study detected unusually high levels of chlorine compounds in the stratosphere, and, in 1987, the instrumented aircraft of NOZE-2 measured enormous increases in chlorine compounds when it entered the polar vortex. These findings, in conjunction with other chemical discoveries, allowed scientists to pinpoint *chlorine* from CFCs as the main cause of the ozone hole.

Even with a decline in ozone-destroying chemicals, the largest Antarctic ozone hole observed to date occurred during September, 2000, and the second largest during September, 2003 (Fig. 1). Apparently, these yearly variations in the size and depth of the ozone hole are mainly due to changes in polar stratospheric temperatures.

In the Northern Hemisphere's polar Arctic, airborne instruments and satellites during the late 1980s and 1990s measured high levels of ozone-destroying chlorine compounds in the stratosphere. By 1997, springtime ozone levels in the Arctic were the lowest ever observed in the region—about 40 percent below average (see Fig. 2). But observations could not detect an ozone hole like the one that forms over the Antarctic.

Apparently, several factors inhibit massive ozone loss in the Arctic. For one thing, in the stratosphere, the circulation of air over the Arctic differs from that over the Antarctic. Then, too, the

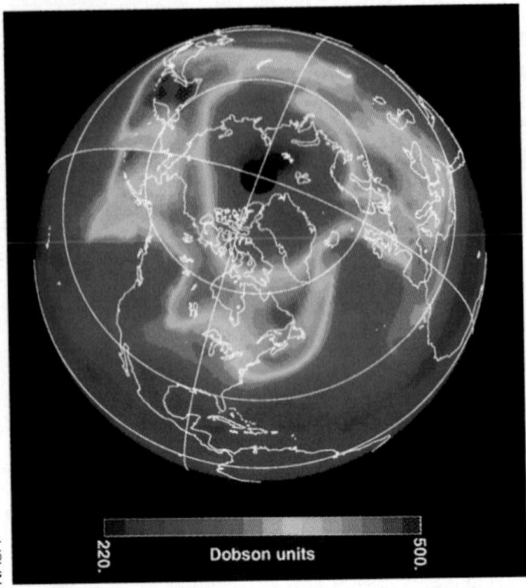

FIGURE 2 Color image of total ozone amounts over the Northern Hemisphere for March 24, 1997. Notice that minimum ozone values (purple shades) appear over a region near the North Pole. The color scale on the bottom of the image shows total ozone values in Dobson units (DU). A Dobson unit is the physical thickness of the ozone layer if it were brought to the earth's surface (500 DU equals 5 mm).

Arctic stratosphere is normally too warm for the widespread development of clouds that help activate chlorine molecules. However, it appears that a very cold Arctic stratosphere, along with ozone-destroying chemicals were responsible for the low readings in 1997. Moreover, during January, 2000, more polar stratospheric clouds formed over the Arctic, and they lasted longer than during any previous year. This situation contributed to significant ozone loss.

Ozone depletion is not just confined to the stratospheric Arctic and Antarctic. For example, in March, 1995, satellite measurements revealed that the United State's ozone levels fell between 15 and 20 percent below the values observed during March, 1979.

We still have much to learn about stratospheric ozone and the processes that both form and destroy it. Presently, atmospheric studies are providing more information so that a more complete assessment of the ozone problem will become available in the future.

FACTORS THAT AFFECT AIR POLLUTION

If you live in a region that periodically experiences smog, you may have noticed that these episodes often occur with clear skies, light winds, and generally warm sunny weather. Although this may be "typical" air pollution weather, it by no means represents the only weather conditions necessary to produce high concentrations of pollutants, as we will see in the following sections.

The Role of the Wind The wind speed plays a role in diluting pollution. When vast quantities of pollutants are spewed into the air, the wind speed determines how quickly the pollutants mix with the surrounding air and, of course, how fast they move away from their source. Strong winds tend to lower the concentration of pollutants by spreading them apart as they move downwind. Moreover, the stronger the wind, the more turbulent the air. Turbulent air produces swirling eddies that dilute the pollutants by mixing them with the cleaner surrounding air. Hence, when the wind dies down, pollutants are not readily dispersed, and they tend to become more concentrated.

The Role of Stability and Inversions Recall from Chapter 5 that atmospheric stability determines the extent to which air will rise. Remember also that an unstable atmosphere favors vertical air currents, whereas a stable atmosphere strongly resists upward vertical motions. Consequently, smoke emitted into a stable atmosphere tends to spread horizontally, rather than mix vertically.

The stability of the atmosphere is determined by the way the air temperature changes with height (the lapse rate). When the measured air temperature decreases rapidly as we move up into the atmosphere, the atmosphere tends to be more unstable. If, however, the measured air temperature either decreases quite slowly as we ascend, or actually increases with height (remember that this is called an *inversion*), the atmosphere is stable. An inversion represents an extremely stable atmosphere where warm air lies above cool air. Any air parcel that attempts to rise into the inversion will, at some point, be cooler and heavier (more dense) than the warmer air surrounding it. Hence, the inversion acts like a lid on vertical air motions.

One type of very stable atmosphere usually exists during the night and early morning hours. If the sky is clear and the winds are light, the air near the ground can become much cooler than the air higher up. Recall that this situation is called a **radiation** (or **surface**) **inversion** (see Chapter 3, p. 59).

Figure 12.11 shows a strong radiation inversion on a clear calm winter night. Notice that within the stable inversion, the smoke from the shorter stack does not rise very high, but spreads out, contaminating the area around it. In the relatively unstable air above the inversion, smoke from the taller stack is able to rise and become dispersed. Since radiation inversions are often rather shallow, it should be apparent why taller chimneys have replaced many of the shorter ones. In fact, taller chimneys disperse pollutants better than shorter ones even in the absence of a surface inversion because the taller chimneys are able to mix pollutants throughout a greater volume of air. Although these taller stacks do improve the air quality in their immediate area, they may also contribute to the acid rain problem by allowing the pollutants to be swept great distances downwind.

As the sun rises and the surface warms, the radiation inversion normally weakens and disappears before noon. By afternoon, the atmosphere is sufficiently unstable so that, with adequate winds, pollutants are able to disperse vertically. The changing atmospheric stability, from stable in the early morning to conditionally unstable in the afternoon, can have a profound effect on the daily concentrations of pollution in certain regions. For example, on a busy city street corner, carbon monoxide levels can be considerably higher in the early morning than in the early afternoon (with the same flow of traffic). Changes in atmospheric stability can also cause smoke plumes from chimneys to change during the course of a day. (Some of these changes are described in the Focus section on p. 332.)

Radiation inversions normally last just a few hours, while **subsidence inversions** may persist for several days

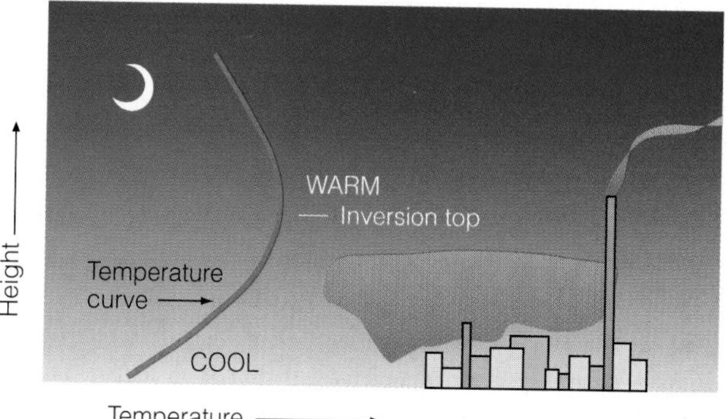

FIGURE 12.11 The smoke from the shorter stack is trapped within the inversion, while the smoke from the taller stack, above the inversion, rises, mixes, and disperses downwind.

FOCUS ON AN OBSERVATION

Smokestack Plumes

We know that the stability of the air (especially near the surface) changes during the course of a day. These changes can influence the pollution near the ground as well as the behavior of smoke leaving a chimney. Figure 3 illustrates different smoke plumes that can develop with adequate wind, but different types of stability.

In Fig. 3, diagram a, it is early morning, the winds are light, and a radiation inversion extends from the surface to well above the height of the smoke stack. In this stable environment, there is little up and down motion, so the smoke spreads horizontally rather than vertically. When viewed from above, the smoke plume resembles the shape of a fan. For this reason, it is referred to as a *fanning smoke plume.*

Later in the morning, the surface air warms quickly and destabilizes as the radiation inversion gradually disappears from the surface upward (Fig. 3, diagram b). However, the air above the chimney is still stable, as indicated by the presence of the inversion. Consequently, vertical motions are confined to the region near the surface. Hence, the smoke mixes downwind, increasing the concentration of pollution at the surface—sometimes to dangerously high levels. This effect is called *fumigation.*

If daytime heating of the ground continues, the depth of atmospheric instability increases. Notice in Fig. 3, diagram c, that the inversion has completely disappeared. Light-to-moderate winds combine with rising and sinking air to cause the smoke to move up and down in a wavy pattern, producing a *looping smoke plume.* Thus, watching smoke plumes provides a clue to the stability of the atmosphere, and knowing the stability yields important information about the dispersion of pollutants.

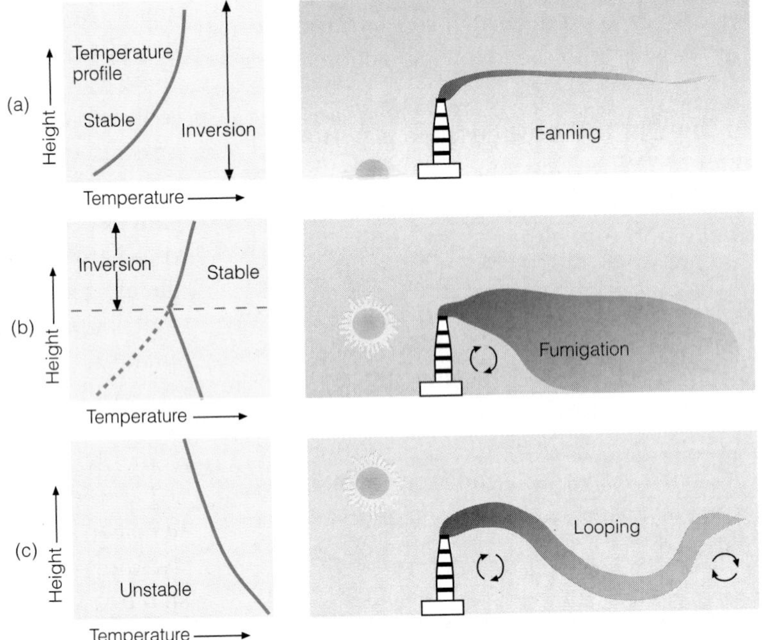

FIGURE 3 As the vertical temperature profile changes during the course of a day (a through c), the pattern of smoke emitted from the stack changes as well.

Of course, other factors influence the dispersion of pollutants from a chimney, including the pollutants' temperature and exit velocity, wind speed and direction, and, of course, the chimney's height. Overall, taller chimneys, greater wind speeds, and higher exit velocities result in a lower concentration of pollutants.

or longer. Subsidence inversions, therefore, are the ones commonly associated with major air pollution episodes. They form as the air above a deep anticyclone slowly sinks (subsides) and warms.*

A typical temperature profile of a subsidence inversion is shown in Fig. 12.12. Notice that in the relatively unstable air beneath the inversion, the pollutants are able to mix vertically up to the inversion base. The stable air of the inversion, however, inhibits vertical mixing and acts like a lid on the pollution below, preventing it from entering into the inversion.

In Fig. 12.12, the region of relatively unstable (well-mixed) air that extends from the surface to the base of the inversion is referred to as the **mixing layer.** The vertical extent of the mixing layer is called the **mixing depth.** Observe that if the inversion rises, the mixing depth increases and the pollutants would be dispersed throughout a greater volume of air; if the inversion lowers, the mixing depth would decrease and the pollutants

*Remember from Chapter 2 that sinking air always warms because it is being compressed by the surrounding air.

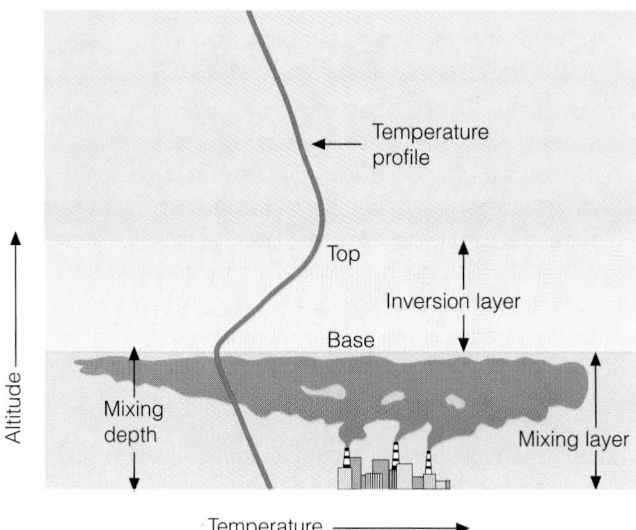

FIGURE 12.12 The inversion layer prevents pollutants from escaping into the air above it. If the inversion lowers, the mixing depth decreases and the pollutants are concentrated within a smaller volume.

FIGURE 12.13 A thick layer of polluted air is trapped in the valley. The top of the polluted air marks the base of a subsidence inversion and the top of the mixing layer.

would become more concentrated, sometimes reaching unhealthy levels. Since the atmosphere tends to be most unstable in the afternoon and most stable in the early morning, we typically find the greatest mixing depth in the afternoon and the most shallow one (if one exists at all) in the early morning. Consequently, during the day, the top of the mixing layer may clearly be visible (see Fig. 12.13). Moreover, during take-off or landing on daylight flights out of large urban areas, the top of the mixing layer may be easily observed.

The position of the semipermanent Pacific high off the coast of California contributes greatly to the air pollution in that region. The Pacific high promotes subsiding air, which warms the air aloft. Surface winds around the high promote upwelling of ocean water (see Chapter 7). Upwelling makes the surface water cool, which, in turn, cools the air above. Warm air aloft coupled with cool, surface (marine) air, together produce a strong and persistent subsidence inversion—one that exists 80 to 90 percent of the time over the city of Los Angeles between June and October, the smoggy months. The pollutants trapped within the cool marine air are occasionally swept eastward by a sea breeze. This action carries smog from the coastal regions into the interior valleys (see Fig. 12.14).

FIGURE 12.14 The leading edge of cool, marine air carries pollutants into Riverside, California.

The Role of Topography The shape of the landscape (topography) plays an important part in trapping pollutants. We know from Chapter 3 that, at night, cold air tends to drain downhill, where it settles into low-lying basins and valleys. The cold air can have several effects: It can strengthen a pre-existing surface inversion, and it can carry pollutants downhill from the surrounding hillsides (see Fig. 12.15).

Valleys prone to pollution are those completely encased by mountains and hills. The surrounding mountains tend to block the prevailing wind. With light winds, and a shallow mixing layer, the poorly ventilated cold valley air can only slosh back and forth like a murky bowl of soup.

Air pollution concentrations in mountain valleys tend to be greatest during the colder months. During the warmer months, daytime heating can warm the sides of the valley to the point that upslope valley winds vent the pollutants upward, like a chimney. Valleys susceptible to stagnant air exist in just about all mountainous regions.

The pollution problem in several large cities is, at least, partly due to topography. For example, the city of Los Angeles is surrounded on three sides by hills and mountains. Cool marine air from off the ocean moves inland and pushes against the hills, which tend to block the air's eastward progress. Unable to rise, the cool air settles in the basin, trapping pollutants from industry and millions of autos. Baked by sunlight, the pollutants become the infamous photochemical smog. By the same token, the "mile high" city of Denver, Colorado, sits in a broad shallow basin that frequently traps both cold air and pollutants.

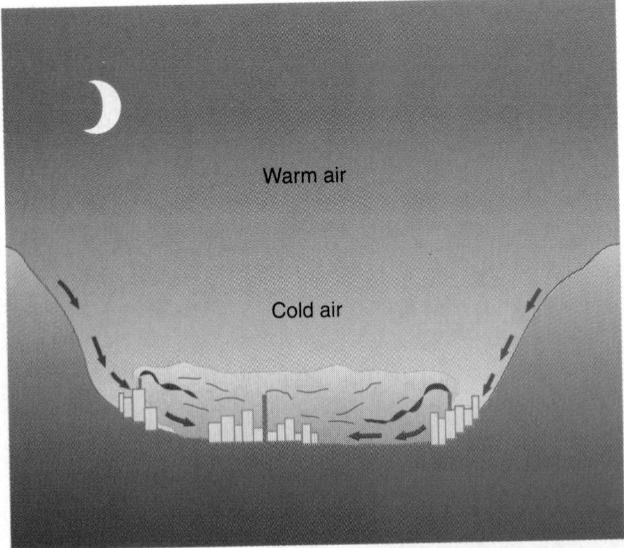

FIGURE 12.15 At night, cold air and pollutants drain downhill and settle in low-lying valleys.

> **DID YOU KNOW?**
>
> Cubatao, Brazil, just may be the most polluted city in the world. Located south of São Paulo, this heavily industrialized area of 100,000 people lies in a coastal valley—known by local residents as "the valley of death." Temperature inversions and stagnant air combine to trap the many pollutants that spew daily into the environment. Recently, nearly one-third of the downtown residents suffered from respiratory disease, and more babies are born deformed there than anywhere else in South America.

Severe Air Pollution Potential The greatest potential for an episode of severe air pollution occurs when all of the factors mentioned in the previous sections come together simultaneously. Ingredients for a major buildup of atmospheric pollution are:

- many sources of air pollution (preferably clustered close together)
- a deep high-pressure area that becomes stationary over a region
- light surface winds that are unable to disperse the pollutants
- a strong subsidence inversion produced by the sinking of air aloft
- a shallow mixing layer with poor ventilation
- a valley where the pollutants can accumulate
- clear skies so that radiational cooling at night will produce a surface inversion, which can cause an even greater buildup of pollutants near the ground
- and, for photochemical smog, adequate sunlight to produce secondary pollutants, such as ozone

Light winds and poor vertical mixing can produce a condition known as **atmospheric stagnation.** When this condition prevails for several days to a week or more, the buildup of pollutants can lead to some of the worst air pollution disasters on record, such as the one in the valley city of Donora, Pennsylvania, where in 1948 seventeen people died within fourteen hours. (Additional information on the Donora disaster is found in the Focus section on the next page.)

FOCUS ON AN OBSERVATION

Five Days in Donora—An Air Pollution Episode

On Tuesday morning, October 26, 1948, a cold surface anticyclone moved over the eastern half of the United States. There was nothing unusual about this high-pressure area; with a central pressure of only 1025 mb (30.27 in.), it was not exceptionally strong (see Fig. 4). Aloft, however, a large blocking-type ridge formed over the region, and the jet stream, which moves the surface pressure features along, was far to the west. Consequently, the surface anticyclone became entrenched over Pennsylvania and remained nearly stationary for five days.

The widely spaced isobars around the high-pressure system produced a weak pressure gradient and generally light winds throughout the area. These light winds, coupled with the gradual sinking of air from aloft, set the stage for a disastrous air pollution episode.

On Tuesday morning, radiation fog gradually settled over the moist ground in Donora, a small town nestled in the Monongahela Valley of western Pennsylvania. Because Donora rests on bottom land, surrounded by rolling hills, its residents were accustomed to fog, but not to what was to follow.

The strong radiational cooling that formed the fog, along with the sinking air of the anticyclone, combined to produce a strong temperature inversion. Light, downslope winds spread cool air and contaminants over Donora from the community's steel mill, zinc smelter, and sulfuric acid plant.

The fog with its burden of pollutants lingered into Wednesday. Cool drainage winds during the night strengthened the inversion and added more effluents to the already filthy air. The dense fog layer blocked sunlight from reaching the ground. With essentially no surface heating, the mixing depth lowered and the pollution became more concentrated. Unable to mix and disperse both horizontally and vertically, the dirty air became confined to a shallow, stagnant layer.

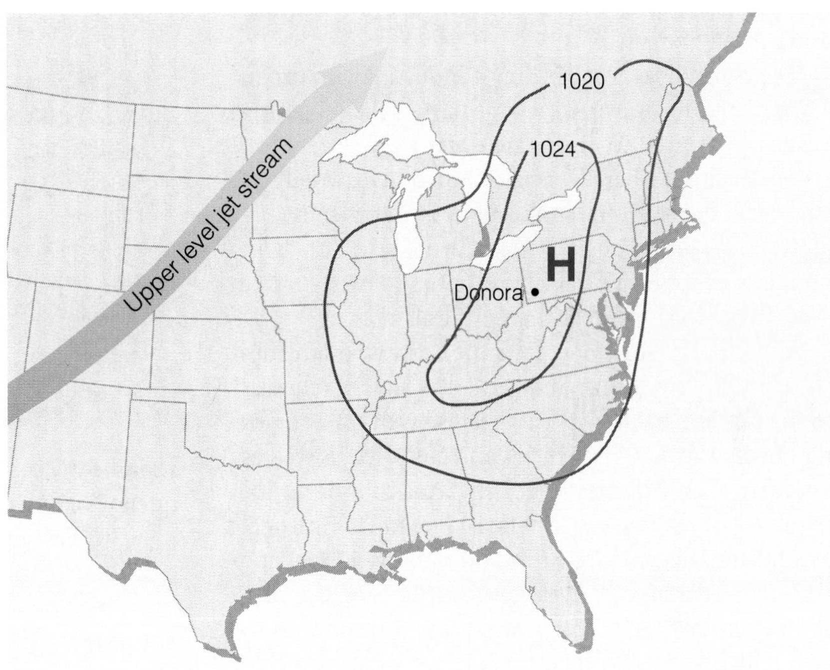

FIGURE 4 Surface weather map that shows a stagnant anticyclone over the eastern United States on October 26, 1948. The heavy arrow represents the position of the jet stream.

Meanwhile, the factories continued to belch impurities into the air (primarily sulfur dioxide and particulate matter) from stacks no higher than 40 m (130 ft) tall. The fog gradually thickened into a moist clot of smoke and water droplets. By Thursday, the visibility had decreased to the point where one could barely see across the street. At the same time, the air had a penetrating, almost sickening, smell of sulfur dioxide. At this point, a large percentage of the population became ill.

The episode reached a climax on Saturday, as 17 deaths were reported. As the death rate mounted, alarm swept through the town. An emergency meeting was called between city officials and factory representatives to see what could be done to cut down on the emission of pollutants.

The light winds and unbreathable air persisted until, on Sunday, an approaching storm generated enough wind to vertically mix the air and disperse the pollutants. A welcome rain then cleaned the air further. All told, the episode had claimed the lives of 22 people. During the five-day period, about half of the area's 14,000 inhabitants experienced some ill effects from the pollution. Most of those affected were older people with a history of cardiac or respiratory disorders.

AIR POLLUTION AND THE URBAN ENVIRONMENT

For more than 100 years, it has been known that cities are generally warmer than surrounding rural areas. This region of city warmth, known as the **urban heat island,** can influence the concentration of air pollution. However, before we look at its influence, let's see how the heat island actually forms.

The urban heat island is due to industrial and urban development. In rural areas, a large part of the incoming solar energy is used to evaporate water from vegetation and soil. In cities, where less vegetation and exposed soil exists, the majority of the sun's energy is absorbed by urban structures and asphalt. Hence, during warm daylight hours, less evaporative cooling in cities allows surface temperatures to rise higher than in rural areas.*

At night, the solar energy (stored as vast quantities of heat in city buildings and roads) is slowly released into the city air. Additional city heat is given off at night (and during the day) by vehicles and factories, as well as by industrial and domestic heating and cooling units. The release of heat energy is retarded by the tall vertical city walls that do not allow infrared radiation to escape as readily as do the relatively level surfaces of the surrounding countryside. The slow release of heat tends to keep nighttime city temperatures higher than those of the faster cooling rural areas. Overall, the heat island is strongest (1) at night when compensating sunlight is absent, (2) during the winter when nights are longer and there is more heat generated in the city, and (3) when the region is dominated by a high-pressure area with light winds, clear skies, and less humid air. Over time, increasing urban heat islands affect climatological temperature records, producing artificial warming in climatic records taken in cities. As we will see in Chapter 14, this warming must be accounted for in interpreting climate change over the past century.

The constant outpouring of pollutants into the environment may influence the climate of a city. Certain particles reflect solar radiation, thereby reducing the sunlight that reaches the surface. Some particles serve as nuclei upon which water and ice form. Water vapor condenses onto these particles when the relative humidity is as low as 70 percent, forming haze that greatly reduces visibility. Moreover, the added nuclei increase the frequency of city fog.*

*The cause of the urban heat island is quite involved. Depending on the location, time of year, and time of day, any or all of the following differences between cities and their surroundings can be important: albedo (reflectivity of the surface), surface roughness, emissions of heat, emissions of moisture, and emissions of particles that affect net radiation and the growth of cloud droplets.

TABLE 12.3
Contrast of the Urban and Rural Environment (Average Conditions)*

CONSTITUENTS	URBAN AREA (CONTRASTED TO RURAL AREA)
Mean pollution level	higher
Mean sunshine reaching the surface	lower
Mean temperature	higher
Mean relative humidity	lower
Mean visibility	lower
Mean wind speed	lower
Mean precipitation	higher
Mean amount of cloudiness	higher
Mean thunderstorm (frequency)	higher

*Values are omitted because they vary greatly depending upon city, size, type of industry, and season of the year.

Studies suggest that precipitation may be greater in cities than in the surrounding countryside. This phenomenon may be due in part to the increased roughness of city terrain, brought on by large structures that cause surface air to slow and gradually converge. This piling-up of air over the city then slowly rises, much like toothpaste does when its tube is squeezed. At the same time, city heat warms the surface air, making it more unstable, which enhances rising air motions, which, in turn, aids in forming clouds and thunderstorms. This process helps explain why both tend to be more frequent over cities. Table 12.3 summarizes the environmental influence of cities by contrasting the urban environment with the rural.

On clear still nights when the heat island is pronounced, a small thermal low-pressure area forms over the city. Sometimes a light breeze—called a **country breeze**—blows from the countryside into the city. If there are major industrial areas along the city's outskirts, pollutants are carried into the heart of town, where they became even more concentrated. Such an event is especially true if an inversion inhibits vertical mixing and dispersion (see Fig. 12.16).

Pollutants from urban areas may even affect the weather downwind from them. In a controversial study conducted at La Porte, Indiana—a city located about 30

*The impact that tiny liquid and solid particles (aerosols) may have on a larger scale is complex and depends upon a number of factors, which are addressed in Chapter 14.

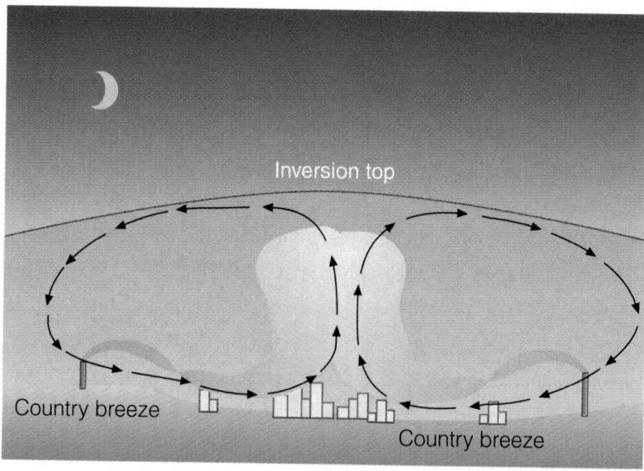

FIGURE 12.16 On a clear, relatively calm night, a weak country breeze carries pollutants from the outskirts into the city, where they concentrate and rise due to the warmth of the city's urban heat island. This effect may produce a pollution (or dust) dome from the suburbs to the center of town.

miles downwind of the industries of south Chicago—scientists suggested that La Porte had experienced a notable increase in annual precipitation since 1925. Because this rise closely followed the increase in steel production, it was suggested that the phenomenon was due to the additional emission of particles or moisture (or both) by industries to the west of La Porte.

A study conducted in St. Louis, Missouri (the *Metro*politan *Me*teorological *Ex*periment, or *METROMEX*), indicated that the average annual precipitation downwind from this city increased by about 10 percent. These increases closely followed industrial development upwind. This study also demonstrated that precipitation amounts were significantly greater on weekdays (when pollution emissions were higher) than on weekends (when pollution emissions were lower). Corroborative findings have been reported for Paris, France, and for other cities as well. However, in areas with marginal humidity to support the formation of clouds and precipitation, studies suggest that the rate of precipitation may actually decrease as excess pollutant particles (nuclei) compete for the available moisture, similar to the effect of overseeding a cloud, discussed in Chapter 5. Moreover, recent studies using satellite data indicate that fine airborne particles, concentrated over an area, can greatly reduce precipitation.

ACID DEPOSITION

Air pollution emitted from industrial areas, especially products of combustion, such as oxides of sulfur and nitrogen, can be carried many kilometers downwind. Either these particles and gases slowly settle to the ground in dry form *(dry deposition)* or they are removed from the air during the formation of cloud particles and then carried to the ground in rain and snow *(wet deposition)*. **Acid rain** and *acid precipitation* are common terms used to describe wet deposition, while **acid deposition** encompasses both dry and wet acidic substances. How, then, do these substances become acidic?

Emissions of sulfur dioxide (SO_2) and oxides of nitrogen may settle on the local landscape, where they transform into acids as they interact with water, especially during the formation of dew or frost. The remaining airborne particles may transform into tiny dilute drops of sulfuric acid (H_2SO_4) and nitric acid (HNO_3) during a complex series of chemical reactions involving sunlight, water vapor, and other gases. These acid particles may then fall slowly to earth, or they may adhere to cloud droplets or to fog droplets, producing **acid fog.** They may even act as nuclei on which the cloud droplets begin to grow. When precipitation occurs in the cloud, it carries the acids to the ground. Because of this, precipitation is becoming increasingly acidic in many parts of the world, especially downwind of major industrial areas.

Airborne studies conducted during the middle 1980s revealed that high concentrations of pollutants that produce acid rain can be carried great distances from their sources. For example, in one study scientists discovered high concentrations of pollutants hundreds of miles off the east coast of North America. It is suspected that they came from industrial East Coast cities. Although most pollutants are washed from the atmosphere during storms, some may be swept over the Atlantic, reaching places like Bermuda and Ireland. Acid rain knows no national boundaries.

Although studies suggest that acid precipitation may be nearly worldwide in distribution, regions noticeably affected are eastern North America, central Europe, and Scandinavia. Sweden contends that most of the sulfur emissions responsible for its acid precipitation are coming from factories in England. In some places, acid precipitation occurs naturally, such as in northern Canada, where natural fires in exposed coal beds produce tremendous quantities of sulfur dioxide. By the same token, acid fog can form by natural means.

Precipitation is naturally somewhat acidic. The carbon dioxide occurring naturally in the air dissolves in precipitation, making it slightly acidic with a pH between 5.0 and 5.6. Consequently, precipitation is considered acidic when its pH is below about 5.0 (see Fig. 12.17). In the northeastern United States, where

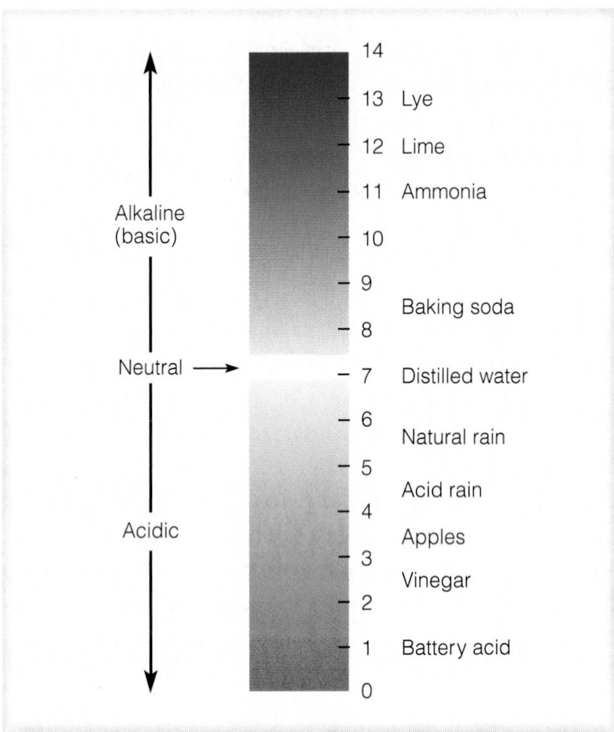

FIGURE 12.17 The pH scale ranges from 0 to 14, with a value of 7 considered neutral. Values greater than 7 are alkaline and below 7 are acidic. The scale is logarithmic, which means that rain with pH 3 is 10 times more acidic than rain with pH 4 and 100 times more acidic than rain with pH 5.

emissions of sulfur dioxide are primarily responsible for the acid precipitation, typical pH values range between 4.0 and 4.5 (see Fig. 12.18). But acid precipitation is not confined to the Northeast; the acidity of precipitation has increased rapidly during the past 20 years in the southeastern states, too. Further west, rainfall acidity also appears to be on the increase. Along the West Coast, the main cause of acid deposition appears to be the oxides of nitrogen released in automobile exhaust. In Los Angeles, acid fog is a more serious problem than acid rain, especially along the coast, where fog is most prevalent. The fog's pH is usually between 4.4 and 4.8, although pH values of 3.0 and below have been measured.

High concentrations of acid deposition can damage plants and water resources (freshwater ecosystems seem to be particularly sensitive to changes in acidity). Concern centers chiefly on areas where interactions with alkaline soil are unable to neutralize the acidic inputs. Studies indicate that thousands of lakes in the United States and Canada are so acidified that entire fish populations may have been adversely affected. In an attempt to reduce acidity, lime (calcium carbonate, $CaCO_3$) is being poured into some lakes. Natural alkaline soil particles can be swept into the air where they neutralize the acid.

About a third of the trees in Germany show signs of a blight that is due, in part, to acid deposition. Appar-

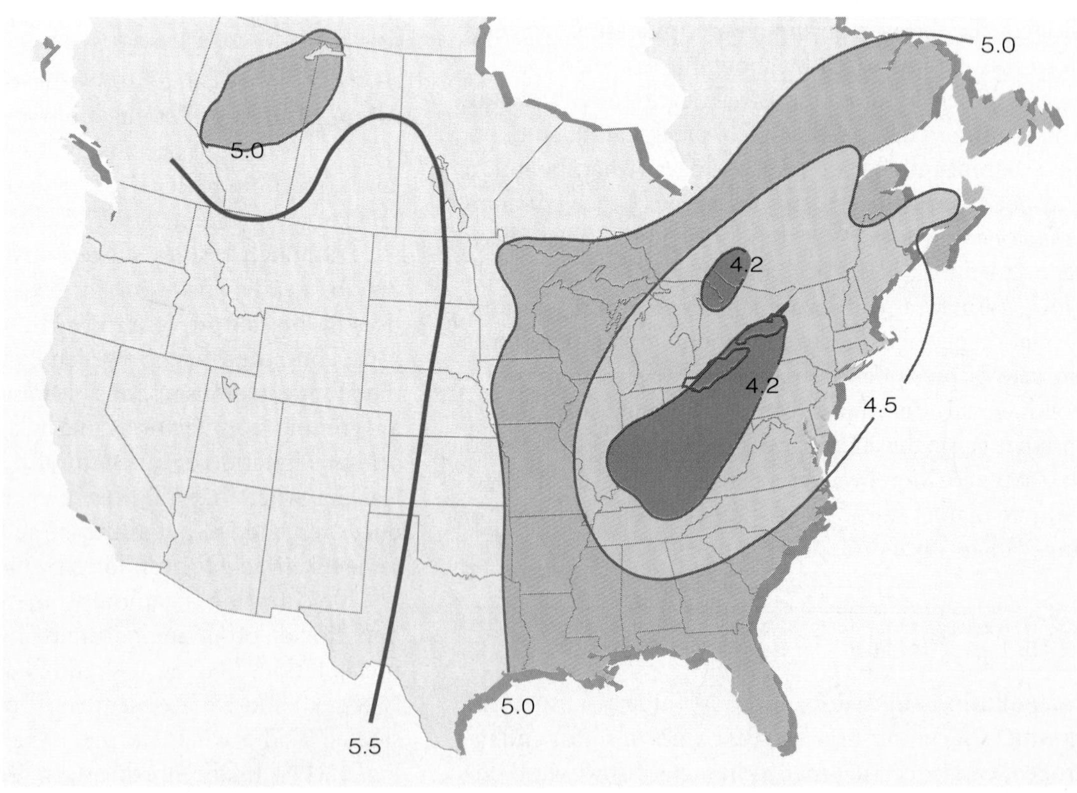

FIGURE 12.18 Annual average value of pH in precipitation weighted by the amount of precipitation in the United States and Canada for 1980.

ently, acidic particles raining down on the forest floor for decades have caused a chemical imbalance in the soil that, in turn, causes serious deficiencies in certain elements necessary for the trees' growth. The trees are thus weakened and become susceptible to insects and drought. The same type of processes may be affecting North American forests, but at a much slower pace, as many forests at higher elevations from southeastern Canada to South Carolina appear to be in serious decline (see Fig. 12.19). Moreover, acid precipitation is a problem in the mountainous West where high mountain lakes and forests seem to be most affected.

Also, acid deposition is eroding the foundations of structures in many cities throughout the world. In Rome, the acidity of rainfall is beginning to disfigure priceless outdoor fountain sculptures and statues. The estimated annual cost of this damage to building surfaces, monuments, and other structures is more than $2 billion.

Control of acid deposition is a difficult political problem because those affected by acid rain can be quite distant from those who cause it. Technology can control sulfur emissions (for example, stack scrubbers and fluidized bed combustion) and nitrogen emissions (catalytic converters on cars), but some people argue the cost is too high. If the United States turns more to coal-fired power plants, which are among the leading sources of sulfur oxide emissions, many scientists believe that the acid deposition problem will become more acute.

© Frederic Georgia/Photo Researchers

FIGURE 12.19 The effects of acid fog in the Great Smoky Mountains of Tennessee.

DID YOU KNOW?

Estimates are that acid rain has severely affected aquatic life in about 10 percent of the lakes and streams in the eastern United States.

In an attempt to better understand acid deposition, the *N*ational *C*enter for *A*tmospheric *R*esearch (NCAR) and the Environmental Protection Agency have been working to develop computer models that better describe the many physical and chemical processes contributing to acid deposition. To deal with the acid deposition problem, the Clean Air Act of 1990 imposed a reduction in the United States' emissions of sulfur dioxide and nitrogen dioxide. Canada has imposed new pollution control standards and set a goal of reducing industrial air pollution by 50 percent.

SUMMARY

In this chapter, we found that air pollution has plagued humanity for centuries. Air pollution problems began when people tried to keep warm by burning wood and coal. These problems worsened during the industrial revolution as coal became the primary fuel for both homes and industry. Even though many American cities do not meet all of the air quality standards set by the federal Clean Air Act of 1990, the air over our large cities is cleaner today than it was 50 years ago due to stricter emission standards and cleaner fuels.

We examined the types and sources of air pollution and found that primary air pollutants enter the atmosphere directly, whereas secondary pollutants form by chemical reactions that involve other pollutants. The secondary pollutant ozone is the main ingredient of photochemical smog—a smog that irritates the eyes and forms in the presence of sunlight. In polluted air, ozone forms during a series of chemical reactions involving nitrogen oxides and hydrocarbons (VOCs). In the stratosphere, ozone is a naturally occurring gas that protects us from the sun's harmful ultraviolet rays. We learned that human-induced gases, such as chlorofluorocarbons, work their way into the stratosphere where they release chlorine that rapidly destroys ozone, especially in polar regions.

We looked at the air quality index and found that a number of areas across the United States still have days considered unhealthy by the standards set by the United States Environmental Protection Agency. We also looked at the main factors affecting air pollution and found that most air pollution episodes occur when the winds are light, skies are clear, the mixing layer is shallow, the atmosphere is stable, and a strong inversion exists. These conditions usually prevail when a high-pressure area stalls over a region.

We observed that, on the average, urban environments tend to be warmer and more polluted than the rural areas that surround them. We saw that pollution from industrial areas can modify environments downwind of them, as oxides of sulfur and nitrogen are swept into the air, where they may transform into acids that fall to the surface. Acid deposition, a serious problem in many regions of the world, knows no national boundaries—the pollution of one country becomes the acid rain of another.

Meteorology Now™ Assess your understanding of this chapter's topics with additional quizzing and tutorials at http://earthscience.brookscole.com/ahrens/ess4e.

KEY TERMS

The following terms are listed in the order they appear in the text. Define each. Doing so will aid you in reviewing the material covered in this chapter.

air pollutants
primary air pollutants
secondary air pollutants
particulate matter
carbon monoxide (CO)
sulfur dioxide (SO_2)
volatile organic compounds (VOCs)
hydrocarbons
nitrogen dioxide (NO_2)
nitric oxide (NO)
smog
photochemical smog
ozone (O_3)
ozone hole
air quality index (AQI)
radiation (surface) inversion
subsidence inversion
mixing layer
mixing depth
atmospheric stagnation
urban heat island
country breeze
acid rain
acid deposition
acid fog

QUESTIONS FOR REVIEW

1. What are some of the main sources of air pollution?
2. How do primary air pollutants differ from secondary air pollutants?
3. List a few of the substances that fall under the category of particulate matter.
4. How does PM-10 particulate matter differ from that called PM-2.5? Which poses the greatest risk to human health?
5. How is particulate matter removed from the atmosphere?
6. Describe the primary sources and some of the health problems associated with each of the following pollutants:
   (a) carbon monoxide (CO)
   (b) sulfur dioxide (SO_2)
   (c) volatile organic compounds (VOCs)
   (d) nitrogen oxides
7. How does London-type smog differ from Los Angeles-type smog?
8. What is photochemical smog? How does it form? What is the main component of photochemical smog?
9. Why is photochemical smog more prevalent during the summer and early fall than during the middle of winter?
10. Why is stratospheric ozone beneficial to life on earth, whereas tropospheric (ground-level) ozone is not?
11. If all the ozone in the stratosphere were destroyed, what possible effects might this have on the earth's inhabitants?
12. According to Fig. 12.9, there is a dramatic drop in the concentration of several pollutants after 1970. What is the reason for this decrease?
13. (a) On the AQI scale, when is a pollutant considered unhealthful?
    (b) On the AQI scale, how would air be described if it had an AQI value of 250 for ozone?
    (c) What would be the general health effects with an AQI value of 250 for ozone? What precautions should a person take with this value?
14. Why is a light wind, rather than a strong wind, more conducive to high concentrations of air pollution?
15. How does atmospheric stability influence the accumulation of air pollutants?
16. Why is it that polluted air and inversions seem to go hand in hand?
17. Major air pollution episodes are mainly associated with radiation inversions or subsidence inversions. Why?
18. Give several reasons why taller smokestacks are better than shorter ones at improving the air quality in their immediate area.
19. How does the mixing depth normally change during the course of a day? As the mixing depth changes, how does it affect the concentration of pollution near the surface?

20. For least-polluting conditions, what would be the best time of day for a farmer to burn agricultural debris? Explain your reasoning.
21. Explain why most severe episodes of air pollution are associated with slow moving or stagnant high-pressure areas.
22. How does topography influence the concentration of pollutants in cities such as Los Angeles and Denver? In mountainous terrain?
23. List the factors that can lead to a major buildup of atmospheric pollution.
24. What is an urban heat island? Is it more strongly developed at night or during the day? Explain.
25. What causes the "country breeze"? Why is it usually more developed at night than during the day? Would it be more easily developed in summer or winter? Explain.
26. How can pollution play a role in influencing the precipitation downwind of certain large industrial complexes?
27. What is acid deposition? Why is acid deposition considered a serious problem in many regions of the world? How does precipitation become acidic?

QUESTIONS FOR THOUGHT AND EXPLORATION

1. Would you expect a fumigation-type smoke plume on a warm, sunny afternoon? Explain.
2. Give a few reasons why, in industrial areas, nighttime pollution levels might be higher than daytime levels.
3. Explain this apparent paradox: High levels of tropospheric (ground-level) ozone are "bad" and we try to reduce them, whereas high levels of stratospheric ozone are "good" and we try to maintain them.
4. A large industrial smokestack located within an urban area emits vast quantities of sulfur dioxide and nitrogen dioxide. Following criticism from local residents that emissions from the stack are contributing to poor air quality in the area, the management raises the height of the stack from 10 m (33 ft) to 100 m (330 ft). Will this increase in stack height change any of the existing air quality problems? Will it create any new problems? Explain.
5. If the sulfuric acid and nitric acid in rainwater are capable of adversely affecting soil, trees, and fish, why doesn't this same acid adversely affect people when they walk in the rain?
6. Which do you feel is likely to be more acidic: acid rain or acid fog? Explain your reasoning.
7. Keep a log of the daily AQI readings in your area and note the pollutants listed in the index. Also, keep a record of the daily weather conditions, such as cloud cover, high temperature for the day, average wind direction and speed, etc. See if there is any relationship between these weather conditions and high AQI readings for certain pollutants.
8. Air Pollution Maps (http://www.epa.gov/air/data/reports.htm): Using the maps of nonattainment areas, (areas where air pollution levels persistently exceed national air quality standards), determine the major pollution problem(s) affecting your area.
9. Air Trajectory Model (http://www.arl.noaa.gov/ready/hysplit4.html): Use an online, interactive air trajectory model to predict the movement of air 48 hours into the future, starting at a location of your choice. Describe the predicted movement. What weather patterns are guiding this movement? How can this model be used to forecast air pollution episodes?

Go to the Brooks/Cole Earth Sciences Resource Center (http://earthscience.brookscole.com) for critical thinking exercises, articles, and additional readings from InfoTrac College Edition, Brooks/Cole's online student library.

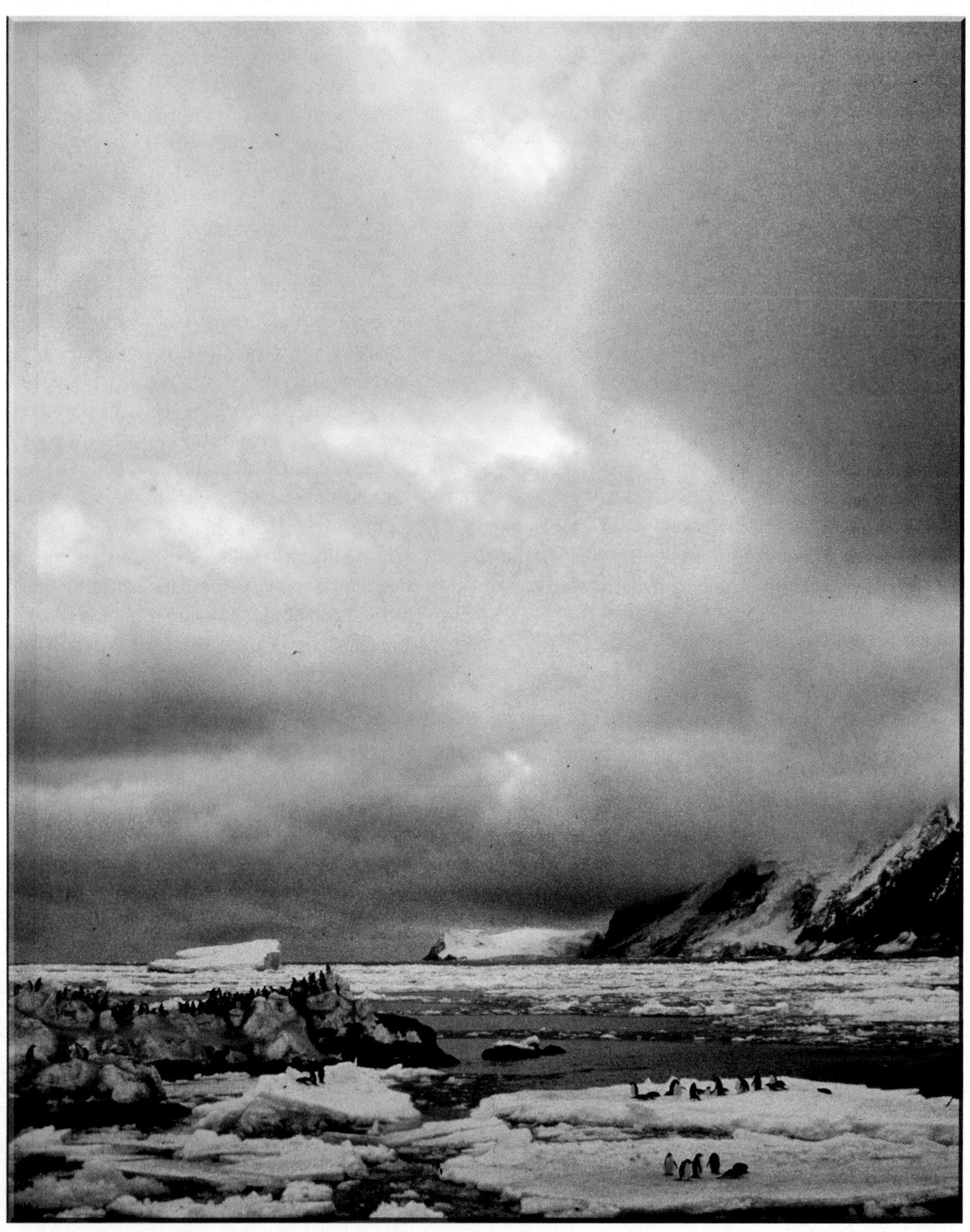

An Antarctic sunset. The climate of this region can be quite brutal, as air temperatures seldom rise much above freezing, and low temperatures may approach −80°C (−112°F).

Meteorology Now™ This icon, appearing throughout the book, indicates an opportunity to explore interactive tutorials, animations, or practice problems available on the MeteorologyNow Web site at http://earthscience.brookscole.com/ahrens/ess4e.

Global Climate

The climate is unbearable. . . . At noon today the highest temperature measured was −33°C. We really feel that it is late in the season. The days are growing shorter, the sun is low and gives no warmth, katabatic winds blow continuously from the south with gales and drifting snow. The inner walls of the tent are like glazed parchment with several millimeters thick ice-armour. . . . Every night several centimeters of frost accumulate on the walls, and each time you inadvertently touch the tent cloth a shower of ice crystals falls down on your face and melts. In the night huge patches of frost from my breath spread around the opening of my sleeping bag and melt in the morning. The shoulder part of the sleeping bag facing the tent-side is permeated with frost and ice, and crackles when I roll up the bag. . . . For several weeks now my fingers have been permanently tender with numb fingertips and blistering at the nails after repeated frostbites. All food is frozen to ice and it takes ages to thaw out everything before being able to eat. At the depot we could not cut the ham, but had to chop it in pieces with a spade. Then we threw ourselves hungrily at the chunks and chewed with the ice crackling between our teeth. You have to be careful with what you put in your mouth. The other day I put a piece of chocolate from an outer pocket directly in my mouth and promptly got frostbite with blistering of the palate.

Ove Wilson (Quoted in David M. Gates, *Man and His Environment*)

CONTENTS

Our opening comes from a report by Norwegian scientists on their encounter with one of nature's cruelest climates—that of Antarctica. Their experience illustrates the profound effect that climate can have on even ordinary events, such as eating a piece of chocolate. Though we may not always think about it, climate profoundly affects nearly everything in the middle latitudes, too. For instance, it influences our housing, clothing, the shape of landscapes, agriculture, how we feel and live, and even where we reside, as most people will choose to live on a sunny hillside rather than in a cold, dark, and foggy river basin. Entire civilizations have flourished in favorable climates and have moved away from, or perished in, unfavorable ones. We learned early in this text that *climate* is the average of the day-to-day weather over a long duration. But the concept of climate is much larger than this, for it encompasses, among other things, the daily and seasonal extremes of weather within specified areas.

When we speak of climate, then, we must be careful to specify the spatial location we are talking about. For example, the Chamber of Commerce of a rural town may boast that its community has mild winters with air temperatures seldom below freezing. This may be true several meters above the ground in an instrument shelter, but near the ground the temperature may drop below freezing on many winter nights. This small climatic region near or on the ground is referred to as a **microclimate.** Because a much greater extreme in daily air temperatures exists near the ground than several meters above, the microclimate for small plants is far more harsh than the thermometer in an instrument shelter would indicate.

When we examine the climate of a small area of the earth's surface, we are looking at the **mesoclimate.** The size of the area may range from a few acres to several square kilometers. Mesoclimate includes regions such as forests, valleys, beaches, and towns. The climate of a much larger area, such as a state or a country, is called **macroclimate.** The climate extending over the entire earth is often referred to as *global climate.*

In this chapter, we will concentrate on the larger scales of climate. We will begin with the factors that regulate global climate; then we will discuss how climates are classified. Finally, we will examine the different types of climate.

A WORLD WITH MANY CLIMATES

The world is rich in climatic types. From the teeming tropical jungles to the frigid polar "wastelands," there seems to be an almost endless variety of climatic regions. The factors that produce the climate in any given place—the **climatic controls**—are the same that produce our day-to-day weather. Briefly, the controls are the

1. intensity of sunshine and its variation with latitude
2. distribution of land and water
3. ocean currents
4. prevailing winds
5. positions of high- and low-pressure areas
6. mountain barriers
7. altitude

We can ascertain the effect these controls have on climate by observing the global patterns of two weather elements—temperature and precipitation.

Global Temperatures Figure 13.1 shows mean annual temperatures for the world. To eliminate the distorting effect of topography, the temperatures are corrected to sea level.* Notice that in both hemispheres the isotherms are oriented east-west, reflecting the fact that locations at the same latitude receive nearly the same amount of solar energy. In addition, the annual solar heat that each latitude receives decreases from low to high latitude; hence, annual temperatures tend to decrease from equatorial toward polar regions.**

The bending of the isotherms along the coastal margins is due in part to the unequal heating and cooling properties of land and water, and to ocean currents and upwelling. For example, along the west coast of North and South America, ocean currents transport cool water equatorward. In addition to this, the wind in both regions blows toward the equator, parallel to the coast. This situation favors upwelling of cold water (see Chapter 7, p. 191), which cools the coastal margins. In the area of the eastern North Atlantic Ocean (north of 40°N), the poleward bending of the isotherms is due to the Gulf Stream and the North Atlantic Drift, which carry warm water northward.

The fact that land masses heat up and cool off more quickly than do large bodies of water means that variations in temperature between summer and winter will be far greater over continental interiors than along the west coastal margins of continents. By the same token,

*This correction is made by adding to each station above sea level an amount of temperature that would correspond to the normal (standard) temperature lapse rate of 6.5°C per 1000 m (3.6°F per 1000 ft).

**Average global temperatures for January and July are given in Figs. 3.8 and 3.9, respectively, on p. 64.

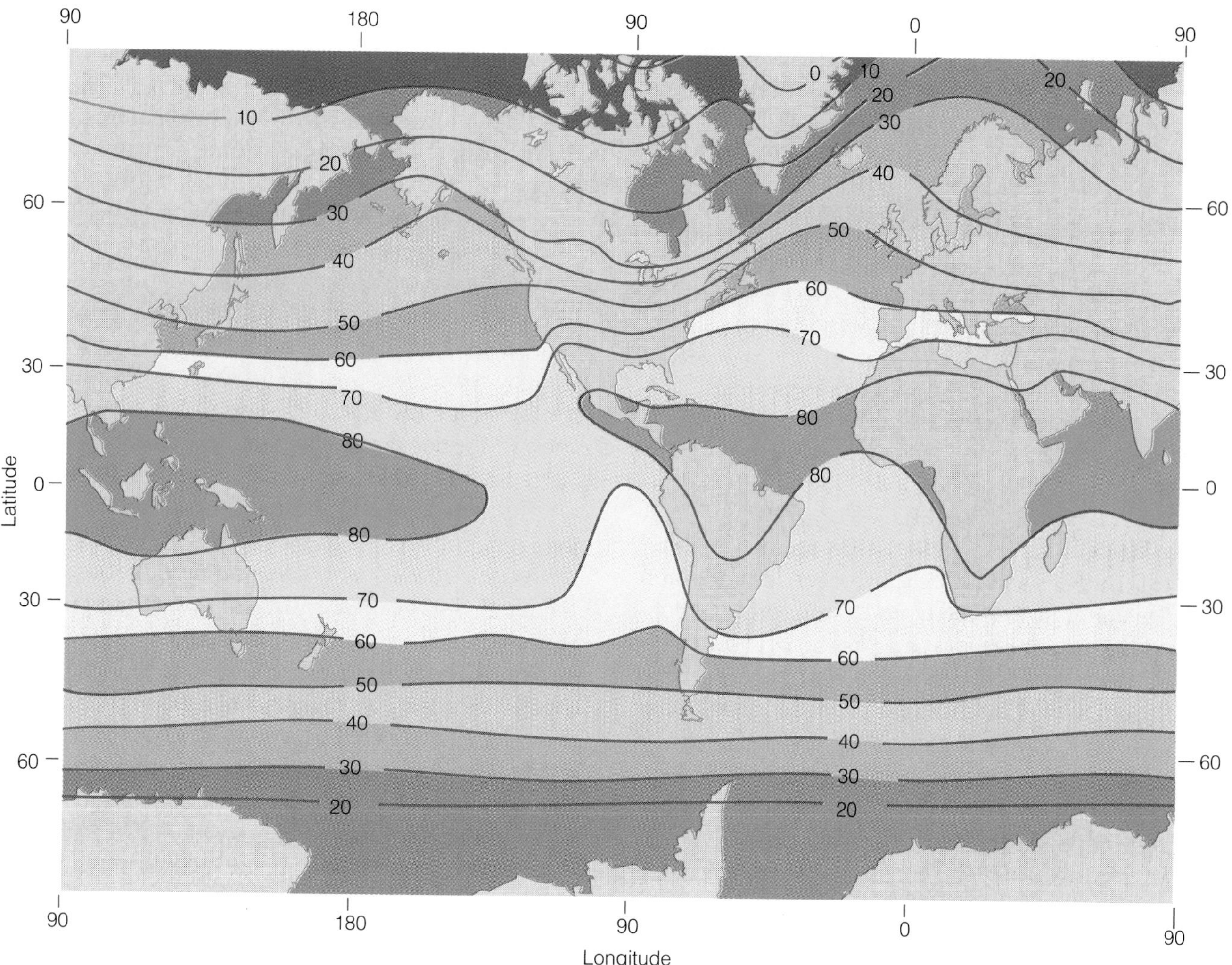

FIGURE 13.1 Average annual sea-level temperatures throughout the world (°F).

the climates of interior continental regions will be more extreme, as they have (on the average) higher summer temperatures and lower winter temperatures than their west-coast counterparts. In fact, west-coast climates are typically quite mild for their latitude.

The highest mean temperatures do not occur in the tropics, but rather in the subtropical deserts of the Northern Hemisphere. Here, the subsiding air associated with the subtropical anticyclones produces generally clear skies and low humidity. In summer, the high sun beating down upon a relatively barren landscape produces scorching heat.

The lowest mean temperatures occur over large land masses at high latitudes. The coldest area of the world is the Antarctic. During part of the year, the sun is below the horizon; when it is above the horizon, it is low in the sky and its rays do not effectively warm the surface. Consequently, the land remains snow- and ice-covered year-round. The snow and ice reflect perhaps 80 percent of the sunlight that reaches the surface. Much of the unreflected solar energy is used to transform the ice and snow into water vapor. The relatively dry air and the Antarctic's high elevation permit rapid radiational cooling during the dark winter months, producing extremely cold surface air. The extremely cold Antarctic helps to explain why, overall, the Southern Hemisphere is cooler than the Northern Hemisphere. Other contributing factors for a cooler Southern Hemisphere include the fact that polar regions of the Southern Hemisphere reflect more incoming sunlight, and the fact that less land area is found in tropical and subtropical areas of the Southern Hemisphere.

Global Precipitation Appendix H, pp. 440–441, shows the worldwide general pattern of annual precipitation,

DID YOU KNOW?

Even "summers" in the Antarctic can be brutal. In 1912, during the Antarctic summer, Robert Scott of Great Britain not only lost the race to the South Pole to Norway's Roald Amundsen, but perished in a blizzard trying to return. Temperature data taken by Scott and his crew showed that the winter of 1912 was unusually cold, with air temperatures remaining below −34°C (−30°F) for nearly a month. These exceptionally low temperatures eroded the men's health and created an increase in frictional drag on the sleds the men were pulling. Just before Scott's death, he wrote in his journal that "no one in the world would have expected the temperatures and surfaces which we encountered at this time of year."

which varies from place to place. There are, however, certain regions that stand out as being wet or dry. For example, equatorial regions are typically wet, while the subtropics and the polar regions are relatively dry. The global distribution of precipitation is closely tied to the general circulation of winds in the atmosphere (Chapter 7) and to the distribution of mountain ranges and high plateaus.

Figure 13.2 shows in simplified form how the general circulation influences the north-to-south distribution of precipitation to be expected on a uniformly water-covered earth. Precipitation is most abundant where the air rises, least abundant where it sinks. Hence, one expects a great deal of precipitation in the tropics and along the polar front, and little near subtropical highs and at the poles. Let's look at this in more detail.

In tropical regions, the trade winds converge along the Intertropical Convergence Zone (ITCZ), producing rising air, towering clouds, and heavy precipitation all year long. Poleward of the equator, near latitude 30°, the sinking air of the subtropical highs produces a "dry belt" around the globe. The Sahara Desert of North Africa is in this region. Here, annual rainfall is exceedingly light and varies considerably from year to year. Because the major wind belts and pressure systems shift with the season—northward in July and southward in January—the area between the rainy tropics and the dry subtropics is influenced by both the ITCZ and the subtropical highs.

In the cold air of the polar regions there is little moisture, so there is little precipitation. Winter storms drop light, powdery snow that remains on the ground for a long time because of the low evaporation rates. In summer, a ridge of high pressure tends to block storm systems that would otherwise travel into the area; hence, precipitation in polar regions is meager in all seasons.

There are exceptions to this idealized pattern. For example, in middle latitudes the migrating position of the subtropical anticyclones also has an effect on the west-to-east distribution of precipitation. The sinking air associated with these systems is more strongly developed on their eastern side. Hence, the air along the east-

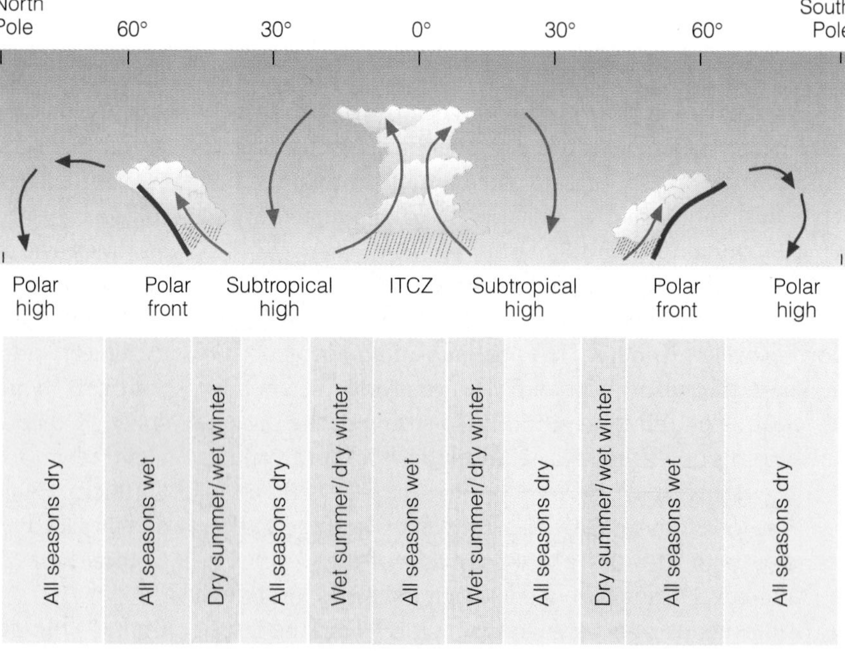

FIGURE 13.2 A vertical cross section along a line running north to south illustrates the main global regions of rising and sinking air and how each region influences precipitation.

ern side of an anticyclone tends to be more stable; it is also drier, as cooler air moves equatorward because of the circulating winds around these systems. In addition, along coastlines, cold upwelling water cools the surface air even more, adding to the air's stability. Consequently, in summer, when the Pacific high moves to a position centered off the California coast, a strong, stable subsidence inversion forms above coastal regions. With the strong inversion and the fact that the anticyclone tends to steer storms to the north, central and southern California areas experience little, if any, rainfall during the summer months.

On the western side of subtropical highs, the air is less stable and more moist, as warmer air moves poleward. In summer, over the North Atlantic, the Bermuda high pumps moist tropical air northward from the Gulf of Mexico into the eastern two-thirds of the United States. The humid air is conditionally unstable to begin with, and by the time it moves over the heated ground, it becomes even more unstable. If conditions are right, the moist air will rise and condense into cumulus clouds, which may build into towering thunderstorms.

In winter, the subtropical North Pacific high moves south, allowing storms traveling across the ocean to penetrate the western states, bringing much needed rainfall to California after a long, dry summer. The Bermuda high also moves south in winter. Across much of the United States, intense winter storms develop and travel eastward, frequently dumping heavy precipitation as they go. Usually, however, the heaviest precipitation is concentrated in the eastern states, as moisture from the Gulf of Mexico moves northward ahead of these systems. Therefore, cities on the plains typically receive more rainfall in summer, and those on the west coast have maximum precipitation in winter, whereas cities in the Midwest and East usually have abundant precipitation all year long. The contrast in seasonal precipitation among a West Coast city (San Francisco), a central plains city (Kansas City), and an eastern city (Baltimore) is clearly shown in Fig. 13.3.

Mountain ranges disrupt the idealized pattern of global precipitation (1) by promoting convection (because their slopes are warmer than the surrounding air) and (2) by forcing air to rise along their windward slopes *(orographic uplift)*. Consequently, the windward

DID YOU KNOW?

Wynoochee Oxbow, Washington, on the Olympic Peninsula, is considered the wettest weather station in the continental United States, with an average rainfall of 366 cm (144 in.)—a total 86 times greater than the average 4.3 cm (1.7 in.) for Death Valley, California.

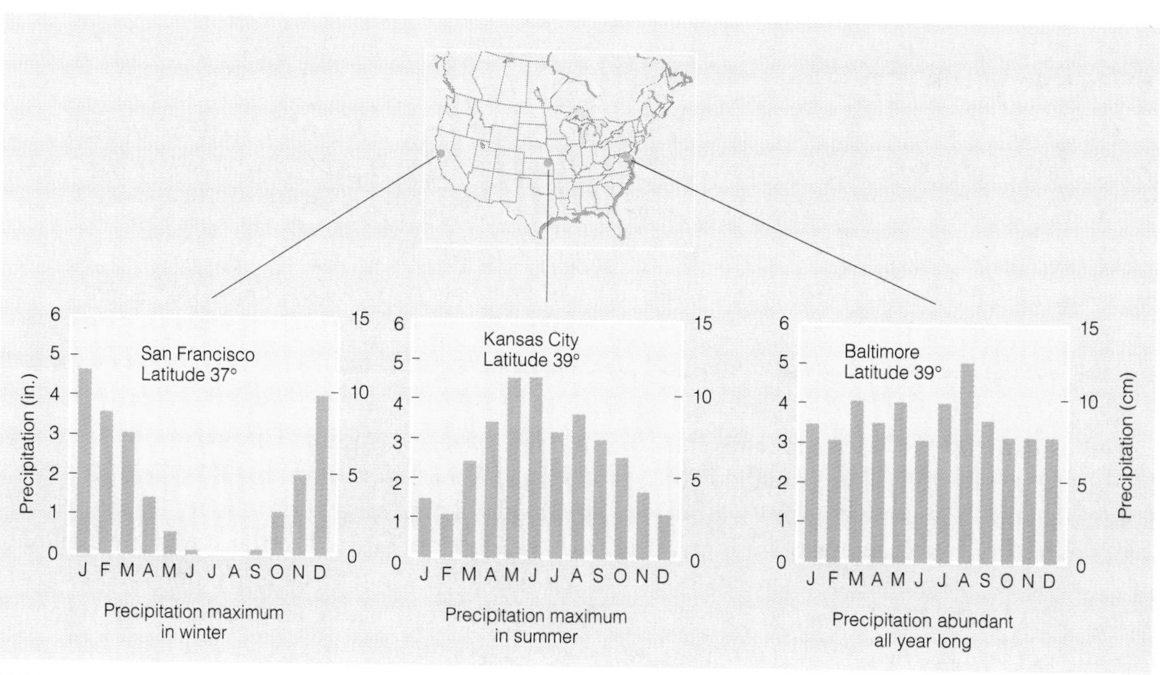

FIGURE 13.3 Variation in annual precipitation for three Northern Hemisphere cities.

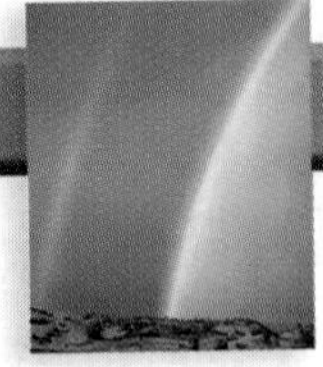

FOCUS ON A SPECIAL TOPIC

Precipitation Extremes

Most of the "rainiest" places in the world are located on the windward side of mountains. For example, Mount Waialeale on the island of Kauai, Hawaii, has the greatest annual average rainfall on record: 1168 cm (460 in.). Cherrapunji, on the crest of the southern slopes of the Khasi Hills in northeastern India, receives an average of 1080 cm (425 in.) of rainfall each year, the majority of which falls during the summer monsoon, between April and October. Cherrapunji, which holds the greatest twelve-month rainfall total of 2647 cm (1042 in.), once received 380 cm (150 in.) of rain in just five days.

Record rainfall amounts are often associated with tropical storms. On the island of La Réunion (about 650 km east of Madagascar in the Indian Ocean), a tropical cyclone dumped 135 cm (53 in.) of rain on Belouve in twelve hours. Heavy rains of short duration often occur with severe thunderstorms that move slowly or stall over a region. On July 4, 1956, 3 cm (1.2 in.) of rain fell from a thunderstorm on Unionville, Maryland, in one minute.

Snowfalls tend to be heavier where cool, moist air rises along the windward slopes of mountains. One of the snowiest places in North America is located at the Paradise Ranger Station in Mt. Rainier National Park, Washington. Situated at an elevation of 1646 m (5400 ft) above sea level, this station receives an average 1575 cm (620 in.) of snow annually. However, a record annual snowfall amount of 2896 cm (1140 in.) was recorded at Mt. Baker ski area during the winter of 1998–1999.

As we noted earlier, the driest regions of the world lie in the frigid polar region, the leeward side of mountains, and in the belt of subtropical high pressure, between 15° and 30° latitude. Arica in northern Chile holds the world record for lowest annual rainfall, 0.08 cm (0.03 in.). In the United States, Death Valley, California, averages only 4.5 cm (1.78 in.) of precipitation annually. Figure 1 gives additional information on world precipitation records.

KEY TO MAP

1	World's greatest annual average rainfall	1168 cm (460 in.)	Mt. Waialeale, Hawaii
2	Greatest 1-month rainfall total	930 cm (366 in.)	Cherrapunji, India, July, 1861
3	Greatest 12-hour rainfall total	135 cm (53 in.)	Belouve, La Réunion Island, February 28, 1964
4	Greatest 24-hour rainfall total in United States	109 cm (43 in.)	Alvin, Texas, July 25, 1979
5	Greatest 42-minute rainfall total	30 cm (12 in.)	Holt, Missouri, June 22, 1947
6	Greatest 1-minute rainfall total in United States	3 cm (1.2 in.)	Unionville, MD, July 4, 1956
7	Lowest annual average rainfall in Northern Hemisphere	3 cm (1.2 in.)	Bataques, Mexico
8	Lowest annual average rainfall in the world	0.08 cm (0.03 in.)	Arica, Chile
9	Greatest annual snowfall in United States	2896 cm (1140 in.)	Mt. Baker ski area, WA,1998
10	Greatest snowfall in 1 month	991 cm (390 in.)	Tamarack, CA, January, 1911
11	Greatest snowfall in 24 hours	193 cm (76 in.)	Silverlake, Boulder, CO April 14–15, 1921
12	Longest period without measurable precipitation in U.S. (993 days)	0.0 cm (0.0 in.)	Bagdad, CA August 1909 to May 1912

side of mountains tends to be "wet." As air descends and warms along the leeward side, there is less likelihood of clouds and precipitation. Thus, the leeward side of mountains tends to be "dry." As Chapter 5 points out, a region on the leeward side of a mountain where precipitation is noticeably less is called a *rain shadow.*

A good example of the rain shadow effect occurs in the northwestern part of Washington State. Situated on the western side at the base of the Olympic Mountains, the Hoh River Valley annually receives an average 380 cm (150 in.) of precipitation. On the eastern (leeward) side of this range, only about 100 km (62 mi) from the Hoh rain forest, the mean annual precipitation is less than 43 cm (17 in.), and irrigation is necessary to grow certain crops. Figure 13.4 shows a classic example of how topography produces several rain shadow effects. (Additional information on precipitation extremes is given in the Focus section above.)

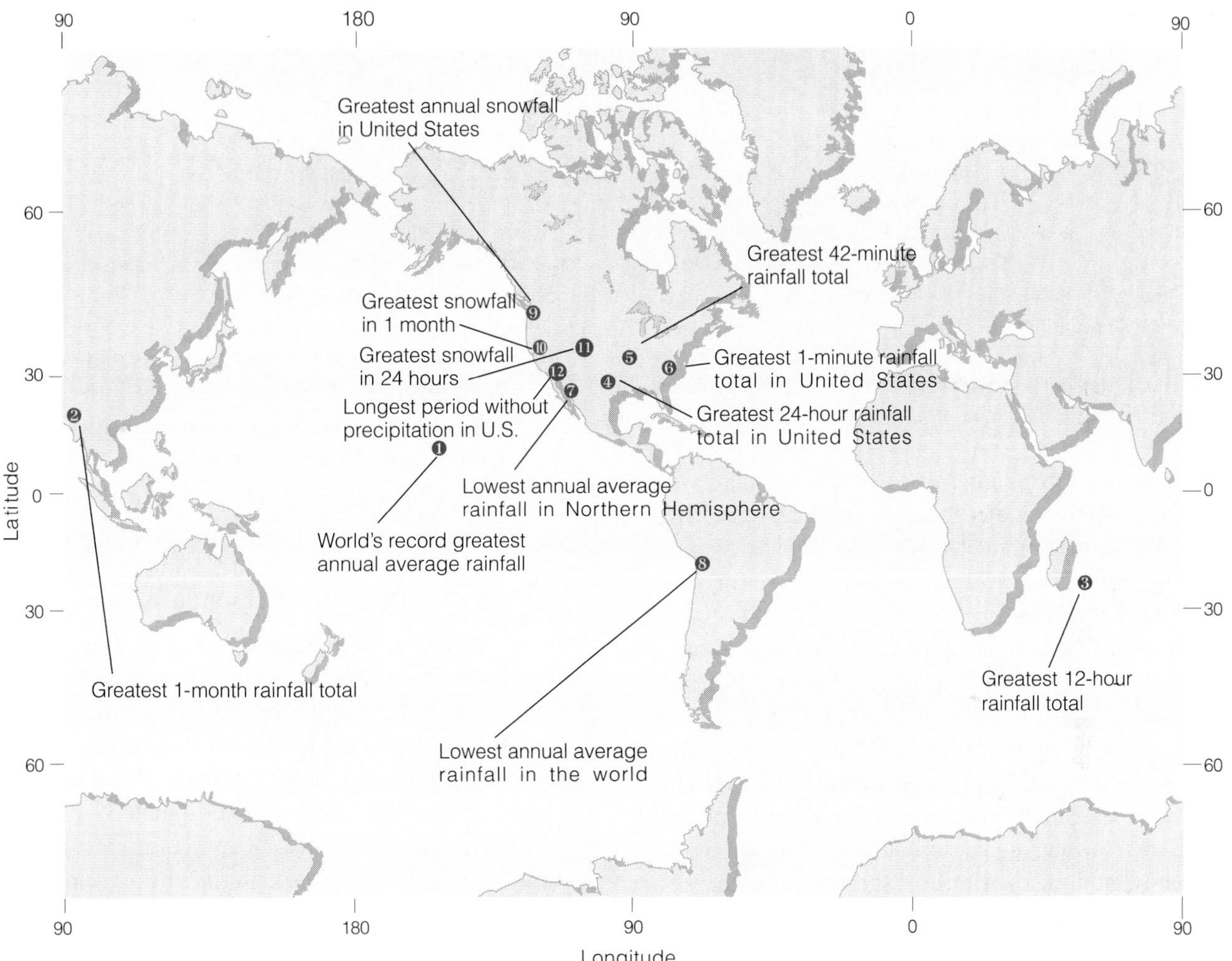

FIGURE 1 Some precipitation records throughout the world.

BRIEF REVIEW

Before going on to the section on climate classification, here is a brief review of some of the facts covered so far:

- The climate controls are the factors that govern the climate of any given region.
- The hottest places on earth tend to occur in the subtropical deserts of the Northern Hemisphere, where clear skies and sinking air, coupled with low humidity and a high summer sun beating down upon a relatively barren landscape, produce extreme heat.
- The coldest places on earth tend to occur in the interior of high-latitude land masses. The coldest areas of the Northern Hemisphere are found in the interior of Siberia and Greenland, whereas the coldest area of the world is the Antarctic.
- The wettest places in the world tend to be located on the windward side of mountains where warm, humid air rises upslope. On the downwind (leeward) side of a mountain there often exists a "dry" region, known as a *rain shadow.*

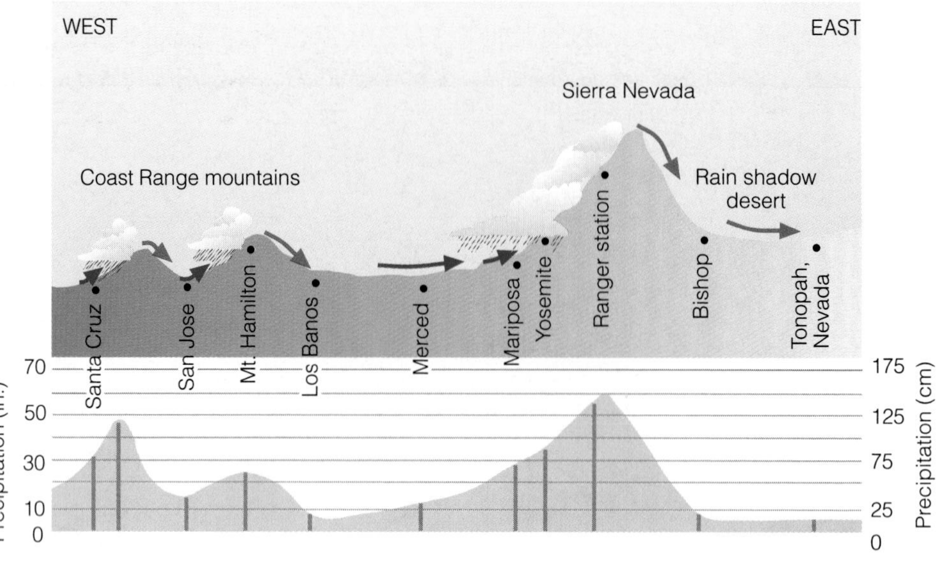

FIGURE 13.4 The effect of topography on average annual precipitation along a line running from the Pacific Ocean through central California into western Nevada.

CLIMATIC CLASSIFICATION—THE KÖPPEN SYSTEM

The climatic controls interact to produce such a wide array of different climates that no two places experience exactly the same climate. However, the similarity of climates within a given area allows us to divide the earth into climatic regions.

A widely used classification of world climates based on the annual and monthly averages of temperature and precipitation was devised by the famous German scientist Waldimir Köppen (1846–1940). Initially published in 1918, the original **Köppen classification system** has since been modified and refined. Faced with the lack of adequate observing stations throughout the world, Köppen related the distribution and type of native vegetation to the various climates. In this way, climatic boundaries could be approximated where no climatological data were available.

Köppen's scheme employs five major climatic types; each type is designated by a capital letter:

A *Tropical moist climates:* All months have an average temperature above 18°C (64°F). Since all months are warm, there is no real winter season.

B *Dry climates:* Deficient precipitation most of the year. Potential evaporation and transpiration exceed precipitation.

C *Moist mid-latitude climates with mild winters:* Warm-to-hot summers with mild winters. The average temperature of the coldest month is below 18°C (64°F) and above –3°C (27°F).

D *Moist mid-latitude climates with severe winters:* Warm summers and cold winters. The average temperature of the warmest month exceeds 10°C (50°F), and the coldest monthly average drops below –3°C (27°F).

E *Polar climates:* Extremely cold winters and summers. The average temperature of the warmest month is below 10°C (50°F). Since all months are cold, there is no real summer season.

Each group contains subregions that describe special regional characteristics, such as seasonal changes in temperature and precipitation. In mountainous country, where rapid changes in elevation bring about sharp changes in climatic type, delineating the climatic regions is impossible. These regions are designated by the letter H, for highland climates. (Köppen's climate classification system, including the criteria for the various subdivisions, is given in Appendix G on p. 439.)

Köppen's system has been criticized primarily because his boundaries (which relate vegetation to monthly temperature and precipitation values) do not correspond to the natural boundaries of each climatic zone. In addition, the Köppen system implies that there is a sharp boundary between climatic zones, when in reality there is a gradual transition.

The Köppen system has been revised several times, most notably by the German climatologist Rudolf Geiger, who worked with Köppen on amending the climatic boundaries of certain regions. A popular modification of the Köppen system was developed by the American climatologist Glenn T. Trewartha, who redefined some of the climatic types and altered the climatic world map by putting more emphasis on the lengths of growing seasons and average summer temperatures.

THE GLOBAL PATTERN OF CLIMATE

Figure 13.5 displays how the major climatic regions of the world are distributed, based mainly on the work of Köppen. We will first examine humid tropical climates in low latitudes and then we'll look at middle-latitude and polar climates. Bear in mind that each climatic region has many subregions of local climatic differences wrought by such factors as topography, elevation, and large bodies of water. Remember, too, that boundaries of climatic regions represent gradual transitions. Thus, the major climatic characteristics of a given region are best observed away from its periphery.

Tropical Moist Climates (Group A)

General characteristics: year-round warm temperatures (all months have a mean temperature above 18°C, or 64°F); abundant rainfall (typical annual average exceeds 150 cm, or 59 in.).

Extent: northward and southward from the equator to about latitude 15° to 25°.

Major types (based on seasonal distribution of rainfall): *tropical wet* (Af), *tropical monsoon* (Am), and *tropical wet and dry* (Aw).

At low elevations near the equator, in particular the Amazon lowland of South America, the Congo River Basin of Africa, and the East Indies from Sumatra to New Guinea, high temperatures and abundant yearly rainfall combine to produce a dense, broadleaf, evergreen forest called a **tropical rain forest.** Here, many different plant species, each adapted to differing light intensity, present a crudely layered appearance of diverse vegetation. In the forest, little sunlight is able to penetrate to the ground through the thick crown cover. As a result, little plant growth is found on the forest floor. However, at the edge of the forest, or where a clearing has been made, abundant sunlight allows for the growth of tangled shrubs and vines, producing an almost impenetrable *jungle* (see Fig. 13.6).

Within the **tropical wet climate*** (Af), seasonal temperature variations are small (normally less than 3°C) because the noon sun is always high and the number of daylight hours is relatively constant. However, there is a greater variation in temperature between day (average high about 32°C) and night (average low about 22°C) than there is between the warmest and coolest months. This is why people remark that winter comes to the tropics at night. The weather here is monotonous and sultry. There is little change in temperature from one day to the next. Furthermore, almost every day, towering cumulus clouds form and produce heavy, localized showers by early afternoon. As evening approaches, the showers usually end and skies clear. Typical annual rainfall totals are greater than 150 cm (59 in.) and, in some cases, especially along the windward side of hills and mountains, the total may exceed 400 cm (157 in.).

The high humidity and cloud cover tend to keep maximum temperatures from reaching extremely high values. In fact, summer afternoon temperatures are normally higher in middle latitudes than here. Nighttime radiational cooling can produce saturation and, hence, a blanket of dew and—occasionally—fog covers the ground.

An example of a station with a tropical wet climate (Af) is Iquitos, Peru (see Fig. 13.7). Located near the equator (latitude 4°S), in the low basin of the upper Amazon River, Iquitos has an average annual temperature of 25°C (77°F), with an annual temperature range of only 2.2°C (4°F). Notice also that the monthly rainfall totals vary more than do the monthly temperatures. This is due primarily to the migrating position of the Intertropical Convergence Zone (ITCZ) and its associated wind-flow patterns. Although monthly precipitation totals vary considerably, the average for each month exceeds 6 cm, and consequently no month is considered deficient of rainfall.

Take a minute and look again at Fig. 13.6. From the photo, one might think that the soil beneath the forest's canopy would be excellent for agriculture. Actually, this is not true. As heavy rain falls on the soil, the water works its way downward, removing nutrients in a process called *leaching.* Strangely enough, many of the nutrients needed to sustain the lush forest actually come from dead trees that decompose. The roots of the living trees absorb this matter before the rains leach it away. When the forests are cleared for agricultural purposes, or for the timber, what is left is a thick red soil called **laterite.** When exposed to the intense sunlight of the tropics, the soil may harden into a bricklike consistency, making cultivation almost impossible.

DID YOU KNOW?

Hot and humid Belem, Brazil—a city situated near the equator with a tropical wet climate—had an all-time record high temperature of 98°F, exactly 2°F *less* than the highest temperature (100°F) ever measured in Prospect Creek, Alaska, a city with a subpolar climate.

*The tropical wet climate is also known as the *tropical rain forest climate.*

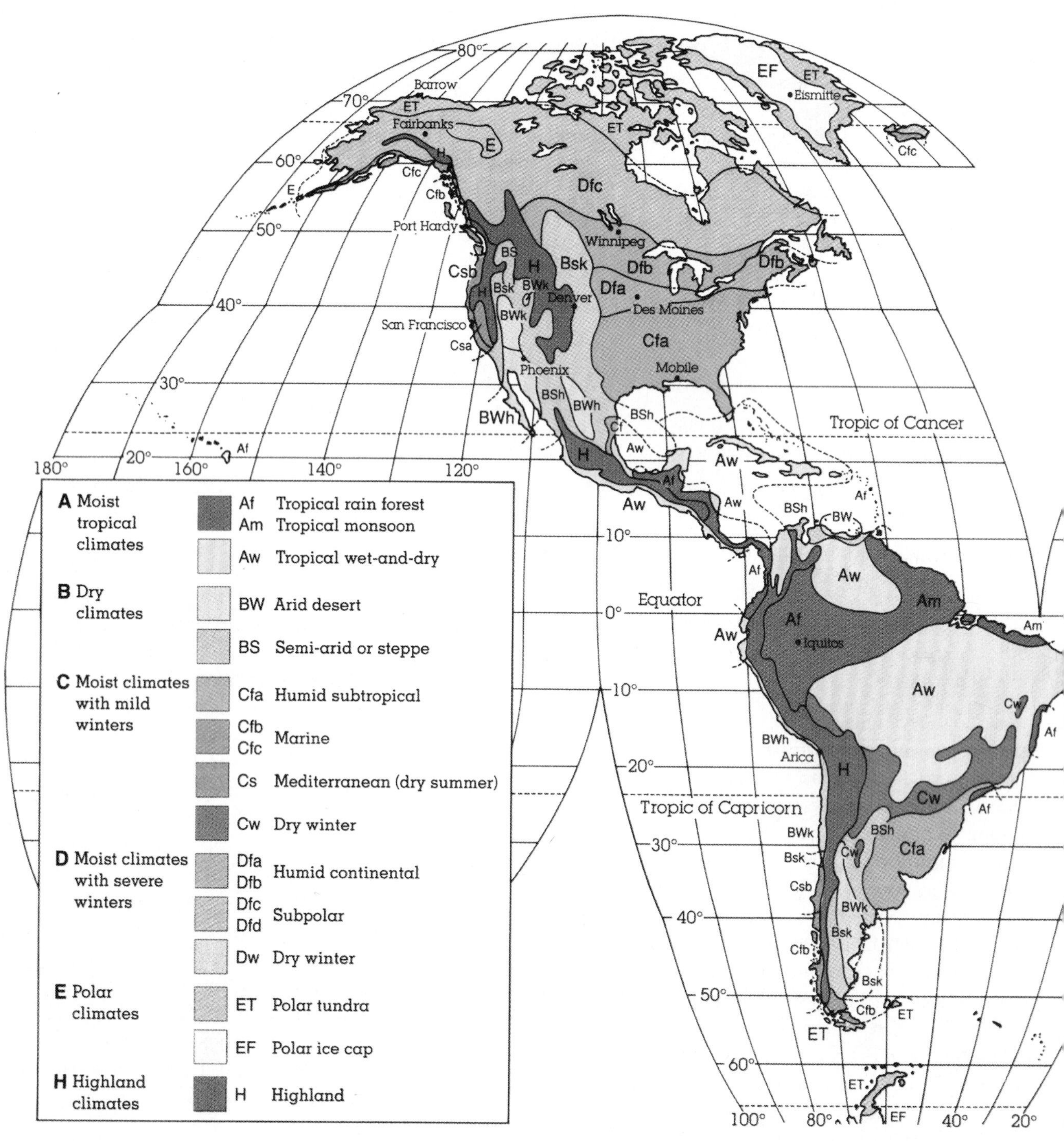

FIGURE 13.5 Worldwide distribution of climatic regions (after Köppen).

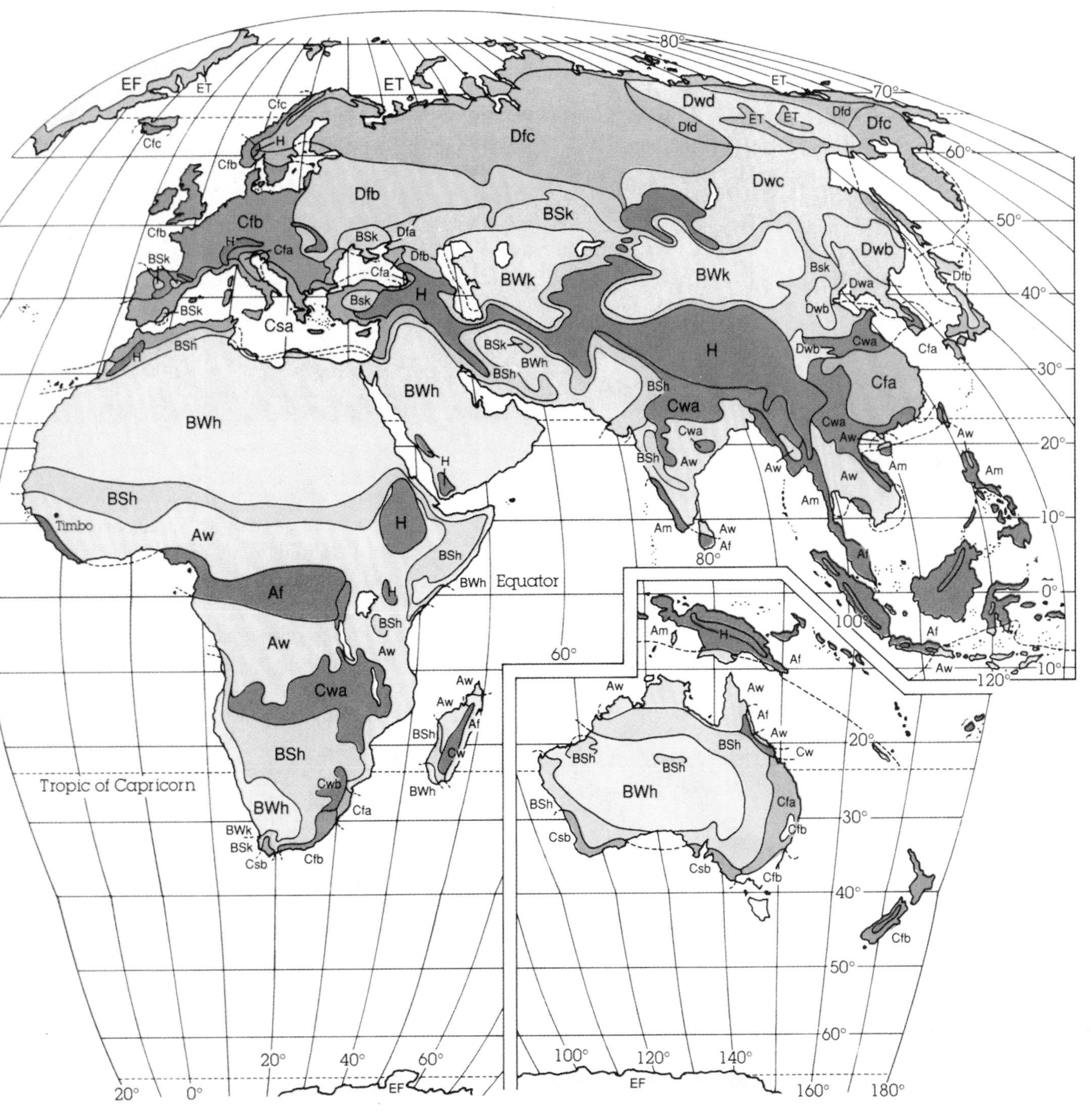

EF
ET
Cfc
Dfc
Dwd
Dfd
Dwc
Dfb
Cfb
BSk
Dfa
Cfa
Csa
H
BWk
Dwb
Dwa
BSh
BWh
Cwa
Aw
Am
Af
Equator
Timbo
Cw
Cwb
BWk
Csb
Tropic of Capricorn
EF
80°
70°
60°
50°
40°
30°
20°
10°
0°
20°
0°
20°
40°
60°
100°
120°
140°
160°
180°

FIGURE 13.6 Tropical rain forest near Iquitos, Peru. (Climatic information for this region is presented in Fig. 13.7.)

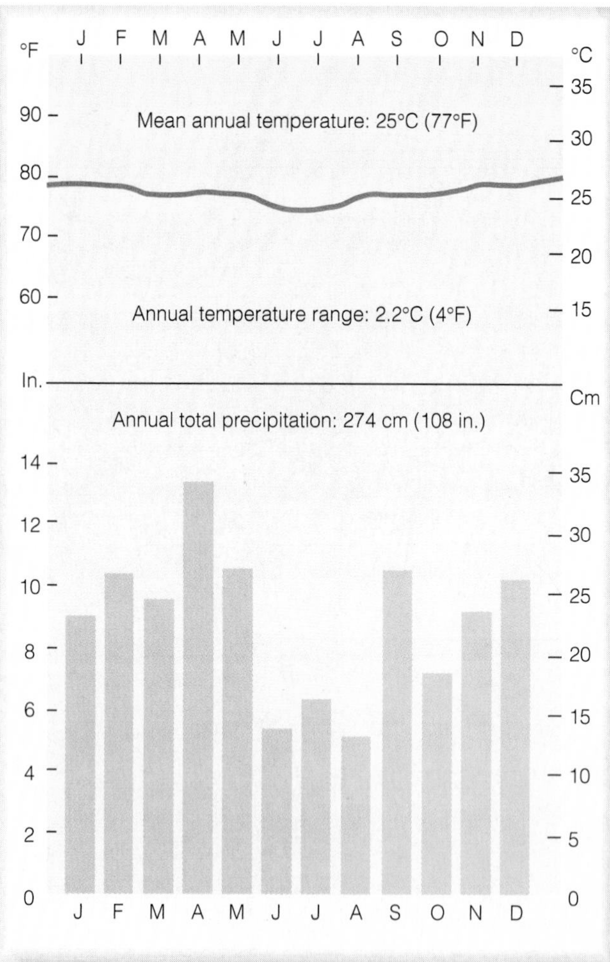

FIGURE 13.7 Temperature and precipitation data for Iquitos, Peru, latitude 4°S. A station with a tropical wet climate (Af). (This type of diagram is called a *climograph.* It shows monthly mean temperatures with a solid red line and monthly mean precipitation with bar graphs.)

Köppen classified tropical wet regions, where the monthly precipitation totals drop below 6 cm for perhaps one or two months, as **tropical monsoon climates** (Am). Here, yearly rainfall totals are similar to those of the tropical wet climate, usually exceeding 150 cm a year. Because the dry season is brief and copious rains fall throughout the rest of the year, there is sufficient soil moisture to maintain the tropical rain forest through the short dry period. Tropical monsoon climates can be seen in Fig. 13.5 along the coasts of Southeast Asia, India, and in northeastern South America.

Poleward of the tropical wet region, total annual rainfall diminishes, and there is a gradual transition from the tropical wet climate to the **tropical wet-and-dry climate** (Aw), where a distinct dry season prevails. Even though the annual precipitation usually exceeds 100 cm, the dry season, where the monthly rainfall is less than 6 cm (2.4 in.), lasts for more than two months. Because tropical rain forests cannot survive this "drought," the jungle gradually gives way to tall, coarse **savanna grass,** scattered with low, drought-resistant deciduous trees (see Fig. 13.8). The dry season occurs during the winter (low sun period), when the region is under the influence of the subtropical highs. In summer, the ITCZ moves poleward, bringing with it heavy precipitation, usually in the form of showers. Rainfall is enhanced by slow moving shallow lows that move through the region.

Tropical wet-and-dry climates not only receive less total rainfall than the tropical wet climates, but the rain that does occur is much less reliable, as the total rainfall often fluctuates widely from one year to the next. In the course of a single year, for example, destructive floods

FIGURE 13.8 Baobob and acacia trees illustrate typical trees of the East African grassland savanna, a region with a tropical wet-and-dry climate (Aw).

may be followed by serious droughts. As with tropical wet regions, the daily range of temperature usually exceeds the annual range, but the climate here is much less monotonous. There is a cool season in winter when the maximum temperature averages 30°C to 32°C (86°F to 90°F). At night, the low humidity and clear skies allow for rapid radiational cooling and, by early morning, minimum temperatures drop to 20°C (68°F) or below.

From Fig. 13.5, pp. 352–353, we can see that the principal areas having a tropical wet-and-dry climate (Aw) are those located in western Central America, in the region both north and south of the Amazon Basin (South America), in southcentral and eastern Africa, in parts of India and Southeast Asia, and in northern Australia. In many areas (especially within India and Southeast Asia), the marked variation in precipitation is associated with the *monsoon*—the seasonal reversal of winds.

As we saw in Chapter 7, the monsoon circulation is due in part to differential heating between land masses and oceans. During winter in the Northern Hemisphere, winds blow outward, away from a cold, shallow high-pressure area centered over continental Siberia. These downslope, relatively dry northeasterly winds from the interior provide India and Southeast Asia with generally fair weather and the dry season. In summer, the wind-flow pattern reverses as air flows into a developing thermal low over the continental interior. The humid air from the water rises and condenses, resulting in heavy rain and the wet season.

An example of a station with a tropical wet-and-dry climate (Aw) is given in Fig. 13.9. Located at latitude 11°N in west Africa, Timbo, Guinea, receives an annual average

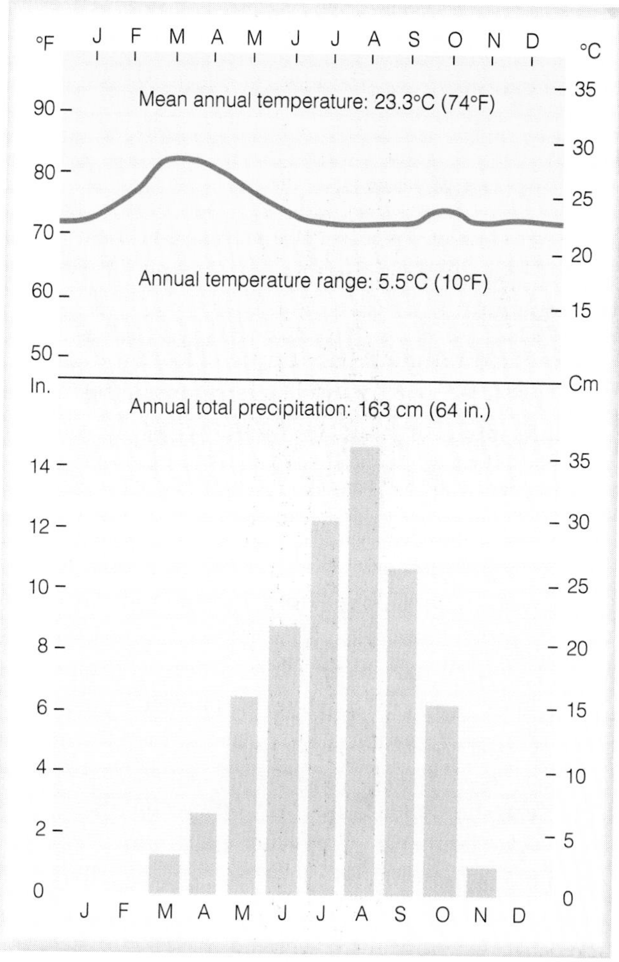

FIGURE 13.9 Climatic data for Timbo, Guinea, latitude 11°N. A station with a tropical wet-and-dry climate (Aw).

163 cm (64 in.) of rainfall. Notice that the rainy season is during the summer when the ITCZ has migrated to its most northern position. Note also that practically no rain falls during the months of December, January, and February, when the region comes under the domination of the subtropical high-pressure area and its sinking air.

The monthly temperature patterns at Timbo are characteristic of most tropical wet-and-dry climates. As spring approaches, the noon sun is slightly higher, and the more intense sunshine produces greater surface heating and higher afternoon temperatures—usually above 32°C (90°F) and occasionally above 38°C (100°F) —creating hot, dry desertlike conditions. After this brief hot season, a persistent cloud cover and the evaporation of rain tends to lower the temperature during the summer. The warm, muggy weather of summer often resembles that of the tropical wet climate (Af). The rainy summer is followed by a warm, relatively dry period, with afternoon temperatures usually climbing above 30°C (86°F).

Poleward of the tropical wet-and-dry climate, the dry season becomes more severe. Clumps of trees are more isolated and the grasses dominate the landscape. When the potential annual water loss through evaporation and transpiration exceeds the annual water gain from precipitation, the climate is described as dry.

Dry Climates (Group B)

General characteristics: deficient precipitation most of the year; potential evaporation and transpiration exceed precipitation.

Extent: the subtropical deserts extend from roughly 20° to 30° latitude in large continental regions of the middle latitudes, often surrounded by mountains.

Major types: arid (BW)—the "true desert"—and semi-arid (BS).

A quick glance at Fig. 13.5, pp. 352–353, reveals that, according to Köppen, the dry regions of the world occupy more land area (about 26 percent) than any other major climatic type. Within these dry regions, a deficiency of water exists. Here, the potential annual loss of water through evaporation is greater than the annual water gained through precipitation. Thus, classifying a climate as dry depends not only on precipitation totals but also on temperature, which greatly influences evaporation. For example, 35 cm (14 in.) of precipitation in a hot climate will support only sparse vegetation, while the same amount of precipitation in northcentral Canada will support a conifer forest. In addition, a region with a low annual rainfall total is more likely to be classified as dry if the majority of precipitation is concentrated during the warm summer months, when evaporation rates are greater.

Precipitation in a dry climate is both meager and irregular. Typically, the lower the average annual rainfall, the greater its variability. For example, a station that reports an annual rainfall of 5 cm (2 in.) may actually measure no rainfall for two years; then, in a single downpour, it may receive 10 cm (4 in.).

The major dry regions of the world can be divided into two primary categories. The first includes the area of the subtropics (between latitudes 15° and 30°), where the sinking air of the subtropical anticyclones produces generally clear skies. The second is found in the continental areas of the middle latitudes. Here, far removed from a source of moisture, areas are deprived of precipitation. Dryness here is often accentuated by mountain ranges that produce a rain shadow effect.

Köppen divided dry climates into two types based on their degree of dryness: the *arid* (BW)* and the *semi-arid,* or steppe (BS). These two climatic types can be divided even further. For example, if the climate is hot and dry with a mean annual temperature above 18°C (64°F), it is either BWh or BSh (the *h* is for *heiss,* meaning hot in German). On the other hand, if the climate is cold (in winter, that is) and dry with a mean annual temperature below 18°C, then it is either BWk or BSk (where the *k* is for *kalt,* meaning cold in German).

The **arid climates** (BW) occupy about 12 percent of the world's land area. From Fig. 13.5, pp. 352–353, we can see that this climatic type is found along the west coast of South America and Africa and over much of the interior of Australia. Notice, also, that a swath of arid climate extends from northwest Africa all the way into central Asia. In North America, the arid climate extends from northern Mexico into the southern interior of the United States and northward along the leeward slopes of the Sierra Nevada. This region includes both the Sonoran and Mojave deserts and the Great Basin.

The southern desert region of North America is dry because it is dominated by the subtropical high most of the year, and winter storm systems tend to weaken before they move into the area. The northern region is in the rain shadow of the Sierra Nevada. These regions are deficient in precipitation all year long, with many stations receiving less than 13 cm (5 in.) annually. As noted earlier, the rain that does fall is spotty, often in the form of scattered summer afternoon showers. Some of these showers can be downpours that change a gentle

*The letter *W* is for *Wüste,* the German word for desert.

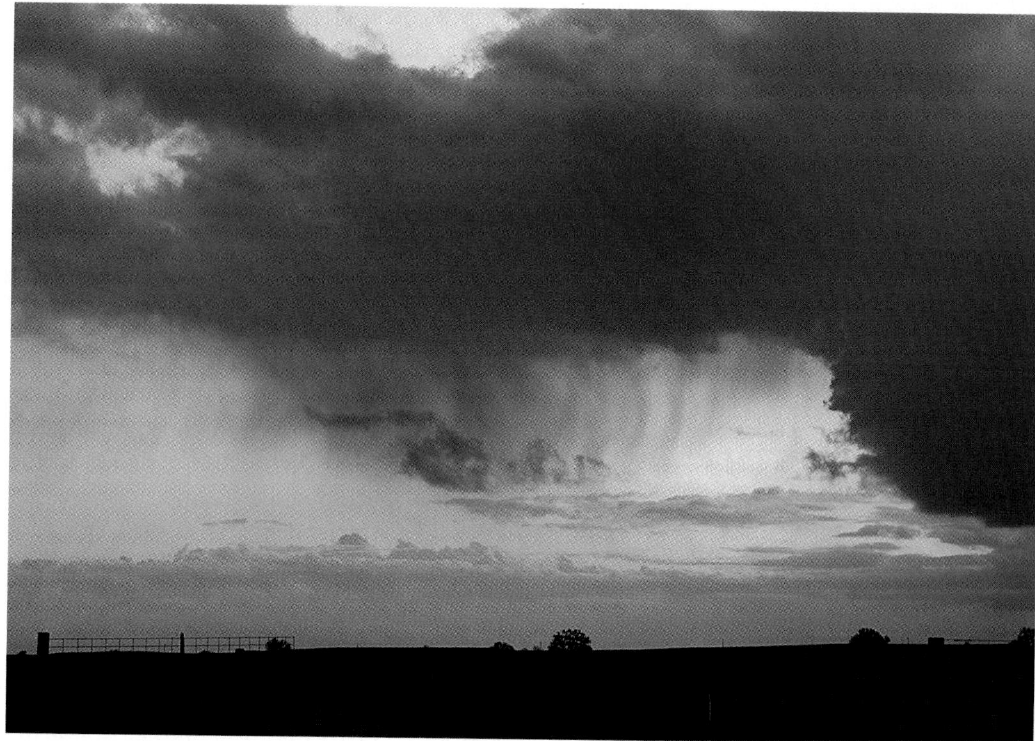

Photo by author

FIGURE 13.10 Rain streamers (virga) are common in dry climates, as falling rain evaporates into the drier air before ever reaching the ground.

gully into a raging torrent of water. More often than not, however, the rain evaporates into the dry air before ever reaching the ground, and the result is rain streamers (virga) dangling beneath the clouds (see Fig. 13.10).

Contrary to popular belief, few deserts are completely without vegetation. Although meager, the vegetation that does exist must depend on the infrequent rains. Thus, most of the native plants are **xerophytes**—those capable of surviving prolonged periods of drought (see Fig. 13.11). Such vegetation includes various forms of cacti and short-lived plants that spring up during the rainy periods.

In low-latitude deserts (BWh), intense sunlight produces scorching heat on the parched landscape. Here, air temperatures are as high as anywhere in the world. Maximum daytime readings during the summer can exceed 50°C (122°F), although 40°C to 45°C (104°F to 113°F) are more common. In the middle of the day, the relative humidity is usually between 5 and 25 percent. At night, the air's relatively low water vapor content allows for rapid radiational cooling. Minimum temperatures often

DID YOU KNOW?

The driest major city in the contiguous United States is Yuma, Arizona. Yuma has a total average annual precipitation of 6.5 cm (2.6 in.)—it rains there only about 17 days a year.

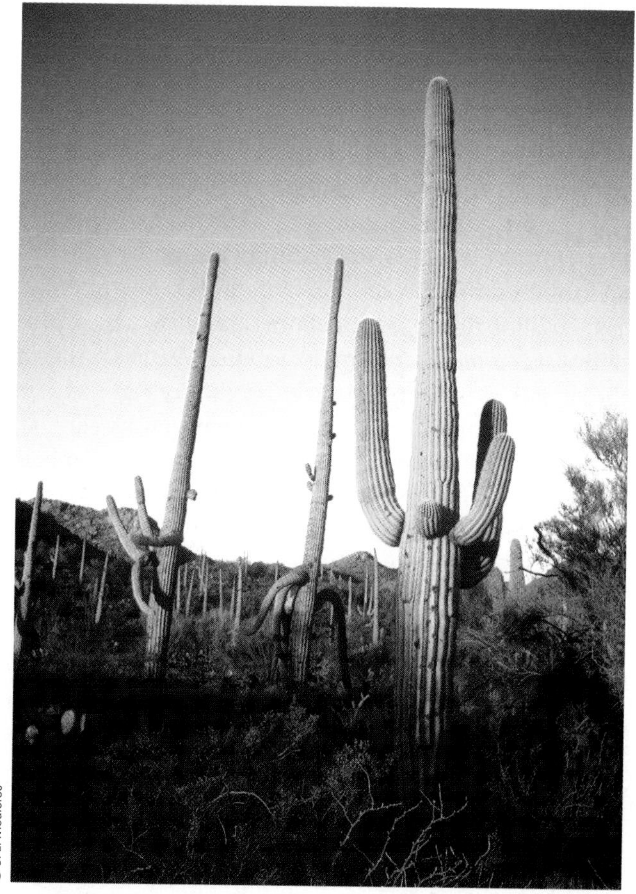

© J. L. Medieros

FIGURE 13.11 Creosote bushes and cactus are typical of the vegetation found in the arid southwestern American deserts (BWh).

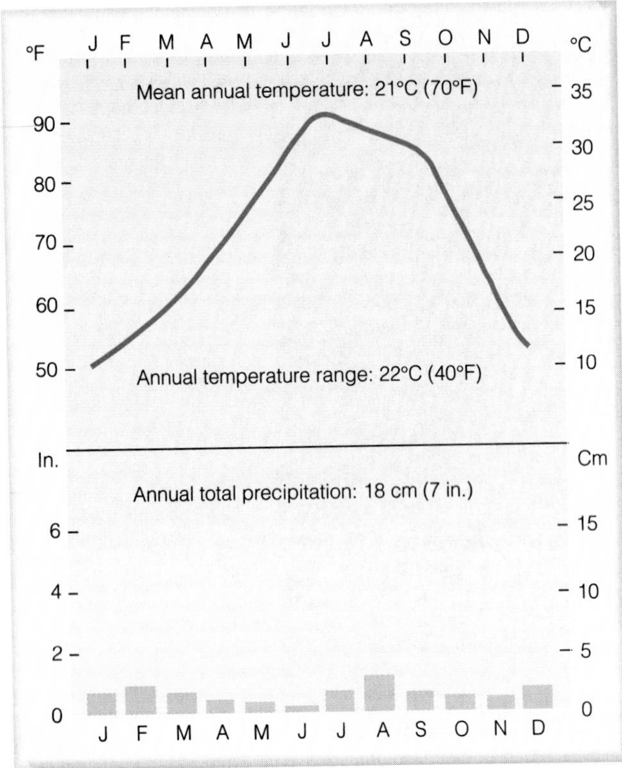

FIGURE 13.12 Climatic data for Phoenix, Arizona, latitude 33.5°N. A station with an arid climate (BWh).

drop below 25°C (77°F). Thus, arid climates have large daily temperature ranges, often between 15°C and 25°C (27°F and 45°F) and occasionally higher.

During the winter, temperatures are more moderate, and minimums may, on occasion, drop below freezing. The variation in temperature from summer to winter produces large annual temperature ranges. We can see this in the climate record for Phoenix, Arizona (see Fig. 13.12), a city in the southwestern United States with a BWh climate. Notice that the average annual temperature in Phoenix is 22°C (72°F), and that the average temperature of the warmest month (July) reaches a sizzling 32°C (90°F). As we would expect, rainfall is meager in all months. There is, however, a slight maximum in July and August. This is due to the summer monsoon, when more humid, southerly winds are likely to sweep over the region and develop into afternoon showers and thunderstorms.

In middle-latitude deserts (BWk), average annual temperatures are lower. Summers are typically warm to hot, with afternoon temperatures frequently reaching 40°C (104°F). Winters are usually extremely cold, with minimum temperatures sometimes dropping below −35°C (−31°F). Many of these deserts lie in the rain shadow of an extensive mountain chain, such as the Sierra Nevada and the Cascade mountains in North America, the Himalayan Mountains in Asia, and the Andes in South America. The meager precipitation that falls comes from an occasional summer shower or a passing mid-latitude cyclonic storm in winter.

Again, refer to Fig. 13.5 and notice that around the margins of the arid regions, where rainfall amounts are greater, the climate gradually changes into **semi-arid** (BS). This region is called **steppe** and typically has short bunch grass, scattered low bushes, trees, or sagebrush (see Fig. 13.13). In North America, this climatic region includes most of the Great Plains, the southern coastal sections of California, and the northern valleys of the Great Basin. As in the arid region, northern areas experience lower winter temperatures and more frequent snowfalls. Annual precipitation is generally between 20 and 40 cm (8 and 16 in.). The climatic record for Denver, Colorado (see Fig. 13.14), exemplifies the semi-arid (BSk) climate.

As average rainfall amounts increase, the climate gradually changes to one that is more humid. Hence, the semi-arid (steppe) climate marks the transition between the arid and the humid climatic regions. (Before reading about moist climates, you may wish to read the Focus section on p. 360 about deserts that experience drizzle but little rainfall.)

Moist Subtropical Mid-Latitude Climates (Group C)

General characteristics: humid with mild winters (i.e., average temperature of the coldest month below 18°C, or 64°F, and above −3°C, or 27°F).

Extent: on the eastern and western regions of most continents, from about 25° to 40° latitude.

Major types: humid subtropical (Cfa), marine (Cfb), and dry-summer subtropical, or Mediterranean (Cs).

The Group C climates of the middle latitudes have distinct summer and winter seasons. Additionally, they have ample precipitation to keep them from being classified as dry. Although winters can be cold, and air tem-

DID YOU KNOW?

Arid Bagdad, California, once went for almost three consecutive years without measurable precipitation, from August 18, 1909, to May 6, 1912.

Photo by author

FIGURE 13.13 Cumulus clouds forming over the steppe grasslands of western North America, a region with a semi-arid climate (BS).

peratures can change appreciably from one day to the next, no month has a mean temperature below −3°C (27°F), for if it did, it would be classified as a D climate—one with severe winters.

The first C climate we will consider is the **humid subtropical climate** (Cfa). Notice in Fig. 13.5, pp. 352–353, that Cfa climates are found principally along the east coasts of continents, roughly between 25° and 40° latitude. They dominate the southeastern section of the United States, as well as eastern China and southern Japan. In the Southern Hemisphere, they are found in southeastern South America and along the southeastern coasts of Africa and Australia.

A trademark of the humid subtropical climate is its hot, muggy summers. This sultry summer weather occurs because Cfa climates are located on the western side of subtropical highs, where maritime tropical air from lower latitudes is swept poleward into these regions. Generally, summer dew-point temperatures are high (often exceeding 23°C, or 73°F) and so is the relative humidity, even during the middle of the day. The high humidity combines with the high air temperature (usually above 32°C, or 90°F) to produce more oppressive conditions than are found in equatorial regions. Summer morning low temperatures often range between 21°C and 27°C (70°F and 81°F). Occasionally, a weak summer cool front will bring temporary relief from the sweltering conditions. However, devastating heat waves, sometimes lasting many weeks, can occur when an upper-level ridge moves over the area.

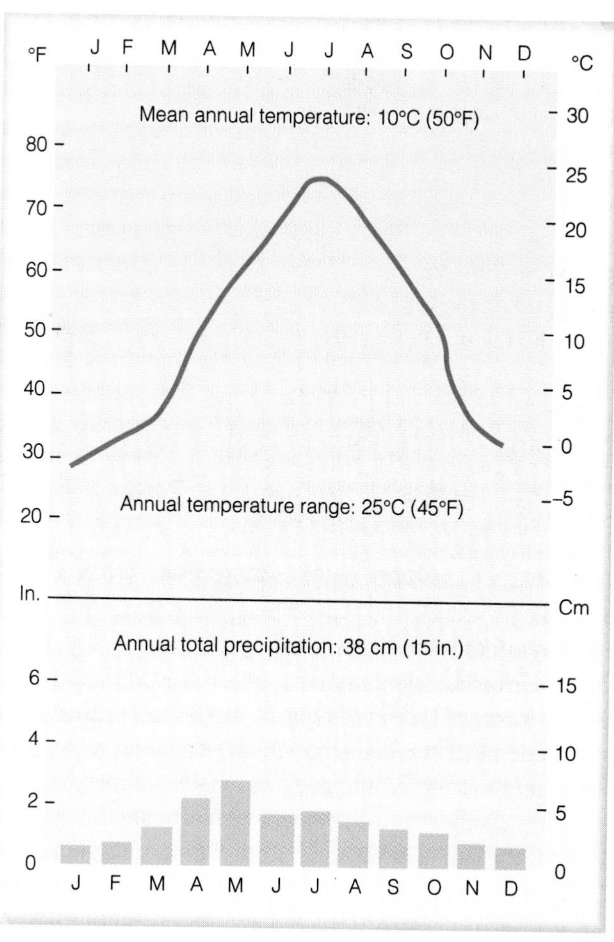

FIGURE 13.14 Climatic data for Denver, Colorado, latitude 40°N. A station with a semi-arid climate (BSk).

FOCUS ON AN OBSERVATION

A Desert with Clouds and Drizzle

We already know that not all deserts are hot. By the same token, not all deserts are sunny. In fact, some coastal deserts experience considerable cloudiness, especially low stratus and fog.

Amazingly, these coastal deserts are some of the driest places on earth. They include the Atacama Desert of Chile and Peru, the coastal Sahara Desert of northwest Africa, the Namib Desert of southwestern Africa, and a portion of the Sonoran Desert in Baja, California (see Fig. 2). On the Atacama Desert, for example, some regions go without measurable rainfall for decades. And Arica, in northern Chile, has an annual rainfall of only 0.08 cm (0.03 in.).

The cause of this aridity is, in part, due to the fact that each region is adjacent to a large body of relatively cool water. Notice in Fig. 2 that these deserts are located along the western coastal margins of continents, where a subtropical high-pressure area causes prevailing winds to move cool water from higher latitudes along the coast. In addition, these winds help to accentuate the water's coldness by initiating *upwelling*—the rising of cold water from lower levels. The combination of these conditions tends to produce coastal water temperatures between 10°C and 15°C (50°F and 59°F), which is quite cool for such low latitudes. As surface air sweeps across the cold water, it is chilled to its dew point, often producing a blanket of fog and low clouds, from which drizzle falls. The drizzle, however, accounts for very little rainfall. In most regions, it is only enough to dampen the streets with a mere trace of precipitation.

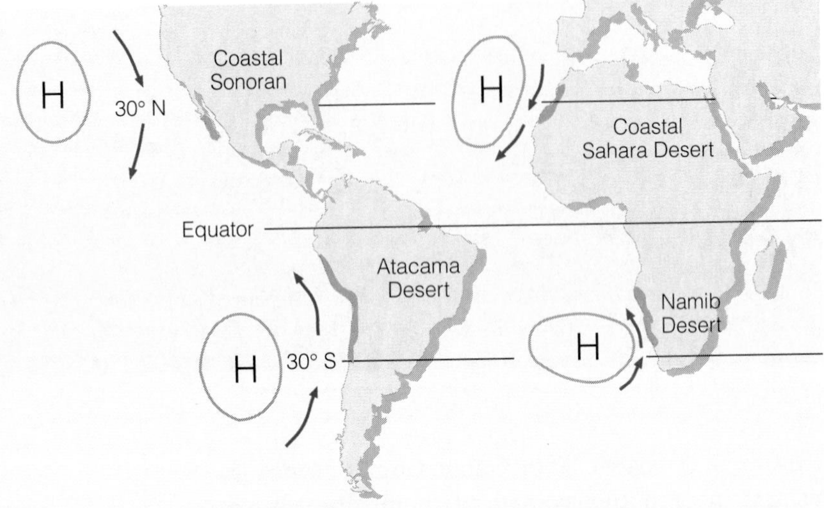

FIGURE 2 Location of coastal deserts that experience frequent fog, drizzle, and low clouds.

As the cool stable air moves inland, it warms, and the water droplets evaporate. Hence, most of the cloudiness and drizzle is found along the immediate coast. Although the relative humidity of this air is high, the dew-point temperature is comparatively low (often near that of the coastal surface water). Inland, further warming causes the air to rise. However, a stable subsidence inversion, associated with the subtropical highs, inhibits vertical motions by capping the rising air, causing it to drift back toward the ocean, where it sinks, completing a rather strong sea breeze circulation. The position of the subtropical highs, which tend to remain almost stationary, plays an additional role by preventing the Intertropical Convergence Zone with its rising, unstable air from entering the region.

And so we have a desert with clouds and drizzle—a desert that owes its existence, in part, to its proximity to rather cold ocean water and, in part, to the position and air motions of a subtropical high.

Winters tend to be relatively mild, especially in the lower latitudes, where air temperatures rarely dip much below freezing. Poleward regions experience winters that are colder and harsher. Here, frost, snow, and ice storms are more common, but heavy snowfalls are rare. Winter weather can be quite changeable, as almost summerlike conditions can give way to cold rain and wind in a matter of hours when a middle-latitude cyclonic storm and its accompanying fronts pass through the region.

Humid subtropical climates experience adequate and fairly well-distributed precipitation throughout the year, with typical annual averages between 80 and 165 cm (31 and 65 in.). In summer, when thunderstorms are common, much of the precipitation falls as afternoon showers. Tropical storms entering the United States and China can substantially add to their summer and autumn rainfall totals. Winter precipitation most often occurs with eastward-trekking middle-latitude cyclonic storms. In the southeastern United States, the abundant rainfall supports a thick pine forest that becomes mixed with oak at higher latitudes. The climate data for Mobile, Alabama, a city with a Cfa climate, is given in Fig. 13.15.

Glance back at Fig. 13.5, pp. 352–353, and observe that C climates extend poleward along the western side of most continents from about latitude 40° to 60°. These regions are dominated by prevailing winds from the ocean that moderate the climate, keeping winters considerably milder than stations located at the same latitude farther inland. In addition to this, summers are quite cool. When the summer season is both short and cool, the climate is designated as Cfc.* Equatorward, where summers are longer (but still cool), the climate is classified as *west coast marine,* or simply **marine,** Cfb.

Where mountains parallel the coastline, such as along the west coasts of North and South America, the marine influence is restricted to narrow belts. Unobstructed by high mountains, prevailing westerly winds pump ocean air over much of western Europe and thus provide this region with a marine climate (Cfb).

During much of the year, marine climates are characterized by low clouds, fog, and drizzle. The ocean's influence produces adequate precipitation in all months, with much of it falling as light or moderate rain associated with maritime polar air masses. Snow does fall, but frequently it turns to slush after only a day or so. In some locations, topography greatly enhances precipitation totals. For example, along the west coast of North America, coastal mountains not only force air upward enhancing precipitation, they also slow the storm's eastward progress, which enables the storm to drop more precipitation on the area.

Along the northwest coast of North America, rainfall amounts decrease in summer. This phenomenon is caused by the northward migration of the subtropical Pacific high, which is located southwest of this region. The summer decrease in rainfall can be seen by examining the climatic record of Port Hardy (see Fig. 13.16), a station situated along the coast of Canada's Vancouver Island. The data illustrate another important characteristic of marine climates: the low annual temperature range for such a high-latitude station. The ocean's influence keeps daily temperature ranges low as well. In this climate type, it rains on many days and when it is not raining, skies are usually overcast. The heavy rains produce a dense forest of Douglas fir.

Moving equatorward of marine climates, the influence of the subtropical highs becomes greater, and the summer dry period more pronounced. Gradually, the climate changes from marine to one of **dry-summer subtropical** (Cs), or **Mediterranean,** because it also borders the coastal areas of the Mediterranean Sea. Along the west coast of North America, Portland, Oregon, because it has rather dry summers, marks the transition between the marine climate and the dry-summer subtropical climate to the south.

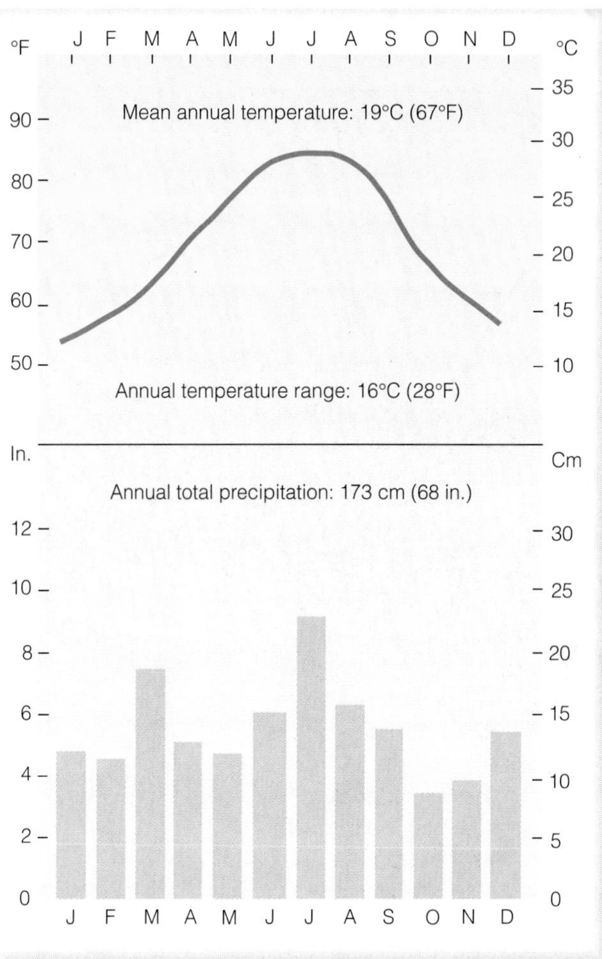

FIGURE 13.15 Climatic data for Mobile, Alabama, latitude 30°N. A station with a humid subtropical climate (Cfa).

The extreme summer aridity of the Mediterranean climate, which in California may exist for five months, is caused by the sinking air of the subtropical highs. In addition, these anticyclones divert summer storm systems poleward. During the winter, when the subtropical highs move equatorward, mid-latitude storms from the ocean frequent the region, bringing with them much needed rainfall. Consequently, Mediterranean climates are characterized by mild, wet winters, and mild-to-hot, dry summers.

DID YOU KNOW?

Marine climates can be very cloudy and wet. Quillayute, Washington—a city with a marine climate—averages 242 cloudy days a year, with rain on 212 days.

*Appendix G, p. 439, details the necessary criteria for each climatic type.

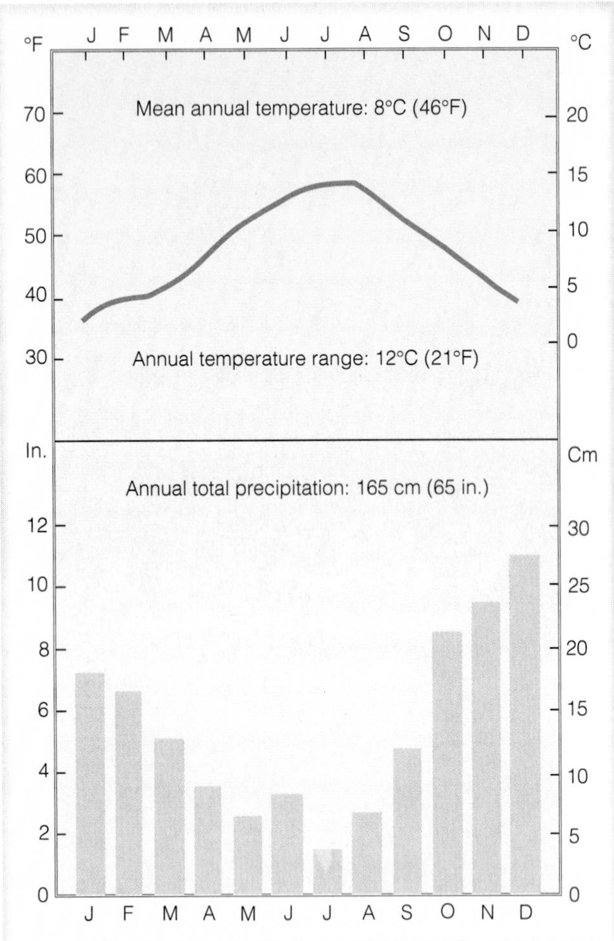

FIGURE 13.16 Climatic data for Port Hardy, Canada, latitude 51°N. A station with a marine climate (Cfb).

Where surface winds parallel the coast, upwelling of cold water helps keep the water itself and the air above it cool all summer long. In these coastal areas, which are often shrouded in low clouds and fog, the climate is called *coastal Mediterranean* (Csb). Here, summer daytime maximum temperatures usually reach about 21°C (70°F), while overnight lows often drop below 15°C (59°F). Inland, away from the ocean's influence, summers are hot and winters are a little cooler than coastal areas. In this *interior Mediterranean climate* (Csa), summer afternoon temperatures usually climb above 34°C (93°F) and occasionally above 40°C (104°F).

Figure 13.17 contrasts the coastal Mediterranean climate of San Francisco, California, with the interior Mediterranean climate of Sacramento, California. While Sacramento is only 130 km (80 mi) inland from San Francisco, Sacramento's average July temperature is 9°C (16°F) higher. As we would expect, Sacramento's annual temperature range is considerably higher, too. Although Sacramento and San Francisco both experience an occasional frost, snow in these areas is a rarity.

In Mediterranean climates, yearly precipitation amounts range between 30 and 90 cm (11 and 35 in.). However, much more precipitation falls on surrounding hillsides and mountains. Because of the summer dryness, the land supports only a scrubby type of low-growing woody plants and trees called *chaparral* (see Fig. 13.18).

At this point, we should note that summers are not as dry along the Mediterranean Sea as they are along the west coast of North America. Moreover, coastal Mediterranean areas are also warmer, due to the lack of upwelling in the Mediterranean Sea.

Before leaving our discussion of C climates, note that when the dry season is in winter, the climate is classified as Cw. Over northern India and portions of China, the relatively dry winters are the result of northerly winds from continental regions circulating southward around the cold Siberian high. Many lower-latitude regions with a Cw climate would be tropical if it were not for the fact they are too high in elevation and, consequently, too cool to be designated as tropical.

DID YOU KNOW?

The warm water of the Gulf Stream helps to keep the average winter temperature in Bergen, Norway (which is located just south of the Arctic Circle at latitude 60°N), about 0.6°C (about 1°F) warmer than the average winter temperature in Philadelphia, Pennsylvania (latitude 40°N).

Moist Continental Climates (Group D)

General characteristics: warm-to-cool summers and cold winters (i.e., average temperature of warmest month exceeds 10°C, or 50°F, and the coldest monthly average drops below −3°C, or 27°F); winters are severe with snowstorms, blustery winds, bitter cold; climate controlled by large continent.

Extent: north of moist subtropical mid-latitude climates.

Major types: humid continental with hot summers (Dfa), humid continental with cool summers (Dfb), and subpolar (Dfc).

The D climates are controlled by large land masses. Therefore, they are found only in the Northern Hemi-

°F J F M A M J J A S O N D °C

Mean annual temperature: 14°C (57°F)

Annual temperature range: 6°C (11°F)

In. Cm

Annual total precipitation: 53 cm (21 in.)

J F M A M J J A S O N D

San Francisco, California, 37°N

Climate: Csb

°F J F M A M J J A S O N D °C

Mean annual temperature: 15°C (59°F)

Annual temperature range: 17°C (30°F)

In. Cm

Annual total precipitation: 41 cm (16 in.)

J F M A M J J A S O N D

Sacramento, California, 38°N

Climate: Csa

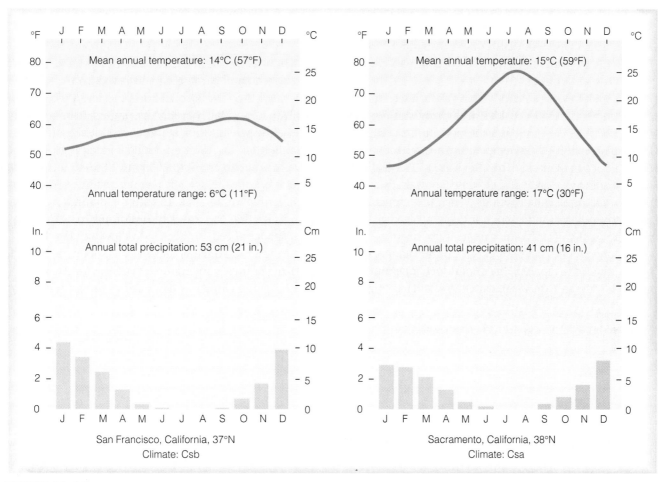

FIGURE 13.17 Comparison of a coastal Mediterranean climate, Csb (San Francisco, at left), with an interior Mediterranean climate, Csa (Sacramento, at right).

Photo by author

FIGURE 13.18 In the Mediterranean-type climates of North America, typical chaparral vegetation includes chamise, manzanita, and foothill pine.

sphere. Look at the climate map, Fig. 13.5, pp. 352–353, and notice that D climates extend across North America and Eurasia, from about latitude 40°N to almost 70°N. In general, they are characterized by cold winters and warm-to-cool summers.

As we know, for a station to have a D climate, the average temperature of its coldest month must dip below −3°C (27°F). This is not an arbitrary number. Köppen found that, in Europe, this temperature marked the southern limit of persistent snow cover in winter.* Hence, D climates experience a great deal of winter snow that stays on the ground for extended periods. When the temperature drops to a point such that no month has an average temperature of 10°C (50°F), the climate is classified as polar (E). Köppen found that the average monthly temperature of 10°C tended to represent the minimum temperature required for tree growth. So no matter how cold it gets in a D climate (and winters can get extremely cold), there is enough summer warmth to support the growth of trees.

There are two basic types of D climates: the **humid continental** (Dfa and Dfb) and the **subpolar** (Dfc). Humid continental climates are observed from about latitude 40°N to 50°N (60°N in Europe). Here, precipitation is adequate and fairly evenly distributed throughout the year, although interior stations experience maximum precipitation in summer. Annual precipitation totals usually range from 50 to 100 cm (20 to 40 in.). Native vegetation in the wetter regions includes forests of spruce, fir, pine, and oak. In autumn, nature's pageantry unveils itself as the leaves of deciduous trees turn brilliant shades of red, orange, and yellow (see Fig. 13.19).

Humid continental climates are subdivided on the basis of summer temperatures. Where summers are long and hot,* the climate is described as *humid continental*

*In North America, studies suggest that an average monthly temperature of 0°C (32°F) or below for the coldest month seems to correspond better to persistent winter snow cover.

*"Hot" means that the average temperature of the warmest month is above 22°C (72°F) and at least four months have a monthly mean temperature above 10°C (50°F).

FIGURE 13.19 The leaves of deciduous trees burst into brilliant color during autumn over the countryside of Adirondack Park, a region with a humid continental climate.

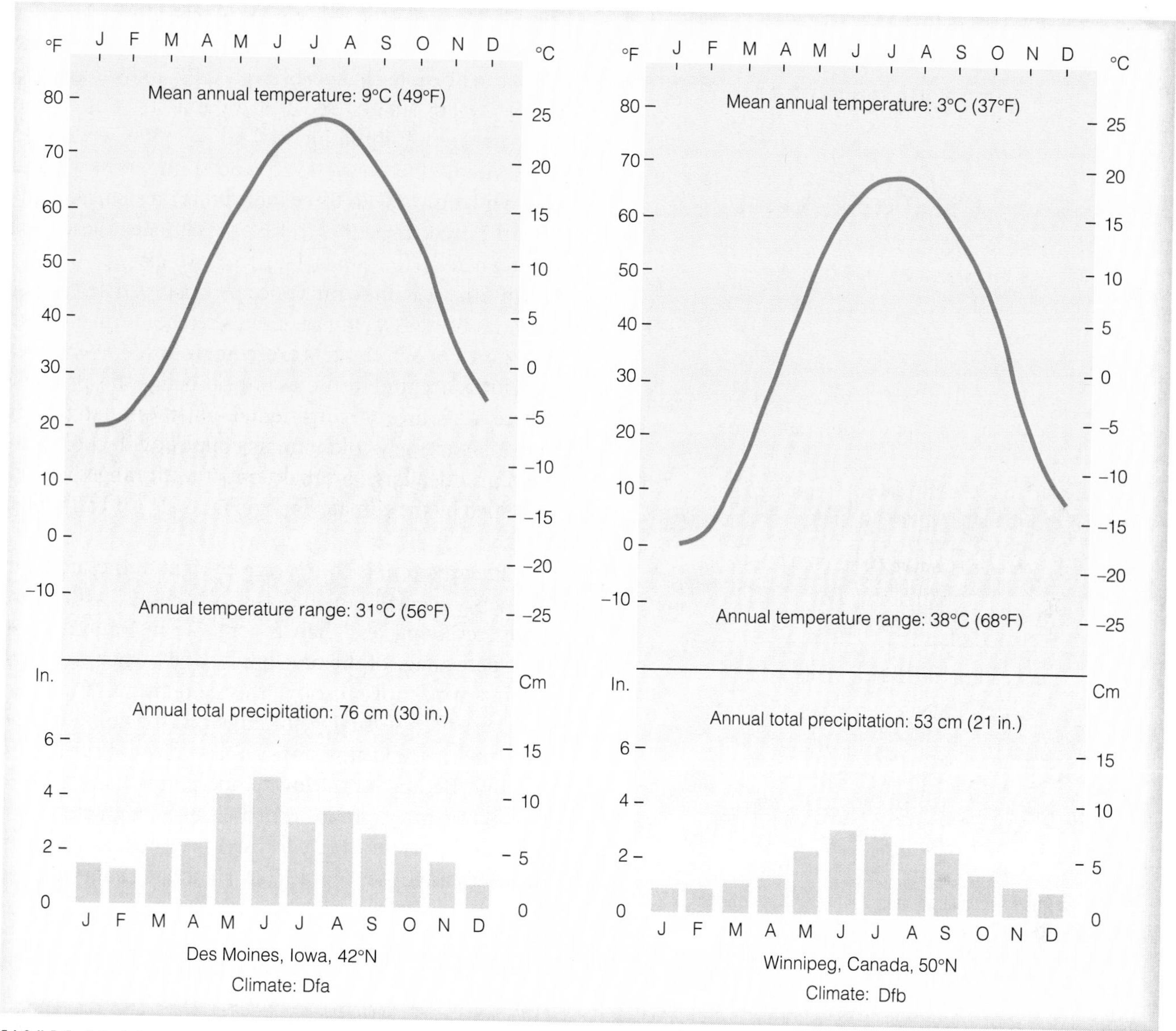

FIGURE 13.20 Comparison of a humid continental hot summer climate, Dfa (Des Moines, at left), with a humid continental cool summer climate, Dfb (Winnipeg, at right).

with hot summers (Dfa). Here summers are often hot and humid, especially in the southern regions. Midday temperatures often exceed 32°C (90°F) and occasionally 40°C (104°F). Summer nights are usually warm and humid, as well. The frost-free season normally lasts from five to six months, long enough to grow a wide variety of crops. Winters tend to be windy, cold, and snowy. Farther north, where summers are shorter and not as hot,* the climate is described as *humid continental with long cool summers* (Dfb). In Dfb climates, summers are not only cooler but much less humid. Temperatures may exceed 35°C (95°F) for a time, but extended hot spells lasting many weeks are rare. The frost-free season is shorter than in the Dfa climate, and normally lasts between three and five months. Winters are long, cold, and windy. It is not uncommon for temperatures to drop below –30°C (–22°F) and stay below –18°C (0°F) for days and sometimes weeks. Autumn is short, with winter often arriving right on the heels of summer. Spring, too, is short, as late spring snowstorms are common, especially in the more northern latitudes.

Figure 13.20 compares the Dfa climate of Des Moines, Iowa, with the Dfb climate of Winnipeg, Canada.

*"*Not as hot*" means that the average temperature of the warmest month is below 22°C (72°F) and at least four months have a monthly mean temperature above 10°C(50°F).

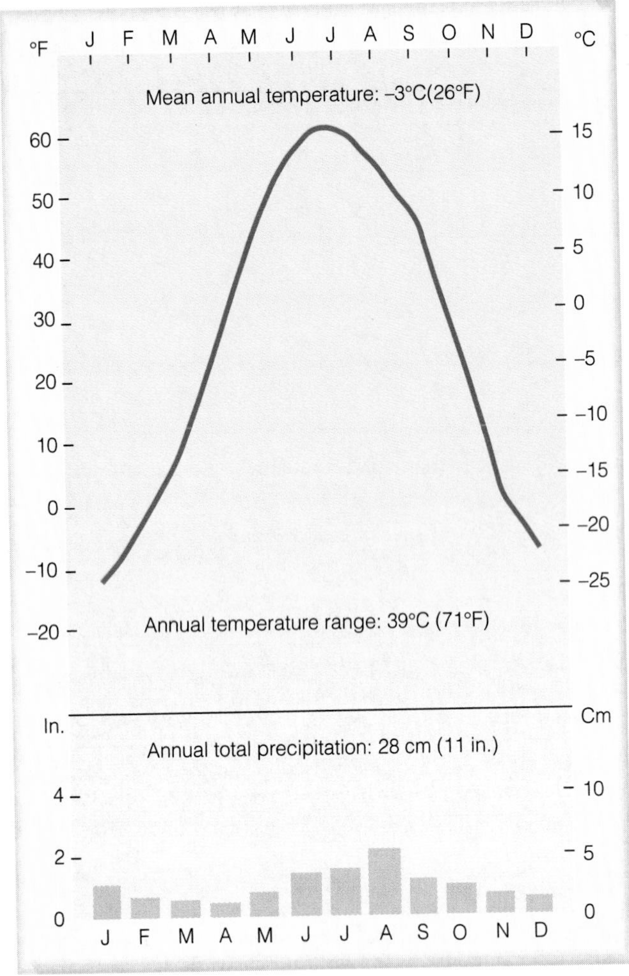

FIGURE 13.21 Climatic data for Fairbanks, Alaska, latitude 65°N. A station with a subpolar climate (Dfc).

Notice that both cities experience a large annual temperature range. This is characteristic of climates located in the northern interior of continents. In fact, as we move poleward, the annual temperature range increases. In Des Moines, it is 31°C (56°F), while 950 km (590 mi) to the north in Winnipeg, it is 38°C (68°F). The summer precipitation maximum expected for these interior continental locations shows up well in Fig. 13.20. Most of the summer rain is in the form of isolated convective showers, although an occasional weak frontal system can produce more widespread precipitation, as can a cluster of thunderstorms—the Mesoscale Convective Complex described in Chapter 10. The weather in both climatic types can be quite changeable, especially in winter, when a brief warm spell is replaced by blustery winds and temperatures plummeting well below −30°C (−22°F).

When winters are severe and summers short and cool, with only one to three months having a mean temperature exceeding 10°C (50°F), the climate is described as *subpolar* (Dfc). From Fig. 13.5, we can see that, in North America, this climate occurs in a broad belt across Canada and Alaska; in Eurasia, it stretches from Norway over much of Siberia. The exceedingly low temperatures of winter account for these areas being the primary source regions for continental polar and arctic air masses. Extremely cold winters coupled with cool summers produce large annual temperature ranges, as exemplified by the climate data in Fig. 13.21 for Fairbanks, Alaska.

Precipitation is comparatively light in the subpolar climates, especially in the interior regions, with most places receiving less than 50 cm (20 in.) annually. A good percentage of the precipitation falls when weak cyclonic storms move through the region in summer. The total snowfall is usually not large but the cold air prevents melting, so snow stays on the ground for months at a time. Because of the low temperatures, there is a low annual rate of evaporation that ensures adequate moisture to support the boreal forests of conifers and birches known as **taiga** (see Fig. 13.22). Hence, the subpolar climate is known also as a *boreal climate* and as a *taiga climate.*

In the taiga region of northern Siberia and Asia, where the average temperature of the coldest month drops to a frigid −38°C (−36°F) or below, the climate is designated Dfd. Where the winters are considered dry, the climate is designated Dwd.

DID YOU KNOW?

The word *boreal* comes from the ancient Greek *Boreas,* meaning "wind from the north."

Polar Climates (Group E)

General characteristics: year-round low temperatures (i.e., average temperature of the warmest month is below 10°C, or 50°F).

Extent: northern coastal areas of North America and Eurasia; Greenland; and Antarctica.

Major types: polar tundra (ET) and polar ice caps (EF).

In the **polar tundra** (ET), the average temperature of the warmest month is below 10°C (50°F), but above freezing. (See Fig. 13.23, the climate data for Barrow, Alaska.) Here, the ground is permanently frozen to

FIGURE 13.22 Coniferous forests (taiga) such as this occur where winter temperatures are low and precipitation is abundant.

depths of hundreds of meters, a condition known as **permafrost.** Summer weather, however, is just warm enough to thaw out the upper meter or so of soil. Hence, during the summer, the tundra turns swampy and muddy. Annual precipitation on the tundra is meager, with most stations receiving less than 20 cm (8 in.). In lower latitudes, this would constitute a desert, but in the cold polar regions evaporation rates are very low and moisture remains adequate. Because of the extremely short growing season, *tundra vegetation* consists of mosses, lichens, dwarf trees, and scattered woody vegetation, fully grown and only several centimeters tall (see Fig. 13.24).

Even though summer days are long, the sun is never very high above the horizon. Additionally, some of the sunlight that reaches the surface is reflected by snow and ice, while some is used to melt the frozen soil. Consequently, in spite of the long hours of daylight, summers are quite cool. The cool summers and the extremely cold winters produce large annual temperature ranges.

When the average temperature for every month drops below freezing, plant growth is impossible, and the region is perpetually covered with snow and ice. This climatic type is known as **polar ice cap** (EF). It occupies the interior ice sheets of Greenland and Antarctica, where the depth of ice in some places measures thousands of meters. In this region, temperatures are never

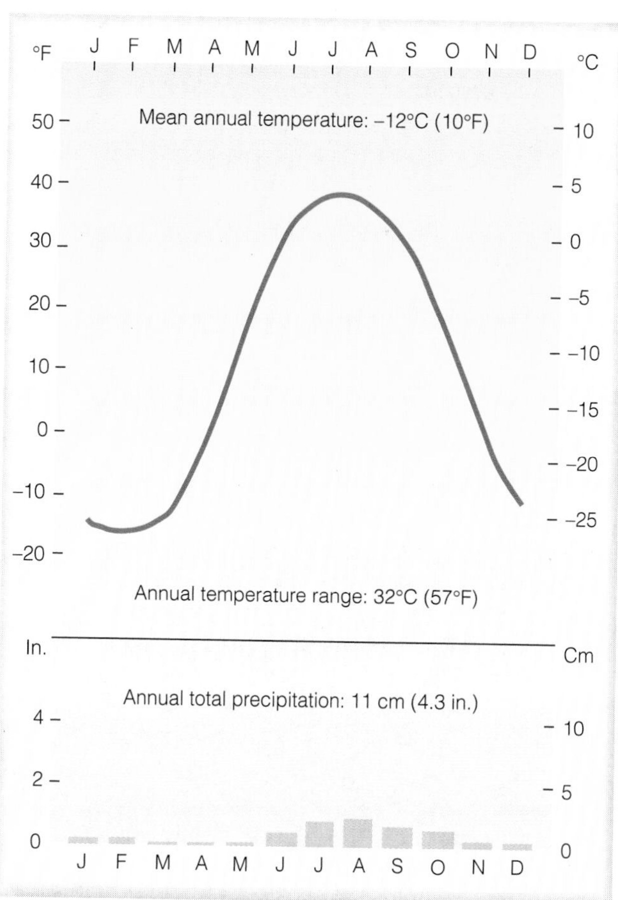

FIGURE 13.23 Climatic data for Barrow, Alaska, latitude 71°N. A station with a polar tundra climate (ET).

FIGURE 13.24 Tundra vegetation in Alaska. This type of tundra is composed mostly of sedges and dwarfed wildflowers that bloom during the brief growing season.

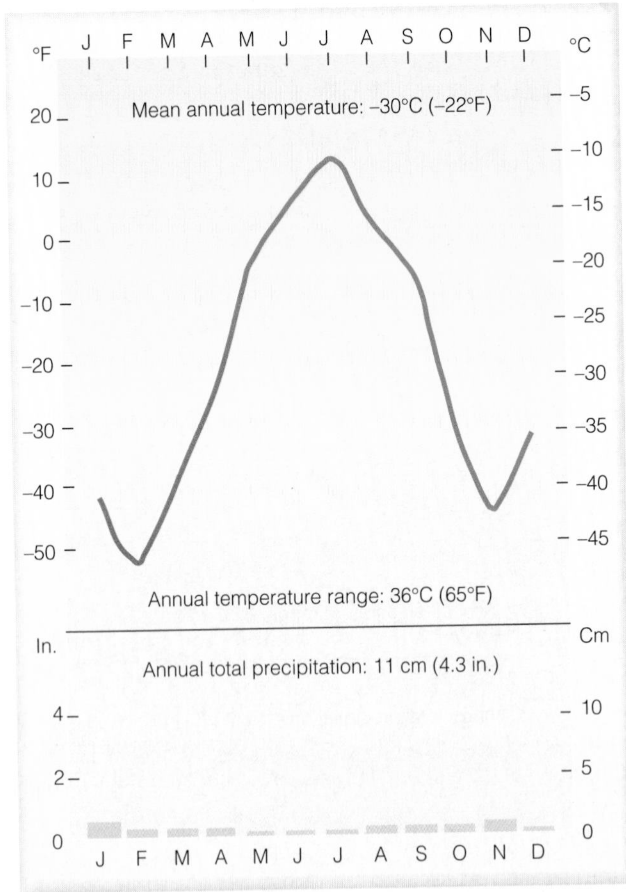

FIGURE 13.25 Climatic data for Eismitte, Greenland, latitude 71°N. Located in the interior of Greenland at an elevation of almost 10,000 feet above sea level. Eismitte has a polar ice cap climate (EF).

much above freezing, even during the middle of "summer." The coldest places in the world are located here. Precipitation is extremely meager with many places receiving less than 10 cm (4 in.) annually. Most precipitation falls as snow during the "warmer" summer. Strong downslope katabatic winds frequently whip the snow about, adding to the climate's harshness. The data in Fig. 13.25 for Eismitte, Greenland, illustrate the severity of an EF climate.

Highland Climates (Group H) It is not necessary to visit the polar regions to experience a polar climate. Because temperature decreases with altitude, climatic changes experienced when climbing 300 m (1000 ft) in elevation are about equivalent in high latitudes to horizontal changes experienced when traveling 300 km (186 mi) northward. (This distance is equal to about 3° latitude.) Therefore, when ascending a high mountain, one can travel through many climatic regions in a relatively short distance.

Figure 13.26 shows how the climate and vegetation change along the western slopes of the central Sierra Nevada. (See Fig. 13.4, p. 350, for the precipitation patterns for this region.) Notice that, at the base of the mountains, the climate and vegetation represent semiarid conditions, while in the foothills the climate becomes Mediterranean and the vegetation changes to chaparral. Higher up, thick fir and pine forests prevail. At still higher elevations, the climate is subpolar and the

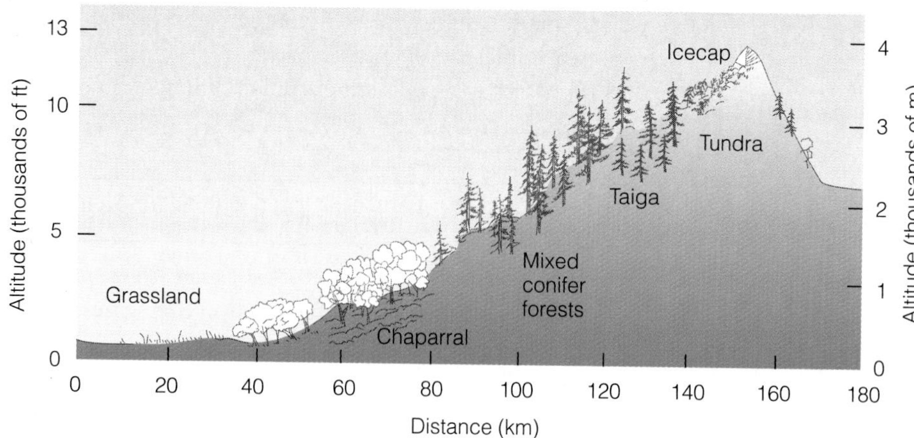

FIGURE 13.26 Vertical view of changing vegetation and climate due to elevation in the central Sierra Nevada.

taiga gives way to dwarf trees and tundra vegetation. Near the summit there are permanent patches of ice and snow, with some small glaciers nestled in protected areas. Hence, in less than 13,000 vertical feet, the climate has changed from semi-arid to polar.

SUMMARY

In this chapter, we examined global temperature and precipitation patterns, as well as the various climatic regions throughout the world. Tropical climates are found in low latitudes, where the noon sun is always high, day and night are of nearly equal length, every month is warm, and no real winter season exists. Some of the rainiest places in the world exist in the tropics, especially where warm, humid air rises upslope along mountain ranges.

Dry climates prevail where potential evaporation and transpiration exceed precipitation. Some deserts, such as the Sahara, are mainly the result of sinking air associated with the subtropical highs, while others, due to the rain shadow effect, are found on the leeward side of mountains. Many deserts form in response to both of these effects.

Middle latitudes are characterized by a distinct winter and summer season. Winters tend to be milder in lower latitudes and more severe in higher latitudes. Along the east coast of some continents, summers tend to be hot and humid as moist air sweeps poleward around the subtropical highs. The air often rises and condenses into afternoon thunderstorms in this humid subtropical climate. The west coasts of many continents tend to be drier, especially in summer, as the combination of cool ocean water and sinking air of the subtropical highs, to a large degree, inhibit the formation of cumuliform clouds.

In the middle of large continents, such as North America and Eurasia, summers are usually wetter than winters. Winter temperatures are generally lower than those experienced in coastal regions. As one moves northward, summers become shorter and winters longer and colder. Polar climates prevail at high latitudes, where winters are severe and there is no real summer. When ascending a high mountain, one can travel through many climatic zones in a relatively short distance.

Meteorology Now™ Assess your understanding of this chapter's topics with additional quizzing and tutorials at http://earthscience.brookscole.com/ahrens/ess4e.

KEY TERMS

The following terms are listed in the order they appear in the text. Define each. Doing so will aid you in reviewing the material covered in this chapter.

microclimate
mesoclimate
macroclimate
climatic controls
Köppen classification system
tropical rain forest
tropical wet climate
laterite
tropical monsoon climate
tropical wet-and-dry climate
savanna grass
arid climate
xerophytes
semi-arid climate
steppe
humid subtropical climate
marine climate
dry-summer subtropical (Mediterranean) climate
humid continental climate
subpolar climate
taiga
polar tundra climate
permafrost
polar ice cap climate

QUESTIONS FOR REVIEW

1. What factors determine the global pattern of precipitation?
2. Explain why, in North America, precipitation typically is a maximum along the West Coast in winter, a maximum on the Central Plains in summer, and fairly evenly distributed between summer and winter along the East Coast.
3. According to Köppen's climatic system (Fig. 13.5, pp. 352–353), what major climatic type is most abundant
   (a) in North America;
   (b) in South America;
   (c) throughout the world?
4. What is the primary factor that makes a dry climate "dry"?
5. What climatic information did Köppen use in classifying climates?
6. In which climatic region would each of the following be observed: tropical rain forest, xerophytes, steppe, taiga, tundra, and savanna?
7. What are the controlling factors (the major climatic controls) that produce the following climatic regions:
   (a) tropical wet and dry;
   (b) Mediterranean;
   (c) marine;
   (d) humid subtropical;
   (e) subpolar;
   (f) polar ice cap?
8. Why are marine climates (Cs) usually found on the west coast of continents?
9. Why are large annual temperature ranges characteristic of D-type climates?
10. Why are D climates found in the Northern Hemisphere but not in the Southern Hemisphere?
11. Explain why a tropical rain forest climate will support a tropical rain forest, while a tropical wet-and-dry climate will not.
12. What is the primary distinction between a Cfa and a Dfa climate?
13. Explain how arid deserts can be found adjacent to oceans.
14. Why did Köppen use the 10°C (50°F) average temperature for July to distinguish between D and E climates?
15. What accounts for the existence of a BWk climate in the western Great Basin of North America?
16. Barrow, Alaska, receives a mere 11 cm (about 4.3 in.) of precipitation annually. Explain why its climate is not classified as arid or semi-arid.
17. Explain why subpolar climates are also known as boreal climates and taiga climates.
18. How did Köppen define a polar climate? How did he define a tropical climate?

QUESTIONS FOR THOUGHT AND EXPLORATION

1. Why do cities east of the Rockies, such as Denver, Colorado, get much more precipitation than cities east of the Sierra Nevada, such as Reno, Nevada?
2. According to the Köppen system of climate classification, which type of climate is found in your area?
3. Los Angeles, Seattle, and Boston are all coastal cities, yet Boston has a continental rather than a marine climate. Explain why.
4. Why are many structures in polar regions built on pilings?
5. Why are summer afternoon temperatures in a humid subtropical climate (Cfa) often higher than in a tropical wet climate (Af)?
6. Why are humid subtropical climates (Cfa) found in regions bounded by 20° and 40° (N or S) latitudes, and nowhere else?
7. In which of the following climate types is virga likely to occur most frequently: humid continental, arid desert, or polar tundra? Explain why.
8. As shown in Figure 13.17, San Francisco and Sacramento, California, have similar mean annual temperatures but different annual temperature ranges. What factors control the annual temperature ranges at these two locations?
9. Why is there a contrast in climate types on either side of the Rocky Mountains, but not on either side of the Appalachian Mountains?
10. Sketch graphs of annual variation of temperature and precipitation for a coastal location, and also for a location in the center of a large continent. Explain any differences in your graphs.
11. On a blank map of the world, roughly outline where Köppen's major climatic regions are located.
12. Over the past 100 years or so the earth has warmed by more than 0.7°C (1.3°F). If this warming should continue over the next 100 years, explain how this rise in temperature might influence the boundary between C and D climates. How would the warming influence the boundary between D and E climates?
13. U.S. Climate Data (http://www.cdc.noaa.gov/USclimate/states.fast.html): Compare graphs of maximum and minimum temperature and precipitation for three

cities in different parts of the United States. Describe the differences in climate.

14. Current Global Temperatures (http://www.ssec.wisc.edu/data/composites.html): Look at the current global pattern of surface temperatures. How does it compare to the average conditions? Are there any areas experiencing significant anomalies (differences from average conditions)? What might be causing these anomalies?

- Go to the Brooks/Cole Earth Sciences Resource Center (http://earthscience.brookscole.com) for critical thinking exercises, articles, and additional readings from InfoTrac College Edition, Brooks/Cole's online student library.

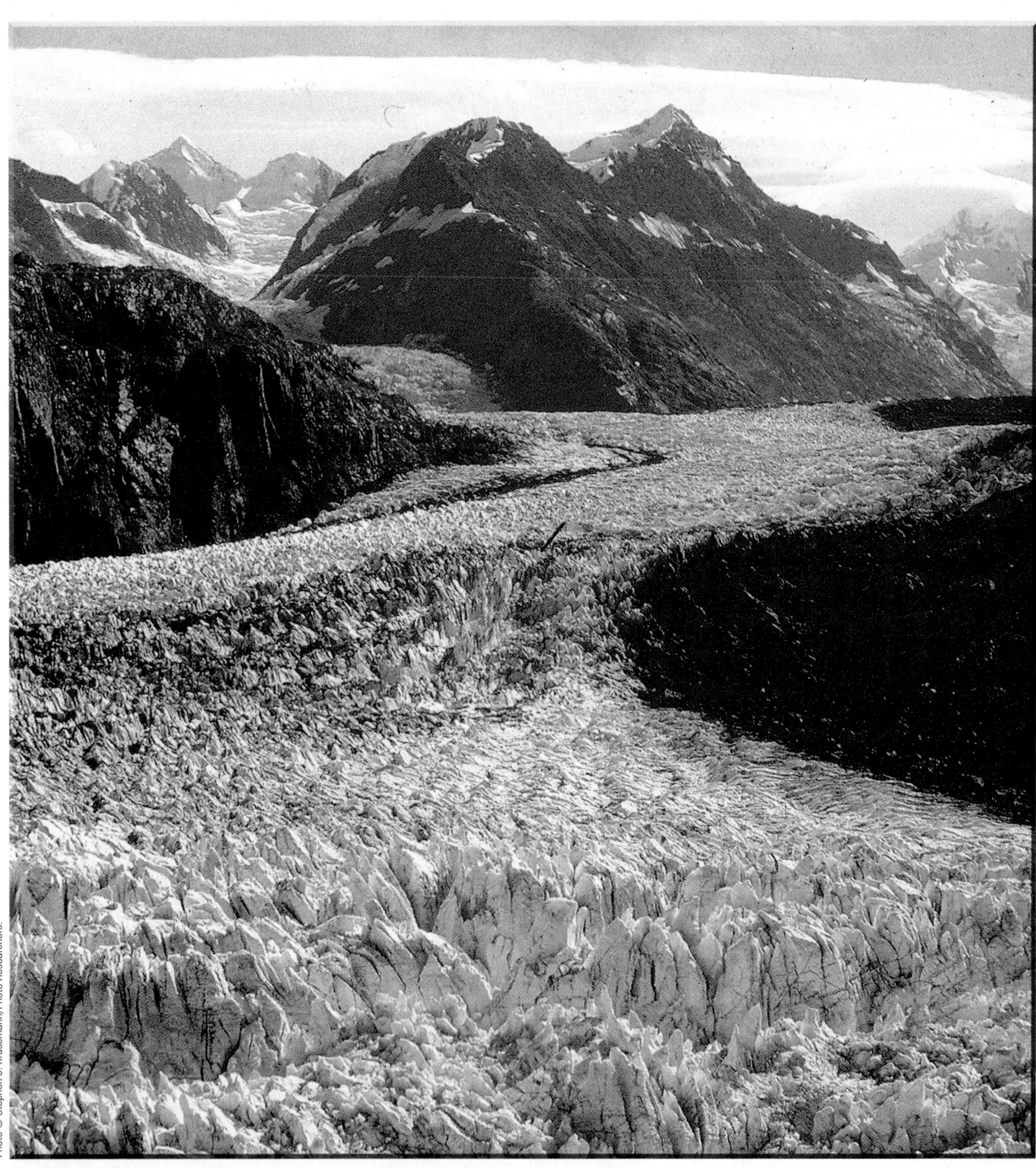

Twenty thousand years ago, glaciers covered three times more land than they do today. Over portions of North America, the ice was thousands of meters thick and extended well into the eastern half of the United States. Today, the ice over North America is gone except for the relatively small valley glaciers that still exist in high mountain valleys.

Meteorology Now™ This icon, appearing throughout the book, indicates an opportunity to explore interactive tutorials, animations, or practice problems available on the MeteorologyNow Web site at http://earthscience.brookscole.com/ahrens/ess4e.

CHAPTER 14

Climate Change

A change in our climate however is taking place very sensibly. Both heats and colds are becoming much more moderate within the memory even of the middle-aged. Snows are less frequent and less deep. They do not often lie, below the mountains, more than one, two, or three days, and very rarely a week. They are remembered to have been formerly frequent, deep, and of long continuance. The elderly inform me the earth used to be covered with snow about three months in every year. The rivers, which then seldom failed to freeze over in the course of the winter, scarcely ever do now. This change has produced an unfortunate fluctuation between heat and cold, in the spring of the year, which is very fatal to fruits. In an interval of twenty-eight years, there was no instance of fruit killed by the frost in the neighborhood of Monticello. The accumulated snows of the winter remaining to be dissolved all together in the spring, produced those overflowings of our rivers, so frequent then, and so rare now.

Thomas Jefferson, *Notes on the State of Virginia,* 1781

CONTENTS

The climate is always changing. Evidence shows that climate has changed in the past, and nothing suggests that it will not continue to change. In Chapter 12, we saw that as the urban environment grows, its climate differs from that of the region around it. Sometimes the difference is striking, as when city nights are warmer than the nights of the outlying rural areas. Other times, the difference is subtle, as when the climate is modified by a layer of sulfur-rich haze that blankets the city. Climate change, in the form of a persistent drought or a delay in the annual monsoon rains, can adversely affect the lives of millions. Even small changes can have an adverse effect when averaged over many years, as when grasslands once used for grazing gradually become uninhabited deserts. In this chapter, first we will investigate how the global climate has changed; then we will examine some theories on why it has changed.

THE EARTH'S CHANGING CLIMATE

Not only is the earth's climate always changing, but a mere 18,000 years ago the earth was in the grip of a cold spell, with *alpine glaciers* extending their ice fingers down river valleys and huge ice sheets *(continental glaciers)* covering vast areas of North America and Europe. The ice measured several kilometers thick and extended as far south as New York and the Ohio River Valley. Perhaps the glaciers advanced 10 times during the last 2 million years, only to retreat. In the warmer periods, between glacier advances, average global temperatures were slightly higher than at present. Hence, some scientists feel that we are still in an ice age, but in the comparatively warmer part of it.

Presently, glaciers cover only about 10 percent of the earth's land surface. Most of this ice is in the Greenland and Antarctic ice sheets. If global temperatures were to rise enough so that all of this ice melted, the level of the ocean would rise about 65 m (213 ft) (see Fig. 14.1). Imagine the catastrophic results: Many major cities (such as New York, Tokyo, and London) would be inundated. Even a rise in global temperature of several degrees Celsius might be enough to raise sea level by about half a meter or so, flooding coastal lowlands.

Determining Past Climates The study of the geological evidence left behind by advancing and retreating glaciers is one factor suggesting that global climate has undergone slow but continuous changes. To reconstruct past climates, scientists must examine and then carefully piece together all the available evidence. Unfortunately, the evidence only gives a general understanding of what the climate was like. For example, fossil pollen of a tun-

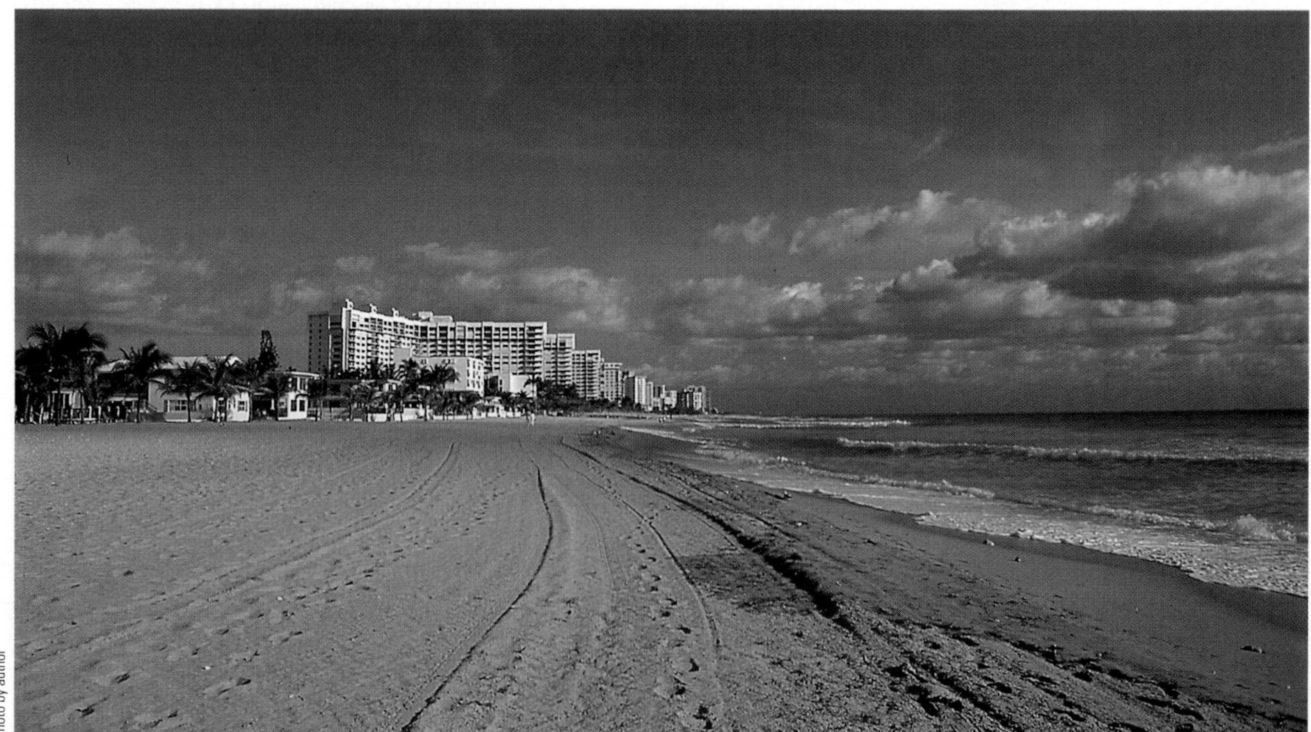

Photo by author

FIGURE 14.1 If all the ice locked up in glaciers and ice sheets were to melt, estimates are that this coastal area of south Florida would be under 65 m (213 ft) of water. Even a relatively small one-meter rise in sea level would threaten half of the world's population with rising seas.

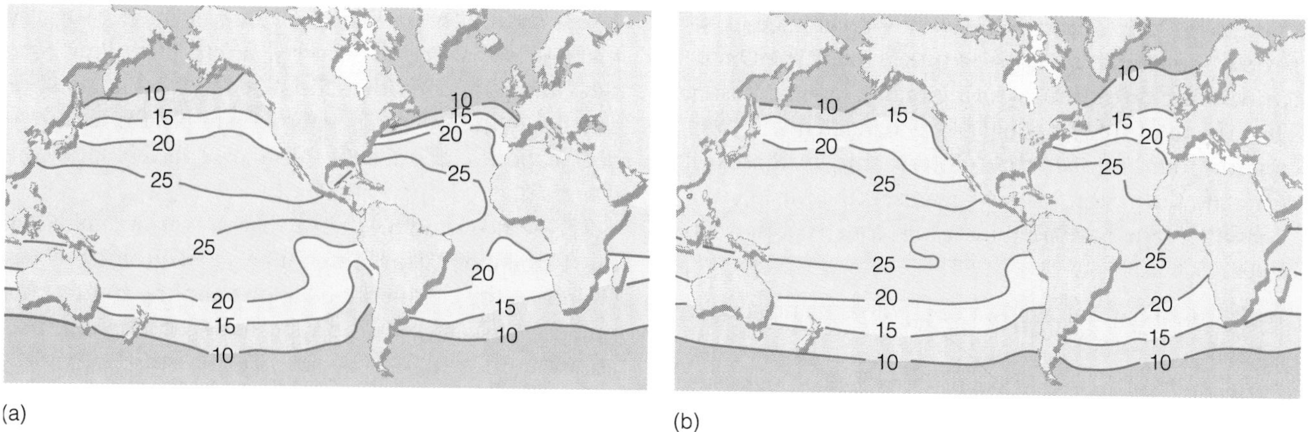

FIGURE 14.2 (a) Sea surface isotherms (°C) during August 18,000 years ago and (b) during August today. Apparently, during the Ice Age (diagram a) the Gulf Stream shifted to a more easterly direction, depriving northern Europe of its warmth and causing a rapid north-to-south ocean surface temperature gradient.

dra plant collected in a layer of sediment in New England and dated to be 12,000 years old suggests that the climate of that region was much colder than it is today.

Other evidence of global climatic change comes from core samples taken from ocean floor sediments and ice from Greenland. A multiuniversity research project known as CLIMAP (*C*limate: *l*ong-range *i*nvestigation *map*ping and *p*rediction) studied the past million years of global climate. Thousands of meters of ocean sediment obtained with a hollow-centered drill were analyzed. The sediment contains the remains of calcium carbonate shells of organisms that once lived near the surface. Because certain organisms live within a narrow range of temperature, the distribution and type of organisms within the sediment indicate the surface water temperature.

In addition, the oxygen-isotope* ratio of these shells provides information about the sequence of glacier advances. For example, most of the oxygen in sea water is composed of 8 protons and 8 neutrons in its nucleus, giving it an atomic weight of 16. However, about one out of every thousand oxygen atoms contains an extra 2 neutrons, giving it an atomic weight of 18. When ocean water evaporates, the heavy oxygen 18 tends to be left behind. Consequently, during periods of glacier advance, the oceans, which contain less water, have a higher concentration of oxygen 18. Since the shells of marine organisms are constructed from the oxygen atoms existing in ocean water, determining the ratio of oxygen 18 to oxygen 16 within these shells yields information about how the climate may have varied in the past. A higher ratio of oxygen 18 to oxygen 16 in the sediment record suggests a colder climate, while a lower ratio suggests a warmer climate. Using data such as these, the CLIMAP project was able to reconstruct the earth's surface ocean temperature for various times during the past (see Fig. 14.2).

Vertical ice cores extracted from ice sheets in Antarctica and Greenland provide additional information on past temperature patterns. Glaciers form over land where temperatures are sufficiently low so that, during the course of a year, more snow falls than will melt. Successive snow accumulations over many years compact the snow, which slowly recrystallizes into ice. Since ice is composed of hydrogen and oxygen, examining the oxygen-isotope ratio in ancient cores provides a past record of temperature trends. Generally, the colder the air when the snow fell, the richer the concentration of oxygen 16 in the core. Moreover, bubbles of ancient air trapped in the ice can be analyzed to determine the past composition of the atmosphere (see Fig. 14.10, p. 386).

Ice cores also record the causes of climate changes. One such cause is deduced from layers of sulfuric acid in the ice. The sulfuric acid originally came from large volcanic explosions that injected huge quantities of sulfur into the stratosphere. The resulting sulfate aerosols eventually fell to the earth in polar regions as acid snow, which was preserved in the ice sheets. The Greenland ice cores also provide a continuous record of sulfur from human sources. Moreover, ice cores at both poles record a beryllium isotope ($^{10}B_e$) that indicates solar activity. Various types of dust collected in the cores indicate whether the climate was arid or wet.

Still other evidence of climatic change comes from the study of annual growth rings of trees, called **dendrochronology.** As a tree grows, it produces a layer of wood cells under its bark. Each year's growth appears as a ring. The changes in thickness of the rings indicate climatic changes that may have taken place from one year

*Isotopes are atoms whose nuclei have the same number of protons but different numbers of neutrons.

to the next. Tree rings are only useful in regions that experience an annual cycle and in trees that are stressed by temperature or moisture during their growing season. The growth of tree rings has been correlated with precipitation and temperature patterns for hundreds of years into the past in various regions of the world.

Other data have been used to reconstruct past climates, such as:

1. records of natural lake-bottom sediment and soil deposits
2. the study of pollen in deep ice caves, soil deposits, and sea sediments
3. certain geologic evidence (ancient coal beds, sand dunes, and fossils) and the change in the water level of closed-basin lakes
4. documents concerning droughts, floods, and crop yields
5. the study of oxygen-isotope ratios of corals
6. dating calcium carbonate layers of stalactites in caves
7. borehole temperature profiles, which can be inverted to give records of past temperature change at the surface
8. deuterium (heavy hydrogen) ratios in ice cores, which indicate temperature changes

Despite all of these data, our knowledge about past climates is still incomplete. Now that we have reviewed *how* the climatologist gains information about the past, let's look at *what* this information reveals.

Climate Through the Ages Throughout much of the earth's history, long before humanity came onto the scene, the global climate was much warmer than now, with the global mean temperature perhaps between 8°C and 15°C warmer than it is today. During most of this time, the polar regions were free of ice. These comparatively warm conditions, however, were interrupted by several periods of glaciation. Geologic evidence suggests that one glacial period occurred about 700 million years ago (m.y.a.) and another about 300 m.y.a. The most recent one—the *Pleistocene epoch* or, simply, the **Ice Age**—began about 2 m.y.a. Let's summarize the climatic conditions that led up to the Pleistocene.

About 65 m.y.a., the earth was warmer than it is now; polar ice caps did not exist. Beginning about 55 m.y.a., the earth entered a long cooling trend. After millions of years, polar ice appeared. As average temperatures continued to lower, the ice grew thicker, and by about 10 m.y.a. a deep blanket of ice covered the Antarctic. Meanwhile, snow and ice began to accumulate in high mountain valleys of the Northern Hemisphere, and alpine, or valley, glaciers soon appeared.

About 2 m.y.a., continental glaciers appeared in the Northern Hemisphere, marking the beginning of the Pleistocene epoch. The Pleistocene, however, was not a period of continuous glaciation but a time when glaciers alternately advanced and retreated (melted back) over large portions of North America and Europe. Between the glacial advances were warmer periods called **interglacial periods,** which lasted for 10,000 years or more.

The most recent North American glaciers reached their maximum thickness and extent about 18,000–22,000 years ago (y.a.). At that time, the sea level was perhaps 120 m (395 ft) lower than it is now. The lower sea level exposed vast areas of land, such as the *Bering land bridge* (a strip of land that connected Siberia to Alaska), which allowed human and animal migration from Asia to North America.

The ice began to retreat about 14,000 y.a. as surface temperatures slowly rose, producing a warm spell (see Fig. 14.3). Then, about 12,700 y.a., the average temperature suddenly dropped and northeastern North America and northern Europe reverted back to glacial conditions.

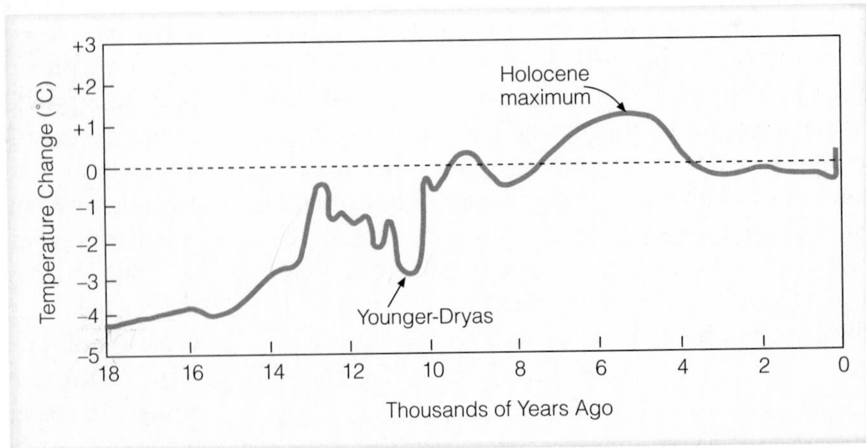

FIGURE 14.3 Average air-temperature variations for the past 18,000 years. These data, which represent temperature records compiled from a variety of sources, only give an approximation of temperature changes. Some regions of the world experienced a cooling and other regions a warming that either preceded or lagged behind the temperature variations shown in the diagram.

About 1000 years later, the cold spell (known as the **Younger-Dryas***) ended abruptly and temperatures rose rapidly in many areas. Beginning about 8000 y.a. the mean temperature dropped by as much as 2°C over central Europe. During this cold period, which was not experienced worldwide, the European alpine timberline fell about 200 m (600 ft). The cold period ended, temperatures began to rise, and by about 6000 y.a. the continental ice sheets over North America were gone. This warm spell during the current interglacial period, or *Holocene epoch,* is sometimes called the *mid-Holocene maximum,* and because this warm period favored the development of plants, it is also known as the *climatic optimum.* About 5000 y.a. a cooling trend set in, during which extensive alpine glaciers returned, but not continental glaciers.

It is interesting to note that ice core data from Greenland reveal that rapid shifts in climate (from ice age conditions to a much warmer state) took place in as little as three years over central Greenland around the end of the Younger-Dryas. The data also reveal that similar rapid shifts in climate occurred several times toward the end of the Ice Age. What could cause such rapid changes in temperature? One possible explanation is given in the Focus section on p. 378.

Climate During the Last 1000 Years Figure 14.4 shows how the average surface air temperature changed in the Northern Hemisphere during the last 1000 years. The data needed to reconstruct the temperature profile in Fig. 14.4 comes from a variety of sources, including tree rings, corals, ice cores, historical records, and thermometers. Notice that about 1000 y.a., the Northern Hemisphere was slightly cooler than average (where average represents the average temperature from 1961 to 1990). However, certain regions in the Northern Hemisphere were warmer than others. For example, during this time vineyards flourished and wine was produced in England, indicating warm, dry summers and the absence of cold springs. It was during the early part of the millennium that Vikings colonized Iceland and Greenland.*

Notice in Fig. 14.4 that the temperature curve shows a relatively warm period during the 11th to the 14th centuries—relatively warm, but still cooler than the 20th century. During this time, the relatively mild climate of Western Europe began to show large variations. For several hundred years the climate grew stormy. Both great floods and great droughts occurred. Extremely cold winters were followed by relatively warm ones. During the cold spells, the English vineyards and the Viking settlements suffered. Europe experienced several famines during the 1300s.

> **DID YOU KNOW?**
>
> About 20,000 years ago, the sea level was so low that there was no English Channel and the River Thames was actually a tributary of the Rhine River.

*This exceptionally cold spell is named after the *Dryas,* an arctic flower.

*This relatively warm, tranquil period of several hundred years over western Europe is sometimes referred to in that region as the *Medieval Climatic Optimum.*

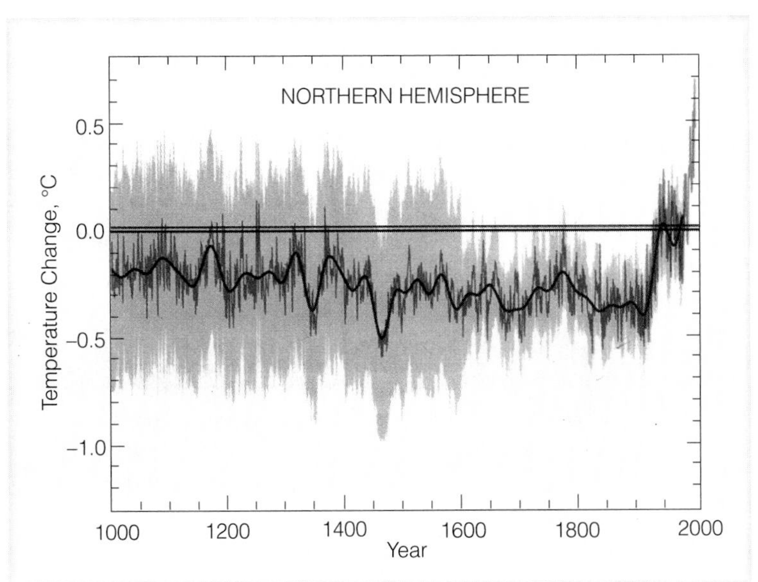

FIGURE 14.4 The average temperature variations over the Northern Hemisphere for the last 1000 years relative to the 1961 to 1990 average (zero line). Yearly temperature data from tree rings, corals, ice cores, and historical records are shown in blue. Yearly temperature data from thermometers are in red. The black line represents a smoothing of the data. (The gray shading represents a statistical 95 percent confidence range in the annual temperature data.) (*Source:* From Climate Change 2001: The Scientific Basis, 2001, by J.T. Houghton, et al. Copyright © 2001 Cambridge University Press. Reprinted with permission of the Intergovernmental Panel on Climate Change.)

FOCUS ON A SPECIAL TOPIC

The Ocean Conveyor Belt and Climate Change

During the last glacial period, the climate around Greenland (and probably other areas of the world) underwent shifts, from ice-age temperatures to much warmer conditions in a matter of years. What could bring about such large fluctuations in temperature over such a short period of time? It now appears that a vast circulation of ocean water, known as the *conveyor belt,* plays a major role in the climate picture.

Figure 1 illustrates the movement of the ocean conveyor belt, or *thermohaline circulation.** The conveyor-like circulation begins in the north Atlantic near Greenland and Iceland, where salty surface water is cooled through contact with cold Arctic air masses. The cold, dense water sinks and flows southward through the deep Atlantic Ocean, around Africa, and into the Indian and Pacific Oceans. In the North Atlantic, the sinking of cold water draws warm water northward from lower latitudes. As this water flows northward, evaporation increases the water's salinity (dissolved salt content) and density. When this salty, dense water reaches the far regions of the North Atlantic, it gradually sinks to great depths. This warm part of the conveyor delivers an incredible amount of tropical heat to the northern Atlantic. During the winter, this heat is transferred to the overlying atmosphere, and evaporation moistens the air. Strong westerly winds then carry this warmth and moisture into northern and western Europe, where it causes winters to be much warmer and wetter than one would normally expect for this latitude.

Ocean sediment records along with ice-core records from Greenland suggest that the giant conveyor belt has switched on and off during the last glacial period. Such events have apparently coincided with rapid changes in climate. For example, when the conveyor belt is strong, winters in northern Europe tend to be wet and relatively mild. However, when the conveyor belt is weak or stops altogether, winters in northern Europe appear to turn much colder. This switching from a period of milder winters to one of severe cold shows up many times in the climate record. One such event—the Younger-Dryas—illustrates how quickly climate can change and how western and northern Europe's climate can cool within a matter of decades, then quickly return back to milder conditions.

Apparently, one mechanism that can switch the conveyor belt off is a massive influx of freshwater. For example, about 11,000 years ago during the Younger-Dryas event, fresh water from a huge glacial lake began to flow down the St. Lawrence River and into the North Atlantic. This massive inflow of freshwater reduced the salinity (and, hence, density) of the surface water to the point that it stopped sinking. The conveyor shut down for about 1000 years during which time severe cold engulfed much of northern Europe. The conveyor belt started up again when the fresh water began to drain down the Mississippi rather than into the North Atlantic. It was during this time that milder conditions returned to northern Europe.

Will increasing levels of CO_2 have an effect on the conveyor belt? Some climate models predict that as CO_2 levels increase, more precipitation will fall over the North Atlantic. This situation reduces the density of the sea water and slows down the conveyor belt. In fact, if CO_2 levels double, computer models predict that the conveyor belt will slow by about 30 percent. If CO_2 levels quadruple, models predict that the conveyor belt will stop and severe cold will return to northern Europe, even though global temperatures will likely increase dramatically.

*Thermohaline circulations are ocean circulations produced by differences in temperature and/or salinity. Changes in ocean water temperature or salinity create changes in water density.

Again look back at Fig. 14.4 and observe that the Northern Hemisphere experienced a slight cooling during the 15th to 19th centuries. This cooling was significant enough in certain areas to allow alpine glaciers to increase in size and advance down river canyons. In many areas in Europe, winters were long and severe; summers, short and wet. The vineyards in England vanished, and farming became impossible in the more northern latitudes. Cut off from the rest of the world by an advancing ice pack, the Viking colony in Greenland perished.

There is no evidence that this cold spell existed worldwide. However, over Europe, this cold period has come to be known as the **Little Ice Age.** During these colder times, one particular year stands out: 1816. In Europe that year, bad weather contributed to a poor wheat crop, and famine spread across the land. In Northern America, unusual blasts of cold arctic air moved through Canada and the northeastern United States between May and September. The cold spells brought heavy snow in June and killing frosts in July and August. In the warmer days that followed each cold snap, farm-

DID YOU KNOW?

During the 1700s, winters were much colder over North America than they are today. Those cold winters, in fact, allowed soldiers during the Revolutionary War to drag cannons from Staten Island to Manhattan across the frozen Upper New York Bay.

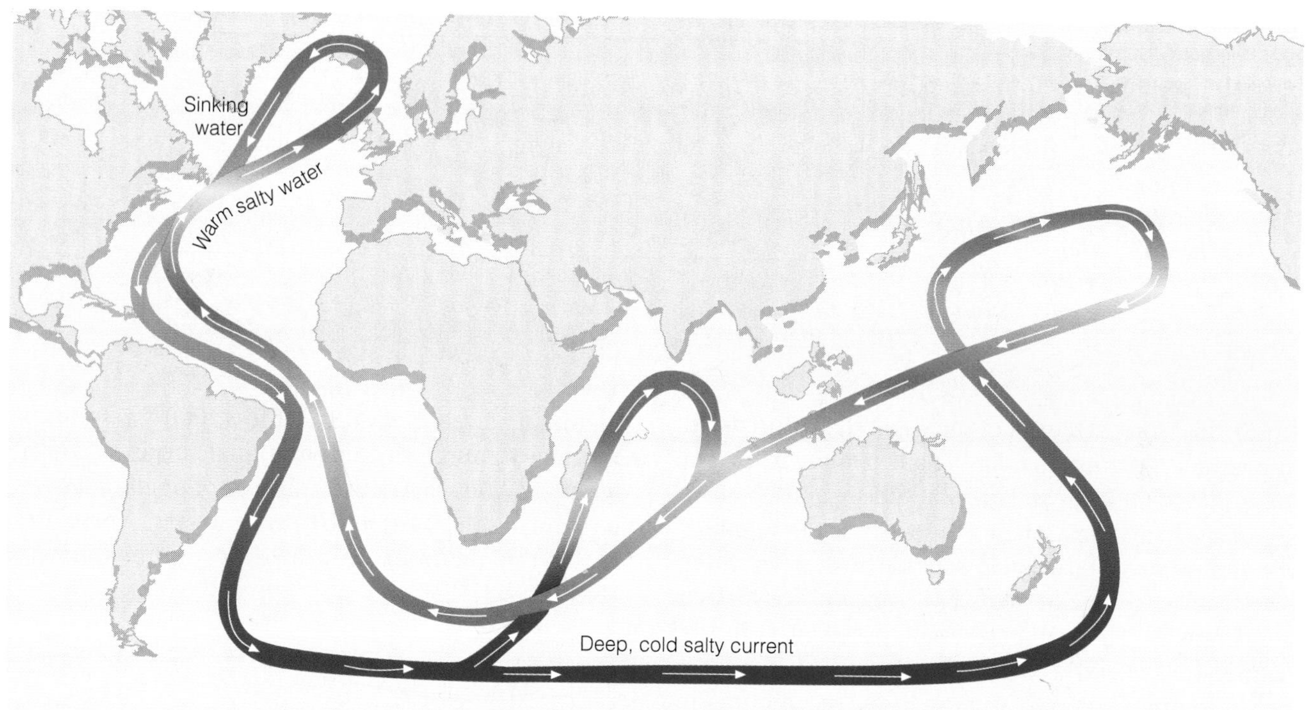

FIGURE 1 The ocean conveyor belt. In the North Atlantic, cold, salty water sinks, drawing warm water northward from lower latitudes. The warm water provides warmth and moisture for the air above, which is then swept into northern Europe by westerly winds that keep the climate of that region milder than one would normally expect. When the conveyor belt stops, winters apparently turn much colder over northern Europe.

ers replanted, only to have another cold outbreak damage the planting. The year 1816 has come to be known as "the year without a summer" or "eighteen hundred and froze-to-death." The unusually cold summer was followed by a bitterly cold winter.

In the early 1900s, the average global surface temperature began to rise (see Fig. 14.5). Notice that, from about 1900 to 1945, the average temperature rose nearly 0.5°C. Following the warmer period, the earth began to cool slightly over the next 25 years or so. In the late 1960s and 1970s, the cooling trend ended over most of the Northern Hemisphere. In the mid-1970s, a warming trend set in that continued into the twenty-first century. It appears, in fact, that over the Northern Hemisphere, the decade of the 1990s was the warmest of the 20th century, with 1998 being the warmest year in over 1000 years.* Moreover, it appears that the increase in average temperature experienced over the Northern Hemisphere during the 20th century is likely to have been the largest increase in temperature of any century during the past 1000 years.

The average warming experienced over the globe, however, has not been uniform. The greatest warming has occurred over the mid-latitude continents in winter and spring, whereas a few areas have not warmed in recent decades, such as areas of the oceans in the Southern Hemisphere and parts of Antarctica. The United States has experienced little warming as compared to the rest of

*The exceptionally warm year of 1998 happened to coincide with a major El Niño warming of the tropical Pacific Ocean. The second warmest year to date, 2003, did not coincide with a major El Niño event.

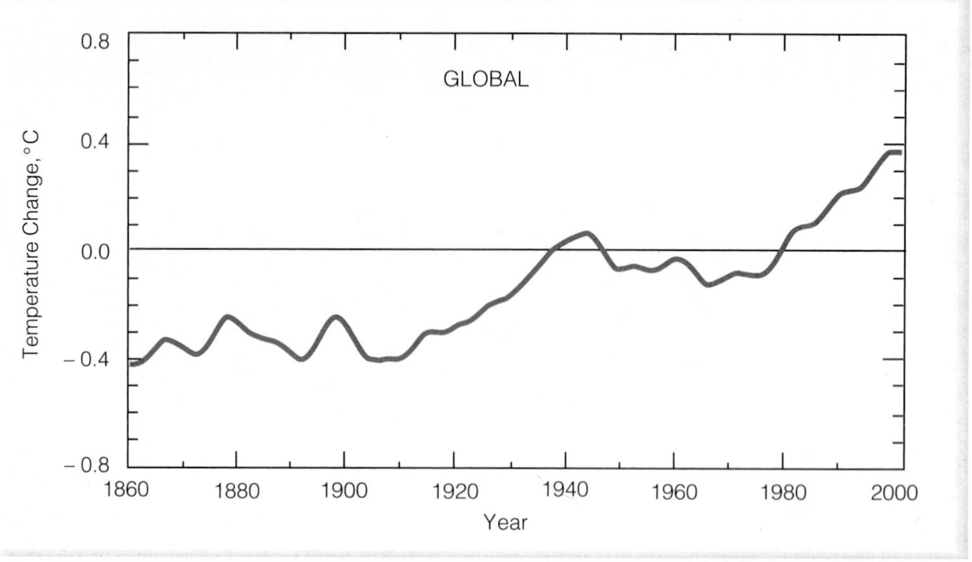

FIGURE 14.5 Average temperature variations over the globe (land and sea) from 1860 to 1999. The zero line represents the average surface temperature from 1961 to 1990. (*Source:* From Climate Change 2001: The Scientific Basis, 2001, by J.T. Houghton, et al. Copyright © 2001 Cambridge University Press. Reprinted with permission of the Intergovernmental Panel on Climate Change.)

the world. Moreover, most of the warming has occurred at night—a situation that has lengthened the frost-free seasons in many mid- and high-latitude regions.

The changes in air temperature shown in Fig. 14.5 are derived from three main sources: air temperatures over land, air temperatures over ocean, and sea surface temperatures. There are, however, uncertainties in the temperature record. For example, during this time period recording stations have moved, and techniques for measuring temperature have varied. Also, marine observing stations are scarce. In addition, urbanization (especially in developed nations) tends to artificially raise average temperatures as cities grow (the urban heat island effect). When urban warming is taken into account, and improved sea surface temperature information is incorporated into the data, the warming over the past 100 years measures about 0.6°C (about 1°F).

A global increase in temperature of 0.6°C may seem small, but in Fig. 14.3, p. 376, we can see that global temperatures have varied no more than 2°C during the past 10,000 years. Consequently, an increase of 0.6°C becomes significant when compared with temperature changes over thousands of years.

Up to this point we have examined the temperature record of the earth's surface and observed that during the past century the earth has been in a warming trend. Climate scientists believe that a good part of the warming is due to an enhanced greenhouse effect caused by increasing levels of greenhouse gases, such as CO_2.* If increasing levels of CO_2 are at least partly responsible for the warming, why did the climate begin to cool after 1940? And what caused the exceptionally cold winters during the 14th and 19th centuries? These are a few of the questions we will address in the following sections.

*The earth's atmospheric greenhouse effect is due mainly to the absorption and emission of infrared radiation by gases, such as water vapor, CO_2, methane, nitrous oxide, and chlorofluorocarbons. Refer back to Chapter 2, p. 35, for additional information on this topic.

POSSIBLE CAUSES OF CLIMATIC CHANGE

Why the earth's climate changes naturally is not totally understood. Many theories attempt to explain the changing climate, but no single theory alone can satisfactorily account for *all* the climatic variations of the past.

Why hasn't the riddle of a fluctuating climate been completely solved? One major problem facing any comprehensive theory is the intricate interrelationship of the elements involved. For example, if temperature changes, many other elements may be altered as well. The interactions among the atmosphere, the oceans, and the ice are extremely complex and the number of possible interactions among these systems is enormous. No climatic element within the system is isolated from the others. With this in mind, we will first investigate how feedback systems work; then we will consider some of the current theories of climatic change.

Climate Change and Feedback Mechanisms In Chapter 2, we learned that the earth-atmosphere system is in a delicate balance between incoming and outgoing energy. If this balance is upset, even slightly, global climate can undergo a series of complicated changes.

Let's assume that the earth-atmosphere system has been disturbed to the point that the earth has entered a slow warming trend. Over the years the temperature slowly rises, and water from the oceans rapidly evaporates into the warmer air. The increased quantity of water vapor absorbs more of the earth's infrared energy, thus strengthening the atmospheric greenhouse effect. This strengthening of the greenhouse effect raises the air temperature even more, which, in turn, allows more water to evaporate into the atmosphere. The greenhouse effect becomes even stronger and the air temperature rises even more. This situation is known as the **water vapor–greenhouse feedback.** (Also called the *water vapor–temperature rise* feedback.) It represents a **positive feedback mechanism** because the initial increase in temperature is reinforced by the other processes. If this feedback were left unchecked, the earth's temperature would increase until the oceans evaporated away. Such a chain reaction is called a *runaway greenhouse effect.*

Another positive feedback mechanism is the **snow-albedo feedback,** where an increase in global surface air temperature might cause snow and ice to melt in polar latitudes. This melting would reduce the albedo (reflectivity) of the surface, allowing more solar energy to reach the surface, which would further raise the temperature.

Helping to counteract the positive feedback mechanisms are **negative feedback mechanisms**—those that tend to weaken the interactions among the variables rather than reinforce them. For example, as the surface warms, it emits more infrared radiation.* This increase in radiant energy from the surface would greatly slow the rise in temperature and help to stabilize the climate. The increase in radiant energy from the surface with increasing surface temperature is the strongest negative feedback in the climate system, and greatly lowers the possibility of a runaway greenhouse effect. Consequently, there is no evidence that a runaway greenhouse effect ever occurred on earth, and it is not very likely that it will occur in the future. (However, for information on the greenhouse effect on the planet Venus, read the Focus section on p. 382.)

Another negative feedback on a warming planet might come from clouds. Suppose, for example, that as the surface warms, more water evaporates from the oceans and global low cloudiness increases. Low clouds tend to reflect a large percentage of incoming sunlight, and with less solar energy to heat the surface, the warming slows.

*Outgoing infrared energy actually increases by an amount proportional to the fourth power of the absolute temperature. Doubling the surface temperature results in 16 times more energy emitted.

DID YOU KNOW?

If all the snow that normally falls over central Canada during the course of one year were to stay on the ground and not melt (even during the summer), it would take nearly 3000 years to build an ice sheet comparable to the one that existed there 18,000 years ago.

All feedback mechanisms work simultaneously and in both directions. Earlier, we saw that the snow-albedo feedback produces a positive feedback on a warming planet, but it produces a positive feedback on a cooling planet as well. For example, suppose the earth were in a slow global cooling trend that lasted for hundreds or even thousands of years. Lower temperatures might allow for a greater snow cover in middle and high latitudes, which would increase the albedo of the surface so that much of the incident sunlight would be reflected back to space. Less sunlight absorbed at the surface might cause a further drop in temperature. This action might further increase the snow cover, lowering the temperature even more.* If left unchecked, the snow-albedo feedback would produce a *runaway ice age* which, of course, is not likely on earth because other feedback mechanisms in the atmospheric system are constantly working to moderate the magnitude of the cooling. In summary, the earth-atmosphere system has a number of checks and balances that help it to counteract tendencies of climate change.

Climate Change, Plate Tectonics, and Mountain Building During the geologic past, the earth's surface has undergone extensive modifications. One involves the slow shifting of the continents and the ocean floors. This motion is explained in the widely accepted **theory of plate tectonics** (formerly called the *theory of continental drift*). According to this theory, the earth's outer shell is composed of huge plates that fit together like pieces of a jigsaw puzzle. The plates, which slide over a partially molten zone below them, move in relation to one another. Continents are embedded in the plates and move along like luggage riding piggyback on a conveyor belt. The rate of motion is extremely slow, only a few centimeters per year.

Besides providing insights into many geological processes, plate tectonics also helps to explain past climates. For example, we find glacial features near sea

*This snow-albedo positive feedback on a cooling planet operates on all time scales, including seasonal temperature cycles.

FOCUS ON A SPECIAL TOPIC

THE GREENHOUSE EFFECT ON VENUS

Our closest planetary neighbor, Venus, is about the same size as Earth. Venus is slightly closer to the sun, so compared to Earth, its average surface temperature should be slightly warmer. However, observations reveal that the surface temperature of Venus is not slightly warmer—it is scorching hot, averaging about 480°C (900°F). The cause for these high temperatures is a positive feedback mechanism that some scientists refer to as a *runaway greenhouse effect.*

Unlike Earth, the atmosphere of Venus is almost entirely CO_2 with minor amounts of other gases such as water vapor, sulfur dioxide, and nitrogen. The CO_2 probably originated in much the same way as it did in the Earth's early atmosphere—through volcanic outgassing of CO_2, water vapor, and hydrogen compounds from the planet's hot interior. As the Earth's atmosphere cooled, however, its water vapor condensed into clouds that produced vast amounts of liquid water, which filled the basins to form the seas. Much of the CO_2 dissolved in the ocean water, and through chemical and biological processes became carbonate rocks. Plants evolved that further removed CO_2 and, during photosynthesis, enriched the Earth's atmosphere with oxygen.

On Venus, the story is different. Being closer to the sun, Venus was warmer. In the warmer air, the water vapor probably did not condense, but remained as a vapor to enhance the greenhouse effect. The lack of oceans on Venus meant that its CO_2 was to remain in its atmosphere. Gradually, the atmosphere became more dense. As infrared energy from the surface tried to penetrate this thick atmosphere, it was absorbed and radiated back. Volcanoes continued to spew CO_2 and water vapor into the atmosphere. More greenhouse gases meant more warming, and the runaway positive feedback mechanism was underway.* Eventually, the outgoing energy from the surface balanced the incoming energy (mainly from the atmosphere), but not until the average surface temperature reached an unbearable 480°C.

We know that, given the checks and balances in our own atmosphere, a runaway greenhouse effect on Earth is not likely. But these extremely high temperatures are not likely on Earth for other reasons, too. For one thing, the atmosphere of Venus is about 96 percent CO_2, whereas the Earth's atmosphere contains only about 0.03 percent CO_2. The Earth has oceans that dissolve CO_2; Venus does not. Moreover, the atmosphere of Venus is about 90 times more dense than that of Earth. While the surface air pressure on Earth is close to 1000 millibars, on Venus the surface pressure is about 90,000 millibars. This thick, dense atmosphere of CO_2 on Venus produces an incredible greenhouse effect.

*On Venus, at some point, energetic rays from the sun probably separated the water vapor into hydrogen and oxygen. The lighter hydrogen more than likely escaped from the hot atmosphere, while the heavier oxygen became trapped in surface rocks and minerals.

level in Africa today, suggesting that the area underwent a period of glaciation hundreds of millions of years ago. Were temperatures at low elevations near the equator ever cold enough to produce ice sheets? Probably not. The ice sheets formed when this land mass was located at a much higher latitude. Over the many millions of years since then, the land has slowly moved to its present position. Along the same line, we can see how the fossil remains of tropical vegetation can be found under layers of ice in polar regions today.

According to plate tectonics, the now existing continents were at one time joined together in a single huge continent, which broke apart. Its pieces slowly moved across the face of the earth, thus changing the distribution of continents and ocean basins, as illustrated in Fig. 14.6. Some scientists feel that, when land masses are concentrated in middle and high latitudes, ice sheets are more likely to form. During these times, there is a greater likelihood that more sunlight will be reflected back into space and that the snow-albedo feedback mechanism mentioned earlier will amplify the cooling.

The various arrangements of the continents may also influence the path of ocean currents, which, in turn, could not only alter the transport of heat from low to high latitudes but could also change both the global wind system and the climate in middle and high latitudes. As an example, suppose that plate movement "pinches off" a rather large body of high-latitude ocean water such that the transport of warm water into the region is cut off. In winter, the surface water would eventually freeze over with ice. This freezing would, in turn, reduce the amount of sensible and latent heat given up to the atmosphere. Furthermore, the ice allows snow to accumulate on top of it, thereby setting up conditions that could lead to even lower temperatures.

There are other mechanisms by which tectonic processes* may influence climate. In Fig. 14.7, notice that the formation of oceanic plates (plates that lie beneath the

*Tectonic processes are large-scale geologic processes that deform the earth's crust.

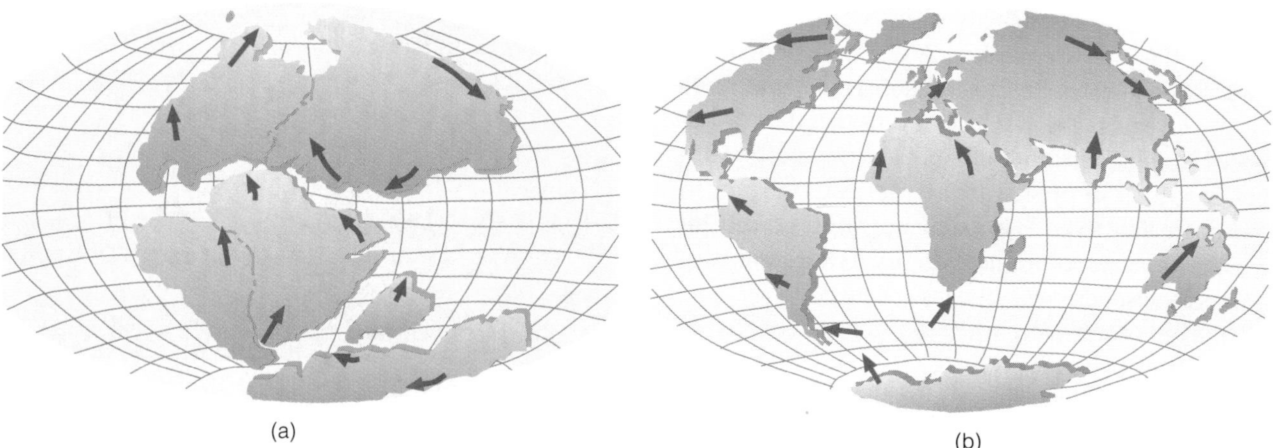

FIGURE 14.6 Geographical distribution of (a) land masses about 180 million years ago, and (b) today. Arrows show the relative direction of continental movement.

ocean) begins at a *ridge,* where dense, molten material from inside the earth wells up to the surface, forming new sea floor material as it hardens. Spreading (on the order of several centimeters a year) takes place at the ridge center, where two oceanic plates move away from one another. When an oceanic plate encounters a lighter continental plate, it responds by diving under it, in a process called *subduction.* Heat and pressure then melt a portion of the subducting rock, which usually consists of volcanic rock and calcium-rich ocean sediment. The molten rock may then gradually work its way to the surface, producing volcanic eruptions that spew water vapor, carbon dioxide, and minor amounts of other gases into the atmosphere. The release of these gases (called *degassing*) usually takes place at other locations as well (for instance, at ridges where new crustal rock is forming.)

Some scientists speculate that climatic change, taking place over millions of years, might be related to the rate at which the plates move and, hence, related to the amount of CO_2 in the air. For example, during times of rapid spreading, a relatively wide ridge forms, causing sea level to rise relative to the continents. At the same time,

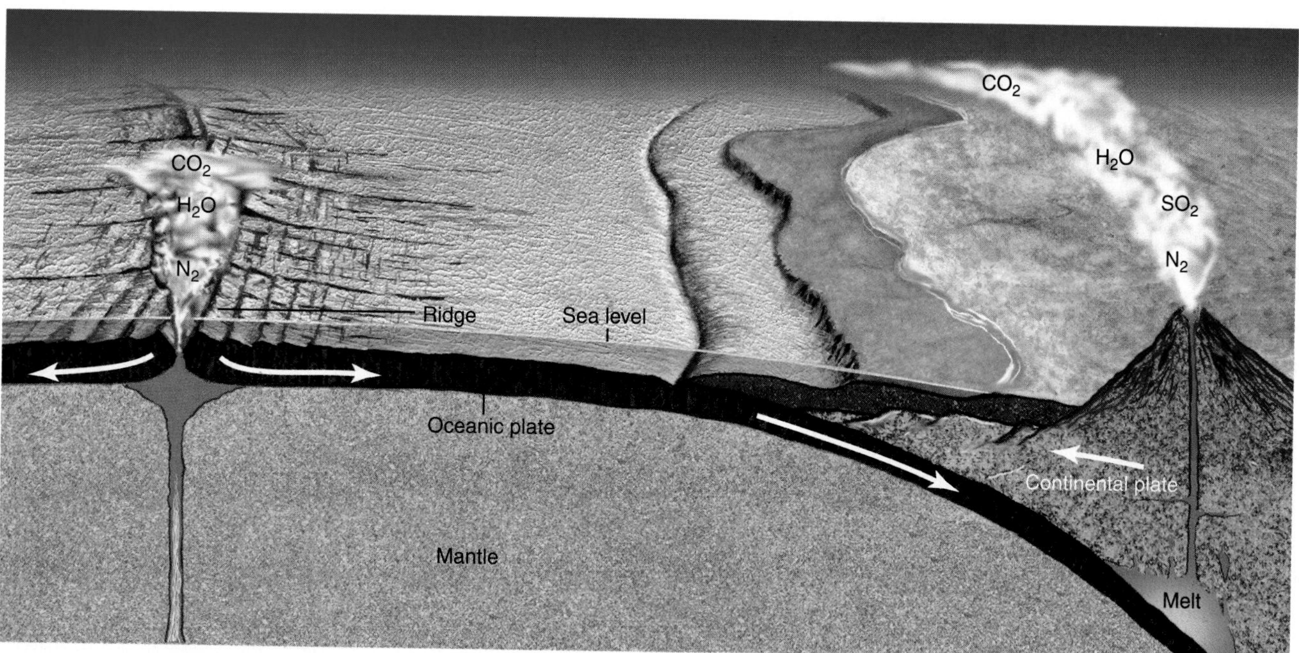

FIGURE 14.7 The earth is composed of a series of moving plates. The rate at which plates move (spread) may influence global climate. During times of rapid spreading, increased volcanic activity may promote global warming by enriching the CO_2 content of the atmosphere.

an increase in volcanic activity vents large quantities of CO_2 into the atmosphere, which enhances the atmospheric greenhouse effect, causing global temperatures to rise. Moreover, a higher sea level means that there is less exposed landmass and, presumably, less chemical weathering of rocks*—a process that removes CO_2 from the atmosphere. However, as global temperatures climb, increasing temperatures promote chemical weathering that removes atmospheric CO_2 at a faster rate.

Millions of years later, when spreading rates decrease, less volcanic activity means less degassing. The changing shape of the underwater ridge causes the sea level to drop relative to the continents, exposing more rocks for chemical attack and the removal of CO_2 from the air. A reduction in CO_2 levels weakens the greenhouse effect, which causes global temperatures to drop. The accumulation of ice and snow over portions of the continents may promote additional cooling by reflecting more sunlight back to space. The cooling, however, will not go unchecked, as lower temperatures retard both the chemical weathering of rocks and the depletion of atmospheric CO_2.

A chain of volcanic mountains forming above a subduction zone may disrupt the airflow over them. By the same token, mountain building that occurs when two continental plates collide (like that which presumably formed the Himalayan mountains and Tibetan highlands) can have a marked influence on global circulation patterns and, hence, on the climate of an entire hemisphere.

Up to now, we have examined how climatic variations can take place over millions of years due to the movement of continents and the associated restructuring of landmasses, mountains, and oceans. We will now turn our attention to variations in the earth's orbit that may account for climatic fluctuations that take place on a time scale of tens of thousands of years.

Climate Change and Variations in the Earth's Orbit

A theory ascribing climatic changes to variations in the earth's orbit is the **Milankovitch theory,** named for the astronomer Milutin Milankovitch, who first proposed the idea in the 1930s. The basic premise of this theory is that, as the earth travels through space, three separate cyclic movements combine to produce variations in the amount of solar energy that falls on the earth.

The first cycle deals with changes in the shape (**eccentricity**) of the earth's orbit as the earth revolves about the sun. Notice in Fig. 14.8 that the earth's orbit changes from being elliptical to being nearly circular. To go from less elliptical to more elliptical and back again takes about 100,000 years. The greater the eccentricity of the orbit (that is, the more eccentric the orbit), the greater the variation in solar energy received by the earth between its closest and farthest approach to the sun.

*Chemical weathering is the process by which rocks decompose.

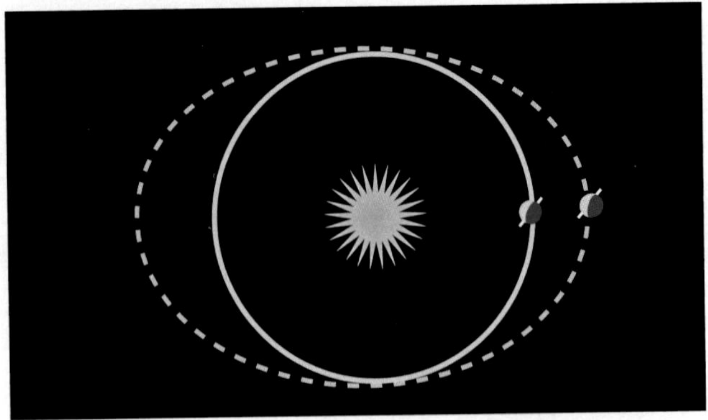

FIGURE 14.8 For the earth's orbit to stretch from nearly a circle (dashed line) to an elliptical orbit (solid line) and back again takes nearly 100,000 years. (Diagram is highly exaggerated and is not to scale.)

Presently, we are in a period of low eccentricity. The earth is closer to the sun in January and farther away in July (see Chapter 2, p. 42). The difference in distance (which only amounts to about 3 percent) is responsible for a nearly 7 percent increase in the solar energy received at the top of the atmosphere from July to January. When the difference in distance is 9 percent (a highly eccentric orbit), the difference in solar energy received will be on the order of 20 percent. In addition, the more eccentric orbit will change the length of seasons in each hemisphere by changing the length of time between the vernal and autumnal equinoxes.*

The second cycle takes into account the fact that, as the earth rotates on its axis, it wobbles like a spinning top. This wobble, known as the **precession** of the earth's axis, occurs in a cycle of about 23,000 years. Presently, the earth is closer to the sun in January and farther away in July. Due to precession, the reverse will be true in about 11,000 years (see Fig. 14.9). In about 23,000 years we will be back to where we are today. This means, of course, that if everything else remains the same, 11,000 years from now seasonal variations in the Northern Hemisphere should be greater than at present. The opposite would be true for the Southern Hemisphere.

The third cycle takes about 41,000 years to complete and relates to the changes in tilt (**obliquity**) of the

*Although rather large percentage changes in solar energy can occur between summer and winter, the globally and annually averaged change in solar energy received by the earth (due to orbital changes) hardly varies at all. It is the distribution of incoming solar energy that changes, not the totals.

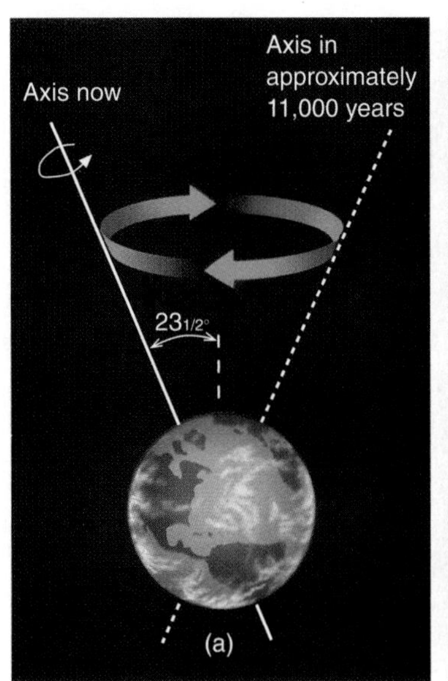

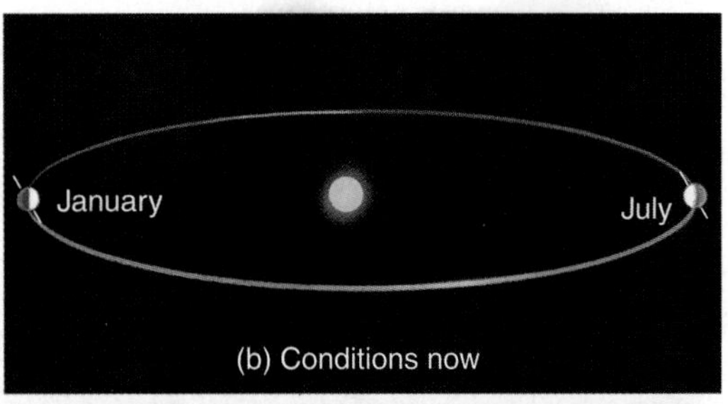

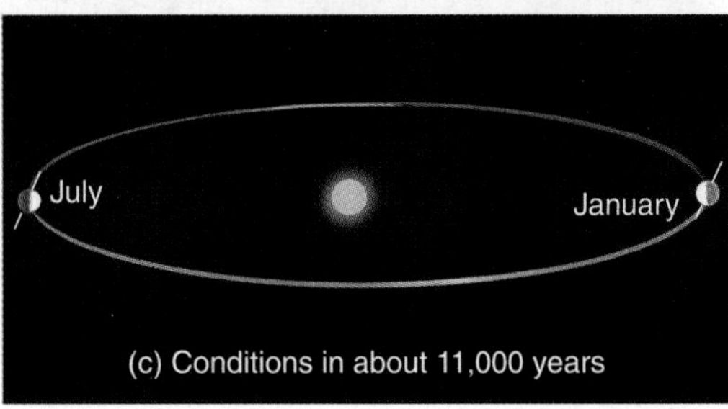

Meteorology Now™

ACTIVE FIGURE 14.9 (a) Like a spinning top, the earth's axis of rotation slowly moves and traces out the path of a cone in space. (b) Presently the earth is closer to the sun in January, when the Northern Hemisphere experiences winter. (c) In about 11,000 years, due to precession, the earth will be closer to the sun in July, when the Northern Hemisphere experiences summer. Watch this Active Figure at http://earthscience.brookscole.com/ahrens/ess4e

earth as it orbits the sun. Presently, the earth's orbital tilt is 23½°, but during the 41,000-year cycle the tilt varies from about 22° to 24½°. The smaller the tilt, the less seasonal variation there is between summer and winter in middle and high latitudes. Thus, winters tend to be milder and summers cooler. During the warmer winters, more snow would probably fall in polar regions due to the air's increased capacity for water vapor. And during the cooler summers, less snow would melt. As a consequence, the periods of smaller tilt would tend to promote the formation of glaciers in high latitudes. In fact, when all of the cycles are taken into account, the present trend should be toward a cooler climate over the Northern Hemisphere.

In summary, the Milankovitch cycles that combine to produce variations in solar radiation received at the earth's surface include:

1. changes in the shape *(eccentricity)* of the earth's orbit about the sun
2. *precession* of the earth's axis of rotation, or wobbling
3. changes in the tilt *(obliquity)* of the earth's axis

In the 1970s, scientists of the CLIMAP project found strong evidence in deep-ocean sediments that variations in climate during the past several hundred thousand years were closely associated with the Milankovitch cycles. More recent studies have strengthened this premise. For example, studies conclude that during the past 800,000 years, ice sheets have peaked about every 100,000 years. This conclusion corresponds naturally to variations in the earth's eccentricity. Superimposed on this situation are smaller ice advances that show up at intervals of about 41,000 years and 23,000 years. It appears, then, that eccentricity is the *forcing factor*—the external cause—for the frequency of glaciation, as it appears to control the severity of the climatic variation.

But orbital changes alone are probably not totally responsible for ice buildup and retreat. Evidence (from trapped air bubbles in the ice sheets of Greenland and Antarctica representing thousands of years of snow accumulation) reveals that CO_2 levels were about 30 percent lower during colder glacial periods than during warmer interglacial periods. Analysis of air bubbles in Antarctic ice cores reveals that methane follows a pattern similar to that of CO_2 (see Fig. 14.10). This knowledge suggests that lower atmospheric CO_2 levels may have had the effect of amplifying the cooling initiated by the orbital changes. Likewise, increasing CO_2 levels at the end of the glacial period may have accounted for the rapid melting of the ice sheets. Just why atmospheric CO_2 levels have varied as glaciers expanded and contracted is not clear, but it appears to be due to changes in biological activity taking place in the oceans.

Perhaps, also, changing levels of CO_2 indicate a shift in ocean circulation patterns. Such shifts, brought on by changes in precipitation and evaporation rates, may alter the distribution of heat energy around the world. Alteration wrought in this manner could, in turn, affect the global circulation of winds, which may explain why alpine glaciers in the Southern Hemisphere expanded and contracted in tune with Northern Hemisphere glaciers during the last ice age, even though the Southern Hemisphere (according to the Milankovitch cycles) was not in an orbital position for glaciation.

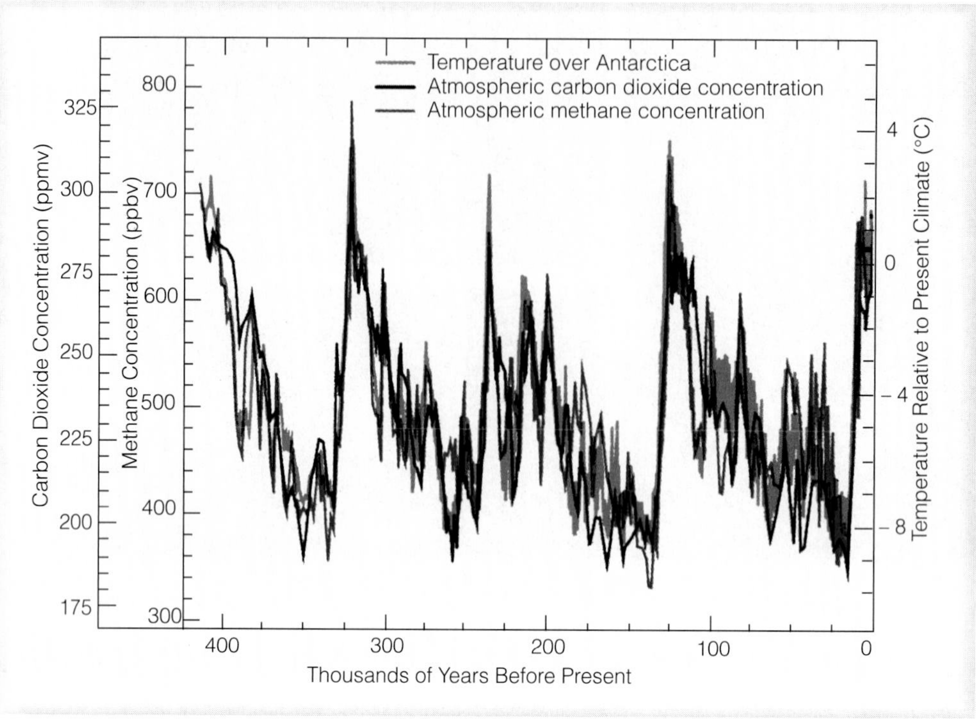

FIGURE 14.10 Variations of temperature (red line, °C), carbon dioxide (black line, ppmv), and methane (blue line, ppbv). Concentrations of gases are derived from air bubbles trapped within the ice sheets of Antarctica and extracted from ice cores. Temperatures are derived from the analysis of oxygen isotopes. (Note: ppmv represents parts per million by volume, and ppbv represents parts per billion by volume.) (*Source:* From Climate Change 2001: The Scientific Basis, 2001, by J.T. Houghton, et al. Copyright © 2001 Cambridge University Press. Reprinted with permission of the Intergovernmental Panel on Climate Change.)

Still other factors may work in conjunction with the earth's orbital changes to explain the temperature variations between glacial and interglacial periods. Some of these are

1. the amount of dust and other aerosols in the atmosphere
2. the reflectivity of the ice sheets
3. the concentration of other trace gases, such as methane
4. the changing characteristics of clouds
5. the rebounding of land, having been depressed by ice

Hence, the Milankovitch cycles, in association with other natural factors, may explain the advance and retreat of ice over periods of 10,000 to 100,000 years. But what caused the Ice Age to begin in the first place? And why have periods of glaciation been so infrequent during geologic time? The Milankovitch theory does not attempt to answer these questions.

Climate Change and Atmospheric Particles Microscopic liquid and solid particles *(aerosols)* that enter the atmosphere from both human-induced (anthropogenic) and natural sources can have an effect on climate. The effect, however, is exceedingly complex and depends upon a number of factors, such as the particle's size, shape, color, chemical composition, and vertical distribution above the surface. In this section, we will first examine aerosols in the lower atmosphere. Then we will examine the effect that volcanic aerosols in the stratosphere have on climate.

Aerosols in the Troposphere Aerosols enter the lower atmosphere in a variety of ways—from factory and auto emissions, agricultural burning, wild land fires, and dust storms. Some particles (such as soil dust and sulfate particles) mainly reflect and scatter incoming sunlight, while others (such as smoky soot) readily absorb sunlight, which warms the air around them. Many aerosols that reduce the amount of sunlight reaching the earth's surface tend to cause net cooling of the surface air during the day. Certain aerosols also selectively absorb and emit infrared energy back to the surface, producing a net warming of the surface air at night. However, the overall net effect of human-induced (anthropogenic) aerosols on climate is to *cool the surface.*

In recent years, the effect of highly reflective **sulfate aerosols** on climate has been extensively researched. In the lower atmosphere, the majority of these particles come from the combustion of sulfur-containing fossil fuels but emissions from smoldering volcanoes can also be a significant source of tropospheric sulfate aerosols. Sulfur pollution, which has more than doubled globally since preindustrial times, enters the atmosphere mainly as sulfur dioxide gas. There, it transforms into tiny sulfate droplets or particles. Since these aerosols usually remain in the atmosphere for only a few days, they do not have time to spread around the globe. Hence, they are not well mixed and their effect is felt mostly over the Northern Hemisphere, especially over polluted regions.

Sulfate aerosols not only scatter incoming sunlight back to space, but they also serve as cloud condensation nuclei. Consequently, they have the potential for altering the physical characteristics of clouds. For example, if the number of sulfate aerosols and, hence, condensation nuclei inside a cloud should increase, the cloud would have to share its available moisture with the added nuclei, a situation that should produce many more (but smaller) cloud droplets. The greater number of droplets would reflect more sunlight and have the effect of brightening the cloud and reducing the amount of sunlight that reaches the surface.

In summary, sulfate aerosols reflect incoming sunlight, which tends to lower the earth's surface temperature during the day. Sulfate aerosols may also modify clouds by increasing their reflectivity. Because sulfate pollution has increased significantly over industrialized areas of eastern Europe, northeastern North America and China, the cooling effect brought on by these particles may explain: (1) why the Northern Hemisphere has warmed less than the Southern Hemisphere during the past several decades, (2) why the United States has experienced little warming compared to the rest of the world, and (3) why most of the global warming has occurred at night and not during the day, especially over polluted areas. Research is still being done, and the overall effect of tropospheric aerosols on the climate system is not totally understood. (Information regarding the possible effect on climate from huge masses of particles being injected into the atmosphere is given in the Focus section on p. 388.)

Volcanic Eruptions and Aerosols in the Stratosphere Volcanic eruptions can have a definitive impact on climate. During volcanic eruptions, fine particles of ash and dust (as well as gases) can be ejected into the stratosphere (see Fig. 14.11). Scientists agree that the volcanic eruptions having the greatest impact on climate are those rich in sulfur gases. These gases, over a period of about two months, combine with water vapor in the presence of sunlight to produce tiny, reflective sulfuric acid particles that grow in size, forming a dense layer of haze. The haze may reside in the stratosphere for several years, absorbing and reflecting back to space a portion of the sun's incoming energy. The absorption of the sun's energy along with the absorption of infrared energy from the earth, warms the stratosphere. The reflection of incoming sunlight by the haze tends to cool the air at the earth's surface, especially in the hemisphere where the eruption occurs.

Two of the largest volcanic eruptions of the 20th century in terms of their sulfur-rich veil, were that of El

FIGURE 14.11 Large volcanic eruptions rich in sulfur can affect climate. As sulfur gases in the stratosphere transform into tiny reflective sulfuric acid particles, they prevent a portion of the sun's energy from reaching the surface. Here, the Philippine volcano Mount Pinatubo erupts during June, 1991.

FOCUS ON A SPECIAL TOPIC

Nuclear Winter, Cold Summers, and Dead Dinosaurs

A number of studies indicate that a nuclear war involving hundreds or thousands of nuclear detonations would drastically modify the earth's climate.

Researchers assume that a nuclear war would raise an enormous pall of thick, sooty smoke from massive fires that would burn for days, even weeks, following an attack. The smoke would drift higher into the atmosphere, where it would be caught in the upper-level westerlies and circle the middle latitudes of the Northern Hemisphere. Unlike soil dust, which mainly scatters and reflects incoming solar radiation, soot particles readily absorb sunlight. Hence, for several weeks after the war, sunlight would virtually be unable to penetrate the smoke layer, bringing darkness or, at best, twilight at midday.

Such reduction in solar energy would cause surface air temperatures over land masses to drop below freezing, even during the summer, resulting in extensive damage to plants and crops and the death of millions (or perhaps billions) of people. The dark, cold, and gloomy conditions that would be brought on by nuclear war are often referred to as *nuclear winter.*

As the lower troposphere cools, the solar energy absorbed by the smoke particles in the upper troposphere would cause this region to warm. The end result would be a strong, stable temperature inversion extending from the surface up into the higher atmosphere. A strong inversion would lead to a number of adverse effects, such as suppressing convection, altering precipitation processes, and causing major changes in the general wind patterns.

The heating of the upper part of the smoke cloud would cause it to rise upward into the stratosphere, where it would then drift southward. Thus, about one-third of the smoke would remain in the atmosphere for a year or longer. The other two-thirds would be washed out in a month or so by precipitation. This smoke lofting, combined with persisting sea ice formed by the initial cooling, would produce climatic change that would remain for several years.

Virtually all research on nuclear winter, including models and analog studies, confirms this gloomy scenario. Observations of forest fires show lower temperatures under the smoke, confirming part of the theory. The implications of nuclear winter are clear: A nuclear war would drastically alter global climate and would devastate our living environment.

Could atmospheric particles and a nuclear winter–type event have contributed to the demise of the dinosaurs? About 65 million years ago, the dinosaurs, along with about half of all plant and animal species on earth, died in a mass extinction. What could cause such a catastrophe?

One popular theory proposes that about 65 million years ago a giant meteorite measuring some 10 km (6 mi) in diameter slammed into the earth at about 44,000 mi/hr (see Fig. 2). The impact (possibly located near the Yucatan Peninsula) sent billions of tons of dust and debris into the upper atmosphere, where such particles circled the globe for months and greatly reduced the sunlight reaching the earth's surface. Reduced sunlight disrupted photosynthesis in plants which, in turn, led to a breakdown in the planet's food chain. Lack of food, as well as cooler conditions brought on by the dust, must have had an adverse effect on life, especially large plant-eating dinosaurs.

Evidence for this catastrophic collision comes from the geologic record, which shows a thin layer of sediment deposited worldwide, about the time the dinosaurs disappeared. The sediment contains iridium, a rare element on earth, but common in certain types of meteorites.

Was what caused this disaster an isolated phenomenon or did other events, such as huge volcanic eruptions, play an additional role in altering the climate? Have such meteorite collisions been more common in the geologic past than was once thought? And what is the likelihood of such an event occurring in the near future? Questions like these are certainly interesting to ponder.

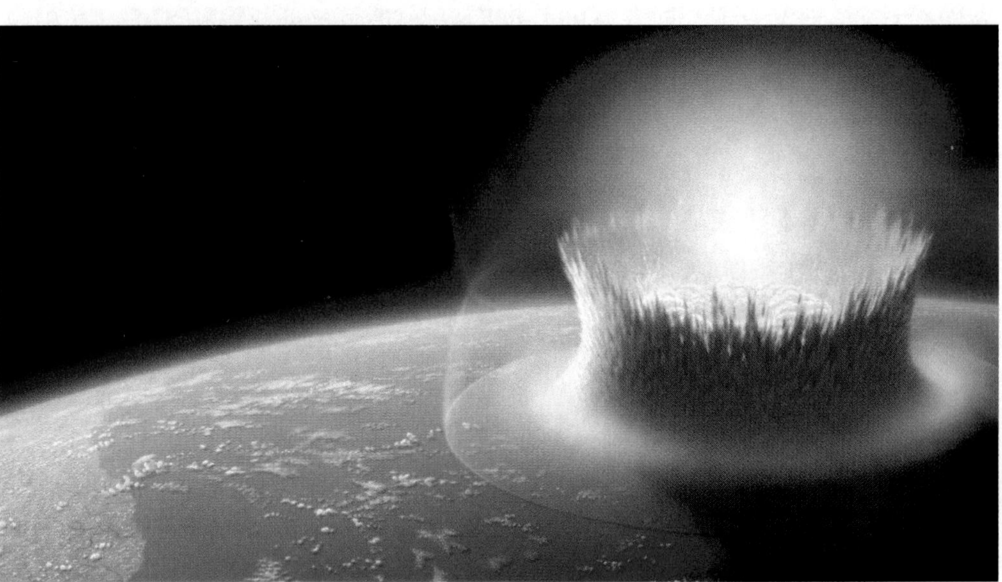

FIGURE 2 Artist interpretation of a giant meteorite striking the earth's surface 65 million years ago, creating a nuclear winter event.

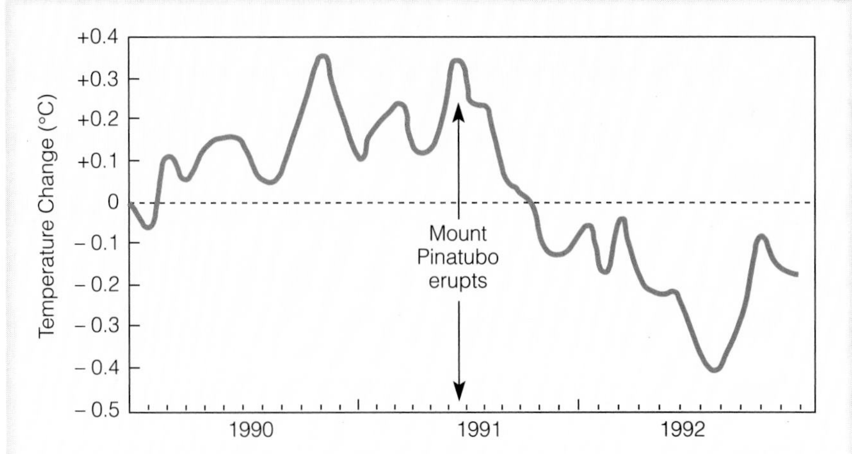

FIGURE 14.12 Changes in average global air temperature from 1990 to 1992. After the eruption of Mount Pinatubo in June, 1991, the average global temperature by July, 1992, decreased by almost 0.5°C (0.9°F) from the 1981 to 1990 average (dashed line).

Chichón in Mexico during April, 1982, and Mount Pinatubo in the Philippines during June, 1991.* Mount Pinatubo ejected an estimated 20 million tons of sulfur dioxide (more than twice that of El Chichón) that gradually worked its way around the globe. For major eruptions such as this one, mathematical models predict that average hemispheric temperatures can drop by about 0.2° to 0.5°C or more for one to three years after the eruption. Model predictions agreed with temperature changes brought on by the Pinatubo eruption, as in early 1992 the mean global surface temperature had decreased by about 0.5°C (see Fig. 14.12). The cooling might even have been greater had the eruption not coincided with a major El Niño event that began in 1990 and lasted until early 1995 (see Chapter 7, p. 192, for information on El Niño).

An infamous cold spell often linked to volcanic activity occurred during the year 1816, "the year without a summer" mentioned earlier. Apparently, a rather stable longwave pattern in the atmosphere produced unseasonably cold summer weather over eastern North America and western Europe. The cold weather followed the massive eruption in 1815 of Mount Tambora in Indonesia. In addition to this, major volcanic eruptions occurred in the four years preceding Tambora. If, indeed, the cold weather pattern was brought on by volcanic eruptions, it was probably an accumulation of several volcanoes loading the stratosphere with particles—particles that probably remained there for several years.

*The eruption of Mount Pinatubo in 1991 was many times greater than that of Mount St. Helens in the Pacific Northwest in 1980. In fact, the largest eruption of Mount St. Helens was a lateral explosion that pulverized a portion of the volcano's north slope. The ensuing dust and ash (and very little sulfur) had virtually no effect on global climate as the volcanic material was confined mostly to the lower atmosphere and fell out quite rapidly over a large area of the northwestern United States.

DID YOU KNOW?

About 100 million years ago, when dinosaurs roamed this planet, the earth's mean surface temperature was between 10°C and 15°C (18°F and 27°F) warmer than it is today, and the concentration of CO_2 in the atmosphere was much higher.

In an attempt to correlate sulfur-rich volcanic eruptions with long-term trends in global climate, scientists are measuring the acidity of annual ice layers in Greenland and Antarctica. Generally, the greater the concentration of sulfuric acid particles in the atmosphere, the greater the acidity of the ice layer. Relatively acidic ice has been uncovered from about A.D. 1350 to about 1700, a time that corresponds to a cooling trend over Europe referred to as the *Little Ice Age.* Such findings suggest that sulfur-rich volcanic eruptions may have played an important role in triggering this comparatively cool period and, perhaps, other cool periods during the geologic past. Moreover, recent core samples taken from the northern Pacific Ocean reveal that volcanic eruptions in the northern Pacific were at least 10 times larger 2.6 million years ago (a time when Northern Hemisphere glaciation began) than previous volcanic events recorded elsewhere in the sediment.

DID YOU KNOW?

The year without a summer (1816) even had its effect on literature. Inspired (or perhaps dismayed) by the cold, gloomy, summer weather along the shores of Lake Geneva, Mary Shelley wrote the novel *Frankenstein.*

Climate Change and Variations in Solar Output

In the past, it was thought that solar energy did not vary by more than a fraction of a percent over many centuries. However, measurements made by sophisticated radiometers aboard satellites suggest that the sun's energy output may vary more than was thought. Moreover, the sun's energy output appears to change slightly with sunspot activity.

Sunspots are huge magnetic storms on the sun that show up as cooler (darker) regions on the sun's surface. They occur in cycles, with the number and size reaching a maximum approximately every 11 years. During periods of maximum sunspots, the sun emits more energy (about 0.1 percent more) than during periods of sunspot minimums (see Fig. 14.13). Evidently, the greater number of bright areas *(faculae)* around the sunspots radiate more energy, which offsets the effect of the dark spots.

It appears that the 11-year sunspot cycle has not always prevailed. Apparently, between 1645 and 1715, during the period known as the **Maunder minimum,*** there were few, if any, sunspots. It is interesting to note that the minimum occurred during a cool spell in the temperature record shown in Fig. 14.4, p. 377. Some scientists suggest that a reduction in the sun's energy output was, in part, responsible for this cold spell.

In an attempt to better understand the sun's behavior, solar researchers are examining stars that are similar in age and mass to our sun. Recent observations suggest that, in some of these stars, energy output may vary by as much as 0.4 percent, leading some scientists to speculate that changes in the sun's brightness might account for part of the global warming during the last century.

The sun's magnetic field varies with sunspot activity and actually reverses every 11 years. Because it takes 22 years to return to its original state, the sun's *magnetic cycle* is 22 years, rather than 11. Some researchers point to the fact that periodic 20-year droughts on the Great Plains of the United States seem to correlate with this 22-year solar cycle. More recently, scientists have found a relationship between the 11-year sunspot cycle and weather patterns across the Northern Hemisphere. It appears that winter warmings might be related to variations in sunspots and to a pattern of reversing stratospheric winds over the tropics.

To sum up, fluctuations in solar output may account for climatic changes over time scales of decades and centuries. To date, many theories have been proposed linking solar variations to climate change, but none has been proven. However, instruments aboard satellites and solar telescopes on the earth are monitoring the sun to observe how its energy output may vary. Because many years of data are needed, it may be some time before we fully understand the relationship between solar activity and climate change on earth.

*This period is named after E. W. Maunder, the British solar astronomer who first discovered the low sunspot period sometime in the late 1880s.

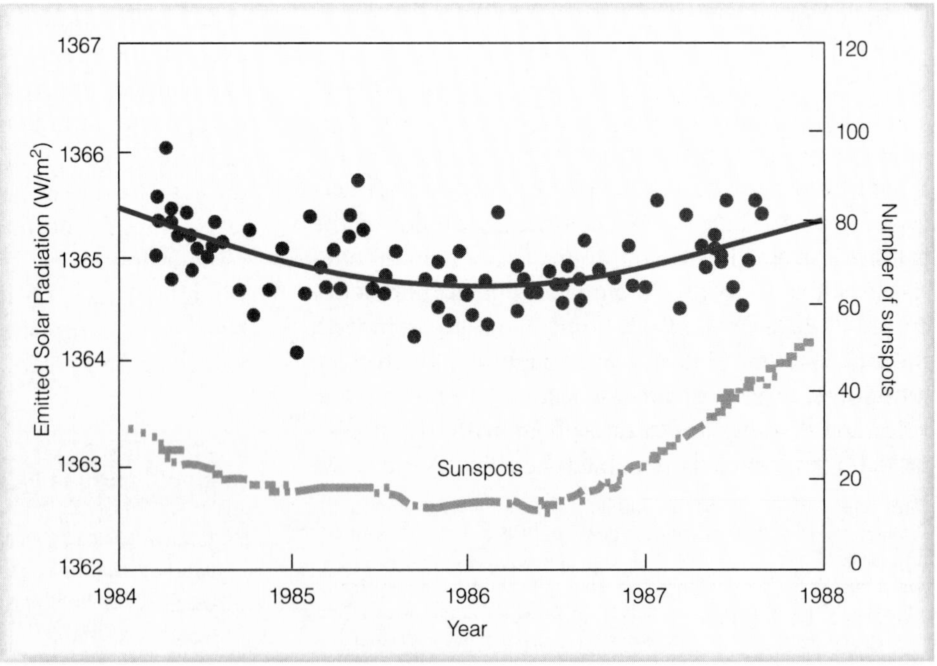

FIGURE 14.13 Changes in solar energy output (upper curve) in watts per square meter as measured by the *Earth Radiation Budget Satellite.* Bottom curve represents the yearly average number of sunspots. As sunspot activity increases from minimum to maximum, the sun's energy output increases by about 0.1 percent.

BRIEF REVIEW

Before going on to the next section, here is a brief review of some of the facts and concepts we covered so far:

- The earth's climate is constantly undergoing change. Evidence suggests that throughout much of the earth's history, the earth's climate was much warmer than it is today.
- The most recent glacial period (or Ice Age) began about 2 million years ago. During this time, glacial advances were interrupted by warmer periods called *interglacial periods.* In North America, glaciers reached their maximum thickness and extent about 18,000 to 22,000 years ago and disappeared completely from North America by about 6000 years ago.
- The Younger-Dryas event represents a time about 12,000 years ago when northeastern North America and northern Europe reverted back to glacial conditions.
- Over the last 100 years or so, the earth's surface temperature has increased by about 0.6°C (about 1°F).
- The shifting of continents, along with volcanic activity and mountain building, are possible causes of climate change.
- The Milankovitch theory (in association with other natural forces) proposes that altering glacial and interglacial episodes during the past 2 million years are the result of small variations in the tilt of the earth's axis and in the geometry of the earth's orbit around the sun.
- Trapped air bubbles in the ice sheets of Greenland and Antarctica reveal that CO_2 levels and methane levels were lower during colder glacial periods and higher during warmer interglacial periods.
- Sulfate aerosols in the troposphere reflect incoming sunlight, which tends to lower the earth's surface temperature during the day. Sulfate aerosols may also modify clouds by increasing the cloud's reflectivity.
- Volcanic eruptions, rich in sulfur, may be responsible for cooler periods that span years and decades in the geologic past.
- Fluctuation in solar output (brightness) may account for climatic changes over time scales of decades and centuries.

In previous sections, we saw how increasing levels of CO_2 may have contributed to changes in global climate spanning thousands and even millions of years. Today, we are undertaking a global scientific experiment by injecting vast quantities of greenhouse gases into our atmosphere without fully understanding the long-term consequences. The next section describes how CO_2 and other trace gases appear to be enhancing the earth's greenhouse effect, producing global warming.

GLOBAL WARMING

We know from Chapter 2 that CO_2 is a greenhouse gas that strongly absorbs infrared radiation and plays a major role in warming the lower atmosphere. We also know that CO_2 has been increasing steadily in the atmosphere, primarily due to the burning of fossil fuel (see Fig. 1.3, p. 5). However, deforestation is also adding to this increase as tropical rain forests are cut down and replaced with plants less efficient in removing CO_2 from the atmosphere. Presently, the annual average of CO_2 is about 375 parts per million (ppm), and present estimates are that if CO_2 levels continue to increase at the same rate that they have been (about 1.5 ppm per year), atmospheric concentrations will rise to between 540 and 970 ppm by the end of this century. To complicate the picture, trace gases such as methane (CH_4), nitrous oxide (N_2O), and chlorofluorocarbons (CFCs), all of which readily absorb infrared radiation, have been increasing in concentration over the past century.* Collectively, the increase in these gases is about equal to CO_2 in its ability to enhance the atmospheric greenhouse effect.

Numerical climate models (mathematical models that simulate climate) predict that by the end of this century increasing concentrations of greenhouse gases will result in a mean global warming of surface air between 1.4° and 5.8°C (between about 2.5° and 10.5°F) above the average surface air temperature of 1990. The newest, most sophisticated models take into account a number of important relationships, including the interactions between the oceans and the atmosphere, the processes by which CO_2 is removed from the atmosphere, and the cooling effect produced by sulfate aerosols in the lower atmosphere. The models also predict that as the air warms, additional water vapor will evaporate from the oceans into the air. The added water vapor (which is the most abundant greenhouse gas) will produce a positive feedback on the climate system by enhancing the atmospheric greenhouse effect and accelerating the temperature rise. (This is the *water vapor–greenhouse feedback* described on p. 381.) Without this feedback produced by the added water vapor, the models predict that the warming will be much less.

The Recent Warming Earlier in this chapter we saw that, since the beginning of the 20th century, the average global surface air temperature has risen about 0.6°C. Is this warming due to increasing greenhouse gases and an enhanced greenhouse effect? Before we can address this question, we need to review a few concepts we learned in Chapter 2.

*Refer back to Chapter 1 and to Table 1.1, p. 3, for additional information on the concentration of these gases.

Radiative Forcing Agents We know from Chapter 2 that our world without water vapor, CO_2, and other greenhouse gases would be a colder world—about 33°C (59°F) colder than at present. With an average surface temperature of about −18°C (0°F), much of the planet would be uninhabitable. In Chapter 2, we also learned that when the rate of the incoming solar energy balances the rate of outgoing infrared energy from the earth's surface and atmosphere, the earth-atmosphere system is in a state of *radiative equilibrium*. Increasing concentrations of greenhouse gases can disturb this equilibrium and are, therefore, referred to as **radiative forcing agents.** The **radiative forcing*** provided by extra CO_2 and other greenhouse gases increased by about 2.43 watts per square meter (2.43 W/m^2) from the middle 1700s to the present, with CO_2 contributing about 1.46 W/m^2, or 60 percent of the increase. So it is very likely that part of the warming during the last century is due to increasing levels of greenhouse gases. But what part does natural climate variability play in global warming? And with levels of CO_2 increasing by more than 25 percent over the last century, why has the observed increase in global temperature been relatively small?

We know that the climate may change due to natural events. For example, changes in the sun's energy output (called *solar irradiance*) and volcanic eruptions rich in sulfur are two major natural radiative forcing agents. Studies show that since the middle 1700s, changes in the sun's energy output may have contributed a small positive forcing (about 0.3 W/m^2) on the climate system, most of which occurred during the first half of the 20th century. On the other hand, volcanic eruptions that inject sulfur-rich particles into the stratosphere produce a negative forcing, which lasts for a few years after the eruption. Because several major eruptions occurred between 1880 and 1920, as well as between 1960 and 1991, the combined change in radiative forcing due to both volcanic activity and solar activity over the past 20 to 40 years appears to be negative. (The combined effect is that of cooling.) Recall that sulfur-rich aerosols near the surface also have a net cooling effect on the climate.

Climate Models and Recent Temperature Trends How, then, does the temperature change observed over the last century compare with temperature changes derived from climate models? Before we look at what climate models reveal, it is important to realize that the interactions between the earth and its atmosphere are so complex that it is difficult to unequivocally *prove* that the warming trend during the past 100 years has been due primarily to increasing concentrations of greenhouse gases. The problem is that any human-induced signal of climate change is superimposed on a background of natural climatic variations ("noise"), such as the El Niño–Southern Oscillation (ENSO) phenomenon (discussed in Chapter 7). Moreover, in the temperature observations, it is difficult to separate a signal from the noise of natural climate variability. However, today's more sophisticated climate models are much better at filtering out this noise while at the same time taking into account those forcing agents that are both natural and human-induced.

Figure 14.14 shows the predicted changes in surface air temperature from 1860 to 2000 made by different climate models using various scenarios (different forcing agents). The gray line presents the actual changes in surface air temperature from 1860 to 2000. Notice that when only increasing levels of greenhouse gases are plugged into the model (yellow line), the model shows a surface temperature increase in excess of 1°C. When greenhouse gases and aerosols are both added to the model (blue line), the increase in surface temperature is much less; in fact, it is less than the temperature increase observed over the past 100 years. However, when greenhouse gases, sulfate aerosols, and changes in solar radiation are *all* added to the model (red line), the projected temperature change and the observed temperature change closely match.

It is climate studies using computer models such as these that have led scientists to conclude that some of the warming during the 20th century is very likely due to increasing levels of greenhouse gases. In fact, the Intergovernmental Panel on Climate Change (IPCC), a committee of over 2000 leading earth scientists, considered the issues of climate change in a report published in 1990 and updated in 1992, in 1995, and again in 2001. The 2001 report, called the IPCC, TAR (Third Assessment Report), states that:

> In the light of new evidence and taking into account the remaining uncertainties, most of the observed warming over the last 50 years is likely to have been due to the increase in greenhouse gas concentrations.

Future Warming—Projections, Questions, and Uncertainties As we saw in an early section, today's climate models project that, due to increasing levels of

---

*Radiative forcing is interpreted as an increase (positive) or a decrease (negative) in net radiant energy observed over an area at the tropopause. All factors being equal, an increase in radiative forcing may induce surface *warming*, whereas a *decrease* may induce surface *cooling*.

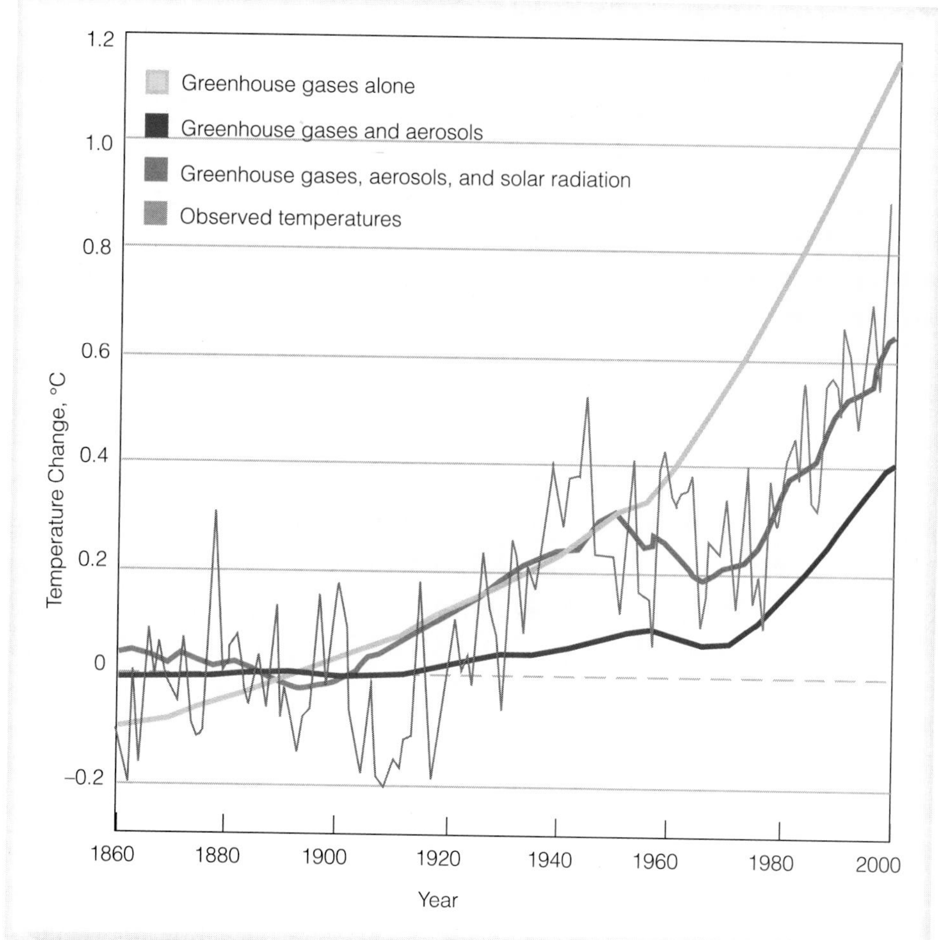

FIGURE 14.14 Projected surface air temperature changes from different climate models. Model input from greenhouse gases only is shown in yellow; input from greenhouse gases plus aerosols is shown in blue; input from greenhouse gases, sulfate aerosols, and solar energy changes is shown in red. The gray line shows observed surface temperatures. The dashed line is the 1880 to 1999 mean temperature. (Redrawn from "The Science of Climate Change" by Tom M. L. Wigley, published by the Pew Center of Global Climate Change.)

greenhouse gases, the surface air temperature will increase between 1.4°C and 5.8°C from the year 1990 to the year 2100 (see Fig. 14.15). Notice that the smallest projected increase in temperature of 1.4°C is still twice that experienced during the 20th century. An increase of 5.8°C would have potentially devastating effects worldwide. Consequently, it is likely that the warming over this century will be much larger than the warming experienced during the 20th century and probably greater than any warming during the past 10,000 years.

There are, however, uncertainties in the climate picture. For example, it is not known how fast CO_2 levels will increase. Currently, the oceans and vegetation on land absorb about half of the CO_2 emitted by human sources. As a result, the oceans play a major role in the climate system, yet the exact effect they will have on rising levels of CO_2 and global warming is not totally clear. Microscopic plants (phytoplankton) extract CO_2 from the atmosphere during photosynthesis and store some of it below the ocean's surface, where they die. Will a warming earth trigger a large blooming of these tiny plants, in effect reducing the rate at which atmospheric CO_2 is increasing? Or would a gradual rise in ocean temperature increase the amount of CO_2 in the air due to the fact that warmer oceans can't hold as much CO_2 as colder ones?

In addition, rising temperatures may alter the way landmasses absorb and emit CO_2. For example, temperatures over the Alaskan tundra have risen dramatically during the past 35 years to the point where more frozen soil melts in summer than it used to. During warmer months, deep layers of decaying peat release CO_2 into the atmosphere. Until recently, this region absorbed more CO_2 than it released. Now, however, much of the tundra acts as a source for CO_2.

At present, deforestation accounts for about one-fourth of the observed increase in atmospheric CO_2. Hence, changes in land use could influence CO_2 concentrations, especially if the practice of deforestation is replaced by reforestation. And it is unknown what future

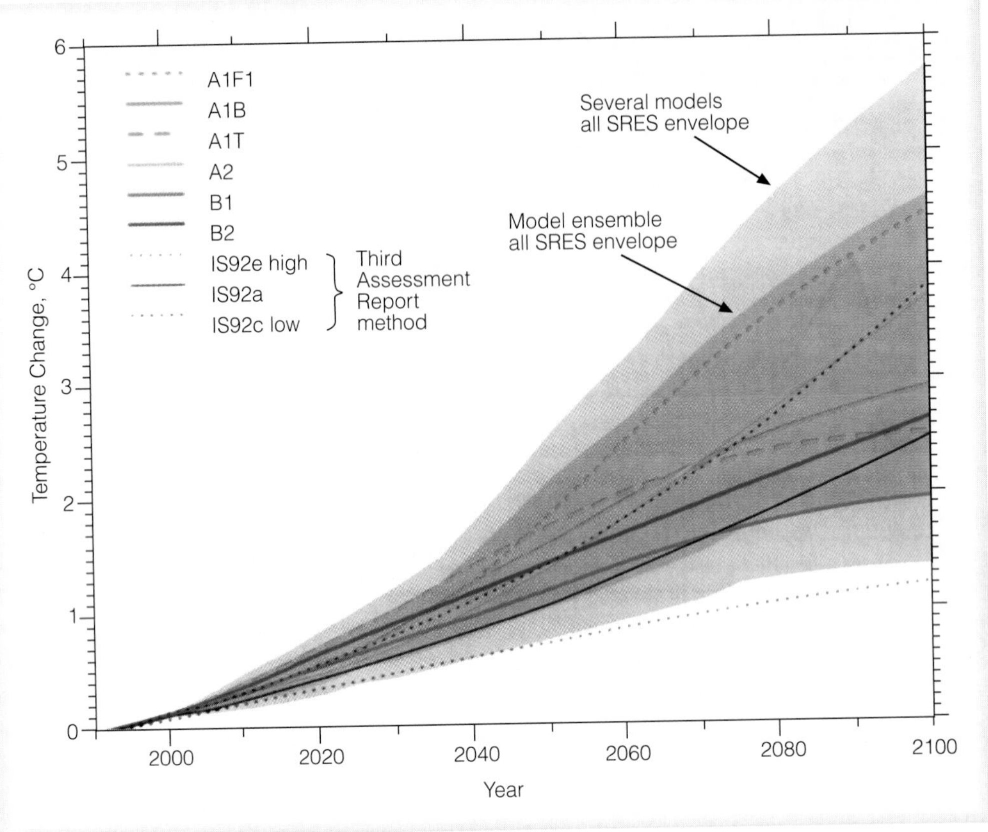

FIGURE 14.15 Global average projected temperature changes (°C) from 1990 to 2100 using climate models with six different scenarios. Each scenario describes how the average temperature will change based on different concentrations of greenhouse gases and various forcing agents. (SRES stands for "IPCC Special Report on Emission Scenarios") (*Source:* From Climate Change 2001: The Scientific Basis, 2001, by J.T. Houghton, et al. Copyright © 2001 Cambridge University Press. Reprinted with permission of the Intergovernmental Panel on Climate Change.)

steps countries will take to limit the emissions of CO_2 from the burning of fossil fuels. Overall, present trends indicate that as concentrations of atmospheric CO_2 increase, the oceans and landmasses will absorb a *decreasing* percentage of this greenhouse gas. In addition, it is not known how quickly greenhouse gases other than CO_2 will increase.

As the atmosphere warms and more water vapor is added to the air, global cloudiness might increase as well. How, then, would clouds—which come in a variety of shapes and sizes and form at different altitudes—affect the climate system? Clouds reflect incoming sunlight back to space, a process that tends to cool the climate, but clouds also absorb infrared radiation from the earth, which tends to warm it. Just how the climate will respond to changes in cloudiness will probably depend on the type of clouds that form and their physical properties, such as liquid water (or ice) content and droplet size distribution. For example, high, thin cirriform clouds (composed mostly of ice) tend to promote a net warming effect: They allow a good deal of sunlight to pass through (which warms the earth's surface), yet because they are cold, they warm the atmosphere by absorbing more infrared radiation from the earth than they emit upward. Low stratified clouds, on the other hand, tend to promote a net cooling effect. Composed mostly of water droplets, they reflect much of the sun's incoming energy, and, because their tops are relatively warm, they radiate away much of the infrared energy they receive from the earth. Satellite data confirm that, overall, clouds presently have a *net cooling effect* on our planet, which means that, without clouds, our atmosphere would be warmer.

Additional clouds in a warmer world would not necessarily have a net cooling effect, however. Their influence on the average surface air temperature would depend on their extent and on whether low or high clouds dominate the climate scene. Consequently, the feedback from clouds could potentially enhance or reduce the warming produced by increasing greenhouse gases. Most models show that as the surface air warms, there will be more convection, more convective-type clouds, and an increase in cirrus clouds. This situation would tend to provide a positive feedback on the climate system, and the effect of clouds on cooling the earth would be diminished (see Fig. 14.16).

Critics of global warming point to the fact that, even though the surface has warmed dramatically over the past two decades, the overall troposphere has not. Since 1979, satellite measurements indicate that, within the troposphere, the air has warmed 0°C to 0.2°C, whereas surface stations during the same period show a warming of 0.25°C to 0.4°C. If an enhanced atmospheric greenhouse effect is in fact causing the surface warming, why hasn't the atmo-sphere warmed in tandem? Although this question has not been totally resolved, one answer may be that perhaps natural events, such as the ocean warming during El Niño and the cooling induced by large volcanic eruptions, may account for part of the temperature differences. Also, it could well be the case that the thinning of ozone in the stratosphere may be partly responsible for a cooler upper troposphere. In fact, studies indicate that, from 1979 to 2000, the depletion of ozone in the stratosphere caused a small negative radiative forcing (and a slight cooling effect). As the ozone layer recovers, it is likely that this negative forcing will diminish.

Meteorology Now™ Click "Temperature Trends" to view the global distribution of future temperature changes according to one climate model.

Possible Consequences of Global Warming If the world warms as predicted by climate computer models, what will the warmer world be like? Climate models predict that land areas will warm more rapidly than the global average, particularly in the northern high latitudes in winter (see Fig. 14.17). In the high latitudes of the Northern Hemisphere, the dark green boreal forests absorb up to three times as much solar energy as the snow-covered tundra. Consequently, the winter temperatures in subarctic regions are, on the average, about 11.5°C (21°F) higher than they would be without trees. If warming allows the boreal forests to expand into the tundra, the forests may accelerate the warming in that region. As the temperature rises, organic matter in the soil should decompose at a faster rate, adding more CO_2 to the air, which might accelerate the warming even more. Moreover, trees that grow in a climate zone defined by temperature may become especially hard hit as rising temperatures place them in an inhospitable environment. In a weakened state, they may become more susceptible to insects and disease.

In a warmer world, enhanced evaporation of water should lead to greater worldwide average precipitation. During the warming of the 20th century, there appears to have been an increase in precipitation by as much as 10 percent over the middle- and high-latitude land areas of the Northern Hemisphere. In contrast, it appears that over subtropical land areas, a decrease in precipitation has occurred. It also appears that there has been an increase in the frequency of heavy precipitation events during the last half of the 20th century.

FIGURE 14.16 Clouds play an important role in the earth's climate system. How they will respond to increasing global temperatures is not totally clear. An increase in global cloudiness could potentially enhance or reduce the warming brought on by increasing greenhouse gases.

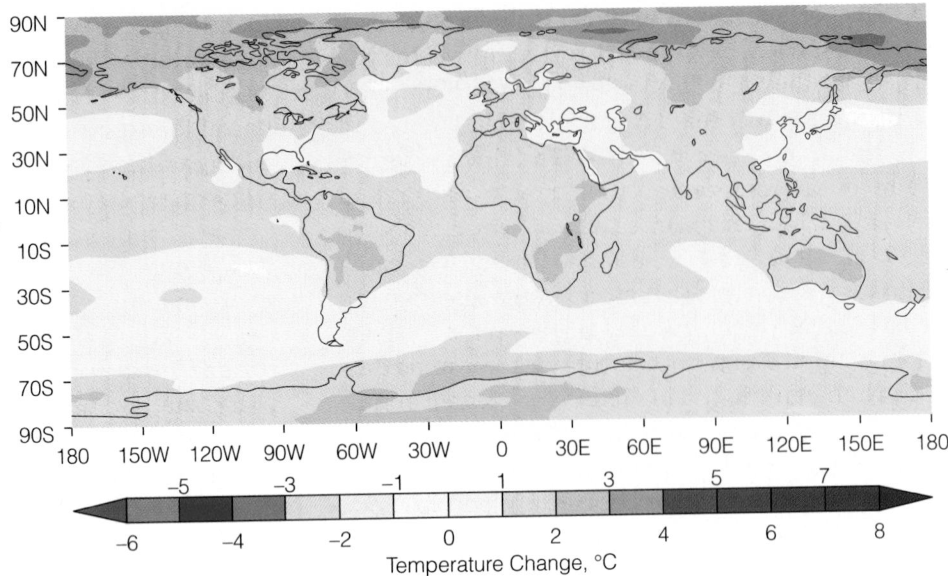

FIGURE 14.17 Projected changes in surface air temperature due to a doubling of CO_2 and human-induced sulfide emissions with an Atmospheric Ocean General Circulation Model (AOGCM). Notice that the greatest warming is projected for the northern polar latitudes. [After F. B. Mitchell et al., "Transient climate response to increasing sulphate aerosols and greenhouse gases," *Nature* (1995) 376: 501–504.]

As the warming continues, some models predict that the jet stream will weaken and global winds will shift from their "normal" position. The shifting upper-level winds might reduce precipitation over certain areas, which, in turn, would put added stress on certain agricultural regions. These same models indicate that more precipitation will fall in winter over higher latitudes, and that precipitation intensity will increase, suggesting a possibility for more extreme rainfall events, such as floods and severe drought. In the mountainous regions of western North America, where much of the precipitation falls in winter, precipitation might fall mainly as rain, causing a decrease in snow-melt runoff that fills the reservoirs during the spring. As the planet warms, total rainfall must increase to balance the increase in evaporation. But at this point, climate models are unable to determine *exactly* how global precipitation patterns will change.

Other consequences of global warming will likely be a rise in sea level as alpine glaciers recede, polar ice melts, and the oceans continue to expand as they slowly warm. During the 20th century, sea level rose between 10 and 20 centimeters, and today's improved climate models estimate that sea level will rise an additional 9 to 88 centimeters by the end of this century, depending, of course, on how much the surface temperature increases. Rising ocean levels might have a damaging influence on coastal ecosystems. In addition, coastal groundwater supplies might become contaminated with saltwater.

How the ice sheets in Antarctica and Greenland respond to global warming is not exactly clear. In polar regions, as elsewhere around the globe, rising temperatures produce complex interactions among temperature, precipitation, and wind patterns. Hence, it is now believed that as temperatures rise in south polar regions, more snow will fall in the warmer (but still cold) air, causing snow and ice to build up over the continent of Antarctica. Over Greenland, where melting snow and ice are expected to exceed any increase in precipitation, the ice sheet is expected to shrink.

Increasing levels of CO_2 in a warmer world might have additional consequences. For example, higher levels of CO_2 might act as a "fertilizer" for some plants, accelerating their growth. Increased plant growth consumes more CO_2, which might retard the increasing rate of CO_2 in the environment. On the other hand, the increased plant growth might force some insects to eat more, resulting in a net loss in vegetation. It is possible that a major increase in CO_2 might upset the balance of nature, with some plant species becoming so dominant that others are eliminated. In tropical areas, where many developing nations are located, the warming may actually decrease crop yield, whereas in cold climates, where

DID YOU KNOW?

Beware before you purchase that ocean-front property. By the end of this century, if the ocean level should rise 50 cm (about 1.6 ft.), ocean shorelines along the east coast of North America could retreat by as much as 750 m, or 2460 ft.

crops are now grown only marginally, the warming effect may actually increase crop yields.

Following are some conclusions about global warming and its future impact on our climate system summarized from the 2001 Third Assessment Report (TAR) of the Intergovernmental Panel on Climate Change (IPCC):

- The concentrations of the greenhouse gases methane, nitrous oxide, and carbon dioxide are all increasing. The present concentration of carbon dioxide has not been exceeded during the past 420,000 years, and it is likely not to have been during the past 20 million years. The current rate of increase is unprecedented for at least the past 20,000 years.
- Emissions of long-lived greenhouse gases have a lasting effect on atmospheric composition, radiative forcing, and climate.
- The global average surface temperature has increased over the 20th century about 0.6°C (about 1°F).
- There is new and stronger evidence that most of the warming observed over the last 50 years is attributable to human activities.
- Global average temperature and sea level are projected to rise under all IPCC SRES scenarios, with the globally averaged surface temperature projected to increase by 1.4°C to 5.8°C during the period 1990 to 2100.
- After greenhouse gas concentrations have stabilized, global average surface temperatures would rise at a rate of only a few tenths of a degree Celsius per century rather than the several degrees Celsius per century projected for the 21st century without stabilization.
- Global average water vapor concentration and precipitation are projected to increase during the 21st century.
- It is very likely that the 20th-century warming has, through thermal expansion of seawater and widespread loss of land ice, contributed significantly to the observed sea-level rise.
- Since 1950 it is very likely that there has been a reduction in the frequency of extreme low temperatures, with a smaller increase in the frequency of extreme high temperatures.
- It is likely that higher maximum temperatures and more hot days, along with higher minimum temperatures and fewer cold days, will occur during this century over nearly all land areas.
- Northern Hemisphere spring and summer sea-ice extent has decreased by about 10 to 15 percent since the late 1950s, and is projected to decrease further.
- It is likely that the warming associated with increasing greenhouse gas concentrations will cause an increase of Asian summer monsoon precipitation variability.
- Current projections show little change or a small increase in amplitude for El Niño events over the next 100 years.
- It is likely that over some tropical ocean, increasing water temperature may cause an increase in tropical cyclone peak winds, along with an increase in tropical cyclone mean and peak precipitation.
- Most models show weakening of the ocean thermohaline circulations, which leads to a reduction of the heat transport into high latitudes. (This phenomenon is the conveyor belt circulation described on pp. 378–379.)

In Perspective Cutting down on the emissions of greenhouse gases and pollutants has several potentially positive benefits. For example, a reduction in greenhouse gas emissions might slow down the enhancement of the earth's greenhouse effect; a reduction in air pollutants might reduce acid rain, diminish haze, and slow the production of photochemical smog. Even if the greenhouse warming proves to be less than what modern climate models project, these measures would certainly benefit humanity.

As we look to the future of a warmer world, any modification of the earth's surface taking place right now could potentially influence the immediate climate of certain regions. For example, studies show that about half the rainfall in the Amazon River Basin is returned to the atmosphere through evaporation and through transpiration from the leaves of trees. Consequently, clearing large areas of tropical rain forests in South America to create open areas for farms and cattle ranges will most likely cause a decrease in evaporative cooling. This decrease, in turn, could lead to a warming in that area of at least several degrees Celsius. In turn, the reflectivity of the deforested area will change. Similar changes in albedo result from the overgrazing and excessive cultivation of grasslands in semi-arid regions, causing an increase in desert conditions (a process known as **desertification**).

Currently, billions of acres of the world's range and cropland, along with the welfare of millions of people,

FOCUS ON A SPECIAL TOPIC

The Sahel—An Example of Climatic Variability and Human Existence

The Sahel is in North Africa, located between about 14° and 18°N latitude (see Fig. 3). Bounded on the north by the dry Sahara and on the south by the grasslands of the Sudan, the Sahel is a semi-arid region of variable rainfall. Precipitation totals may exceed 50 cm (20 in.) in the southern portion while in the north, rainfall is scanty. Yearly rainfall amounts are also variable as a year with adequate rainfall can be followed by a dry one.

During the winter, the Sahel is dry, but, as summer approaches, the Intertropical Convergence Zone (ITCZ) with its rain usually moves into the region. The inhabitants of the Sahel are mostly nomadic people who migrate to find grazing land for their cattle and goats. In the early and middle 1960s, adequate rainfall led to improved pasture lands; herds grew larger and so did the population. However, in 1968, the annual rains did not reach as far north as usual, marking the beginning of a series of dry years and a severe drought.

The decrease in rainfall, along with overgrazing, turned thousands of square kilometers of pasture into barren wasteland. By 1973, when the severe drought reached its climax, rainfall totals were 50 percent of the long-term average, and perhaps 50 percent of the cattle and goats had died. The Sahara Desert had migrated southward into the northern fringes of the region, and a great famine had taken the lives of more than 100,000 people.

Although low rainfall years have been followed by wetter ones, relatively dry conditions have persisted over the region for the past 30 years or so. The wetter years of the 1950s and 1960s appear to be due to the northward displacement of the ITCZ. The drier years, however, appear to be more related to the intensity of rain that falls during the so-called rainy season. But what causes the lack of intense rain? Some scientists feel that this situation is due to a *biogeophysical feedback mechanism* wherein less rainfall and reduced vegetation cover modify the surface and promote a positive feedback relationship: Surface changes act to reduce convective activity, which in turn promotes or reinforces the dry conditions. As an example, when the vegetation is removed from the surface (perhaps through overgrazing or excessive cultivation), the surface albedo (reflectivity) increases, and the surface temperature drops. But studies show that less vegetation cover does not always result in a higher albedo.

Since the mid-1970s the Sahara Desert has not progressively migrated southward into the Sahel. In fact, during dry years, the desert does migrate southward, but in wet years, it retreats. By the same token, vegetation cover throughout the Sahel is more extensive during the wetter years. Consequently, desertification is not presently overtaking the Sahel, nor is the albedo of the region showing much year-to-year change.

So the question remains: Why did the Sahel go from a period of abundant rainfall in the 1950s and early 1960s to relatively dry conditions since then? Was there a large change in the surface albedo brought on by reduced vegetation? Without adequate satellite imagery during those years, it is impossible to tell. Does this relatively dry spell indicate a long-term fluctuation in climate, or will the wetter years of the 1950s return? And if global temperatures continue to rise, how will precipitation patterns change? At present, we have no answers.

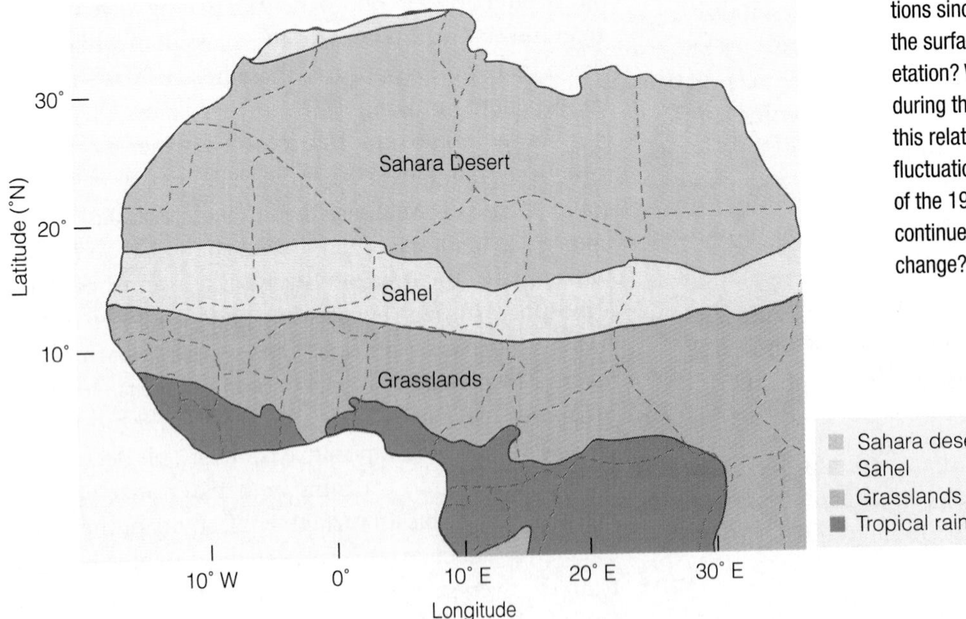

FIGURE 3 The semi-arid Sahel of North Africa is bounded by the Sahara Desert to the north and grasslands to the south.

are affected by desertification. Annually, millions of acres are reduced to a state of near or complete uselessness. The main cause is overgrazing, although overcultivation, poor irrigation practices, and deforestation also play a role. The effect this will have on climate, as surface albedos increase and more dust is swept into the air, is uncertain. (For a look at how a modified surface influences the inhabitants of a region in Africa, read the Focus section on p. 398.)

SUMMARY

In this chapter, we considered some of the many ways the earth's climate can be changed. First, we saw that the earth's climate has undergone considerable change during the geologic past. Some of the evidence for a changing climate comes from tree rings (dendrochronology), chemical analysis of oxygen isotopes in ice cores and fossil shells, and geologic evidence left behind by advancing and retreating glaciers. The evidence from these suggest that, throughout much of the geologic past (before humanity arrived on the scene), the earth was much warmer than it is today. There were cooler periods, however, during which glaciers advanced over large sections of North America and Europe.

We examined some of the possible causes of climate change, noting that the problem is extremely complex, as a change in one variable in the climate system almost immediately changes other variables. One theory suggests that the shifting of the continents, along with volcanic activity and mountain building, may account for variations in climate that take place over millions of years.

The Milankovitch theory proposes that alternating glacial and interglacial episodes during the past 2 million years are the result of small variations in the tilt of the earth's axis and in the geometry of the earth's orbit around the sun. Another theory suggests that certain cooler periods in the geologic past may have been caused by volcanic eruptions rich in sulfur. Still another theory postulates that climatic variations on earth might be due to variations in the sun's energy output.

We looked at temperature trends over the past 100 years and found that, over this span of time, the earth has warmed by about 0.6°C (about 1°F). It is likely that most of the warming during the last 50 years is due to increasing concentrations of greenhouse gases. Sophisticated climate models project that, as levels of CO_2 and other greenhouse gases continue to increase, the earth will warm by between 1.4°C and 5.8°C from 1990 to the end of this century. The models also predict that, as the earth warms, there will be a global increase in atmospheric water vapor, an increase in global precipitation, and a rise in sea level.

Meteorology Now™ Assess your understanding of this chapter's topics with additional quizzing and tutorials at http://earthscience.brookscole.com/ahrens/ess4e.

KEY TERMS

The following terms are listed in the order they appear in the text. Define each. Doing so will aid you in reviewing the material covered in this chapter.

dendrochronology
Ice Age
interglacial period
Younger-Dryas (event)
Little Ice Age
water vapor–greenhouse feedback
positive feedback mechanism
snow-albedo feedback
negative feedback mechanism
theory of plate tectonics
Milankovitch theory
eccentricity
precession
obliquity
sulfate aerosols
Maunder minimum
radiative forcing agents
radiative forcing
desertification

QUESTIONS FOR REVIEW

1. What methods do scientists use to determine climate conditions that have occurred in the past?
2. Explain how the changing climate influenced the formation of the Bering land bridge.
3. How does today's average global temperature compare with the average temperature during most of the past 1000 years?
4. What is the Younger-Dryas episode? When did it occur?

5. How does a positive feedback mechanism differ from a negative feedback mechanism? Is the water vapor–greenhouse feedback considered positive or negative? Explain.

6. How does the theory of plate tectonics explain climate change over periods of millions of years?

7. Describe the Milankovitch theory of climatic change by explaining how each of the three cycles alters the amount of solar energy reaching the earth.

8. Given the analysis of air bubbles trapped in polar ice during the past 160,000 years, were CO_2 levels generally higher or lower during colder glacial periods?

9. How do sulfate aerosols in the lower atmosphere affect surface air temperatures during the day?

10. Volcanic eruptions rich in sulfur warm the stratosphere. Do they warm or cool the earth's surface? Explain.

11. Explain how variations in the sun's energy output might influence global climate.

12. Climate models predict that increasing levels of CO_2 will cause the mean global surface temperature to rise by as much as 5.8°C by the year 2100. What other greenhouse gas must also increase in concentration in order for this condition to occur?

13. In Fig. 14.14, p. 393, explain why the actual rise in surface air temperature (gray line) is much less than the projected rise in air temperature due to increasing levels of greenhouse gases (yellow line).

14. Describe some of the natural radiative forcing agents and their effect on climate.

15. (a) Describe how clouds influence the climate system.

    (b) Which clouds would tend to promote surface cooling: high clouds or low clouds?

16. Even though CO_2 concentrations have risen dramatically over the past 100 years, how do scientists explain the fact that global temperatures have risen by only 0.6°C?

17. Explain how the ocean's conveyor belt circulation works. How does the conveyor belt appear to influence the climate of northern Europe?

18. Why do climate scientists now believe that at least part of the warming experienced during the 20th century was due to increasing levels of greenhouse gases?

19. List some of the consequences that global warming might have on the atmosphere and its inhabitants.

20. Is CO_2 the only greenhouse gas we should be concerned with for climate change? If not, what are the other gases?

QUESTIONS FOR THOUGHT AND EXPLORATION

1. Ice cores extracted from Greenland and Antarctica have yielded valuable information on climate changes during the past few hundred thousand years. What do you feel might be some of the limitations in using ice core information to evaluate past climate changes?

2. When glaciation was at a maximum (about 18,000 years ago), was global precipitation greater or less than at present? Explain your reasoning.

3. Consider the following climate change scenario. Warming global temperatures increase saturation vapor pressures over the ocean. As more water evaporates, increasing quantities of water vapor build up in the troposphere. More clouds form as the water vapor condenses. The clouds increase the albedo, resulting in decreased amounts of solar radiation reaching the earth's surface. Is this scenario plausible? What type(s) of feedback(s) is/are involved?

4. Are ice ages in the Northern Hemisphere more likely when the tilt of the earth is at a maximum or a minimum? Explain.

5. Are ice ages in the Northern Hemisphere more likely when the sun is closest to the earth during summer or during winter? Explain.

6. The oceans are a major sink (absorber) of CO_2. According to one hypothesis, continued global warming will result in less CO_2 being dissolved in the oceans. Under this scenario, would you expect the earth to warm or to cool further? Explain your reasoning.

7. Global Climate Change Data (http://cdiac.esd.ornl.gov/trends/trends.htm): Compare graphs of temperature trends for the Northern Hemisphere, the Southern Hemisphere, and the globe. Compare and contrast these trends. Which hemisphere has a trend that is most similar to the global trend?

8. **8.** Paleoclimate (http://www.ngdc.noaa.gov/paleo/education.html): What is known about past climates? How does the climate change of the past 100 years compare to climate changes that have occurred in the past? When was the last glaciation? Do you think it could happen again?

Go to the Brooks/Cole Earth Sciences Resource Center (http://earthscience.brookscole.com) for critical thinking exercises, articles, and additional readings from InfoTrac College Edition, Brooks/Cole's online student library.

Sunlight bending through ice crystals in cirriform clouds produces bands of color called sundogs, or parhelia, on both sides of the sun on this cold winter day in Minnesota.

MeteorologyNow™ This icon, appearing throughout the book, indicates an opportunity to explore interactive tutorials, animations, or practice problems available on the MeteorologyNow Web site at http://earthscience.brookscole.com/ahrens/ess4e.

Light, Color, and Atmospheric Optics

CONTENTS

The sky is clear, the weather cold, and the year, 1818. Near Baffin Island in Canada, a ship with full sails enters unknown waters. On board are the English brothers James and John Ross, who are hoping to find the elusive "Northwest Passage," the waterway linking the Atlantic and Pacific oceans. On this morning, however, their hopes would be dashed, for directly in front of the vessel, blocking their path, is a huge towering mountain range. Disappointed, they turn back and report that the Northwest Passage does not exist. About seventy-five years later Admiral Perry met the same barrier and called it "Crocker land." What type of treasures did this mountain conceal—gold, silver, precious gems? The curiosity of explorers from all over the world had been aroused. Speculation was the rule, until, in 1913, the American Museum of Natural History commissioned Donald MacMillan to lead an expedition to solve the mystery of Crocker land. At first, the journey was disappointing. Where Perry had seen mountains, MacMillan saw only vast stretches of open water. Finally, ahead of his ship was Crocker land, but it was more than two hundred miles farther west from where Perry had encountered it. MacMillan sailed on as far as possible. Then he dropped anchor and set out on foot with a small crew of men. As the team moved toward the mountains, the mountains seemed to move away from them. If they stood still, the mountains stood still; if they started walking, the mountains receded again. Puzzled, they trekked onward over the glittering snow-fields until huge mountains surrounded them on three sides. At last the riches of Crocker land would be theirs. But in the next instant the sun disappeared below the horizon and, as if by magic, the mountains dissolved into the cold arctic twilight. Dumbfounded, the men looked around only to see ice in all directions—not a mountain was in sight. There they were, the victims of one of nature's greatest practical jokes, for Crocker land was a mirage.

The sky is full of visual events. Optical illusions (mirages) can appear as towering mountains or wet roadways. In clear weather, the sky can appear blue, while the horizon appears milky white. Sunrises and sunsets can fill the sky with brilliant shades of pink, red, orange, and purple. At night, the sky is black, except for the light from the stars, planets, and the moon. The moon's size and color seem to vary during the night, and the stars twinkle. To understand what we see in the sky, we will take a closer look at sunlight, examining how it interacts with the atmosphere to produce an array of atmospheric visuals.

WHITE AND COLORS

We know from Chapter 2 that nearly half of the solar radiation that reaches the atmosphere is in the form of visible light. As sunlight enters the atmosphere, it is either absorbed, reflected, scattered, or transmitted on through. How objects at the surface respond to this energy depends on their general nature (color, density, composition) and the wavelength of light that strikes them. How do we see? Why do we see various colors? What kind of visual effects do we observe because of the interaction between light and matter? In particular, what can we *see* when light interacts with our atmosphere?

We perceive light because electromagnetic waves stimulate antenna-like nerve endings in the retina of the human eye. These antennae are of two types—*rods* and *cones.* The rods respond to all wavelengths of visible light and give us the ability to distinguish light from dark. If people possessed rod-type receptors only, then only black and white vision would be possible. The cones respond to specific wavelengths of visible light. The cones fire an impulse through the nervous system to the brain, and we perceive this impulse as the sensation of color. (Color blindness is caused by missing or malfunctioning cones.) Wavelengths of radiation shorter than those of visible light do not stimulate color vision in humans.

White light is perceived when all visible wavelengths strike the cones of the eye with nearly equal intensity.* Because the sun radiates almost half of its energy as visible light, all visible wavelengths from the midday sun reach the cones, and the sun usually appears white. A star that is cooler than our sun radiates most of its energy at slightly longer wavelengths; therefore, it appears redder. On the other hand, a star much hotter than our sun radiates more energy at shorter wavelengths and thus appears bluer. A star whose temperature is about the same as the sun's appears white.

Objects that are not hot enough to produce radiation at visible wavelengths can still have color. Everyday objects we see as red are those that absorb all visible radiation except red. The red light is reflected from the object to our eyes. Blue objects have blue light returning from them, since they absorb all visible wavelengths except blue. Some surfaces absorb all visible wavelengths and reflect no light at all. Since no radiation strikes the rods or cones, these surfaces appear black. Therefore, when we see colors, we know that light must be reaching our eyes.

*Recall from Chapter 2 that visible white light is a combination of waves with different wavelengths. The wavelength of visible light in decreasing order are: red (longest), orange, yellow, green, blue, and violet (shortest).

WHITE CLOUDS AND SCATTERED LIGHT

One exciting feature of the atmosphere can be experienced when we watch the underside of a puffy, growing cumulus cloud change color from white to dark gray or black. When we see this change happen, our first thought is usually, "It's going to rain." Why is the cloud initially white? Why does it change color? To answer these questions, let's examine the concept of *scattering.*

When sunlight bounces off a surface at the same angle at which it strikes the surface, we say that the light is **reflected,** and call this phenomenon *reflection.* There are various constituents of the atmosphere, however, that tend to deflect sunlight from its path and send it out in all directions. We know from Chapter 2 that radiation reflected in this way is said to be **scattered.** (Scattered light is also called *diffuse light.*) During the scattering process, no energy is gained or lost and, therefore, no temperature changes occur. In the atmosphere, scattering is usually caused by small objects, such as air molecules, fine particles of dust, water molecules, and some pollutants. Just as the ball in a pinball machine bounces off the pins in many directions, so solar radiation is knocked about by small particles in the atmosphere.

Typical cloud droplets are large enough to effectively scatter all wavelengths of visible radiation more or less equally. Clouds, even small ones, are optically thick, meaning that very little unscattered light gets through them. These same clouds are poor absorbers of sunlight. Hence, when we look at a cloud, it appears white because countless cloud droplets scatter all wavelengths of visible sunlight in all directions (see Fig. 15.1).

As a cloud grows larger and taller, more sunlight is reflected from it and less light can penetrate all the way through it (see Fig. 15.2). In fact, relatively little light penetrates a cloud whose thickness is 1000 m (3300 ft).

DID YOU KNOW?

Cumulus clouds are usually white in appearance, whereas thunderstorms, especially severe thunderstorms that form over the Great Plains, may appear green. The green color was once thought to be reflected light from green ground foliage or scattered light from falling hail. But green thunderstorms have been observed above various colored surfaces and without hail. The green color may be due to reddish sunlight (especially at sunset) penetrating the storm, then being scattered by cloud particles composed of water and ice. With much of the red light removed, the scattered light casts the underside of the cloud as a faint greenish hue.

Since little sunlight reaches the underside of the cloud, little light is scattered, and the cloud base appears dark. At the same time, if droplets near the cloud base grow larger, they become less effective scatterers and better absorbers. As a result, the meager amount of visible light that does reach this part of the cloud is absorbed rather than scattered, which makes the cloud appear even darker. These same cloud droplets may even grow large and heavy enough to fall to the earth as rain. From a casual observation of clouds, we know that dark, threatening ones frequently produce rain. Now we know why they appear so dark.

BLUE SKIES AND HAZY DAYS

The sky appears blue because light that stimulates the sensation of blue color is reaching the retina of the eye. How does this happen?

Individual air molecules are much smaller than cloud droplets—their diameters are small even when compared with the wavelength of visible light. Each air molecule of oxygen and nitrogen is a *selective scatterer* in that each scatters shorter waves of visible light much more effectively than longer waves.

As sunlight enters the atmosphere, the shorter visible wavelengths of violet, blue, and green are scattered more by atmospheric gases than are the longer wavelengths of yellow, orange, and especially red. (Violet light is scattered about 16 times more than red light.) As we view the sky, the scattered waves of violet, blue, and green strike the eye from all directions. Because our eyes are more sensitive to blue light, these waves, viewed together, produce the sensation of blue coming from all around us (see Fig. 15.3). Therefore, when we look at the sky it appears blue (see Fig. 15.4). (Earth, by the way, is not the only planet with a colorful sky. On Mars, dust in the air turns the sky red at midday and purple at sunset.)

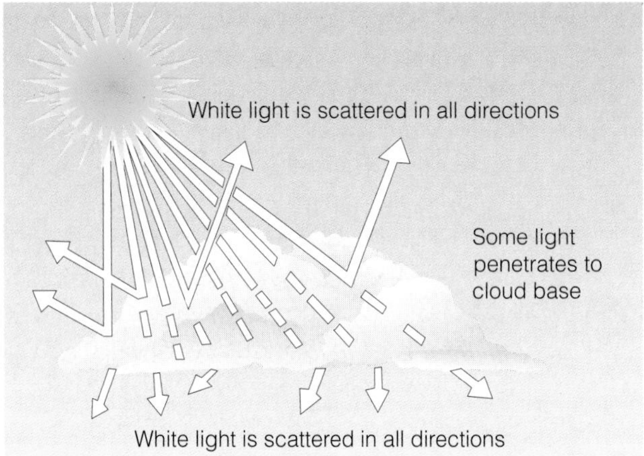

FIGURE 15.1 Since tiny cloud droplets scatter visible light in all directions, light from many billions of droplets turns a cloud white.

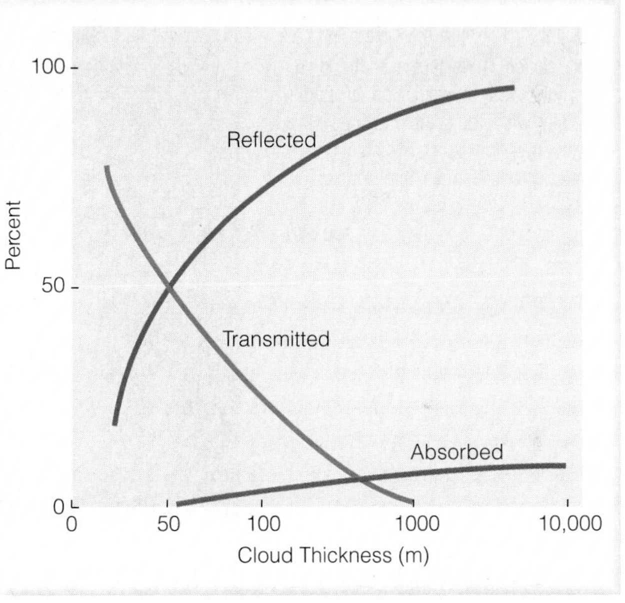

FIGURE 15.2 Average percent of radiation reflected, absorbed, and transmitted by clouds of various thickness.

The selective scattering of blue light by air molecules and very small particles can make distant mountains appear blue, such as the Blue Ridge Mountains of Virginia (see Fig. 15.5) and the Blue Mountains of Australia. In some places, a *blue haze* may cover the landscape, even in areas far removed from human contamination. Although its cause is still controversial, the blue haze appears to be the result of a particular process. Extremely tiny particles (hydrocarbons called *terpenes*) are released by vegetation to combine chemically with small

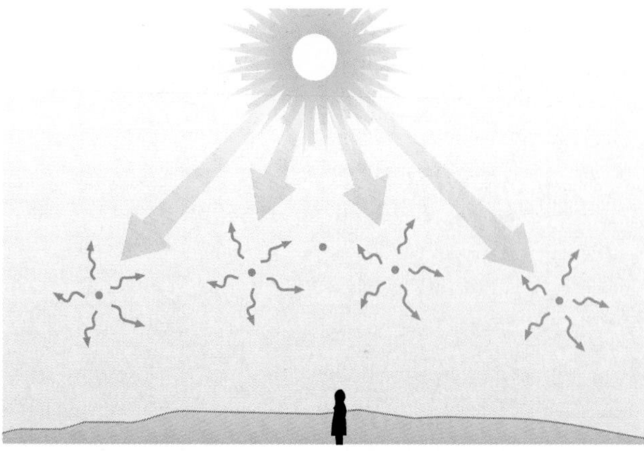

FIGURE 15.3 The sky appears blue because billions of air molecules selectively scatter the shorter wavelengths of visible light more effectively than the longer ones. This causes us to see blue light coming from all directions.

amounts of ozone. This reaction produces extremely tiny particles that selectively scatter blue light.

When small particles, such as fine dust and salt, become suspended in the atmosphere, the color of the sky begins to change from blue to milky white. Although these particles are small, they are large enough to scatter all wavelengths of visible light fairly evenly in all directions. When our eyes are bombarded by all wavelengths of visible light, the sky appears milky white, the visibility lowers, and we call the day "hazy." If the humidity is high enough, soluble particles (nuclei) will "pick up" water vapor and grow into haze particles. Thus, the color of the sky gives us a hint about how much material is suspended in the air: the more particles, the more scattering, and the whiter the sky becomes. On top of a high

FIGURE 15.4 Blue skies and white clouds. The selective scattering of blue light by air molecules produces the blue sky, while the scattering of all wavelengths of visible light by liquid cloud droplets produces the white clouds.

Photo by author

Photo by author

FIGURE 15.5 The Blue Ridge Mountains in Virginia. The blue haze is caused by the scattering of blue light by extremely small particles—smaller than the wavelengths of visible light. Notice that the scattered blue light causes the most distant mountains to become almost indistinguishable from the sky.

NCAR

FIGURE 15.6 The scattering of sunlight by dust and haze produces these white bands of crepuscular rays.

mountain, when we are above many of these haze particles, the sky usually appears a deep blue.

Haze can scatter light from the rising or setting sun so that we see bright lightbeams, or **crepuscular rays,** radiating across the sky. A similar effect occurs when the sun shines through a break in a layer of clouds (see Fig. 15.6). Dust, tiny water droplets, or haze in the air beneath the clouds scatter sunlight, making that region of the sky appear bright with rays. Because these rays seem to reach downward from clouds, some people will remark that the "sun is drawing up water." In England, this same phenomenon is referred to as "Jacob's ladder." No matter what name these sunbeams go by, it is the scattering of sunlight by particles in the atmosphere that makes them visible.

RED SUNS AND BLUE MOONS

At midday, the sun seems a brilliant white, while at sunset it usually appears to be yellow, orange, or red. At noon, when the sun is high in the sky, light from the sun is most intense—all wavelengths of visible light are able to reach the eye with about equal intensity, and the sun appears white. (Looking directly at the sun, especially during this time of day, can cause irreparable damage to the eye. Normally, we get only glimpses or impressions of the sun out of the corner of our eye.)

Near sunrise or sunset, however, the rays coming directly from the sun strike the atmosphere at a low angle. They must pass through much more atmosphere than at any other time during the day. (When the sun is 4° above the horizon, sunlight must pass through an atmosphere more than 12 times thicker than when the sun is directly overhead.) By the time sunlight has penetrated this large amount of air, most of the shorter waves of visible light have been scattered away by the air molecules. Just about the only waves from a setting sun that make it through the atmosphere on a fairly direct path are the yellow, orange, and red. Upon reaching the eye, these waves produce a sunset that is bright yellow-orange (see Fig. 15.7).

Bright, yellow-orange sunsets only occur when the atmosphere is fairly clean, such as after a recent rain. If the atmosphere contains many fine particles whose diameters are a little larger than air molecules, slightly longer (yellow) waves also would be scattered away. Only orange and red waves would penetrate through to the eye, and the sun would appear red-orange. When the atmosphere becomes loaded with particles, only the

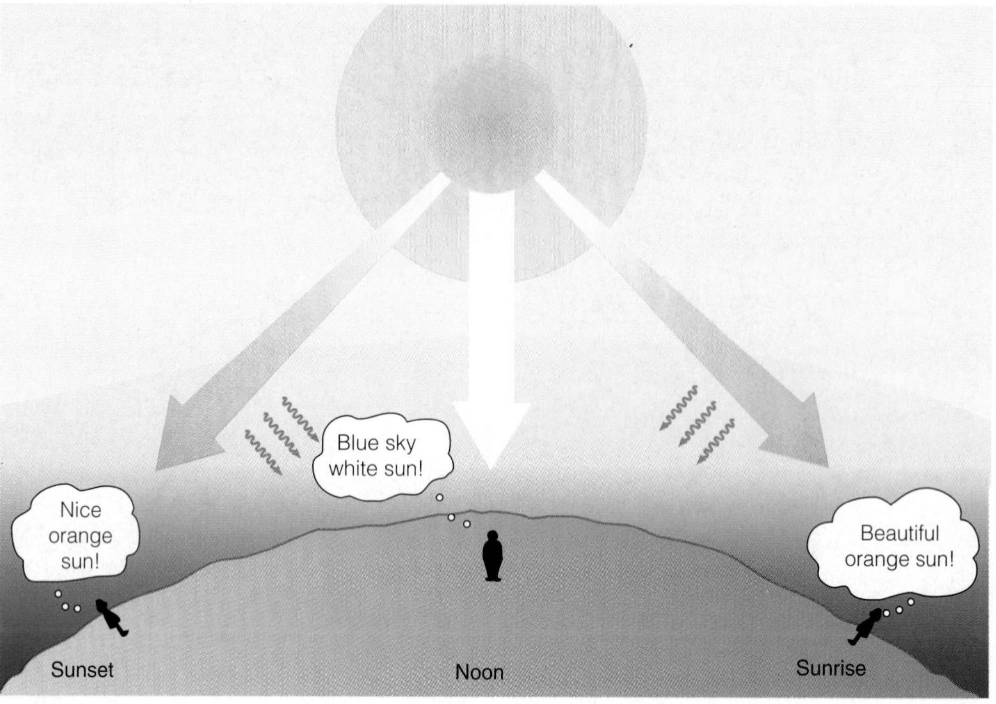

FIGURE 15.7 Because of the selective scattering by a thick section of atmosphere, the sun at sunrise and sunset appears either yellow, orange, or red. At noon, it is usually white.

longest red wavelengths are able to penetrate the atmosphere, and we see a red sun.

Natural events may produce red sunrises and sunsets. Over the oceans, for example, the scattering characteristics of small, suspended salt particles and water vapor are responsible for the brilliant red suns observed from the beach (see Fig. 15.8). Moreover, major volcanic eruptions send vast amounts of dust and ash high into the atmosphere. These fine particles, moved by the winds aloft, circle the globe, producing beautiful sunrises and sunsets for months and even years. There were beautiful ruddy sunsets in many parts of the Northern Hemisphere after the the eruption of the Mexican volcano El Chichón in 1982 and the Philippine volcano Mount Pinatubo in 1991.

Photo by author

FIGURE 15.8 Red sunset near the coast of Iceland. The reflection of sunlight off the slightly rough water is producing a glitter path.

Occasionally, the atmosphere becomes so laden with dust, smoke, and pollutants that even red waves are unable to pierce the filthy air. An eerie effect then occurs. Because no visible waves enter the eye, the sun literally disappears before it reaches the horizon.

The scattering of light by large quantities of atmospheric particles can cause some rather unusual sights. If the volcanic ash, dust, smoke particles, or pollutants are roughly uniform in size, they can selectively scatter the sun's rays. Even at noon, various colored suns have appeared: orange suns, green suns, and even blue suns. For blue suns to appear, the size of the suspended particles must be similar to the wavelength of visible light. When these particles are present they tend to scatter red light more than blue, which causes a bluing of the sun and a reddening of the sky. Although rare, the same phenomenon can happen to moonlight, making the moon appear blue; thus, the expression "once in a blue moon."

In summary, the scattering of light by small particles in the atmosphere causes many familiar effects: white clouds, blue skies, hazy skies, crepuscular rays, and colorful sunsets. In the absence of any scattering, we would simply see a white sun against a black sky—not an attractive alternative.

TWINKLING, TWILIGHT, AND THE GREEN FLASH

Light that passes through a substance is said to be *transmitted.* Upon entering a denser substance, transmitted light slows in speed. If it enters the substance at an angle, the light's path also bends. This bending is called **refraction.** The amount of refraction depends primarily on two factors: the density of the material and the angle at which the light enters the material.

Refraction can be demonstrated in a darkened room by shining a flashlight into a beaker of water (see Fig. 15.9). If the light is held directly above the water so that the beam strikes the surface of the water straight on, no bending occurs. But, if the light enters the water at some angle, it bends toward the *normal,* which is the dashed line in the diagram running perpendicular to the air-water boundary. (The normal is simply a line that intersects any surface at a right angle. We use it as a reference to see how much bending occurs as light enters and leaves various substances.) A small mirror on the bottom of the beaker reflects the light upward. This reflected light bends away from the normal as it re-enters the air. We can summarize these observations as follows: *Light that travels from a less-dense to a more-dense medium loses speed and bends toward the normal, whereas light that enters a less-dense medium increases in speed and bends away from the normal.*

The refraction of light within the atmosphere causes a variety of visual effects. At night, for example, the light from the stars that we see directly above us is not bent, but starlight that enters the earth's atmosphere at an angle is bent. In fact, a star whose light enters the atmosphere just above the horizon has more atmosphere to penetrate and is thus refracted the most. As we can see in Fig. 15.10, the bending is toward the normal as the light enters the more-dense atmosphere. By the time this "bent" starlight reaches our eyes, the star appears to be higher than it actually is because our eyes cannot detect that the light path is bent. We see light coming from a particular direction and interpret the star to be in that direction. So, the next time you take a midnight stroll, point to any star near the horizon and remember: this is where the star appears to be. To point to the star's true position, you would have to lower your arm just a tiny bit.

As starlight enters the atmosphere, it often passes through regions of differing air density. Each of these regions deflects and bends the tiny beam of starlight, constantly changing the apparent position of the star. This causes the star to appear to *twinkle* or flicker, a condition known as **scintillation.** Planets, being much closer to us, appear larger, and usually do not twinkle because their size is greater than the angle at which their light deviates as it penetrates the atmosphere. Planets sometimes twinkle, however, when they are near the horizon, where the bending of their light is greatest.

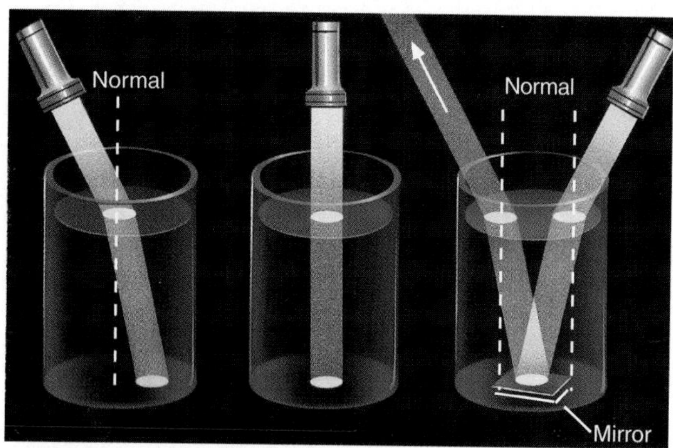

FIGURE 15.9 The behavior of light as it enters and leaves a more-dense substance, such as water.

The refraction of light by the atmosphere has some other interesting consequences. For example, the atmosphere gradually bends the rays from a rising or setting

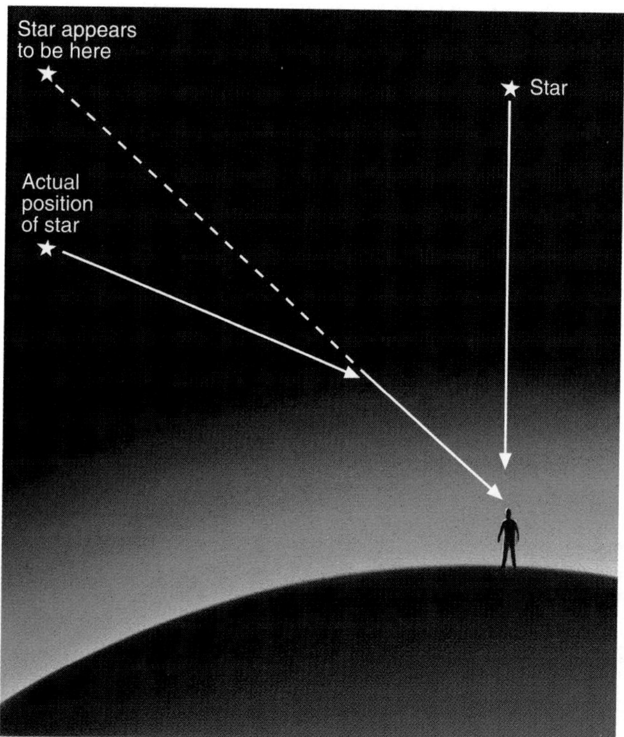

FIGURE 15.10 Due to the bending of starlight by the atmosphere, stars not directly overhead appear to be higher than they really are.

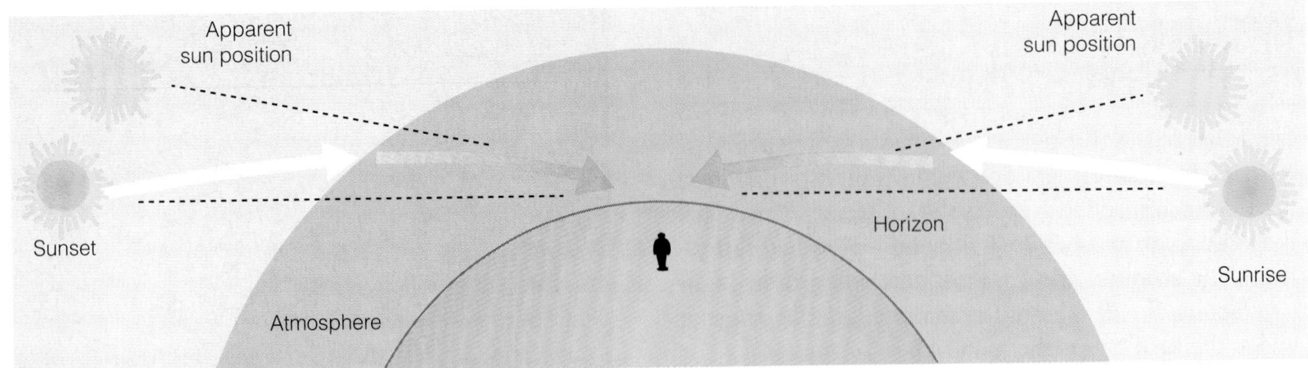

FIGURE 15.11 The bending of sunlight by the atmosphere causes the sun to rise about two minutes earlier, and set about two minutes later, than it would otherwise.

sun or moon. Because light rays from the lower part of the sun (or moon) are bent more than those from the upper part, the sun appears to flatten out on the horizon, taking on an elliptical shape. (The sun in Fig. 15.13 shows this effect.) Also, since light is bent most on the horizon, the sun and moon both appear to be higher than they really are. Consequently, they both rise about two minutes earlier and set about two minutes later than they would if there were no atmosphere (see Fig. 15.11).

You may have noticed that on clear days the sky is often bright for some time after the sun sets. The atmosphere refracts and scatters sunlight to our eyes, even though the sun itself has disappeared from our view. **Twilight** is the name given to the time after sunset (and immediately before sunrise) when the sky remains illuminated and allows outdoor activities to continue without artificial lighting.

The length of twilight depends on season and latitude. During the summer in middle latitudes, twilight adds about thirty minutes of light to each morning and evening for outdoor activities. The duration of twilight increases with increasing latitude, especially in summer. At high latitudes during the summer, morning and evening twilight may converge, producing a *white night*—a night long twilight.

The color of twilight can change when particles of volcanic ash and dust are present in the upper atmosphere. For example, after the volcanic eruptions of El Chichón in 1982 and Mount Pinatubo in 1991, brilliant red twilights occurred over parts of North America as sulfur-rich volcanic particles scattered red light from the setting (and rising) sun (see Fig. 15.12).

In general, without the atmosphere, there would be no refraction or scattering, and the sun would rise later and set earlier than it now does. Instead of twilight, darkness would arrive immediately when the sun disappeared below the horizon. Imagine the number of sandlot baseball games that would be called because of instant darkness.

Occasionally, a flash of green light—called the **green flash**—may be seen near the upper rim of a rising or setting sun (see Fig. 15.13). Remember from our earlier discussion that, when the sun is near the horizon, its light must penetrate a thick section of atmosphere. This thick atmosphere refracts sunlight, with purple and blue light bending the most, and red light the least. Because of this bending, more blue light should appear along the top of the sun. But because the atmosphere selectively scatters blue light, very little reaches us, and we see green light instead.

Usually, the green light is too faint to see with the human eye. However, under certain atmospheric conditions, such as when the surface air is very hot or when an upper-level inversion exists, the green light is magnified by the atmosphere. When this happens, a momentary flash of green light appears, often just before the sun disappears from view.

The flash usually lasts about a second, although in polar regions it can last longer. Here, the sun slowly changes in elevation and the flash may exist for many minutes. Members of Admiral Byrd's expedition in the south polar region reported seeing the green flash for 35 minutes in September as the sun slowly rose above the horizon, marking the end of the long winter.

DID YOU KNOW?

An old Scottish legend says that if you see the green flash you will not err in matters of love.

Photo by author

FIGURE 15.12 Bright twilight over California produced by the sulfur-rich particles from the volcano Mt. Pinatubo during September, 1992.

© Pekka Parviainen

FIGURE 15.13 The very light green on the upper rim of the sun is the green flash. Also, observe how the atmosphere makes the sun appear to flatten on the horizon into an elliptical shape.

BRIEF REVIEW

Up to this point, we have examined how light can interact with our atmosphere. Before going on, here is a review of some of the important concepts and facts we have covered:

- When light is scattered, it is sent in all directions—forward, sideways, and backward.
- White clouds, blue skies, hazy skies, crepuscular rays, and colorful sunsets are the result of sunlight being scattered.
- The bending of light as it travels through regions of differing density is called *refraction.*
- As light travels from a less-dense substance (such as outer space) and enters a more-dense substance at an angle (such as our atmosphere), the light bends downward, toward the normal. This effect causes stars, the moon, and the sun to appear just a tiny bit higher than they actually are.

THE MIRAGE: SEEING IS NOT BELIEVING

In the atmosphere, when an object appears to be displaced from its true position, we call this phenomenon a **mirage.** A mirage is not a figment of the imagination—our minds are not playing tricks on us, but the atmosphere is.

Atmospheric mirages are created by light passing through and being bent by air layers of different densities. Such changes in air density are usually caused by sharp changes in air temperature. The greater the rate of temperature change, the greater the light rays are bent. For example, on a warm, sunny day, black road surfaces absorb a great deal of solar energy and become very hot. Air in contact with these hot surfaces warms by conduction and, because air is a poor thermal conductor, we find much cooler air only a few meters higher. On hot days, these road surfaces often appear wet (see Fig. 15.14). Such "puddles" disappear as we approach them, and advancing cars seem to swim in them. Yet, we know the road is dry. The apparent wet pavement above a road is the result of blue skylight refracting up into our eyes as it travels through air of different densities. A similar type of mirage occurs in deserts during the hot summer. Many thirsty travelers have been disappointed to find that what appeared to be a water hole was in actuality hot desert sand.

Sometimes, these "watery" surfaces appear to *shimmer.* The shimmering results as rising and sinking air near the ground constantly change the air density. As light moves through these regions, its path also changes, causing the shimmering effect.

When the air near the ground is much warmer than the air above, objects may not only appear to be lower than they really are, but also (often) inverted. These mirages are called **inferior** (lower) **mirages.** The tree in Fig. 15.15 certainly doesn't grow upside down. So why does it look that way? It appears to be inverted because light reflected from the top of the tree moves outward in all directions. Rays that enter the hot, less-dense air above the sand are refracted upward, entering the eye from below. The brain is fooled into thinking that these rays came from below the ground, which makes the tree appear upside down. Some light from the top of the tree

Photo by author

FIGURE 15.14 The road in the photo appears wet because blue skylight is bending up into the camera as the light passes through air of different densities.

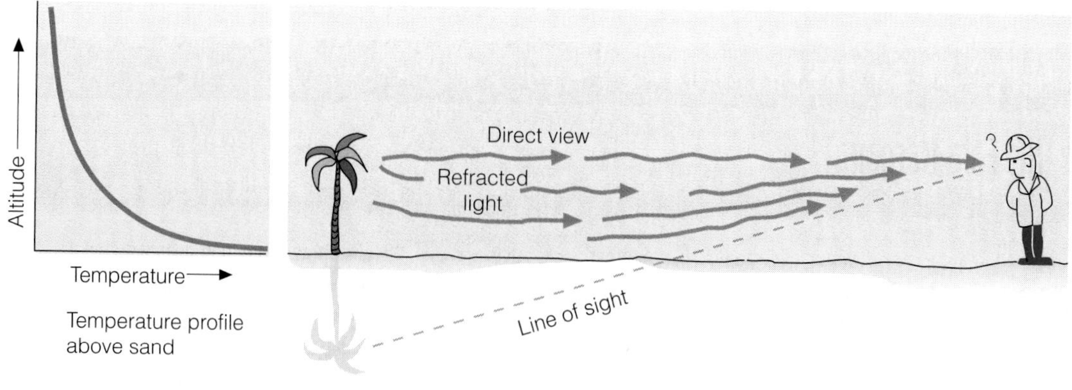

FIGURE 15.15
Inferior mirage over hot desert sand.

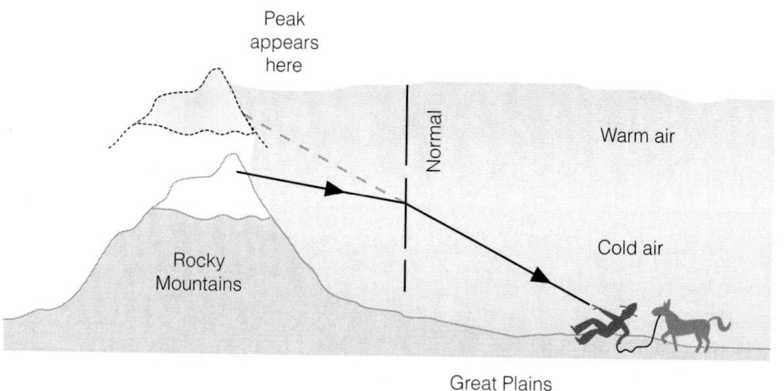

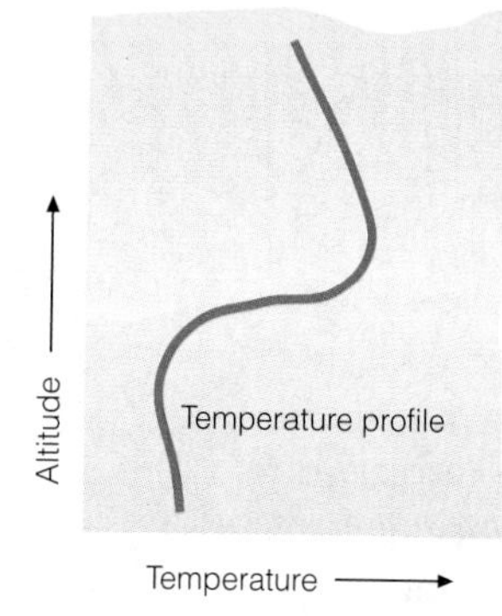

FIGURE 15.16
The formation of a superior mirage. When cold air lies close to the surface with warm air aloft, light from distant mountains is refracted toward the normal as it enters the cold air. This causes an observer on the ground to see mountains higher and closer than they really are.

travels directly toward the eye through air of nearly constant density and, therefore, bends very little. These rays reach the eye "straight-on," and the tree appears upright. Hence, off in the distance, we see a tree and its upside-down image beneath it. (Some of the trees in Fig. 15.14 show this effect.)

The atmosphere can play optical jokes on us in extremely cold areas, too. In polar regions, air next to a snow surface can be much colder than the air many meters above. Because the air in this cold layer is very dense, light from distant objects entering it bends toward the normal in such a way that the objects can appear to be shifted upward. This phenomenon is called a **superior** (upward) **mirage.** Figure 15.16 shows the conditions favorable for a superior mirage. (A special type of mirage, the *Fata Morgana,* is described in the Focus section on p. 414.)

DID YOU KNOW?

On December 14, 1890, a mirage—lasting several hours—gave the residents of Saint Vincent, Minnesota, a clear view of cattle nearly 13 km (8 mi) away.

HALOS, SUNDOGS, AND SUN PILLARS

A ring of light encircling and extending outward from the sun or moon is called a **halo.** Such a display is produced when sunlight or moonlight is refracted as it passes through ice crystals. Hence, the presence of a halo indicates that *cirriform clouds* are present. (Cirriform clouds are described more completely in Chapter 6.)

The most common type of halo is the 22° halo—a ring of light 22° from the sun or moon.* Such a halo forms when tiny suspended column-type ice crystals (with diameters less than 20 micrometers) become randomly oriented as air molecules constantly bump against them. The refraction of light rays through these crystals forms a halo like the one shown in Fig. 15.17. Less common is the 46° halo, which forms in a similar fashion to the 22° halo (see Fig. 15.18). With the 46° halo, however, the light is refracted through column-type ice crystals that have diameters in a narrow range between about 15 and 25 micrometers.

Occasionally, a bright arc of light may be seen at the top of a 22° halo (see Fig. 15.19). Since the arc is tangent

*Extend your arm and spread your fingers apart. An angle of 22° is about the distance from the tip of the thumb to the tip of the little finger.

FOCUS ON AN OBSERVATION

The Fata Morgana

A special type of superior mirage is the *Fata Morgana,* a mirage that transforms a fairly uniform horizon into one of vertical walls and columns with spires (see Fig. 1). According to legend, *Fata Morgana* (Italian for fairy Morgan) was the half-sister of King Arthur. Morgan, who was said to live in a crystal palace beneath the water, had magical powers that could build fantastic castles out of thin air. Looking across the Straits of Messina (between Italy and Sicily), residents of Reggio, Italy, on occasion would see buildings, castles, and sometimes whole cities appear, only to vanish again in minutes. The *Fata Morgana* is observed where the air temperature increases with height above the surface, slowly at first, then more rapidly, then slowly again. Consequently, mirages like the *Fata Morgana* are frequently seen where warm air rests above a cold surface, such as above large bodies of water and in polar regions.

FIGURE 1 The *Fata Morgana* mirage over water. The mirage is the result of refraction—light from small islands and ships is bent in such a way as to make them appear to rise vertically above the water.

FIGURE 15.17 A 22° halo around the sun, produced by the refraction of sunlight through ice crystals.

to the halo, it is called a **tangent arc.** Apparently, the arc forms as large six-sided (hexagonal) pencil-shaped ice crystals fall with their long axes horizontal to the ground. Refraction of sunlight through the ice crystals produces the bright arc of light.

A halo is usually seen as a bright, white ring, but there are refraction effects that can cause it to have color. To understand this, we must first examine refraction more closely.

When white light passes through a glass prism, it is refracted and split into a spectrum of visible colors (see Fig. 15.20). Each wavelength of light is slowed by the glass, but each is slowed a little differently. Because longer wavelengths (red) slow the least and shorter wavelengths (violet) slow the most, red light bends the least, and violet light bends the most. The breaking up of white light by "selective" refraction is called **dispersion.**

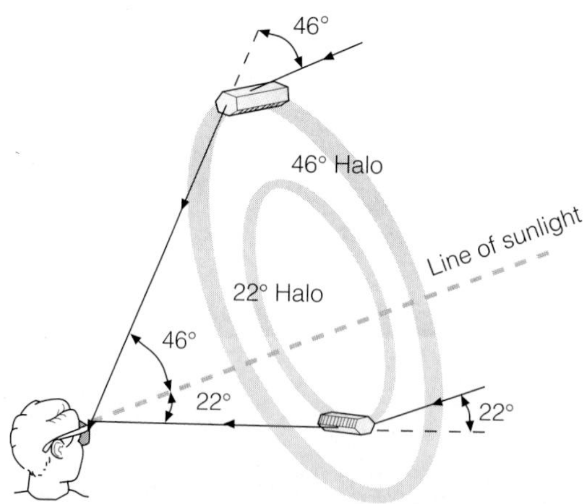

FIGURE 15.18 The formation of a 22° and a 46° halo with column-type ice crystals.

As light passes through ice crystals, dispersion causes red light to be on the inside of the halo and blue light on the outside.

When hexagonal platelike ice crystals are present in the air, they tend to fall slowly and orient themselves horizontally (see Fig. 15.21). (The horizontal orientation of these ice crystals prevents a ring halo.) In this position, the ice crystals act as small prisms, refracting and dispersing sunlight that passes through them. If the sun is near the horizon in such a configuration that it, ice crystals, and observer are all in the same horizontal plane, the observer will see a pair of brightly colored spots, one on either side of the sun. These colored spots are called **sundogs,** *mock suns,* or **parhelia**—meaning "with the sun" (see Fig. 15.22).* The colors usually grade from red (bent least) on the inside closest to the sun to blue (bent more) on the outside.

Whereas sundogs, tangent arcs, and halos are caused by *refraction* of sunlight *through* ice crystals, **sun pillars** are caused by *reflection* of sunlight *off* ice crystals. Sun pillars appear most often at sunrise or sunset as a vertical shaft of light extending upward or downward from the sun (see Fig. 15.23). Pillars may form as hexagonal platelike ice crystals fall with their flat bases oriented horizontally. As the tiny crystals fall in still air, they tilt from side to side like a falling leaf. This motion allows sunlight to reflect off the tipped surfaces of the crystals, producing a relatively bright area in the sky above or below the sun. Pillars may also form as sunlight reflects off hexagonal pencil-shaped ice crystals that fall with their long axes oriented horizontally. As these crystals fall, they can rotate about their horizontal axes, pro-

*Also look at the opening photo of this chapter on p. 402.

Photo by author

FIGURE 15.19 Halo with an upper tangent arc.

FIGURE 15.20 Refraction and dispersion of light through a glass prism.

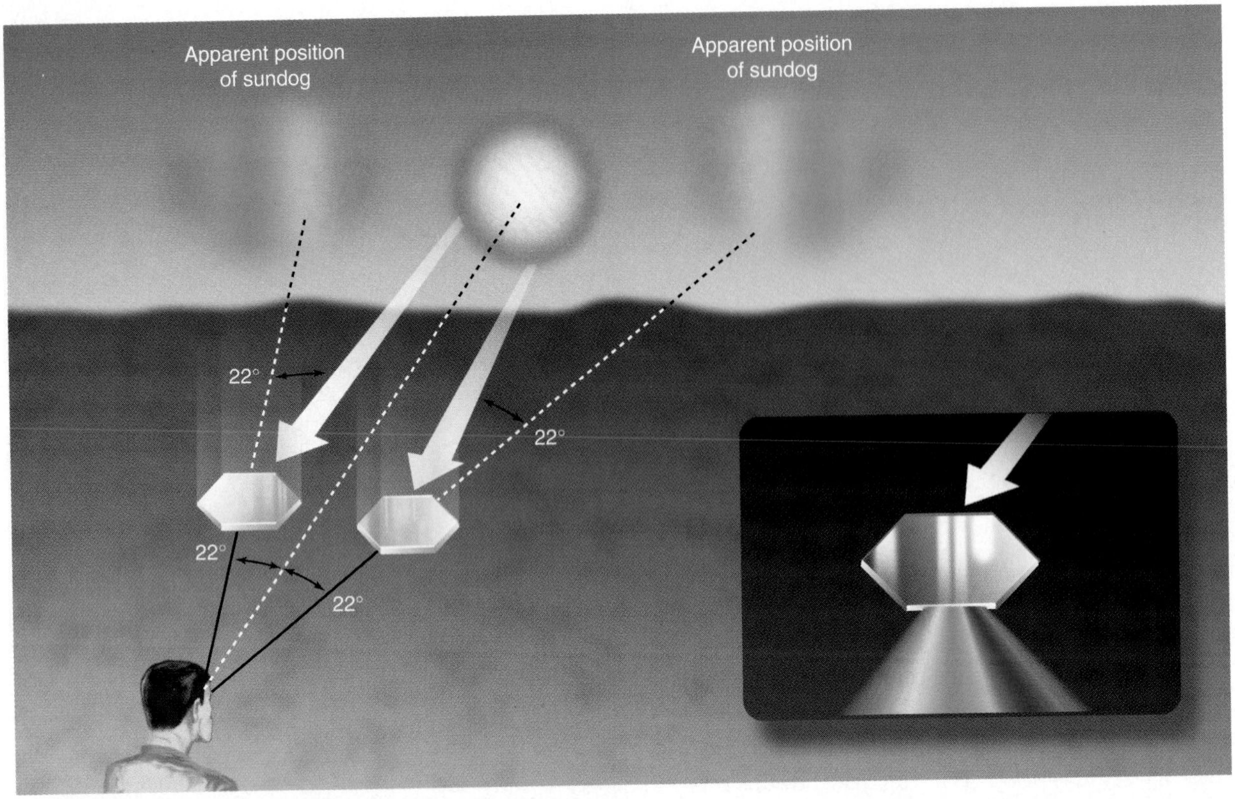

Apparent position of sundog

Apparent position of sundog

22°

22°

22°

22°

FIGURE 15.21 Platelike ice crystals falling with their flat surfaces parallel to the earth produce sundogs.

FIGURE 15.22 The bright areas on each side of the sun are sundogs.

FIGURE 15.23 A sun pillar produced by the reflection of sunlight off ice crystals.

ducing many orientations that reflect sunlight. So, look for sun pillars when the sun is low on the horizon and cirriform (ice crystal) clouds are present.

Meteorology Now™ Click "Atmosphere Optics" to see examples of atmospheric optical phenomena.

RAINBOWS

Now we come to one of the most spectacular light shows observed on the earth—the rainbow. **Rainbows** occur when rain is falling in one part of the sky, and the sun is shining in another. (Rainbows also may form by the sprays from waterfalls and water sprinklers.) To see the rainbow, we must face the falling rain with the sun at our backs. Look at Fig. 15.24 closely and note that, when we see a rainbow in the evening, we are facing east toward the rain shower. Behind us—in the west—it is clear. Because clouds tend to move from west to east in middle latitudes, the clear skies in the west suggest that the showers will give way to clearing. However, when we see a rainbow in the morning, we are facing west, toward the rain shower. It is a good bet that the clouds and showers will move toward us and it will rain soon. These observations explain why the following weather rhyme became popular:

> Rainbow in morning, sailors take warning
> Rainbow at night, a sailor's delight.*

When we look at a rainbow we are looking at sunlight that has entered the falling drops, and, in effect, has been redirected back toward our eyes. Exactly how this process happens requires some discussion.

As sunlight enters a raindrop, it slows and bends, with violet light refracting the most and red light the least (see Fig. 15.25). Although most of this light passes right on through the drop and is not seen by us, some of it strikes the backside of the drop at such an angle that it is reflected within the drop. The angle at which this occurs is called the *critical angle.* For water, this angle is 48°. Light that strikes the back of a raindrop at an angle exceeding the critical angle bounces off the back of the drop and is *internally reflected* toward our eyes (see Fig. 15.25a). Because each light ray bends differently from the rest, each ray emerges from the drop at a slightly different angle. For red light, the angle is 42° from the beam of sunlight; for violet light, it is 40° (see Fig. 15.25b). The light leaving the drop is, therefore, dispersed into a spectrum of colors from red to violet. Since we see only a single color from each drop, it takes a myriad of raindrops (each refracting and reflecting light back to our eyes at slightly different angles) to produce the brilliant colors of a *primary rainbow.*

Figure 15.25b might lead us erroneously to believe that red light should be at the bottom of the bow and violet at the top. A more careful observation of the behavior of light leaving two drops (Fig. 15.26) shows us why the reverse is true. When violet light from the *lower drop* reaches an observer's eye, red light from the same drop is incident elsewhere, toward the waist. Notice that red light reaches the observer's eye from the *higher drop.* Because the color red comes from higher drops and the color violet from lower drops, the colors of a primary rainbow change from red on the outside (top) to violet on the inside (bottom).

*This rhyme is often used with the words "red sky" in the place of rainbow. The red sky makes sense when we consider that it is the result of red light from a rising or setting sun being reflected from the underside of clouds above us. In the morning, a red sky indicates that it is clear to the east and cloudy to the west. A red sky in the evening suggests the opposite.

DID YOU KNOW?

During the summer, rainbows are occasionally seen after a thunderstorm. Because of this fact, the Shoshone Indians viewed the rainbow as a giant serpent that would sometimes rub its back on the icy sky and hurl pieces of ice (hail) to the ground.

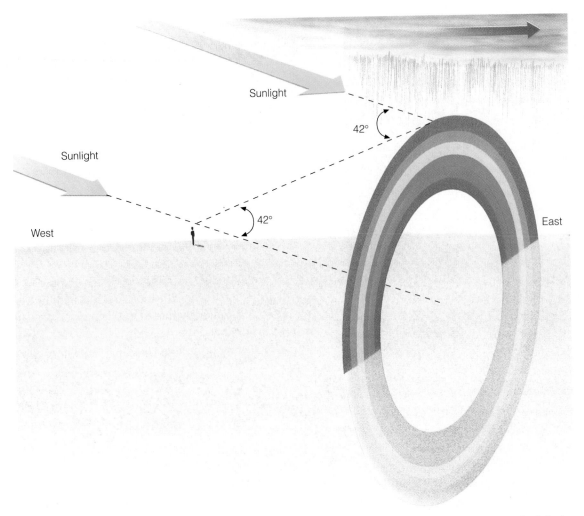

FIGURE 15.24 When you observe a rainbow, the sun is always to your back. In middle latitudes, a rainbow in the evening indicates that clearing weather is ahead.

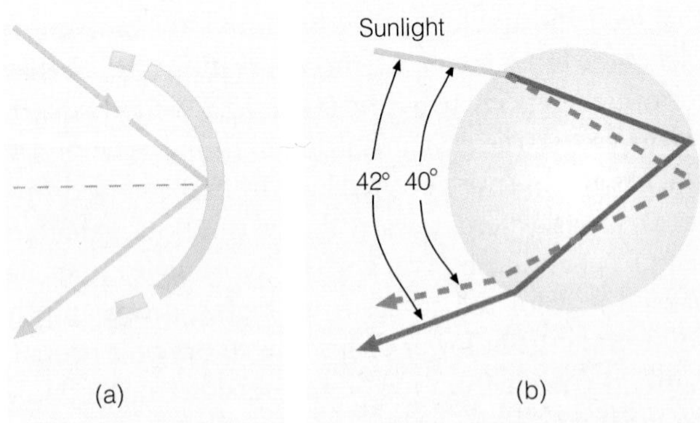

FIGURE 15.25 Sunlight internally reflected and dispersed by a raindrop. (a) The light ray is internally reflected only when it strikes the backside of the drop at an angle greater than the critical angle for water. (b) Refraction of the light as it enters the drop causes the point of reflection (on the back of the drop) to be different for each color. Hence, the colors are separated from each other when the light emerges from the raindrop.

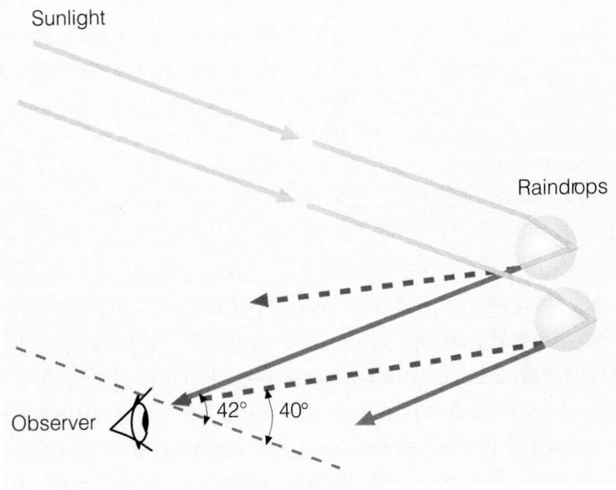

FIGURE 15.26 The formation of a primary rainbow. The observer sees red light from the upper drop and violet light from the lower drop.

Photo by author

FIGURE 15.27 A primary and a secondary rainbow.

Frequently, a larger, second (secondary) rainbow with its colors reversed can be seen above the primary bow (see Fig. 15.27). Usually this secondary bow is much fainter than the primary one. The *secondary bow* is caused when sunlight enters the raindrops at an angle that allows the light to make two internal reflections in each drop. Each reflection weakens the light intensity and makes the bow dimmer. Figure 15.28 shows that the color reversals—with red now at the bottom and violet on top—are due to the way the light emerges from each drop after going through two internal reflections.

As you look at a rainbow, keep in mind that only one ray of light is able to enter your eye from each drop. Every time you move, whether it be up, down, or sideways, the rainbow moves with you. The reason why this happens is that, with every movement, light from different raindrops enters your eye. The bow you see is not exactly the same rainbow that the person standing next to you sees. In effect, each of us has a personal rainbow to ponder and enjoy!

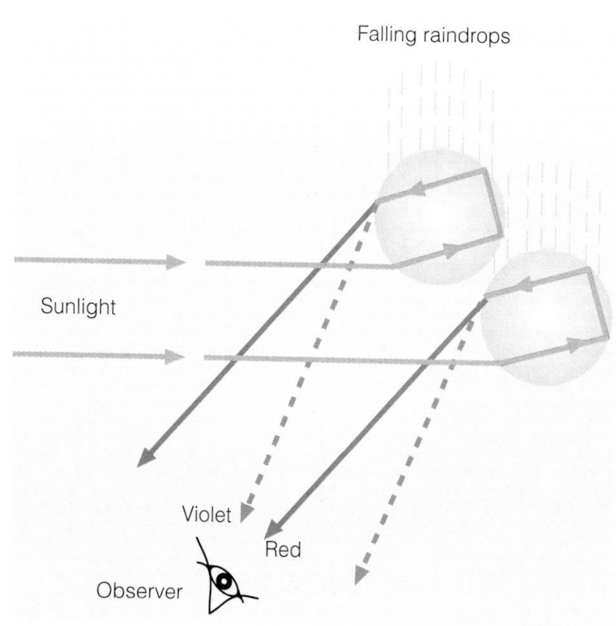

FIGURE 15.28 Two internal reflections are responsible for the weaker, secondary rainbow. Notice that the eye sees violet light from the upper drop and red light from the lower drop.

CORONAS AND CLOUD IRIDESCENCE

When the moon is seen through a thin veil of clouds composed of tiny spherical water droplets, a bright ring of light, called a **corona** (meaning crown), may appear to rest on the moon (see Fig. 15.29). The same effect can occur with the sun, but, due to the sun's brightness, it is usually difficult to see.

The corona is due to **diffraction**—the bending of light as it passes *around* objects. To understand the corona, imagine water waves moving around a small stone in a pond. As the waves spread around the stone, the trough of one wave may meet the crest of another wave. This situation results in the waves canceling each other, thus producing calm water. Where two crests come together, they produce a much larger wave. The same thing happens when light passes around tiny cloud droplets. Where light waves come together, we see bright light; where they cancel each other, we see darkness. Sometimes, the corona appears white, with alternating bands of light and dark. On other occasions, the rings have color (see Fig. 15.30).

The colors appear when the cloud droplets (or any kind of small particles, such as volcanic ash) are of uniform size. Because the amount of bending due to diffraction depends upon the wavelength of light, the shorter wavelength blue light appears on the inside of a ring, while the longer wavelength red light appears on the outside. These colors may repeat over and over, becoming fainter as each ring is farther from the moon or sun. Clouds that have recently formed (such as thin altostratus and altocumulus) are the best corona producers.

© Elizabeth Beaver Burnett

FIGURE 15.30 Corona around the sun. This type of corona, called *Bishop's ring,* is the result of diffraction of sunlight by tiny volcanic particles emitted from the volcano El Chichón in 1982.

When different size droplets exist within a cloud, the corona becomes distorted and irregular. Sometimes the cloud exhibits patches of color, often pastel shades of pink, blue, or green. These bright areas produced by diffraction are called **iridescence** (see Fig. 15.31). Cloud iridescence is most often seen within 20° of the sun and is often associated with thin clouds, such as cirrocumulus and altocumulus. (Additional optical phenomena—the glory and the *Heiligenschein*—are described in the Focus section on p. 421.)

Photo by author

FIGURE 15.29 The corona around the moon results from the diffraction of light by tiny liquid cloud droplets of uniform size.

© Pekka Parviainen

FIGURE 15.31 Cloud iridescence.

FOCUS ON AN OBSERVATION

Glories and the Heiligenschein

When an aircraft flies above a cloud layer composed of tiny water droplets, a set of colored rings called the *glory* may appear around the shadow of the aircraft (see Fig. 2). The same effect can happen when you stand with your back to the sun and look into a cloud or fog bank, as a bright ring of light may be seen around the shadow of your head. In this case, the glory is called the *brocken bow,* after the Brocken Mountains in Germany, where it is particularly common.

For the glory and the brocken bow to occur, the sun must be to your back, so that sunlight can be returned to your eye from the water droplets. Sunlight that enters the small water droplet along its edge is refracted, then reflected off the backside of the droplet. The light then exits at the other side of the droplet, being refracted once again (see Fig. 3). The colorful rings may be due to the various angles at which different colors leave the droplet.

On a clear morning with dew on the grass, stand facing the dew with your back to the sun and observe that, around the shadow of your head, is a bright area—the *Heiligenschein* (German for halo). The *Heiligenschein* forms when sunlight, which falls on nearly spherical dew drops, is focused and reflected back toward the sun along nearly the same path that it took originally. The light, however, does not travel along the exact path; it actually spreads out just enough to be seen as bright white light around the shadow of your head on a dew-covered lawn (see Fig. 4).

FIGURE 2 The series of rings surrounding the shadow of the aircraft is called the glory.

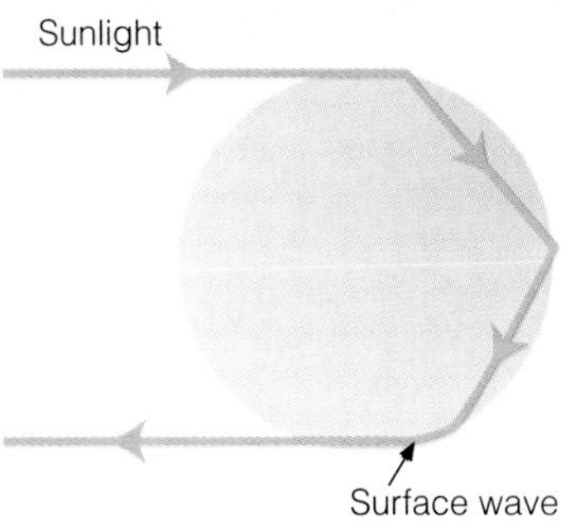

FIGURE 3 Light that produces the glory follows this path in a water droplet.

FIGURE 4 The *Heiligenschein* is the ring of light around the shadow of the observer's head.

SUMMARY

The scattering of sunlight in the atmosphere can produce a variety of atmospheric visuals, from hazy days and blue skies to crepuscular rays and blue moons. Refraction of light by the atmosphere causes stars near the horizon to appear higher than they really are. It also causes the sun and moon to rise earlier and set later than they otherwise would. Under certain atmospheric conditions, the amplification of green light near the upper rim of a rising or setting sun produces the elusive green flash.

Mirages form when refraction of light displaces objects from their true positions. Inferior mirages cause objects to appear lower than they really are, while superior mirages displace objects upward.

Halos and sundogs form from the refraction of light through ice crystals. Sun pillars are the result of sunlight reflecting off gently falling ice crystals. Diffraction of light produces coronas and cloud iridescence. The refraction, reflection, and dispersion of light in raindrops create a rainbow. To see a rainbow, the sun must be to your back, and rain must be falling in front of you.

Meteorology Now™ Assess your understanding of this chapter's topics with additional quizzing and tutorials at http://earthscience.brookscole.com/ahrens/ess4e.

KEY TERMS

The following terms are listed in the order they appear in the text. Define each. Doing so will aid you in reviewing the material covered in this chapter.

reflected light
scattered light
crepuscular rays
refraction (of light)
scintillation
twilight
green flash
mirage
inferior mirage
superior mirage
halo
tangent arc
dispersion (of light)
sundog (or parhelia)
sun pillar
rainbow
corona
diffraction
iridescence

QUESTIONS FOR REVIEW

1. Why are cumulus clouds normally white?
2. Why do the undersides of building cumulus clouds frequently change color from white to dark gray or even black?
3. Explain why the sky is blue during the day and black at night.
4. What can make a setting (or rising) sun appear red?
5. If the earth had no atmosphere, what would be the color of the daytime sky?
6. Explain why the horizon sky appears white on a hazy day.
7. What process (refraction or scattering) produces crepuscular rays?
8. Why do stars "twinkle"?
9. How does refraction of light differ from reflection of light?
10. How long does twilight last on the moon? (Hint: The moon has no atmosphere.)
11. At what time of day would you expect to observe the green flash?
12. How does light bend as it enters a more-dense substance at an angle? How does it bend upon leaving the more-dense substance? Make a sketch to illustrate your answer.
13. On a clear, dry, warm day, why do dark road surfaces frequently appear wet?
14. What atmospheric conditions are necessary for an inferior mirage? A superior mirage?
15. (a) Describe how a halo forms.
    (b) How is the formation of a halo different from that of a sundog?
16. Would you expect to see a ringed halo if the sky contains a few wispy cirrus clouds? Explain.
17. What process is believed to be mainly responsible for the formation of sun pillars: refraction, reflection, or scattering?
18. Explain why this rhyme makes sense:
    Rainbow in morning, joggers take warning.
    Rainbow at night (evening), jogger's delight.
19. Why can a rainbow only be observed if the sun is at the observer's back?
20. Why are secondary rainbows much dimmer than primary rainbows?
21. How would you distinguish a corona from a halo?

QUESTIONS FOR THOUGHT AND EXPLORATION

1. You are flying in a jet airliner at an altitude of 18,000 feet, and looking out the window you see stratocumulus clouds below. Under what circumstances would the tops of these clouds *not* be white? Explain.
2. Explain why the notion that "the sky is blue because of reflected light from the oceans" is false.
3. Why does smoke rising from a cigarette appear blue, yet appears white when blown from the mouth?
4. If there were no atmosphere surrounding the earth, what color would the sky be at sunrise? At sunset? What color would the sun be at noon? At sunrise? At sunset?
5. Explain why, on a cloudless day, the sky will usually appear milky white before it rains and a deeper blue after it rains.
6. Why are rainbows seldom observed at noon?
7. During the day, clouds are white and the sky is blue. Why then, during a full moon, do cumulus clouds appear faintly white, while the sky does not appear blue?
8. The Green Flash (http://mintaka.sdsu.edu/GF/): Using your own words, explain the principle behind the green flash. Where and when are you most likely to see it? How might you photograph it? Is it always green?

9. Choose a 3-day period in which to observe the sky 5 times each day. Record in a notebook the number of times you see halos, crepuscular rays, coronas, cloud iridescence, sun dogs, rainbows, and other phenomena.

- Go to the Brooks/Cole Earth Sciences Resource Center (http://earthscience.brookscole.com) for critical thinking exercises, articles, and additional readings from InfoTrac College Edition, Brooks/Cole's online student library.

APPENDIX A

Units, Conversions, Abbreviations, and Equations

LENGTH

1 kilometer (km)	= 1000 m
	= 3281 ft
	= 0.62 mi
1 mile (mi)	= 5280 ft
	= 1609 mi
	= 1.61 km
1 meter (m)	= 100 cm
	= 3.28 ft
	= 39.37 in.
1 foot (ft)	= 12 in.
	= 30.48 cm
	= 0.305 m
1 centimeter (cm)	= 0.39 in.
	= 0.01 m
	= 10 mm
1 inch (in.)	= 2.54 cm
	= 0.08 ft
1 millimeter (mm)	= 0.1 cm
	= 0.001 m
	= 0.039 in.
1 micrometer (μm)	= 0.0001 cm
	= 0.000001 m
1 degree latitude	= 111 km
	= 60 nautical mi
	= 69 statute mi

AREA

1 square centimeter (cm^2)	= 0.15 $in.^2$
1 square inch ($in.^2$)	= 6.45 cm^2
1 square meter (m^2)	= 10.76 ft^2
1 square foot (ft^2)	= 0.09 m^2

VOLUME

1 cubic centimeter (cm^3)	= 0.06 $in.^3$
1 cubic inch ($in.^3$)	= 16.39 cm^3
1 liter (l)	= 1000 cm^3
	= 0.264 gallon (gal) U.S.

SPEED

1 knot	= 1 nautical mi/hr
	= 1.15 statute mi/hr
	= 0.51 m/sec
	= 1.85 km/hr
1 mile per hour (mi/hr)	= 0.87 knots
	= 0.45 m/sec
	= 1.61 km/hr
1 kilometer per hour (km/hr)	= 0.54 knots
	= 0.62 mi/hr
	= 0.28 m/sec
1 meter per second (m/sec)	= 1.94 knots
	= 2.24 mi/hr
	= 3.60 km/hr

FORCE

1 dyne	= 1 gram centimeter per second per second
	= 2.2481 × 10^{-6} pound (lb)
1 newton (N)	= 1 kilogram meter per second per second
	= 10^5 dynes
	= 0.2248 lb

MASS

1 gram (g)	= 0.035 ounce
	= 0.002 lb
1 kilogram (kg)	= 1000 g
	= 2.2 lb

ENERGY

1 erg	= 1 dyne per cm
	= 2.388×10^{-8} cal
1 joule (J)	= 1 newton meter
	= 0.239 cal
	= 10^{7} erg
1 calorie (cal)	= 4.186 J
	= 4.186×10^{7} erg

PRESSURE

1 millibar (mb)	= 1000 dynes/cm^2
	= 0.75 millimeter of mercury (mm Hg)
	= 0.02953 inch of mercury (in. Hg)
	= 0.01450 pound per square inch (lb/in.^{2})
	= 100 pascals (Pa)
1 standard atmosphere	= 1013.25 mb
	= 760 mm Hg
	= 29.92 in. Hg
	= 14.7 lb/in.^{2}
1 inch of mercury	= 33.865 mb
1 millimeter of mercury	= 1.3332 mb
1 pascal	= 0.01 mb
	= 1 N/m^2
1 hectopascal (hPa)	= 1 mb
1 kilopascal (kPa)	= 10 mb

POWER

1 watt (W)	= 1 J/sec
	= 14.3353 cal/min
1 cal/min	= 0.06973 W
1 horse power (hp)	= 746 W

POWERS OF TEN

Prefix			
nano	one-billionth	= 10^{-9}	= 0.000000001
micro	one-millionth	= 10^{-6}	= 0.000001
milli	one-thousandth	= 10^{-3}	= 0.001
centi	one-hundredth	= 10^{-2}	= 0.01
deci	one-tenth	= 10^{-1}	= 0.1
hecto	one hundred	= 10^{2}	= 100
kilo	one thousand	= 10^{3}	= 1000
mega	one million	= 10^{6}	= 1,000,000
giga	one billion	= 10^{9}	= 1,000,000,000

TEMPERATURE

$$°C = \frac{5}{9}(°F - 32)$$

To convert degrees Fahrenheit (°F) to degrees Celsius (°C): Subtract 32 degrees from °F, then divide by 1.8.

To convert degrees Celsius (°C) to degrees Fahrenheit (°F): Multiply °C by 1.8, then add 32 degrees.

To convert degrees Celsius (°C) to Kelvins (K): Add 273 to Celsius temperature, as

$$K = °C + 273.$$

F = C x 1.8 + 32
F = (1.8C) + 32

TABLE A.1 Temperature Conversions

°F	°C	°F	°C	°F	°C	°F	°C	°F	°C	°F	°C	°F	°C	°F	°C
–40	–40	–20	–28.9	0	–17.8	20	–6.7	40	4.4	60	15.6	80	26.7	100	37.8
–39	–39.4	–19	–28.3	1	–17.2	21	–6.1	41	5.0	61	16.1	81	27.2	101	38.3
–38	–38.9	–18	–27.8	2	–16.7	22	–5.6	42	5.6	62	16.8	82	27.8	102	38.9
–37	–38.3	–17	–27.2	3	–16.1	23	–5.0	43	6.1	63	17.2	83	28.3	103	39.4
–36	–37.8	–16	–26.7	4	–15.6	24	–4.4	44	6.7	64	17.8	84	28.9	104	40.0
–35	–37.2	–15	–26.1	5	–15.0	25	–3.9	45	7.2	65	18.3	85	29.4	105	40.6
–34	–36.7	–14	–25.6	6	–14.4	26	–3.3	46	7.8	66	18.9	86	30.0	106	41.1
–33	–36.1	–13	–25.0	7	–13.9	27	–2.8	47	8.3	67	19.4	87	30.6	107	41.7
–32	–35.6	–12	–24.4	8	–13.3	28	–2.2	48	8.9	68	20.0	88	31.1	108	42.2
–31	–35.0	–11	–23.9	9	–12.8	29	–1.7	49	9.4	69	20.6	89	31.7	109	42.8
–30	–34.4	–10	–23.3	10	–12.2	30	–1.1	50	10.0	70	21.1	90	32.2	110	43.3
–29	–33.9	–9	–22.8	11	–11.7	31	–0.6	51	10.6	71	21.7	91	32.8	111	43.9
–28	–33.3	–8	–22.2	12	–11.1	32	0.0	52	11.1	72	22.2	92	33.3	112	44.4
–27	–32.8	–7	–21.7	13	–10.6	33	0.6	53	11.7	73	22.8	93	33.9	113	45.0
–26	–32.2	–6	–21.1	14	–10.0	34	1.1	54	12.2	74	23.3	94	34.4	114	45.6
–25	–31.7	–5	–20.6	15	–9.4	35	1.7	55	12.8	75	23.9	95	35.0	115	46.1
–24	–31.1	–4	–20.0	16	–8.9	36	2.2	56	13.3	76	24.4	96	35.6	116	46.7
–23	–30.6	–3	–19.4	17	–8.3	37	2.8	57	13.9	77	25.0	97	36.1	117	47.2
–22	–30.0	–2	–18.9	18	–7.8	38	3.3	58	14.4	78	25.6	98	36.7	118	47.8
–21	–29.4	–1	–18.3	19	–7.2	39	3.9	59	15.0	79	26.1	99	37.2	119	48.3

TABLE A.2 SI Units* and Their Symbols

QUANTITY	NAME	UNITS	SYMBOL
length	meter	m	m
mass	kilogram	kg	kg
time	second	sec	sec
temperature	Kelvin	K	K
density	kilogram per cubic meter	kg/m^3	kg/m^3
speed	meter per second	m/sec	m/sec
force	newton	$m \mid kg/sec^2$	N
pressure	pascal	N/m^2	Pa
energy	joule	N · m	J
power	watt	J/sec	W

*SI stands for Système International, which is the international system of units and symbols.

Equations and Constants

GAS LAW (EQUATION OF STATE)

The relationship among air pressure, air density, and air temperature can be expressed by

Pressure = density × temperature × constant.

This relationship, often called the gas law (or equation of state), can be expressed in symbolic form as:

$$p = \rho RT$$

where p is air pressure, ρ is air density, R is a constant, and T is air temperature.

UNITS/CONSTANTS

p = pressure in N/m^2 (SI)
ρ = density (kg/m^3)
T = temperature (K)
R = 287 J/kg • K (SI) or
$R = 2.87 \times 10^6$ erg/g • K

STEFAN-BOLTZMANN LAW

The Stefan-Boltzmann law is a law of radiation. It states that all objects with temperatures above absolute zero emit radiation at a rate proportional to the fourth power of their absolute temperature. It is expressed mathematically as:

$$E = \sigma T^4$$

where E is the maximum rate of radiation emitted each second per unit surface area, T is the object's surface temperature, and σ is a constant.

UNITS/CONSTANTS

E = radiation emitted in W/m^2 (SI)
$\sigma = 5.67 \times 10^{-8}\ W/m^2 \cdot K^4$ (SI) or
$\sigma = 5.67 \times 10^{-5}\ erg/cm^2 \cdot K^4 \cdot sec$
T = temperature (K)

WIEN'S LAW

Wien's law (or Wien's displacement law) relates an object's maximum emitted wavelength of radiation to the object's temperature. It states that the wavelength of maximum emitted radiation by an object is inversely proportional to the object's absolute temperature. In symbolic form, it is written as:

$$\lambda_{max} = \frac{w}{T}$$

where λ_{max} is the wavelength at which maximum radiation emission occurs, T is the object's temperature, and w is a constant.

UNITS/CONSTANTS

λ_{max} = wavelength (micrometers)
w = 0.2897 μm K
T = temperature (K)

GEOSTROPHIC WIND EQUATION

The geostrophic wind equation gives an approximation of the wind speed above the level of friction, where the wind blows parallel to the isobars or contours. The equation is expressed mathematically as:

$$V_g = \frac{1}{2\Omega \sin\phi\rho} \frac{\Delta p}{d}$$

where V_g is the geostrophic wind, Ω is a constant (twice the earth's angular spin), $\sin\phi$ is a trigonometric function that takes into account the variation of latitude (ϕ), ρ is the air density, Δp is the pressure difference between two places on the map some horizontal distance (d) apart.

UNITS/CONSTANTS

V_g	=	geostrophic wind (m/sec)
Ω	=	7.29×10^{-5} radian*/sec
ϕ	=	latitude
ρ	=	air density (kg/m^3)
d	=	distance (m)
Δp	=	pressure difference (newton/m^2)

*2π radians equal 360°.

HYDROSTATIC EQUATION

The hydrostatic equation relates to how quickly the air pressure decreases in a column of air above the surface. The equation tells us that the rate at which the air pressure decreases with height is equal to the air density times the acceleration of gravity. In symbolic form, it is written as:

$$\frac{\Delta p}{\Delta z} = -\rho g$$

where Δp is the decrease in pressure along a small change in height Δz, ρ is the air density, and g is the force of gravity.

UNITS/CONSTANTS

Δp	=	pressure difference (newton/m^2)
Δz	=	change in height (m)
ρ	=	air density (kg/m^3)
g	=	force of gravity (9.8m/sec^2)

RELATIVE HUMIDITY

The relative humidity of the air can be expressed as:

$$RH = \frac{e}{e_s} \times 100\%.$$

To determine e and e_s, when the air temperature and dew-point temperature are known, consult Table B.1. Simply read the value adjacent to the air temperature and obtain e_s; read the value adjacent to the dew-point temperature and obtain e.

UNITS/CONSTANTS

e	=	actual vapor pressure (millibars)
e_s	=	saturation vapor pressure (millibars)
RH	=	relative humidity (percent)

TABLE B.1 Saturation Vapor Pressure over Water for Various Air Temperatures

AIR TEMPERATURE (°C)	(°F)	SATURATION VAPOR PRESSURE (MB)	AIR TEMPERATURE (°C)	(°F)	SATURATION VAPOR PRESSURE (MB)
−18	(0)	1.5	18	(65)	21.0
−15	(5)	1.9	21	(70)	25.0
−12	(10)	2.4	24	(75)	29.6
−9	(15)	3.0	27	(80)	35.0
−7	(20)	3.7	29	(85)	41.0
−4	(25)	4.6	32	(90)	48.1
−1	(30)	5.6	35	(95)	56.2
2	(35)	6.9	38	(100)	65.6
4	(40)	8.4	41	(105)	76.2
7	(45)	10.2	43	(110)	87.8
10	(50)	12.3	46	(115)	101.4
13	(55)	14.8	49	(120)	116.8
16	(60)	17.7	52	(125)	134.2

APPENDIX C

Weather Symbols and the Station Model

SIMPLIFIED SURFACE-STATION MODEL

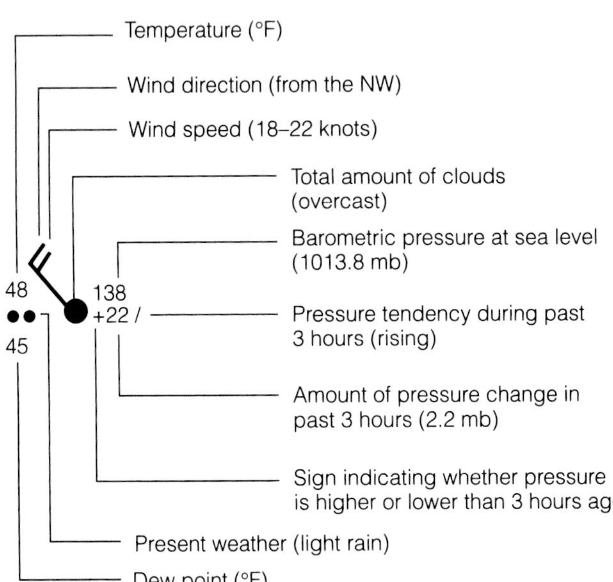

UPPER-AIR MODEL (500 MB)

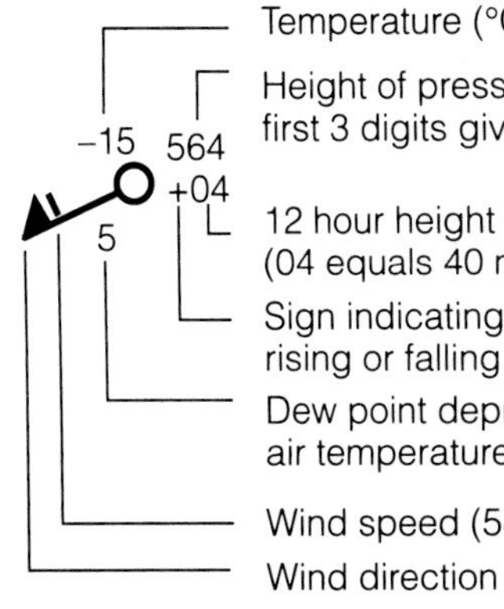

CLOUD COVERAGE

Clear

 1/8

 Scattered

 3/8

 4/8

 5/8

 Broken

 7/8

 Overcast

 Obscured

 Missing

COMMON WEATHER SYMBOLS

Light rain

 Moderate rain

 Heavy rain

 Light snow

 Moderate snow

 Heavy snow

 Light drizzle

Ice pellets (sleet)

 Freezing rain

Freezing drizzle

Rain shower

Snow shower

Showers of hail

Drifting or blowing snow

Dust storm

Fog

Haze

Smoke

Thunderstorm

Hurricane

WIND ENTRIES

	Miles (statute) per hour	Knots	Kilometers per Hour
	Calm	Calm	Calm
	1–2	1–2	1–3
	3–8	3–7	4–13
	9–14	8–12	14–19
	15–20	13–17	20–32
	21–25	18–22	33–40
	26–31	23–27	41–50
	32–37	28–32	51–60
	38–43	33–37	61–69
	44–49	38–42	70–79
	50–54	43–47	80–87
	55–60	48–52	88–96
	61–66	53–57	97–106
	67–71	58–62	107–114
	72–77	63–67	115–124
	78–83	68–72	125–134
	84–89	73–77	135–143
	119–123	103–107	144–198

PRESSURE TENDENCY

- Rising, then falling
- Rising, then steady; or rising, then rising more slowly
- Rising steadily or unsteadily
- Falling or steady, then rising; or rising, then rising more quickly

(Barometer now higher than 3 hours ago)

- Steady, same as 3 hours ago
- Falling, then rising, same or lower than 3 hours ago
- Falling, then steady; or falling, then falling more slowly
- Falling steadily, or unsteadily
- Steady or rising, then falling; or falling, then falling more quickly

(Barometer now lower than 3 hours ago)

FRONT SYMBOLS

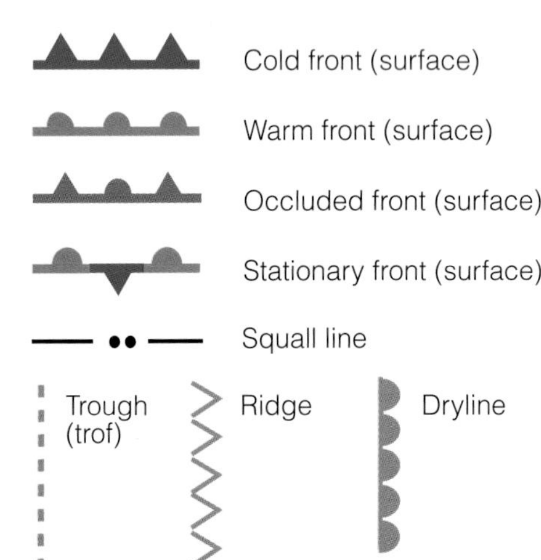

Humidity and Dew-Point Tables (Psychromatic Tables)

To obtain the dew point (or relative humidity), simply read down the temperature column and then over to the wet-bulb depression. For example, in Table D.1, a temperature of 10°C with a wet-bulb depression of 3°C produces a dew-point temperature of 4°C. (Dew-point temperature and relative humidity readings are appropriate for pressures near 1000 mb.)

TABLE D.1 Dew-Point Temperature (°C)

Air (Dry-Bulb) Temperature (°C)	WET-BULB DEPRESSION (DRY-BULB TEMPERATURE MINUS WET-BULB TEMPERATURE) (°C)															
	0.5	*1.0*	*1.5*	*2.0*	*2.5*	*3.0*	*3.5*	*4.0*	*4.5*	*5.0*	*7.5*	*10.0*	*12.5*	*15.0*	*17.5*	*20.0*
−20	−25	−33														
−17.5	−21	−27	−38													
−15	−19	−23	−28													
−12.5	−15	−18	−22	−29												
−10	−12	−14	−18	−21	−27	−36										
−7.5	−9	−11	−14	−17	−20	−26	−34									
−5	−7	−8	−10	−13	−16	−19	−24	−31								
−2.5	−4	−6	−7	−9	−11	−14	−17	−22	−28	−41						
0	−1	−3	−4	−6	−8	−10	−12	−15	−19	−24						
2.5	1	0	−1	−3	−4	−6	−8	−10	−13	−16						
5	4	3	2	0	−1	−3	−4	−6	−8	−10	−48					
7.5	6	6	4	3	2	1	−1	−2	−4	−6	−22					
10	9	8	7	6	5	4	2	1	0	−2	−13					
12.5	12	11	10	9	8	7	6	4	3	2	−7	−28				
15	14	13	12	12	11	10	9	8	7	5	−2	−14				
17.5	17	16	15	14	13	12	12	11	10	8	2	−7	−35			
20	19	18	18	17	16	15	14	14	13	12	6	−1	−15			
22.5	22	21	20	20	19	18	17	16	16	15	10	3	−6	−38		
25	24	24	23	22	21	21	20	19	18	18	13	7	0	−14		
27.5	27	26	26	25	24	23	23	22	21	20	16	11	5	−5	−32	
30	29	29	28	27	27	26	25	25	24	23	19	14	9	2	−11	
32.5	32	31	31	30	29	29	28	27	26	26	22	18	13	7	−2	
35	34	34	33	32	32	31	31	30	29	28	25	21	16	11	4	
37.5	37	36	36	35	34	34	33	32	32	31	28	24	20	15	9	0
40	39	39	38	38	37	36	36	35	34	34	30	27	23	18	13	6
42.5	42	41	41	40	40	39	38	38	37	36	33	30	26	22	17	11
45	44	44	43	43	42	42	41	40	40	39	36	33	29	25	21	15
47.5	47	46	46	45	45	44	44	43	42	42	39	35	32	28	24	19
50	49	49	48	48	47	47	46	45	45	44	41	38	35	31	28	23

TABLE D.2 Relative Humidity (Percent)

Air (Dry-Bulb) Temperature (°C)	WET-BULB DEPRESSION (DRY-BULB TEMPERATURE MINUS WET-BULB TEMPERATURE) (°C)																	
	0.5	*1.0*	*1.5*	*2.0*	*2.5*	*3.0*	*3.5*	*4.0*	*4.5*	*5.0*	*7.5*	*10.0*	*12.5*	*15.0*	*17.5*	*20.0*	*22.5*	*25.0*
−20	70	41	11															
−17.5	75	51	26	2														
−15	79	58	38	18														
−12.5	82	65	47	30	13													
−10	85	69	54	39	24	10												
−7.5	87	73	60	48	35	22	10											
−5	88	77	66	54	43	32	21	11	1									
−2.5	90	80	70	60	50	42	37	22	12	3								
0	91	82	73	65	56	47	39	31	23	15								
2.5	92	84	76	68	61	53	46	38	31	24								
5	93	86	78	71	65	58	51	45	38	32	1							
7.5	93	87	80	74	68	62	56	50	44	38	11							
10	94	88	82	76	71	65	60	54	49	44	19							
12.5	94	89	84	78	73	68	63	58	53	48	25	4						
15	95	90	85	80	75	70	66	61	57	52	31	12						
17.5	95	90	86	81	77	72	68	64	60	55	36	18	2					
20	95	91	87	82	78	74	70	66	62	58	40	24	8					
22.5	96	92	87	83	80	76	72	68	64	61	44	28	14	1				
25	96	92	88	84	81	77	73	70	66	63	47	32	19	7				
27.5	96	92	89	85	82	78	75	71	68	65	50	36	23	12	1			
30	96	93	89	86	82	79	76	73	70	67	52	39	27	16	6			
32.5	97	93	90	86	83	80	77	74	71	68	54	42	30	20	11	1		
35	97	93	90	87	84	81	78	75	72	69	56	44	33	23	14	6		
37.5	97	94	91	87	85	82	79	76	73	70	58	46	36	26	18	10	3	
40	97	94	91	88	85	82	79	77	74	72	59	48	38	29	21	13	6	
42.5	97	94	91	88	86	83	80	78	75	72	61	50	40	31	23	16	9	2
45	97	94	91	89	86	83	81	78	76	73	62	51	42	33	26	18	12	6
47.5	97	94	92	89	86	84	81	79	76	74	63	53	44	35	28	21	15	9
50	97	95	92	89	87	84	82	79	77	75	64	54	45	37	30	23	17	11

TABLE D.3 Dew-Point Temperature (°F)

Air (Dry-Bulb) Temperature (°F)	WET-BULB DEPRESSION (DRY-BULB TEMPERATURE MINUS WET-BULB TEMPERATURE) (°F)																						
	1	*2*	*3*	*4*	*5*	*6*	*7*	*8*	*9*	*10*	*11*	*12*	*13*	*14*	*15*	*16*	*17*	*18*	*19*	*20*	*25*	*30*	*35*	*40*
0	−7	−20																						
5	−1	−9	−24																					
10	5	−2	−10	−27																				
15	11	6	0	−9	−26																			
20	16	12	8	2	−7	−21																		
25	22	19	15	10	5	−3	−15	−51																
30	27	25	21	18	14	8	2	−7	−25															
35	33	30	28	25	21	17	13	7	0	−11														
40	38	35	33	30	28	25	21	18	13	7	−1	−14												
45	43	41	38	36	34	31	28	25	22	18	13	7	−1	−14										
50	48	46	44	42	40	37	34	32	29	26	22	18	13	8	0	−13								
55	53	51	50	48	45	43	41	38	36	33	30	27	24	20	15	9	1	−12						
60	58	57	55	53	51	49	47	45	43	40	38	35	32	29	25	21	17	11	4	−8				
65	63	62	60	59	57	55	53	51	49	47	45	42	40	37	34	31	27	24	19	14				
70	69	67	65	64	62	61	59	57	55	53	51	49	47	44	42	39	36	33	30	26	−11			
75	74	72	71	69	68	66	64	63	61	59	57	55	54	51	49	47	44	42	39	36	15			
80	79	77	76	74	73	72	70	68	67	65	63	62	60	58	56	54	52	50	47	44	28	−7		
85	84	82	81	80	78	77	75	74	72	71	69	68	66	64	62	61	59	57	54	52	39	19		
90	89	87	86	85	83	82	81	79	78	76	75	73	72	70	69	67	65	63	61	59	48	32		
95	94	93	91	90	89	87	86	85	83	81	80	79	78	76	74	73	71	70	68	66	56	43	24	
100	99	98	96	95	94	93	91	90	89	87	86	85	83	82	80	79	77	76	74	72	63	52	37	12
105	104	103	101	100	99	98	96	95	94	93	91	90	89	87	86	84	83	82	80	78	70	61	48	30
110	109	108	106	105	104	103	102	100	99	98	97	95	94	93	91	90	89	87	86	84	77	68	57	43
115	114	113	112	110	109	108	107	106	104	103	102	101	99	98	97	96	94	93	92	90	83	75	65	54
120	119	118	117	115	114	113	112	111	110	108	107	106	105	104	102	101	100	98	97	96	89	81	73	63

TABLE D.4 Relative Humidity (Percent)

Air (Dry-Bulb) Temperature (°F)	WET-BULB DEPRESSION (DRY-BULB TEMPERATURE MINUS WET-BULB TEMPERATURE) (°F)																							
	1	2	3	4	5	6	7	8	9	10	11	12	13	14	15	16	17	18	19	20	25	30	35	40
0	67	31	1																					
5	73	46	20																					
10	78	56	34	13																				
15	82	64	46	29	11																			
20	85	70	55	40	26	12																		
25	87	74	62	49	37	25	13	1																
30	89	78	67	56	46	36	26	16	6															
35	91	81	72	63	54	45	36	27	19	10	2													
40	92	83	75	68	60	52	45	37	29	22	15	7												
45	93	86	78	71	64	57	51	44	38	31	25	18	12	6										
50	93	87	80	74	67	61	55	49	43	38	32	27	21	16	10	5								
55	94	88	82	76	70	65	59	54	49	43	38	33	28	23	19	14	9	5						
60	94	89	83	78	73	68	63	58	53	48	43	39	34	30	26	21	17	13	9	5				
65	95	90	85	80	75	70	66	61	56	52	48	44	39	35	31	27	24	20	16	12				
70	95	90	86	81	77	72	68	64	59	55	51	48	44	40	36	33	29	25	22	19	3			
75	96	91	86	82	78	74	70	66	62	58	54	51	47	44	40	37	34	30	27	24	9			
80	96	91	87	83	79	75	72	68	64	61	57	54	50	47	44	41	38	35	32	29	15	3		
85	96	92	88	84	80	76	73	69	66	62	59	56	52	49	46	43	41	38	35	32	20	8		
90	96	92	89	85	81	78	74	71	68	65	61	58	55	52	49	47	44	41	39	36	24	13	3	
95	96	93	89	85	82	79	75	72	69	66	63	60	57	54	51	49	46	43	41	38	27	17	7	1
100	96	93	89	86	83	80	77	73	70	68	65	62	59	56	54	51	49	46	44	41	30	21	12	4
105	97	93	90	87	83	80	77	74	71	69	66	63	60	58	55	53	50	48	46	43	33	23	15	7
110	97	93	90	87	84	81	78	75	73	70	67	65	62	60	57	55	52	50	48	46	36	26	18	11
115	97	94	91	88	85	82	79	76	74	71	68	66	63	61	58	56	54	52	49	47	37	28	21	13
120	97	94	91	88	85	82	80	77	74	72	69	67	65	62	60	58	55	53	51	49	40	31	23	17

Standard Atmosphere

ALTITUDE				PRESSURE	TEMPERATURE		DENSITY
meters	*feet*	*kilometers*	*miles*	*millibars*	°C	°F	kg/m^3
0	0	0.0	0.0	1013.25	15.0	(59.0)	1.225
500	1,640	0.5	0.3	954.61	11.8	(53.2)	1.167
1,000	3,280	1.0	0.6	898.76	8.5	(47.3)	1.112
1,500	4,921	1.5	0.9	845.59	5.3	(41.5)	1.058
2,000	6,562	2.0	1.2	795.01	2.0	(35.6)	1.007
2,500	8,202	2.5	1.5	746.91	–1.2	(29.8)	0.957
3,000	9,842	3.0	1.9	701.21	–4.5	(23.9)	0.909
3,500	11,483	3.5	2.2	657.80	–7.7	(18.1)	0.863
4,000	13,123	4.0	2.5	616.60	–11.0	(12.2)	0.819
4,500	14,764	4.5	2.8	577.52	–14.2	(6.4)	0.777
5,000	16,404	5.0	3.1	540.48	–17.5	(0.5)	0.736
5,500	18,045	5.5	3.4	505.39	–20.7	(–5.3)	0.697
6,000	19,685	6.0	3.7	472.17	–24.0	(–11.2)	0.660
6,500	21,325	6.5	4.0	440.75	–27.2	(–17.0)	0.624
7,000	22,965	7.0	4.3	411.05	–30.4	(–22.7)	0.590
7,500	24,606	7.5	4.7	382.99	–33.7	(–28.7)	0.557
8,000	26,247	8.0	5.0	356.51	–36.9	(–34.4)	0.526
8,500	27,887	8.5	5.3	331.54	–40.2	(–40.4)	0.496
9,000	29,528	9.0	5.6	308.00	–43.4	(–46.1)	0.467
9,500	31,168	9.5	5.9	285.84	–46.6	(–51.9)	0.440
10,000	32,808	10.0	6.2	264.99	–49.9	(–57.8)	0.413
11,000	36,089	11.0	6.8	226.99	–56.4	(–69.5)	0.365
12,000	39,370	12.0	7.5	193.99	–56.5	(–69.7)	0.312
13,000	42,651	13.0	8.1	165.79	–56.5	(–69.7)	0.267
14,000	45,932	14.0	8.7	141.70	–56.5	(–69.7)	0.228
15,000	49,213	15.0	9.3	121.11	–56.5	(–69.7)	0.195
16,000	52,493	16.0	9.9	103.52	–56.5	(–69.7)	0.166
17,000	55,774	17.0	10.6	88.497	–56.5	(–69.7)	0.142
18,000	59,055	18.0	11.2	75.652	–56.5	(–69.7)	0.122
19,000	62,336	19.0	11.8	64.674	–56.5	(–69.7)	0.104
20,000	65,617	20.0	12.4	55.293	–56.5	(–69.7)	0.089
25,000	82,021	25.0	15.5	25.492	–51.6	(–60.9)	0.040
30,000	98,425	30.0	18.6	11.970	–46.6	(–51.9)	0.018
35,000	114,829	35.0	21.7	5.746	–36.6	(–33.9)	0.008
40,000	131,234	40.0	24.9	2.871	–22.8	(–9.0)	0.004
45,000	147,638	45.0	28.0	1.491	–9.0	(15.8)	0.002
50,000	164,042	50.0	31.1	0.798	–2.5	(27.5)	0.001
60,000	196,850	60.0	37.3	0.220	–26.1	(–15.0)	0.0003
70,000	229,659	70.0	43.5	0.052	–53.6	(–64.5)	0.00008
80,000	262,467	80.0	49.7	0.010	–74.5	(–102.1)	0.00002

APPENDIX F

Beaufort Wind Scale (Over Land)

TABLE F.1 Estimating Wind Speed from Surface Observation

BEAUFORT NUMBER	DESCRIPTION	WIND SPEED MI/HR	KNOTS	KM/HR	OBSERVATIONS
0	Calm	0–1	0–1	0–2	Smoke rises vertically
1	Light air	1–3	1–3	2–6	Direction of wind shown by drifting smoke, but not by wind vanes
2	Slight breeze	4–7	4–6	7–11	Wind felt on face; leaves rustle; wind vanes moved by wind; flags stir
3	Gentle breeze	8–12	7–10	12–19	Leaves and small twigs move; wind will extend light flag
4	Moderate breeze	13–18	11–16	20–29	Wind raises dust and loose paper; small branches move; flags flap
5	Fresh breeze	19–24	17–21	30–39	Small trees with leaves begin to sway; flags ripple
6	Strong breeze	25–31	22–27	40–50	Large tree branches in motion; whistling heard in telegraph wires; umbrellas used with difficulty
7	High wind	32–38	28–33	51–61	Whole trees in motion; inconvenience felt walking against wind; flags extend
8	Gale	39–46	34–40	62–74	Wind breaks twigs off trees; walking is difficult
9	Strong gale	47–54	41–47	75–87	Slight structural damage occurs (signs and antennas blown down)
10	Whole gale	55–63	48–55	88–101	Trees uprooted; considerable damage occurs
11	Storm	64–74	56–64	102–119	Winds produce widespread damage
12	Hurricane	≥ 75	≥ 65	≥ 120	Winds produce extensive damage

Köppen's Climatic Classification System

TABLE G.1 Köppen's Climatic Classification System

Letter Symbol 1st	2nd	3rd	Climatic Characteristics	Criteria
A			Humid tropical	All months have an average temperature of 18°C (64°F) or higher
	f		Tropical wet (rain forest)	Wet all seasons; all months have at least 6 cm (2.4 in.) of rainfall
	w		Tropical wet and dry (savanna)	Winter dry season; rainfall in driest month is less than 6 cm (2.4 in.) and less than 10 – *P*/25 (*P* is mean annual rainfall in cm)
	m		Tropical monsoon	Short dry season; rainfall in driest month is less than 6 cm (2.4 in.) but equal to or greater than 10 – *P*/25.
B			Dry	Potential evaporation and transpiration exceed precipitation. The dry/humid boundary is defined by the following formulas: $p = 2t + 28$ when 70% or more of rain falls in warmer 6 months (dry winter) $p = 2t$ when 70% or more of rain falls in cooler 6 months (dry summer) $p = 2t + 14$ when neither half year has 70% or more of rain (p is the mean annual precipitation in cm and t is the mean annual temperature in °C)*
	S		Semi-arid (steppe)	The BS/BW boundary is exactly ½ the dry/humid boundary
	W		Arid (desert)	
		h	Hot and dry	Mean annual temperature is 18°C (64°F) or higher
		k	Cool and dry	Mean annual temperature is below 18°C (64°F)
C			Moist with mild winters	Average temperature of coolest month is below 18°C (64°F) and above –3°C (27°F)
	w		Dry winters	Average rainfall of wettest summer month at least 10 times as much as in driest winter month
	s		Dry summers	Average rainfall of driest summer month less than 4 cm (1.6 in.); average rainfall of wettest winter month at least 3 times as much as in driest summer month
	f		Wet all seasons	Criteria for w and s cannot be met
		a	Summers long and hot	Average temperature of warmest month above 22°C (72°F); at least 4 months with average above 10°C (50°F)
		b	Summers long and cool	Average temperature of all months below 22°C (72°F); at least 4 months with average above 10°C (50°F)
		c	Summers short and cool	Average temperature of all months below 22°C (72°F); 1 to 3 months with average above 10°C (50°F)
D			Moist with cold winters	Average temperature of coldest month is –3°C (27°F) or below; average temperature of warmest month is greater than 10°C (50°F)
	w		Dry winters	Same as under C
	s		Dry summers	Same as under C
	f		Wet all seasons	Same as under C
		a	Summers long and hot	Same as under C
		b	Summers long and cool	Same as under C
		c	Summers short and cool	Same as under C
		d	Summers short and cool; winters severe	Average temperature of coldest month is –38°C (–36°F) or below
E			Polar climates	Average temperature of warmest month is below 10°C (50°F)
	T		Tundra	Average temperature of warmest month is greater than 0°C (32°F) but less than 10°C (50°F)
	F		Ice cap	Average temperature of warmest month is 0°C (32°F) or below

*The dry/humid boundary is defined in English units as: $p = 0.44t - 3$ (dry winter); $p = 0.44t - 14$ (dry summer); and $p = 0.44t - 8.6$ (rainfall evenly distributed). Where p is mean annual rainfall in inches and t is mean annual temperature in °F.

Annual Global Pattern of Precipitation

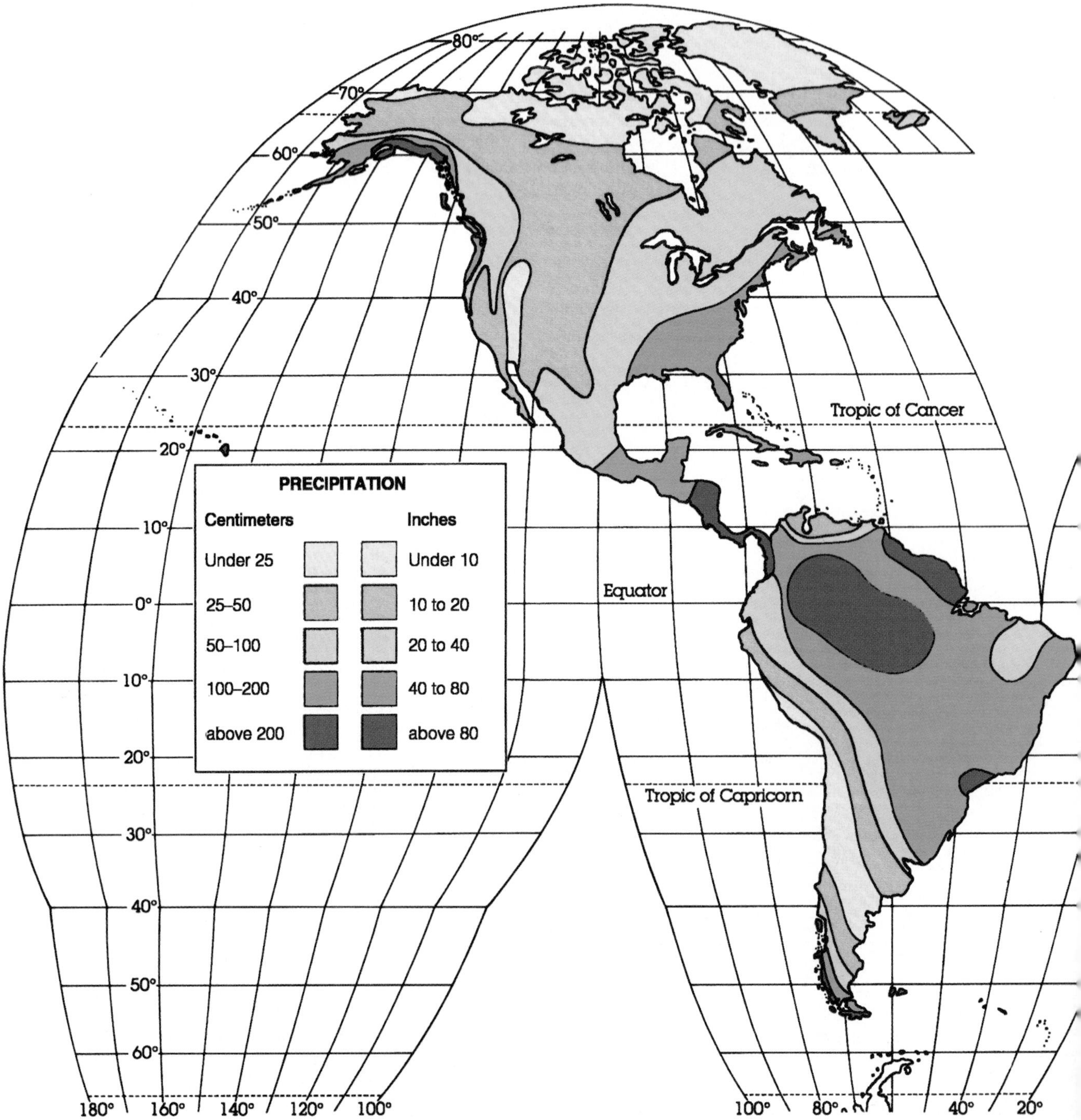

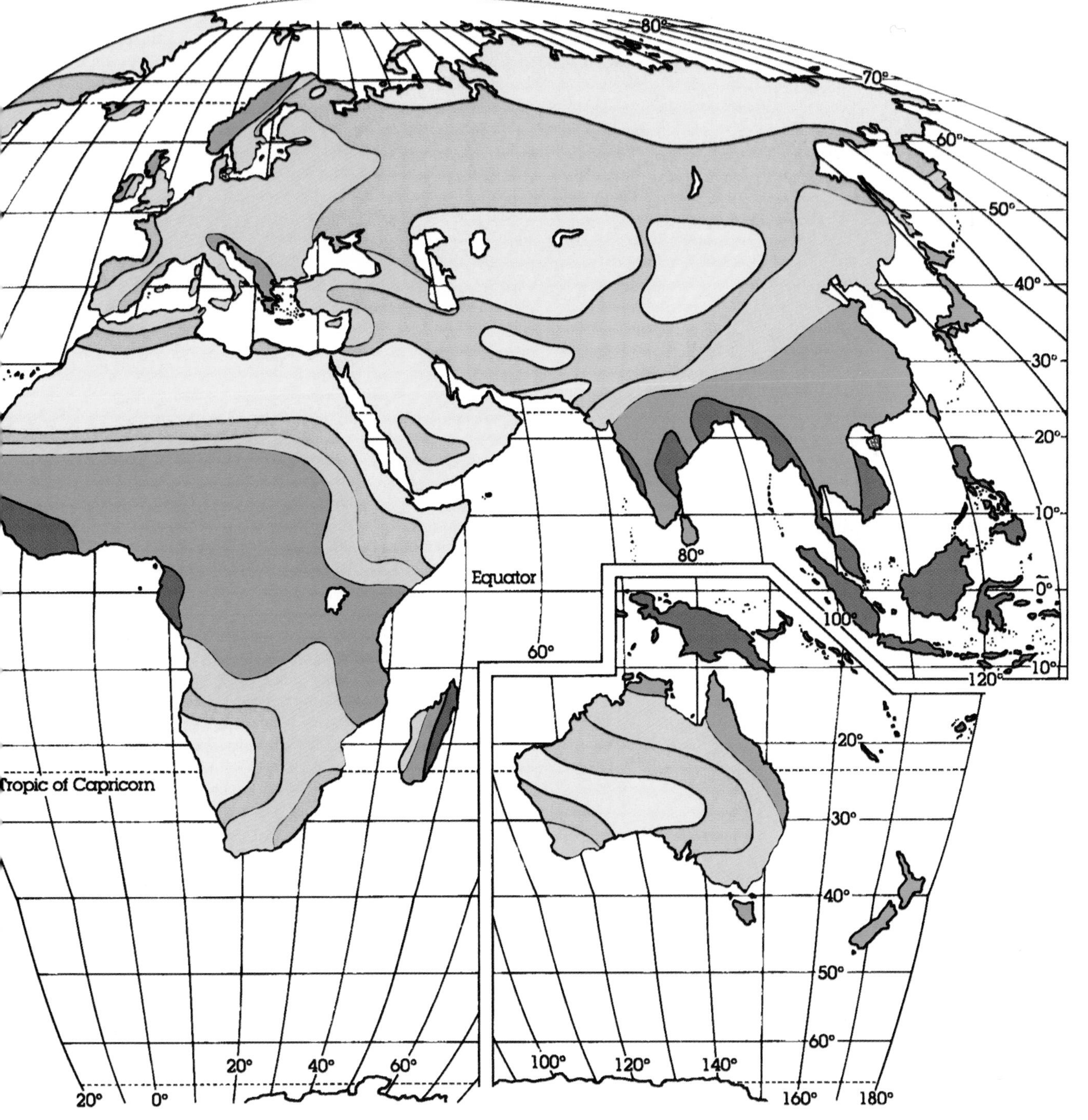

80°
70°
60°
50°
40°
30°
20°
10°
0°
10°
120°
100°
80°
60°
Equator
Tropic of Capricorn
20°
30°
40°
50°
60°
20°
0°
20°
40°
60°
100°
120°
140°
160°
180°

Additional Reading Material

PERIODICALS

Selected nontechnical periodicals that contain articles on weather and climate.

Bulletin of the American Meteorological Society. Monthly. The American Meteorological Society, 45 Beacon St., Boston, MA 02108.

Meteorological Magazine. Monthly. British Meteorological Office, British Information Services, 845 Third Avenue, New York, NY.

National Weather Digest. Quarterly. National Weather Association, 4400 Stamp Road, Room 404, Marlow Heights, MD 20031. (Deals mainly with weather forecasting.)

Weather. Monthly. Royal Meteorological Society, James Glaisher House, Grenville Place, Bracknell, Berkshire, England.

Weatherwise. Bimonthly. Heldref Publications, 4000 Albermarle St., N.W., Washington, DC 20016.

SELECTED TECHNICAL PERIODICALS

EOS—Transaction of the American Geophysical Union. American Geophysical Union (AGS), Washington, DC.

Journal of Applied Meteorology. American Meteorological Society (AMS), Boston, MA.

Journal of Atmospheric and Oceanic Technology. AMS, Boston, MA.

Journal of Atmospheric Science. AMS, Boston, MA.

Journal of Climate. AMS, Boston, MA.

Journal of Geophysical Research. American Geophysical Union, Washington, DC.

Monthly Weather Review. AMS, Boston, MA.

Weather and Forecasting. AMS, Boston, MA.

Additional periodicals that frequently contain articles of meteorological interest.

American Scientist. Bimonthly. Sigma Xi, the Scientific Research Society, Inc., New Haven, CT.

Science. Weekly. American Association for the Advancement of Science, Washington, DC.

Scientific American. Monthly. Scientific American, Inc., New York, NY.

Smithsonian. Monthly. The Smithsonian Association, Washington, DC.

BOOKS

The titles listed below may be drawn upon for additional information. Many are written at the introductory level. Those that are more advanced are marked with an asterisk.

Ahrens, C. Donald. *Meteorology Today* (7th ed.), Brooks/Cole Publishing Co., Pacific Grove, CA, 2003.

Anthes, R. A. *Tropical Cyclones: Their Evolution, Structure, and Effect,* American Meteorological Society, Boston, MA, 1982.

Arya, Pal S. *Air Pollution Meteorology,* Oxford University Press, New York, 1998.

Battan, Louis J. *Harvesting the Clouds,* Anchor Books, Doubleday and Co., Garden City, NY, 1961.

——. *The Unclean Sky,* Anchor Books, Doubleday and Co., Garden City, NY, 1966.

Bigg, Grant R. *The Oceans and Climate,* Cambridge University Press, New York, 1996.

*Bluestein, Howard B. *Synoptic-Dynamic Meteorology in Midlatitudes. Vol. 1: Principles of Kinematics and Dynamics,* Oxford University Press, New York, 1992.

——. *Synoptic-Dynamic Meteorology in Midlatitudes. Vol. II: Observations and Theory of Weather Systems,* Oxford University Press, New York, 1993.

——. *Tornado Alley: Monster Storms of the Great Plains,* Oxford University Press, New York, 1999.

Bohren, Craig F. *Clouds in a Glass of Beer: Simple Experiments in Atmospheric Physics,* Wiley, New York, 1987.

——. *What Light Through Yonder Window Breaks?,* Wiley, New York, 1991.

Boubel, Richard W., et al., *Fundamentals of Air Pollution* (3rd ed.), Academic Press, New York, 1994.

Burgess, Eric, and Douglass Torr. *Into the Thermosphere: The Atmosphere Explorers,* National Aeronautics and Space Administration, Washington, DC, 1987.

Burroughs, William J. *Watching the World's Weather,* Cambridge University Press, New York, 1991.

——. *Climate Revealed,* Cambridge University Press, Cambridge, England, 1999.

Carlson, Toby N. *Mid-Latitude Weather Systems,* American Meteorological Society, Boston, MA, 1998.

*Cotton, W. R., and R. A. Anthes. *Storm and Cloud Dynamics,* Academic Press, New York, 1989.

Cotton, William R. *Storms,* ASTeR Press, Fort Collins, CO, 1990.

Cotton, William R., and Roger A. Pielke. *Human Impacts on Weather and Climate,* Cambridge University Press, New York, 1995.

De Blij, H. J. *Nature on the Rampage,* Smithsonian Books, Washington, DC, 1994.

Eagleman, Joe R. *Air Pollution Meteorology,* Trimedia Publishing Co., Lenexa, KS, 1991.

Elsner, James B., and A. Biral Kara. *Hurricanes of the North Atlantic,* Oxford University Press, New York, 1999.

Elsom, Derek M. *Atmospheric Pollution: A Global Problem* (2nd ed.), Blackwell Publishers, Oxford, England, 1992.

Encyclopedia of Climate and Weather, Vol. 1 and Vol. 2, Stephen H. Schneider, Ed., Oxford University Press, New York, 1996.

Energy and Climate Change. Report of the DOE Multi-Laboratory Climate Change Committee, Lewis Publishers, Chelsea, MI, 1991.

England, Gary A. *Weathering the Storm,* University of Oklahoma Press, Norman, OK, 1996.

Firor, John. *The Changing Atmosphere: A Global Challenge,* Yale University Press, New Haven, CT, 1990.

*Fleagle, Robert G., and Joost A. Businger. *An Introduction to Atmospheric Physics* (2nd ed.), Academic Press, New York, 1980.

Fujita, T. T. *The Downburst—Microburst and Macroburst,* University of Chicago Press, Chicago, 1985.

Glossary of Meteorology. Todd S. Glickman, Managing Ed., American Meteorological Society, Boston, MA, 2000.

Glossary of Weather and Climate. Ira W. Geer, Ed., American Meteorological Society, Boston, MA, 1996.

*Graedel, T. E. and Paul J. Crutzen. *Atmospheric Change: An Earth System Perspective,* W. H. Freeman, New York, 1993.

Graedel, Thomas E., and Paul J. Crutzen. *Atmosphere, Climate, and Change,* W. H. Freeman, New York, 1995.

Greenler, Robert. *Rainbows, Halos and Glories,* Cambridge University Press, New York, 1980.

*Grotjahn, Richard. *Global Atmospheric Circulations: Observations and Theories,* Oxford University Press, Oxford, England, 1993.

*Hobbs, Peter V. *Basic Physical Chemistry for Atmospheric Sciences,* Cambridge University Press, New York, 1995.

Houghton, David D. *Handbook of Applied Meteorology,* Wiley (through AMS), Boston, MA, 1985.

Houghton, J. T., et al. *Climate Change 2001: The Scientific Basis,* Cambridge University Press, Cambridge, England, 2001.

——. *Climate Change 1995. The Science of Climate Change,* Cambridge University Press, Cambridge, England, 1996.

Hoyt, Douglas V. and Kenneth H. Schatten. *The Role of the Sun in Climate Change,* Oxford University Press, New York, 1997.

Imbrie, John, and K. P. Imbrie. *Ice Ages: Solving the Mystery,* Enslow Publishers, Short Hills, NJ, 1979.

International Cloud Atlas. World Meteorological Organization, Geneva, Switzerland, 1987.

James, Bruce P., et al., Eds. *Climate Change 1995. Economic and Social Dimensions of Climate Change,* Cambridge University Press, Cambridge, England, *1996.*

*Karoly, David J., and Dayton G. Vincent, Eds. *Meteorology of the Southern Hemisphere,* American Meteorological Society, Boston, MA, 1998.

Keen, Richard A. *Skywatch: The Western Weather Guide.* Fulcrum Incorporated, Golden, CO, 1987.

——. *Skywatch East: A Weather Guide,* Fulcrum Incorporated, Golden, CO, 1992.

Kessler, Edwin. *Thunderstorm Morphology and Dynamics* (2nd ed.), University of Oklahoma Press, Norman, OK, 1986.

Kocin, Paul J., and L. W. Uccellini. *Snowstorms along the Northeastern Coast of the United States: 1955 to 1985,* American Meteorological Society, Boston, MA, 1990.

*Ludlam, F. H. *Clouds and Storms: The Behavior and Effects of Water in the Atmosphere,* Pennsylvania State University Press (through AMS), Boston, MA, 1980.

Ludlum, D. M. *The Audubon Society Field Guide to North American Weather,* Alfred A. Knopf, New York, 1991.

——. *The Country Journal New England Weather Book,* Houghton Mifflin, Boston, MA, 1976.

——. *Early American Winters: 1604–1870,* American Meteorological Society, Boston, MA, 1967.

Lynch, David K., and William Livingston. *Color and Light in Nature,* Cambridge University Press, New York, 1995.

Mason, B. J. *Acid Rain: Its Causes and Its Effects on Inland Waters,* Oxford University Press, New York, 1992.

Meinel, Aden, and Marjorie Meinel. *Sunsets, Twilights and Evening Skies.* Cambridge University Press, New York, 1983.

National Academy Press. *Acid Deposition—Atmospheric Processes in Eastern North America,* Washington, DC, 1983.

——. *The Effects on the Atmosphere of a Major Nuclear Exchange,* Washington, DC, 1985.

——. *Ozone Depletion, Greenhouse Gases, and Climate Change,* Washington, DC, 1989.

Nelson, Mike. *The Colorado Weather Book,* Westcliff Publishers, Englewood, CO, 1999.

*Palmen, E., and C. W. Newton. *Atmospheric Circulation Systems,* Academic Press, New York, 1969.

Prospects for Future Climate. Special US/USSR Report on Climate and Climate Change, Lewis Publishers, Chelsea, MI, 1990.

*Ray, Peter S., Ed. *Mesoscale Meteorology and Forecasting,* American Meteorological Society, Boston, MA, 1986.

Righter, Robert W. *Wind Energy in America,* University of Oklahoma Press, Norman, OK, 1996.

*Rogers, R. R. *A Short Course in Cloud Physics* (3rd ed.), Pergamon Press, Oxford, England, 1989.

Schaeter, Vincent J., and John A. Day. *A Field Guide to the Atmosphere,* Houghton Mifflin (through AMS), Boston, MA, 1981.

——. *Peterson First Guide to Clouds and Weather,* Houghton Mifflin, Boston, MA, 1991.

Scorer, Richard S., and Arjen Verkaik. *Spacious Skies,* David and Charles Publishers, London, 1989.

Simpson, Robert H. and Herbert Riehl. *The Hurricane and Its Impact,* Louisiana State University Press, Baton Rouge, LA, 1981.

*Stull, Roland B. *Meteorology Today for Scientists and Engineers (2nd Ed.),* Brooks/Cole Publishing Co., Pacific Grove, CA, 2000.

Van Andel, Tjeerd H. *New Views on an Old Planet—A History of Global Change,* Cambridge University Press, Cambridge, England, 1994.

*Watson, Robert T., et al., Eds. *Climate Change 1995. Impacts, Adaptations and Mitigation of Climate Change: Scientific-Technical Analyses,* Cambridge University Press, Cambridge, England, 1996.

Williams, Jack. *The USA Today Weather Book,* Vintage Books, Random House, New York, 1992.

Glossary

Absolute humidity The mass of water vapor in a given volume of air. It represents the density of water vapor in the air.
Absolute zero A temperature reading of –273°C, –460°F, or 0K. Theoretically, there is no molecular motion at this temperature.
Absolutely stable atmosphere An atmospheric condition that exists when the environmental lapse rate is less than the moist adiabatic rate. This results in a lifted parcel of air being colder than the air around it.
Absolutely unstable atmosphere An atmospheric condition that exists when the environmental lapse rate is greater than the dry adiabatic rate. This results in a lifted parcel of air being warmer than the air around it.
Accretion The growth of a precipitation particle by the collision of an ice crystal or snowflake with a supercooled liquid droplet that freezes upon impact.
Acid deposition The depositing of acidic particles (usually sulfuric acid and nitric acid) at the earth's surface. Acid deposition occurs in dry form (*dry deposition*) or wet form (*wet deposition*). Acid rain and acid precipitation often denote wet deposition. (*See* Acid rain.)
Acid fog *See* Acid rain.
Acid rain Cloud droplets or raindrops combining with gaseous pollutants, such as oxides of sulfur and nitrogen, to make falling rain (or snow) acidic—pH less than 5.0. If fog droplets combine with such pollutants it becomes *acid fog.*
Actual Vapor Pressure *See* Vapor pressure.
Adiabatic process A process that takes place without a transfer of heat between the system (such as an air parcel) and its surroundings. In an adiabatic process, compression always results in warming, and expansion results in cooling.
Advection The horizontal transfer of any atmospheric property by the wind.
Advection fog Occurs when warm, moist air moves over a cold surface and the air cools to below its dew point.
Advection-radiation fog Fog that forms as relatively warm moist air moves over a colder surface that cooled mainly by radiational cooling.
Aerosols Tiny suspended solid particles (dust, smoke, etc.) or liquid droplets that enter the atmosphere from either natural or human (anthropogenic) sources, such as the burning of fossil fuels. Sulfur-containing fossil fuels, such as coal, produce *sulfate aerosols.*
Aerovane A wind instrument that indicates or records both wind speed and wind direction. Also called a *skyvane.*
Aggregation The clustering together of ice crystals to form snowflakes.
Air density *See* Density.
Air mass A large body of air that has similar horizontal temperature and moisture characteristics.
Air-mass thunderstorm *See* Ordinary thunderstorm.
Air-mass weather A persistent type of weather that may last for several days (up to a week or more). It occurs when an area comes under the influence of a particular air mass.
Air parcel *See* Parcel of air.
Air pollutants Solid, liquid, or gaseous airborne substances that occur in concentrations high enough to threaten the health of people and animals, to harm vegetation and structures, or to toxify a given environment.
Air pressure (atmospheric pressure) The pressure exerted by the mass of air above a given point, usually expressed in millibars (mb), inches of mercury (Hg) or in hectopascals (hPa).
Air quality index (AQI) An index of air quality that provides daily air pollution concentrations. Intervals on the scale relate to potential health effects.
Albedo The percent of radiation returning from a surface compared to that which strikes it.
Aleutian low The subpolar low-pressure area that is centered near the Aleutian Islands on charts that show mean sea-level pressure.
Altimeter An instrument that indicates the altitude of an object above a fixed level. Pressure altimeters use an aneroid barometer with a scale graduated in altitude instead of pressure.
Altocumulus A middle cloud, usually white or gray. Often occurs in layers or patches with wavy, rounded masses or rolls.

Altostratus A middle cloud composed of gray or bluish sheets or layers of uniform appearance. In the thinner regions, the sun or moon usually appears dimly visible.
Analogue forecasting method A forecast made by comparison of past large-scale synoptic weather patterns that resemble a given (usually current) situation in its essential characteristics.
Analysis The drawing and interpretation of the patterns of various weather elements on a surface or upper-air chart.
Anemometer An instrument designed to measure wind speed.
Aneroid barometer An instrument designed to measure atmospheric pressure. It contains no liquid.
Annual range of temperature The difference between the warmest and coldest months at any given location.
Anticyclone An area of high atmospheric pressure around which the wind blows clockwise in the Northern Hemisphere and counterclockwise in the Southern Hemisphere. Also called a *high.*
Apparent temperature What the air temperature "feels like" for various combinations of air temperature and relative humidity.
Arid climate An extremely dry climate—drier than the semi-arid climate. Often referred to as a "true desert" climate.
ASOS Acronym for *A*utomated *S*urface *O*bserving *S*ystems. A system designed to provide continuous information of wind, temperature, pressure, cloud base height, and runway visibility at selected airports.
Atmosphere The envelope of gases that surround a planet and are held to it by the planet's gravitational attraction. The earth's atmosphere is mainly nitrogen and oxygen.
Atmospheric boundary layer The layer of air from the earth's surface usually up to about 1 km (3300 ft) where the wind is influenced by friction of the earth's surface and objects on it. Also called the *planetary boundary layer* and the *friction layer.*
Atmospheric greenhouse effect The warming of an atmosphere by its absorbing and emitting infrared radiation while allowing shortwave radiation to pass on through. The gases mainly responsible for the earth's atmospheric greenhouse effect are water vapor and carbon dioxide. Also called the *greenhouse effect.*
Atmospheric models Simulation of the atmosphere's behavior by mathematical equations or by physical models.
Atmospheric stagnation A condition of light winds and poor vertical mixing that can lead to a high concentration of pollutants. Air stagnations are most often associated with fair weather, an inversion, and the sinking air of a high-pressure area.
Atmospheric window The wavelength range between 8 and 11 μm in which little absorption of infrared radiation takes place.
Aurora Glowing light display in the nighttime sky caused by excited gases in the upper atmosphere giving off light. In the Northern Hemisphere it is called the *aurora borealis* (northern lights); in the Southern Hemisphere, the *aurora australis* (southern lights).
Autumnal equinox The equinox at which the sun approaches the Southern Hemisphere and passes directly over the equator. Occurs around September 23.
AWIPS Acronym for *A*dvanced *W*eather *I*nteractive *P*rocessing *S*ystem. New computerized system that integrates and processes data received at a Weather Forecasting Office from NEXRAD, ASOS, and analysis and guidance products prepared by NMC.

Back-door cold front A cold front moving south or southwest along the Atlantic seaboard of the United States.
Backing wind A wind that changes direction in a counterclockwise sense (e.g., north to northwest to west).
Ball lightning A rare form of lightning that may consist of a reddish, luminous ball of electricity or charged air.
Barograph A recording barometer.
Barometer An instrument that measures atmospheric pressure. The two most common barometers are the *mercury barometer* and the *aneroid barometer.*
Bergeron process *See* Ice-crystal process.
Bermuda high *See* Subtropical high.
Billow clouds Broad, nearly parallel lines of wavelike clouds oriented at right angles to the wind.
Bimetallic thermometer A temperature-measuring device usually consisting of two dissimilar metals that expand and contract differentially as the temperature changes.
Blackbody A hypothetical object that absorbs all of the radiation that strikes it. It also emits radiation at a maximum rate for its given temperature.
Blizzard A severe weather condition characterized by low temperatures and strong winds (greater than 35 mi/hr) bearing a great amount of snow either falling or blowing. When these conditions continue after the falling snow has ended, it is termed a *ground blizzard.*
Boulder winds Fast-flowing, local downslope winds that may attain speeds of 100 knots or more. They are especially strong along the eastern foothills of the Rocky Mountains near Boulder, Colorado.
Boundary layer *See* Atmospheric boundary layer.
Brocken bow A bright ring of light seen around the shadow of an observer's head as the observer peers into a cloud or fog bank. Formed by *diffraction* of light.
Buoyant force (buoyancy) The upward force exerted upon an air parcel (or any object) by virtue of the density (mainly temperature) difference between the parcel and that of the surrounding air.
Buys-Ballot's law A law describing the relationship between the wind direction and the pressure distribution. In the Northern Hemisphere, if you stand with your back to the surface wind, then turn clockwise about 30°, lower pressure will be to your left. In the Southern Hemisphere, stand with your back to the surface wind, then turn counterclockwise about 30°, lower pressure will be to your right.

California current The ocean current that flows southward along the west coast of the United States from about Washington to Baja, California.

Cap cloud *See* Pileus cloud.

Carbon dioxide (CO_2) A colorless, odorless gas whose concentration is about 0.037 percent (375 ppm) in a volume of air near sea level. It is a selective absorber of infrared radiation and, consequently, it is important in the earth's atmospheric greenhouse effect. Solid CO_2 is called *dry ice.*

Carbon monoxide (CO) A colorless, odorless, toxic gas that forms during the incomplete combustion of carbon-containing fuels.

Celsius scale A temperature scale where zero is assigned to the temperature where water freezes and 100 to the temperature where water boils (at sea level).

Centripetal acceleration The inward-directed acceleration on a particle moving in a curved path.

Centripetal force The radial force required to keep an object moving in a circular path. It is directed toward the center of that curved path.

Chaos The property describing a system that exhibits erratic behavior in that very small changes in the initial state of the system rapidly lead to large and apparently unpredictable changes sometime in the future.

Chinook wall cloud A bank of clouds over the Rocky Mountains that signifies the approach of a chinook.

Chinook wind A warm, dry wind on the eastern side of the Rocky Mountains. In the Alps, the wind is called a *foehn.*

Chlorofluorocarbons (CFCs) Compounds consisting of methane (CH_4) or ethane (C_2H_6) with some or all of the hydrogen replaced by chlorine or fluorine. Used in fire extinguishers, as refrigerants, as solvents for cleaning electronic microcircuits, and as propellants. CFCs contribute to the atmospheric greenhouse effect and destroy ozone in the stratosphere.

Cirrocumulus A high cloud that appears as a white patch of clouds without shadows. It consists of very small elements in the form of grains or ripples.

Cirrostratus High, thin, sheetlike clouds, composed of ice crystals. They frequently cover the entire sky and often produce a halo.

Cirrus A high cloud composed of ice crystals in the form of thin, white, featherlike clouds in patches, filaments, or narrow bands.

Clear air turbulence (CAT) Turbulence encountered by aircraft flying through cloudless skies. Thermals, wind shear, and jet streams can each be a factor in producing CAT.

Clear ice A layer of ice that appears transparent because of its homogeneous structure and small number and size of air pockets.

Climate The accumulation of daily and seasonal weather events over a long period of time.

Climatic controls The relatively permanent factors that govern the general nature of the climate of a region.

Climatic optimum *See* Mid-Holocene maximum.

Climatological forecast A weather forecast, usually a month or more in the future, which is based upon the climate of a region rather than upon current weather conditions.

Cloud A visible aggregate of tiny water droplets and/or ice crystals in the atmosphere above the earth's surface.

Cloudburst Any sudden and heavy rain shower.

Cloud seeding The introduction of artificial substances (usually silver iodide or dry ice) into a cloud for the purpose of either modifying its development or increasing its precipitation.

Coalescence The merging of cloud droplets into a single larger droplet.

Cold fog *See* Supercooled cloud.

Cold front A transition zone where a cold air mass advances and replaces a warm air mass.

Cold occlusion *See* Occluded front.

Cold wave A rapid fall in temperature within 24 hours that often requires increased protection for agriculture, industry, commerce, and human activities.

Comma cloud A band of organized cumuliform clouds that looks like a comma on a satellite photograph.

Computer enhancement A process where the temperatures of radiating surfaces are assigned different shades of gray (or different colors) on an infrared picture. This allows specific features to be more clearly delineated.

Condensation The process by which water vapor becomes a liquid.

Condensation level The level above the surface marking the base of a cumuliform cloud.

Condensation nuclei Also called *cloud condensation nuclei.* Tiny particles upon whose surfaces condensation of water vapor begins in the atmosphere.

Conditionally unstable atmosphere An atmospheric condition that exists when the environmental lapse rate is less than the dry adiabatic rate but greater than the moist adiabatic rate. Also called *conditional instability.*

Conduction The transfer of heat by molecular activity from one substance to another, or through a substance. Transfer is always from warmer to colder regions.

Constant-height chart (constant-level chart) A chart showing variables, such as pressure, temperature, and wind, at a specific altitude above sea level. Variation in horizontal pressure is depicted by isobars. The most common constant-height chart is the surface chart, which is also called the *sea-level chart* or *surface weather map.*

Constant-pressure chart (isobaric chart) A chart showing variables, such as temperature and wind, on a constant-pressure surface. Variations in height are usually shown by lines of equal height (contour lines).

Contact freezing The process by which contact with a nucleus such as an ice crystal causes supercooled liquid droplets to change into ice.

Continental arctic air mass An air mass characterized by extremely low temperatures and very dry air.

Continental polar air mass An air mass characterized by low temperatures and dry air. Not as cold as arctic air masses.

Continental tropical air mass An air mass characterized by high temperatures and low humidity.

Contour line A line that connects points of equal elevation above a reference level, most often sea level.

Contrail (condensation trail) A cloudlike streamer frequently seen forming behind aircraft flying in clear, cold, humid air.

Controls of temperature The main factors that cause variations in temperature from one place to another.

Convection Motions in a fluid that result in the transport and mixing of the fluid's properties. In meteorology, convection usually refers to atmospheric motions that are predominantly vertical, such as rising air currents due to surface heating. The rising of heated surface air and the sinking of cooler air aloft is often called *free convection.* (Compare with *forced convection.*)

Convergence An atmospheric condition that exists when the winds cause a horizontal net inflow of air into a specified region.

Cooling degree-day A form of degree-day used in estimating the amount of energy necessary to reduce the effective temperature of warm air. A cooling degree-day is a day on which the average temperature is one degree above a desired base temperature.

Coriolis force An apparent force observed on any free-moving object in a rotating system. On the earth, this deflective force results from the earth's rotation and causes moving particles (including the wind) to deflect to the right in the Northern Hemisphere and to the left in the Southern Hemisphere.

Corona (optic) A series of colored rings concentrically surrounding the disk of the sun or moon. Smaller than the halo, the corona is often caused by the diffraction of light around small water droplets of uniform size.

Country breeze A light breeze that blows into a city from the surrounding countryside. It is best observed on clear nights when the urban heat island is most pronounced.

Crepuscular rays Alternating light and dark bands of light that appear to fan out from the sun's position, usually at twilight.

Cumulonimbus An exceptionally dense and vertically developed cloud, often with a top in the shape of an anvil. The cloud is frequently accompanied by heavy showers, lightning, thunder, and sometimes hail. It is also known as a *thunderstorm cloud.*

Cumulus A cloud in the form of individual, detached domes or towers that are usually dense and well defined. It has a flat base with a bulging upper part that often resembles cauliflower. Cumulus clouds of fair weather are called *cumulus humilis.* Those that exhibit much vertical growth are called *cumulus congestus* or *towering cumulus.*

Cumulus stage The initial stage in the development of an ordinary thunderstorm in which rising, warm, humid air develops into a cumulus cloud.

Cut-off low A cold upper-level low that has become displaced out of the basic westerly flow and lies to the south of this flow.

Cyclogenesis The development or strengthening of middle latitude (extratropical) cyclones.

Cyclone An area of low pressure around which the winds blow counterclockwise in the Northern Hemisphere and clockwise in the Southern Hemisphere.

Daily range of temperature The difference between the maximum and minimum temperatures for any given day.

Dart leader The discharge of electrons that proceeds intermittently toward the ground along the same ionized channel taken by the initial lightning stroke.

Dendrochronology The analysis of the annual growth rings of trees as a means of interpreting past climatic conditions.

Density The ratio of the mass of a substance to the volume occupied by it. Air density is usually expressed as g/cm^3 or kg/m^3.

Deposition A process that occurs in subfreezing air when water vapor changes directly to ice without becoming a liquid first.

Derecho Strong, damaging, straight-line winds associated with a cluster of severe thunderstorms that most often form in the evening or at night.

Desertification A general increase in the desert conditions of a region.

Dew Water that has condensed onto objects near the ground when their temperatures have fallen below the dew point of the surface air.

Dew cell An instrument used to determine the dew-point temperature.

Dew point (dew-point temperature) The temperature to which air must be cooled (at constant pressure and constant water vapor content) for saturation to occur.

Dew-point hygrometer An instrument that determines the dew-point temperature of the air.

Diffraction The bending of light around objects, such as cloud and fog droplets, producing fringes of light and dark or colored bands.

Dispersion The separation of white light into its different component wavelengths.

Dissipating stage The final stage in the development of an ordinary thunderstorm when downdrafts exist throughout the cumulonimbus cloud.

Divergence An atmospheric condition that exists when the winds cause a horizontal net outflow of air from a specific region.

Doldrums The region near the equator that is characterized by low pressure and light, shifting winds.

Doppler lidar The use of light beams to determine the velocity of objects such as dust and falling rain by taking into account the *Doppler shift.*

Doppler radar A radar that determines the velocity of falling precipitation either toward or away from the radar unit by taking into account the *Doppler shift.*

Doppler shift (effect) The change in the frequency of waves that occurs when the emitter or the observer is moving toward or away from the other.

Downburst A severe localized downdraft that can be experienced beneath a severe thunderstorm. (Compare *Microburst* and *Macroburst.*)

Drizzle Small water drops between 0.2 and 0.5 mm in diameter that fall slowly and reduce visibility more than light rain.

Drought A period of abnormally dry weather sufficiently long enough to cause serious effects on agriculture and other activities in the affected area.

Dry adiabatic rate The rate of change of temperature in a rising or descending unsaturated air parcel. The rate of adiabatic cooling or warming is about 10°C per 1000 m (5.5°F per 1000 ft).

Dry-bulb temperature The air temperature measured by the dry-bulb thermometer of a psychrometer.

Dry climate A climate deficient in precipitation where annual potential evaporation and transpiration exceed precipitation.

Dry haze *See* Haze.

Dryline A boundary that separates warm, dry air from warm, moist air. It usually represents a zone of instability along which thunderstorms form.

Dry-summer subtropical climate A climate characterized by mild, wet winters and warm to hot, dry summers. Typically located between 30 and 45 degrees latitude on the western side of continents. Also called *Mediterranean climate.*

Dust devil (or whirlwind) A small but rapidly rotating wind made visible by the dust, sand, and debris it picks up from the surface. It develops best on clear, dry, hot afternoons.

Easterly wave A migratory wavelike disturbance in the tropical easterlies. Easterly waves occasionally intensify into tropical cyclones. They are also called *tropical waves.*

Eccentricity (of the earth's orbit) The deviation of the earth's orbit from elliptical to nearly circular.

Eddy A small volume of air (or any fluid) that behaves differently from the larger flow in which it exists.

Ekman spiral An idealized description of the way the wind-driven ocean currents vary with depth. In the atmosphere it represents the way the winds vary from the surface up through the friction layer or planetary boundary layer.

Ekman transport Net surface water transport due to the Ekman spiral. In the Northern Hemisphere the transport is 90° to the right of the surface wind direction.

Electrical hygrometer *See* Hygrometer.

Electrical thermometers Thermometers that use elements that convert energy from one form to another (transducers). Common electrical thermometers include the electrical resistance thermometer, thermocouple, and thermistor.

Electromagnetic waves *See* Radiant energy.

El Niño An extensive ocean warming that begins along the coast of Peru and Ecuador and extends westward over the tropical Pacific. Major El Niño events, or strong El Niños, occur once every 2 to 7 years as a current of nutrient-poor tropical water moves southward along the west coast of South America.

Energy The property of a system that generally enables it to do work. Some forms of energy are kinetic, radiant, potential, chemical, electric, and magnetic.

Ensemble forecasting A forecasting technique that entails running several forecast models, each beginning with slightly different weather information. The forecaster's level of confidence is based on how well the models agree (or disagree) at the end of some specified time.

ENSO (El Niño/southern oscillation) A condition in the tropical Pacific whereby the reversal of surface air pressure at opposite ends of the Pacific Ocean induces westerly winds, a strengthening of the equatorial countercurrent, and extensive ocean warming.

Entrainment The mixing of environmental air into a preexisting air current or cloud so that the environmental air becomes part of the current or cloud.

Environmental lapse rate The rate of decrease of air temperature with elevation. It is most often measured with a radiosonde.

Evaporation The process by which a liquid changes into a gas.

Evaporation (mixing) fog Fog produced when sufficient water vapor is added to the air by evaporation, and the moist air mixes with relatively drier air. The two common types are *steam fog,* which forms when cold air moves over warm water, and *frontal fog,* which forms as warm raindrops evaporate in a cool air mass.

Exosphere The outermost portion of the atmosphere.

Extratropical cyclone A cyclonic storm that most often forms along a front in middle and high latitudes. Also called a *middle-latitude cyclonic storm,* a *depression,* and a *low.* It is not a tropical storm or hurricane.

Eye A region in the center of a hurricane (tropical storm) where the winds are light and skies are clear to partly cloudy.

Eye wall A wall of dense thunderstorms that surrounds the eye of a hurricane.

Fahrenheit scale A temperature scale where 32 is assigned to the temperature at which water freezes and 212 to the temperature at which water boils (at sea level).

Fallstreaks Falling ice crystals that evaporate before reaching the ground.

Fall wind A strong, cold katabatic wind that blows downslope off snow-covered plateaus.

Fata Morgana A complex mirage that is characterized by objects being distorted in such a way as to appear as castlelike features.

Feedback mechanism A process whereby an initial change in an atmospheric process will tend to either reinforce the process (*positive feedback*) or weaken the process (*negative feedback*).

Ferrel cell The name given to the middle-latitude cell in the 3-cell model of the general circulation.

Flash flood A flood that rises and falls quite rapidly with little or no advance warning, usually as the result of intense rainfall over a relatively small area.
Foehn *See* Chinook wind.
Fog A cloud with its base at the earth's surface.
Forced convection On a small scale, a form of mechanical stirring taking place when twisting eddies of air are able to mix hot surface air with the cooler air above. On a larger scale, it can be induced by the lifting of warm air along a front (*frontal uplift*) or along a topographic barrier (*orographic uplift*).
Free convection *See* Convection.
Freeze A condition occurring over a widespread area when the surface air temperature remains below freezing for a sufficient time to damage certain agricultural crops. A freeze most often occurs as cold air is advected into a region, causing freezing conditions to exist in a deep layer of surface air. Also called *advection frost.*
Freezing rain and freezing drizzle Rain or drizzle that falls in liquid form and then freezes upon striking a cold object or ground. Both can produce a coating of ice on objects which is called *glaze.*
Friction layer The atmospheric layer near the surface usually extending up to about 1 km (3300 ft) where the wind is influenced by friction of the earth's surface and objects on it. Also called the *atmospheric boundary layer* and *planetary boundary layer.*
Front The transition zone between two distinct air masses.
Frontal fog *See* Evaporation fog.
Frontal thunderstorms Thunderstorms that form in response to forced convection (forced lifting) along a front. Most go through a cycle similar to those of ordinary thunderstorms.
Frontal wave A wavelike deformation along a front in the lower levels of the atmosphere. Those that develop into storms are termed *unstable waves,* while those that do not are called *stable waves.*
Frost (also called **hoarfrost**) A covering of ice produced by deposition on exposed surfaces when the air temperature falls below the frost point.
Frostbite The partial freezing of exposed parts of the body, causing injury to the skin and sometimes to deeper tissues.
Frost point The temperature at which the air becomes saturated with respect to ice when cooled at constant pressure and constant water vapor content.
Frozen dew The transformation of liquid dew into tiny beads of ice when the air temperature drops below freezing.
Fujita scale A scale developed by T. Theodore Fujita for classifying tornadoes according to the damage they cause and their rotational wind speed.
Funnel cloud A tornado whose circulation has not reached the ground. Often appears as a rotating conelike cloud that extends downward from the base of a thunderstorm.

Gas law The thermodynamic law applied to a perfect gas that relates the pressure of the gas to its density and absolute temperature.
General circulation of the atmosphere Large-scale atmospheric motions over the entire earth.
Geostationary satellite A satellite that orbits the earth at the same rate that the earth rotates and thus remains over a fixed place above the equator.
Geostrophic wind A theoretical horizontal wind blowing in a straight path, parallel to the isobars or contours, at a constant speed. The geostrophic wind results when the Coriolis force exactly balances the horizontal pressure gradient force.
Glaciated cloud A cloud or portion of a cloud where only ice crystals exist.
Global climate Climate of the entire globe.
Global scale The largest scale of atmospheric motion. Also called the *planetary scale.*
Global warming Increasing global surface air temperatures that show up in the climate record. The term *global warming* is usually attributed to human activities, such as increasing concentrations of greenhouse gases.
Glory Colored rings that appear around the shadow of an object.
Gradient wind A theoretical wind that blows parallel to curved isobars or contours.
Graupel Ice particles between 2 and 5 mm in diameter that form in a cloud often by the process of accretion. Snowflakes that become rounded pellets due to riming are called *graupel* or *snow pellets.*
Green flash A small green color that occasionally appears on the upper part of the sun as it rises or sets.
Greenhouse effect *See* Atmospheric greenhouse effect.
Ground fog *See* Radiation fog.
Growing degree-day A form of the degree-day used as a guide for crop planting and for estimating crop maturity dates.
Gulf stream A warm, swift, narrow ocean current flowing along the east coast of the United States.
Gust front A boundary that separates a cold downdraft of a thunderstorm from warm, humid surface air. On the surface its passage resembles that of a cold front.
Gustnado A relatively weak tornado associated with a thunderstorm's outflow. It most often forms along the gust front.

Haboob A dust or sandstorm that forms as cold downdrafts from a thunderstorm turbulently lift dust and sand into the air.
Hadley cell A thermal circulation proposed by George Hadley to explain the movement of the trade winds. It consists of rising air near the equator and sinking air near 30° latitude.
Hailstones Transparent or partially opaque particles of ice that range in size from that of a pea to that of golf balls.
Hailstreak The accumulation of hail at the earth's surface along a relatively long (10 km), narrow (2 km) band.
Hair hygrometer *See* Hygrometer.
Halos Rings or arcs that encircle the sun or moon when seen through an ice crystal cloud or a sky filled with falling ice crystals. Halos are produced by refraction of light.
Haze Fine dry or wet dust or salt particles dispersed through a portion of the atmosphere. Individually these are not visible but

cumulatively they will diminish visibility. *Dry haze* particles are very small, on the order of 0.1 μm. *Wet haze* particles are larger.

Heat A form of energy transferred between systems by virtue of their temperature differences.

Heat capacity The ratio of the heat absorbed (or released) by a system to the corresponding temperature rise (or fall).

Heat index (HI) An index that combines air temperature and relative humidity to determine an apparent temperature—how hot it actually feels.

Heating degree-day A form of the degree-day used as an index for fuel consumption.

Heat lightning Distant lightning that illuminates the sky but is too far away for its thunder to be heard.

Heatstroke A physical condition induced by a person's overexposure to high air temperatures, especially when accompanied by high humidity.

Hectopascal Abbreviated hPa. One hectopascal is equal to 100 Newtons/m^2, or 1 millibar.

Heiligenschein A faint white ring surrounding the shadow of an observer's head on a dew-covered lawn.

Heterosphere The region of the atmosphere above about 85 km where the composition of the air varies with height.

High *See* Anticyclone.

High inversion fog A fog that lifts above the surface but does not completely dissipate because of a strong inversion (usually subsidence) that exists above the fog layer.

Homosphere The region of the atmosphere below about 85 km where the composition of the air remains fairly constant.

Hook echo The shape of an echo on a Doppler radar screen that indicates the possible presence of a tornado.

Horse latitudes The belt of latitude at about 30° to 35° where winds are predominantly light and the weather is hot and dry.

Humid continental climate A climate characterized by severe winters and mild to warm summers with adequate annual precipitation. Typically located over large continental areas in the Northern Hemisphere between about 40° and 70° latitude.

Humidity A general term that refers to the air's water vapor content. (*See* Relative humidity.)

Humid subtropical climate A climate characterized by hot muggy summers, cool to cold winters, and abundant precipitation throughout the year.

Hurricane A tropical cyclone having winds in excess of 64 knots (74 mi/hr).

Hurricane warning A warning given when it is likely that a hurricane will strike an area within 24 hours.

Hurricane watch A hurricane watch indicates that a hurricane poses a threat to an area (often within several days) and residents of the watch area should be prepared.

Hydrocarbons Chemical compounds composed of only hydrogen and carbon—they are included under the general term volatile organic compounds (VOCs).

Hydrologic cycle A model that illustrates the movement and exchange of water among the earth, atmosphere, and oceans.

Hydrostatic equation An equation that states that the rate at which the air pressure decreases with height is equal to the air density times the acceleration of gravity. The equation relates to how quickly the air pressure decreases in a column of air.

Hydrostatic equilibrium The state of the atmosphere when there is a balance between the vertical pressure gradient force and the downward pull of gravity.

Hygrometer An instrument designed to measure the air's water vapor content. The sensing part of the instrument can be hair (*hair hygrometer*), a plate coated with carbon (*electrical hygrometer*), or an infrared sensor (*infrared hygrometer*).

Hygroscopic The ability to accelerate the condensation of water vapor. Usually used to describe condensation nuclei that have an affinity for water vapor.

Hypothermia The deterioration in one's mental and physical condition brought on by a rapid lowering of human body temperature.

Hypoxia A condition experienced by humans when the brain does not receive sufficient oxygen.

Ice Age *See* Pleistocene epoch.

Ice-crystal (Bergeron) process A process that produces precipitation. The process involves tiny ice crystals in a supercooled cloud growing larger at the expense of the surrounding liquid droplets. Also called the *Bergeron process.*

Ice fog A type of fog that forms at very low temperatures, composed of tiny suspended ice particles.

Icelandic low The subpolar low-pressure area that is centered near Iceland on charts that show mean sea-level pressure.

Ice nuclei Particles that act as nuclei for the formation of ice crystals in the atmosphere.

Ice pellets *See* Sleet.

Indian summer An unseasonably warm spell with clear skies near the middle of autumn. Usually follows a substantial period of cool weather.

Inferior mirage *See* Mirage.

Infrared radiation Electromagnetic radiation with wavelengths between about 0.7 and 1000 μm. This radiation is longer than visible radiation but shorter than microwave radiation.

Infrared radiometer An instrument designed to measure the intensity of infrared radiation emitted by an object. Also called *infrared sensor.*

Insolation The *in*coming *sol*ar radi*ation* that reaches the earth and the atmosphere.

Instrument shelter A boxlike, often wooden, structure designed to protect weather instruments from direct sunshine and precipitation.

Interglacial period A time interval of relatively mild climate during the Ice Age when continental ice sheets were absent or limited in extent to Greenland and the Antarctic.

Intertropical convergence zone (ITCZ) The boundary zone separating the northeast trade winds of the Northern

Hemisphere from the southeast trade winds of the Southern Hemisphere.

Inversion An increase in air temperature with height.

Ion An electrically charged atom, molecule, or particle.

Ionosphere An electrified region of the upper atmosphere where fairly large concentrations of ions and free electrons exist.

Iridescence Brilliant spots or borders of colors, most often red and green, observed in clouds up to about 30° from the sun.

Isobar A line connecting points of equal pressure.

Isobaric map *See* Constant-pressure chart.

Isobaric surface A surface along which the atmospheric pressure is everywhere equal.

Isotach A line connecting points of equal wind speed.

Isotherm A line connecting points of equal temperature.

Isothermal layer A layer where the air temperature is constant with increasing altitude. In an isothermal layer, the air temperature lapse rate is zero.

Jet maximum *See* Jet streak.

Jet streak A region of high wind speed that moves through the axis of a jet stream. Also called *jet maximum.*

Jet stream Relatively strong winds concentrated within a narrow band in the atmosphere.

Katabatic (fall) wind Any wind blowing downslope. It is usually cold.

Kelvin A unit of temperature. A Kelvin is denoted by K and 1 K equals 1°C. Zero Kelvin is absolute zero, or –273.15°C.

Kelvin scale A temperature scale with zero degrees equal to the theoretical temperature at which all molecular motion ceases. Also called the *absolute scale.* The units are sometimes called "degrees Kelvin"; however, the correct SI terminology is "Kelvins," abbreviated K.

Kinetic energy The energy within a body that is a result of its motion.

Kirchhoff's law A law that states: good absorbers of a given wavelength of radiation are also good emitters of that wavelength.

Knot A unit of speed equal to 1 nautical mile per hour. One knot equals 1.15 mi/hr.

Köppen classification system A system for classifying climates developed by W. Köppen that is based mainly on annual and monthly averages of temperature and precipitation.

Lake breeze A wind blowing onshore from the surface of a lake.

Lake-effect snows Localized snowstorms that form on the downwind side of a lake. Such storms are common in late fall and early winter near the Great Lakes as cold, dry air picks up moisture and warmth from the unfrozen bodies of water.

Land breeze A coastal breeze that blows from land to sea, usually at night.

Landspout Relatively weak nonsupercell tornado that originates with a cumliform cloud in its growth stage and with a cloud that does not contain a mid-level mesocyclone. Its spin originates near the surface. Landspouts often look like waterspouts over land.

La Niña A condition where the central and eastern tropical Pacific Ocean turns cooler than normal.

Lapse rate The rate at which an atmospheric variable (usually temperature) decreases with height. (*See* Environmental lapse rate.)

Latent heat The heat that is either released or absorbed by a unit mass of a substance when it undergoes a change of state, such as during evaporation, condensation, or sublimation.

Laterite A soil formed under tropical conditions where heavy rainfall leaches soluble minerals from the soil. This leaching leaves the soil hard and poor for growing crops.

Lee-side low Storm systems (extratropical cyclones) that form on the downwind (lee) side of a mountain chain. In the United States lee-side lows frequently form on the eastern side of the Rockies and Sierra Nevada mountains.

Lenticular cloud A cloud in the shape of a lens.

Level of free convection The level in the atmosphere at which a lifted air parcel becomes warmer than its surroundings in a conditionally unstable atmosphere.

Lidar An instrument that uses a laser to generate intense pulses that are reflected from atmospheric particles of dust and smoke. Lidars have been used to determine the amount of particles in the atmosphere as well as particle movement that has been converted into wind speed. Lidar means *li*ght *d*etection *a*nd *r*anging.

Lightning A visible electrical discharge produced by thunderstorms.

Liquid-in-glass thermometer *See* Thermometer.

Little Ice Age The period from about 1550 to 1850 when average temperatures over Europe were lower.

Local winds Winds that tend to blow over a relatively small area; often due to regional effects, such as mountain barriers, large bodies of water, local pressure differences, and other influences.

Long-range forecast Generally used to describe a weather forecast that extends beyond about 8.5 days into the future.

Longwave radiation A term most often used to describe the infrared energy emitted by the earth and the atmosphere.

Longwaves in the westerlies A wave in the upper level of the westerlies characterized by a long length (thousands of kilometers) and significant amplitude. Also called *Rossby waves.*

Low *See* Extratropical cyclone.

Low-level jet streams Jet streams that typically form near the earth's surface below an altitude of about 2 km and usually attain speeds of less than 60 knots.

Macroburst A strong downdraft (*downburst*) greater than 4 km wide that can occur beneath thunderstorms. A downburst less than 4 km across is called a *microburst.*

Macroclimate The general climate of a large area, such as a country.

Macroscale The normal meteorological synoptic scale for obtaining weather information. It can cover an area ranging from the size of a continent to the entire globe.

Mammatus clouds Clouds that look like pouches hanging from the underside of a cloud.

Marine climate A climate controlled largely by the ocean. The ocean's influence keeps winters relatively mild and summers cool.

Maritime air Moist air whose characteristics were developed over an extensive body of water.

Maritime polar air mass An air mass characterized by low temperatures and high humidity.

Maritime tropical air mass An air mass characterized by high temperatures and high humidity.

Mature thunderstorm The second stage in the three-stage cycle of an ordinary thunderstorm. This mature stage is characterized by heavy showers, lightning, thunder, and violent vertical motions inside cumulonimbus clouds.

Maunder minimum A period from about 1645 to 1715 when few, if any, sunspots were observed.

Maximum thermometer A thermometer with a small constriction just above the bulb. It is designed to measure the maximum air temperature.

Mean annual temperature The average temperature at any given location for the entire year.

Mean daily temperature The average of the highest and lowest temperature for a 24-hour period.

Mediterranean climate *See* Dry-summer subtropical climate.

Medium-range forecast Generally used to describe a weather forecast that extends from about 3 to 8.5 days into the future.

Mercury barometer A type of barometer that uses mercury to measure atmospheric pressure. The height of the mercury column is a measure of atmospheric pressure.

Meridional flow A type of atmospheric circulation pattern in which the north-south component of the wind is pronounced.

Mesoclimate The climate of an area ranging in size from a few acres to several square kilometers.

Mesocyclone A vertical column of cyclonically rotating air within a severe thunderstorm.

Mesohigh A relatively small area of high atmospheric pressure that forms beneath a thunderstorm.

Mesopause The top of the mesosphere. The boundary between the mesosphere and the thermosphere, usually near 85 km.

Mesoscale The scale of meteorological phenomena that range in size from a few km to about 100 km. It includes local winds, thunderstorms, and tornadoes.

Mesoscale convective complex (MCC) A large organized convective weather system comprised of a number of individual thunderstorms. The size of an MCC can be 1000 times larger than an individual air-mass thunderstorm.

Mesoscale convective system (MCS) A large cloud system that represents an ensemble of thunderstorms that form by convection, and produce precipitation over a wide area.

Mesosphere The atmospheric layer between the stratosphere and the thermosphere. Located at an average elevation between 50 and 80 km above the earth's surface.

Meteogram A chart that shows how one or more weather variables has changed at a station over a given period of time or how the variables are likely to change with time.

Meteorology The study of the atmosphere and atmospheric phenomena as well as the atmosphere's interaction with the earth's surface, oceans, and life in general.

Microburst A strong localized downdraft (downburst) less than 4 km wide that occurs beneath thunderstorms. A strong downburst greater than 4 km across is called a *macroburst.*

Microclimate The climate structure of the air space near the surface of the earth.

Micrometer (μm) A unit of length equal to one-millionth of a meter.

Microscale The smallest scale of atmospheric motions.

Mid-Holocene maximum A warm period in geologic history (about 5000 to 6000 years ago) that favored the development of plants.

Middle latitudes The region of the world typically described as being between 30° and 50° latitude.

Middle latitude cyclone *See* Extratropical cyclone.

Milankovitch theory A theory proposed by Milutin Milankovitch in the 1930s suggesting that changes in the earth's orbit were responsible for variations in solar energy reaching the earth's surface and climatic changes.

Millibar (mb) A unit for expressing atmospheric pressure. Sea-level pressure is normally close to 1013 mb.

Minimum thermometer A thermometer designed to measure the minimum air temperature during a desired time period.

Mini-swirls Small whirling eddies perhaps 30 to 100 m in diameter that form in a region of strong wind shear of a hurricane's eye wall. Same as *spin-up vortices.*

Mirage A refraction phenomenon that makes an object appear to be displaced from its true position. When an object appears higher than it actually is, it is called a *superior mirage.* When an object appears lower than it actually is, it is an *inferior mirage.*

Mixing depth The vertical extent of the mixing layer.

Mixing layer The unstable atmospheric layer that extends from the surface up to the base of an inversion. Within this layer, the air is well stirred.

Mixing ratio The ratio of the mass of water vapor in a given volume of air to the mass of dry air.

Moist adiabatic rate The rate of change of temperature in a rising or descending saturated air parcel. The rate of cooling or warming varies but a common value of 6°C per 1000 m (3.3°F per 1000 ft) is used.

Molecule A collection of atoms held together by chemical forces.

Monsoon wind system A wind system that reverses direction between winter and summer. Usually the wind blows from land to sea in winter and from sea to land in summer.

Mountain and valley breeze A local wind system of a mountain valley that blows downhill (*mountain breeze*) at night and uphill (*valley breeze*) during the day.
Multicell storms Thunderstorms often in a line, each of which may be in a different stage of its life cycle.

Nacreous clouds Clouds of unknown composition that have a soft, pearly luster and that form at altitudes about 25 to 30 km above the earth's surface. They are also called *mother-of-pearl clouds.*
Negative feedback mechanism *See* Feedback mechanism.
Neutral stability (neutrally stable atmosphere) An atmospheric condition that exists in dry air when the environmental lapse rate equals the dry adiabatic rate. In saturated air the environmental lapse rate equals the moist adiabatic rate.
NEXRAD An acronym for *Ne*xt Generation Weather *Rad*ar. The main component of NEXRAD is the WSR 88-D, Doppler radar.
Nimbostratus A dark, gray cloud characterized by more or less continuously falling precipitation. It is rarely accompanied by lightning, thunder, or hail.
Nitric oxide (NO) A colorless gas produced by natural bacterial action in soil and by combustion processes at high temperatures. In polluted air, nitric oxide can react with ozone and hydrocarbons to form other substances. In this manner, it acts as an agent in the production of photochemical smog.
Nitrogen (N_2) A colorless and odorless gas that occupies about 78 percent of dry air in the lower atmosphere.
Nitrogen dioxide (NO_2) A reddish-brown gas, produced by natural bacterial action in soil and by combustion processes at high temperatures. In the presence of sunlight, it breaks down into nitric oxide and atomic oxygen. In polluted air, nitrogen dioxide acts as an agent in the production of photochemical smog.
Nitrogen oxides (NO_X) Gases produced by natural processes and by combustion processes at high temperatures. In polluted air, nitric oxide (NO) and nitrogen dioxide (NO_2) are the most abundant oxides of nitrogen, and both act as agents for the production of photochemical smog.
Noctilucent clouds Wavy, thin, bluish-white clouds that are best seen at twilight in polar latitudes. They form at altitudes about 80 to 90 km above the surface.
Nocturnal inversion *See* Radiation inversion.
Nonsupercell tornado A tornado that occurs with a cloud that is often in its growing stage and one that does not contain a mid-level mesocyclone or wall cloud. Landspouts and gustanadoes are examples of nonsupercell tornadoes.
Northeaster A name given to a strong, steady wind from the northeast that is accompanied by rain and inclement weather. It often develops when a storm system moves northeastward along the coast of North America. Also called Nor'easter.
Northern lights *See* Aurora.
Nowcasting Short-term weather forecasts varying from minutes up to a few hours.
Nuclear winter The dark, cold, and gloomy conditions that presumably would be brought on by nuclear war.
Numerical weather prediction (NWP) Forecasting the weather based upon the solutions of mathematical equations by high-speed computers.

Obliquity (of the earth's axis) The tilt of the earth's axis. It represents the angle from the perpendicular to the plane of the earth's orbit.
Occluded front (occlusion) A complex frontal system that ideally forms when a cold front overtakes a warm front. When the air behind the front is colder than the air ahead of it, the front is called a *cold occlusion.* When the air behind the front is milder than the air ahead of it, it is called a *warm occlusion.*
Offshore wind A breeze that blows from the land out over the water. Opposite of an onshore wind.
Onshore wind A breeze that blows from the water onto the land. Opposite of an offshore wind.
Open wave The stage of development of a wave cyclone (mid-latitude cyclonic storm) where a cold front and a warm front exist, but no occluded front. The center of lowest pressure in the wave is located at the junction of the two fronts.
Orchard heaters Oil heaters placed in orchards that generate heat and promote convective circulations to protect fruit trees from damaging low temperatures. Also called *smudge pots.*
Ordinary thunderstorm (formally called *air-mass thunderstorm*) A thunderstorm produced by local convection within a conditionally unstable air mass. It often forms in the afternoon and does not reach the intensity of a severe thunderstorm.
Orographic uplift The lifting of air over a topographic barrier. Clouds that form in this lifting process are called *orographic clouds.*
Outgassing The release of gases dissolved in hot, molten rock.
Overrunning A condition that occurs when air moves up and over another layer of air.
Oxygen (O_2) A colorless and odorless gas that occupies about 21 percent of dry air in the lower atmosphere.
Ozone (O_3) An almost colorless gaseous form of oxygen with an odor similar to weak chlorine. The highest natural concentration is found in the stratosphere where it is known as *stratospheric ozone.* It also forms in polluted air near the surface where it is the main ingredient of photochemical smog. Here, it is called *tropospheric ozone.*
Ozone hole A sharp drop in stratospheric ozone concentration observed over the Antarctic during the spring.

Pacific decadal oscillation (PDO) A reversal in ocean surface temperatures that occurs every 20 to 30 years over the northern Pacific Ocean.
Pacific high *See* Subtropical high.
Parcel of air An imaginary small body of air a few meters wide that is used to explain the behavior of air.
Parhelia *See* Sundog.
Particulate matter Solid particles or liquid droplets that are small enough to remain suspended in the air. Also called *aerosols.*

Pattern recognition An analogue method of forecasting where the forecaster uses prior weather events (or similar weather map conditions) to make a forecast.
Permafrost A layer of soil beneath the earth's surface that remains frozen throughout the year.
Persistence forecast A forecast that the future weather condition will be the same as the present condition.
Photochemical smog *See* Smog.
Photodissociation The splitting of a molecule by a photon.
Photon A discrete quantity of energy that can be thought of as a packet of electromagnetic radiation traveling at the speed of light.
Pileus cloud A smooth cloud in the form of a cap. Occurs above, or is attached to, the top of a cumuliform cloud. Also called a *cap cloud.*
Planetary boundary layer *See* Atmospheric boundary layer.
Planetary scale The largest scale of atmospheric motion. Sometimes called the *global scale.*
Plate tectonics The theory that the earth's surface down to about 100 km is divided into a number of plates that move relative to one another across the surface of the earth. Once referred to as continental drift.
Pleistocene Epoch (or Ice Age) The most recent period of extensive continental glaciation that saw large portions of North America and Europe covered with ice. It began about 2 million years ago and ended about 10,000 years ago.
Polar easterlies A shallow body of easterly winds located at high latitudes poleward of the subpolar low.
Polar front A semipermanent, semicontinuous front that separates tropical air masses from polar air masses.
Polar front jet stream (polar jet) The jet stream that is associated with the polar front in middle and high latitudes. It is usually located at altitudes between 9 and 12 km.
Polar front theory A theory developed by a group of Scandinavian meteorologists that explains the formation, development, and overall life history of cyclonic storms that form along the polar front.
Polar ice cap climate A climate characterized by extreme cold, as every month has an average temperature below freezing.
Polar low An area of low pressure that forms over polar water behind (poleward of) the main polar front.
Polar orbiting satellite A satellite whose orbit closely parallels the earth's meridian lines and thus crosses the polar regions on each orbit.
Polar tundra climate A climate characterized by extremely cold winters and cool summers, as the average temperature of the warmest month climbs above freezing but remains below 10°C (50°F).
Pollutants Any gaseous, chemical, or organic matter that contaminates the atmosphere, soil, or water.
Pollutant standards index (PSI) *See* Air quality index.
Positive feedback mechanism *See* Feedback mechanism.
Potential energy The energy that a body possesses by virtue of its position with respect to other bodies in the field of gravity.
Precession (of the earth's axis of rotation) The wobble of the earth's axis of rotation that traces out the path of a cone over a period of about 23,000 years.
Precipitation Any form of water particles—liquid or solid—that falls from the atmosphere and reaches the ground.
Pressure The force per unit area. *See also* Air pressure.
Pressure gradient The rate of decrease of pressure per unit of horizontal distance. On the same chart, when the isobars are close together, the pressure gradient is steep. When the isobars are far apart, the pressure gradient is weak.
Pressure gradient force (PGF) The force due to differences in pressure within the atmosphere that causes air to move and, hence, the wind to blow. It is directly proportional to the pressure gradient.
Pressure tendency The rate of change of atmospheric pressure within a specified period of time, most often three hours. Same as *barometric tendency.*
Prevailing westerlies The dominant westerly winds that blow in middle latitudes on the poleward side of the subtropical high-pressure areas. Also called *westerlies.*
Prevailing wind The wind direction most frequently observed during a given period.
Primary air pollutants Air pollutants that enter the atmosphere directly.
Probability forecast A forecast of the probability of occurrence of one or more of a mutually exclusive set of weather conditions.
Prognostic chart (prog) A chart showing expected or forecasted conditions, such as pressure patterns, frontal positions, contour height patterns, and so on.
Psychrometer An instrument used to measure the water vapor content of the air. It consists of two thermometers (dry bulb and wet bulb). After whirling the instrument, the dew point and relative humidity can be obtained with the aid of tables.

Radar An electronic instrument used to detect objects (such as falling precipitation) by their ability to reflect and scatter microwaves back to a receiver. (*See also* Doppler radar.)
Radiant energy (radiation) Energy propagated in the form of electromagnetic waves. These waves do not need molecules to propagate them, and in a vacuum they travel at nearly 300,000 km per sec (186,000 mi per sec).
Radiational cooling The process by which the earth's surface and adjacent air cool by emitting infrared radiation.
Radiation fog Fog produced over land when radiational cooling reduces the air temperature to or below its dew point. It is also known as *ground fog* and *valley fog.*
Radiation inversion An increase in temperature with height due to radiational cooling of the earth's surface. Also called a *nocturnal inversion.*
Radiative equilibrium temperature The temperature achieved when an object, behaving as a blackbody, is absorbing and emitting radiation at equal rates.
Radiative forcing An increase (positive) or a decrease (negative) in net radiant energy observed over an area at the

tropopause. An increase in radiative forcing may induce surface warming, whereas a decrease may induce surface cooling.
Radiative forcing agent Any factor (such as increasing greenhouse gases and variations in solar output) that can change the balance between incoming energy from the sun and outgoing energy from the earth and the atmosphere.
Radiometer *See* Infrared radiometer.
Radiosonde A balloon-borne instrument that measures and transmits pressure, temperature, and humidity to a ground-based receiving station.
Rain Precipitation in the form of liquid water drops that have diameters greater than that of drizzle.
Rainbow An arc of concentric colored bands that spans a section of the sky when rain is present and the sun is positioned at the observer's back.
Rain gauge An instrument designed to measure the amount of rain that falls during a given time interval.
Rain shadow The region on the leeside of a mountain where the precipitation is noticeably less than on the windward side.
Rawinsonde observation A radiosonde observation that includes wind data.
Reflection The process whereby a surface turns back a portion of the radiation that strikes it.
Refraction The bending of light as it passes from one medium to another.
Relative humidity The ratio of the amount of water vapor in the air compared to the amount required for saturation (at a particular temperature and pressure). The ratio of the air's actual vapor pressure to its saturation vapor pressure.
Return stroke The luminous lightning stroke that propagates upward from the earth to the base of a cloud.
Ridge An elongated area of high atmospheric pressure.
Rime A white or milky granular deposit of ice formed by the rapid freezing of supercooled water drops as they come in contact with an object in below-freezing air.
Riming *See* Accretion.
Roll cloud A dense, cylindrical, elongated cloud that appears to slowly spin about a horizontal axis behind the leading edge of a thunderstorm's gust front.
Rotor cloud A turbulent cumuliform type of cloud that forms on the leeward side of large mountain ranges. The air in the cloud rotates about an axis parallel to the range.
Rotors Turbulent eddies that form downwind of a mountain chain, creating hazardous flying conditions.

Saffir-Simpson scale A scale relating a hurricane's central pressure and winds to the possible damage it is capable of inflicting.
St. Elmo's fire A bright electric discharge that is projected from objects (usually pointed) when they are in a strong electric field, such as during a thunderstorm.
Santa Ana wind A warm, dry wind that blows into southern California from the east off the elevated desert plateau. Its warmth is derived from compressional heating.
Saturation (of air) An atmospheric condition whereby the level of water vapor is the maximum possible at the existing temperature and pressure.
Saturation vapor pressure The maximum amount of water vapor necessary to keep moist air in equilibrium with a surface of pure water or ice. It represents the maximum amount of water vapor that the air can hold at any given temperature and pressure. (*See* Equilibrium vapor pressure.)
Savanna A tropical or subtropical region of grassland and drought-resistant vegetation. Typically found in tropical wet-and-dry climates.
Scales of motion The hierarchy of atmospheric circulations from tiny gusts to giant storms.
Scattering The process by which small particles in the atmosphere deflect radiation from its path into different directions.
Scintillation The apparent twinkling of a star due to its light passing through regions of differing air densities in the atmosphere.
Sea breeze A coastal local wind that blows from the ocean onto the land. The leading edge of the breeze is termed a *sea breeze front.*
Sea-level pressure The atmospheric pressure at mean sea level.
Secondary air pollutants Pollutants that form when a chemical reaction occurs between a primary air pollutant and some other component of air. Tropospheric ozone is a secondary air pollutant.
Selective absorbers Substances such as water vapor, carbon dioxide, clouds, and snow that absorb radiation only at particular wavelengths.
Semi-arid climate A dry climate where potential evaporation and transpiration exceed precipitation. Not as dry as the arid climate. Typical vegetation is short grass.
Semipermanent highs and lows Areas of high pressure (anticyclones) and low pressure (extratropical cyclones) that tend to persist at a particular latitude belt throughout the year. In the Northern Hemisphere, typically they shift slightly northward in summer and slightly southward in winter.
Sensible heat The heat we can feel and measure with a thermometer.
Sensible temperature The sensation of temperature that the human body feels in contrast to the actual temperature of the environment as measured with a thermometer.
Severe thunderstorms Intense thunderstorms capable of producing heavy showers, flash floods, hail, strong and gusty surface winds, and tornadoes.
Shear *See* Wind shear.
Sheet lightning Occurs when the lightning flash is not seen but the flash causes the cloud (or clouds) to appear as a diffuse luminous white sheet.
Shelf cloud A dense, arch-shaped, ominous-looking cloud that often forms along the leading edge of a thunderstorm's gust front, especially when stable air rises up and over cooler air at the surface. Also called an *arcus cloud.*

Short-range forecast Generally used to describe a weather forecast that extends from about 6 hours to a few days into the future.

Shortwave (in the atmosphere) A small wave that moves around longwaves in the same direction as the air flow in the middle and upper troposphere. Shortwaves are also called *shortwave troughs.*

Shortwave radiation A term most often used to describe the radiant energy emitted from the sun, in the visible and near ultraviolet wavelengths.

Shower Intermittent precipitation from a cumuliform cloud, usually of short duration but often heavy.

Siberian high A strong, shallow area of high pressure that forms over Siberia in winter.

Sleet A type of precipitation consisting of transparent pellets of ice 5 mm or less in diameter. Same as *ice pellets.*

Smog Originally *smog* meant a mixture of smoke and fog. Today, *smog* means air that has restricted visibility due to pollution, or pollution formed in the presence of sunlight—*photochemical smog.*

Smog front (also smoke front) The leading edge of a sea breeze that is contaminated with smoke or pollutants.

Snow A solid form of precipitation composed of ice crystals in complex hexagonal form.

Snow-albedo feedback A positive feedback whereby increasing surface air temperatures enhance the melting of snow and ice in polar latitudes. This reduces the earth's albedo and allows more sunlight to reach the surface, which causes the air temperature to rise even more.

Snowflake An aggregate of ice crystals that falls from a cloud.

Snow flurries Light showers of snow that fall intermittently.

Snow grains Precipitation in the form of very small, opaque grains of ice. The solid equivalent of drizzle.

Snow pellets White, opaque, approximately round ice particles between 2 and 5 mm in diameter that form in a cloud either from the sticking together of ice crystals or from the process of accretion. Also called *graupel.*

Snow squall (shower) An intermittent heavy shower of snow that greatly reduces visibility.

Solar constant The rate at which solar energy is received on a surface at the outer edge of the atmosphere perpendicular to the sun's rays when the earth is at a mean distance from the sun. The value of the solar constant is about two calories per square centimeter per minute or about 1376 W/m^2 in the SI system of measurement.

Solar wind An outflow of charged particles from the sun that escapes the sun's outer atmosphere at high speed.

Sonic boom A loud explosive-like sound caused by a shock wave eminating from an aircraft (or any object) traveling at or above the speed of sound.

Sounding An upper-air observation, such as a radiosonde observation. A vertical profile of an atmospheric variable such as temperature or winds.

Source regions Regions where air masses originate and acquire their properties of temperature and moisture.

Southern oscillation The reversal of surface air pressure at opposite ends of the tropical Pacific Ocean that occur during major El Niño events.

Specific heat The ratio of the heat absorbed (or released) by the unit mass of the system to the corresponding temperature rise (or fall).

Specific humidity The ratio of the mass of water vapor in a given parcel to the total mass of air in the parcel.

Spin-up vortices Small whirling eddies perhaps 30 to 100 m in diameter that form in a region of strong wind shear in a hurricane's eye wall. Same as *mini-swirls.*

Squall line A line of thunderstorms that form along a cold front or out ahead of it.

Stable air *See* Absolutely stable atmosphere.

Standard atmosphere A hypothetical vertical distribution of atmospheric temperature, pressure, and density in which the air is assumed to obey the gas law and the hydrostatic equation. The lapse rate of temperature in the troposphere is taken as 6.5°C/1000 m or 3.6°F/1000 ft.

Standard atmospheric pressure A pressure of 1013.25 millibars (mb), 29.92 inches of mercury (Hg), 760 millimeters (mm) of mercury, 14.7 pounds per square inch (lb/in.^{2}), or 1013.25 hectopascals (hPa).

Standard rain gauge A nonrecording rain gauge with an 8-inch diameter collector funnel and a tube that amplifies rainfall by tenfold.

Stationary front A front that is nearly stationary with winds blowing almost parallel and from opposite directions on each side of the front.

Station pressure The actual air pressure computed at the observing station.

Statistical forecast A forecast based on a mathematical/statistical examination of data that represents the past observed behavior of the forecasted weather element.

Steady-state forecast A weather prediction based on the past movement of surface weather systems. It assumes that the systems will move in the same direction and at approximately the same speed as they have been moving. Also called *trend forecasting.*

Steam fog *See* Evaporation (mixing) fog.

Stefan-Boltzmann law A law of radiation which states that the amount of radiant energy emitted from a unit surface area of an object (ideally a blackbody) is proportional to the fourth power of the object's absolute temperature.

Steppe An area of grass-covered, treeless plains that has a semi-arid climate.

Stepped leader An initial discharge of electrons that proceeds intermittently toward the ground in a series of steps in a cloud-to-ground lightning stroke.

Storm surge An abnormal rise of the sea along a shore; primarily due to the winds of a storm, especially a hurricane.

Stratocumulus A low cloud, predominantly stratiform, with low, lumpy, rounded masses, often with blue sky between them.

Stratosphere The layer of the atmosphere above the troposphere and below the mesosphere (between 10 km and 50 km), generally characterized by an increase in temperature with height.

Stratospheric polar night jet A jet stream that forms near the top of the stratosphere over polar latitudes during the winter months.

Stratus A low, gray cloud layer with a rather uniform base whose precipitation is most commonly drizzle.

Streamline A line that shows the wind flow pattern.

Sublimation The process whereby ice changes directly into water vapor without melting.

Subpolar climate A climate observed in the Northern Hemisphere that borders the polar climate. It is characterized by severely cold winters and short, cool summers. Also known as *taiga climate* and *boreal climate*.

Subpolar low A belt of low pressure located between 50° and 70° latitude. In the Northern Hemisphere, this "belt" consists of the *Aleutian low* in the North Pacific and the *Icelandic low* in the North Atlantic. In the Southern Hemisphere, it exists around the periphery of the Antarctic continent.

Subsidence The slow sinking of air, usually associated with high-pressure areas.

Subsidence inversion A temperature inversion produced by compressional warming—the adiabatic warming of a layer of sinking air.

Subtropical high A semipermanent high in the subtropical high-pressure belt centered near 30° latitude. The *Bermuda high* is located over the Atlantic Ocean off the east coast of North America. The *Pacific high* is located off the west coast of North America.

Subtropical jet stream The jet stream typically found between 20° and 30° latitude at altitudes between 12 and 14 km.

Suction vortices Small, rapidly rotating whirls perhaps 10 m in diameter that are found within large tornadoes.

Sulfate aerosols *See* Aerosols.

Sulfur dioxide (SO_2) A colorless gas that forms primarily in the burning of sulfur-containing fossil fuels.

Summer solstice Approximately June 21 in the Northern Hemisphere when the sun is highest in the sky and directly overhead at latitude $23\frac{1}{2}$°N, the Tropic of Cancer.

Sundog A colored luminous spot produced by refraction of light through ice crystals that appears on either side of the sun. Also called *parhelia*.

Sun pillar A vertical streak of light extending above (or below) the sun. It is produced by the reflection of sunlight off ice crystals.

Sunspots Relatively cooler areas on the sun's surface. They represent regions of an extremely high magnetic field.

Supercell storm An enormous severe thunderstorm that consists primarily of a single rotating updraft. Its organized internal structure allows that storm to maintain itself for several hours. The storm can produce large hail and dangerous tornadoes.

Supercell tornadoes Tornadoes that occur within supercell thunderstorms that contain well-developed mid-level mesocyclones.

Supercooled cloud (or cloud droplets) A cloud composed of liquid droplets at temperatures below 0°C (32°F). When the cloud is on the ground it is called *supercooled fog* or *cold fog*.

Superior mirage *See* Mirage.

Supersaturation A condition whereby the atmosphere contains more water vapor than is needed to produce saturation with respect to a flat surface of pure water or ice, and the relative humidity is greater than 100 percent.

Super typhoon A tropical cyclone (typhoon) in the western Pacific that has sustained winds of 130 knots or greater.

Surface inversion *See* Radiation inversion.

Surface map A map that shows the distribution of sea-level pressure with isobars and weather phenomena. Also called a *surface chart*.

Synoptic scale The typical weather map scale that shows features such as high- and low-pressure areas and fronts over a distance spanning a continent. Also called the *cyclonic scale*.

Taiga (boreal forest) The open northern part of the coniferous forest. Taiga also refers to subpolar climate.

Tangent arc An arc of light tangent to a halo. It forms by refraction of light through ice crystals.

Tcu An abbreviation sometimes used to denote a towering cumulus cloud (cumulus congestus).

Teleconnections A linkage between weather changes occurring in widely separated regions of the world.

Temperature The degree of hotness or coldness of a substance as measured by a thermometer. It is also a measure of the average speed or kinetic energy of the atoms and molecules in a substance.

Temperature inversion An increase in air temperature with height, often simply called an *inversion*.

Thermal A small, rising parcel of warm air produced when the earth's surface is heated unevenly.

Thermal belts Horizontal zones of vegetation found along hillsides that are primarily the result of vertical temperature variations.

Thermal circulations Air flow resulting primarily from the heating and cooling of air.

Thermal lows and thermal highs Areas of low and high pressure that are shallow in vertical extent and are produced primarily by surface temperatures.

Thermal tides Atmospheric pressure variations due to the uneven heating of the atmosphere by the sun.

Thermograph An instrument that measures and records air temperature.

Thermometer An instrument for measuring temperature. The most common are liquid-in-glass, which have a sealed glass tube attached to a glass bulb filled with liquid.

Thermosphere The atmospheric layer above the mesosphere (above about 85 km) where the temperature increases rapidly with height.
Thunder The sound due to rapidly expanding gases along the channel of a lightning discharge.
Thunderstorm A local storm produced by cumulonimbus clouds. Always accompanied by lightning and thunder.
Tornado An intense, rotating column of air that often protrudes from a cumulonimbus cloud in the shape of a funnel or a rope whose circulation is present on the ground. (*See* Funnel cloud.)
Tornado outbreak A series of tornadoes that forms within a particular region—a region that may include several states. Often associated with widespread damage and destruction.
Tornado vortex signature (TVS) An image of a tornado on the Doppler radar screen that shows up as a small region of rapidly changing wind directions inside a mesocyclone.
Tornado warning A warning issued when a tornado has actually been observed either visually or on a radar screen. It is also issued when the formation of tornadoes is imminent.
Tornado watch A forecast issued to alert the public that tornadoes may develop within a specified area.
Trace (of precipitation) An amount of precipitation less than 0.01 in. (0.025 cm).
Trade wind inversion A temperature inversion frequently found in the subtropics over the eastern portions of the tropical oceans.
Trade winds The winds that occupy most of the tropics and blow from the subtropical highs to the equatorial low.
Transpiration The process by which water in plants is transferred as water vapor to the atmosphere.
Tropical cyclone The general term for storms (cyclones) that form over warm tropical oceans.
Tropical depression A mass of thunderstorms and clouds generally with a cyclonic wind circulation of between 20 and 34 knots.
Tropical disturbance An organized mass of thunderstorms with a slight cyclonic wind circulation of less than 20 knots.
Tropical easterly jet A jet stream that forms on the equatorward side of the subtropical highs near 15 km.
Tropical monsoon climate A tropical climate with a brief dry period of perhaps one or two months.
Tropical rain forest A type of forest consisting mainly of lofty trees and a dense undergrowth near the ground.
Tropical storm Organized thunderstorms with a cyclonic wind circulation between 35 and 64 knots.
Tropical wave A migratory wavelike disturbance in the tropical easterlies. Tropical waves occasionally intensify into tropical cyclones. They are also called *easterly waves.*
Tropical wet-and-dry climate A tropical climate poleward of the tropical wet climate where a distinct dry season occurs, often lasting for two months or more.
Tropical wet climate A tropical climate with sufficient rainfall to produce a dense tropical rain forest.
Tropopause The boundary between the troposphere and the stratosphere.
Troposphere The layer of the atmosphere extending from the earth's surface up to the tropopause (about 10 km above the ground).
Trough An elongated area of low atmospheric pressure.
Turbulence Any irregular or disturbed flow in the atmosphere that produces gusts and eddies.
Twilight The time at the beginning of the day immediately before sunrise and at the end of the day after sunset when the sky remains illuminated.
Typhoon A hurricane that forms in the western Pacific Ocean.

Ultraviolet (UV) radiation Electromagnetic radiation with wavelengths longer than X-rays but shorter than visible light.
Unstable air *See* Absolutely unstable atmosphere.
Upslope fog Fog formed as moist, stable air flows upward over a topographic barrier.
Upslope precipitation Precipitation that forms due to moist, stable air gradually rising along an elevated plain. Upslope precipitation is common over the western Great Plains, especially east of the Rocky Mountains.
Upwelling The rising of water (usually cold) toward the surface from the deeper regions of a body of water.
Urban heat island The increased air temperatures in urban areas as contrasted to the cooler surrounding rural areas.

Valley breeze *See* Mountain breeze.
Valley fog *See* Radiation fog.
Vapor pressure The pressure exerted by the water vapor molecules in a given volume of air.
Veering wind The wind that changes direction in a clockwise sense—north to northeast to east, and so on.
Vernal equinox The equinox at which the sun approaches the Northern Hemisphere and passes directly over the equator. Occurs around March 20.
Very short-range forecast Generally used to describe a weather forecast that is made for up to a few hours (usually less than 6 hours) into the future.
Virga Precipitation that falls from a cloud but evaporates before reaching the ground. (*See* Fallstreaks.)
Visible radiation (light) Radiation with a wavelength between 0.4 and 0.7 μm. This region of the electromagnetic spectrum is called the visible region.
Visible region *See* Visible radiation.
Visibility The greatest distance an observer can see and identify prominent objects.
Volatile organic compounds (VOCs) A class of organic compounds that are released into the atmosphere from sources such as motor vehicles, paints, and solvents. VOCs (which include hydrocarbons) contribute to the production of secondary pollutants, such as ozone.

Wall cloud An area of rotating clouds that extends beneath a severe thunderstorm and from which a funnel cloud may appear. Also called a *collar cloud* and *pedestal cloud.*

Warm-core low A low-pressure area that is warmer at its center than at its periphery. Tropical cyclones exhibit this temperature pattern.

Warm front A front that moves in such a way that warm air replaces cold air.

Warm occlusion *See* Occluded front.

Warm sector The region of warm air within a wave cyclone that lies between a retreating warm front and an advancing cold front.

Water equivalent The depth of water that would result from the melting of a snow sample. Typically about 10 inches of snow will melt to 1 inch of water, producing a water equivalent of 10 to 1.

Waterspout A column of rotating wind over water that has characteristics of a dust devil and tornado.

Water vapor Water in a vapor (gaseous) form. Also called *moisture.*

Water vapor–greenhouse effect feedback A positive feedback whereby increasing surface air temperatures cause an increase in the evaporation of water from the oceans. Increasing concentrations of atmospheric water vapor enhance the greenhouse effect, which causes the surface air temperature to rise even more.

Watt (W) The unit of power in SI units where 1 watt is equivalent to 1 joule per second.

Wave cyclone An extratropical cyclone that forms and moves along a front. The circulation of winds about the cyclone tends to produce a wavelike deformation on the front.

Wavelength The distance between successive crests, troughs, or identical parts of a wave.

Weather The condition of the atmosphere at any particular time and place.

Weather elements The elements of *air temperature, air pressure, humidity, clouds, precipitation, visibility,* and *wind* that determine the present state of the atmosphere, the weather.

Weather type forecasting A forecasting method where weather patterns are categorized into similar groups or types.

Weather types Certain weather patterns categorized into similar groups. Used as an aid in weather prediction.

Weather warning A forecast indicating that hazardous weather is either imminent or actually occurring within the specified forecast area.

Weather watch A forecast indicating that atmospheric conditions are favorable for hazardous weather to occur over a particular region during a specified time period.

Westerlies The dominant westerly winds that blow in the middle latitudes on the poleward side of the subtropical high-pressure areas.

Wet-bulb depression The difference in degrees between the air temperature (dry-bulb temperature) and the wet-bulb temperature.

Wet-bulb temperature The lowest temperature that can be obtained by evaporating water into the air.

Whirlwinds *See* Dust devils.

Wien's law A law of radiation which states that the wavelength of maximum emitted radiation by an object (ideally a blackbody) is inversely proportional to the object's absolute temperature.

Wind Air in motion relative to the earth's surface.

Wind-chill index The cooling effect of any combination of temperature and wind, expressed as the loss of body heat. Also called *wind-chill factor.*

Wind direction The direction *from which* the wind is blowing.

Wind machines Fans placed in orchards for the purpose of mixing cold surface air with warmer air above.

Wind profiler A Doppler radar capable of measuring the turbulent eddies that move with the wind. Because of this, it is able to provide a vertical picture of wind speed and wind direction.

Wind rose A diagram that shows the percent of time that the wind blows from different directions at a given location over a given time.

Wind shear The rate of change of wind speed or wind direction over a given distance.

Wind vane An instrument used to indicate wind direction.

Windward side The side of an object facing into the wind.

Winter solstice Approximately December 21 in the Northern Hemisphere when the sun is lowest in the sky and directly overhead at latitude $23\frac{1}{2}$°S, the Tropic of Capricorn.

Xerophytes Drought-resistant vegetation.

Younger-Dryas event A cold episode that took place about 11,000 years ago, when average temperatures dropped suddenly and portions of the Northern Hemisphere reverted back to glacial conditions.

Zonal wind flow A wind that has a predominate west-to-east component.

Index

A

D

E

F

G

R

V

W

X

Y

Z

Credits

Chapter 1
Fig. 1.3 Data courtesy of NOAA.
Fig. 1.4 © David Weintraub/Photo Researchers
Fig. 1.13 © Judy Champlin.
Fig. 1.14 © Warren Faidley/Weatherstock.
Fig. 1.15 AP/Wide World.
Fig. 1.16 © 1993 C. Doswell

Chapter 2
Opener © Brad Perks.
Fig. 2 © Lindsey Martin.
Fig. 2.18 © Michael Orton/Stone.
Fig. 2.28 © Larry Ulrich/Stone.
Fig. 4 © Leland Bobbe/Stone.

Chapter 3
Opener Photos by author.
Fig. 3.7 © Ted Benson/Modesto Bee.
Figs. 3.12 and 3.13 Data from U.S. Department of Commerce.
Tables 3.2 and 3.3 Data courtesy of NOAA/National Weather Service.

Chapter 4
Opener © Chad Ehlers/Stone.
Fig. 4.7 Data from several sources.
Fig. 4.9 NOAA/National Weather Service.
Fig. 4.16 © Russell D. Curtis/Photo Researchers.
Fig. 4.33 National Center for Atmospheric Research/University Corporation for Atmospheric Research/National Science Foundation.
Figs. 4.34 and 4.35 © Pekka Parviainen.

Chapter 5
Opener © Ed Darack.
Fig. 5.14 © Jeff Smith.
Fig. 5.25 © Scott Cunazine/Photo Researchers.
Fig. 5.27 National Center for Atmospheric Research/University Corporation for Atmospheric Research/National Science Foundation.
Fig. 5.28 © George Champlin.
Fig. 5.34 Weather Graphics Courtesy of AccuWeather, Inc. 385 Science Park Road, State College, PA 16803 (814)237-0309 © 1999

Chapter 7
Fig. 2 National Center for Atmospheric Research/University Corporation for Atmospheric Research/National Science Foundation.
Fig. 7.14 Courtesy of Sherwood B. Idso.
Fig. 7.30 JISAO, University of Washington, obtained via http://www.tao. atmos.washington. edu/pdo. Used with permission.

Chapter 9
Fig. 9.4 NOAA/National Weather Service.
Fig. 9.12 Courtesy of NOAA.

Chapter 10
Opener © G. Kaufman.
Fig. 10.3 © Howard B. Bluestein.
Fig. 10.5 © Warren Faidley/WeatherStock.
Fig. 10.7 © Richard Picanso.
Fig. 10.8 © Howard B. Bluestein.
Fig. 10.16 © Wide World Photos.
Fig. 10.17 After *Thunderstorms,* Vol. 2, U.S. Government Printing Office, Washington, D.C., 1982.
Fig. 10.18 Data courtesy of NOAA.
Fig. 10.21 Photo © Richard Lee Kaylin.
Fig. 10.24 Global Atmospherics, Inc.
Fig. 2 Photo by Johnny Autery.
Fig. 10.25 Mary K. Hurley.
Fig. 10.26 Data courtesy of NOAA.
Fig. 10.29 © Wade Balzer/Weatherstock.
Fig. 10.35 © Howard B. Bluestein.
Fig. 10.36 Photo courtesy of James Tyler.
Fig. 10.37 Figure modified after Wakimoto and Wilson
Fig. 10.39 National Center for Atmospheric Research/University Corporation for Atmospheric Research/National Science Foundation.
Figs. 10.40 Joseph H. Golden, NOAA.

Chapter 11
Fig. 11.3 Modified after National Weather Service.
Fig. 11.4 © 1998 WSI Corporation. Intellicast.com.
Fig. 11.16 Photo © Weather Catalog/Media Services, Inc.
Tables 11.1, 11.2, and 11.3 NOAA/National Hurricane Center.

Chapter 12
Opener © B. & C. Alexander, Photo Researchers.
Fig. 12.3 Courtesy John Day and the University of Colorado Health Sciences Center.
Fig. 12.7 Data courtesy of NOAA.
Fig. 12.14 Jim Edwards, Riverside Press Enterprise.
Fig. 12.18 After United States and Canada Work Group No. 2, 1982.
Fig. 12.19 © Frederica Georgia/Photo Researchers.

Chapter 13
Opener © B. & C. Alexander, Photo Researchers
Fig. 13.6 © Frans Lanting, Minden Pictures.
Fig. 13.19 © Carr Clifton, Minden Pictures.
Fig. 13.24 © Michio Hoshino, Minden Pictures.

Chapter 14
Opener © Stephen J. Krasemann/Photo Researchers.
Fig. 14.2 Modified from *CLIMAP*, 1976.
Fig. 14.3 Modified from several sources, including J. T. Houghton, et al., *Climate Change 2001: The Scientific Basis.* Cambridge University Press, Cambridge, England, 2001.
Fig. 14.11 © Gamma Liaison.
Fig. 2 © Don Davis
Fig. 14.12 Data courtesy of John Christy, University of Alabama, Huntsville, and R. Spencer, NASA Marshall Space Flight Center.
Fig. 14.13 From V. Ramanathan, B. R. Barstrom, and E. F. Harrison, "Climate and the earth's radiation budget," *Physics Today* (May, 1989), Fig. 5, p. 27.
Fig. 14.14 © Frans Lanting, Minden Pictures.
Fig. 14.15 After F. B. Mitchell, et al., "Transient climate response to increasing sulphate aerosols and greenhouse gases." *Nature* (1995) 376: 501–504.
Fig. 1 © Chris Butler/SPL/Photo Researchers.

Chapter 15
Opener © 2002 STAR TRIBUNE/Minneapolis-St. Paul.
Fig.15.6 National Center for Atmospheric Research/University Corporation for Atmospheric Research/National Science Foundation.
Fig. 15.15 © Pekka Parviainen.
Fig. 15.20 Science Source/Photo Researchers.
Fig. 1 © Pekka Parviainen.
Fig. 15.22 © Norbert Rosing/Getty.
Fig. 15.31 © Pekka Parviainen.
Fig. 2 © H. Michael Mogil.

Cirrus High ice cryst
they are observed at alt

Altocumulus Middle
ice crystals at elevation
(6500 ft and 23,000 ft)
hand at arm's length, t
of your thumbnail.)

Cumulus Small, puffy clouds with relatively flat bases and limited vertical growth.

Cumulus congestus Cumulus clouds that show extensive vertical development. They may form in rows, such as these, or as individual clouds that appear as a head of cauliflower. Showery precipitation may fall from cumulus congestus.

Cumulus The sprouting clouds with pronounced vertical growth are *cumulus congestus*. The clouds with much smaller vertical extent are called *cumulus humilis*. The very small ragged-looking clouds are called *cumulus fractus*.

Cumulonimbus This cumuliform cloud may develop vertically to great heights, often with a sprouting anvil-shaped top. It produces heavy showers of rain or snow, lightning and thunder, and strong, gusty surface winds.

Mammatus Downward moving air causes these clouds to hang from the base of another cloud—most often a cumulonimbus or an altostratus.

Lenticular These lens-shaped clouds often form in waves that develop downwind of a mountain. They normally remain in one place as the air rushes through them.